Modern Simulation and Modeling

WILEY SERIES IN PROBABILITY AND STATISTICS
APPLIED PROBABILITY AND STATISTICS SECTION

Established by WALTER A. SHEWHART and SAMUEL S. WILKS

A complete list of the titles in this series appears at the end of this volume.

Modern Simulation and Modeling

REUVEN Y. RUBINSTEIN
Technion, Israel Institute of Technology
Haifa, Israel

BENJAMIN MELAMED
Rutgers University
New Brunswick, New Jersey

A Wiley-Interscience Publication
JOHN WILEY & SONS, INC.
New York · Chichester · Weinheim · Brisbane · Singapore · Toronto

This book is printed on acid-free paper. ♾

Published simultaneously in Canada.

Library of Congress Cataloging-in-Publication Data:

Rubinstein, Reuven Y.
Modern simulation and modeling / Reuven Y. Rubinstein, Benjamin Melamed.
p. cm. – (Wiley series in probability and statistics. Applied probability and statistics)
"A Wiley-Interscience publication."
Includes index.
ISBN 0-471-17077-1 (cloth : alk. paper)
1. Simulation methods. 2. Discrete-time systems. I. Melamed, Benjamin. II. Title. III. Series.
T57.62.R83 1998
003′.83–dc21 97-11869

Printed in the United States of America.

10 9 8 7 6 5 4 3

To My Beloved Son Eitan Jacob Simon Melamed—The Model Son

and

To Eitan Rubinstein—My Beloved Son

Contents

Preface

This book is about modeling and simulation of *discrete-event systems* (DES). Examples of DES are computer-communication networks, flexible manufacturing systems, PERT (project evaluation and review technique) networks and flow networks, including queueing systems. These systems are typically driven by the occurrence of discrete events and their state changes in time. In view of the complex interaction among such discrete events, the corresponding systems are often studied via stochastic simulation. Numerous textbooks on performance evaluation of complex DES have been written during the past two decades. Simulation is now one of the most widely used techniques, pervading both theory and practice, in management, science, and engineering.

This book is divided into two parts: Part I treats *conventional simulation* and consist of Chapters 1–4; Part II addresses *modern simulation* and consist of Chapters 5–9. Chapters 1–5 are based on an undergraduate course on simulation, while Chapters 6–9 are based on a graduate course on DES; both have been taught by Reuven Rubinstein at Technion—Israel Institute of Technology for the past 15 and 5 years, respectively. Each chapter contains exercises, with more advanced ones marked by an asterisk. Sections containing advanced material are also marked by asterisks.

Although in some sections the treatment is theoretical, emphasis is placed throughout on concepts rather than on mathematical completeness. We assume that the reader has a basic knowledge of probability, statistics, and optimization. More sophisticated concepts are either explained or referred to external literature.

Chapter 1 describes fundamental concepts such as systems, models, and simulation. In addition to the terminology, some examples and ideas pertaining to discrete-event simulation are presented, and one of its most important ingredients, the simulation clock, is discussed.

Chapter 2 discusses basic techniques for generating pseudorandom numbers, random variables (variates), and stochastic processes on a computer. The chapter also discusses stochastic sequences from a versatile class, called TES, designed to capture both first-order and second-order statistics of empirical records, and in particular, traffic *burstiness*.

Chapter 3 deals with output analysis of DES, particularly with the statistical analysis of simulated time series. It covers performance estimation of *discrete-event static systems* (DESS) and *discrete-event dynamic systems* (DEDS). In the latter case, we distinguish between *finite-horizon* and *steady-state* simulation. The fundamental difference between DESS and DEDS is that DESS do not evolve in time, whereas DEDS do. An example of DESS is a stochastic PERT network, and an example of DEDS is a queueing network.

Chapter 4 provides a basic treatment of variance reduction techniques, such as antithetic and common variates, control variates, conditional Monte Carlo, importance sampling, and stratified sampling. Variance reduction can be viewed as a means of using some available information about the system so as to obtain more accurate estimators of its performance.

Part II of the book deals with sensitivity analysis and optimization of DES. Consider, for example:

1. **Manufacturing Systems.** Here, the performance measure might be the average waiting time of an item to be processed at several workstations, given some schedule and route. The sensitivity and decision parameters might be the average rates at which the workstations process the item. A typical issue of interest might be the minimization of the average make-span, consisting of the processing time and delay time, subject to certain constraints (e.g., cost).
2. **Stochastic PERT Networks.** Here, the performance measure might be the mean shortest path. One might be interested in minimizing the mean shortest path in the network with respect to various parameters, such as activity durations on network links, subject to some constraints.

A central topic in Part II is the *score function* (SF) method, which we view as a component of *modern simulation*. We shall show that the SF method allows us to evaluate, *simultaneously from a single sample path* (simulation experiment), not only the performance (the mean value of some process in equilibrium as a function of some vector of decision parameters) and all its sensitivities (gradient, Hessian, etc. with respect to the same parameters), but also to solve an entire optimization problem, assuming that the vector of decision parameters comes from a parametric family of distributions. We shall also show that to this end, we do not have to know explicitly the sample performance (realization process), nor need we assume its differentiability with respect to the decision parameters; all we need is a *single sample* from the performance (output) process, differentiability of the underlying parametric family of distributions with respect to its vector of decision parameters, and some mild regularity conditions relating to the interchangeability of the expectation and differentiation operators. The SF method allows us to perform sensitivity analysis and optimization with hundreds of decision parameters, *based on a single sample path* (*simulation run*)! SF algorithms and procedures have been implemented in a simulation package called QNSO (queueing network stabilizer and optimizer), and they can be readily reimplemented in any discrete-event simulation language. The extra computation time consumed by SF computer code is on the order of 10–50% of the main simulation execution time.

Chapters 5 and 6 treat sensitivity analysis and stochastic optimization of DESS and DEDS with respect to *distributional* parameters, such as the service rate in a queueing model, while Chapter 7 treats sensitivity analysis with respect to *structural* parameters of DEDS, such as the (s, S) parameters in an $s - S$ policy inventory model. Chapter 7 presents two techniques, called *push-out* and *push-in*, respectively. These terms derive from the fact that in the former case we push out the structural parameters from the original sample performance function into an auxiliary pdf via a suitable transformation, and then apply the SF method; in the latter case, we operate the other way around, namely, we first push in (via a suitable transformation) the distributional parameters into the sample performance function, and then differentiate the resulting (auxiliary) sample performance function. Conditions will be discussed under which such transformations are computationally efficient. It will also be shown that the *infinitesimal perturbation analysis* (IPA) method, introduced by Ho and his co-workers, corresponds to the push-in technique; the latter can be also viewed as a dual of the push-out technique. The chapter also discusses how to combine efficiently the crude Monte Carlo method with the SF method.

Chapter 8 is devoted to the response-surface methodology. In particular, it is shown how to estimate (evaluate) an entire response surface of a DES, such as the expected waiting time and its derivatives in a queueing network, from a single simulation run.

Finally, Chapter 9 deals with rare-event probability estimation for complex DES, particularly queueing networks. Here we present a framework for the time complexity of Monte Carlo estimators and, in particular, define the concepts of polynomial and exponential time estimators. We use modern notions such as the *optimal exponential change of measure* (OECM) and *bottleneck module*, and derive polynomial estimators for classical ruin problem, the $GI/G/1$ queue and its extensions. We also discuss the robustness (stability) of the rare-event estimators.

The first author wishes to express his indebtedness to Søren Asmussen, John Harris, Dimitri Lieber, Vladimir Kalashnikov, Vladimir Katkovnik, Vladimir Kriman, Arkadi Nemirovski, Georg Pflug, Sergei Porotsky, Perwez Shahabuddin, Zhang Shiyng, and Alexander Shapiro for a number of valuable suggestions on an earlier draft of this book. Many thanks are due to Eliesar Goldberg of the Technion for his initial editing of some parts of the manuscript, and Lillian Bluestein for her excellent typing of the bulk of the manuscript.

Part of this book was written during Reuven Rubinstein's visits at Tilburg University. The hospitality extended to him is greatly appreciated.

REUVEN RUBINSTEIN
BENJAMIN MELAMED

Technion, Haifa, Israel
Rutgers University, New Brunswick, New Jersey
April 1997

Acronyms and Symbols

cdf	cumulative distribution function
CLT	central limit theorem
CMC	crude Monte Carlo
CSF	conditional score function
$\stackrel{D}{\rightarrow}$	convergence in distribution
DEDS	discrete-event dynamic systems
DES	discrete-event system
DESS	discrete-event static system
DSF	decomposable score function
FIFO	first in/first out
iid	independent identically distributed
IS	importance sampling
IPA	infinitesimal perturbation analysis
LCRV	linear control random variables
LIFO	last in/last out
LLN	law of large numbers
LR	likelihood ratio
PA	perturbation analysis
PERT	project evaluation and review technique
pdf	probability density function
QNSO	queueing network stabilizer and optimizer
RBM	regulated Brownian motion
RSF	regenerative score function
SF	score function
TSF	truncated score function
w.p.1:	with probability 1
$\sim$	distributed as
$f(x \mid A)$	conditional pdf (conditioned on event A)
$F(x \mid A)$	conditional cdf (conditioned on event A).

Modern Simulation and Modeling

PART I

Conventional Simulation

Monte Carlo simulation (simulation, for short) came to the fore in the first few decades following World War II. The flowering of simulation as a theoretical discipline and practical tool manifested itself in a concomitant growth of the corresponding research and practitioner community. Today, simulation is in widespread use in countless academic institutions and industrial sites, and it has become a standard topic of teaching and research, with its own dedicated scientific journals and professional conferences.

Just as probability was historically motivated by gambling (Cardano and Pascal are credited with pioneering contributions (Feller, 1958, Vol. I, Section II.10)), the origins of simulation appear to lie in experiments involving chance (e.g., Buffon's needle experiment and other anecdotal evidence in Feller (1958, Vol. II, Section II.8). As such, simulation introduced an alternative paradigm to traditional analytical modeling with random elements. In both cases a well-defined model is presented, but the "solution method" is drastically different. The analytical approach employs strictly mathematical tools to compute various quantities of interest in relatively simple models. In contrast, the simulation approach merely generates possible histories (sample path realizations) and then calculates statistics from them. For example, to compute performance measures of a queueing system in equilibrium, such as the queue length distribution, the analytical approach calls for the solution of certain equilibrium equations, whereas the simulation approach generates sufficiently long queueing histories, and calculates the corresponding time averages. For the two approaches to lead to consistent results in this particular example, one relies on the presumed ergodicity of the system under study to ensure that time averages converge to the corresponding "phase averages." The different tools employed by the two approaches (mathematics versus statistics) are the source of their corresponding strengths and weaknesses: Analytical methods yield *exact* solutions but typically can handle only relatively simple models, whereas simulation can be applied to far more complex models but only yields statistical *estimates*, subject to experimental error. Nevertheless, therein lies the great advantage of simulation, namely, the fact that sample path realizations can always be computed, in principle, though subject to the physical constraints of computing time and available storage. Thus, a simulation environment serves as an in vitro laboratory that can afford the analyst enormous flexibility.

From a practical point of view, simulation has nowadays become the tool of choice for the analysis of most complex systems. The advent of powerful computer systems and graphical displays has provided the enabling technology for the explosive growth and dissemination of simulation technologies. Furthermore, the diffusion of simulation technology to a broad range of technology areas, such as manufacturing, telecommunications, and computer systems, has been accompanied by vigorous efforts to develop both general-purpose and special-purpose simulation languages and software environments.

This book is wholly devoted to discrete-event simulation—the most widespread paradigm of the simulation discipline. It covers probabilistic and statistical aspects of simulation; implementation issues, such as programming languages or parallel/distributed simulation, are outside its scope. The book is divided into two parts: Part I treats *conventional simulation*, while Part II covers more recent developments that we term *modern simulation*. Roughly speaking, conventional simulation pertains to the fundamentals of discrete-event simulation, largely established before the 1980s, while modern simulation pertains to "smart techniques" dealing with sensitivity analysis and stochastic optimization of computer simulation models.

Chapter 1 starts out with the notions of systems, models, and discrete-event simulation. As its name suggests, discrete-event simulation views both time and state as discrete (rather than continuous) in the sense that its state trajectories are step functions. Thus, the prevailing state in a discrete-event simulation is always considered to be constant over some interval, and state transitions take place at discrete points in time from one discrete state to another. Consequently, the notions of *state transitions* (or *state jumps*) and *holding time* (or *sojourn time*) in a state are fundamental to discrete-event simulation. Philosophically, a discrete-event simulation simplifies reality by assuming that something "fundamentally new" occurs only at state transitions, whereas nothing "fundamentally new" or "unforeseen" takes place in between transitions. State transitions in a discrete-event simulation are triggered by *events*. In practice, events are fragments of computer code associated with an execution time; these are kept in chronological order in an *event list*, maintained by the simulation program. A *simulation clock* variable provides the notion of time. From a high-level viewpoint, a discrete-event simulation consists of a single repeatedly executed loop that first dequeues the current most imminent event from the event list, increments the simulation clock to the associated event time, and finally executes that event (in the course of which new events may be added to the event list).

Typically, a discrete-event simulation incorporates some randomness. From a high-level viewpoint, this randomness manifests itself in just two aspects: Either state transitions are random or the holding times are random or, most commonly, both are random. If a holding time is zero, then a succession of state transitions takes place instantaneously; the new simulation state will be well-defined, provided the number of instantaneous transitions is finite. At the applied level, the modeler generates random quantities (e.g., interarrival times and service times of customers in a queue) from prescribed distributions or probability laws. Chapter 2 covers computer generation of random variables and stochastic processes.

A large part of discrete-event simulation consists of so-called *output analysis*, namely, the collection of statistics from simulation runs, the formation of statistical estimates and confidence intervals for quantities of interest, and the performance of statistical tests in order to make decisions, based on statistical inference. Statistics collection and estimator formation are normally handled as part of event execution, while statistical inference is normally deferred until the end of the simulation or even relegated to separate software outside it. Chapter 3 covers basic output analysis techniques, including the *regenerative method* and *batch-means method.*

The smaller the variance of the statistical (output) estimators, the better (a smaller variance gives rise to narrower confidence intervals for a fixed confidence level). Conversely, if the confidence intervals are unsatisfactorily wide, the analyst has two choices. The first choice is to reduce the variance by collecting additional observations. Naturally, this requires more or longer simulation runs with an attendant computational cost, typically in terms of increased computing time. An alternative approach is to change the experimental conditions and include more information about the system in such a way that the estimator is still valid, but the resulting variance is much smaller for a comparable computational effort. In that case, we say that *variance reduction* was attained. Chapter 4 explains basic variance reduction techniques.

CHAPTER 1

Systems, Models, and Simulation

1.1 INTRODUCTION

1.1.1 Discrete-Event Systems and Simulation

This book is about simulation and modeling of *discrete-event systems* (DES). All too often, real-life DES are too complex to model analytically, requiring the analyst to resort to statistical computer simulation models in order to approximate numerically their desired characteristics. Simulation has now become one of the most widespread modeling approaches in operations research, management science, and engineering. Furthermore, the sustained growth in size and complexity of emerging real-world systems (e.g., high-speed communications networks) will undoubtedly ensure that the popularity of computer simulation will continue to grow.

This chapter is organized as follows. Section 1.2 describes basic concepts, such as systems, models, simulation, and Monte Carlo methods. Finally, Section 1.3 deals with discrete-event simulation, and in particular, with one of its most important ingredients—the simulation clock.

1.2 SYSTEMS, MODELS, AND SIMULATION

This section describes the concepts of systems, models, and simulation. By a *system*, we mean a set of related entities, sometimes called *components* or *elements*. For instance, a hospital may be considered as a system, with doctors, nurses, and patients as elements. The elements possess certain characteristics or *attributes* that take on logical or numerical values. In our example, an attribute might be the number of beds, the number of X-ray machines, skill level, and so on. Typically, the activities of individual components interact in time. These activities cause changes in the system's state. For example, the state of a hospital's waiting room might be described by the number of patients waiting for a doctor. When a patient arrives at the hospital or leaves it, the system jumps to a new state.

We shall be solely concerned with discrete-event systems, to wit, those systems in which the state variables change instantaneously through jumps at discrete points in time; DES are qualitatively distinct from *continuous systems* (CS), where state variables change continuously with respect to time. Examples of a discrete and a continuous sytem are, respectively, a bank serving customers and a car moving on a freeway. In the former case, the number of waiting customers is a piecewise-constant state variable that changes only when either a new customer arrives at the bank or a customer finishes to transact his business and departs from the bank; in the latter case, the car velocity is a state variable that can change continuously in time.

The first step in studying a system is to build a model from which we can obtain predictions concerning the behavior of the system under study. The importance of models and model building has been discussed frequently. An apt quotation appears in Rosenbluth and Wiener (1945) who wrote:

> No substantial part of the universe is so simple that it can be grasped and controlled without abstraction. Abstraction consists in replacing the part of the universe under consideration by a model of similar but simpler structure. Models ... are thus a central necessity of scientific procedure.

By a *model*, we mean an abstraction of some real system that can be used to obtain predictions and formulate control strategies. In particular, models are used to analyze one or more changes in various aspects of the modeled system that may affect other aspects of the same system.

In order to be useful, a model must necessarily incorporate elements of two conflicting attributes—realism and simplicity. On the one hand, the model should serve as a reasonably close approximation to the real system and incorporate most of the important aspects of the real system. On the other hand, the model must not be overly complex so as to preclude its understanding and manipulation. Being a formalism, a model is necessarily an abstraction.

One may think that the more detailed the model is, the better it captures reality. However, adding excessive detail to a model tends to make the solution more difficult; at best it tends to convert the method for solving the problem at hand from an analytical to an approximate numerical one. Often, it is not even necessary for the model to capture all system characteristics. All we require is that there be a high *correlation* between model predictions and real-life system performance. To ascertain whether this requirement is satisfied or not, it is important to test the model's validity.

Usually, we begin testing a model by reexamining the formulation of the problem and uncovering possible flaws. Another check on the validity of a model is to ascertain that all mathematical expressions are dimensionally consistent. A third useful test consists of varying input parameters and checking that the output from the model behaves in a plausible manner. The fourth test is the so-called *retrospective* test. It involves using historical data to reconstruct the past and then determining how well the resulting solution would have performed, if it had been used. Comparing the effectiveness of this hypothetical performance with what actually happened then indicates how well the model predicts "reality." However, a disadvantage of retrospective testing is that it uses the same data as the model. Unless the past is a representative replica of the future, it is better not to resort to this test at all. A crucial step in building an optimization-oriented model is the construction of the objective function, formulated as a mathematical function of the decision variables.

Having constructed some model for the problem under consideration, the next step is to derive a solution from this model. To this end, both *analytical* and *numerical* solution methods may be invoked. An analytical solution is usually obtained directly from its mathematical representation in the form of formulas. A numerical solution is generally an approximation via a suitable approximation procedure.

This book deals with numerical solution/estimation methods obtained via computer simulation. More precisely, we use stochastic simulation, which includes some randomness in the underlying model, rather than deterministic simulation (see below).

Computer simulation has long served as an important tool in a wide variety of disciplines, such as supersonic jet flight, telephone communications systems, wind tunnel testing, large-scale battle management (e.g., to evaluate defensive or offensive weapon systems), or maintenance operations (e.g., to determine the optimal size of repair crews), to mention a few. Although simulation is still sometimes viewed as a "method of last resort" to be employed when "everything else has failed," recent advances in simulation methodologies, software availability, sensitivity analysis, and stochastic optimization have combined to make simulation one of the most widely accepted and practiced tools in system analysis and operations research.

Naylor et al. (1966) define simulation as follows:

> Simulation is a numerical technique for conducting experiments on a digital computer, which involves certain types of mathematical and logical models that describe the behavior of business or economic system (or some component thereof) over extended periods of real time.

This definition is extremely broad, encompassing such diverse and seemingly unrelated enterprises as economic and financial modeling, wind tunnel testing of aircraft, war games, and business management games.

The motivation for simulation modeling is nicely put by Naylor et al. (1966):

> The fundamental rationale for using simulation is man's unceasing quest for knowledge about the future. This search for knowledge and the desire to predict the future are as old as the history of mankind. But prior to the seventeenth century the pursuit of predictive power was limited almost entirely to the purely deductive methods of such philosophers as Plato, Aristotle, Euclid, and others.

The following list of typical situations should give the reader some idea on where simulation would be an appropriate tool.

- The system may be so complex that a formulation in terms of simple mathematical equations may be impossible. Most economic systems fall into this category. For example, it is often virtually impossible to describe the operation of a business firm, an industry, or an economy in terms of a few simple equations. Another class of problems that leads to similar difficulties is that of large-scale complex queueing systems. Simulation has been an extremely effective tool for dealing with problems of this type.
- Even if a mathematical model can be formulated, which captures the behavior of some system of interest, it may not be possible to obtain a solution to the model by straightforward analytical techniques. Again, economic systems and complex queueing systems exemplify this type of difficulty.
- Simulation may be used as a pedagogical device for teaching both students and practitioners basic skills in system analysis, statistical analysis, and decision making. Among the disciplines in which simulation has been used successfully for this purpose are business administration, economics, medicine, and law.
- The formal exercise of designing a computer simulation model may be more valuable than the actual simulation itself. The knowledge obtained in designing a simulation study serves to crystallize the analyst's thinking and often suggests changes in the system being simulated. The effects of these changes can then be tested via simulation before implementing them in the actual system.
- Simulation can yield valuable insight into the problem of identifying which variables are important and which have negligible effect on the system, and can shed light on how these variables interact (see Chapters 5 and 6).
- Simulation can be used to experiment with new scenarios, so as to glean insight into system behavior under new circumstances.
- Simulation provides an *in-vitro* lab, allowing the analyst to discover better controls of the system under study.
- Simulation makes it possible to study dynamic systems in either real, compressed, or expanded time horizons.

As a modeling methodology, simulation modeling is by no means ideal. Some of its shortcomings and various caveats are:

- Simulation provides *statistical estimates* rather than the *exact* characteristics and performance measures of the model. Thus, simulation results are subject to uncertainty and contain "experimental errors."
- Simulation modeling is typically time-consuming and consequently expensive in terms of analyst time.

- Simulation results, no matter how precise, accurate, and impressive, provide consistently useful information about the actual system, *only* if the model is a "valid" representation of the system under study.

Computer simulation models can be classified in several ways:

1. *Static Versus Dynamic Models.* Static models are those that do not evolve in time, and therefore do not represent the passage of time. In contrast, dynamic models represent systems that evolve over time (e.g., traffic light operation).
2. *Deterministic Versus Stochastic Models.* If a simulation contains *only* deterministic (i.e., nonrandom) components, it is called *deterministic.* In a deterministic model, all mathematical and logical relationships between the elements (variables) are fixed in advance and not subject to uncertainty. A typical example is a complicated and analytically unsolvable system of standard differential equations, describing, say, a chemical reaction. In contrast, a model with at least one random input variable is called a *stochastic* model. Most queueing and inventory systems are modeled stochastically.
3. *Continuous Versus Discrete Simulation Models. Discrete* and *continuous* models are defined in the same way as discrete and continuous systems. Note that decisions on whether to use a discrete or continuous model for a given system depends on the objectives of the study. An example of a continuous simulation is a mathematical model aiming to calculate a numerical solution for a system of differential equations. Queueing network simulations are predominantly discrete simulations.

This book treats, in the main, discrete, dynamic, and stochastic simulation models, collectively called *discrete-event dynamic models* (DEDM), with the exception of Chapter 5 which addresses *discrete-event static models* (DESM). A DESM typically deals with evaluating (estimating) complex multidimensional integrals via the Monte Carlo method.

The term "Monte Carlo" was used by von Neumann and Ulam during World War II as a code word for the secret work at Los Alamos on problems related to the atomic bomb. That work involved simulation of random neutron diffusion in nuclear materials.

One of the earliest problems connected with the Monte Carlo method is Buffon's needle problem. This famous problem can be stated as follows. A needle of length l units is thrown randomly onto a floor composed of parallel planks of equal width of d units, where $d > l$. What is the probability that the needle, once it comes to rest, will cross (or touch) a crack separating the planks on the floor? It can be shown that the probability of the needle hitting a crack is $P = 2l/\pi d$, which can be estimated as the ratio of the number of throws hitting a crack to the total number of throws.

At the turn of the century, the Monte Carlo method was used to examine the Boltzmann equation. In 1908, one of the famous statistician's students used the Monte Carlo method for estimating the correlation coefficient in his t distribution.

For more details on the Monte Carlo method, see Ermakov (1976), Kalos and Wittlock (1986), and Rubinstein (1981).

It should be pointed out that recent advances in computer technology have enabled the use of *parallel* or *distributed* simulation, that is, the execution of discrete-event simulation on multiple linked (networked) computers operating simultaneously in a cooperative manner. Such an environment allows simultaneous "distribution" of different computing tasks among the individual processors, thus reducing the overall simulation time. For more details on this interesting and fast-developing area, see Heidelberger (1988) and Misra (1986) and references therein.

1.3 DISCRETE-EVENT SIMULATION

Discrete-event simulation deals with discrete-event dynamic systems (DEDS) modeling, and computes the evolution of their trajectories over time. Over an infinite time horizon, these systems change their state only at a countable number of time points. (In practical simulations, the number of state changes is, of course, finite). State changes are triggered by the execution of simulation events occurring at the corresponding time points; an *event* is a collection of attributes (values, flags, etc.), chief among which is the *event occurrence time* and an associated algorithm to execute state changes.

Example 1.3.1. Consider a single-server queue, say, an information desk in a hotel. Assume that we are interested in estimating the average number of customers waiting for information. The state variables for such a DES model might be the time of arrival of each customer and the status of the server (busy or idle). The arrival time of a customer is needed to compute the number of customers in the queue, which is incremented by one upon each customer arrival. The status of the server is needed to determine whether the newly arrived customer can be served immediately or must join the end of the queue. It is readily seen that there are two types of events in this simple system: customer arrival at the system and customer departure from it, immediately following service completion. Clearly, both are events, since they cause the number of customers to be incremented and decremented by one, respectively.

1.3.1 Simulation Clock and Event List

Because of their dynamic nature, DEDS require a time-keeping mechanism to *advance the simulated time* from one event to another, as the simulation unfolds in time. The variable recording the current simulation time is called the *simulation clock*. To keep track of events, the simulation maintains a list of all pending events. This list is called the *event list*, and its task is to maintain all pending events in chronological order, that is, events are inserted into it ordered by their time of occurrence. In particular, the most imminent event is always located at the head of the event list.

Initially, the simulation clock is set to zero, and the initial event(s) are loaded into the event list (chronologically ordered). Next, the most imminent event is unloaded from the event list for execution, and the simulation clock is advanced to its occurrence time. In the course of executing the current event, the state of the system is updated, and future events are typically generated and loaded into the event list. The process of unloading events from the event list, advancing the simulation clock, and executing the next most imminent event terminates when some specified stopping condition is met, say, as soon as a prescribed number of customers depart from the system.

This approach for advancing the simulation time is called the *next-event time advance* approach. The following example illustrates this approach for the single-server queueing model of Example 1.3.1.

Example 1.3.2. Define the following random variables (variates):

t_K is the clock time of arrival of customer K (by convention, $K_0 = 0$ is not a customer arrival time).

$A_K = t_K - t_{K-1}$ is the interarrival time, separating the arrivals of customer $(K-1)$ and customer K.

S_K is the service time of customer K; during this time, the server is unavailable to serve other customers.

D_K is the delay (sojourn time) of customer K in the system. (D_K equals the delay of customer K in the queue plus the service time S_K.)

$t'_K = t_K + D_K$ is the completion time, that is, the simulation clock time when customer K completed service and departed from the system.

E_K is the time of occurrence of the Kth event (by convention $E_0 = 0$ does not correspond to an event time).

V_t is the virtual time in the system (time to serve all customers) at time t.

Assume that the interarrival times, $A_1, A_2, \ldots$, and the service times, $S_1, S_2, \ldots$, are random variables, generated from some given cumulative distribution functions (cdf's) F_1 and F_2, respectively. Assume further that at time $t_0 = E_0 = 0$ the server is idle, so that at time t_1 the first customer arrives for service at an empty system (see Figure 1.1). The event list is initialized by an arrival event with occurrence time $t_1 = A_1$, and t_1 is determined by generating a random variable A_1 from F_1. This event is unloaded for execution, and the simulation clock is advanced from 0 to the time of the first event $E_1 = t_1$. Since the first customer arrives at an idle system, his delay in the system is $D_1 = S_1$, and the status of the server at time t_1 changes from idle to busy. Clearly, a service completion event will be scheduled at time $t'_1 = D_1$ and the next arrival event will take place at time $t_2 = t_1 + A_2$. If $t_2 < t'_1$, as depicted in Figure 1.1, the simulation clock is advanced from time E_1 to the next arrival event, that is, we set $E_2 = t_2$. Note that the number of customers in the system would be then incremented from 1 to 2. Since the customer arriving at time t_2 finds the server busy, we record t_2 and compute $t_3 = t_2 + A_3$, the time of the third arrival.

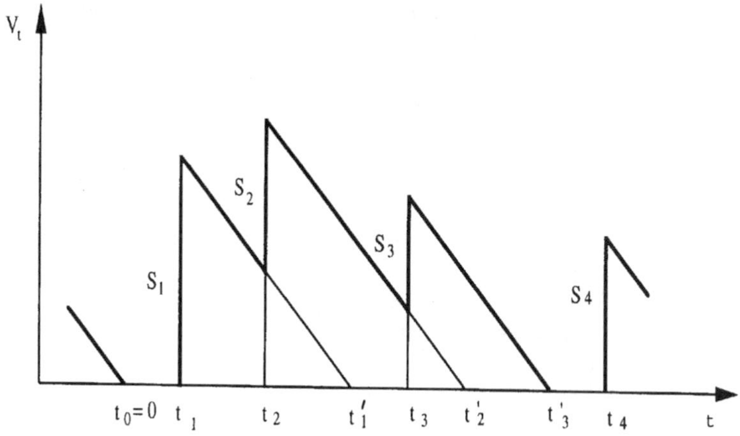

Figure 1.1. Sample path realization of the virtual time process, $\{V_t\}$, in a queueing system.

If $t_3 > t'_1$, as depicted in Figure 1.1, we advance the simulation clock from E_2 to the time of the next event, $E_3 = t'_1$. Executing this event means that the second customer completes his service and departs from the system, while the first customer in the queue (the second customer to arrive) begins service at the same time. At this point, we generate S_2 from F_2, calculate the sojourn time D_2 as $D_2 = t'_1 - t_2 + S_2$ and decrement the number of customers in the system from 2 to 1. If $t_3 < t'_2$, as in Figure 1.1, the simulation clock is advanced from E_3 to the time of the next event, $E_4 = t_3$, and so forth. In Figure 1.1, we terminate our "simulation" after executing the event at time $E_7 = t_4$, which takes place when the fifth customer arrives at an idle server. The event time $E_7 = t_4$ follows event times $E_5 = t'_2$ and $E_6 = t'_3$.

Note that the event time $E_6 = t'_3$ corresponds to the time the server becomes idle again, having completed a busy period of length $t'_3 - t_1$. The time between the onset of two consecutive busy periods is called a *cycle*. Clearly, each cycle is composed of a busy period and an idle period in succession. In our case (see Figure 1.1):

$$\text{cycle} = \text{busy period} + \text{idle period} = (t_4 - t_1) = (t'_3 - t_1) + (t_4 - t'_3).$$

Figure 1.1 actually depicts a sample path of the *virtual waiting time* (unfinished work), with V_t defined as the time to complete service for all customers present in the queueing system at time t. The evolution of the process $\{V_t\}$ is characterized by upward jumps (customer arrivals) followed by downward linear decreases (work depletion). Figure 1.2 depicts the associated sample path of the number of customers, N_t, present in the queueing system at time t. Sample paths of the process $\{N_t\}$ are piecewise constant with upward and downward unit jumps representing customer arrivals and departures, respectively.

More details on discrete-event simulation are found in Bratley et al. (1987), Banks and Garson (1984), Kleijnen (1987, 1991), Kleijnen and Van Groenendaal (1992), Law and Kelton (1991), Pegden (1989), Pollatschek (1996), Pritsker (1992), and Yakovitz (1977).

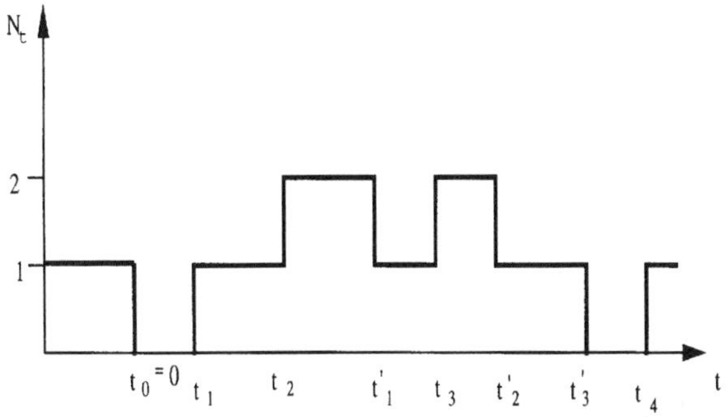

Figure 1.2. Sample path realization of the number of customers process, $\{N_t\}$, in a queueing system.

REFERENCES

Banks, J. and Carson, J. S. (1984). *Discrete-Event System Simulation*, Prentice-Hall, Englewood Cliffs, NJ.

Bratley, P., Fox, B. L., and Schrage, L. E. (1987). *A Guide to Simulation*, 2nd ed., Springer-Verlag, New York.

Ermakov, J. M. (1976). *Monte Carlo Method and Related Questions*, Nauka, Moscow (in Russian).

Feller, W. (1958). *An Introduction to Probability Theory and Its Application*, Vol. 1, Wiley, New York.

Heidelberger, P. (1988). "Discrete event simulations and parallel processing: Statistical properties," *SIAM J. Statist. Comput.*, **9**, 1114–1132.

Kleijnen, J. P. C. (1987). *Statistical Tools for Simulation Practitioners*, Marcel Dekker, New York.

Kleijnen, J. P. C. (1991). "Sensitivity analysis of simulation experiments: Tutorial on regression analysis and statistical design," Tilburg University, Tilburg, The Netherlands.

Kleijnen, J. P. C. and Van Groenendaal, W. (1992). *Simulation: A Statistical Perspective*, Wiley, Chichester.

Law, A. M. and Kelton, W. D. (1991). *Simulation Modeling and Analysis*, McGraw-Hill, New York.

Misra, J. (1986). "Distributed discrete-event simulation," *Computing Surveys*, **18**, 39–65.

Naylor, T. J., Balintfy, J. L., Burdick, D. S., and Chu, K. (1966). *Computer Simulation Techniques*, Wiley, New York.

Pegden, C. D. (1989). *Introduction to SIMAN* (January 1989 version), Systems Modeling Corp., Sewickley, PA.

Pollatschek, M. A. (1996). *Programming Discrete Simulations* R&D Books, Lawrence, KS.

Pritsker, A. A. B. (1992). *Introduction to Simulation and SLAM II*, 3rd ed., Systems Publishing, West Lafayette, IN.

Rosenbluth, A. and Wiener, N. (1945). "The role of models in science," *Phil. Sci.*, **XII**(4), 316–321.

Rubinstein, R. Y. (1981). *Simulation and the Monte Carlo Method*, Wiley, New York.

Yakowitz, S. J. (1977). *Computational Probability and Simulation*, Addison-Wesley, Reading, MA.

CHAPTER 2

Random Numbers, Variates, and Stochastic Process Generation

This chapter deals with computer generation of pseudorandom numbers (random numbers, for short) and random variables (variates, for short). In this book, the terms random variables and variates are largely interchangeable; however, the former term connotes the formal concept, while the latter is used mainly in simulation context as the computer-generated realization of the former. Section 2.1 presents a brief discussion on generation of variate sequences, whose marginals are uniformly distributed on the interval (0, 1); such random numbers are available on most computers. Section 2.2 deals with generation methods of one-dimensional variates from prescribed distributions: the *inverse-transform* method, the *composition* method, and the *acceptance-rejection* method. Section 2.3 deals with the multidimensional case. In particular, Section 2.3.1 discusses cases where the one-dimensional inverse-transform method can be extended to a multidimensional version. Section 2.3.2 deals with the multidimensional version of the acceptance-rejection method for generating random vectors from arbitrary distributions, while Section 2.3.3 describes random vector generation from multinormal distributions. Sections 2.4 and 2.5 describe specific generation methods for continuous and discrete variates, respectively, while Section 2.6 deals with generation of stochastic processes. Section 2.7 describes generation of points, uniformly distributed over different regions. Finally, Section 2.8 deals with generation of autocorrelated sequences; it describes a versatile class of stochastic sequences, called TES processes, and input analysis based on them.

2.1 RANDOM NUMBER GENERATION

In a typical stochastic simulation, randomness is introduced into simulation models via sequences $U_1, U_2, \ldots,$ of independent random variables, uniformly distributed on the interval (0, 1). This chapter details how such random numbers are generated and discusses some of the generator properties. These random numbers are used to generate samples of random variables from specified distributions (Devroye, 1986) in order to characterize system behavior. For example, for a queueing model, one needs to specify the interarrival and service time distributions in order to drive the simulation over time; and for a stochastic inventory model, one needs to specify the distribution of random demands.

In the early days of simulation, randomness was generated by *manual* techniques, such as coin flipping, dice rolling, card shuffling, and roulette spinning. Later on, *physical devices*, such as noise diodes and Geiger counters, were attached to computers for the same purpose. The prevailing belief held that *only* mechanical or electronic devices could produce "truly" random sequences. Although mechanical devices are still widely used in gambling and lotteries, these methods were abandoned by the computer simulation community for several reasons: Mechanical methods are too slow for general use; the generated sequences cannot be reproduced; and, most importantly, it has been found that the generated numbers exhibit both bias and dependence.

As computer simulation is becoming ever more widespread, increasing attention has been paid to random number generators (RNG), which are suitable for computer implementation. Perhaps the earliest such method is due to von Neumann who suggested that an n-digit number be squared and that the middle n digits of the result be used as the next number in the sequence. Most RNGs are quite fast, can readily reproduce a given sequence of random numbers, and require little storage space. In fact, the RNGs used in practice produce *deterministic* sequences. Still, they capture certain statistical properties, mimicking *true* random sequences. For this reason, the RNGs are called *pseudorandom*. In practice, however, it is sufficient that the joint distributions of a *pseudorandom sequence* $(U_{i+1}, \ldots, U_{i+k})$, $i = 1, 2, \ldots,$ be close to uniformity over $(0, 1)^k$, say, for $k \leq 6$ (see Ripley, 1983).

The most common methods for generating pseudorandom sequences use the so-called *linear congruential generators*, introduced in Lehmer (1951). These generate a deterministic sequence of numbers by means of a recursive formula that is based on calculating the modulo-m residues of a linear transformation, for some integer m. More precisely, congruential generators are based on a fundamental congruence relationship, given by

$$X_{i+1} = aX_i + c \pmod{m}, \tag{2.1.1}$$

where the initial value, X_0, is called the *seed*, a and m are positive integers, and c is a nonnegative integer.

Note that applying the modulo-m operator in (2.1.1) means that $aX_i + c$ is divided by m, and the remainder is taken as the value of X_{i+1}. That is, each X_i can only assume a value from the set $\{0, 1, \ldots, m-1\}$, and the quantities

$$U_i = \frac{X_i}{m}, \tag{2.1.2}$$

called *pseudorandom numbers*, constitute approximations to the true sequence of uniform random variables. In the special case where $c = 0$, formula (2.1.1) simply reduces to

$$X_i = aX_i \pmod{m}. \tag{2.1.3}$$

Such a generator is called a *multiplicative congruential generator* to distinguish it from (2.1.1), which is called a *mixed congruential generator* (as it involves both a multiplicative and an additive term).

A recursive scheme of the form (2.1.1) or (2.1.3) is called a *congruential method*; the integers a, c, and m are called the *multiplier*, the *increment*, and the *modulus*, respectively. Sequences thus generated are called pseudorandom sequences.

It follows from (2.1.1) and (2.1.2) that the sequence $\{U_i\}$ is completely deterministic, provided the four values X_0, a, c, and m are fixed. Although the sequence $\{U_i\}$ is completely deterministic, we may accept it as a random sequence if it appears to be "sufficiently random" in the sense that the generated numbers, $U_1, U_2, \ldots$, appear to be uniformly distributed and statistically independent. It is readily seen that an arbitrary choice of X_0, a, c and m will not lead to a pseudorandom sequence with good statistical properties. In fact, number theory has been used to show that only a few combinations of these produce satisfactory results.

Note also that the sequence $\{X_i\}$ will repeat itself in at most m steps and will therefore be periodic with period not exceeding m. For example, let $a = c = X_0 = 3$ and $m = 5$; then the sequence obtained from the recursive formula $X_{i+1} = 3X_i + 3 \pmod 5$ is $X_i = 3, 2, 4, 0, 3$. Indeed, the period of the generator in the above example is 4, while $m = 5$.

Because of the deterministic character of the sequence, the entire sequence recurs as soon as any number is repeated. This is referred to as "getting into a loop," that is, there is a cycle of numbers that is repeated endlessly. Knuth (1981) shows that all sequences of the form $X_{i+1} = f(X_i)$ get into a loop. One wants, of course, to choose m as large as possible to ensure a sufficiently long sequence of distinct numbers in a cycle. To do so, let p be the period of the sequence. (When p equals its maximum m, we say that the random number generator has a *full period*.) It is shown in Knuth (1981) that the generator defined in (2.1.1) has a full period, m, if and only if:

1. c is *relatively prime* to m, that is, c and m have no common divisors greater than one.
2. $a \equiv 1 \pmod g$ for every prime factor g of m.
3. $a \equiv 1 \pmod 4$ if m is a multiple of 4.

(Readers not acquainted with the congruence relation $\equiv$, should read $=$ instead.) Condition 1 means that the greatest common divisor of c and m is unity. Condition 2 means that $a = g\lfloor a/g \rfloor + 1$. Let g be a prime factor of m; then letting

$k = \lfloor a/g \rfloor$, one has

$$a = 1 + gk. \tag{2.1.4}$$

Condition 3 means that

$$a = 1 + 4\lfloor a/4 \rfloor \tag{2.1.5}$$

if $m/4$ is an integer.

In computer implementations, m is selected as a large prime number that can be accommodated by the computer word size. For example, in a binary 32-bit word computer, statistically acceptable generators can be obtained by choosing $m = 2^{31} - 1$ and $a = 7^5$, provided the first bit is a sign bit. A 64-bit word computer or even a 48-bit word computer will naturally yield better statistical results.

Formulas (2.1.1), (2.1.2), and (2.1.3) can be readily extended to pseudorandom vector generation. For example, the multidimensional version of (2.1.3) and (2.1.2) can be written as

$$\boldsymbol{X}_{i+1} = \boldsymbol{A}\boldsymbol{X}_i \pmod{\boldsymbol{M}}, \tag{2.1.6}$$

and

$$\boldsymbol{U}_{i+1} = \boldsymbol{M}^{-1}\boldsymbol{X}_i, \tag{2.1.7}$$

respectively, where $\boldsymbol{A}$ is a nonsingular $n \times n$ matrix, $\boldsymbol{M}$ is a vector of constants of the same dimensionality as the vector $\boldsymbol{X}$, and $\boldsymbol{M}^{-1}\boldsymbol{X}_i$ is an n-dimensional vector with components $M_1^{-1}X_1, \ldots, M_n^{-1}X_n$. The properties of pseudorandom vector generators and the choice of their parameters, $\boldsymbol{M}$ and $\boldsymbol{A}$, are discussed in Section 10.1 of Niederreiter (1992).

Beside the linear congruential generators, other classes have been proposed, so as to achieve longer periods and better statistical properties (see L'Ecuyer, 1990).

A number of statistical tests of randomness for pseudorandom sequences, such as the serial test, run test, spectral test, chi-squared goodness-of-fit test, and the Kolmogorov-Smirnov test, have been proposed and discussed in the literature on random number generation (e.g., Law and Kelton, 1991).

Most computer languages already contain a built-in pseudorandom number generator. The user is typically requested only to input the initial seed, X_0, and upon invocation, the RNG produces a sequence of independent uniform (0, 1) variates. Instead of exploring the theoretical and practical aspects of pseudorandom number generation [see Law and Kelton (1991), L'Ecuyer and Tezuka (1991), L'Ecuyer (1990), Niederreiter (1991, 1992), Knuth (1981) and references therein], we assume the availability of a "black box," capable of producing pseudorandom numbers on demand.

2.2 VARIATE GENERATION FROM PRESCRIBED DISTRIBUTIONS

This section treats generation of one-dimensional variates from a prescribed distribution. We consider here the inverse-transform method, the composition method, and the acceptance-rejection method.

In the algorithms outlined below, the keyword "return" signals that the requisite random number has been obtained. The algorithm terminates implicitly if only a single observation is desired; otherwise, it is understood to loop back to the beginning to compute more random numbers. The notation $X \sim F$ means that the variate X has distribution F.

2.2.1 Inverse-Transform Method

Let X be a variate with cumulative distribution function (cdf) $F(x)$. Since $F(x)$ is a nondecreasing function, the inverse function $F^{-1}(y)$ may be defined as

$$F^{-1}(y) = \inf\{x\colon F(x) \geq y\}, \qquad 0 \leq y \leq 1. \tag{2.2.1}$$

(Readers not acquainted with the notion of inf (infimum) should read min).

It is easy to prove that if U is uniformly distributed over the interval [0, 1] [denoted, hereafter, by $\mathcal{U}(0, 1)$], then (Figure 2.1)

$$X = F^{-1}(U) \tag{2.2.2}$$

has cdf $F(x)$.

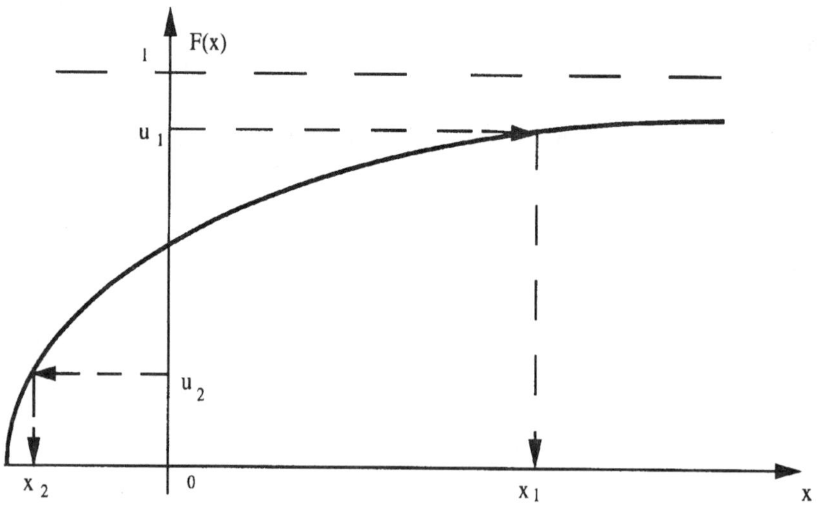

Figure 2.1. Inverse-transform method.

In view of the fact that F is invertible and $P(U \leq u) = u$, we readily obtain

$$P(X \leq x) = P[F^{-1}(U) \leq x] = P[U \leq F(x)] = F(x). \tag{2.2.3}$$

Thus, to generate a value, say x, of a random variate X with cdf $F(x)$, first sample a value, say u, of a uniform variate, U, compute $F^{-1}(u)$, and set it equal to x.

Figure 2.1 illustrates the inverse-transform method given by the following algorithm.

Algorithm 2.2.1 Inverse-Transform Method

1. *Generate U from $\mathscr{U}(0, 1)$.*
2. *Return $X = F^{-1}(U)$.*

Example 2.2.1. Generate a variate from the probability density function (pdf)

$$f(x) = \begin{cases} 2x, & 0 \leq x \leq 1 \\ 0, & \text{otherwise.} \end{cases} \tag{2.2.4}$$

The cdf is

$$F(x) = \begin{cases} 0, & x < 0 \\ \displaystyle\int_0^x 2x\, dx = x^2, & 0 \leq x \leq 1 \\ 1, & x > 1. \end{cases}$$

Applying (2.2.2), we have

$$X = F^{-1}(U) = U^{1/2}, \qquad 0 \leq u \leq 1.$$

Therefore, to generate a variate X from the pdf (2.2.4), first generate a variate U from $\mathscr{U}(0, 1)$, and then take its square root.

Example 2.2.2. Generate variates from the order statistics $Y_n = \max(X_1, \ldots, X_n)$ and $Y_1 = \min(X_1, \ldots, X_n)$, respectively, where $X_1, \ldots, X_n$ are independent and identically distributed (iid) variates with cdf $F(x)$.

The distributions of Y_n and Y_1 are

$$F_n(y) = [F(y)]^n$$

and

$$F_1(y) = 1 - [1 - F(y)]^n,$$

respectively. Applying (2.2.2), we get

$$Y_n = F^{-1}(U^{1/n})$$

and, since $1 - U$ is also from $\mathscr{U}(0, 1)$,

$$Y_1 = F^{-1}(1 - U^{1/n}).$$

In the special case where $X = U$ we have

$$Y_n = U^{1/n}$$

and

$$Y_1 = 1 - U^{1/n}.$$

Example 2.2.3 Generating Variates from a Discrete Distribution. Let X be a discrete variate with probability mass function (pmf)

$$p(x_i) = P(X = x_i), \qquad i = 1, 2, \ldots,$$

and cdf

$$F(x_k) = \sum_{x_i \le x_k} p(x_i), \qquad k = 1, 2, \ldots,$$

respectively.

The algorithm for generating a variate from $F(x)$ (see Figure 2.2) can be written as:

Algorithm 2.2.2 Inverse-Transform Method for a Discrete Distribution

1. *Generate* $U \sim \mathscr{U}(0, 1)$.
2. *Find the smallest positive integer, k, such that* $U \le F(x_k)$ *and return* $X = x_k$.

Much of the execution time in Algorithm 2.2.2 is spent in making the comparisons of step 2. This time can be reduced by using efficient search techniques (see Devroye, 1986).

Example 2.2.4 Generating Variates from an Empirical Distribution. Assume that our data can be grouped to fall into n adjacent intervals $[a_0, a_1)$,

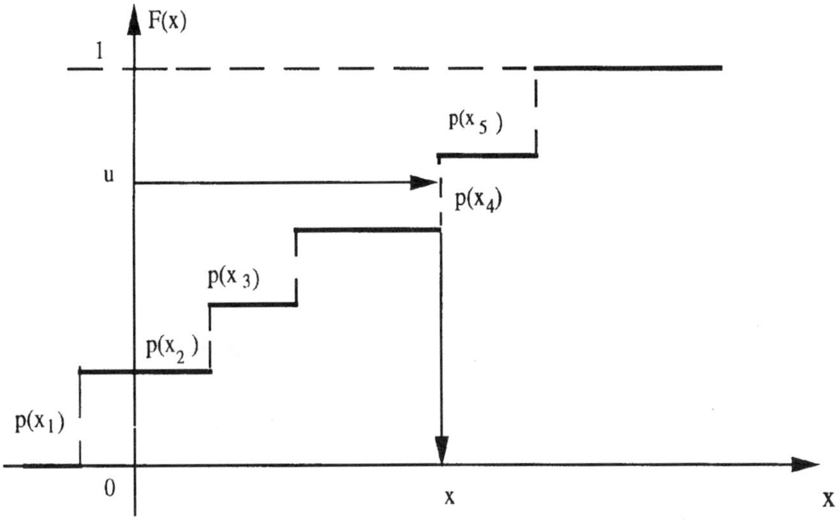

Figure 2.2. Inverse-transform method for a discrete variate.

$[a_1, a_2), \ldots, [a_{n-1}, a_n]$ and define the following piecewise-linear empirical distribution function:

$$F(x) = \begin{cases} 0, & x < a_0 \\ F(a_{i-1}) + \dfrac{x - a_{i-1}}{a_i - a_{i-1}}[F(a_i) - F(a_{i-1})], & a_{i-1} \leq x < a_i \text{ for some } 1 \leq i \leq n \\ 1, & x > a_n, \end{cases}$$

where $F(a_0) = 0$, $F(a_i) = \sum_{j=1}^{i} r_j/r$, $i = 1, 2, \ldots, n$, r_i is the number of observations in the ith interval $[a_{i-1}, a_i)$ and $r = \sum_{i=1}^{n} r_i$.

The algorithm for generating a variate from an empirical distribution $F(x)$ (see Figure 2.3) can be written as:

Algorithm 2.2.3 Inverse-Transform Method for an Empirical Distribution (Histogram)

1. *Generate* $U \sim \mathscr{U}(0, 1)$.
2. *Find the smallest positive integer* k $(0 \leq k \leq n - 1)$, *such that* $U \leq F(x_k)$, *and return*

$$X = F^{-1}(U) = a_k + [U - F(a_k)](a_{k+1} - a_k)/[F(a_{k+1}) - F(a_k)].$$

In general, the inverse-transform method requires that the underlying cdf, $F(x)$, exist in a form for which the corresponding inverse function $F^{-1}(\cdot)$ can be found analytically or algorithmically. Applicable distributions are, for example, exponential, uniform, Weibull, logistic, and Cauchy. Unfortunately, for many

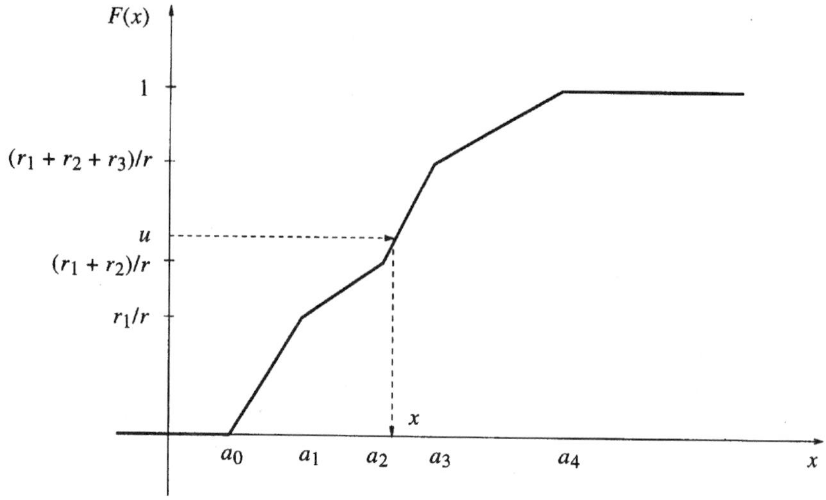

Figure 2.3. Inverse-transform method for an empirical distribution.

probability distributions, it is either impossible or difficult to find the inverse transform, that is, to solve

$$F(x) = \int_{-\infty}^{x} f(t)\, dt = u$$

with respect to x. Even in the case when F^{-1} exists in an explicit form, the inverse-transform method may not necessarily be the most efficient variate generation method (see Devroye, 1986).

*2.2.2 Composition Method

This method assumes that a cdf, $F(x)$, can be expressed as a probabilistic mixture (convex combination) of distributions, H_i, namely

$$F(x) = \sum_{i=1}^{n} p_i H_i(x), \tag{2.2.5}$$

where

$$0 < p_i \leq 1, \qquad \sum_{i=1}^{n} p_i = 1.$$

Let X_i have distribution function H_i, for $1 \leq i \leq n$. Then the random variable X with cdf $F(x)$ can be represented as

$$X = \begin{cases} X_1 \text{ with probability } p_1, \\ \quad\vdots \\ X_n \text{ with probability } p_n. \end{cases}$$

It is readily seen that in order to generate a variate X from $F(x)$, we must first generate a discrete variate Y, with discrete pdf $P(Y = i) = p_i$, $i = 1, \ldots, n$, and then generate the requisite variate, X; equivalently, given $Y=i$, X has distribution function $H(x|i) = P(X \leq x|Y = i)$. To see that, note that using conditioning, we can rewrite (2.2.5) as

$$F(x) = \sum_{i=1}^{n} P(X \leq x|Y = i)P(Y = i).$$

This approach is called the composition method. The corresponding algorithm is straightforward.

Algorithm 2.2.4 Composition Method

1. *Generate a variate (positive integer) Y from the discrete pdf:*

$$P(Y = i) = p_i, \qquad i = 1, \ldots, n.$$

2. *Return X from the cdf $H(x|i)$.*

The reader should show that X has a cdf given by 2.2.5.

The composition method can be readily extended to $n = \infty$ and to the case where Y is a continuous variate and the pdf of X can be expressed as

$$f(x) = \int h(x|y)p(y)\, dy.$$

Example 2.2.5. Generate a variate from

$$f(x) = n \int_1^{\infty} y^{-n} e^{-yx}\, dy.$$

In this case, we define the pdf's $h(x|y)$ and $p(y)$ as

$$h(x|y) = ye^{-yx}, \qquad x \geq 0$$

and

$$p(y) = \frac{n}{y^{n+1}}, \qquad 1 \leq y \leq \infty, \quad n \geq 1,$$

respectively. According to Algorithm 2.2.4, we first generate a variate Y from $p(y)$. Once this Y is determined, we next generate the desired variate X from the conditional pdf $h(x|y) = ye^{-yx}$, $x \geq 0$.

In summary, in order to generate a variate from a prescribed pdf, $f(x)$, we need to perform the following steps:

1. Generate independent variates U_1 and U_2 from $\mathcal{U}(0, 1)$.
2. Generate a variate Y from the pdf $p(y)$, that is, take $Y = F_Y^{-1}(U_1) = U_1^{-1/n}$. where F_Y is the cdf of Y.
3. Generate a variate X from the conditional cdf $H(x|y)$; equivalently, X is the solution of $U_2 = H^{-1}(X|Y)$ which equals $X = -(1/Y)\ln U_2$ [recall that if U is from $\mathcal{U}(0, 1)$, then so is $1 - U$].

2.2.3 Acceptance-Rejection Method

The inverse-transform and the composition methods are direct methods in the sense that they deal directly with the cdf of the variate to be generated. The acceptance-rejection method (ARM), to be presented next, is an indirect method, due to John von Neumann. It may be appealed to when the above-mentioned direct methods either fail or turn out to be computationally inefficient. An ARM, however, calls for specifying a function ϕ that *majorizes* the original pdf, $f(x)$.

To carry out the acceptance-rejection method, we represent $f(x)$ as

$$f(x) = Ch(x)g(x), \tag{2.2.6}$$

where $C \geq 1$, $h(x)$ is a pdf, $\phi(x) = Ch(x)$ majorizes the pdf $f(x)$ [i.e., $Ch(x) \geq f(x)$ for all x], and $0 < g(x) = f(x)/\phi(x) \leq 1$ (see Figure 2.4). Now, generate two variates, U from $\mathcal{U}(0, 1)$ and Y from $h(y)$, and test the inequality $U \leq g(Y)$. If the inequality holds, then we accept Y as the requisite variate from $f(x)$; otherwise, we reject the pair (U, Y) and try again until successful. The acceptance-rejection algorithm can be written as follows.

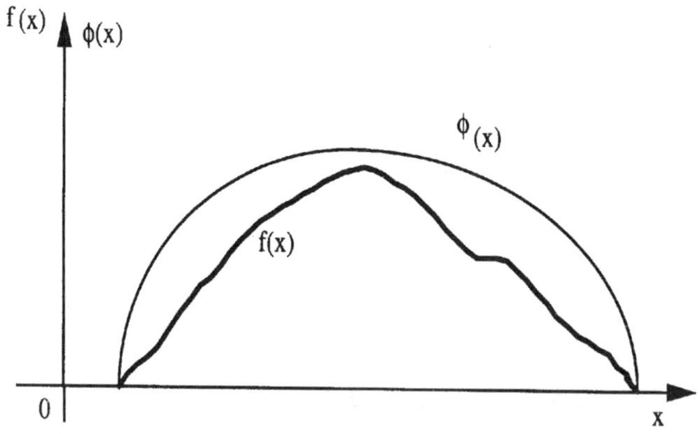

Figure 2.4. Acceptance-rejection method.

Algorithm 2.2.5 Acceptance-Rejection Method

1. *Generate* U *from* $\mathscr{U}(0, 1)$.
2. *Generate* Y *from* $h(y)$, *independent of* U.
3. *If* $U \leq g(Y)$, *return* $X = Y$. *Otherwise, go to step 1.*

Example 2.2.6 (Example 2.2.1 continued). We shall show how to generate a variate from the pdf

$$f(x) = \begin{cases} 2x, & 0 \leq x \leq 1 \\ 0, & \text{otherwise} \end{cases}$$

using the acceptance-rejection method. For simplicity, take $h(y) = 1$, $(0 \leq y \leq 1)$ and $C = 2$. In this case, $g(x) = \frac{1}{2}f(x) = x$, and Algorithm 2.2.5 becomes:

1. Generate U from $\mathscr{U}(0, 1)$.
2. Generate Y from $\mathscr{U}(0, 1)$ independent on U.
3. If $U \leq Y$, return $X = Y$. Otherwise, go to step 1.

Note that this example is merely illustrative, since the inverse-transform method handles it efficiently.

The theoretical basis of ARM is provided by the following theorem:

Theorem 2.2.1. *Let* X *be a random variable distributed according to the pdf* $f(x)$. *Assume that* $f(x)$ *can be represented as*

$$f(x) = Ch(x)g(x),$$

where $C \geq 1$, $h(x)$ *is a pdf, and* $0 < g(x) \leq 1$. *Let* U *and* Y *be distributed according to* $\mathscr{U}(0, 1)$ *and* $h(y)$, respectively. *Then*

$$h(x|U \leq g(Y)) = f(x), \tag{2.2.7}$$

where $h(x|U \leq g(Y)) = (\partial/\partial x)P(Y \leq x|U \leq g(Y))$.

Proof. By Bayes's formula

$$h(x|U \leq g(Y)) = \frac{P(U \leq g(Y)|Y = x)h(x)}{P(U \leq g(Y))}. \tag{2.2.8}$$

Direct computations yield

$$P(U \leq g(Y)|Y = x) = P(U \leq g(x)) = g(x) \tag{2.2.9}$$

and

$$P(U \le g(Y)) = \int P(U \le g(Y) | Y = x) h(x)\, dx$$
$$= \int g(x) h(x)\, dx = \int \frac{f(x)}{C}\, dx = \frac{1}{C}. \tag{2.2.10}$$

Upon substitution of (2.2.9) and (2.2.10) into (2.2.8), we obtain

$$h(x | U \le g(Y)) = C h(x) g(x) = f(x).$$

The efficiency of the acceptance-rejection method is determined by the probability $p = P(U \le g(Y))$ [see (2.2.10)]. Since the trials are independent, the probability of success in each trial is $p = 1/C$. The number of trials, N, before a successful pair (U, Y) occurs has a geometric distribution,

$$P(N = n) = p(1 - p)^{n-1}, \qquad n = 1, \ldots, \tag{2.2.11}$$

with the expected number of trials equal to $1/p = C$.

For this method to be of practical interest, the following criteria must be used in selecting $h(x)$:

1. It should be easy to generate a variate from $h(x)$.
2. The efficiency, $1/C$, of the procedure should be large, that is, C should be close to 1 [which occurs when $h(x)$ is close to $f(x)$].

The maximal efficiency is attained when $f(x) = \phi(x)$, for all x. In this case, $1/C = C = 1$, $g(x) = 1$, $h(x) = f(x)$ and the acceptance-rejection method is obviated, since generating a variate from $f(x)$ is the same as generating it from $h(x)$.

Example 2.2.7. Generate a variate from

$$f(x) = \frac{2}{\pi R^2} \sqrt{R^2 - x^2}, \qquad -R \le x \le R.$$

Take $h(x) = 1/2R(-R < x < R)$ and $C = 4/\pi$. Then the inequality $f(x) \le 2/\pi R = C h(x) = \phi(x)$ holds, $g(x) = f(x)/\phi(x) = \sqrt{1 - (x/R)^2}$, and Algorithm 2.2.5 becomes:

1. Generate two independent variates U_1 and U_2 from $\mathscr{U}(0, 1)$.

2. Use U_2 to generate Y from $h(y)$ via the inverse-transform method, namely, $Y = (2U_2 - 1)R$, and calculate

$$g(Y) = \frac{f(Y)}{\phi(Y)} = \sqrt{1 - (2U_2 - 1)^2}.$$

3. If $U_1 \le g(Y)$, which is equivalent to $(2U_2 - 1)^2 \le 1 - U_1^2$, then return $X = Y = (2U_2 - 1)R$; otherwise go to step 1.

It can be shown that the expected number of trials is $C = 4/\pi$, and the efficiency is $1/C = \pi/4 = 0.785$.

2.3 RANDOM VECTOR GENERATION

2.3.1 Vector Inverse-Transform Method

Suppose that we wish to generate a random vector $\boldsymbol{X} = (X_1, \ldots, X_n)$ from a given cdf $F(\boldsymbol{x})$. We distinguish between the following two cases: (1) the variates $X_1, \ldots, X_n$ are *independent*, and (2) the variates $X_1, \ldots, X_n$ are *dependent.*

1. For independent variates $X_1, \ldots, X_n$, the joint pdf, $f(\boldsymbol{x})$, is

$$f(\boldsymbol{x}) = f(x_1, \ldots, x_n) = \prod_{i=1}^{n} f_i(x_i), \tag{2.3.1}$$

where $f_i(x_i)$ is the marginal pdf of the variate X_i. It is easy to see that in order to generate the random vector $\boldsymbol{X} = (X_1, \ldots, X_n)$ from cdf $F(\boldsymbol{x})$, we can apply the inverse-transform method

$$X_i = F^{-1}(U_i), \qquad i = 1, \ldots, n, \tag{2.3.2}$$

for each variate separately.

Example 2.3.1. Let X_i, $1 \le x_i \le n$, be independent variates with the pdf's

$$f_i(x_i) = \begin{cases} \dfrac{1}{b_i - a_i}, & a_i \le x_i \le b_i, \quad i = 1, \ldots, n \\ 0, & \text{otherwise.} \end{cases}$$

To generate the random vector $\boldsymbol{X} = (X_1, \ldots, X_n)$ with the joint pdf

$$f(x_1, \ldots, x_n) = \begin{cases} \displaystyle\prod_{i=1}^{n} \frac{1}{(b_i - a_i)}, & (x_1, \ldots, x_n) \in D \\ 0, & \text{otherwise} \end{cases}$$

where $D = \{(x_1, \ldots, x_n)\colon a_i \le x_i \le b_i, \quad i = 1, \ldots, n\}$, we apply the inverse-transform formula (2.3.2) and get $X_i = a_i + (b_i - a_i)U_i, \quad i = 1, \ldots, n$, where $U_1, \ldots, U_n$ are iid from $\mathcal{U}(0, 1)$.

2. For dependent variates $X_1, \ldots, X_n$, represent the joint pdf $f(x)$ as

$$f(x_1, \ldots, x_n) = f_1(x_1)f_2(x_2|x_1)\cdots f_n(x_n|x_1, \ldots, x_{n-1}), \tag{2.3.3}$$

where $f_1(x_1)$ is the marginal pdf of X_1, and $f_k(x_k|x_1, \ldots, x_{k-1})$ is the conditional pdf of X_k given $X_1 = x_1, X_2 = x_2, \ldots, X_{k-1} = x_{k-1}$. Let $U_1, \ldots, U_n$ be iid variates from $\mathcal{U}(0, 1)$. It is readily seen that the vector $\boldsymbol{X} = (X_1, \ldots, X_n)$, obtained from the solution of the following system of equations

$$\begin{aligned} F_1(X_1) &= U_1 \\ F_2(X_2|X_1) &= U_2 \\ &\vdots \\ F_n(X_n|X_1, \ldots, X_{n-1}) &= U_n. \end{aligned} \tag{2.3.4}$$

is distributed according to the joint cdf, $F(x)$.

The algorithm for generating random vectors from (2.3.3) consists of just two steps:

Algorithm 2.3.1 Vector Inverse-Transform Method

1. *Generate n iid variates,* $U_1, \ldots, U_n$, *from* $\mathcal{U}(0, 1)$.
2. *Solve the system of Eqs. (2.3.4) with respect to* $\boldsymbol{X} = (X_1, \ldots, X_n)$.

The applicability of this method is quite limited since it requires knowledge of both marginal cdf's and conditional cdf's. This knowledge is typically unavailable in complex discrete-event systems.

2.3.2 Vector Acceptance-Rejection Method

Algorithm 2.2.5 and Theorem 2.2.1 are directly applicable to the multidimensional case. We need only bear in mind that $\boldsymbol{Y}$ (see step 2 of algorithm 2.2.5) becomes an n-dimensional random vector rather than a (unidimensional) random variable. Consequently, we need a convenient way of generating $\boldsymbol{Y}$ from the multidimensional pdf $h(\boldsymbol{y})$, say, by using the vector inverse-transform method. The following examples demonstrate the vector version of the acceptance-rejection method.

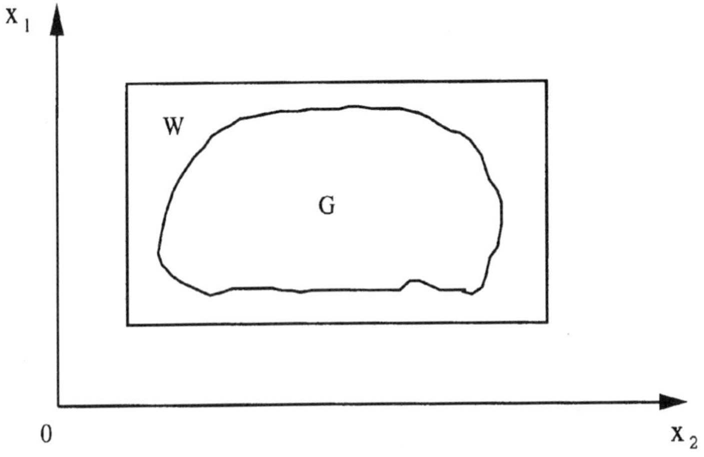

Figure 2.5. Vector acceptance-rejection method.

Example 2.3.2. Generate a random vector, uniformly distributed over the "irregular" two-dimensional region G (see Figure 2.5). The corresponding algorithm is straightforward.

1. Generate a random vector, $\boldsymbol{Y}$, uniformly distributed in W, where W is a "regular" region (multidimensional hypercube, hyper-rectangle, hypersphere, hyperellipsoid, etc.).
2. If $\boldsymbol{Y} \in G$, accept $\boldsymbol{Y}$ as a random vector, uniformly distributed over G; otherwise go to step 1.

As a special case, let G be the n-dimensional unit sphere, and let W be the n-dimensional hypercube, $\{-1 \leq x_i \leq 1\}_{i=1}^n$.

To generate a random vector, uniformly distributed over the surface of the n-dimensional unit sphere, we generate a random vector, uniformly distributed over the n-dimensional hypercube, $\{-1 \leq x_i \leq 1\}_{i=1}^n$, and then accept or reject the sample $(X_1, \ldots, X_n)$, depending on whether the point $(X_1, \ldots, X_n)$ is inside or outside the n-dimensional sphere. The corresponding algorithm is as follows:

Algorithm 2.3.2

1. Generate $U_1, \ldots, U_n$ as iid variates from $\mathcal{U}(0, 1)$.
2. Set $X_1 \leftarrow 1 - 2U_1, \ldots, X_n \leftarrow 1 - 2U_n$, and $Y^2 \leftarrow \sum_{i=1}^n X_i^2$.
3. If $Y^2 < 1$, accept $\boldsymbol{Z} = (Z_1, \ldots, Z_n)$, where $Z_i = X_i/Y$, $i = 1, \ldots, n$, as the desired vector; otherwise go to step 1.

Remark 2.3.1. To generate a random vector, uniformly distributed over the surface and interior of an n-dimensional unit sphere, we need only rewrite step 3 in

Algorithm 2.3.2 as:

3′. *If* $Y^2 \leq 1$, *accept* $Y = (Y_1, \ldots, Y_n)$ *as the desired vector.*

The efficiency of the vector acceptance-rejection method is equal to the ratio

$$\frac{1}{C} = \frac{\text{volume of the sphere}}{\text{volume of the hypercube}} = \frac{1}{n2^{n-1}} \frac{\pi^{n/2}}{\Gamma(n/2)},$$

where the volume of the unit sphere and hypercube are $\pi^{n/2}/[(n/2)\Gamma(n/2)]$ and 2^n, respectively. For even n ($n = 2m$)

$$\frac{1}{C} = \frac{\pi^m}{m!2^{2m}} = \frac{1}{m!}\left(\frac{\pi}{2}\right)^m 2^{-m} \to 0 \text{ as } m \to \infty.$$

In other words, the acceptance-rejection method grows inefficient in n and is asymptotically useless.

*2.3.3 Generating Variates from the Multinormal Distribution

A random vector $\boldsymbol{X} = (X_1, \ldots, X_n)$ has a multinormal distribution, denoted by $N(\boldsymbol{\mu}, \boldsymbol{\Sigma})$, if its pdf is given by

$$f(x) = \frac{1}{(2\pi)^{n/2}|\boldsymbol{\Sigma}|^{1/2}} \exp[-\tfrac{1}{2}(x - \boldsymbol{\mu})'\boldsymbol{\Sigma}^{-1}(x-)], \tag{2.3.5}$$

where $\boldsymbol{\mu} = (\mu_1, \ldots, \mu_n)$ is the mean vector; $\boldsymbol{\Sigma}$ is the covariance $(n \times n)$ matrix

$$\boldsymbol{\Sigma} = \begin{Vmatrix} \sigma_{11} & \sigma_{12} & \cdots & \sigma_{1n} \\ \sigma_{21} & \sigma_{22} & \cdots & \sigma_{2n} \\ \vdots & \vdots & & \vdots \\ \sigma_{n1} & \sigma_{n2} & \cdots & \sigma_{nn} \end{Vmatrix}; \tag{2.3.6}$$

prime denotes the transpose operator; $|\boldsymbol{\Sigma}|$ denotes the determinant of $\boldsymbol{\Sigma}$, and $\boldsymbol{\Sigma}^{-1}$ denotes the inverse matrix of $\boldsymbol{\Sigma}$.

Since $\boldsymbol{\Sigma}$ is positive definite and symmetric, there exists a unique lower triangular matrix,

$$C = \begin{Vmatrix} C_{11} & 0 & \cdots & 0 \\ C_{21} & C_{22} & \cdots & 0 \\ \vdots & \vdots & & \vdots \\ C_{n1} & C_{n2} & \cdots & C_{nn} \end{Vmatrix}, \tag{2.3.7}$$

such that

$$\boldsymbol{\Sigma} = \boldsymbol{CC}'. \tag{2.3.8}$$

Thus, the vector $\boldsymbol{X}$ can be represented as

$$\boldsymbol{X} = \boldsymbol{CZ} + \mu, \tag{2.3.9}$$

where $\boldsymbol{Z} = (Z_1, \ldots, Z_n)$ is a normal vector with zero mean and covariance matrix equal to the identity matrix; consequently, all components Z_i, $i = 1, \ldots, n$, of $\boldsymbol{Z}$ are marginally distributed according to the standard normal distribution, $N(0, 1)$.

In order to obtain $\boldsymbol{C}$ satisfying $\boldsymbol{\Sigma} = \boldsymbol{CC}'$, the so-called *square root method* may be used. This method computes the elements of C via a set of recursive equations as follows. From (2.3.9)

$$X_1 = C_{11}Z_1 + \mu_1. \tag{2.3.10}$$

Therefore, $\operatorname{var}[X_1] = \sigma_{11} = C_{11}^2$ and $C_{11} = \sigma_{11}^{1/2}$. Proceeding with the second component of (2.3.9), we obtain

$$X_2 = C_{21}Z_1 + C_{22}Z_2 + \mu_2 \tag{2.3.11}$$

and

$$\sigma_{22} = \operatorname{var}[X_2] = \operatorname{var}[C_{21}Z_1 + C_{22}Z_2] = C_{21}^2 + C_{22}^2. \tag{2.3.12}$$

Further, from (2.3.10) and (2.3.11),

$$\sigma_{12} = \mathbb{E}\{(X_1 - \mu_1)(X_2 - \mu_2)\} = \mathbb{E}\{C_{11}Z_1(C_{21}Z_1 + C_{22}Z_2)\} = C_{11}C_{21}. \tag{2.3.13}$$

Hence, from (2.3.12) and (2.3.13), and the symmetry of Σ,

$$C_{21} = \frac{\sigma_{12}}{C_{11}} = \frac{\sigma_{12}}{\sigma_{11}^{1/2}} \tag{2.3.14}$$

$$C_{22} = \left(\sigma_{22} - \frac{\sigma_{21}^2}{\sigma_{11}}\right)^{1/2}. \tag{2.3.15}$$

Generally, the C_{ij} can be found from the recursive formula

$$C_{ij} = \frac{\sigma_{ij} - \sum_{k=1}^{j-1} C_{ik}C_{jk}}{\left(\sigma_{jj} - \sum_{k=1}^{j-1} C_{jk}^2\right)^{1/2}}, \tag{2.3.16}$$

where, by convention,

$$\sum_{k=1}^{0} C_{ik}C_{jk} = 0, \qquad 1 \le j \le i \le n,$$

are empty sums that evaluate to zero.

The following algorithm describes the generation of a multinormal variate vector.

Algorithm 2.3.3 Generation of Multinormal Vectors

1. *Generate* $Z_1, \ldots, Z_n$ *as iid variates from* $N(0, 1)$, *(see Algorithm 2.4.6).*
2. *Set*

$$C_{ij} \leftarrow \frac{\sigma_{ij} - \sum_{k=1}^{j-1} C_{ik} C_{jk}}{\left(\sigma_{jj} - \sum_{k=1}^{j-1} C_{jk}^2\right)^{1/2}}.$$

3. *Return* $\mathbf{X} = \mathbf{C}\mathbf{Z} + \boldsymbol{\mu}$.

2.4 GENERATING CONTINUOUS VARIATES

Sections 2.4 and 2.5 present algorithms for generating variates from commonly used continuous and discrete distributions. Of the numerous algorithms available (e.g., Devroye, 1986), we have tried to select those that are reasonably efficient and relatively simple to implement.

2.4.1 Exponential Distribution

A random variable X has an exponential distribution, if its pdf is of the form

$$f(x) = \lambda e^{-\lambda x}, \quad x > 0, \lambda > 0. \tag{2.4.1}$$

An exponential distribution is denoted by $\mathrm{Exp}(\lambda)$, where λ is called the *rate* parameter. The following algorithm generates exponential variates via the inverse-transform method.

Algorithm 2.4.1 Generation of an Exponential Variate

1. *Generate* $U \sim \mathcal{U}(0, 1)$.
2. *Return* $X = (-1/\lambda) \ln U$ *as a variate from* $\mathrm{Exp}(\lambda)$.

There are many alternative procedures for generating variates from the exponential distribution. The interested reader is referred to Marsaglia (1961), and Rubinstein (1981).

2.4.2 Gamma Distribution

A random variable X has a gamma distribution if its pdf is of the form

$$f(x) = \frac{x^{\beta-1}\lambda^{\beta}e^{-\lambda x}}{\Gamma(\beta)}, \qquad x > 0, \beta > 0, \lambda > 0. \tag{2.4.2}$$

The gamma distribution is denoted by $G(\lambda, \beta)$, where λ and β are called the *scale* and *shape* parameters, respectively. Since the cdf for gamma distributions does not generally exist in explicit form, the inverse-transform method cannot always be applied. Alternative methods are then needed.

Assume first that β is a positive integer, say $\beta = m$. In this case, X has the well-known Erlang distribution, denoted by Erl(m, λ), which is frequently used in simulation. It can be represented as sum of iid exponential variates Y_i, that is, $X = \sum_{i=1}^{m} Y_i$, where the Y_i's are iid exponential variates, each with mean $1/\lambda$. From Algorithm 2.4.1, $Y_i = (-1/\lambda)\ln U_i$, whence

$$X = -\frac{1}{\lambda}\sum_{i=1}^{m} \ln U_i = -\frac{1}{\lambda}\ln \prod_{i=1}^{m} U_i. \tag{2.4.3}$$

Equation (2.4.3) suggests the following straightforward generation algorithm.

Algorithm 2.4.2 Generation of an Erlang Variate

1. *Generate iid variates* $U_1, \ldots, U_m$ *from* $\mathscr{U}(0, 1)$.
2. *Return* $X = (-1/\lambda)\ln \prod_{i=1}^{m} U_i$.

It is readily seen that the execution time of the algorithm is proportional to m. Since the Erlang distribution is a special case of the gamma family, one might look for alternatives when m is large. There are many efficient procedures for generating a variate from the gamma distribution, most of which are based on the acceptance-rejection method [see Tadikamalla and Johnson (1981)].

The next proposition, due to Jöhnk (1964), gives rise to an algorithm for generating a gamma variate, $G(\lambda, \beta)$, $0 < \beta < 1$, without resorting to acceptance-rejection. It requires, however, variate generation from the beta distribution.

Proposition 2.4.1. *Let Y and Z be independent random variables from the beta distribution* $Be(\delta, 1-\delta)$ *for* $0 < \delta < 1$ [*see (2.4.4) below*] and Exp(1) distribution, *respectively. Then the variate* $X = \lambda^{-1}YZ$ *has the gamma distribution* $G(\lambda, \delta)$.

Proof. The proof is left as an exercise.

Hint: To prove the proposition, consider the transformation

$$\begin{cases} V = Z \\ X = \lambda^{-1}YZ \end{cases}.$$

Next, find the Jacobian of the transformation

$$\begin{cases} Z = V \\ Y = \lambda X/Z \end{cases}$$

and the joint pdf $f_{V,X}(\cdot)$ of the random vector (V, X). Finally, compute the marginal pdf for X by integrating the joint pdf, $f_{V,X}(v, x)$, with respect to v.

Algorithm 2.4.3 Generation of a Gamma Variate

1. *Generate two independent variates, Y and Z, from* $\mathrm{Be}(\delta, 1-\delta)$, $0 < \delta < 1$ *(see below)*, and $\mathrm{Exp}(1)$, *respectively.*
2. *Return $X = \lambda^{-1} YZ$ as a variate from $G(\lambda, \delta)$.*

To generate a gamma variate from $G(\lambda, \beta)$ with $\beta > 1$, use the fact that the gamma family is closed under addition (of independent gamma variates) to represent β as $\beta = m + \delta$, where m is a positive integer and $0 < \delta < 1$, and write

$$X = \lambda^{-1}(W + YZ),$$

where $W \sim \mathrm{Erl}(m, 1)$, and Y and Z are as in Algorithm 2.4.3. It follows that $X \sim G(\lambda, \beta)$.

2.4.3 Beta Distribution

A random variable X has the beta distribution if its pdf is of the form

$$f(x) = \frac{\Gamma(\alpha+\beta)}{\Gamma(\alpha)\Gamma(\beta)} x^{\alpha-1}(1-x)^{\beta-1}, \qquad 0 \le x \le 1, \alpha > 0, \beta > 0. \tag{2.4.4}$$

A beta distribution is denoted by $\mathrm{Be}(\alpha, \beta)$. Note that when $\alpha = 1$ and $\beta = 1$, the beta density $\mathrm{Be}(1, 1)$ coincides with the $\mathscr{U}(0, 1)$ density.

Consider first the case when either α or β equals 1. For example, for $\beta = 1$, the $\mathrm{Be}(\alpha, 1)$ pdf is

$$f(x) = \alpha x^{\alpha-1}, \qquad 0 \le x \le 1,$$

and the corresponding cdf becomes

$$F(x) = x^{\alpha}, \qquad 0 \le x \le 1.$$

Thus, a variate $X \sim \mathrm{Be}(\alpha, 1)$ can be generated by the inverse-transform method.

A general procedure for generating a variate from $\text{Be}(\alpha, \beta)$ is based on the fact that if $Y_1 \sim G(\alpha, 1)$, $Y_2 \sim G(\beta, 1)$, and Y_1 and Y_2 are independent, then

$$X = \frac{Y_1}{Y_1 + Y_2}$$

is distributed $\text{Be}(\alpha, \beta)$. The reader is requested to prove this assertion. The corresponding algorithm is as follows.

Algorithm 2.2.4 Generation of a Beta Variate, General Case

1. *Generate independent variates,* $Y_1 \sim G(\alpha, 1)$ *and* $Y_2 \sim G(\beta, 1)$.
2. *Return* $X = Y_1/(Y_1 + Y_2)$ *as a variate from* $\text{Be}(\alpha, \beta)$.

For integer α and β, another method may be used, based on the theory of order statistics. Let $U_1, \ldots, U_{\alpha+\beta-1}$ be random variables from $\mathcal{U}(0, 1)$. Then the αth-order statistic, $U_{(\alpha)}$, has distribution $\text{Be}(\alpha, \beta)$, (see, e.g., Wilks, 1962). The resulting algorithm is extremely simple.

Algorithm 2.4.5 Generation of a Beta Variate with Integer Parameters α and β

1. *Generate* $(\alpha + \beta - 1)$ *iid variates* $U_1, \ldots, U_{\alpha+\beta-1}$ *from* $\mathcal{U}(0, 1)$.
2. *Return the* α*th-order statistic,* $U_{(\alpha)}$, *as a variate from* $\text{Be}(\alpha, \beta)$.

It can be shown that the total number of comparisons needed to find $U_{(\alpha)}$ is $(\alpha/2)(\alpha + 2\beta - 1)$, so that this procedure loses efficiency for large α and β.

2.4.4 Normal Distribution

A random variable X has a normal distribution if its pdf is of the form

$$f(x) = \frac{1}{\sigma\sqrt{2\pi}} \exp\left[-\frac{(x-\mu)^2}{2\sigma^2}\right], \qquad -\infty < x < \infty. \tag{2.4.5}$$

The normal distribution is denoted by $N(\mu, \sigma^2)$, where μ is the mean and σ^2 is the variance.

Since the normal cdf does not have an explicit representation, the inverse-transform method cannot be applied, and some other procedures must be devised instead. We consider only generation from $N(0, 1)$ (standard normal variates) since any random variable $X \sim N(\mu, \sigma^2)$ can be represented as $X = \mu + \sigma Z$, where Z is from $N(0, 1)$. One of the earliest methods for generating variates from $N(0, 1)$ is due to Box and Muller (1958). It is based on the fact that if U_1 and U_2 are independent random variables from $\mathcal{U}(0, 1)$, then the random variables

$$\begin{aligned} Z_1 &= (-2 \ln U_1)^{1/2} \cos(2\pi U_2) \\ Z_2 &= (-2 \ln U_1)^{1/2} \sin(2\pi U_2) \end{aligned} \tag{2.4.6}$$

are independent, from the standard normal distribution. To see that, rewrite the system (2.4.6) as

$$\begin{aligned} Z_1 &= (2V)^{1/2} \cos(2\pi U) \\ Z_2 &= (2V)^{1/2} \sin(2\pi U), \end{aligned} \tag{2.4.7}$$

where $V \sim \mathrm{Exp}(1)$ and $U_2 = U$. It follows from (2.4.7) that

$$Z_1^2 + Z_2^2 = 2V \quad \text{and} \quad \frac{Z_2}{Z_1} = \tan(2\pi U).$$

The Jacobian of the transformation $(U, V) \to (Z_1, Z_2)$ is

$$J = \begin{vmatrix} \dfrac{\partial u}{\partial z_1} & \dfrac{\partial u}{\partial z_2} \\ \dfrac{\partial v}{\partial z_1} & \dfrac{\partial v}{\partial z_2} \end{vmatrix} = \begin{vmatrix} \dfrac{-z_2 \cos^2 2\pi u}{2\pi z_1^2} & \dfrac{\cos^2 2\pi u}{2\pi z_1} \\ z_1 & z_2 \end{vmatrix}$$

$$= \begin{vmatrix} \dfrac{-z_2}{4\pi v} & \dfrac{z_1}{4\pi v} \\ z_1 & z_2 \end{vmatrix} = -\frac{1}{4\pi v}(z_2^2 + z_1^2) = -\frac{1}{2\pi}$$

and

$$f_{Z_1,Z_2}(z_1, z_2) = f_{U,V}(u, v)|J| = \frac{1}{2\pi} \exp\left(-\frac{z_1^2 + z_2^2}{2}\right). \tag{2.4.8}$$

The last formula represents the joint pdf of two independent standard normal variates.

Algorithm 2.4.6 Generation of a Normal Variate, Box and Muller Approach

1. *Generate two independent variates, U_1 and U_2, from $\mathcal{U}(0, 1)$.*
2. *Return two independent standard normal variates, Z_1 and Z_2, by substituting U_1 and U_2 into the system of Eqs. (2.4.6).*

An alternative generation method from $N(0, 1)$ is based on the acceptance-rejection method. First, note that in order to generate a variate Y from $N(0, 1)$, one can first generate a nonnegative variate X from the pdf

$$f(x) = \sqrt{\frac{2}{\pi}} e^{-x^2/2}, \qquad x \geq 0, \tag{2.4.9}$$

and then assign to X a random sign. The validity of this procedure follows from the symmetry of the standard normal distribution about zero.

To generate a variate X from (2.4.9), recall Algorithm 2.2.5 and write $f(x)$ as

$$f(x) = Ch(x)g(x),$$

where

$$C = \left(\frac{2e}{\pi}\right)^{1/2} \tag{2.4.10}$$

$$h(x) = e^{-x} \tag{2.4.11}$$

$$g(x) = \exp\left[-\frac{(x-1)^2}{2}\right]. \tag{2.4.12}$$

The efficiency of this method is $\sqrt{\pi/2e} \simeq 0.76$.

The acceptance condition, $U \leq g(Y)$, can be written as

$$U \leq \exp[-(Y-1)^2/2], \tag{2.4.13}$$

which is equivalent to

$$-\ln U \geq \frac{(Y-1)^2}{2}, \tag{2.4.14}$$

where Y is from Exp(1). Since $-\ln U$ is also from Exp(1), the last inequality can be written as

$$V_1 \geq \frac{(V_2-1)^2}{2}, \tag{2.4.15}$$

where both $V_1 = -\ln U$ and $V_2 = Y$ are from Exp(1).

Algorithm 2.4.7 Generation of a Normal Variate, Acceptance-Rejection Approach

1. *Generate independent variates,* V_1 *and* V_2*, from* Exp(1).

2. *If* $V_1 < (V_2 - 1)^2/2$, *go to step 1.*
3. *Generate* U_1 *from* $\mathcal{U}(0, 1)$.
4. *If* $U_1 \leq 0.5$, *return* $Z = V_1$; *otherwise return* $Z = -V_1$.

2.5 GENERATING DISCRETE VARIATES

The most common procedure for generating discrete variates is the inverse-transform method [see (2.2.1) and Example 2.2.3]. An alternative approach for generating arbitrary discrete variates with a *finite* range of values is the so-called *alias method*, due to Walker (1977). The alias method is based on the fact that any n-point pdf

$$P(X = x_i) = p_i, \qquad i = 1, \ldots, n$$

can be represented as an equally weighted mixture of $n - 1$ discrete pdf's, $g^{(k)}$, $k = 1, \ldots, n - 1$, each having at most *two* nonzero components. That is, any n-point pdf f can be represented as

$$f = \frac{1}{n-1} \sum_{k=1}^{n-1} g^{(k)} \tag{2.5.1}$$

for suitably defined 2-point pdf's $g^{(k)}$, $k = 1, \ldots, n$ (see Walker, 1977).

The alias method is rather general and efficient but requires some initial setup and extra storage for the $n - 1$ pdf's, $g^{(k)}$.

Lemma 2.5.1. *For any n-point pdf* $f(x_i)$, $i = 1, \ldots, n$:

(a) There exists an index i, such that $p_i < 1/(n - 1)$; $i = 1, \ldots, n$.
(b) Given this index i, there exists an index $j, j \neq i$, *such that* $p_i + p_j \geq 1/(n - 1)$.

Proof. The proof of this lemma is left as an exercise.

With Lemma 2.5.1 in hand, a procedure for computing these simple pdf's, $g^{(k)}$, $k = 1, \ldots, n - 1$, can be devised such that (2.5.1) holds for an arbitrary n-point pdf, f. For more details, see Walker (1977) or Law and Kelton (1991).

The rest of this section presents several algorithms for generating variates from commonly used discrete distributions.

2.5.1 Bernoulli Distribution

A random variable X has a Bernoulli distribution, if its pdf is of the form

$$f(x) = p^x (1 - p)^{1-x}, \qquad x = 0, 1, \quad 0 < p < 1. \tag{2.5.2}$$

A Bernoulli distribution is denoted by Ber(p), where p is the probability of success. A Bernoulli variate returns the outcome of a binary trial ($0 =$ failure, $1 =$ success). Applying the inverse-transform method, one readily obtains the following generation algorithm.

Algorithm 2.5.1 Generation of a Bernoulli Variate

1. *Generate* $U \sim \mathscr{U}(0, 1)$.
2. *If* $U \leq p$, *return* $X = 1$; *otherwise return* $X = 0$.

2.5.2 Binomial Distribution

A random variable has a binomial distribution if its pdf is of the form

$$f(x) = \binom{n}{x} p^x (1-p)^{n-x}, \qquad x = 0, 1, \ldots, n, \quad 0 < p < 1. \tag{2.5.3}$$

A binomial distribution is denoted by Bin(n, p), and the corresponding binomial variate is the sum of n iid Bernoulli trials, each with parameter p. The simplest generation algorithm can be written as follows.

Algorithm 2.5.2 Generation of a Binomial Variate

1. *Generate iid variates* $Y_1, \ldots, Y_n$ *from* Ber(p).
2. *Return* $X = \sum_{i=1}^{n} Y_i$ *as a variate from* Bin(n, p).

Since the execution time of Algorithm 2.5.2 is proportional to n, one is motivated to use alternative methods for large n. For example, one may consider the normal distribution as an approximation to the binomial. In particular, as n increases, the distribution of

$$Z = \frac{X - np + \frac{1}{2}}{[np(1-p)]^{1/2}} \tag{2.5.4}$$

approaches $N(0, 1)$.

To obtain a binomial variate, we generate Z from $N(0, 1)$, solve (2.5.4) for X, and truncate to the nearest nonnegative integer, that is,

$$X = \max\{0, \lfloor -0.5 + np + Z(np(1-p))^{1/2} \rfloor\}, \tag{2.5.5}$$

where $\lfloor \alpha \rfloor$ denotes the integer part of α. One should consider replacing the binomial with the approximating normal variate for $np > 10$ with $p \geq \frac{1}{2}$, and for $n(1-p) > 10$ with $p < \frac{1}{2}$.

It is worthwhile to note that if $Y \sim \text{Bin}(n, p)$, then $n - Y \sim \text{Bin}(n, 1-p)$. Hence, to enhance efficiency, one may elect to generate X from $\text{Bin}(n, p)$ according to

$$X = \begin{cases} Y_1 \sim \text{Bin}(n, p) & \text{if } p \leq \frac{1}{2} \\ n - Y_2,\, Y_2 \sim \text{Bin}(n, 1-p) & \text{if } p > \frac{1}{2}. \end{cases}$$

2.5.3 Geometric Distribution

A random variable has a geometric distribution, if its pdf is of the form

$$f(x) = p(1-p)^x, \qquad x = 0, 1, \ldots, \qquad 0 < p < 1. \tag{2.5.6}$$

A geometric distribution is denoted by $\text{Ge}(p)$, and the corresponding geometric variate is the number of trials separating successes in an iid sequence of Bernoulli trials.

We now present an algorithm that is based on the relationship between the exponential and geometric distributions. Let $Y \sim \text{Exp}(\beta)$, whence

$$P(x < Y \leq x+1) = \frac{1}{\beta} \int_x^{x+1} e^{-y/\beta} dy = e^{-x/\beta}(1 - e^{-1/\beta}). \tag{2.5.7}$$

Thus for integer $x \geq 0$, Eq. (2.5.7) is the geometric distribution $\text{Ge}(p = 1 - e^{-1/\beta})$.

In particular, for $\beta = -1/\ln(1-p)$, Eq. (2.5.7) is identical to (2.5.6). Therefore, since $P(Y \leq x) = 1 - e^{-x/\beta}$,

$$X = \left\lfloor \frac{\ln U}{\ln(1-p)} \right\rfloor = \left\lfloor -\frac{V}{\ln(1-p)} \right\rfloor, \tag{2.5.8}$$

where $V = -\ln(U)$ is a standard exponential variate implying that X is from $\text{Ge}(p)$. Hence, to generate a variate from $\text{Ge}(p)$, we first generate a variate from the exponential distribution with $\beta = -1/\ln(1-p)$ and then truncate the obtained value to the nearest integer.

Algorithm 2.5.3 Generation of a Geometric Variate

1. *Generate* $V \sim \text{Exp}[-1/\ln(1-p)]$.
2. *Return* $X = \lfloor -V/\ln(1-p) \rfloor$ *as a variate from* $\text{Ge}(p)$.

2.5.4 Poisson Distribution

A random variable X has a Poisson distribution, if its pdf is of the form

$$f(n) = \frac{\lambda^n e^{-\lambda}}{n!}, \qquad n = 0, 1, \ldots; \quad \lambda > 0. \tag{2.5.9}$$

A Poisson distribution is denoted by $P(\lambda)$, where λ is the *rate* parameter. A Poisson variate, X, is the maximal number of iid exponential variates (with common parameter λ) whose sum does not exceed one.

The following algorithm is based on the relationship between the $P(\lambda)$ and $\mathrm{Exp}(\lambda)$ distributions. Write

$$\sum_{i=0}^{i} Y_j \le 1 \le \sum_{i=0}^{i+1} Y_j, \tag{2.5.10}$$

where Y_j, $j = 0, 1, \ldots, i+1$, are iid variates from $\mathrm{Exp}(\lambda)$. Let the Y_j be viewed as intervals, laid end to end, and let the end point of each interval be interpreted as an "event"(e.g., a customer arrival, say, at a queue). Then in this setting Eq. (2.5.10) means that the number of events by time 1, denoted $X(1)$, is Poisson distributed (with mean λ) and can be written as

$$X(1) = \max\left\{n : \sum_{j=1}^{n} Y_j \le 1\right\}. \tag{2.5.11}$$

Further, since $Y_j = (-1/\lambda)\ln U_n$, we can rewrite (2.5.11) as

$$\begin{aligned} X &= \max\left\{n : \sum_{j=1}^{n} -\ln U_j \le \lambda\right\} \\ &= \max\left\{n : \ln\left(\prod_{j=1}^{n} U_j\right) \ge -\lambda\right\} \\ &= \max\left\{n : \prod_{j=1}^{n} U_j \ge e^{-\lambda}\right\}. \end{aligned} \tag{2.5.12}$$

To see that the random variable X has the desired (Poisson) distribution, note again that $-\ln U_j$ is exponential with rate 1. Interpreting $-\ln U_j$, $j \ge 1$, as the interarrival times of a Poisson process with rate 1, we have that X is equal to the number of events by time λ. Thus, X is Poisson distributed with mean λ.

Algorithm 2.5.4 Generating a Poisson Variate

1. *Set* $n \leftarrow 1$ *and* $a \leftarrow 1$.
2. *Generate an independent* $U_n \sim \mathcal{U}(0, 1)$ *and set* $a \leftarrow aU_n$.
3. *If* $a \ge e^{-\lambda}$, *then set* $n \leftarrow n + 1$ *and go to step 2.*
4. *Otherwise, return* $X = n - 1$, *as a variate from* $P(\lambda)$.

It is readily seen that for large λ, this algorithm becomes slow ($e^{-\lambda}$ is small for large λ, and more random numbers, U_j, are required to satisfy $\prod_{j=1}^{n} U_j < e^{-\lambda}$).

Alternative approaches use the inverse-transform method with an efficient search (see Atkinson, 1979a,b, and Devroye, 1981), or the alias method.

2.6 GENERATING POISSON PROCESSES

The rest of this chapter is concerned with generating variate sequences. For example, in order to simulate a renewal process over the interval [0, t], one generates a sequence of iid variates of random length

$$N(t) = \min\left\{n: \sum_{i=1}^{n} X_i > t\right\}. \tag{2.6.1}$$

Clearly, if X_i, $i = 1, 2, \ldots,$ represent the times between arrivals, then (2.6.1) gives rise to $N - 1$ events which occur at times $X_1, X_1 + X_2, \ldots, X_1 + \cdots + X_{N(t)-1}$, within the interval [0, t].

This section treats the generation of stationary and nonstationary Poisson processes.

2.6.1 Generating a Stationary Poisson Process

In the special case where $X_i \sim \text{Exp}(\lambda)$, the process $\{N(t)\}$ is a stationary Poisson process with rate λ. Thus, the $X_i = (-1/\lambda)\ln U_i$, $i = 1, \ldots, n$, are the exponentially distributed time intervals between the $(i - 1)$st, and the ith arrival events, and the time point of the jth arrival is $A_n = \sum_{i=1}^{j} X_i$.

The algorithm for generating a Poisson process $\{N(t)\}$ over the interval [0, T] is given below.

Algorithm 2.6.1 Generation of a Stationary Poisson Process

1. *Set* $t \leftarrow 0$ *and* $n \leftarrow 0$.
2. *Set* $n \leftarrow n + 1$.
3. *Generate an independent variate* $U_n \sim \mathscr{U}(0, 1)$.
4. *Set* $t \leftarrow t - (1/\lambda)\ln U_n$ *and declare an arrival event.*
5. *If* $t > T$, *stop; otherwise go to step 2.*

Note that t is the variable recording the time of the most recent arrival event (by convention, $t = 0$ is a fictitious arrival time).

2.6.2 Generating a Nonstationary Poisson Process

Thinning is a simple method, due to Lewis and Shedler (1979), for generating nonstationary Poisson processes with rate function $\lambda(t)$. For an alternative method, see Law and Kelton (1991) and Ross (1990).

The method first generates a stationary Poisson process with rate $\lambda = \max_{0 \leq t \leq T} \lambda(t)$, which is assumed to be finite. Each arrival event of the stationary Poisson process is rejected (thinned) with probability $1 - \lambda(A_n)/\lambda$, where A_n is the (random) arrival time of the nth event. The surviving events define the desired nonstationary Poisson process.

The algorithm for generating a non-stationary Poisson process $\{N(t)\}$ over the interval $[0, T]$ is similar to the previous algorithm.

Algorithm 2.6.2 Generating a Nonstationary Poisson Process

1. *Set* $t \leftarrow 0$ *and* $n \leftarrow 0$.
2. *Set* $n \leftarrow n + 1$.
3. *Generate an independent variate* $U_n \sim \mathcal{U}(0, 1)$.
4. *Set* $t \leftarrow t - (1/\lambda) \ln U_n$.
5. *If* $t > T$, *stop*; *otherwise, continue.*
6. *Generate an independent variate* $V_n \sim \mathcal{U}(0, 1)$.
7. *If* $V_n \leq \lambda(t)/\lambda$, *declare an arrival event and go to step 2.*

*2.7 GENERATING RANDOM VECTORS, UNIFORMLY DISTRIBUTED OVER VARIOUS REGIONS

This section deals with several algorithms for generating random vectors, uniformly distributed over the surface and interior of (a) a simplex, (b) a hypersphere, and (c) a hyperellipsoid. Most of the material is taken from Rubinstein (1982). For a landmark paper on generating vectors, uniformly distributed over bounded regions, see Smith (1984); and for efficient generation methods of random vectors over the surface and interior of open regions with arbitrary distributions, see Borovkov (1994) and Claude et al. (1993).

2.7.1 Generating Random Vectors over a Simplex

The pdf of a random vector X, uniformly distributed over the interior of an n-dimensional simplex,

$$S_n = \left\{(x_1, \ldots, x_n): x_i \geq 0, i = 1, \ldots, n, \sum_{i=1}^{n} x_i \leq 1\right\}, \tag{2.7.1}$$

is

$$f(x) = \begin{cases} n! & \text{if } x = (x_1, \ldots, x_n) \in S_n, \\ 0 & \text{otherwise.} \end{cases} \tag{2.7.2}$$

This density is a special case of the density

$$f(x) = \frac{\Gamma(v_1 + \cdots + v_{n+1})}{\Gamma(v_1) \cdots \Gamma(v_{n+1})} x_1^{v_1 - 1} \cdots x_n^{v_n - 1} (1 - x_1 - \cdots - x_n)^{v_{n+1} - 1}, \tag{2.7.3}$$

with parameters $v_1, \ldots, v_n, v_{n+1}$. A density of the form (2.7.3) is called a *Dirichlet density*, and the corresponding *Dirichlet distribution* is denoted by $D(v_1, \ldots, v_n, v_{n+1})$ (see Wilks, 1962). Indeed, for $v_1 = v_n = v_{n+1} = 1$, Eq. (2.7.3) reduces to Eq. (2.7.2), and the corresponding distribution is denoted by $D(1, \ldots, 1, 1)$. It is also known (see, e.g., Wilks, 1962) that if Y_i, $i = 1, \ldots, n+1$, are independent random variables from gamma distributions with shape parameter $v_i > 0$ and scale parameter 1, namely,

$$f_{Y_i}(y) = \begin{cases} \dfrac{y^{n-1} e^{-y}}{\Gamma(v_i)} & y > 0 \\ 0 & \text{otherwise,} \end{cases} \tag{2.7.4}$$

then the (normalized) random vector,

$$\boldsymbol{X} = (X_1, \ldots, X_n) = \left(\frac{Y_1}{\sum_{i=1}^{n+1} Y_i}, \ldots, \frac{Y_n}{\sum_{i=1}^{n+1} Y_i} \right), \tag{2.7.5}$$

is distributed $D(v_1, \ldots, v_n, v_{n+1})$. Observing that for the special case $v_i = 1$, $i = 1, \ldots, n+1$, each Y_i is distributed Exp(1), the algorithm for generating the desired random vector $\boldsymbol{X}$ can be written as follows:

Algorithm 2.7.1 Generating a Random Vector over a Simplex (Exponential Approach)

1. *Generate* $n+1$ *variates* $Y_1, \ldots, Y_n, Y_{n+1}$, *each from* Exp(1).
2. *Apply formula (2.7.5) and return the vector* $(X_1, \ldots, X_n)$.

It is known (e.g., Rubinstein, 1982) that if we increase the dimension in (2.7.5) from n to $n+1$, namely,

$$\boldsymbol{X} = (X_1, \ldots, X_{n+1}) = \left(\frac{Y_1}{\sum_{i=1}^{n+1} Y_i}, \ldots, \frac{Y_{n+1}}{\sum_{i=1}^{n+1} Y_i} \right), \tag{2.7.6}$$

then the resultant $\boldsymbol{X} = (X_1, \ldots, X_{n+1})$ is uniformly distributed over the surface of the $(n+1)$-dimensional simplex,

$$S_{n+1} = \left\{ (x_1, \ldots, x_{n+1}) : x_i \geq 0, i = 1, \ldots, n+1, \sum_{i=1}^{n+1} x_k = 1 \right\},$$

whereas the original vector, $\boldsymbol{X} = (X_1, \ldots, X_n)$, is uniformly distributed over the interior of the n-dimensional simplex considered in (2.7.5).

An alternative procedure for generating random vectors, uniformly distributed over the interior of an n-dimensional simplex, is based on the following result. Let $Y_{(1)}, \ldots, Y_{(n)}$ be the order statistics where the variates $Y, \ldots, Y_n$ are iid with a continuous common cdf $F_Y(y)$. Then the random vector $\boldsymbol{X} = (X_1, \ldots, X_n)$, defined by

$$\begin{aligned} X_1 &= F_Y(Y_{(1)}), \\ X_2 &= F_Y(Y_{(2)}) - F_Y(Y_{(1)}), \\ &\vdots \\ X_n &= F_Y(Y_{(n)}) - F_Y(Y_{(n-1)}), \end{aligned} \tag{2.7.7}$$

is distributed $D(1, \ldots, 1, 1)$ (see Wilks, 1962). Observe that in the special case that the random variables $Y_j = U_j$ are distributed $\mathcal{U}(0, 1)$, Eq. (2.7.7) can be rewritten as

$$\begin{aligned} X_1 &= U_{(1)}, \\ X_2 &= U_{(2)} - U_{(1)}, \\ &\vdots \\ X_n &= U_{(n)} - U_{(n-1)}. \end{aligned} \tag{2.7.8}$$

The following is an alternative algorithm for generating a random vector from $D(1, \ldots, 1, 1)$.

Algorithm 2.7.2 Generating a Vector over a Simplex (Order Statistics Approach)

1. *Generate n independent variates $U_1, \ldots, U_n$ from $\mathcal{U}(0, 1)$.*
2. *Permute $U_1, \ldots, U_n$ into the order statistics $U_{(1)}, \ldots, U_{(n)}$.*
3. *Apply formula (2.7.8), and return the vector $\boldsymbol{X} = (X_1, \ldots, X_n)$.*

Again, if we increase the dimension in (2.7.8) from n to $n+1$, namely,

$$\begin{aligned} X_1 &= U_{(1)}, \\ X_2 &= U_{(2)} - U_{(1)}, \\ &\vdots \\ X_n &= U_{(n)} - U_{(n-1)}, \\ X_{n+1} &= 1 - U_{(n)}, \end{aligned} \tag{2.7.9}$$

then the resulting vector, $\boldsymbol{X} = (X_1, \ldots, X_{n+1})$, is uniformly distributed over the surface of the $(n+1)$-dimensional simplex.

Note that both Algorithms 2.7.1 and 2.7.2 are suitable for generating random vectors, uniformly distributed over an n-dimensional simplex with vertices

$$\boldsymbol{x}_0 = (0, 0, \ldots, 0), \boldsymbol{x}_1 = (1, 0, \ldots, 0), \ldots, \boldsymbol{x}_n = (0, 0, \ldots, 1).$$

Consider now the generation of random vectors, uniformly distributed over an n-dimensional simplex with arbitrary vertices, say, $\boldsymbol{z}_0, \boldsymbol{z}_1, \ldots, \boldsymbol{z}_n$, where $\boldsymbol{z}_i = (z_{i1}, \ldots, z_{in})$, $i = 0, 1, \ldots, n$. It is readily seen that if $\boldsymbol{X}$ is uniformly distributed over the simplex (2.7.1), then the linear transformation

$$\boldsymbol{Z} = \boldsymbol{C}\boldsymbol{X} + \boldsymbol{z}_0$$

with the matrix

$$\boldsymbol{C} = \begin{bmatrix} z_{11} - z_{01} & z_{12} - z_{02}, \ldots, & z_{1n} - z_{0n} \\ z_{21} - z_{01} & z_{22} - z_{02}, \ldots, & z_{2n} - z_{0n} \\ \vdots & \vdots & \vdots \\ z_{n1} - z_{01} & z_{n2} - z_{02}, \ldots, & z_{nn} - z_{0n} \end{bmatrix}$$

preserves uniformity, that is, the vector $\boldsymbol{Z}$ will be uniformly distributed on the simplex with vertices $\boldsymbol{z}_0, \boldsymbol{z}_1, \ldots, \boldsymbol{z}_n$.

2.7.2 Generating Random Vectors, Uniformly Distributed over a Unit Hypersphere

The density function of a random vector, uniformly distributed over the surface of an n-dimensional sphere of radius r, is

$$f_Y(y) = \begin{cases} \dfrac{\Gamma(n/2)}{2\pi^{n/2} r^{n-1}} & y = \left\{ (y_1, \ldots, y_n) \colon \sum_{i=1}^{n} y_i^2 = r^2 \right\}, \\ 0 & \text{otherwise.} \end{cases} \tag{2.7.10}$$

Formula (2.7.10) follows from the fact that the surface "size" of the n-dimensional sphere of radius r is

$$\frac{2\pi^{n/2} r^{n-1}}{\Gamma(n/2)}.$$

Before proceeding with the algorithms for generating points, uniformly distributed over the surface and interior of the n-dimensional sphere, it should be noted that any *surface* algorithm (see below) can be converted into an *interior* algorithm and vice versa. Indeed, let $\boldsymbol{Y} = (Y_1, \ldots, Y_n)$ be a vector, uniformly distributed over the surface of an n-dimensional sphere, and let $U \sim \mathcal{U}(0, 1)$. Then the vector $\boldsymbol{X} = (X_1, \ldots, X_n)$, with components $X_i = U^{1/n} Y_i$, is uniformly distributed over the interior of that hypersphere. Conversely, if $\boldsymbol{X} = (X_1, \ldots, X_n)$ is a vector, uniformly distributed over the interior of the n-dimensional sphere, then the vector $\boldsymbol{Y} = (Y_1, \ldots, Y_n)$ with components $Y_i = (\sum_{i=1}^n X_i^2)^{-1/2} X_i$, is distributed uniformly over the surface of that hypersphere. This holds, since $\boldsymbol{Y}$ is the projection of interior sphere points onto the surface of the sphere.

The algorithm for generating a vector, uniformly distributed over the surface of an n-dimensional unit sphere, is based on von Neumann's acceptance-rejection method.

Algorithm 2.7.3 Generating Random Vectors over the Surface of a Unit Hypersphere (Acceptance-Rejection Approach)

1. *Generate iid variates* $U_i \sim \mathcal{U}(0, 1)$, $i = 1, \ldots, n$.
2. *Set* $Z_i \leftarrow 1 - 2U_i$, $i = 1, \ldots, n$, *and* $Y^2 \leftarrow \sum_{i=1}^n Z_i^2$.
3. *If* $Y^2 \geq 1$, *then go to step 1.*
4. *Otherwise, if* $Y^2 < 1$, *return* $\boldsymbol{X} = (X_1, \ldots, X_n)$, *where* $X_i = Z_i / Y$, $i = 1, \ldots, n$.

To obtain a point, uniformly distributed over the interior of an n-dimensional unit sphere, we need only replace step 4 in Algorithm 2.7.3 by

4′. *Otherwise, if* $Y^2 < 1$, *return* $\boldsymbol{Y} = (Y_1, \ldots, Y_n)$, *all other steps remaining the same.*

The main advantage of the acceptance-rejection method is its simplicity. Its main disadvantage is that the number of trials needed to generate a hypersphere point increases explosively in n (see Section 4.2). For this reason, it can be recommended only for low dimensions ($n \leq 5$).

Another algorithm for generating a vector, uniformly distributed over the surface of a unit hypersphere, is based on the following result.

Theorem 2.7.1. *Let* $X_1, \ldots, X_n$ *be iid variates from* $N(0, 1)$, *and define* $Z = (\sum_{i=1}^n X_i^2)^{1/2} = V$. *Then the vector*

$$\boldsymbol{Y} = (Y_1, \ldots, Y_n) = \left(\frac{X_1}{Z}, \ldots, \frac{X_n}{Z}\right) \tag{2.7.11}$$

is distributed uniformly over the surface of the unit hypersphere $\{(y_1, \ldots, y_n): \sum_{i=1}^n y_i^2 = 1\}$.

Proof. Consider the transformation $(x_1, \ldots, x_n, z) \to (y_1, \ldots, y_n, v)$, given by

$$\begin{cases} y_i(x_1, \ldots, x_n, z) = \dfrac{x_i}{z}, & i = 1, \ldots, n \\ v(x_1, \ldots, x_n, z) = z. \end{cases} \tag{2.7.12}$$

Its Jacobian is $J = v^n$, since the inverse transformation is

$$\begin{cases} x_i(y_1, \ldots, y_n, v) = vy_i, & i = 1, \ldots, n \\ z(y_1, \ldots, y_n, v) = v. \end{cases}$$

But

$$f_{X_1,\ldots,X_n,Z}(x_1, \ldots, x_n, z) = \frac{1}{(2\pi)^{n/2}} e^{-(1/2)z^2} I\left(z, \left(\sum x_i^2\right)^{1/2}\right),$$

where

$$I(x, y) = \begin{cases} 1, & \text{if } x = y \\ 0, & \text{otherwise.} \end{cases}$$

This implies

$$f_{Y_1,\ldots,Y_n,V}(y_1, \ldots, y_n, v) = v^n \frac{1}{(2\pi)^{n/2}} \exp\left(-\sum_{i=1}^n \frac{(y_i v)^2}{2}\right) I\left(v, \left(\sum_{i=1}^n (y_i v)^2\right)^{1/2}\right). \tag{2.7.13}$$

In view of

$$\sum_{i=1}^n y_i^2 = \sum_{i=1}^n \left(\frac{x_i}{z^2}\right)^2 = \frac{1}{z^2} \sum_{i=1}^n x_i^2 = 1,$$

we have $I(v, (\sum_{i=1}^n (y_i v)^2)^{1/2}) = 1$, so that (2.7.13) can be further simplified to

$$\begin{aligned} f_{Y_1,\ldots,Y_n,V}(y_1, \ldots, y_n, v) &= v^n \frac{1}{(2\pi)^{n/2}} \exp\left(-\frac{v^2}{2} \sum_{i=1}^n y_i^2\right) \\ &= v^n \frac{1}{(2\pi)^{n/2}} e^{-v^2/2}. \end{aligned} \tag{2.7.14}$$

It follows from (2.7.13) and (2.7.14) that $(Y_1, \ldots, Y_n)$ and V are independent, and that the vector $(Y_1, \ldots, Y_n)$ is uniformly distributed over the surface of the unit sphere, $\{(y_1, \ldots, y_n): \sum_{i=1}^n y_i^2 = 1\}$.

Theorem 2.7.1 motivates the following alternative generation algorithm.

Algorithm 2.7.4 Generating Random Vectors over the Surface of a Unit Hypersphere [Box and Muller (1958)]

1. *Generate n iid variates* $X_i, \ i = 1, \ldots, n$, *from* $N(0, 1)$.
2. *Compute* $Z = (\sum_{i=1}^n X_i^2)^{1/2}$.
3. *Return* $\mathbf{Y} = (Y_1, \ldots, Y_n)$, *where* $Y_i = X_i/Z, \ i = 1, \ldots, n$.

An alternative recursive procedure is as follows. First select a point on the surface of the one-dimensional sphere ($x_1 = 1$ or $x_1 = -1$), and then continue recursively until the required dimensionability is reached: Given that a point $(x_1, x_2, \ldots, x_n)$ on the surface of an n-dimensional sphere has already been generated, the point on the $(n+1)$-dimensional sphere is selected as

$$(Rx_1, Rx_2, \ldots, Rx_n, S\sqrt{1 - R^2}),$$

where S assumes the values $+1$ and -1 with probability $\frac{1}{2}$ each, and R has the pdf

$$f_R(r) = \begin{cases} k\dfrac{r^{n-1}}{\sqrt{1 - r^2}} & 0 \le r \le 1, k = \left(\displaystyle\int_0^1 \frac{r^{n-1}}{\sqrt{1 - r^2}} dr\right)^{-1}, \\ 0 & \text{otherwise.} \end{cases}$$

Consider now the polar transformation

$$\begin{aligned} X_1 &= R \cos\varphi_1 \cos\varphi_2 \cdots \cos\varphi_{n-2} \cos\varphi_{n-1} \\ X_2 &= R \cos\varphi_1 \cos\varphi_2 \cdots \cos\varphi_{n-2} \sin\varphi_{n-1} \\ X_3 &= R \cos\varphi_1 \cos\varphi_2 \cdots \sin\varphi_{n-2} \\ &\vdots \\ X_{n-1} &= R \cos\varphi_1 \sin\varphi_2 \\ X_n &= R \sin\varphi_1 \end{aligned} \tag{2.7.15}$$

where $-\frac{1}{2}\pi \le \varphi_i \le \frac{1}{2}\pi, \ i = 1, \ldots, n-2, \ 0 \le \varphi_{n-1} \le 2\pi$, and $0 \le R \le 1$ has pdf f_R; furthermore, R and $\varphi_1, \ldots, \varphi_{n-1}$ are mutually independent variates. The following algorithm generates points, uniformly distributed over the interior of the hypersphere of radius r.

Algorithm 2.7.5 Generating Random Vectors over the Interior of a Hypersphere (Polar Transformation Approach)

1. *Set* $R \leftarrow U_1^{1/n}$, *where* $U_1 \sim \mathcal{U}(0, 1)$.
2. *Set* $\varphi_{n-1} \leftarrow 2\pi U_2$, *where* $U_2 \sim \mathcal{U}(0, 1)$ *is independent of* U_1.
3. *Generate independent random-angle variates* φ_i, $i = 1, \ldots, n-1$, *from the pdf*

$$f_i(x) = c_i \cos^{n-i-1} x, \quad \textit{where } c_i^{-1} = \int_{-\pi/2}^{\pi/2} \cos^{n-i-1} x \, dx.$$

4. *Apply formula (2.7.15) and return* $\boldsymbol{X}$ *as the requisite uniform random vector.*

It is readily seen that if $\boldsymbol{X} = (X_1, \ldots, X_n)$ is a vector, uniformly distributed over the interior (surface) of the n-dimensional unit hypersphere, then the vector $\boldsymbol{Y} = (Y_1, \ldots, Y_n)$ with coordinates $Y_i = rX_i + b_i$, $i = 1, \ldots, n$, is distributed uniformly over the interior (surface) of the sphere with radius r centered at point $\boldsymbol{b} = (b_1, \ldots, b_n)$.

2.7.3 Generating Random Vectors, Uniformly Distributed over a Hyperellipsoid

The equation of a hyperellipsoid, centered at the origin, can be written as

$$\boldsymbol{X}'\boldsymbol{\Sigma}\boldsymbol{X} = r^2, \tag{2.7.16}$$

where $\boldsymbol{\Sigma}$ is a positive definite and symmetric $(n \times n)$ matrix. The special case, where $\boldsymbol{\Sigma} = \boldsymbol{I}$ (identity matrix), corresponds to a hypersphere of radius r. Since $\boldsymbol{\Sigma}$ is positive definite and symmetric, there exists a unique lower triangular matrix

$$\boldsymbol{C} = \begin{bmatrix} c_{11} & 0 & \cdots & 0 \\ c_{21} & c_{22} & \cdots & 0 \\ \vdots & \vdots & \vdots & \vdots \\ c_{n1} & c_{n2} & \cdots & c_{nn} \end{bmatrix},$$

such that $\boldsymbol{\Sigma} = \boldsymbol{C}\boldsymbol{C}'$. It can be shown that if the vector $\boldsymbol{X}$ is uniformly distributed over the surface or interior of an n-dimensional sphere of radius r, then the vector $\boldsymbol{Y} = \boldsymbol{C}\boldsymbol{X}$ is uniformly distributed over the surface or interior of a hyperellipsoid [see (2.7.16)]. The corresponding generation algorithm is given below.

Algorithm 2.7.6 Generating Random Vectors over the Surface or Interior of Hyperellipsoids

1. *Generate* $\boldsymbol{X} = (X_1, \ldots, X_n)$, *uniformly distributed over the hypersphere of radius r, centered at the origin.*
2. *Calculate the matrix* $\boldsymbol{C}$, *satisfying* $\boldsymbol{\Sigma} = \boldsymbol{CC}'$.
3. *Return* $\boldsymbol{Y} = \boldsymbol{CX}$ *as the requisite uniform random vector.*

The elements c_{ij} of the matrix $\boldsymbol{C}$ can be calculated from the recursive formula

$$c_{ij} = \frac{\sigma_{ij} - \sum_{k=1}^{j-1} c_{ik}c_{jk}}{\left(\sigma_{jj} - \sum_{k=1}^{j-1} c_{jk}^2\right)^{1/2}},$$

where $\sum_{k=1}^{0} c_{ik}c_{jk} = 0, \quad 1 \le j \le i \le n$, are empty sums, and the σ_{ij} are elements of $\boldsymbol{\Sigma}$.

*2.8 INPUT ANALYSIS USING TES SEQUENCES

This section describes a versatile modeling methodology, called TES (*transform-expand-sample*), specifically designed to fit models to autocorrelated empirical records with general marginal distribution (histogram) (Melamed, 1991; Jagerman and Melamed, 1992a,b; Melamed, 1993). In simulation parlance, TES is an *input analysis* method. Essentially, TES is a modulo-1 reduction of a simple linear autoregressive scheme, followed by additional transformations. But what sets TES apart from traditional models, such as linear autoregression (AR), is that TES strives to capture the statistical signature of empirical records in terms of their first-order *and* second-order properties, *simultaneously*. More specifically, given a stationary empirical time series, the TES modeling methodology aims to fit both the empirical marginal distribution (histogram) as well as the empirical autocorrelation function, the latter being a standard proxy for temporal dependence. A more precise statement of the TES modeling requirements will be given later.

Autocorrelated stochastic processes abound in real-world phenomena and applications. Typical examples are autocorrelated job arrivals to a manufacturing shop or compressed video traffic in a telecommunications network (Lee et al., 1992, Melamed et al., 1992a). However, analytical models generally attempt to minimize temporal dependence assumptions in model description in order to simplify model specifications and facilitate their mathematical solutions. Thus, the bulk of queueing theory consists of $GI/GI/m$-type models, where *GI* stands for *general independent* interarrival and service times. Additionally, the imposition of parametric assumptions on the corresponding marginal distributions often yields elegant mathematical solutions. In queueing, the "memoryless" exponential distribution occupies a special position, as evidenced by the popularity of the simple $M/M/m$ queue and networks thereof.

On the other hand it must be recognized that the presence of temporal dependence in real-life systems does not automatically preclude simple models that assume their absence. The pertinent question to ask concerns the modeling risk attendant to ignoring dependence: Does it entail a substantial error in predicting performance as compared, say, to renewal models? Common sense suggests that the prediction error would depend on the "magnitude" of real-life dependence being ignored in the proposed model. Queueing systems provide instructive examples. A number of studies (Fendick et al., 1989; Heffes and Lucantoni, 1986; Livny et al., 1993; Patuwo et al., 1993) have shown that when autocorrelated customer interarrival times drive a queuing system, the resulting performance measures are much worse than those corresponding to renewal traffic (which lacks dependence among interarrival times). In particular, positively autocorrelated interarrivals manifest themselves as "bursty" traffic. A little reflection reveals that bursty traffic should give rise to higher mean waiting times, since the effect of burstiness is that "everybody arrives at once," thereby exacerbating waiting times. What can be startling is the magnitude of this effect. For example, Livny et al. (1993) shows that strongly autocorrelated interarrival and/or service times can easily yield orders-of-magnitude discrepancies from predicted mean waiting times that ignore autocorrelations, that is, when compared to renewal interarrival/service times with the same marginal distribution. This effect is observed in all traffic conditions including light traffic, but is greatly exacerbated in heavy traffic.

The TES modeling methodology is motivated by the importance of capturing both the histogram and leading autocorrelations of empirical data. To set up the problem formally, we introduce some notation. Let $\{Y_n\}_{n=0}^{N-1}$ be an empirical data sequence (record) of size N, sampled from an unknown real-valued, discrete-time stationary time series. Further, let $\{X_n\}_{n=0}^{\infty}$ denote a stationary stochastic sequence, purporting to fit the empirical data (a "model"); the X_n have common density function f_X, with mean μ_X and variance $0 < \sigma_X < \infty$. The marginal density of $\{Y_n\}$ is modeled by the empirical histogram density

$$\hat{h}_Y(y) = \sum_{j=1}^{J} 1_{[l_j, r_j)}(y)\frac{\hat{p}_j}{w_j}, \qquad -\infty < y < \infty, \tag{2.8.1}$$

where $1_A(x)$ denotes the indicator function of set A, J is the number of histogram cells, $[l_j, r_j)$ is the support of cell j with width $w_j = r_j - l_j > 0$, and $\hat{p}_j$ is the probability of cell j (the relative frequency of observations that fell in the interval $[l_j, r_j)$). The hat symbol is used as a reminder that the corresponding quantity is a statistical estimator. Note that the empirical marginal density (2.8.1) is modeled as a step-function density, that is, a probabilistic mixture of J uniform densities, such that cell j corresponds to a uniform density over $[l_j, r_j)$ with mixing probability $\hat{p}_j$. This modeling devise has two major advantages. First, step functions are mathematically very simple and, second, they incur no practical loss of generality as they can approximate measurable functions arbitrarily closely.

The autocorrelation function of $\{X_n\}$ is

$$\rho_X(\tau) = \frac{E[X_n X_{n+\tau}] - \mu_X^2}{\sigma_X^2}, \qquad \tau \geq 1, \tag{2.8.2}$$

while $\hat{\rho}_Y(\tau)$, $1 \leq \tau \ll N$, is some estimator of the autocorrelation function of $\{Y_n\}$, based on (2.8.2).

The TES modeling philosophy considers a candidate model $\{X_n\}$ (for the empirical record $\{Y_n\}$) "good" if it satisfies the following goodness-of-fit requirements:

1. The marginal density function of the model precisely matches its empirical counterpart, namely, $f_X \equiv \hat{h}_Y$.
2. The autocorrelation function of the model approximates its empirical counterpart, namely, $\rho_X(\tau) \simeq \hat{\rho}_Y(\tau)$ for all leading lags (often defined as the range $1 \leq \tau \leq \sqrt{N}$, for "small" N).
3. Sample paths generated by Monte Carlo simulation of the model "resemble" the empirical data.

The first two requirements are quantitative goodness-of-fit criteria, which call for first-order and second-order empirical statistics to be adequately captured. These are precise and well-defined conditions, although the particular metric of goodness is left up to the analyst. In contrast, the third requirement is qualitative and cannot be defined with mathematical rigor, being inherently subjective. Still, it constitutes a common-sense check on model goodness, which serves to increase the confidence in the model. Note carefully that qualitative "resemblance" is not a substitute for quantitative fitting, but more like a supplementary criterion: Among all models that fit the empirical histogram and autocorrelation function, one would select the model whose Monte Carlo realizations "resemble" most the empirical data.

The main advantage of TES modeling stems from the fact that the TES approach addresses itself directly to the three modeling requirements above, and consequently, a successful TES model is a priori more accurate than models subject to less stringent modeling requirements. Furthermore, the versatility of the class of TES processes keeps the chance of success comfortably high. TES sequences give rise to a host of qualitatively diverse sample path behaviors, encompassing cyclical and nondirectional behaviors, as well as probabilistic structures that range from (conditional) determinism to iid. The marginal distributions of TES sequences are unconstrained, and their autocorrelation functions admit a variety of functional forms (monotone, oscillating, alternating, and others). All in all, the TES modeling methodology enjoys a considerable latitude in approximating the empirical autocorrelation function, even as it maintains the matching of the empirical marginal distribution.

The main disadvantage of TES modeling is that it is not entirely algorithmic in nature. Rather, it consists of a heuristic search (explained later) over a large parametric space. Implementing the TES methodology requires software support with extensive graphics for visually assessing the goodness of the current model. Fortunately, the autocorrelation function of a TES model can be computed from fast numerical formulas without requiring simulation. This is important in interactive modeling since the numerical computations are much faster than the corresponding simulation-based statistical calculations. A software environment, called TEStool, which supports this methodology has been constructed (Geist and Melamed, 1992; Melamed et al., 1992b; Hill and Melamed, 1995) and extensively used to model a variety of empirical time series (Lee et al., 1992; Melamed et al., 1992a; Melamed and Hill, 1995).

2.8.1 TES Processes

Let $\lfloor x \rfloor = \max\{n \text{ integer}: n \le x\}$ be the integral part of x, and define the modulo-1 operator $\langle \cdot \rangle$ to be the fractional part of x, namely, $\langle x \rangle = x - \lfloor x \rfloor$. Let $\{V_n\}$ be a sequence of iid random variables with a common density function f_V. Further, let U_0 be uniform on [0, 1) and independent of each element of the sequence $\{V_n.\}$ The random variables V_n are referred to as *innovations*.

The construction of TES processes involves two stochastic sequences in time lockstep, called *background* and *foreground* processes (sequences), respectively, and related by suitable transformations. The background sequence is auxiliary, whereas the corresponding foreground sequence is the target model.

Background TES processes are stationary and fall in two classes, denoted by TES$^+$ and TES$^-$, which consist of stochastic sequences $\{U_n^+\}_{n=1}^{\infty}$ and $\{U_n^-\}_{n=0}^{\infty}$, respectively. The definition of $\{U_n^+\}$ is given recursively by

$$U_n^+ = \begin{cases} U_0, & n = 0 \\ \langle U_{n-1}^+ V_n \rangle, & n > 0, \end{cases} \tag{2.8.3}$$

and the sequence $\{U_n^-\}$ is defined in terms of $\{U_n^+\}$ by

$$U_n^- = \begin{cases} U_n^+, & n \text{ even} \\ 1 - U_n^+, & n \text{ odd}. \end{cases} \tag{2.8.4}$$

The plus and minus superscripts are suggestive of the fact that background TES processes can realize any lag-1 autocorrelation (see Melamed, 1991); TES$^+$ sequences cover the positive range [0, 1], while TES$^-$ sequences cover the negative range [−1, 0]. From now on, we shall consistently append a plus or minus superscript to other mathematical objects associated with those classes, unless the distinction is immaterial; in particular, Eqs. (2.8.3) and (2.8.4) imply that their associated autocorrelation functions are related by $\rho_U^-(\tau) = (-1)^\tau \rho_U^+(\tau)$, $\tau \ge 0$. It can also be shown (Jagerman and Melamed, 1992a) that (2.8.3) and (2.8.4) each give rise to stationary Markovian sequences; however, while $\{U_n^+\}$ has stationary transition probabilities, $\{U_n^-\}$ has a nonstationary transition structure due to its dependence on whether the time index is even or odd. Most importantly, the marginal distributions of both $\{U_n^+\}$ and $\{U_n^-\}$ are always uniform on [0, 1), *regardless* of the innovation sequence chosen.

Foreground TES processes, $\{X_n^+\}$ and $\{X_n^-\}$, are obtained from Eqs. (2.8.3) and (2.8.4), respectively, by

$$X_n^+ = D(U_n^+), \tag{2.8.5}$$

$$X_n^- = D(U_n^-), \tag{2.8.6}$$

where the transformation D is called a *distortion*. The fact that all sequences, $\{U_n^+\}$ and $\{U_n^-\}$, have uniform marginals on [0, 1) allows us to generate random sequences with essentially arbitrary marginals by using the *inverse-transform* method (see Section 2.2.1); for a given distribution function F, take $D = F^{-1}$ (recall Algorithm 2.2.1). For a general distortion D, the resultant sequences $\{X_n^+\}$ and $\{X_n^-\}$ may or may not be Markovian, depending on D.

Recall that the empirical histogram density, $\hat{h}_Y$ of the form (2.8.1), is a step function. Hence, the corresponding distribution function, $\hat{H}_Y$, is a piecewise-linear function. By the monotonicity of $\hat{H}_Y$, its inverse function, $\hat{H}_Y^{-1}$, can be defined as the (piecewise-linear) function

$$\hat{H}_Y^{-1} = \sum_{j=1}^{J} 1_{[\hat{C}_{j-1}, \hat{C}_j)}(x) \left[l_j + (x - \hat{C}_{j-1}) \frac{w_j}{\hat{p}_j} \right], \tag{2.8.7}$$

where $\{\hat{C}_j\}_{j=0}^J$ is the cumulative distribution of $\{\hat{p}_j\}_{j=1}^J$, that is, $\hat{C}_j = \sum_{i=1}^j \hat{p}_i$ ($\hat{C}_0 = 0$ and $\hat{C}_J = 1$). An inverse-transform method of the form (2.8.7) is called a *histogram inversion*. Clearly, the method guarantees that for any background TES sequence $\{U_n\}$, the associated foreground sequence $\{\hat{H}_Y^{-1}(U_n)\}$ has marginal density $\hat{h}_Y$. In a similar vein, it is advantageous, in practice, to restrict the scope of innovation densities under consideration to the subclass of step-function densities, namely, probabilistic mixtures of uniform densities whose support is contained in the unit circle. Such densities have the form

$$f_V(x) = \sum_{k=1}^{K} 1_{[L_k, R_k)}(x) \frac{P_k}{R_k - L_k}, \tag{2.8.8}$$

where K is the number of steps, step k has support $[L_k, R_k)$ satisfying $-0.5 \le L_k < R_k < 0.5$, and $0 < P_k \le 1$ is the probability of step k ($\sum_{k=1}^K P_k = 1$). For convenience, we also require that distinct steps do not overlap (i.e., $R_k \le L_{k+1}$ for all $1 \le k \le K - 1$). In addition to the usual advantages of simplicity and generality, step-function densities are also easy to represent and manipulate graphically. As will be seen later on, this is key to the heuristic search approach to TES modeling to be discussed in Section 2.8.2.

A Monte Carlo simulation of (2.8.5) and (2.8.6) can always provide an estimate of the associated autocorrelation functions, ρ_X^+ and ρ_X^-, respectively, provided a sufficient sample size is generated. This approach, however, can be costly in terms of computation time, especially when high autocorrelations necessitate large sample sizes for adequate statistical reliability. Thus, a numerical approach is preferable. Computationally efficient formulas for the autocorrelation functions of $\{X_n^+\}$ and $\{X_n^-\}$ are available for general innovation densities and general distortions (Jagerman and Melamed, 1992a,b). For a given lag τ, the corresponding autocorrelation functions are given, respectively, by

$$\rho_X^+(\tau) = \frac{2}{\sigma_X^2} \sum_{\nu=1}^{\infty} \Re[\tilde{f}_V^\tau(i2\pi\nu)] \left|\tilde{D}(i2\pi\nu)\right|^2 \tag{2.8.9}$$

and

$$\rho_X^-(\tau) = \begin{cases} \rho_X^+(\tau), & \tau \text{ even} \\ \dfrac{2}{\sigma_X^2} \displaystyle\sum_{\nu=1}^{\infty} \Re[\tilde{f}_V^\tau(i2\pi\nu)] \Re\left[\tilde{D}^2(i2\pi\nu)\right], & \tau \text{ odd,} \end{cases} \tag{2.8.10}$$

where σ_X^2 is the common variance of $\{X_n^+\}$ and $\{X_n^-\}$, $\tilde{f}_V$ and $\tilde{D}$ denote the Laplace transform of f_V and D respectively, $\Re[z]$ denotes the real part of a complex number z, and $i = \sqrt{-1}$. It is worth noting that the effects of the innovation and distortion are conveniently separated in (2.8.9) and (2.8.10). More importantly, the series converge rapidly.

Intuitively, the modulo-1 arithmetic used in defining TES processes in Eqs. (2.8.3) and (2.8.4) has a simple geometrical interpretation as a random walk on the unit circle (circumference 1), with random step size V_n. This is illustrated in Figure 2.6. Here the unit circle lies at the bottom, in the two-dimensional plane, with the current TES variate U_n in the north/north-easterly sector. A step-function innovation density f_V was erected over the unit circle, with values in the third dimension (the "up" direction). The origin for the support of f_V was set at the current TES variate U_n, indicating that the support of f_V is the interval $[-0.5, 0.5)$. To obtain the next TES$^+$ variate, U_{n+1}, just sample a value on the unit circle from the innovation density f_V. Because the origin of f_V was set at U_n, this procedure implies $U_{n+1} = \langle U_n + V_n \rangle$, in agreement with Eq. (2.8.3). To map a TES$^-$ method into a geometrical interpretation in Figure 2.6, simply use alternately f_V for even indices n and its antithetic counterpart (corresponding to sampling $U_{n+1}^- = 1 - U_n^+$) for odd indices n, as prescribed by Eq. (2.8.4).

Figure 2.6 provides some insight into the qualitative nature of TES sequences as a function of the innovation density. It can be readily seen that the way the innovation mass is distributed around the origin determines whether or not the random walk will have a drift around the unit circle. For example, a clockwise drift will ensue when the probability mass to the "right" of the origin exceeds the mass to its "left," resulting in sample paths with a (random) cyclical appearance. By the same token, a symmetric density about the origin gives rise to a driftless random

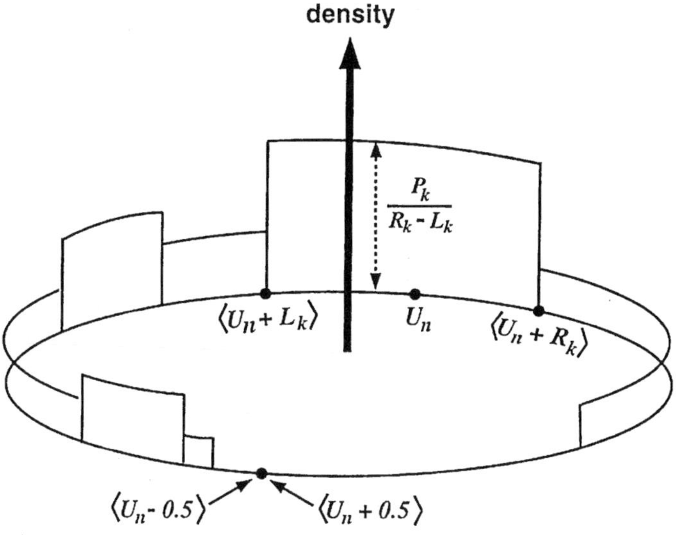

Figure 2.6. Background TES sequences with step-function innovations.

walk. Furthermore, when the innovation density mass has a small support about the origin, then successive TES variates must lie in relatively close proximity on the unit circle. This results in higher autocorrelations as compared to innovation densities with a larger support.

Figure 2.6 also explains why the sample paths of both background and foreground TES sequences often exhibit a visual "discontinuity" whenever the background sequence "crosses" point 0 on the unit circle in either direction (Melamed, 1991; Jagerman and Melamed, 1992a,b). In order to "smooth" TES sample paths, one can employ the family of so-called *stitching transformations*, S_ξ, parameterized by $0 \le \xi \le 1$, and defined for all $y \in [0, 1)$ by

$$S_\xi(y) = \begin{cases} y/\xi, & 0 \le y < \xi \\ (1-y)/(1-\xi), & \xi \le y < 1. \end{cases} \tag{2.8.11}$$

Processes of the form $\{S_\xi(U_n^+)\}$ and $\{S_\xi(U_n^-)\}$ are called *stitched* TES processes. Note that for $0 < \xi < 1$, the S_ξ are continuous on the unit circle; in particular, $S_\xi(0) = S_\xi(1)$, which intuitively explains why stitching transformations have a "smoothing" effect on sample paths. It can be shown (Melamed, 1991) that all S_ξ, $0 \le \xi \le 1$, preserve uniformity; consequently, stitched TES processes can still be distorted to arbitrary marginals via the inverse-transform method (Algorithm 2.2.1). A commonly used distortion is

$$D_{Y,\xi}(x) = \hat{H}_Y^{-1}(S_\xi(x)), \qquad 0 \le x < 1, \tag{2.8.12}$$

which applies stitching and histogram inversion in succession.

A detailed discussion of the qualitative behavior of TES sample paths and autocorrelation functions as a function of the innovation density and the stitching parameter may be found in Melamed (1991) and Jagerman and Melamed (1992a,b). The spectral functions of TES processes were studied in Jagerman and Melamed (1994).

2.8.2 TES Modeling Methodology

The TES modeling methodology is essentially a heuristic search for a pair (ξ, f_V) of a stitching parameter and an innovation density. It is best understood in terms of the typical user interaction with a visual TES modeling software environment such as TEStool (Geist and Melamed, 1992; Melamed et al., 1992b; Hill and Melamed, 1995).

Figure 2.7 displays a typical TEStool screen at the end of a modeling session. The screen is subdivided into four canvas areas, designed to display various types of graphics and controlled by buttons and menu selections. The empirical data modeled here consisted of (random) frame bit rates (sizes) of compressed video (Melamed and Sengupta, 1992). The upper-left canvas plots the empirical sample path of compressed frame bit rates (round markers); superimposed on it is a sample path generated by a Monte Carlo simulation of a TES model (diamond markers). Similarly, the upper-right canvas contains the corresponding histograms, while the lower-left canvas contains the empirical autocorrelation function and its numerically computed TES model counterpart. The TES model itself is displayed graphically in the lower-right canvas, which contains a joint specification of a TES sign, stitching parameter and an innovation density.

A typical modeling scenario would unfold as follows. The empirical data is read in and its empirical histogram (2.8.1) is calculated (with a user-supplied number of cells). The empirical histogram then completely determines a histogram inversion (2.8.7), which together with a choice of $\xi \in [0, 1]$, determine, in turn, a composite distortion of the form (2.8.12). Next, the empirical autocorrelation function is estimated from the empirical sample path. The core activity of TES modeling is a heuristic search for a suitable stitching parameter and innovation density. The modeler searches through stitching parameters in the range [0, 1], and innovation densities in the space of step-function densities (2.8.8) whose support is contained in $[-0.5, 0.5)$, as displayed in the lower-right canvas of Figure 2.7. Note that the innovation-density graph depicted there is a two-dimensional representation corresponding to Figure 2.6, where the unit circle was severed at the origin and then unraveled and shifted to the interval $[-0.5, 0.5)$; this can be done without loss of generality, since modulo-1 arithmetic works with any innovation-density support of unit length. At the same time, the user also looks for models whose simulated sample paths "resemble" their empirical counterparts. Recall that $\{X_n^+\}$ and $\{X_n^-\}$ always have the requisite distribution $\hat{H}_Y$ *regardless* of the innovation sequence selected. However, different innovation densities give rise to different stochastic processes with different autocorrelation functions, thereby providing a large degree of freedom in fitting a temporal dependence structure. To summarize, the TES approach decomposes the selection of the marginal distribution and autocorrelation function, rendering them largely orthogonal choices.

The TEStool approach to implementing the heuristic search is to cast it into a visual interactive activity, guided by visual feedback, in the spirit of an arcade game. The user uses the mouse to choose a TES class by pressing a button and to select a stitching parameter from a slider. Taking advantage of the fact that step functions can be visually represented as nonoverlapping rectangles, the user manipulates the mouse and its buttons so as to visually create, delete, resize, and relocate innovation density steps in a natural way: the mouse is used to "grab" rectangles, to "stretch" and "shrink" them horizontally and vertically, and to "drag" them about horizontally. Each such operation results in a change of a TES model specification, which triggers in turn a recomputation and redisplay of a TES autocorrelation graph and/or simulation of a sample path or histogram. Graphs of TES statistics are then superimposed on the corresponding empirical counterparts for visual comparison. Visual feedback guides the search process—the goal being to bring the TES autocorrelation graph as close as possible to the corresponding empirical target while simultaneously judging the "qualitative similarity" of the model-generated sample path to the empirical data [recall that matching the empirical histogram is guaranteed in any TES model utilizing the composite distortion (2.8.12), and need not be checked by the modeler]. The fast computations of Eqs. (2.8.9) and (2.8.10) permit a real-time interaction.

Figure 2.7 exhibits excellent agreement between empirical and TES statistics: The empirical histogram and leading empirical autocorrelations are well approximated by their TES model counterparts, and a marked "similarity" is evident in the functional form of the empirical and model sample paths. Thus, Figure 2.7 can be deemed to be a successful outcome of TES modeling, in line with the three afore-stipulated modeling requirements.

A more algorithmic TES modeling approach has been devised and implemented in TEStool (see Jelenkovic and Melamed, 1995). The algorithm first carries out a brute-force search over a subspace of step-function innovation densities and various stitching parameters; recall that the distortion is completely determined by the empirical record and user-supplied histogram parameters. Of those, the algorithm selects the best n combinations of pairs (f_V, ξ), in the sense that the associated TES model autocorrelation functions have the smallest sum of squared deviations. The algorithm then performs nonlinear optimization on each candidate model to further reduce the aforementioned sum of squared deviations. Finally, the analyst peruses the results and selects from the n optimized candidate models the one whose Monte Carlo sample paths bear the "most resemblance" to the empirical record, in addition to having a small sum of squared deviations. Preliminary experience suggests that algorithmic TES modeling produces better and faster results than its heuristic counterpart; see Jelenkovic and Melamed (1995).

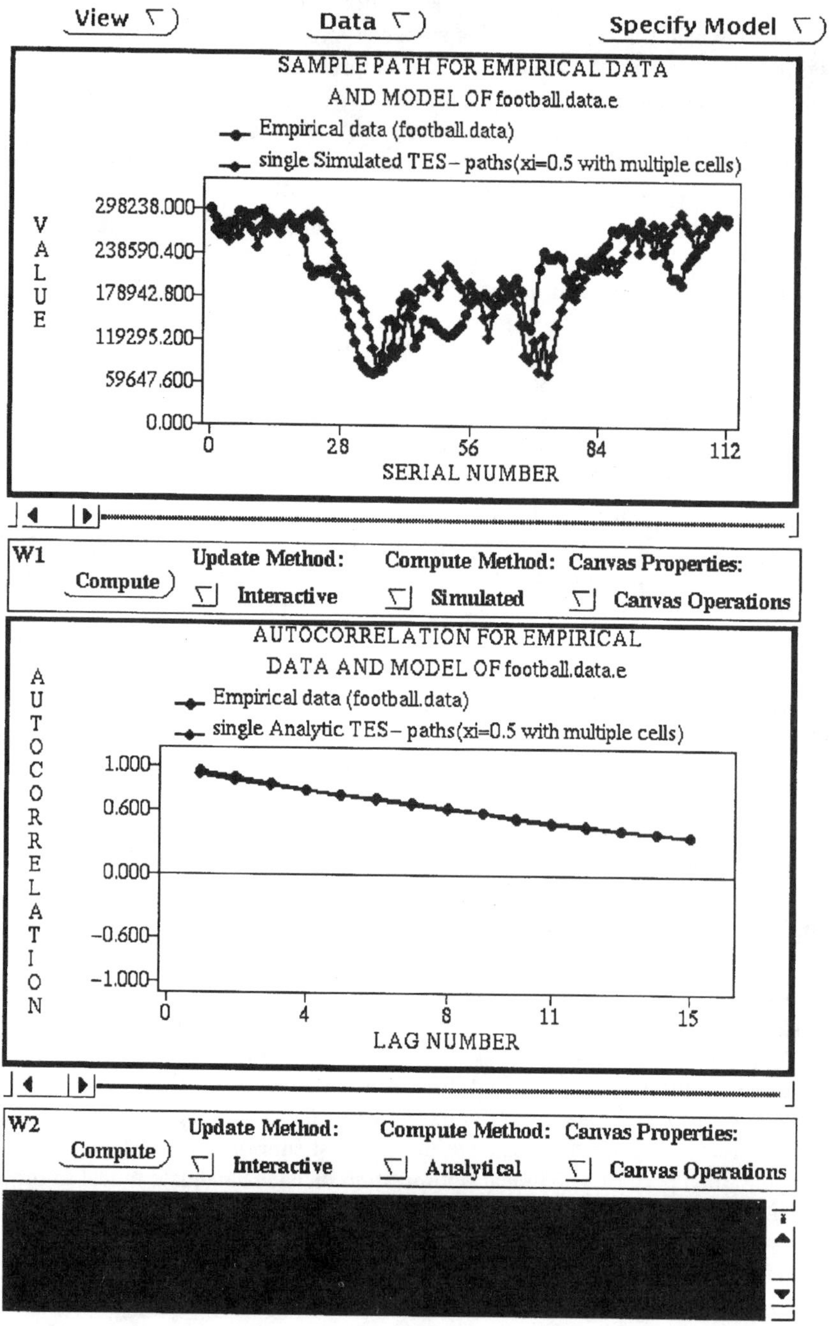

Figure 2.7. Screen of a TEStool model (lefthand side canvases).

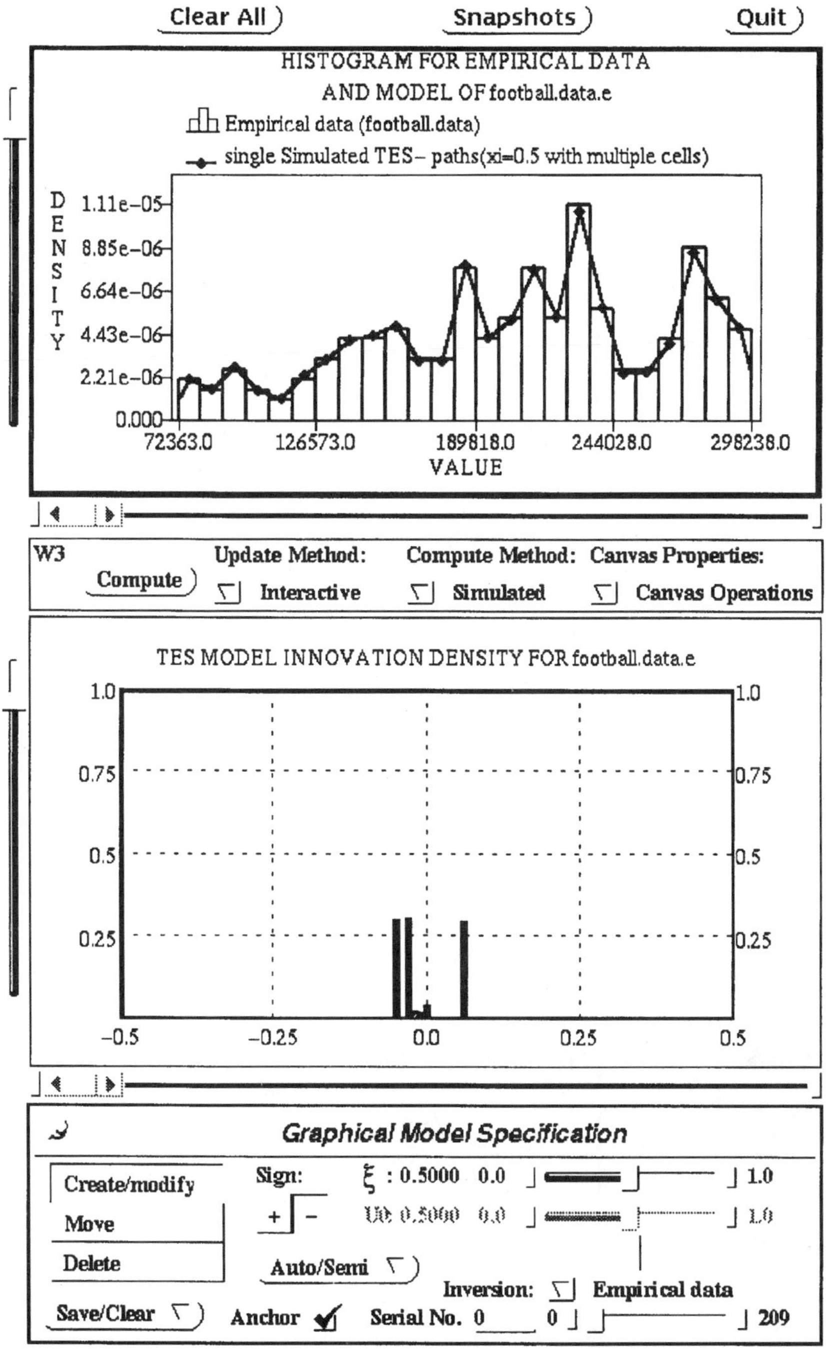

Figure 2.7. *Continued* (righthand side canvases).

2.9 EXERCISES

1. Apply the inverse-transform method to variate generation from a Laplace distribution (shifted two-sided exponential distribution)

$$f(x) = \frac{1}{2\beta}\exp\left[\frac{-|x-\theta|}{\beta}\right], \qquad -\infty < x < \infty, \quad \beta > 0.$$

***2.** Apply the inverse-transform method to variate generation from the extreme value distribution

$$f(x) = \frac{1}{\sigma}\exp\left[-\frac{1}{\sigma}(x-\mu) - \exp\left(\frac{-(x-\mu)}{\sigma}\right)\right], \qquad -\infty < x < \infty.$$

3. Consider the triangular variate, with the pdf

$$f(x) = \begin{cases} 0 & \text{if } x < 2a \text{ or } x \geq 2b \\ \dfrac{x-2a}{(b-a)^2} & \text{if } 2a \leq x < a+b \\ \dfrac{2b-x}{(b-a)^2} & \text{if } a+b \leq x < 2b, \end{cases}$$

and cdf

$$F(x) = \begin{cases} 0 & \text{if } x < 2a \\ \dfrac{(x-2a)^2}{2(b-a)^2} & \text{if } 2a \leq x < a+b \\ 1 - \dfrac{(2b-x)^2}{2(b-a)^2} & \text{if } a+b \leq x < 2b \\ 1 & \text{if } x \geq 2b. \end{cases}$$

Show that applying the inverse-transform method yields

$$X = \begin{cases} 2a + (b-a)\sqrt{2U} & \text{if } 0 \leq U < 0.5 \\ 2b + (a-b)\sqrt{2(1-U)} & \text{if } 0.5 \leq U < 1. \end{cases}$$

***4.** Let

$$f(x) = \begin{cases} C_i x & x_{i-1} \leq x < x_i,\ i = 1, \ldots, n \\ 0 & \text{otherwise} \end{cases}$$

$$x_0 = a, \quad x_n = b, \quad C_i \geq 0, \quad a \geq 0.$$

Using the inverse-transform method, show that

$$X = \left[x_{i-1}^2 + \frac{2(U - F_{i-1})}{C_i}\right]^{1/2}$$

where $F_i = \sum_{j=1}^{i} \int_{x_{j-1}}^{x_j} C_j x \, dx$. Describe an algorithm for variate generation from $f(x)$.

5. Apply the inverse-transform method to generate a variate from the following pdf:

$$f(x) = \begin{cases} \dfrac{1}{n+1} & x = 0, 1, \ldots, n \\ 0 & \text{otherwise.} \end{cases}$$

6. Devise an acceptance-rejection algorithm for generating a variate from the pdf

$$f(x) = \sqrt{\frac{2}{\pi}} e^{-x^2/2}, \qquad x \geq 0,$$

using the representation $f(x) = Ch(x, \beta)g(x)$, where

$$h(x, \beta) = \frac{1}{\beta} e^{-x/\beta}, \qquad x > 0, \ \beta > 0,$$

for fixed β.

7. Devise an algorithm for generating a variate from the pdf

$$f(x) = ke^{-x}, \qquad 0 \leq x \leq a$$

using
(a) the inverse-transform method
(b) the acceptance-rejection method with

$$f(x) = Ch(x)g(x), \qquad \text{where } h(x) = \lambda e^{-\lambda x}, x > 0.$$

Find the efficiency of the acceptance-rejection method for the cases $a = 1$, $a \to \infty$, and $a \to 0$.

***8.** Give an acceptance-rejection algorithm for generating a variate from the pdf

$$f(x) = \begin{cases} C_i x & x_{i-1} \leq x < x_i, i = 1, \ldots, n \\ 0 & \text{otherwise,} \end{cases}$$

where $x_0 = 0$, $x_n = 1$, $C_i \geq 0$, $a \geq 0$. Represent $f(x)$ as

$$f(x) = Ch(x)g(x),$$

where

$$h(x) = 2x, \qquad 0 < x < 1.$$

Calculate the efficiency of the algorithm for the case

$$C_1 = C_2 = \cdots = C_n, \qquad x_i = \frac{i}{n}, \quad i = 1, \ldots, n.$$

9. Present an inverse-transform algorithm for generating a variate from the piecewise-constant pdf

$$f(x) = \begin{cases} C_i & x_{i-1} \leq x \leq x_i, i = 1, 2, \ldots, n \\ 0 & \text{otherwise,} \end{cases}$$

where $C_i \geq 0$ and $a = x_0 < x_1 < \cdots < x_{n-1} < x_n = b$.

10. Let the random variable X have pdf

$$f(x) = \begin{cases} \frac{1}{4} & 0 < x < 1 \\ \frac{1}{4} + x & 1 \leq x \leq 2. \end{cases}$$

Generate a variate from $f(x)$, using
(a) the inverse-transform method
(b) the acceptance-rejection method.
In case (b), represent $f(x)$ as

$$f(x) = Ch(x)g(x), \qquad \text{where} \quad h(x) = \tfrac{1}{2}, 0 \leq x \leq 2.$$

11. Let the random variable X have pdf

$$f(x) = \begin{cases} \frac{1}{2}x & 0 < x < 1 \\ \frac{1}{2} & 1 \leq x \leq \frac{5}{2}. \end{cases}$$

Generate a variate from $f(x)$, using
(a) the inverse-transform method
(b) the acceptance-rejection method.

In case (b), represent $f(x)$ as

$$f(x) = Ch(x)g(x), \quad \text{where } h(x) = \tfrac{8}{25}x,\ 0 \le x \le \tfrac{5}{2}.$$

12. Let the random variable X have pdf

$$f(x) = kp(1-p)^x, \quad x = 0, 1, \ldots, n, \quad k \ge 1.$$

Generate a variate from $f(x)$, using

(a) the inverse-transform method

(b) the acceptance-rejection method.

In case (b), represent $f(x)$ as

$$f(x) = Ch(x)g(x), \quad \text{where} \quad h(x) = p(1-p)^x, \quad x \ge 0.$$

Find the efficiency of the acceptance-rejection method for $n = 1$ and $n = \infty$.

13. Prove the validity of the composition algorithm (2.2.4).

14. Generate a variate $Y = \min_{i=1,\ldots,m} \max_{j=1,\ldots,r} \{X_{ij}\}$, assuming that the variates X_{ij}, $i = 1, \ldots, m$; $j = 1, \ldots, r$, are independent and identically distributed with common cdf $F(x)$. Apply both

(a) the inverse-transform method

(b) the acceptance-rejection method.

Hint: Use the results of the distribution of order statistics in Example 2.2.2.

***15.** Prove Lemma 2.5.1 of the alias method.

REFERENCES

Atkinson, A. C. (1979a). "A family of switching algorithms for the computer generation of Beta random variables," *Biometrika*, **66**, 141–145.

Atkinson, A. C. (1979b). "The computer generation of Poisson random variables," *Appl. Statist.*, **28**, 29–35.

Borovkov, S. (1994). "On simulation of random vectors with given densities in regions and on their boundaries," *J. Appl. Prob.*, **31**, 205–220.

Box, G. E. P. and Muller, M. E. (1958). "A note on the generation of random normal deviates," *Ann. Math. Stat.*, **29**, 610–611.

Claude, J. P., Bélisle, H., Romeijn, E., and Smith, R. L. (1993). "Hit-and-run algorithm for generating multivariate distributions," *Math. Oper. Res.*, **18**(2), 255–266.

Devroye, L. (1981). "The computer generation of Poisson random variables," *Computing*, **26**, 197–207.

Devroye, L. (1986). *Non-Uniform Random Variate Generation*, Springer, New-York.

Fendick, K. W., Saksena, V. R., and Whitt, W. (1989). "Dependence in packet queues," *IEEE Trans. Commun.*, **37**, 1173–1183.

Geist D. and Melamed, B. (1992). "TEStool: An environment for visual interactive modeling of autocorrelated traffic," *Proceedings of ICC '92*, Chicago, pp. 1285–1289.

Heffes, H. and Lucantoni, D. M. (1986). "A Markov modulated characterization of packetized voice and data traffic and related statistical multiplexer performance," *IEEE J. Selected Areas Commun.*, **SAC-4**, 856–868.

Hill, J. R. and Melamed, B. (1995). "TEStool: A visual interactive environment for modeling autocorrelated time series," *Performance Eval.*, **24**, 3–22.

Jagerman, D. L. and Melamed, B. (1992a). '"The transition and autocorrelation structure of TES processes, part I: general theory," *Stochastic Models*, **8**(2), 193–219.

Jagerman, D. L. and Melamed, B. (1992b). "The transition and autocorrelation structure of TES processes, part II: special cases," *Stochastic Models*, **8**(3), 499–527.

Jagerman, D. and Melamed, B. (1994). "The spectral structure of TES processes," *Stochastic Models*, **10**(3), 599–618.

Jelenkovic, P. and Melamed, B. (1995). "Algorithmic modeling of TES processes," *IEEE Trans. Autom. Con.*, **40**(7), 1305–1312.

Jöhnk, M. D. (1964). "Erzeugung von Betaverteilten und Gammaverteilten Zufallszahlen," *Metrika*, **8**, 5–15.

Law, A. M. and Kelton, W. D. (1991). *Simulation Modeling and Analysis*, 2nd Ed. McGraw Hill, New York.

Knuth, D. E. (1981). *The Art of Computer Programming, Vol. 2: Seminumerical Algorithms*, 2nd ed., Addison-Wesley, Reading, MA.

L'Ecuyer, P. (1990). "Random numbers for simulation," *Commun. ACM*, **33**(10), 85–97.

L'Ecuyer, P. and Tezuka, S. (1991). "Structural properties for two classes of combined random number generators," *Math. Comput.*, **57**, 735–746.

Lee, D. -S., Melamed, B., Reibman, A., and Sengupta, B. (1992). "TES modeling for analysis of a video multiplexor," *Performance Eval.*, **16**, 21–34.

Lehmer, D. H. (1951). "Mathematical methods in large-scale computing units," *Ann. Comp. Lab. Harvard Univ.*, **26**, 141–146.

Lewis, P. A. W. and Shedler, G. S. (1979). "Simulation of nonhomogeneous Poisson process by thinning," *Nav. Res. Logist. Quart.*, **26**, 403–413.

Livny, M., Melamed, B., and Tsiolis, A. K. (1993). "The impact of autocorrelation on queuing systems," *Manage. Sci.*, **39**(3), 322–339.

Melamed, B. (1991). "TES: A class of methods for generating autocorrelated uniform variates," *ORSA J. Comput.*, **3**, 317–329.

Melamed, B. (1993). "An overview of TES processes and modeling methodology," in *Performance Evaluation of Computer and Communications Systems*, (L. Donatiello and R. Nelson, Eds.), Lecture Notes in Computer Science, Springer-Verlag, pp. 359–393.

Melamed, B. and Sengupta, B. (1992). "TES modeling of video traffic," *IEICE Trans. Commun.*, **E75-B**(12), 1291–1300.

Melamed, B., Raychaudhuri, D., Sengupta, B., and Zdepski, J. (1992a). "TES-based traffic modeling for performance evaluation of integrated networks," *Proceedings of INFOCOM '92*, Florence, Italy, pp. 75–84.

Melamed, B., Goldsman, D., and Hill, J. R. (1992b). "The TES methodology: Nonparametric modeling of stationary time series," *Proceedings of the 1992 Winter Simulation Conference*, Arlington, Virginia, pp. 1291–1300.

Melamed, B. and Hill, J. R. (1995). "A survey of TES modeling applications," *Simulation*, **64**(6), 353–370.

Niederreiter, H. (1991). "Recent trends in random number and random vector generation," *Ann. Oper. Res.*, **31**, 323–345.

Niederreiter, H. (1992). *Random Number Generation and Quasi-Monte Carlo Methods*, SIAM, Philadelphia.

Patuwo, B. E., Disney, R. L., and McNickle, D. C. (1993). "The effect of correlated arrivals on queues," *IIE Trans.*, **25**(3), 105–110.

Ripley, B. D. (1983). "Computer generation of random variables: A tutorial," *Int. Statist. Rev.*, **51**, 301–319.

Rubinstein, R. Y. (1981). *Simulation and the Monte Carlo Method*, Wiley, New York.

Rubinstein R. Y. (1982). "Generating random vectors uniformly distributed inside and on the surface of different regions," *Euro. J. OR*, **10**, 205–209.

Smith, R. L. (1984). "Efficient Monte Carlo procedures for generating points uniformly distributed over bounded regions," *Oper. Res.*, **32**, 1296–1308.

Tadikamalla, P. R. and Johnson, M. E. (1981). "A Complete guide to gamma variate generation," *Am. J. Math. Manage. Sci.*, **1**, 78–95.

Walker, A. J. (1977). "An efficient method for generating discrete random variables with general distributions," *Assoc. Comput. Mach. Trans. Math. Software*, **3**, 253–256.

CHAPTER 3

Output Analysis of Discrete-Event Systems via Simulation

3.1 INTRODUCTION

The purpose of this chapter is to present a statistical approach to analyzing output data obtained via simulation. Section 3.2 treats statistical analysis of output data from static models, while the rest of the chapter treats statistical analysis of output data from dynamic models, such as queueing models. In particular, Sections 3.3 and 3.4 discuss output analysis for finite-horizon and steady-state simulation. Section 3.5 deals with planning queueing simulations, so as to predict how long the simulation should be run in order to obtain valid confidence intervals for such parameters as the expected number of customers and the expected waiting time in the system. It is shown that the required sample size (simulation time) increases rapidly (nonlinearly) with the traffic intensity, ρ, and that when ρ approaches unity, simulation becomes effectively useless. Recall that $\rho = \mathbb{E}\{X\}/\mathbb{E}\{Y\}$, where X and Y denote the service time and the interarrival time random variables, respectively. Finally, Sections 3.6 and 3.7 present the batch-means method and the regenerative method for estimating steady-state performance measures.

3.2 STATIC MODELS

This section deals with output analysis of a discrete-event static system (DESS). Let ℓ be some expected performance, and assume that ℓ can be represented as

$$\ell = \mathbb{E}\{L(\boldsymbol{Y})\} = \int L(\boldsymbol{y})\, dF(\boldsymbol{y}), \tag{3.2.1}$$

where $F(\boldsymbol{y})$ is the cumulative distribution function (cdf) of the random vector $\boldsymbol{Y}$. The random variable $L(\boldsymbol{Y})$ is called the *sample performance function*, while independent simulation runs yielding $\boldsymbol{Y}_1, \ldots, \boldsymbol{Y}_N$ from the same probability law are called *replications*. As examples of such DESS, consider the following:

1. *Reliability Systems.* The mean lifetime of a coherent reliability system can be written as

$$\ell = \mathbb{E}\left\{ \max_{j=1,\ldots,p} \min_{i \in \mathcal{L}_j} Y_i \right\}. \tag{3.2.2}$$

   Here, $\mathcal{L}_j$ is the jth complete path from a source to a sink; p is the number of complete paths; and Y_i, $i = 1, \ldots, m$, are the durations (lifetimes) of the components, where Y_i has cdf $F_i(y_i)$.

2. *Stochastic Shortest Path.* The mean shortest path (minimal project duration) in a stochastic PERT (project evaluation and review technique) network can be written as

$$\ell = \mathbb{E}\left\{ \min_{j=1,\ldots,p} \sum_{i \in \mathcal{L}_j} Y_i \right\}, \tag{3.2.3}$$

   where $\mathcal{L}_j$, Y_i and p are similar to their counterparts in (3.2.2).

The statistical analysis of a DESS is straightforward. Indeed, an unbiased estimator and a $(1 - \alpha)100\%$ confidence interval for ℓ can be written as

$$\bar{\ell}_N = \frac{1}{N} \sum_{i=1}^{N} L(\boldsymbol{Y}_i) \tag{3.2.4}$$

and

$$(\bar{\ell}_N \pm z_{1-\alpha/2} \sigma N^{-1/2}), \tag{3.2.5}$$

respectively [the notation $(a \pm b)$ is shorthand for the interval $(a - b, a + b)$]. Here, N is the sample size, σ^2 is the variance of the $L(\boldsymbol{Y}_i)$, $i = 1, \ldots, N$, and $\Phi(z_{1-\alpha}) = 1 - \alpha$, Φ being the standard normal cdf. Typically σ^2 is unknown and must be replaced by the (unbiased) sample variance estimator

$$S^2 = \frac{1}{N-1} \sum_{i=1}^{N} (L(\boldsymbol{Y}_i) - \bar{\ell}_N)^2. \tag{3.2.6}$$

The following algorithm estimates the system performance, $\ell = \mathbb{E}\{L(\boldsymbol{Y})\}$, and calculates the associated confidence interval.

Algorithm 3.2.1

1. *Perform N replications,* $\boldsymbol{Y}_1, \ldots, \boldsymbol{Y}_N$, *for the underlying model, and calculate* $L(\boldsymbol{Y}_i)$, $i = 1, \ldots, N$.
2. *Calculate a point estimator and a confidence interval for* ℓ *from (3.2.4) and (3.2.5), respectively.*

3.3 FINITE-HORIZON SIMULATION OF DISCRETE-EVENT DYNAMIC SYSTEMS (DEDS)

Let $\{L_t : t > 0\}$ be the underlying stochastic process, and suppose we wish to estimate the expected value of $X_t = \varphi(L_t)$, where φ is a real-valued measurable function on the state space of $\{L_t\}$. For example, if $\varphi(x) = x^2$, then $\mathbb{E}\{X_t\} = \mathbb{E}\{L_t^2\}$. If $\varphi(x) = I_{(-\infty,a)}(x)$ is the indicator function of the interval $(-\infty, a)$, then $\mathbb{E}\{X_t\} = \mathbb{E}\{I_{(-\infty,a)}(L_t)\} = P\{L_t < a\}$.

Before proceeding further, we make a distinction between the *finite-horizon* and *steady-state* simulation regimes. In the finite-horizon regime, which is referred to as *terminating* simulation, measurements of system performance are defined relatively to the interval of simulation time $[0, T]$, where T is the occurrence time of a specific termination event, say A. Note that the event A, as well as the initial state of the model, must be specified before the simulation begins. Moreover, T may be a random variable, say, the inauguration time of a new busy cycle. In the steady-state regime, which is also referred to as *long-run* simulation, performance measures are defined to be limiting measures, as the time horizon (simulation length) tends to infinity.

Special care must be taken when making inferences concerning steady-state performance. The reason is that the output data are typically autocorrelated, and consequently, classical statistical analysis, based on independent observations, is not applicable. For some stochastic models, only the finite-horizon simulation is feasible, since the steady-state regime either does not exist or the finite-horizon period is so long that the steady-state analysis is computationally prohibitive (e.g., Law and Kelton, 1991).

The following illustrative example offers further insight into finite-horizon and steady-state simulation. Suppose that the performance variate, L_t, represents the sojourn time of customer t in a stable $GI/G/c/b$ queue, where GI, G, c, and b in $GI/G/c/b$ stand for the interarrival time distribution, the service time distribution, the number of servers, and the buffer size, respectively; GI means that the interarrival times have a general (G) distribution and are independent (I). Let

$$F_{t,m}(x) = P\{\varphi(L_t) \le x\} \tag{3.3.1}$$

be the distribution function of the variate $\varphi(L_t)$, given the initial state $\mathcal{L}(0) = m$ (m customers are initially present). $F_{t,m}$ is called the *finite-horizon distribution* of $\varphi(L_t)$, given that $\mathcal{L}(0) = m$.

We say that the process $\{L_t\}$ *settles into steady state* (equivalently, that *steady state exists*), if for all m

$$\lim_{t\to\infty} F_{t,m}(x) = F(x) \equiv P\{\varphi(L_\infty) \le x\}. \tag{3.3.2}$$

In other words, "steady state" implies that as $t \to \infty$, the transient cdf, $F_{t,m}(x)$ (which generally depends on t and m), approaches a steady-state cdf, $F(x)$, which *does not depend* on the initial state, m. In particular, (3.3.2) assumes that the underlying stochastic process, $\{L_t\}$, converges in distribution to a proper limit, L_∞, independently of the event $\{\mathscr{L}(0) = m\}$. Thus, the operational meaning of steady state is that after "some" period of time, the transient cdf, $F_{t,m}(x)$, of the process $\{\varphi(L_t)\}$ approaches its limiting (steady-state) cdf, $F(x)$; this does not mean, however, that the realizations of $\{L_t\}$ generated from the simulation run become independent nor constant at any point in time. The exact distributions (transient and steady state) are usually available only for simple Markovian models (see, e.g., Gross and Harris, 1985, and Abate and Whitt, 1987a,b,c). In particular, for queueing models of the $GI/G/1$ type, neither the distributions (transient and steady state) nor even the associated moments are available via analytical methods. For performance analysis of such models, one must resort to simulation. Figure 3.1 illustrates a typical behavior of the virtual waiting time process, $\{L_t\}$, and expected delay, $\ell_t = \mathbb{E}\{L_t\}$, for the $M/M/1$ queue as a function of time.

Let $\{L_t\}$ be a discrete-time process and suppose we wish to estimate the expected value,

$$\ell(k, m) = \mathbb{E}\left\{\frac{1}{k}\sum_{t=1}^{k} L_t\right\}, \tag{3.3.3}$$

as a function of the number of observations, k, conditioned on the initial state $\mathscr{L}(0) = m$; if $\{L_t\}$ is a continuous-time process, then the sum $\sum_{t=1}^{k} L_t$ is replaced by $\int_0^k L_t\, dt$. With these definitions, it is readily seen that if $\{L_t\}$ is the delay process in a stable $GI/G/1$ queue, then $\ell(k, m)$ represents the average delay of the first k customers in the system, given $\mathscr{L}(0) = m$. Similarly, if $\{L_t\}$ is the queue length process, then $\ell(k, m)$ represents the average number of customers in the system during the time interval $(0, k]$, given $\mathscr{L}(0) = m$.

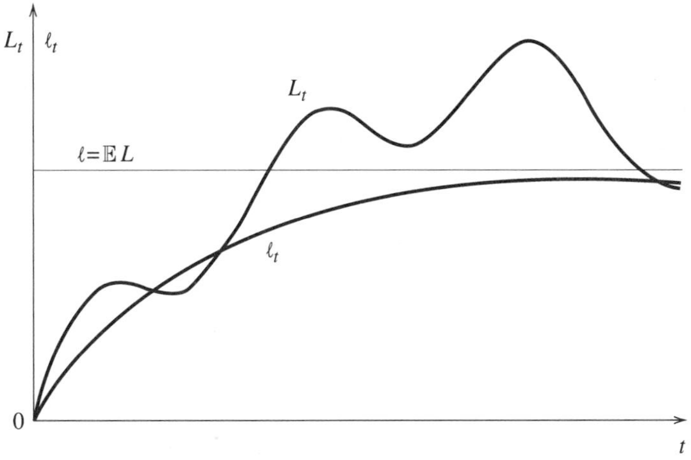

Figure 3.1. Typical behavior of the delay process, $\{L_t\}$, and expected delay, $\ell_t = \mathbb{E}\{L_t\}$, for the $M/M/1$ queue starting with an empty system.

Assume now that N-independent replications were performed, each starting at state $\mathscr{L}(0)=m$. Then the point estimator and the $(1-\alpha)100\%$ confidence interval for $\ell(k, m)$ can be written, similarly to the static case, as

$$\bar{\ell}_N(k, m) = \frac{1}{N}\sum_{i=1}^{N} X_i \tag{3.3.4}$$

and

$$\left(\bar{\ell}_N(k, m) \pm z_{1-\alpha/2} S N^{-1/2}\right), \tag{3.3.5}$$

respectively, where $X_i = k^{-1}\sum_{t=1}^{k} L_{ti}$, L_{ti} is the tth observation from the ith replication, $\Phi(z_{1-\alpha}) = 1-\alpha$, Φ being the standard normal cdf, and S^2 is the sample variance (3.2.6), calculated for $\{X_i\}$.

The algorithm for estimating the finite-horizon performance, $\ell(k, m)$, is similar to that of static models.

Algorithm 3.3.1

1. *Perform N-independent replications, each of length k, starting each replication from the initial state* $\mathscr{L}(0) = m$.
2. *Calculate the point estimator and the confidence interval of* $\ell(k, m)$ *from (3.3.4) and (3.3.5), respectively.*

3.4 STEADY-STATE SIMULATION OF DEDS

Suppose that the steady-state cdf, $F(x)$, of the sample performance process, $\{\varphi(L_t)\}$, exists, and we wish to estimate the steady-state expected value of $\{\varphi(L_t)\}$, say, the expected steady-state delay of customers in the $GI/G/1$ queue, given that the initial state of the system is $\mathscr{L}(0)=m$. Recall from (3.3.2) that the steady-state distribution, $F(\mathbf{x})$ of $\varphi(L_t)$, does not depend on the initial state, $\mathscr{L}(0)=m$. Consequently, $\ell = \mathbb{E}\{\varphi(L_\infty)\}$.

Denote $X_t = \varphi(L_t)$ and let $X_1, \ldots, X_T$ be a sample taken in the steady-state regime. Then, ℓ is estimated as either

$$\bar{\ell}_T = \frac{1}{T}\sum_{t=1}^{T} X_t \quad \text{or} \quad \bar{\ell}_T = \frac{1}{T}\int_0^T X_t \, dt,$$

respectively, depending on whether $\{L_t\}$ is a discrete-time or continuous-time process.

The standard statistical analysis for $\bar{\ell}_T$ in the steady-state regime is based on the central limit theorem (CLT), as $T \to \infty$; that is, one assumes that $T^{1/2}(\bar{\ell}_T - \ell)$

converges in distribution to the normal distribution with mean 0 and variance σ^2. Assume that the CLT holds, namely,

$$T^{1/2}(\bar{\ell}_T - \ell) \Rightarrow N(0, \sigma^2) \quad \text{as } T \to \infty,$$

where $\Rightarrow$ denotes weak convergence (convergence in distribution) (see, e.g., Karlin and Taylor, 1975). Assuming $\varphi(L_t) \Rightarrow \varphi(L_\infty)$, it is valid to conclude that for large T, $\bar{\ell}_T$ is approximately normally distributed with mean $\ell = \mathbb{E}\{\varphi(L_\infty)\}$ and variance σ^2/T, where σ^2/T is called the *asymptotic variance* of the sample mean $\bar{\ell}_T$.

It is well known (e.g., Karlin and Taylor, 1975) that if $\{L_t\}$ is a discrete-time stationary process, then

$$\sigma^2 = \lim_{T\to\infty} T \operatorname{var} \bar{\ell}_T = \lim_{T\to\infty} \sum_{s=-T}^{T} \left(1 - \frac{|s|}{T}\right) R(s), \tag{3.4.1}$$

where

$$R(s) = \operatorname{cov}\{(L_t, L_{t+s})\} = \mathbb{E}\{L_t, L_{t+s}\} - \ell^2$$

is the autocovariance function of $\{L_t\}$. Similarly, if $\{L_t\}$ is a continuous-time process, the sum in (3.4.1) is replaced with the corresponding integral, while all other data remain the same. It follows from (3.4.1) that if L_t and L_{t+s}, $s \neq 0$, are uncorrelated, then $\sigma^2 = \operatorname{var}\{L_t\} = R(0) > 0$; and if L_t and L_{t+s} are "perfectly" correlated [i.e. $\operatorname{cov}\{L_t, L_{t+s}\} = \operatorname{var}\{L_t\} = R(0)$], then $\sigma^2 = \lim_{T\to\infty} T \operatorname{var}\{\bar{\ell}_T\} = \infty$.

Most applications of interest fall in between these two extremes. Note also that for typical steady-state queueing processes (e.g., sojourn times or number of customers in the queue), we have:

1. The autocovariance function, $R(s)$, is nonnegative and decays in time, provided some regularity conditions hold (e.g., Reynolds, 1975, Whitt, 1991).
2. The rate of decay of $R(s)$ typically decreases, as the traffic intensity, ρ, increases.

From now on we routinely omit the subscript ∞ in the steady-state regime to economize on notation. Accordingly, consider confidence intervals for $\ell = \mathbb{E}\{\varphi(L)\}$. Based on the normal approximation for $\bar{\ell}_T$, a $(1-\alpha)100\%$ confidence interval for ℓ can be written as

$$J = J(\alpha, T) = \left(\bar{\ell}_T \pm z_{1-\alpha/2} \frac{\sigma}{T^{1/2}}\right), \tag{3.4.2}$$

where $\Phi(z_{1-\alpha}) = 1 - \alpha$, and Φ denotes the standard normal cdf. The *absolute width* of the confidence interval, J, in (3.4.2) is

$$w_a(\alpha, T) = 2z_{1-\alpha/2}\sigma/T^{1/2}, \tag{3.4.3}$$

while the corresponding *relative width* is

$$w_r(\alpha, T) = \frac{w_a(\alpha, T)}{|\ell|}. \tag{3.4.4}$$

The absolute and relative widths above may be used as stopping rules (criteria) to control the length of a simulation run. For a specified absolute width Δ, and precision level $1 - \alpha$, the *requisite simulation run length* for the absolute width is

$$T_a(\Delta, \alpha) = \frac{4\sigma^2 z_{1-\alpha/2}^2}{\Delta^2}. \tag{3.4.5}$$

Similarly, the corresponding requisite simulation run length for the relative width is

$$T_r(\Delta, \alpha) = \frac{T_a(\Delta, \alpha)}{\ell^2} = \frac{4\sigma^2 z_{1-\alpha/2}^2}{\ell^2 \Delta^2}. \tag{3.4.6}$$

Unless stated otherwise, the relative width criterion will be implicitly used.

From (3.4.6) it follows that the requisite run length, subject to the relative width criterion, is proportional to σ^2/ℓ^2; the latter is called the (relative width) *run-length ratio* (see Whitt, 1989a). It is often helpful to decompose the run-length ratio, σ^2/ℓ^2, into

$$\frac{\sigma^2}{\ell^2} = \frac{\sigma^2}{\mathrm{var}\{L\}} \cdot \frac{\mathrm{var}\{L\}}{\ell^2}, \tag{3.4.7}$$

where, following the terminology of Whitt (1989a), $\sigma^2/\mathrm{var}\{L\}$ is called the *correlation factor* (because it describes the effect of correlation over time), and $\mathrm{var}\{L\}/\ell^2$ is the *squared coefficient of variation*, which is a normalized measure of variability of the random variable $L = L_\infty$. Note that (3.4.1) implies that typically $\sigma^2/\mathrm{var}\{L\} \geq 1$, with equality (to unity) holding when L_t and L_{t+s}, $s \neq 0$, are uncorrelated. Note also that $\mathrm{var}\{L\}/\ell^2 \geq 0$, with equality (to zero) holding when L is deterministic and $\ell^2 > 0$. It is shown in Asmussen (1992) and Whitt (1989b) that in typical situations, both σ^2/ℓ^2 and that $\sigma^2/\mathrm{var}\{L\}$ exceed unity and both increase rather rapidly in ρ, especially when ρ approaches unity [see, e.g. (3.4.8) and also Whitt (1992)]. The reason is that typical output processes are positively autocorrelated and the autocorrelation increases rapidly in ρ. To demonstrate this phenomenon quantitatively, consider the following example, borrowed from Whitt (1989a).

Example 3.4.1 $M/M/1$ Queue. Let $\{L_t\}$ be the queue length process (including the customer in service) in a stable $M/M/1$ queue. Assume for simplicity that the service rate is $\mu = 1$, so that the traffic intensity is $\rho = \lambda/\mu = \lambda$. Then (see Whitt, 1989a)

$$\mathbb{E}\{L\} = \frac{\rho}{1-\rho}, \qquad \text{var}\{L\} = \frac{\rho}{(1-\rho)^2}, \qquad \sigma^2 = \frac{2\rho(1+\rho)}{(1-\rho)^4}. \tag{3.4.8}$$

The run-length ratio, the correlation factor, and the squared coefficient of variation become

$$\frac{\sigma^2}{\ell^2} = \frac{2(1+\rho)}{\rho(1-\rho)^2}, \qquad \frac{\sigma^2}{\text{var}\{L\}} = \frac{2(1+\rho)}{(1-\rho)^2}, \qquad \frac{\text{var}\{L\}}{\ell^2} = \frac{1}{\rho}, \tag{3.4.9}$$

respectively. Furthermore, as $\rho \to 1$, both σ^2/ℓ^2 and $\sigma^2/\text{var}\{L\}$ tend to infinity at a rate of $O(1-\rho)^{-2}$.

3.5 PLANNING QUEUEING SIMULATIONS

In a typical simulation, neither $\text{var}\{L\}$ nor $\ell \equiv \mathbb{E}\{L\}$ are available. Still, one would like to plan a simulation and, in particular, estimate how long the simulation must be run so as to obtain good confidence intervals.

To fix the ideas, consider the $M/M/1$ queue. Substituting (3.4.8) in (3.4.5) and (3.4.6), the estimated simulation run length under the absolute and relative width criteria are:

$$T_a(\Delta, \alpha) = \frac{8\rho(1+\rho)\, z_{1-\alpha/2}^2}{(1-\rho)^4 \Delta^2} \tag{3.5.1}$$

and

$$T_r(\Delta, \alpha) = \frac{8(1+\rho)\, z_{1-\alpha/2}^2}{\rho(1-\rho)^2 \Delta^2}, \tag{3.5.2}$$

respectively. It follows from (3.5.1) and (3.5.2) that as $\rho \to 0$, T_a tends to zero while T_r tends to infinity. However, as $\rho \to 1$, both T_a and T_r tend to infinity, but at different rates.

As an example, set $\Delta = 0.05$ and $z_{1-\alpha/2} = 1.96$ (which corresponds to $1 - \alpha = 0.95$), and suppose that we wish to estimate $\ell = \mathbb{E}\{L\}$ for $\rho = 0.8$. From (3.4.8),

$$\ell = 4, \qquad \sigma^2 = 1800,$$

and from (3.5.1) and (3.5.2),

$$T_a(0.05, 0.05) = 7200\left(\frac{1.96}{0.05}\right)^2 = 1.1 \times 10^7,$$

$$T_r(0.05, 0.05) = 450\left(\frac{1.96}{0.05}\right)^2 = 6.9 \times 10^5.$$

To attain a width of $\Delta = 0.005$ instead of $\Delta = 0.05$, one would have to multiply these T_a and T_r by a factor of 100.

Table 3.5.1 displays T_a and T_r as functions of ρ for $\Delta = 0.05$ and $z_{1-\alpha/2} = 1.96$. In particular, Table 3.5.1 demonstrates that the simulations lengths, T_a and T_r, increase very rapidly in ρ, and that for $\rho \geq 0.95$ (heavy traffic regime), the simulation-based approach becomes virtually useless, since it calls for a huge simulation interval (e.g., $T_a = 3.6 \times 10^9$ and $T_r = 1.0 \times 10^7$ for $\rho = 0.95$). In this case, one is better off using analytical approximations, based on heavy traffic limit theorems and the theory of Brownian motion. Using an analytical approach, Asmussen (1992) and Whitt (1989b) show that for typical output processes, such as waiting time and queue length processes,

$$\sigma^2 = O\left[(1-\rho)^{-4}\right] \quad \text{and} \quad \frac{\sigma^2}{\ell^2} = O\left[(1-\rho)^{-2}\right], \tag{3.5.3}$$

as $\rho \to 1$; thus, as in the heavy traffic regime, T_a and T_r are proportional to $(1-\rho)^{-4}$ and $(1-\rho)^{-2}$, respectively. Notice that these results conform to formula (3.5.1) and (3.5.2) for the $M/M/1$ queue, as $\rho \to 1$.

The next two sections describe two important methods for estimating steady-state system parameters: the *batch-means* method and the *regenerative* method. Other methods, such as the *replication/deletion* method, the *spectral* method (Heidelberger and Welch, 1981), and the *standardized time series* method (Schruben, 1982, 1983), may be found in Law and Kelton (1991).

Table 3.5.1 Requisite Length of Simulation Run for Queue-Length Processes in the $M/M/1$ Queue

ρ	$(1-\rho)^{-4}$	$(1-\rho)^{-2}$	T_a	T_r
0.3	4.1	2.0	$2.0 \cdot 10^4$	$1.1 \cdot 10^5$
0.5	16	4	$1.5 \cdot 10^5$	$1.5 \cdot 10^5$
0.7	123	11.1	$1.8 \cdot 10^6$	$3.3 \cdot 10^5$
0.8	625	25	$1.1 \cdot 10^7$	$6.9 \cdot 10^5$
0.9	10^4	10^2	$2.1 \cdot 10^8$	$2.6 \cdot 10^6$
0.95	$1.6 \cdot 10^5$	4.10^2	$3.6 \cdot 10^9$	$1.0 \cdot 10^7$
0.99	10^8	10^4	$2.4 \cdot 10^{13}$	$2.5 \cdot 10^9$

3.6 BATCH-MEANS METHOD

The *batch-means* method is most widely used by simulation practitioners to estimate steady-state parameters from a single simulation run of length, say, M. The initial K observations, corresponding to a finite-horizon simulation, are deleted, and the remaining $M - K$ observations are divided into N batches, each of length

$$T = \frac{M - K}{N}.$$

The deletion serves to eliminate or reduce the initial bias, so that the remaining observations are statistically more "typical" of steady state. The expected steady-state performance, $\ell = \mathbb{E}\{L\}$, is estimated by

$$\tilde{\ell}_N = \frac{1}{NT} \sum_{i=1}^{N} \sum_{t=1}^{T} L_{ti}. \tag{3.6.1}$$

where L_{ti} is the ith observation from the tth batch.

Assume, for simplicity, that $\{L_t\}$ is a discrete-time process, and let $X_i = T^{-1} \sum_{t=1}^{T} L_{ti}$ and $X_j = T^{-1} \sum_{t=1}^{T} L_{tj}$ be the sample means from the ith and jth batches, respectively. Then, for $i \neq j$, the variates X_i and X_j are approximately independent, provided the batches are sufficiently large, say, $T \approx T_r$, where $T_r = T_r(\rho)$ is given in Table 3.5.1. An approximate confidence interval for ℓ can then be written as

$$\tilde{\ell}_N \pm z_{1-\alpha/2} \frac{S}{N^{1/2}}, \tag{3.6.2}$$

where

$$S^2 = \frac{1}{N - 1} \sum_{i=1}^{N} (X_i - \tilde{\ell}_N)^2 \tag{3.6.3}$$

is the sample variance of σ^2 in (3.4.2). Schmeiser (1982) and Whitt (1991) recommend a selection of the number of batches, N, in the range 20–30. [See also the discussion in Glynn and Whitt (1991) on consistent estimation of the asymptotic variance via the batch-means method as the run length, M, increases].

Recall that the deletion of the K initial observations should render the remaining $M - K$ observations a "typical" steady-state path. The following example gives some idea of how K might be selected. Let $\{L_t\}$ be the queue length process (not including the customer in service) in an $M/M/1$ queue, and assume that we start the simulation at time zero with an empty queue. It is shown in Abate and Whitt (1987a,b) that in order to be within 1% of the steady-state mean, the length of the initial portion, K, to be deleted should be on the order of $8/(\mu(1 - \rho)^2)$, where $1/\mu$

is the expected service time. Thus, for $\rho = 0.5$, 0.8, 0.9, and 0.95, K equals 32, 200, 800, and 3200 expected service times, respectively. Looking up the values of the batch size T_r in Table 3.5.1 (using the relative width criterion), the ratios T_r/K are 53, 137.5, 168.8, and 185 (for $\rho = 0.5$, 0.8, 0.9, and 0.95, respectively). For more details on the transient behavior of queueing models, see Abate and Whitt (1987a,b,c) and Whitt (1991, 1992).

The batch-means algorithm is sketched below.

Algorithm 3.6.1 Batch-Means Method

1. *Make a single simulation run of length M, and delete K initial observations corresponding to a finite-horizon simulation.*
2. *Divide the remaining $M - K$ observations into N batches, each of length*

$$T = \frac{M - K}{N}.$$

3. *Calculate the point estimator and the confidence interval for ℓ from (3.6.1) and (3.6.2), respectively.*

Remark 3.6.1 Replication-Deletion Method. The batch-means method should be compared and contrasted with the replication-deletion method; the latter calls for N-independent runs rather than a single simulation run as in the former. From each replication, one deletes K initial observations corresponding to the finite-horizon simulation, and then calculates the point estimator and the confidence interval for ℓ via (3.6.1) and (3.6.2), respectively, exactly as in the batch-means approach. Note that the confidence interval obtained with the replication-deletion method is unbiased, whereas the one obtained from the batch-means method is slightly biased. However, the former requires deletion from *each* replication, as compared to *a single* deletion in the latter. For more details on the replication-deletion method, see Law and Kelton (1991) and Glynn and Whitt (1991).

3.7 REGENERATIVE METHOD

Roughly speaking, a stochastic process $\{L_t\}$ is called *regenerative* if there exist random time points $0 = T_0 < T_1 < T_2, \ldots$ such that at each such time point, the process "restarts" probabilistically. More precisely, the process $\{L_t\}$ can be split into independent identically distributed (iid) probabilistic replicas, called *regenerative cycles*, of iid random lengths $\tau_i = T_i - T_{i-1},\ i = 1, 2, \ldots$.

Example 3.7.1 The $GI/G/1$ Queue. A classic example is the $GI/G/1$ queue with regeneration times $0 = T_0 < T_1 < T_2, \ldots$ corresponding to customer arrivals at an empty system. At each such random time, T_i, the queue length process starts afresh, independently of the past. We then say that the queue length process *regenerates* itself. The regeneration times, T_i, are so-called *stopping times* (see, e.g., Asmussen, 1987). More precisely, a random variable T taking values in $[0, \infty)$ is said to be a stopping time for a stochastic process $\{X_t\}$, if for every $t \geq 0$, the occurrence of the event $\{T \leq t\}$ can be determined from the history $\{X_s: s \leq t\}$ of the process up until time t. Technically, the event $\{T \leq t\}$ belongs to the σ algebra $\{X_s: s \leq t\}$, for all $t \geq 0$.

Figure 3.2 and Table 3.7.1 illustrate a typical realization (sample path) of the process $\{L_t\}$, where L_t is the number of customers at time t in a $GI/G/1$ queue.

Example 3.7.2 $S - s$ Policy Inventory Model. Consider a continuous review, single-commodity inventory model supplying external demands and receiving stock from a production facility. When demand occurs, it is either filled or back-ordered (to be satisfied by delayed deliveries). At time t, the *net inventory* (on-hand inventory minus back orders) is N_t, and the *inventory position* (on-hand inventory plus on-order inventory minus back orders) is L_t. The control policy is an (s, S) policy that operates on the inventory position. Specifically, whenever $N_t < s$, an order of size $S - N_t$ is placed; otherwise, no action is taken. Nominally, the magnitude of the order $S - N_t$ at the time it is placed brings the inventory position up to S; however, the order $S - N_t$ arrives r times units later (r is called the *lead* time). Clearly, $L_t = N_t$, if $r = 0$.

Let Y_{1n} and Y_{2n} be the size of the nth demand and the length of the nth interdemand time, respectively. Under the back-order policy and the assumptions that the sequences $\{Y_{1n}: n = 1, 2, \ldots\}$ and $\{Y_{2n},\ n = 1, 2, \ldots\}$ are independent, the inventory position process $\{L_t: t \geq 0\}$ is regenerative (see, e.g., Sahin, 1989). Each time an order is placed, the inventory position is raised to S, and the process regenerates. Figure 3.3 illustrate two typical sample path: one for the inventory position process $\{L_t: t \geq 0\}$ (purposely not completed) and another for the inventory on-hand process $\{N_t: t \geq 0\}$. It is readily seen that the sample path of $\{L_t: t \geq 0\}$ contains four regenerative cycles, while the sample path of $\{N_t: t \geq 0\}$ contains a single regenerative cycle, which occurs after the third lead time during which no demand was placed.

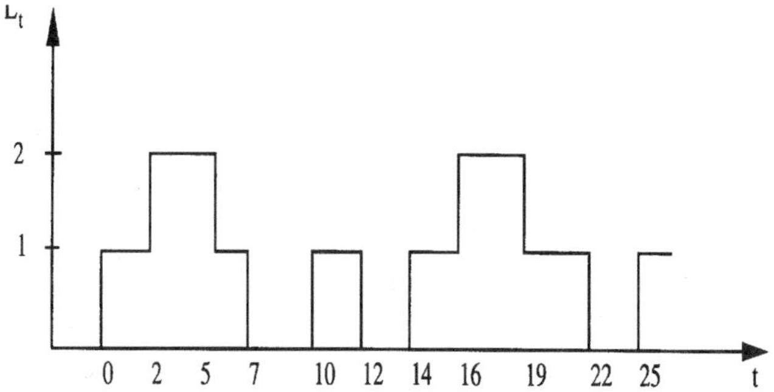

Figure 3.2. Graph of a sample path realization of the number of customers process, $\{L_t\}$, in a $GI/G/1$ queue.

Table 3.7.1 Sample Path Realization of Number of Customers Process, $\{L_t\}$, in a $GI/G/1$ Queue

Cycle	t	L_t
1	0–2	1
	2–5	2
	5–7	1
	7–10	0
2	10–12	1
	12–14	0
3	14–16	1
	16–19	2
	19–22	1
	22–25	0

A third example of a regenerative process is an irreducible and positive recurrent (ergodic) Markov chain, $\{L_t\}$, with a countable state space. In this case, regeneration takes place every time the process enters a fixed state, j. More formally (see Asmussen, 1987), a stochastic process $\{L_t\}$ is regenerative if there exists a sequence of stopping times $T_0, T_1, T_2, \ldots$, such that:

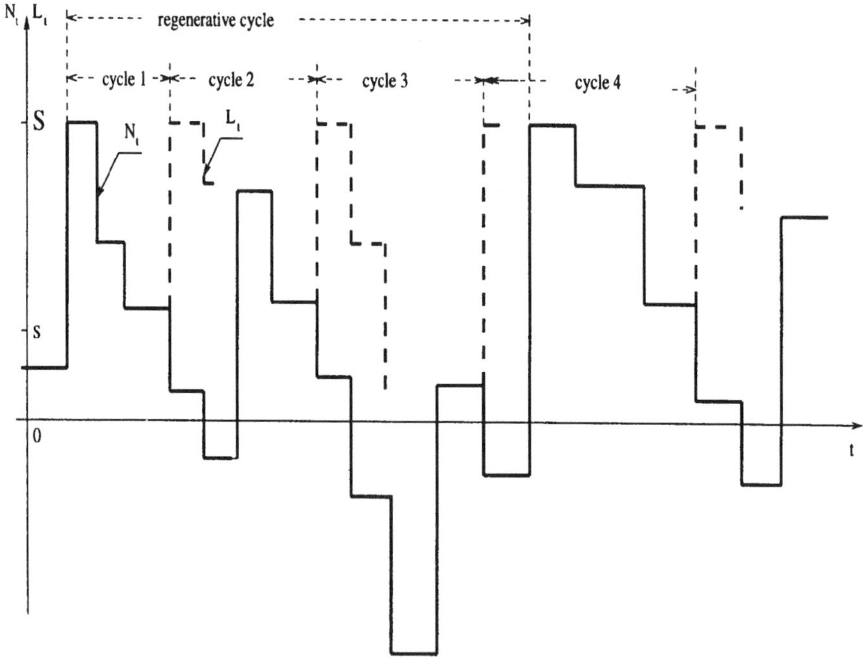

Figure 3.3. Graph of a sample path realization of the processes inventory position, $\{L_t\}$, and inventory on hand, $\{N_t\}$.

1. The sequence of random times, $\{T_i: i \geq 0\}$, is a renewal process, that is, the random variables $\tau_{i+1} = T_{i+1} - T_i$, $i = 0, 1, \ldots$, (with $T_0 = 0$ by convention) are iid.
2. For any indices $l, m \in \{0, 1, \ldots\}$ and time points $t_1, \ldots, t_l > 0$, the random vectors, $\{L_{t_1}, \ldots, L_{t_l}\}$ and $\{L_{t_1+T_m}, \ldots, L_{t_l+T_m}\}$, are identically distributed, and the process $\{L_t: t < T_m\}$ is independent of the process $\{L_{T_m+t}: t \geq 0\}$.

The following properties of regenerative processes (see, e.g., Asmussen, 1987) will be needed later on.

1. If $\{L_t\}$ is regenerative, then the process $\{\varphi(L_t)\}$ is regenerative as well, for any real-valued measurable function φ.
2. Under mild regularity conditions, the process $\{\varphi(L_t)\}$ has a limiting steady-state distribution in the sense that there exists a random variable L, such that

$$\lim_{t\to\infty} P\{\varphi(L_t) \leq x\} = P\{\varphi(L) \leq x\} = F(x).$$

   That is, $\varphi(L_t) \Rightarrow \varphi(L)$.
3. Consider a cycle $L_1, \ldots, L_\tau$ of a regenerative process $\{L_t\}$. Assume that $\mathbb{E}\{\tau\} < \infty$ and, depending on whether $\{L_t\}$ is a discrete-time or continuous-time process, define

$$X = \sum_{t=1}^{\tau} \varphi(L_t)$$

   or

$$X = \int_0^\tau \varphi(L_t)\, dt,$$

   respectively. Then under mild conditions, the steady-state expected value, $\ell = \mathbb{E}\{\varphi(L)\}$, exists and is given by (see, e.g., Asmussen, 1987)

$$\ell = \mathbb{E}\{\varphi(L)\} = \frac{\mathbb{E}\{X\}}{\mathbb{E}\{\tau\}}. \tag{3.7.1}$$

   We remark that X can be viewed as the reward accrued during a cycle of length τ, and ℓ as the ratio of the expected reward, accrued during a cycle, to the expected length of the cycle.
4. (X_i, τ_i), $i = 1, 2, \ldots$, is a sequence of iid random vectors that satisfy the relations

$$X_i = \sum_{t=T_{i-1}+1}^{T_i} \varphi(L_t) \quad \text{or} \quad X_i = \int_{T_{i-1}}^{T_i} \varphi(L_t)\, dt,$$

respectively, depending on whether $\{L_t\}$ is a discrete-time or continuous- time process.

Note that property (1) states that the behavior patterns of the system (or any measurable function thereof), during distinct cycles, are statistically independent and identically distributed, while property (4) asserts that rewards and cycle lengths are jointly iid for distinct cycles.

Formula (3.7.1) is fundamental to regenerative simulation. For typical non-Markovian queueing models, the quantity ℓ (the steady-state expected performance) is unknown and must be evaluated either via analytical approximations (see, e.g., Whitt, 1983) or via simulation, with regenerative simulation constituting an important special case. Regenerative simulation was first introduced by Polyak (1970a,b) and later developed by Crane and Iglehart (1974a,b) and Iglehart (1975, 1976). In particular, given an iid sequence of two-dimensional random vectors $\{X_i, \tau_i\}$, $i = 1, \ldots,$ obtained from a regenerative simulation, one can compute point estimates and confidence intervals for the unknown quantity, ℓ, based on (3.7.1).

To obtain a point estimate of ℓ, one generates N regenerative cycles, calculates the iid sequence of two-dimensional random vectors $\{X_i, \tau_i\}$, $i = 1, \ldots, N$, and finally estimates ℓ by the estimator

$$\bar{\ell}_N = \frac{\bar{X}}{\bar{\tau}}, \tag{3.7.2}$$

where $\bar{X} = N^{-1} \sum_{i=1}^{N} X_i$ and $\bar{\tau} = N^{-1} \sum_{i=1}^{N} \tau_i$. By the law of large numbers (LLN), $\bar{X}$ and $\bar{\tau}$ are strongly consistent estimators for $\mathbb{E}\{X\}$ and $\mathbb{E}\{\tau\}$, respectively, whence $\bar{\ell}$ is a strongly consistent estimator for ℓ.

The *advantages* of the regenerative simulation method are:

1. No initial deletion of transients is necessary.
2. No prior parameters are needed (e.g., number of batches).
3. It is asymptotically exact.
4. It is easy to understand and implement.

The *disadvantages* of the regenerative simulation method are:

1. For many practical cases, the output process, $\{L_t\}$, is either nonregenerative or it is difficult to identify its regeneration points. Moreover, in complex systems (e.g., large queueing networks), checking for the occurrence of regeneration points could be computationally expensive.
2. The estimator $\bar{\ell}_N$ is biased, that is, for finite N, $\mathbb{E}\{\bar{\ell}_N\} \neq \ell$.
3. The regenerative cycles may be very long.
4. Convergence in CLT may be rather slow.

To derive confidence intervals for ℓ, let $Z_i = X_i - \ell\tau_i$. It is readily seen that the Z_i are iid variates, since so are the random vectors (X_i, τ_i).

Denoting

$$\bar{X} = \frac{1}{N}\sum_{i=1}^{N} X_i \quad \text{and} \quad \bar{\tau} = \frac{1}{N}\sum_{i=1}^{N} \tau_i,$$

the CLT ensures that

$$\frac{N^{1/2}(\bar{X} - \ell\bar{\tau})}{\sigma} \Rightarrow N(0, 1) \text{ as } N \to \infty, \tag{3.7.3}$$

and

$$\sigma^2 = \text{var}\{Z\} = \text{var}\{X\} - 2\ell \text{ cov}\{X, \tau\} + \ell^2 \text{var}\{\tau\}. \tag{3.7.4}$$

On division by $\bar{\tau}$, Eq. (3.7.3) becomes

$$\frac{N^{1/2}(\bar{\ell} - \ell)}{\sigma/\bar{\tau}} \Rightarrow N(0, 1) \text{ as } N \to \infty. \tag{3.7.5}$$

The asymptotic $(1 - \alpha)100\%$ confidence interval for $\ell = E\{X\}/E\{\tau\}$ is

$$J = \left(\bar{\ell}_N \pm \frac{z_{1-\alpha/2} S}{\bar{\tau} N^{1/2}}\right). \tag{3.7.6}$$

Here, S^2 is the unbiased sample estimator of σ^2 [see (3.7.4)], namely,

$$S^2 = S_{11} - 2\bar{\ell}_N S_{12} + \bar{\ell}_N^2 S_{22}, \tag{3.7.7}$$

where

$$S_{11} = \frac{1}{N-1}\sum_{i=1}^{N} (X_i - \bar{X})^2, \qquad S_{22} = \frac{1}{N-1}\sum_{i=1}^{N} (\tau_i - \bar{\tau})^2$$

are the sample variances for X and τ, respectively, and

$$S_{12} = \frac{1}{N-1}\sum_{i=1}^{N} (X_i - \bar{X})(\tau_i - \bar{\tau})$$

is the sample covariance of X and τ. The algorithm for estimating the $(1 - \alpha)100\%$ confidence interval for ℓ is sketched below.

Algorithm 3.7.1 Regenerative Simulation Method

1. Simulate N regenerative cycles of the process $\{L_t\}$.

2. *Compute the sequence* $\{(X_i, \tau_i)\}$, $i = 1, \ldots, N$.
3. *Calculate the point estimator* $\bar{\ell}_N$ *and the confidence interval of* ℓ *from (3.7.2) and (3.7.6), respectively.*

Example 3.7.3 (Example 3.7.1 continued). Consider Figure 3.2, which depicts a sample path of the number of customers, L_t, in a $GI/G/1$ queue during the period $0 \le t \le 25$.

Figure 3.2 and Table 3.7.1 reveal three complete cycles with the following pairs: $(X_1, \tau_1) = (10, 10)$, $(X_2, \tau_2) = (2, 4)$, and $(X_3, \tau_3) = (11, 11)$. The resultant statistics are $\bar{\ell}_N = 23/25$, $S_{11} = 21$, $S_{22} = 14.35$, $S_{12} = 4.65$, $S^2 = 38.925$, and $J = 23/25 \pm 0.847$.

Example 3.7.4 (Example 3.7.2 continued). Let $\{L_t : t \ge 0\}$ be the inventory position process in the continuous review inventory model. Table 3.7.2 presents a sample path of $\{L_t : t \ge 0\}$ with $s = 1$, $S = 5$, and $r = 2$. Based on the data of Table 3.7.2 we wish to derive the point estimator and the 95% confidence interval for the steady-state quantity $\ell = P\{L < 4\}$, namely, the probability that the inventory position is less than 4. Recall that $\ell = P\{L < x\} = \mathbb{E}[I_{\{L<x\}}]$, where $I_{\{L<x\}}$ is the indicator of the event $\{L < x\}$.

Table 3.7.2 reveals three complete cycles with the following pairs: $(X_1, \tau_1) = (2, 4)$, $(X_2, \tau_2) = (2, 5)$, $(X_3, \tau_3) = (2, 3)$, where $X_i = \sum_{t=1}^{\tau_i} I_{\{L_{ti} < 4\}}$. The resulting statistics are $\bar{\ell}_N = \frac{1}{2}$, $S_{11} = 0$, $S_{22} = 1$, $S_{12} = 0$, $S^2 = \frac{1}{4}$, and $J = \frac{1}{2} \pm \frac{1}{7}$.

Let us now apply the results of Section 3.5 to the batch-means and regenerative methods. To this end, consider the $M/M/1$ queue with $\rho = 0.8$ and refer to Table 3.5.1. It follows from the results above that regardless of the simulation method used (batch-means or regenerative), the requisite simulation lengths, T_a and T_r, remain $T_a = 1.1 \times 10^7$ and $T_r = 6.9 \times 10^5$, for a confidence interval of width $\Delta = 0.05$, and similarly for other statistics. That is, both the batch-means and regenerative methods provide us with information on how to partition the predetermined simulation lengths, T_a and T_r, into independent pieces (batches and

Table 3.7.2 Sample Path Realization of Inventory Position Process, $\{L_t\}$, with $s = 1$, $S = 5$, and $r = 2$

Cycle	t	L_t
	1–2	5
1	2–4	4
	4–5	2
2	5–8	5
	8–10	3
3	10–11	5
	11–13	3

regenerative cycles), but do not provide the requisite lengths T_a and T_r themselves. These can be estimated, using the following two-stage procedure:

1. Make a pilot run of the system under study. Find the sample variance S^2, using either (3.6.3) or (3.7.7) (batch-means or regenerative estimators, respectively).
2. Replace σ^2 in (3.4.5) and (3.4.6) by the sample variance S^2, found in stage 1, and estimate T_a and T_r accordingly.

3.8 EXERCISES

1. Let $\{L_t\}$ be the queue length process (including the customer in the service) in a stable $M/M/1$ queue. Assume, for simplicity, that the service rate is $\mu = 1$ and the traffic intensity is $\rho = \lambda/\mu = \lambda = 0.6$.

(a) Find the correlation coefficient and the squared coefficient of variation [see (3.4.9)].

(b) Find the 95% confidence interval for $\ell = \mathbb{E}\{L\}$ and the absolute and the relative widths of the confidence interval, corresponding to a sample size (number of simulated customers) of $T = 10^4$.

(c) Find the estimated lengths, $T_a(\Delta, \alpha)$ and $T_r(\Delta, \alpha)$, of the simulation run, for $\Delta = 0$, 1 and 95% confidence interval.

(d) Repeat (a), (b), and (c) for $\rho = 0.8$, and discuss (a), (b) and (c) for $\rho \to 1$.

***2.** Prove Formula (3.4.1).

3. Table 3.8.1 displays a realization (sample path) of the steady-state waiting time process, $\{L_t\}$, in a $GI/G/1$ queue over six regenerative cycles with a total of $t = 15$ customers. Find the point estimator, $\bar{\ell}_N$, and the 95% confidence intervals for $\ell = \mathbb{E}\{L\}$, using (i) the regenerative method and (ii) the batch-means method. Assume that each batch consists of five customers.

Table 3.8.1 Realization of Waiting Time Process, $\{L_t\}$, in a $GI/G/1$ Queue

t	1	2	3	4	5	6	7	8	9	10	11	12	13	14	15
L_t	0	3	0	1	2	1	0	2	0	1	0	1	0	2	0

4. Table 3.8.2 displays three sample paths (realizations) of the process $\{L_{ti}\}$, $i = 1, 2, 3$, of the number of customers (including the customer in service) in a $GI/G/1$ queue, during the interval [0, 23]. Each realization starts with an empty system.

(a) Find the point estimator, $\bar{\ell}_N$, and the 95% confidence interval for the expected number of customers (i) over the interval [0, 10] and (ii) over the interval [15, 23].

Table 3.8.2 Realization of Number of Customers, $\{L_t\}$, in a Steady-State $GI/G/1$ Queue

Path 1		Path 2		Path 3	
t	L_{t1}	t	L_{t2}	t	L_{t3}
0–2	1	0–3	1	0–1	1
2–3	2	3–5	2	1–5	0
3–8	1	5–8	3	5–6	1
8–10	0	8–10	2	6–7	2
10–11	1	10–11	1	7–11	1
11–13	0	11–12	0	11–12	0
13–16	1	12–16	1	12–13	1
16–18	0	16–20	2	13–18	0
18–21	1	20–21	1	18–20	1
21–23	2	21–23	0	20–23	2

(b) Find the point estimator, $\bar{\ell}_N$, and the 95% confidence interval for the expected sojourn time of a customer in the $GI/G/1$ queue (i) for the first four customers and (ii) for customers 2 and 3.

(c) Repeat (b) for the expected waiting time of a customer in a $GI/G/1$ queue.

5. Consider the following sample performance:

$$L(\boldsymbol{Y}) = \min\{(Y_1 + Y_2), (Y_1 + Y_5 + Y_4), (Y_3 + Y_4)\}.$$

Assume that the variates Y_i, $i = 1, \ldots, 5$ are iid with common distribution

(a) $G(\lambda_i, \beta_i)$, where $\lambda_i = i, \beta_i = i$,

(b) $\mathrm{Ber}(p_i)$, where $p_i = 1/2i$.

Run a computer simulation with $N = 1000$ replications, and find point estimators and 95% confidence intervals for $\ell(\boldsymbol{v}) = \mathbb{E}\{L(\boldsymbol{Y})\}$.

6. Run a computer simulation of an $M/M/1$ queue with $\rho = 0.5$ ($\lambda = 1$), starting with an empty system. Find point estimators and confidence intervals in both the finite-horizon and steady-state regimes, as follows:

(a) For the finite-horizon regime, use $N = 50$ replications and estimate the expected waiting time in the system of the first 25 customers.

(b) For the steady-state regime, take a sample of 10,000 customers and estimate the expected number of customers in the system. Apply both the regenerative and the batch-means methods. For the batch-means method, delete the first $K = 100$ customers and use $N = 30$ batches.

(c) In both the regenerative and the batch-means methods, find the requisite sample size (number of customers to simulate) which ensures a relative width of the confidence interval, not exceeding 5%.

7. Run a computer simulation of an $M/M/1$ queue with $\rho = 0.5$ ($\lambda = 1$) in the finite-horizon regime, starting with an empty system. Use $N = 500$ replications and find point estimators and confidence intervals for the expected average delay in the queue for customers $21, \ldots, 70$.
Hint: the point estimator and confidence interval required are for the following parameter:

$$\ell(k, m) = \mathbb{E}\left\{\frac{1}{k}\sum_{t=r+1}^{r+k} L_t\right\},$$

where $r = 20$ and $k = 50$.

8. Run a computer simulation of 1000 regenerative cycles of a continuous review $S - s$ policy inventory model (see Example 3.7.2) with Poisson distribution $Y_1 \sim P(\lambda = 2)$, $Y_2 \sim \text{Exp}(\lambda = 1)$, $s = 1$, $S = 6$, and lead time $r = 2$, starting at $L_0 = N_0 = 4$. Find point estimators and confidence intervals for the quantity $\ell = P\{2 \le L \le 4\}$, where $\{L_t\}$ is the steady-state inventory position process.

REFERENCES

Abate, J. and Whitt, W. (1987a). "Transient behavior of regulated Brownian motion, I: Starting at the origin," *Adv. Appl. Prob.*, **19**, 560–598.

Abate, J. and Whitt, W. (1987b). "Transient behavior of regulated Brownian motion, II: Non-zero initial conditions," *Adv. Appl. Prob.*, **19**, 599–631.

Abate, J. and Whitt, W. (1987c). "Transient behavior of the M/M/1 queue: Starting at the origin," *Queueing Syst.*, **2**, 41–65.

Asmussen, S. (1987). *Applied Probability and Queues*, Wiley, New York.

Asmussen, S. (1992). "Queueing simulation in heavy traffic," *Math. Opns. Res.*, **17**, 84–111.

Crane, M. A. and Iglehart, D. L. (1974a). "Simulating stable stochastic systems, I: General multiserver queues," *J. Assoc. Comput. Mach.*, **21**, 103–113.

Crane, M. A. and Iglehart, D. L. (1974b). "Simulating stable stochastic systems, II: Markov chains," *J. Assoc. Comput. Mach.*, **21**, 114–123.

Glynn, P. W. and Whitt, W. (1991). "Estimating the asymptotic variance with batch means," *Oper. Res. Lett.*, **10**, 431– 435.

Gross, D. and Harris, C. M. (1985). *Fundamentals of Queueing Theory*, 2nd ed., Wiley, New York.

Heidelberger, P. and Welch. P. D. (1981). "A spectral method for confidence interval generation and run length control in simulations," *Commun. Assoc. Comput. Mach.*, **24**, 233– 245.

Iglehart, D. L. (1975). "Simulating stable stochastic systems, V: Comparison of ratio estimators," *Naval Res. Logist. Quart.*, **22**, 553–565.

Iglehart, D. L. (1976). "Simulating stable stochastic systems, VI: Quantile estimation," *J. Assoc. Comp. Mach.*, **23**, 347–360.

Karlin, S. and Taylor, H. J. (1975). *A First Course in Stochastic Processes*, Academic, New York.

Law, A. M. and Kelton, W. D. (1991). *Simulation Modeling and Analysis*, McGraw-Hill, New York.

Polyak, D. G. (1970a). "Precision of statistical simulation of queueing systems," *Eng. Cybern. (NY)*, **1**, 72–80.

Polyak, D. G. (1970b). "Increasing the accuracy of simulation queueing systems," *Eng. Cybern. (NY)*, **4**, 687–695.

Reynolds, J. F. (1975). "The covariance structure of queues and related stochastic processes—a survey of recent work," *Adv. Appl. Prof.*, **7**, 383–415.

Sahin, I. (1989). *Regenerative Inventory Systems. Operating Characteristics and Optimization*, Springer-Verlag, New York.

Schmeiser, B. W. (1982). "Batch size effects in the analysis of simulation output," *Oper. Res.*, **30**, 556–568.

Schruben, L. W. (1982). "Detecting initialization bias in simulation output," *Oper. Res.*, **30**, 569–590.

Schruben, L. W. (1983). "Confidence interval estimation using standardized time series," *Oper. Res.*, **31**, 1090–1108.

Whitt, W. (1983). "The queueing network analyzer," *BSTJ*, **62**(9), 2779–2813.

Whitt, W. (1989a). "Simulation run length planning," *Proceedings of the 1989 Winter Simulation Conference*, Washington, DC, pp. 106-112.

Whitt, W. (1989b). "Planning queueing simulations," *Manage. Sci.*, **35**(11), 1341–1366.

Whitt, W. (1991). "The efficiency of one long run versus independent replications in steady-state simulation," *Manag. Sci.*, **37**(6), 645–666.

Whitt, W. (1992). "Asymptotic formulas for Markov processes with applications to simulation," *Oper. Res.*, **40**, 279–291.

CHAPTER 4

Variance Reduction Techniques

4.1 INTRODUCTION

This chapter treats basic theoretical and practical aspects of *variance reduction* techniques. Variance reduction can be viewed as a means of utilizing known information about the model in order to obtain more accurate estimators of its performance. A variance reduction technique essentially transforms the underlying simulation model into a related one, the latter permitting more accurate estimation of the parameters of interest. In fact, variance reduction cannot be achieved without prior knowledge of the system. At the other extreme, a complete knowledge of the system implies zero variance and obviates the need for simulation. Generally, the more we know about the system, the more effective the variance reduction. One way of gaining this knowledge (information) is through an initial crude simulation of the model. Results from this (first stage) simulation can then be used to formulate variance reduction techniques that will subsequently improve the accuracy of the estimators in the second simulation stage. Variance reduction techniques include the *common and antithetic variates methods*, *control variates*, *conditional Monte Carlo*, *stratified sampling and importance sampling*. For a comprehensive study of variance reduction techniques see Fishman (1996).

All variance reduction techniques are inherently model-dependent and their successful application is viewed more as art than science. Also, one should realize that most variance reduction techniques typically reduce the variance by a constant factor. The exception to this rule is the importance sampling approach, which can lead to dramatic variance reduction in many nontrivial cases of interest. Chapter 9 describes a clever use of importance sampling utilizing exponential change of measure, which makes it possible to obtain polynomial-time algorithms for rare-event probability estimation in queueing networks. Sections 4.2 through 4.6 treat common and antithetic variates, control variates, conditional Monte Carlo, importance sampling, and stratified sampling, respectively.

4.2 COMMON AND ANTITHETIC VARIATES

To motivate the use of common and antithetic variates in simulation, consider the following simple example. Let X and Y be variates with known cumulative distribution functions (cdf's), F_1 and F_2, respectively. Wanted is an estimator of the expected value of their difference, $\mathbb{E}\{X - Y\}$, with minimal variance. Since

$$\text{var}\{X - Y\} = \text{var}\{X\} + \text{var}\{Y\} - 2\,\text{cov}\{X, Y\}, \tag{4.2.1}$$

and since the marginal cdf's of X and Y have been prescribed, it follows that the variance of $X - Y$ is minimized by maximizing the covariance in (4.2.1). Assume that both X and Y can be generated by the inverse-transform method, so that

$$\begin{aligned} X &= F_1^{-1}(U_1) = \inf\{x\colon F_1(x) \geq U_1\}, \\ Y &= F_2^{-1}(U_2) = \inf\{y\colon F_2(y) \geq U_2\}, \end{aligned} \tag{4.2.2}$$

where U_1 and U_2 are uniformly distributed on (0, 1).

Definition. We say that *common random variables* (CRVs) are used, if $U_2 = U_1$, and that *antithetic random variables* (ARVs) are used, if $U_2 = 1 - U_1$ [recall that $U_1 \sim \mathcal{U}(0, 1)$ implies $U_2 = 1 - U_1 \sim \mathcal{U}(0, 1)$].

Since both F_1^{-1} and F_2^{-1} are monotonic nondecreasing functions of U, it is readily seen that using CRVs implies

$$\text{cov}\{F_1^{-1}(U),\ F_2^{-1}(U)\} \geq 0,$$

and consequently, variance reduction is achieved in the sense that the estimator

$$F_1^{-1}(U) - F_2^{-1}(U) \text{ of } \mathbb{E}\{X - Y\}$$

has a smaller variance than the *crude Monte Carlo* (CMC) estimator

$$X - Y = F_1^{-1}(U_1) - F_2^{-1}(U_2).$$

Furthermore, it is well known (see, e.g., Whitt, 1976) that using CRVs does, in fact, *maximize* $\text{cov}\{X, Y\}$, so that $\text{var}\{X - Y\}$ is *minimized* as well. Similarly, $\text{var}\{X + Y\}$ is *minimized* when ARVs are used.

Consider now minimal-variance estimation of $\mathbb{E}\{L_1(X) - L_2(Y)\}$, where X and Y are unidimensional variates with known marginal cdf's, F_1 and F_2, respectively, and the functions L_1 and L_2 are real-valued monotone functions. Mathematically, the problem can be formulated as follows:

Within the set $\mathscr{F}_2$ of all two-dimensional joint cdf's of random variable pairs, (X_1, X_2), find a two-dimensional distribution function, F^*, so as to

$$\textit{minimize } \operatorname{var}\{L_1(X) - L_2(Y)\} \tag{4.2.3}$$

subject to the prescribed marginal cdf's, F_1 and F_2.

This problem has been solved in Gal et al. (1984), which proved that if L_1 and L_2 are *monotonic in the same direction*, then

$$\min_{F \in \mathscr{F}_2} \operatorname{var}\{L_1(X) - L_2(Y)\} = \operatorname{var}\{L_1[F_1^{-1}(U)] - L_2[F_2^{-1}(U)]\}. \tag{4.2.4}$$

From (4.2.4) it follows that the use of CRVs (i.e., $U_1 = U_2 = U$) leads again to optimal variance reduction. The proof of (4.2.4) uses the fact that if $L(u)$ is a monotonic function, then $L(F^{-1}(U))$ is monotonic as well, since $F^{-1}(u)$ is. By symmetry, if L_1 and L_2 are *monotonic in the opposite direction*, then the use of ARVs (i.e., $U_2 = 1 - U_1$) is optimal for the variance minimization problem (4.2.3).

Consider now minimal-variance estimation of $\mathbb{E}\{L_1(\boldsymbol{X}) - L_2(\boldsymbol{Y})\}$, where $\boldsymbol{X}$ and $\boldsymbol{Y}$ are n-dimensional random vectors, and the functions L_1 and L_2 are real-valued and monotone in each component of $\boldsymbol{X}$ and $\boldsymbol{Y}$. The variance minimization problem (4.2.3) extends to multidimensional cases in a natural way:

Within the set of all 2n-dimensional joint cdf's $\boldsymbol{F} \in \mathscr{F}_{2n}$ of random-vector pairs, $(\boldsymbol{X}, \boldsymbol{Y})$, where each random vector has independent components, find a 2n-dimensional distribution function $\boldsymbol{F}^* \in \mathscr{F}_{2n}$ so as to

$$\textit{minimize } \operatorname{var}\{L_1(\boldsymbol{X}) - L_2(\boldsymbol{Y}))\}, \tag{4.2.5}$$

subject to the prescribed n-dimensional distributions

$$F_{\mathbf{1}}(x_1, \ldots, x_n) = \prod_{i=1}^{n} F_{1i}(x_i) \quad \text{and} \quad F_{\mathbf{2}}(y_1, \ldots, y_n) = \prod_{i=1}^{n} F_{2i}(y_i),$$

where

$$\boldsymbol{X} = F_{\mathbf{1}}^{-1}(\boldsymbol{U}_1) = \{F_{11}^{-1}(U_{11}), \ldots, F_{1n}^{-1}(U_{1n})\},$$

and

$$\boldsymbol{Y} = F_{\mathbf{2}}^{-1}(\boldsymbol{U}_2) = \{F_{21}^{-1}(U_{21}), \ldots, F_{2n}^{-1}(U_{2n})\}.$$

[The bold subscripts indicate multivariate distributions.]

Assume in addition that within the set of all cdf's $\boldsymbol{F} \in \mathscr{F}_{2n}$ on $\mathbb{R}^{2n}$, we permit dependence *only between like components* of the vectors $\boldsymbol{X}$ and $\boldsymbol{Y}$ (components with the same indices). The solution of the corresponding minimal-variance problem was given in Rubinstein et al. (1985) which shows that if L_1 and L_2 are monotonic in the same direction in each component of the vectors $\boldsymbol{X} = (X_1, \ldots, X_n)$ and $\boldsymbol{Y} = (Y_1, \ldots, Y_n)$, respectively, and if dependence is permitted only between like components, then

$$\min_{F \in \mathscr{F}_{2n}} \text{var}\{L_1(\boldsymbol{X}) - L_2(\boldsymbol{Y})\} = \text{var}\{L_1(\boldsymbol{F}_1^{-1}(\boldsymbol{U})) - L_2(\boldsymbol{F}_2^{-1}(\boldsymbol{U}))\} \tag{4.2.6}$$

is attained as the solution of (4.2.5) when $\boldsymbol{U}_1 = \boldsymbol{U}_2 = \boldsymbol{U}$ (here and elsewhere, vector equality is componentwise). In other words, the vector of CRVs, $\boldsymbol{U}$, is optimal again. If L_1 and L_2 are monotonic in opposite directions, then $\boldsymbol{U}_2 = \mathbf{1} - \boldsymbol{U}_1$ (the vector of ARVs) is optimal, where $\mathbf{1}$ is a vector of 1's. Finally, if L_1 and L_2 are monotonic increasing with respect to some components and monotonic decreasing with respect to the others, then one can find a proper combination of common and antithetic variates, which is again optimal for the problem (4.2.5).

In many applications, the components of the random vectors $\boldsymbol{X}$ and $\boldsymbol{Y}$ are either dependent or the sample performance functions, $L_1(\boldsymbol{x})$ and $L_2(\boldsymbol{y})$, are not monotonic (or both). The use of CRVs (ARVs) for the case of dependent components of $\boldsymbol{X}$ and $\boldsymbol{Y}$ and strictly monotonic functions, L_1 and L_2, was described in Rubinstein et al. (1985), while the use of CRVs (ARVs) for piecewise-monotonic functions, L_1 and L_2, was treated in Rubinstein (1986). For additional references on common and antithetic variates, see Glasserman and Yao (1992) and Kleijnen (1974, 1976).

Several typical applications will next motivate the use of CRVs and ARVs vis-à-vis the program (4.2.4). Recall that L_1 and L_2 are monotonic functions and that the components of the random vectors $\boldsymbol{X}$ and $\boldsymbol{Y}$ are independent.

Example 4.2.1 Stochastic Shortest Path. Here the sample performance, $L(\boldsymbol{X})$, can be written as

$$L(\boldsymbol{X}) = \min_{j=1,\ldots,m} \sum_{i \in \mathscr{L}_j} X_i, \tag{4.2.7}$$

where $\mathscr{L}_j$ is the jth complete path from the source to the sink of the network; X_i, $i = 1, \ldots, n$, are the durations of the activities (links); and m is the number of complete paths in the network. Note that $L(\boldsymbol{x})$ is nondecreasing in each component of the vector $\boldsymbol{x}$.

Example 4.2.2 Reliability Systems. Here the sample performance (coherent life function), $L(\boldsymbol{X})$, can be written as

$$L(\boldsymbol{X}) = \max_{j=1,\ldots,m} \min_{i \in \mathscr{L}_j} X_i, \tag{4.2.8}$$

where $\mathscr{L}_j$ and X_i are analogous to their counterparts in (4.2.7). Note again that a coherent life function, $L(\boldsymbol{x})$, is nondecreasing in each component of the vector $\boldsymbol{x}$.

Motivated by the examples above, suppose one seeks to estimate

$$\ell = \mathbb{E}\{L(\boldsymbol{X})\},$$

where $\boldsymbol{X}$ is a random vector with independent components and the sample functions, $L(\boldsymbol{X})$ in (4.2.7) and (4.2.8), are monotonic functions in each component of $\boldsymbol{X}$.

An unbiased estimator of ℓ is the CMC estimator, given by

$$\bar{\ell}_N = \frac{1}{N}\sum_{i=1}^{N} L(\boldsymbol{X}_i), \qquad (4.2.9)$$

where $\boldsymbol{X}_1, \ldots, \boldsymbol{X}_N$ is an independent identically distributed (iid) sample from the cdf $F(\boldsymbol{x})$.

An alternative unbiased estimator of ℓ, for N even, is

$$\bar{\ell}_N^{(a)} = \frac{1}{N}\sum_{i=1}^{N/2}\{L(\boldsymbol{X}_i) + L(\boldsymbol{X}_i^{(a)})\}, \qquad (4.2.10)$$

where $\boldsymbol{X}_i = F^{-1}(\boldsymbol{U}_i)$ and $\boldsymbol{X}_i^{(a)} = F^{-1}(\mathbf{1} - \boldsymbol{U}_i)$ are generated via (4.2.2). The estimator $\bar{\ell}_N^{(a)}$ is called the *antithetic estimator* of ℓ.

Since $L(\boldsymbol{X}) + L(\boldsymbol{X}^{(a)})$ is a particular case of $L_1(\boldsymbol{X}) - L_2(\boldsymbol{Y})$ in (4.2.6) [with $L_2(\boldsymbol{Y})$ replaced by $-L(\boldsymbol{X}^{(a)})$], one immediately obtains $\text{var}\{\bar{\ell}_N^{(a)}\} \le \text{var}\{\bar{\ell}_N\}$. That is, the ARV estimator, $\bar{\ell}_N^{(a)}$, is more accurate than the CMC estimator, $\bar{\ell}_N$.

Assume now that one seeks to estimate the expected value of the difference of a pair of functions $L_1(\boldsymbol{X})$ and $L_2(\boldsymbol{Y})$, that is,

$$\ell = \mathbb{E}\{L_1(\boldsymbol{X}) - L_2(\boldsymbol{Y})\}, \qquad (\boldsymbol{X}, \boldsymbol{Y}) \in \mathbb{R}^{2n}.$$

The CMC estimator of ℓ is

$$\bar{\ell}_N = \frac{1}{N}\sum_{i=1}^{N}\{L_1(\boldsymbol{X}_i) - L_2(\boldsymbol{Y}_i)\}, \qquad (4.2.11)$$

while the CRV estimator of ℓ is

$$\bar{\ell}_N^{(c)} = \frac{1}{N}\sum_{i=1}^{N}\{L_1(\boldsymbol{X}_i) - L_2(\boldsymbol{Y}_i^{(c)})\}, \qquad (4.2.12)$$

where $\boldsymbol{X}_i = F_1^{-1}(\boldsymbol{U}_i)$ and $\boldsymbol{Y}_i^{(c)} = F_2^{-1}(\boldsymbol{U}_i)$, that is, $\boldsymbol{X}_i$ and $\boldsymbol{Y}_i^{(c)}$ are generated by using the same uniform random vector $\boldsymbol{U}_i$.

Again, from (4.2.6), $\text{var}\{\bar{\ell}_N^{(c)}\} \le \text{var}\{\bar{\ell}_N\}$, so that the CRV estimator, $\bar{\ell}_N^{(c)}$, is more accurate than the CMC estimator, $\bar{\ell}_N$, provided that both $\boldsymbol{X}$ and $\boldsymbol{Y}$ have independent components, and both L_1 and L_2 are monotonic functions in the same direction.

Let $\bar{\ell}_N(1)$ and $\bar{\ell}_N(2)$ be two estimators of ℓ. To compare $\bar{\ell}_N(1)$ with $\bar{\ell}_N(2)$, define an efficiency measure,

$$\varepsilon = \frac{t_1 \cdot \mathrm{var}\{\bar{\ell}_N(1)\}}{t_2 \cdot \mathrm{var}\{\bar{\ell}_N(2)\}}, \tag{4.2.13}$$

where t_1 and t_2 are the CPU times consumed in calculating the estimators $\bar{\ell}_N(1)$ and $\bar{\ell}_N(2)$, respectively. To fix the ideas, let t_1 and t_2 be the CPU times required to calculate the CMC and the ARV estimators, respectively. Since the ARV estimator, $\bar{\ell}_N^{(a)}$, needs only *half* as many random numbers as its CMC counterpart, $\bar{\ell}_N$, it is readily seen that $t_2 \leq t_1$. Ignoring this advantage of $\bar{\ell}_N^{(a)}$, the efficiency measure (4.2.13) reduces to

$$\varepsilon = \frac{\mathrm{var}\{\bar{\ell}_N(1)\}}{\mathrm{var}\{\bar{\ell}_N(2)\}}. \tag{4.2.14}$$

Table 4.2.1 displays the efficiency ε of the estimator $\bar{\ell}_N^{(a)}$ for various combinations of λ_X in $\mathrm{Exp}(\lambda_X)$ and λ_Y in $\mathrm{Exp}(\lambda_Y)$, resulting from estimating the expected shortest path ℓ for the bridge network in Figure 4.1. Note that for a bridge topology, (4.2.7) and (4.2.8) become

$$L(\boldsymbol{X}) = \min\{X_1 + X_2,\ X_1 + X_5 + X_4,\ X_3 + X_4\} \tag{4.2.15}$$

and

$$L(\boldsymbol{X}) = \max\{\min(X_1, X_2),\ \min(X_1, X_5, X_4),\ \min(X_3, X_4)\}, \tag{4.2.16}$$

respectively.

Assume that the components of the five-dimensional vectors $\boldsymbol{X}$ and $\boldsymbol{Y}$ are independent and that each of their components is distributed exponentially with parameters λ_{1i} and λ_{2i}, $i = 1, \ldots, 5$, respectively. Table 4.2.1 was obtained from simulation runs of 500 replications each, and the results are self-explanatory.

Table 4.2.1 Efficiency ε of the Antithetic Estimator for Various Combinations of λ_X and λ_Y, for the Bridge Network in Figure 4.1

λ_X	$\lambda_Y=1$	$\lambda_Y=2$	$\lambda_Y=3$	$\lambda_Y=4$	$\lambda_Y=5$
1	3.81	1.34	1.01	1.19	1.21
2	1.93	2.95	1.72	1.59	1.26
3	1.20	3.42	4.29	1.64	1.39
4	1.06	1.89	5.01	2.90	2.28
5	1.01	1.52	2.62	5.57	3.24

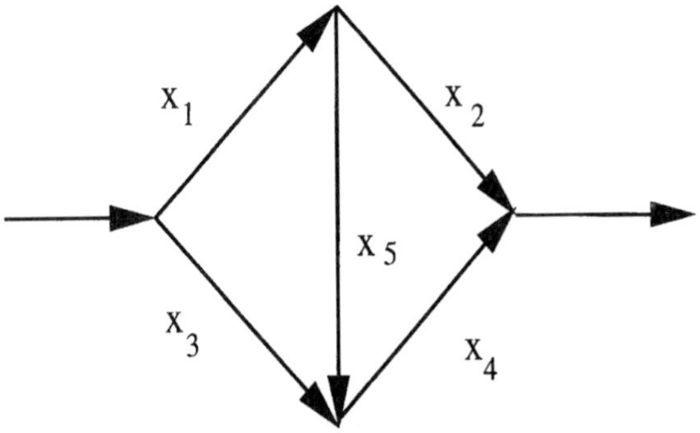

Figure 4.1. Bridge network.

Example 4.2.3 *GI*/*G*/1 Queue. Consider Lindley's equation (see Gross and Harris, 1985)

$$L_{t+1} = \max\{0, L_t + U_t\}, \qquad L_0 = 0 \tag{4.2.17}$$

for the waiting time of the $(t+1)$st customer in a $GI/G/1$ queue. Here $U_j = Y_{1j} - Y_{2(j+1)}$; Y_{1j} and Y_{2j} are the service and the interarrival times of the jth customer, respectively; $Y_{2j} = A_j - A_{j-1}$ for $j \geq 2$, $Y_{2j} = 0$ for $j = 1$; and A_j is the arrival time of the jth customer. Since $L_t = L(Y_{1t}, Y_{2t})$ is monotonic in each component of the two-dimensional vector, $\boldsymbol{X}_t = \{Y_{1t}, Y_{2t}\}$, one can obtain variance reduction relative to the CMC method by using the ARV estimator (4.2.10).

4.3 CONTROL VARIATES

The control variates method is one of the most widely used variance reduction techniques. A sampling of its applications may be found in Asmussen and Rubinstein (1994), Cheng and Feast (1980), Kleijnen (1974), Lavenberg et al. (1982), Lavenberg and Welch (1981), Rubinstein and Marcus (1985), and Wilson (1984).

Consider first the one-dimensional case. Let X be an unbiased estimator of μ, to be estimated from a simulation run. A random variable, C, is called a *control variate* for X, if it is correlated with X and its expectation, r, is known. The control variate C is used to construct an unbiased estimator of μ with a variance smaller than that of X. This estimator,

$$X(\alpha) = X - \alpha(C - r), \tag{4.3.1}$$

where α is a scalar parameter, is called a *linear control variate.* It follows immediately from (4.3.1) that the variance of $X(\alpha)$ is given by

$$\text{var}\{X(\alpha)\} = \text{var}\{X\} - 2\alpha\,\text{cov}\{X, C\} + \alpha^2\,\text{var}\{C\}.$$

Consequently, the value α^* that minimizes $\text{var}\{X(\alpha)\}$ is

$$\alpha^* = \frac{\text{cov}\{X, C\}}{\text{var}\{C\}}, \tag{4.3.2}$$

and the corresponding minimal variance evaluates to

$$\text{var}\{X(\alpha^*)\} = (1 - \rho_{XC}^2)\text{var}\{X\}, \tag{4.3.3}$$

where ρ_{XC} denotes the correlation coefficient of X and C. Notice that the larger $|\rho_{XC}|$, the greater the variance reduction.

Formulas (4.3.1)–(4.3.3) can be easily extended to the case of multiple control variates. Indeed, let $\boldsymbol{C} = (C_1, \ldots, C_m)$ be a (column) vector of m control variates with known mean vector $\boldsymbol{r} = \mathbb{E}\{\boldsymbol{C}\} = (r_1, \ldots, r_m)$, where $r_i = \mathbb{E}\{C_i\}$. Then the extended version of (4.3.1) can be written as

$$X(\alpha) = X - \boldsymbol{\alpha}'(\boldsymbol{C} - \boldsymbol{r}), \tag{4.3.4}$$

where $\boldsymbol{\alpha}$ is an m-dimensional column vector of parameters. It is not difficult to see that the column vector $\boldsymbol{\alpha}^*$ that minimizes $\text{var}\{X(\boldsymbol{\alpha})\}$ is given by

$$\boldsymbol{\alpha}^* = \boldsymbol{\Sigma}_{CC}^{-1}\boldsymbol{\sigma}_{XC}, \tag{4.3.5}$$

where $\boldsymbol{\Sigma}_{CC}$ denotes the $m \times m$ covariance matrix of $\boldsymbol{C}$ and $\boldsymbol{\sigma}_{XC}$ denotes the $m \times 1$ vector whose ith component is the covariance of X and C_i, $i = 1, \ldots, m$. The corresponding minimal variance evaluates to

$$\text{var}\{X(\boldsymbol{\alpha}^*)\} = (1 - R_{XC}^2)\text{var}\{X\}, \tag{4.3.6}$$

where

$$R_{XC}^2 = (\boldsymbol{\sigma}_{XC})'\boldsymbol{\Sigma}_{CC}^{-1}\boldsymbol{\sigma}_{XC}/\text{var}\{X\}$$

is the square of the multiple correlation coefficient of X and $\boldsymbol{C}$. Again the larger $|R_{XC}|$, the greater the variance reduction.

The following examples illustrate applications of control variate methods.

Example 4.3.1 Reliability Bridge Structure. Consider the reliability model (4.2.16) for the bridge structure in Figure 4.1. Assume that all components X_i, $i = 1, \ldots, 5$ are iid exponentially distributed. For control variates, we can use some random variables associated with subnetworks of the full bridge network. Examples of such control variates (control subnetworks) are

$$C_1 = \min(X_1, X_2)$$
$$C_2 = \min(X_1, X_5, X_4)$$
$$C_3 = \min(X_3, X_4).$$

Example 4.3.2 Lindley's Equation. Consider Lindley's equation (4.2.17) for the waiting time process L_t in a the $GI/G/1$ queue.

As a control variate, consider

$$X_t = L_t - \alpha(C_t - r), \tag{4.3.7}$$

where

$$C_t = C_{t-1} - U_{t-1}, \qquad C_0 = 0,$$

and $r = \mathbb{E}\{C_t\} = t\,\mathbb{E}\{U_t\}$. Since both $\{L_t\}$ and $\{C_t\}$ are stochastic processes, $\{X_t\}$ is referred to as the *control random process*.

Example 4.3.3. Consider the estimation of the expected steady-state performance $\ell = \mathbb{E}\{L_\infty\}$ in a queueing network. As a control random process, consider again

$$X_t = L_t - \alpha(C_t - r),$$

where $\{L_t\}$ might be the sojourn time process in the original network, and $\{C_t\}$ might be the sojourn time process in an auxiliary Markovian network that must be run in time lockstep with the original network. In order to produce high correlations between the two processes, $\{L_t\}$ and $\{C_t\}$, it is desirable that both networks have similar topologies and similar loads. In addition, they must use a common stream of random numbers for generating the input variates. Expressions for the expected steady-state performance $r = \mathbb{E}\{C\}$ (such as the expected sojourn time in a Markovian network) may be found in Gross and Harris (1985).

4.4 CONDITIONAL MONTE CARLO

Let

$$\ell = \mathbb{E}\{L(\boldsymbol{Y})\} = \int L(\boldsymbol{y})f(\boldsymbol{y})\,d\boldsymbol{y} \tag{4.4.1}$$

be some expected performance measure of a computer simulation model, where $\boldsymbol{Y}$ is the input random variable (vector) with a probability density function (pdf) $f(\boldsymbol{y})$, and $L(\boldsymbol{Y})$ is the sample performance measure (output random variable). Suppose that there is a second random variable, Y, such that the conditional expectation $\mathbb{E}\{L|Y = y\}$ can be analytically computed. Since

$$\mathbb{E}\{\mathbb{E}\{L|Y\}\} = \mathbb{E}\{L\} = \ell, \tag{4.4.2}$$

it follows that $\mathbb{E}\{L|Y\}$ is an unbiased estimator of ℓ. Furthermore, it will be shown that

$$\text{var}\{\mathbb{E}\{L|Y\}\} \le \text{var}\{L\}, \tag{4.4.3}$$

so that observing the random variable $\mathbb{E}\{L|Y\}$, instead of L, leads to variance reduction. Thus, conditioning *always* leads to variance reduction.

To prove (4.4.3), one uses the well-known fact that var$\{L\}$ can be written as

$$\text{var}\{L\} = \mathbb{E}\{\text{var}\{L|Y\}\} + \text{var}\{\mathbb{E}\{L|Y\}\}. \tag{4.4.4}$$

Since both terms on the right-hand side are nonnegative, (4.4.3) immediately follows. The conditional Monte Carlo algorithm is given next.

Algorithm 4.4.1 Conditional MC

1. *Generate a sample* $Y_1, \ldots, Y_N$ *from* $f(\mathbf{y})$.
2. *Use an analytical formula to calculate* $\mathbb{E}\{L|Y_i = y_i\}, \ i = 1, \ldots, N$.
3. *Estimate* $\ell = \mathbb{E}\{L\}$ *by*

$$\bar{\ell}_N^c = \frac{1}{N}\sum_{i=1}^{N} \mathbb{E}\{L|Y_i = y_i\}. \tag{4.4.5}$$

Note that Algorithm 4.4.1 requires that a random variable Y be found, such that $\mathbb{E}\{L|Y = y\}$ is known analytically for all y. Moreover, for Algorithm 4.4.1 to be of practical use, the following conditions must be met:

1. Y should be easy to generate.
2. $\mathbb{E}\{L|Y = y\}$ should be readily computable for all values of y.
3. $\mathbb{E}\{\text{var}\{L|Y\}\}$ should be large relative to var$\{\mathbb{E}\{L|Y\}\}$.

Example 4.4.1. Consider the shortest path model (4.2.7). Let each Y_i be a Bernoulli random variable with parameter p_i, that is,

$$P(Y_i = y) = p_i(1 - p_i)^{1-y}, \qquad y = 0, 1.$$

In order to achieve variance reduction, estimate $\ell = \mathbb{E}\{L\}$ by conditioning via the estimator:

$$\bar{\ell}_N^c = \frac{1}{N}\sum_{i=1}^{N} \mathbb{E}\{L(Y_{i1}, \ldots, Y_{im})|Y_{i1}, \ldots, Y_{i(j-1)}, Y_{i(j+1)}, \ldots, Y_{im}\}. \tag{4.4.6}$$

Note that $L(\mathbf{Y})$ is conditioned in (4.4.1) on all components of the vector $\mathbf{Y} = (Y_1, \ldots, Y_m)$ except for the jth component. It is not difficult to see that for given $Y_k, \ k \neq j$, the random variable

$$\mathbb{E}\{L(Y_1, \ldots, Y_m)|Y_1, \ldots, Y_{j-1}, Y_{j+1}, \ldots, Y_m\}$$

takes on one of the following three values:

$$\mathbb{E}\{L(\boldsymbol{Y}|\cdot)\} = \begin{cases} 1 & \text{the system is in order} \\ & \text{even if component } j \text{ fails;} \\ 0 & \text{the system fails} \\ & \text{even if component } j \text{ is in order;} \\ p_j & \text{the system is in order} \\ & \text{if component } j \text{ is in order;} \\ & \text{otherwise, the system fails.} \end{cases}$$

Example 4.4.2. Consider the estimation of

$$\ell = P(L_R \le x) = \mathbb{E}\{I_{\{L_R \le x\}}\},$$

for

$$L_R = \sum_{i=1}^{R} X_i,$$

where R is a random variable with some prescribed distribution and the X_i are iid, $X_i \sim F(x)$.

Let F^r denote the cdf of the random variable L_r, for fixed $R = r$. Noting that

$$F^r = P\left(\sum_{i=1}^{r} X_i \le x\right) = \mathbb{E}\left\{F\left(x - \sum_{i=2}^{r} X_i\right)\right\}$$

we obtain

$$\ell = \mathbb{E}_R\{\mathbb{E}\{I_{\{L_R \le x\}}|R\}\} = \mathbb{E}_R\left\{\mathbb{E}\left\{F\left(x - \sum_{i=2}^{R} X_i\right)\right\}\right\}.$$

[The subscript R in $\mathbb{E}_R$ indicates expectation with respect to distribution of R.]

As an estimator of ℓ based on conditioning, we can take

$$\bar{\ell}_N^c = \frac{1}{N}\left\{\sum_{j=1}^{N} F\left(x - \sum_{i=2}^{R_j} X_{ij}\right)\right\}. \tag{4.4.7}$$

An important variant of (4.4.7), where F has a *heavy-tailed* distribution, is considered in Asmussen and Binswanger (1995). For additional interesting applications of conditioning, see Burt and Garman (1971), Law (1975), Law and Kelton (1991), and Minh (1989).

4.5 IMPORTANCE SAMPLING

Consider the expected performance (4.4.1). The CMC estimator of $\mathscr{L}$ is

$$\bar{\ell}_N = \frac{1}{N}\sum_{i=1}^{N} L(\boldsymbol{Y}_i), \tag{4.5.1}$$

where $\boldsymbol{Y}_1, \ldots, \boldsymbol{Y}_N$ is a random sample from the pdf $f(\boldsymbol{y})$.

Let $G(\boldsymbol{y})$ be a probability measure (distribution) such that $dG(\boldsymbol{y}) = g(\boldsymbol{y})\,d\boldsymbol{y}$, $g(\boldsymbol{y})$ being a pdf [for $\boldsymbol{y} = (y_1, \ldots, y_n) \in \mathbb{R}^n$, $d\boldsymbol{y} = \prod_{i=1}^{n} dy_i$]. For any real function h, denote its support by

$$\text{supp}\{h\} = \{y\colon h(y) \neq 0\}.$$

Assume that $g(\boldsymbol{y})$ dominates $f(\boldsymbol{y})$ in the sense of absolute continuity, namely, $g \in \mathscr{D}_f$, where

$$\mathscr{D}_f = \{\text{density } g\colon \text{supp}\{f\} \subset \text{supp}\{g\}\}. \tag{4.5.2}$$

Using any $g \in \mathscr{D}_f$, one can represent ℓ as

$$\ell = \int L(\mathbf{z})\,\frac{f(\mathbf{z})}{g(\mathbf{z})}\,g(\mathbf{z})\,d\mathbf{z} = \mathbb{E}_g\left\{L(\boldsymbol{Z})\,\frac{f(\boldsymbol{Z})}{g(\boldsymbol{Z})}\right\}, \tag{4.5.3}$$

where the subscript g in $\mathbb{E}_g\{L(\boldsymbol{Y})\}$ indicates that the expectation is taken with respect to the pdf g. A comparison of (4.4.1) and (4.5.3) shows that ℓ can be calculated alternatively via a change of measure (from f to g), provided that $L(\boldsymbol{y})$ is changed to $L(\mathbf{z})[f(\mathbf{z})/g(\mathbf{z})]$.

An alternative to the CMC estimator, $\bar{\ell}_N$, is the importance estimator

$$\bar{\ell}^i_N = \frac{1}{N}\sum_{i=1}^{N} L(\boldsymbol{Z}_i)W(\boldsymbol{Z}_i), \tag{4.5.4}$$

where $W(\mathbf{z}) = f(\mathbf{z})/g(\mathbf{z})$ is called the *likelihood ratio* (LR), and $\boldsymbol{Z}_1, \ldots, \boldsymbol{Z}_N$ is a random sample from $g(\mathbf{z})$. A comparison of (4.5.3) and (4.4.1) reveals that the likelihood ratios $W(\boldsymbol{Z}_i)$ can be interpreted as "correction factors" necessitated by the change of measure from f to g. An added advantage of the estimator $\bar{\ell}^i_N$ is that once g is selected, one can estimate the performance $\ell = \ell(v)$ *simultaneously, for different values of* v, *from a single simulation run*. For more details see Chapters 6–8.

Consider now the problem of minimizing the variance of $\bar{\ell}_N^i$ with respect to pdf's $g \in \mathscr{D}_f$, which dominate f, that is,

$$\min_{g \in \mathscr{D}_f} \operatorname{var}_g \left\{ L(\boldsymbol{Z}) \frac{f(\boldsymbol{Z})}{g(\boldsymbol{Z})} \right\}. \tag{4.5.5}$$

It is known (see, e.g., Rubinstein, 1981) that the minimizing $g = g^*$ for program (4.5.5) is given by

$$g^*(\mathbf{z}) = \frac{|L(\mathbf{z})| f(\mathbf{z})}{\int |L(\mathbf{z})| f(\mathbf{z})\, d\mathbf{z}}. \tag{4.5.6}$$

Note that if $L(\mathbf{z}) \geq 0$, then g^* is just

$$g^*(\mathbf{z}) = \frac{L(\mathbf{z}) f(\mathbf{z})}{\ell}, \tag{4.5.7}$$

whence $\bar{\ell}_N^i = \ell$, implying $\operatorname{var}\{\bar{\ell}_N^i\} = 0$. The density $g^*(\mathbf{z})$ as per (4.5.6) and (4.5.7) is called the *importance sampling density*.

Unfortunately, the optimal importance sampling pdf is generally difficult to find. The main difficulty lies in the fact that knowledge of g^* implies knowledge of ℓ. But ℓ is precisely the quantity one wishes to estimate! The situation is then circular. Worse still, the analytical expression for the sample performance L is unknown in most simulations. To overcome this difficulty, one can make a pilot run with the underlying model, obtain a sample $L(\boldsymbol{Y}_1), \ldots, L(\boldsymbol{Y}_N)$, and then use it to estimate (approximate) the importance sampling pdf g^*. It is important to note that sampling from such an artificially constructed pdf g^* might be complicated and time-consuming, especially when g^* is a high-dimensional pdf.

Now, suppose that $g(\boldsymbol{y}) = f(\boldsymbol{y}, \boldsymbol{v}_0)$, that is, $g(\boldsymbol{y})$ belongs to the same parametric family as the distributions $f(\boldsymbol{y}) = f(\boldsymbol{y}, \boldsymbol{v})$, $\boldsymbol{v} \in \boldsymbol{V}$. The parameter vector $\boldsymbol{v}_0$ ($\boldsymbol{v}_0 \neq \boldsymbol{v}$) is called the *reference parameter*. In this case, the factors $W(\boldsymbol{Z}_i)$ in (4.5.4) become just

$$W(\boldsymbol{Z}_i) = \frac{f(\boldsymbol{Z}_i, \boldsymbol{v})}{f(\boldsymbol{Z}_i, \boldsymbol{v}_0)}.$$

In order to achieve variance reduction, one may consider replacing program (4.5.5) by a *simpler* one of the form

$$\min_{\boldsymbol{v}_0} \operatorname{var}_{\boldsymbol{v}_0} \left\{ L(\boldsymbol{Z}) \frac{f(\boldsymbol{Z}, \boldsymbol{v})}{f(\boldsymbol{Z}, \boldsymbol{v}_0)} \right\}. \tag{4.5.8}$$

Let $\boldsymbol{v}_0^*$ be the optimal solution of program (4.5.8). It turns out that for many interesting cases, (see Chapters 6–9), substantial variance reduction is achievable when using the dominating pdf, $f(\mathbf{y}, \mathbf{v}_0^*)$, instead of the importance sampling pdf, $g^*(\mathbf{z})$ of (4.5.6).

Additional discussions and applications of importance sampling may be found in Kleijnen (1974), Asmussen (1985, 1987, 1990), Ermakov (1976), Glynn (1992), Hammersley and Handscomb (1964), Siegmund (1976), Wilson (1984) and Wilson and Pritsker (1984a,b).

4.6 STRATIFIED SAMPLING

In stratified sampling, the integration region D in

$$\ell = \mathbb{E}_f\{L(\mathbf{Y})\} = \int_D L(\mathbf{y})f(\mathbf{y})\,d\mathbf{y}$$

is partitioned into m-disjoint subregions D_i, $i = 1, 2, \ldots, m$, called *strata* (hence the term "stratified"). More precisely, $D = \bigcup_{i=1}^m D_i$, $D_i \cap D_j = \varnothing$, $i \neq j$, where $\varnothing$ is the empty set.

The idea underlying stratified sampling is similar to that of importance sampling. Observations (samples) are collected over subregions D_i that are relatively "important"; however, variance reduction is achieved by "favoring" important subsets D_i, rather than by "changing" the underlying pdf over the entire region D.

To this end, represent ℓ as

$$\ell = \sum_{i=1}^m \ell_i, \tag{4.6.1}$$

where

$$\ell_i = \int_{D_i} L(\mathbf{y})f(\mathbf{y})\,d\mathbf{y}. \tag{4.6.2}$$

Next, rewrite (4.6.2) as

$$\begin{aligned}\ell_i &= \int_{D_i} L(\mathbf{y})f(\mathbf{y})\,d\mathbf{y}\,\frac{p_i}{p_i} = p_i \int_{D_i} L_i(\mathbf{y})\,\frac{f(\mathbf{y})}{p_i}\,d\mathbf{y} \\ &= p_i \mathbb{E}_{f_i}\{L_i(\mathbf{Y})\},\end{aligned} \tag{4.6.3}$$

where

$$p_i = \int_{D_i} f(y)\, dy,$$

$$f_i(y) = \begin{cases} f(y)/p_i & y \in \boldsymbol{D}_i, \\ 0 & \text{otherwise.} \end{cases}$$

$$L_i(y) = \begin{cases} L(y) & y \in \boldsymbol{D}_i \\ 0 & \text{otherwise.} \end{cases}$$

Recall that $\mathbb{E}_{f_i}$ in the last equality of (4.6.3) denotes expectation with respect to the pdf f_i.

With the aid of (4.6.1)–(4.6.3), represent $\ell = \mathbb{E}_f\{L\}$ as

$$\ell = \sum_{i=1}^{m} p_i \mathbb{E}_{f_i}\{L_i(\boldsymbol{Y})\}. \tag{4.6.4}$$

Assuming that $p_i,\ i = 1, \ldots, m$, are available analytically, the *stratified sampling estimator* (SSE) of ℓ, based on a random sample $\{Y_{ij}\},\ i = 1, \ldots, m,\ j = 1, \ldots, N_i$, is the estimator

$$\bar{\ell}_N^s = \sum_{i=1}^{m} \frac{p_i}{N_i} \sum_{j=1}^{N_i} L_i(\boldsymbol{Y}_{ij}), \tag{4.6.5}$$

where N_i is the sample size assigned to subregion D_i, such that

$$\sum_{i=1}^{m} N_i = N, \tag{4.6.6}$$

and for $i = 1, \ldots, m$ and $j = 1, \ldots, N_i$, $\boldsymbol{Y}_{ij}$ is the jth observation from subregion D_i. It is readily seen that

$$\text{var}\{\bar{\ell}_N^s\} = \sum_{i=1}^{m} \frac{p_i^2 \sigma_i^2}{N_i}, \tag{4.6.7}$$

where

$$\sigma_i^2 = \text{var}_{f_i}\{L(\boldsymbol{Y})\}.$$

Once the subsets $D_1, \ldots, D_m$ are selected, the next step is to stratify the sample, that is, to determine the number of samples, N_i, to be assigned to each subregion, D_i. The following theorem provides an optimal stratification rule.

Theorem 4.6.1 Stratified Sampling. *For a given partition* $\bigcup_{i=1}^{m} D_i = D$, *the optimal value and the optimal solution of the mathematical program*

$$\min_{\{N_1,\ldots,N_m\}} \left\{ \operatorname{var}\{\bar{\ell}_N^s\} = \sum_{i=1}^{m} \frac{p_i^2}{N_i} \sigma_i^2 \right\}, \tag{4.6.8}$$

subject to

$$\sum_{i=1}^{m} N_i = N,$$

are

$$N_i^* = N \frac{p_i \sigma_i}{\sum_{j=1}^{m} p_j \sigma_j} \tag{4.6.9}$$

and

$$\operatorname{var}\{\bar{\ell}_N^{*s}\} = \frac{1}{N} \left[\sum_{i=1}^{m} p_i \sigma_i \right]^2, \tag{4.6.10}$$

respectively.

Proof. The theorem is readily proved using Lagrange multipliers and is left as an exercise to the reader.

Theorem 4.6.1 asserts that for any given partition $\{D_i\}$ of D, the minimal variance of $\bar{\ell}_N^s$ is attained for sample sizes N_i that are proportional to $p_i\sigma_i$. However, while the probabilities p_i can be computed, the variances σ_i are usually unknown. In practice, one would estimate the σ_i from "pilot" runs and then proceed to estimate the optimal sample sizes, N_i^*, from (4.6.9).

A simple stratification procedure, which can achieve variance reduction without requiring knowledge of σ_i^2 and $L(\boldsymbol{Y})$, is presented below.

Proposition 4.6.1. *Let the sample sizes* N_i *be proportional to* p_i, *that is,* $N_i = p_i N$, $i = 1, \ldots, m$. *Then*

$$\operatorname{var}\{\bar{\ell}_N^s\} \le \operatorname{var}\{\bar{\ell}_N\}.$$

Proof. Substituting $N_i = p_i N$ in (4.6.7) yields

$$\operatorname{var}\{\bar{\ell}_N^s\} = \frac{1}{N} \sum_{i=1}^{m} p_i \sigma_i^2. \tag{4.6.11}$$

From the Cauchy-Schwarz inequality we have

$$\begin{aligned}\ell^2 &= \left(\sum_{i=1}^{m} \ell_i\right)^2 = \left[\sum_{i=1}^{m} \frac{\ell_i}{\sqrt{p_i}} \sqrt{p_i}\right]^2 \\ &\le \sum_{i=1}^{m} \frac{\ell_i^2}{p_i} \sum_{j=1}^{m} p_j = \sum_{i=1}^{m} \frac{\ell_i^2}{p_i}.\end{aligned} \tag{4.6.12}$$

But, from (4.6.11) and (4.6.3),

$$N \operatorname{var}\{\bar{\ell}_N^s\} = \sum_{i=1}^{m} p_i \sigma_i^2 = \int_D L^2(y) f(y)\, d\mathbf{y} - \sum_{i=1}^{m} \frac{\ell_i^2}{p_i}. \tag{4.6.13}$$

Combining (4.6.13) and (4.6.12) yields the inequality

$$\sum_{i=1}^{m} p_i \sigma_i^2 \le \int_D L^2(y) f(y)\, d\mathbf{y} - \ell^2 = N \operatorname{var}\{\bar{\ell}_N\}, \tag{4.6.14}$$

from which the result immediately follows.

Proposition 4.6.1 states that the estimator $\bar{\ell}_N^s$ is more accurate than the CMC estimator $\bar{\ell}_N$. It effects stratification by favoring those subregions, D_i, whose weights, $p_i = \int_{D_i} f(y)\, dy$, are largest. Intuitively, this cannot, in general, be an optimal assignment, since information on σ_i^2 and $L(\boldsymbol{Y})$ is ignored.

In the special case of equal weights ($p_i = 1/m$ and $N_i = N/m$), the estimator (4.6.5) reduces to

$$\bar{\ell}_N^s = \frac{1}{N} \sum_{i=1}^{m} \sum_{j=1}^{N/m} L(\boldsymbol{Y}_{ij}), \tag{4.6.15}$$

and the method is called the *systematic sampling method*.

4.7 EXERCISES

1. Assume that the expected performance can be written as $\ell = \sum_{i=1}^{m} a_i \ell_i$, where $\ell_i = \int L_i(\boldsymbol{x})\, d\boldsymbol{x}$, and the a_i, $i = 1, \ldots, m$, are known coefficients. An unbiased estimator of ℓ is

$$\bar{\ell} = \sum_{i=1}^{m} a_i \frac{L_i(\boldsymbol{X})}{g(\boldsymbol{X})},$$

where $\boldsymbol{X} \sim g(\boldsymbol{x})$.

(a) Prove that the optimal solution and the optimal value of the program

$$\min_{g(x)} \operatorname{var}\left\{\sum_{i=1}^{m} a_i \frac{L_i(\boldsymbol{X})}{g(\boldsymbol{X})}\right\}$$

is given by

$$g^*(\boldsymbol{x}) = |Q(\boldsymbol{x})| \Big/ \int |Q(\boldsymbol{x})|\, d\boldsymbol{x},$$

and

$$\operatorname{var}_{g^*}\{\bar{\ell}\} = \left(\int |Q(\boldsymbol{x})|\, d\boldsymbol{x}\right)^2 - \ell^2 = \left(\int \left|\sum_{i=1}^{m} a_i L_i(\boldsymbol{x})\right| d\boldsymbol{x}\right)^2 - \left(\int a_i L_i(\boldsymbol{x})\, d\boldsymbol{x}\right)^2,$$

respectively, where $Q(\boldsymbol{x}) = \sum_{i=1}^{n} a_i L_i(\boldsymbol{x})$.

(b) Prove that if the components of the vector $\boldsymbol{X} = (X^{(1)}, \ldots, X^{(m)})$ are independent, then g^* and $\operatorname{var}_{g^*}\{\bar{\ell}\}$ reduce to

$$g^*(\boldsymbol{x}) = \frac{1}{\ell}\left(\sum_{i=1}^{n} a_i^2 L_i^2(\boldsymbol{x})\right)^{1/2}$$

and

$$\operatorname{var}_{g^*}\{\bar{\ell}\} = \theta^2 - \sum_{i=1}^{n} a_i^2 \ell_i^2,$$

respectively, where

$$\theta = \int \left(\sum_{i=1}^{n} a_i^2 L_i^2(\boldsymbol{x})\right)^{1/2} d\boldsymbol{x}$$

(see Evans, 1963).

2. Consider the integral

$$\ell = \int_a^b L(x) f(x)\, dx = \mathbb{E}_f\{L(X)\}.$$

Prove that if $L(x)$ is monotonic in x and $X_i \sim \mathscr{U}(a, b)$, then

$$\operatorname{var}\{\bar{\ell}_N^{(a)}\} \le \tfrac{1}{2} \operatorname{var}\{\bar{\ell}_N\},$$

where

$$\bar{\ell}_N = \frac{1}{N}\sum_{i=1}^{N} L(X_i)$$

and

$$\bar{\ell}_N^{(a)} = \frac{1}{2N}\sum_{i=1}^{N}\{L(X_i) + L(b + a - X_i)\}.$$

In other words, using ARVs is more accurate than using CMCs.

3. Consider the stratified sampling estimator (4.6.5). Assume that the number of strata is $m = 2$, and choose equal sample sizes, $N_1 = N_2 = N/2$. Prove that if $p_1 = \frac{3}{4}$ and $p_2 = \frac{1}{4}$, instead of $p_1 = p_2 = \frac{1}{2}$ [see Proposition (4.6.1)], then for any performance function $L(\boldsymbol{x})$ and pdf $f(\boldsymbol{x})$, the SSE estimator is worse than the CMC estimator; that is, $\text{var}\{\bar{\ell}_N^s\} \geq \text{var}\{\bar{\ell}_N\}$ (see Ermakov, 1976).

4. The Hit-or-Miss Method. Suppose that the sample performance function, L, is bounded on the interval $[0, b]$, say, $0 \leq L(x) \leq c$ for $x \in [0, b]$. Define an estimator of $\ell = \mathbb{E}\{L\}$ by

$$\bar{\ell}_N^h = \frac{1}{N}\sum_{j=1}^{N} I_j,$$

where

$$I_j = \begin{cases} 1 & \text{if } Y_j < L(X_j) \\ 0 & \text{otherwise} \end{cases}$$

and $\{(X_j, Y_j) : j = 1, \ldots, N\}$, is a sequence of points uniformly distributed over the rectangle *ocdb* (see Figure 4.2). The estimator $\bar{\ell}_N^h$ is called the *hit-or-miss estimator*, since a point (X, Y), is accepted or rejected depending on whether that point falls inside or outside of the shaded area, respectively, of Figure 4.2.

Show that the hit-or-miss estimator has a larger variance than the CMC estimator.

5. Run a simulation of the bridge network of Figure 4.1 and estimate the performance $\ell = \mathbb{E}\{L\}$ of the reliability model from 1000 independent replications. Use the vector $\boldsymbol{C} = (C_1, C_2, C_3)$ of Example 4.3.1 as the vector of control variates, assuming that $X_i \sim \text{Exp}(1)$, $i = 1, \ldots, 5$.

6. Prove the importance sampling formula (4.5.6).

7. Prove the stratified sampling Theorem 4.6.1.

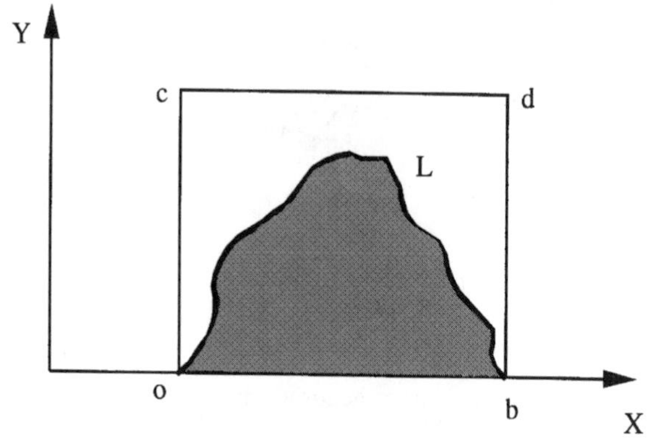

Figure 4.2. Hit-or-miss method.

REFERENCES

Asmussen, S. (1985). "Conjugate processes and the simulation of ruin problems," *Stoch. Proc. Appl.*, **20**, 213–229.

Asmussen, S. (1987). *Applied Probability and Queues*, Wiley, New York.

Asmussen, S. (1990). "Exponential families and regression in the Monte Carlo study of queues and random walks," *Ann. Statist.*, **18**(4), 1851–1867.

Asmussen, S. and Binswanger, K. (1995). "Simulation of ruin probabilities for subsexponential claims," ETH Zurich and University of Lund, unpublished.

Asmussen, S. and Rubinstein, R. Y. (1995). "Complexity properties of steady-state rare events simulation in queueing models," in *Advances in Queueing* (J. Dshalalow, ed.), CRC Press, pp. 429–462.

Burt, J. M., Jr. and Garman, M. B (1971). Conditional Monte Carlo: A simulation technique for stochastic network analysis, *Manag. Sci.*, **18**, 207–217.

Cheng, R. C. H. and Feast, G. M. (1980). "Control variables with known mean and variance," *J. Operl. Res. Soc.*, **31**, 51–56.

Ermakov, S. M. (1976). *Monte Carlo Method and Related Questions*, Nauka, Moscow, in Russian.

Evans, O. H. (1963). "Applied multiplex sampling," *Technometrics*, **5**(3), 341–359.

Fishman, G. S. (1996) *Monte Carlo: Concepts, Algorithms and Aplications*, Springer, New York.

Gal, S., Rubinstein, R. Y., and Ziv, A. (1984). "On the optimality and efficiency of common random numbers," *Math. Comput. Simul.*, **26**, 502–512.

Glasserman, P. and Yao, D. D. (1992). "Some guidance and guaranties for common random numbers," *Manag. Sci.*, **38**, 884–908.

Gross, D. and Harris, C. (1985). *Fundamentals of Queueing Theory*, Wiley, New York.

Hammersley, J. M. and Handscomb, D. C. (1964). *Monte Carlo Methods*, Wiley, New York.

Kleijnen, J. P. C. (1974). *Statistical Techniques in Simulation*, Pt. I, Marcel Dekker, New York.

Kleijnen, J. P. C. (1976). "Analysis of simulation with common random numbers: A note on Heikes et al.," *Simuletter*, **11**, 7–13.

Lavenberg, S. S., Moeller, T. L., and Welch, P. D. (1982). "Statistical results on control variables with application to queueing network simulation," *Oper. Res.*, **30**, 182–202.

Lavenberg, S. S. and Welch, P. D. (1981). "A perspective on the use of control variables to increase the efficiency of Monte Carlo simulations," *Manag. Sci.*, **27**, 322–335.

Law, A. M. (1975). "Efficient estimators for simulated queueing systems," *Manag. Sci.*, **22**, 30–41.

Law, A. M. and Kelton, W. D. (1991). *Simulation Modeling and Analysis*, McGraw-Hill, New York.

Minh, D. L. (1989). "A variant of the conditional expectation variance reduction technique and its application to the simulation of the GI/G/1 queues," *Manag. Sci.*, **35**, 1334–1340.

Rubinstein, R. Y. and Marcus, R. (1985). "Efficiency of multivariate control variates in Monte Carlo simulation," *Oper. Res.*, **33**, 661–667.

Rubinstein, R. Y. (1986). *Monte Carlo Optimization Simulation and Sensitivity of Queueing Networks*, Wiley, New York.

Rubinstein, R. Y. (1981). *Simulation and the Monte Carlo Method*, Wiley, New York.

Rubinstein, R. Y., Samorodnitsky, G., and Shaked, M. (1985). "Antithetic variates, multivariate dependence and simulation of complex stochastic systems," *Manag. Sci.*, **31**, 66–77.

Siegmund, D. (1976). "Importance sampling in the Monte Carlo study of sequential tests," *Ann. Statist.*, **4**, 673–684.

Whitt, W. (1976). "Bivariate distributions with given marginals" *Annals of Stat.*, **4**, 1280–1289.

Wilson, J. R. (1984). "Variance reduction techniques for digital simulation," *Am. J. Math. Manag. Sci.*, **4**, 277–312.

Wilson, J. R. and Pritsker, A. A. B. (1984a). "Variance reduction in queueing simulation using generalized concomitant variables," *J. Statist. Comput. Simul.*, **19**, 129–153.

Wilson, J. R. and Pritsker, A. A. B. (1984b). "Experimental evaluation of variance reduction techniques for queueing simulation using generalized concomitant variables," *Manag. Sci.*, **30**, 1459–1472.

PART II

Modern Simulation

Stochastic systems abound in real-world applications. Examples include traffic systems, flexible manufacturing systems, computer communications systems, inventory systems, production lines, coherent lifetime systems, PERT (project evaluation and review technique) networks, and flow networks. Most of these systems can be modeled in terms of discrete events whose occurrence causes the system to transition from one state to another. Recall that such systems are called *discrete-event systems* (DES). For most real-world DES, analytical methods are not available and they must be studied via simulation.

There are many good references (e.g., Elishakoff, 1983, Law and Kelton, 1991, Whitt, 1989) dealing with simlation-based performance evaluation. In designing, analyzing, and operating such complex DES, one is interested, however, not only in *performance evaluation* but also in *sensitivity analysis* and *optimization* (SAO). Sensitivity analysis is concerned with evaluating sensitivities (gradients, Hessians, etc.) of performance measures with respect to parameters of interest. It provides guidance for design and operational decisions, and plays a pivotal role in identifying the most significant system parameters, as well as bottleneck subsystems. Optimization is concerned with a DES as a whole; in particular, it makes use of sensitivity analysis to find the optimal solution with respect to parameters of interest.

To illustrate, consider the following examples:

1. *Traffic Light Systems.* Here, the performance measure might be a vehicle's average delay as it proceeds from a given origin to a given destination, or the average number of vehicles waiting for a green light at a given intersection. The sensitivity and decision parameters might be the average rate at which vehicles arrive at intersections, and the rate of light changes from green to red. Some performance issues of interest are:

a. What will a vehicle's average delay be, if the interarrival rate at a given intersection were to increase (decrease), say, by 10–50%? What would be the corresponding impact of adding one or more traffic lights to the system?

b. Which parameters are most significant in causing bottlenecks (high congestion in the system), and how can these bottlenecks be prevented or removed most effectively?

c. How can the average delay in the system be minimized, subject to certain constraints?

2. *Queueing Networks.* Here, a performance measure might be the average waiting time of an item to be processed at several workstations, given some schedule and route. The sensitivity and decision parameters might be the average rate at which the workstations process items. A typical performance issue of interest might be the minimization of the average make-span, consisting of the combined processing time and delay time in the system, subject to certain constraints (e.g., cost).

3. *Stochastic Networks.* One might wish to employ sensitivity analysis in order to minimize the mean shortest path in the network with respect, say, to network link

parameters, subject to certain constraints. PERT networks and flow networks are common examples. In the former, input and output variables may represent activity durations and the minimum project duration, respectively. In the latter, they may represent flow capacities and maximal flow capacities.

Consider the following minimization problem:

$$(P_0) = \begin{cases} \text{minimize } \ell_0(\boldsymbol{v}) = \mathbb{E}_{\boldsymbol{v}}\{L_0(\boldsymbol{Y})\}, & \boldsymbol{v} \in V, \\ \text{subject to } \ell_j(\boldsymbol{v}) = \mathbb{E}_{\boldsymbol{v}}\{L_j(\boldsymbol{Y})\} \leq 0, & j = 1, \ldots, k, \\ \qquad\qquad\; \ell_j(\boldsymbol{v}) = \mathbb{E}_{\boldsymbol{v}}\{L_j(\boldsymbol{Y})\} = 0, & j = k+1, \ldots, M, \end{cases} \tag{II.1}$$

where $L_j(\boldsymbol{Y})$ is the jth sample performance, driven by an input vector $\boldsymbol{Y} \in \mathbb{R}^m$ from cumulative distribution function (cdf) $F(\boldsymbol{y}, \boldsymbol{v})$, and $\boldsymbol{v}$ is a decision parameter vector belonging to some parameter set $V \subset \mathbb{R}^n$; the subscript $\boldsymbol{v}$ in $\mathbb{E}_{\boldsymbol{v}}\{L\}$ is shorthand notation for expectation taken with respect to $F(\boldsymbol{y}, \boldsymbol{v})$.

When the objective function $\ell_0(\boldsymbol{v})$ and the constraint functions $\ell_j(\boldsymbol{v})$ are available analytically, then (P_0) becomes a standard nonlinear programming problem, which can be solved either analytically or numerically by standard nonlinear programming techniques (see, e.g., Avriel, 1976). For example, the Markovian queueing system optimization, treated in Kleinrock (1975) falls within this domain. Here, however, it will be assumed that the objective function and some of the constraints functions in (P_0) are not available analytically (typically, due to the complexity of the underlying DES), so that one must resort to stochastic optimization methods, particularly Monte Carlo optimization.

Monte Carlo methods for sensitivity analysis and optimization of DES are typically associated with traditional statistical experimental design, based on finite-difference estimation of the gradient and Hessian via the crude Monte Carlo (CMC) method. This is a time-consuming and inaccurate procedure. For example, with n design variables, CMC requires at least $(n + 1)$ simulations to estimate the gradient.

Two new fundamental approaches to sensitivity analysis and optimization of DES have emerged in the past few decades. The *score function* (SF) approach dates back to the late sixties (see Aleksandrov et al., 1968; Asmussen and Rubinstein, 1992, 1993; Glynn, 1990; L'Ecuyer, 1990; Mikhailov, 1967, 1995; Miller, 1967; Kreimer, 1984; Pflug, 1996; Reiman and Weiss, 1989; Rubinstein, 1969, 1976; Rubinstein and Shapiro, 1993; and Shapiro, 1996). The *infinitesimal perturbation analysis* (IPA) approach followed in the late seventies (see Fu, 1994; Glasserman, 1991; Ho et al., 1979; L'Ecuyer, 1990; Wardi, 1990). Both approaches permit estimation of *all* sensitivities (gradients, Hessians, etc.) from a *single simulation run* (experiment), in the course of evaluating performance (output) measures. However, technically they are very different. The IPA approach involves *perturbation and differentiation of the sample performance with respect to the decision vector*, $\boldsymbol{v}$, and relies on the interchangeability of the differentiation and expectation operators. Such interchangeability conditions can only be ensured for a moderate class of problems. On the other hand, the SF approach is associated with such classical concepts as *efficient score, Fisher information, Radon-Nikodym derivative, and likelihood ratio*; it involves differentiation of the underlying probability density function (pdf) $f(y, \boldsymbol{v}) = dF(y, \boldsymbol{v})/dy$ and some mild regularity conditions. The SF method can handle sensitivity analysis and optimization with hundreds of decision parameters. The SF method has been implemented in a simulation package called QNSO (Queueing Network Stabilizer and Optimizer), though SF algorithms can be readily implemented in any existing discrete-event simulation language. The typical extra computation time required by SF is about 10–50% of the simulation execution time.

The rest of this book is concerned mainly with the SF method. Its goal is to demonstrate its efficacy and efficiency in performing sensitivity analysis and optimization of complex DES. It will be shown that when simulating (stochastic) models (such as *non-Markovian* queueing models), one can perform sensitivity analysis and optimization similar to those performed for analytical models (such as *Markovian* queueing systems). This feasibility is due to the fact that experiments performed on a digital computer are largely computational and not pure statistical experiments. Moreover, as soon as the pseudorandom numbers are generated, one can exploit known analytical information on the system and wield complete control on its simulation. How to use this information and how to control the system in an efficient way will be discussed later on. In a manner of speaking, the original "black- box" system is converted into an "almost white-box" system (the term "almost" stands for the fact that SAO is performed in a stochastic rather than a deterministic sense). It will also be shown that the SF method does not require explicit knowledge of the sample performance (output process), nor does it assume its differentiability. In a nutshell, the SF method requires a *single simulation run* yielding the output process, differentiability of the underlying parametric family of distributions with respect to its vector of parameters, and some mild regularity conditions relating to the interchangeability of the expectation and differentiation operators. It appears that the SF method is the only optimization method to date capable of yielding the optimal solution of an entire optimization problem of the form (P_0) from a single simulation run.

The SF method in simulation context appears to have been discovered and rediscovered independently, starting in the late sixties. The earlier work on SF appeared in Mikhailov (1967), Miller (1967), Aleksandrov et al. (1968), and Rubinstein (1969). Related references in the seventies and early eighties are Polyak (1971), Rubinstein et al. (1972), Rubinstein (1976), Ermakov and Mikhailov (1982), Rubinstein and Kreimer (1983), and Kreimer's (1984) Ph.D. thesis. In the mid eighties, the SF method was rediscovered by Glynn, and independently by Reiman and Weiss who called it the *likelihood-ratio* method [see Glynn (1990) and Reiman and Weiss (1989), and references therein]. Since then, the SF method has evolved over the past decade or so and is now reaching maturity. In its original form, the SF method (see, e.g., Rubinstein, 1976) had certain limitations that will be discussed later on. In particular, it was *only* suitable for those sensitivity and optimization problems where the sample performance, L, *does not depend* on a parameter vector $\boldsymbol{v}$ that parameterizes a family of pdf's $f(y, \boldsymbol{v})$, $\boldsymbol{v} \in V$, whose efficient score, $\nabla \log f(\boldsymbol{Y}, \boldsymbol{v})$, exists and has finite variance. Subsequent modifications, such as the methods of *decomposable score function* (DSF), *truncated score function* (TSF) (see Chapter 6), *conditional score function* (see McLeish and Rollans, 1992) and the *push-out* method (see Chapter 7) have extended the SF method's applicability to a rather broad class of problems.

We now proceed to overview the main features of the SF method. Consider the extended version of the program (P_0), allowing the sample performance, L, to *depend* on a parameter vector, $\boldsymbol{v}$; more precisely, we express the constraint functions $\ell_j(\boldsymbol{v})$ in the program (P_0) as

$$\ell_j(\boldsymbol{v}) = \mathbb{E}_{\boldsymbol{v}_1}\{L_j(\boldsymbol{Y}, \boldsymbol{v}_2)\} = \int L_j(\boldsymbol{y}, \boldsymbol{v}_2)\, dF(\boldsymbol{y}, \boldsymbol{v}_1), \qquad j = 0, 1, \ldots, M. \tag{II.2}$$

Here $\boldsymbol{v}_1$ and $\boldsymbol{v}_2$ are called the *distributional* and the *structural* parameter vectors, respectively, and the subscript $\boldsymbol{v}_1$ in $\mathbb{E}_{\boldsymbol{v}_1}\{L\}$ is shorthand notation for expectation taken with respect to the pdf $f(\boldsymbol{y}, \boldsymbol{v}_1)$. Note that f depends on the parameter vector $\boldsymbol{v}_1$ but not on $\boldsymbol{v}_2$, whereas L_j depends on $\boldsymbol{v}_2$ but not on $\boldsymbol{v}_1$. Examples of $\boldsymbol{v}_1$ and $\boldsymbol{v}_2$ are the service rate and the buffer size in a queueing model, respectively. SAO will be separately considered for (a) distributional system parameters and (b) structural system parameters, which are the subject of Chapters 5 and 6 and Chapter 7, respectively.

DISTRIBUTIONAL PARAMETERS

The crucial step in performing SAO with respect to the distributional parameters involves the choice of a "good" simulation probability measure, also called an *importance sampling* (IS) pdf (see Section 4.5). It will be shown that for many interesting applications, one can obtain nice statistical properties by choosing the IS density from the same parametric family of densities as $f(\boldsymbol{y}, \boldsymbol{v})$, that is, one can choose an IS density of the form $f(\boldsymbol{y}, \boldsymbol{v}_0)$, where $\boldsymbol{v}_0$ is called the *reference parameter*. Simply put, the problem of choosing a good IS density reduces to that of choosing a good reference parameter $\boldsymbol{v}_0$. Once an IS pdf $f(\boldsymbol{y}, \boldsymbol{v}_0)$ is selected, one proceeds as follows: The underlying system is simulated under the IS density $f(\boldsymbol{y}, \boldsymbol{v}_0)$ from which one derives *stochastic counterparts* of the entire response surface $\ell(\boldsymbol{v})$ [see (II.2)], the associated sensitivities $\nabla^k \ell(\boldsymbol{v})$, $k \geq 1$, and the deterministic program (P_0). Finally, the stochastic counterpart of the program (P_0) is solved by standard optimization methods.

We call the system corresponding to the IS pdf, $f(\boldsymbol{y}, \boldsymbol{v}_0)$, the *simulated reference system*. This reference system is selected to have appropriate statistical "goodness" properties pertaining to the unknown parameters associated with SAO, such as the optimal solution, say $\boldsymbol{v}_0^*$, of the program (P_0). Chapters 5 and 6 provide some rules of thumb on how to choose a good reference system such that performance measures at other parameter values may be well estimated from the same single simulation run. Both the technical details and the intuition underlying them will be explored and explained.

To illustrate this point, consider a stable queueing model (traffic intensity $\rho < 1$) and suppose we wish to perform SAO, say, to minimize the mean waiting time in a $GI/G/1$ queue, with respect to the service rate, v. Assume also that ρ must belong to some interval (ρ_1, ρ_2). It will be shown that in this case, the reference service rate, v_0, should correspond to the *highest traffic intensity* among all traffic intensities $\rho \in (\rho_1, \rho_2)$, associated with the feasible set V of decision parameters v. Clearly, v_0 corresponds to ρ_2, and similarly for other decision parameters. In other words, the corresponding rule of thumb is to make a single simulation run with $\rho_0 = \rho_2$, and then perform SAO for other scenarios, $\rho \in (\rho_1, \rho_2)$. Thus, the reference system (with $\rho_0 = \rho_2$) is "knowledgeable enough" to deduce from a simulation run of a $GI/G/1$ queue with, say, $\rho_0 = 0.8$, what would happen if ρ were to be set at other values, say, $\rho = 0.7$, 0.6, 0.5, 0.4, 0.3.

STRUCTURAL PARAMETERS

This is the more difficult case of the two, since in contrast to the pdf $f(y, \boldsymbol{v}_1)$, which is typically smooth and differentiable (with respect to $\boldsymbol{v}_2$), the sample performance, $L(y, \boldsymbol{v}_2)$, typically is not differentiable with respect to $\boldsymbol{v}_2$. Here we present two transformation techniques, called *push-out* and *push-in* respectively. These terms derive from the fact that in the former case we push out the parameter vector $\boldsymbol{v}_2$ from the original sample performance, $L(\boldsymbol{Y}, \boldsymbol{v}_2)$, into an auxiliary pdf via a suitable transformation, and then apply the standard SF method for SAO; in the latter case, we operate the other way around, namely, we first push in (via a suitable transformation) the parameter vector $\boldsymbol{v}_1$ into the sample performance $L(\boldsymbol{Y}, \boldsymbol{v}_2)$, and then differentiate the resulting (auxiliary) sample performance with respect to $\boldsymbol{v} = (\boldsymbol{v}_1, \boldsymbol{v}_2)$. Conditions will be discussed under which such transformations are useful. In particular, it will be shown that the IPA method, pioneered by Ho and his co-workers, corresponds to the push-in technique; the latter can be viewed as a dual of the push-out technique. It will also be shown that as soon as the vector $\boldsymbol{v}_2$ is pushed out into an auxiliary pdf, the sample performance, L, does *not* depend on $\boldsymbol{v}_2$; consequently, the program (P_0) with the functions $\ell_j(\boldsymbol{v})$, given in (II.2), becomes smooth and differentiable, and thus, solvable by the SF method (see Chapters 5–7). Note that although the push-out method is model-dependent, it allows one to perform SAO with respect to many interesting structural parameters $\boldsymbol{v}_2$. Finally several rules of thumb will be provided on how to use the push-out method, with emphasis on cases where $\boldsymbol{v}_2$ is either a location or scale parameter, or an ingredient of a recursive stochastic equation, such as Lindley's recursive equation (see Kleinrock, 1975).

We remark, at this juncture, that the foundations of SAO with respect to distributional parameters is approaching maturity, while the state-of-the-art, regarding the structural parameters, is still far from satisfactory, though active research in this area is in progress.

Estimation of rare-event probabilities is important for many modern systems, such as coherent reliability systems, computer systems, and telecommunications networks, and is the subject of Chapter 9. Importance sampling plays here a crucial role since under the original probability measures, estimation of rare events is practically impossible. As in earlier chapters, it will be assumed that the IS density is from the same *parametric family* as the original density $f(y, \mathbf{v})$, that is, the IS density is of the form $f(y, \mathbf{v}_0)$. We then minimize the variance of the obtained (parametric) IS estimator, with respect to the parameter $\mathbf{v}_0$. To this end, it will be shown that for static systems and reasonably simple queueing models, such as the $GI/G/1$ queue, the optimal choice of the parameters, say $\mathbf{v}_0^*$ (which minimize the variance of the IS estimator), leads to accurate (polynomial time) rare-event estimators, compared to inaccurate (exponential time) estimators obtained under $f(y, \mathbf{v})$. Moreover, it turns out that for complex DES, both static and dynamic, there is no need to parameterize the entire set of input densities, $f(y, \mathbf{v})$. In fact, it will be shown that for many interesting applications, it suffices to parameterize (change the probability measure) only over a small subset of $\mathbf{v}$. This subset, say $\mathbf{v}_{0b} \in V_b$, is associated with the notion of the so-called *bottleneck cut* and *bottleneck queueing module*, for static and queueing networks, respectively. The size of a vector $\mathbf{v}_{0b}$ is typically on the order of tens or less, while the size of the entire vector $\mathbf{v}$ (and $\mathbf{v}_0$) might be on the order of hundreds. In short, it will be demonstrated that rare-event probabilities can be estimated by highly accurate (polynomial time) estimators via an IS pdf, which differs from the original pdf by a small subset of parameters.

REFERENCES

Aleksandrov, V. M., Sysoyev, V. I., and Shemeneva, V. V. (1968). "Stochastic optimization," *Eng. Cybern.*, **5**, 11–16.

Asmussen, S. and Rubinstein, R. Y. (1992). "The efficiency and heavy traffic properties of the score function method in sensitivity analysis of queueing models," *Adv. Appl. Prob.*, **24**(1), 172–201.

Asmussen, S. and Rubinstein, R. Y. (1993). "Response surface estimation and sensitivity analysis via the efficient change of measure," *Stoch. Models*, **9**, 313–339.

Avriel, M. (1976). *Nonlinear Programming Analysis and Methods*, Prentice-Hall, Englewood Cliffs, NJ.

Elishakoff, I. (1983). *Probability Methods in the Theory of Structures*, Wiley, New York.

Ermakov, C. M. and Mikhailov, G. A. (1982). *Statistical Modeling*, Nauka, Moscow (in Russian).

Fu, C. M. (1994). "Optimization in simulation: A review," *Ann. Oper. Res.*, **53**, 99–247.

Glasserman, P. (1991). *Gradient Estimation via Perturbation Analysis*, Kluwer, Norwell, Mass.

Glynn, P. W. (1990). "Likelihood ratio gradient estimation for stochastic systems," *Commun. ACM*, **33**(10), 75–84.

Ho, Y. C., Eyler, M. A., and Chien, T. T. (1979). "A gradient technique for general buffer strorage design in a serial production line," *Int. J. Product. Res.*, **17**(6), 557–580.

Kleinrock, L. (1975). *Queueing Systems*, Vols. I and II, Wiley, New York.

Kreimer, J. (1984). "Stochastic optimization—an adaptive approach," D.Sc. Thesis, Technion, Haifa, Israel.

Law, A. M. and Kelton, W. D. (1991). *Simulation Modeling and Analysis*, McGraw-Hill, New York.

L'Ecuyer, P. (1990). "A unified view of the IPA, SF, and LR gradient estimation techniques," *Manage. Sci.*, **36**, 1364–1384.

McLeish, D. L. and Rollans, S. (1992). "Conditioning for variance reduction in estimating the sensitivity of simulations," *Ann. of Oper. Res.*, **39**, 157–173.

Mikhailov, G. A. (1967). "Calculation of system parameter derivatives of functionals of the solutions to the transport equations," *Zh. Vychisyl. Mat. Mat. Fiz.*, **7**, 915 (see also the English translation, *J. Comput. Math. Math. Phys.*, 1967).

Mikhailov, G. A. (1995). *New Monte Carlo Methods with Estimating Derivatives*, VSP BV, Zeist, The Netherlands.

Miller, L. B. (1967). "Monte Carlo analysis of reactivity coefficients in fast reactors: General theory and applications," ANL-7307 (TID-4500), Argonne National Laboratory, IL.

Polyak, U. G. (1971). *Probabilistic Modeling on Digital Computes*, Sovjet Radio, Moscow (in Russian).

Pflug, G. Ch. (1995). *Simulation and Optimization: The Interface Between Simulation and Optimization*, Kluwer.

Reiman, M. I. and Weiss, A. (1989). "Sensitivity analysis for simulations via likelihood rations," *Oper. Res.*, **37**(5), 830–844.

Rubinstein, R. Y. (1969). "Some problems in Monte Carlo Optimization," PhD Thesis, Riga, Latvia (in Russian).

Rubinstein, R. Y. (1976). "A Monte Carlo method for estimating the gradient in a stochastic network," Manuscript, Technion, Haifa, Israel.

Rubinstein, R. Y. (1986). *Monte Carlo Optimization Simulation and Sensitivity of Queueing Network*, Wiley, New York.

Rubinstein, R. Y. and Kreimer, J. (1983). "About one Monte Carlo method for solving linear equations," *Math. Comput. Simulat.*, **XXV**, 321–334.

Rubinstein, R. Y. and Shapiro, A. (1993). *Discrete Event Systems: Sensitivity Analysis and Stochastic Optimization via the Score Function Method*, Wiley, New York.

Rubinstein, R. Y., Moroz, P., and Kotlar (1972). "Combination of Monte Carlo methods and stochastic approximation for search of extremum," *Quest. Cybern.*, **42**, 107–114 (in Russian).

Shapiro A. (1996). "Simulation based optimization-convergence analysis and statistical inference," *Stoch. Models*, **12**(3), 425–454.

Whitt, W. (1989). "Planning queueing simulations," *Manage. Sci.*, **35**(11), 1341–1366.

CHAPTER 5

Sensitivity Analysis and Optimization of Discrete-Event Static Systems (DESS)

5.1 INTRODUCTION

This chapter deals with sensitivity analysis and optimization of DESS with expected performance

$$\ell(\boldsymbol{v}) = \mathbb{E}_{v}\{L(\boldsymbol{Y})\} = \int L(\boldsymbol{y})\, dF(\boldsymbol{y}, \boldsymbol{v}), \tag{5.1.1}$$

where the cdf $F(\boldsymbol{y}, \boldsymbol{v})$ of the random vector $\boldsymbol{Y}$ in $\ell(\boldsymbol{v}) = \mathbb{E}_{v}\{L(\boldsymbol{Y})\}$ is assumed to belong to a family of absolutely continuous distributions, that is, $dF(\boldsymbol{y}, \boldsymbol{v}) = f(\boldsymbol{y}, \boldsymbol{v})\, d\boldsymbol{y}$, $f(\boldsymbol{y}, \boldsymbol{v})$ being the corresponding pdf. The treatment of the case where $F(\boldsymbol{y}, \boldsymbol{v})$ belongs to a family of discrete or mixed distributions is similar. Examples of DESS possessing these properties are the reliability and stochastic shortest-path systems given in (4.2.8) and (4.2.7), respectively. The analyst might be interested in estimating the sensitivities (gradients, Hessians, etc.) of the mean lifetime of the coherent system (4.2.8) with respect to the expected life of its components, and similarly for the stochastic shortest-path system (4.2.7). One might also be interested in maximizing the mean lifetime and minimizing the mean shortest path in models (4.2.8) and (4.2.7), respectively, subject to certain constraints. The following example is associated with optimization of the stochactic shortest-path model.

Example 5.1.1

$$(P_0) = \begin{cases} \text{minimize} & \ell_0(\boldsymbol{v}), \qquad \boldsymbol{v} \in V, \\ \text{subject to} & \ell_1(\boldsymbol{v}) \le 0, \\ & \ell_2(\boldsymbol{v}) \le 0, \end{cases} \tag{5.1.2}$$

where $V = \mathbb{R}^n_+$, $\ell_0(\boldsymbol{v})$ coincides with $\ell(\boldsymbol{v}) = \mathbb{E}\{L(\boldsymbol{Y})\}$ in (4.2.7) for which

$$L(\boldsymbol{Y}) = \min_{j=1,\ldots,p} \sum_{i \in \mathscr{L}_j} Y_i,$$

and

$$\ell_1(\boldsymbol{v}) = P\{L(\boldsymbol{Y}) > x\} - b_1 = \mathbb{E}\{I_{\{L(\boldsymbol{Y})>x\}}\} - b_1,$$
$$\ell_2(\boldsymbol{v}) = \sum_{j=1}^{n} c_j v_j - b_2.$$

Here I_A is the indicator of the event A, ℓ_1 is the "stochastic" constraint, ℓ_2 is the "deterministic" constraint, c_j is the cost-per-unit increase in $v_j = \mathbb{E}\{Y_j\}$, and the b_i, $i = 1, 2$, are given constants.

The remainder of this chapter is organized as follows. Section 5.2 deals with sensitivity analysis of $\ell(\boldsymbol{v})$ [estimation of the gradients $\nabla\ell(\boldsymbol{v})$, Hessians $\nabla^2\ell(\boldsymbol{v})$, etc.]. The SF method will be introduced and shown to be key in estimating *all* sensitivities *simultaneously* from a *single* simulation experiment, by expressing the sensitivities $\nabla^k\ell(\boldsymbol{v})$, $k = 1, 2, \ldots$, as expectations with respect to the same probability measure (distribution) F. Section 5.3 deals with the what-if problem, which can be formulated as follows: "*What* will be the expected performance, $\ell(\boldsymbol{v})$, and the associated sensitivities, $\nabla^k\ell(\boldsymbol{v})$, $k = 1, 2, \ldots$, *if* we perturb some (all) of the parameters of the vector $\boldsymbol{v}$?" The SF method will be shown to resolve this problem in the sense that it permits estimation of the performance $\ell(\boldsymbol{v})$ at *several* values (scenarios) of $\boldsymbol{v}$ (say, $v_1, \ldots, v_r$), from a *single* simulation experiment. This section also discusses conditions under which the SF method is sure to perform well, and shows that the what-if problem is directly associated with the so-called *response-surface methodology*, which deals with the estimation of entire response surfaces (functions) $\ell(\boldsymbol{v})$ and $\nabla^k\ell(\boldsymbol{v})$, $k = 1, 2, \ldots$ *from a single simulation run*. Additional details on the response-surface methodology will be given in Chapter 8. In Section 5.4 it is shown how the SF method optimizes a DESS, and in particular, how it evaluates the optimal solution of the problem (P_0) from a *single* simulation run. The method is based on a change of measure and on the construction of stochastic counterparts for the conventional deterministic optimization procedures. The section also discusses the statistical properties of the SF estimators and exemplifies them through numerical results.

5.2 SENSITIVITY ANALYSIS OF SYSTEM PERFORMANCE

Let G be a probability measure (distribution) on $\mathbb{R}^m$ with corresponding pdf $g(y)$, so that $dG(y) = g(y)\,dy$. Suppose that for every permissible value of the parameter vector, $\boldsymbol{v}$, the support of $f(\boldsymbol{y}, \boldsymbol{v})$ is contained in the support of $g(\boldsymbol{y})$, that is (see also Section 4.5),

$$\text{supp}\{f(\boldsymbol{y}, \boldsymbol{v})\} \subset \text{supp}\{g(\boldsymbol{y})\}, \qquad \boldsymbol{v} \in V. \tag{5.2.1}$$

[Recall that supp$\{g(\boldsymbol{y})\}$ is the set of those values of $\boldsymbol{y}$, such that $g(\boldsymbol{y}) > 0$]. In this case, we can represent

$$\ell(\boldsymbol{v}) = \mathbb{E}_{\boldsymbol{v}}\{L(\boldsymbol{Y})\}$$

alternatively as

$$\begin{aligned} \ell(\boldsymbol{v}) &= \int L(\boldsymbol{y}) f(\boldsymbol{y}, \boldsymbol{v})\, d\boldsymbol{y} = \int L(\boldsymbol{y}) W(\boldsymbol{y}, \boldsymbol{v})\, dG(\boldsymbol{y}) \\ &= \mathbb{E}_g\{L(\boldsymbol{Z}) W(\boldsymbol{Z}, \boldsymbol{v})\}, \end{aligned} \tag{5.2.2}$$

where

$$W(\mathbf{z}, \boldsymbol{v}) = f(\mathbf{z}, \boldsymbol{v})/g(\mathbf{z}) \quad \text{and} \quad \boldsymbol{Z} \sim g(\mathbf{z}).$$

(By definition, zero divided by zero is zero.) Notice that the function $W(\boldsymbol{y}, \boldsymbol{v})$ is well defined for all $\boldsymbol{v} \in V$ due to condition (5.2.1). In the statistical literature, $W(\boldsymbol{Z}, \boldsymbol{v})$ is called the *likelihood ratio* or *the Radon-Nikodym derivative.* A pdf $g(\boldsymbol{y})$, satisfying condition (5.2.1), will be called a *dominating* pdf. For more details see Rubinstein and Shapiro (1993).

It is important to note that the original expectation, $\ell(\boldsymbol{v}) = \mathbb{E}_{\boldsymbol{v}}\{L(\boldsymbol{Y})\}$, is taken with respect to the underlying pdf, $f(\boldsymbol{y}, \boldsymbol{v})$, while the one on the right-hand side of (5.2.2) is taken with respect to the dominating pdf, $g(\boldsymbol{y})$. Thus, changing the probability density from $f(\boldsymbol{y}, \boldsymbol{v})$ to $g(\boldsymbol{y})$ allows us to express the performance measure, $\ell(\boldsymbol{v})$, as an expectation with respect to $g(\boldsymbol{y})$ and to estimate it accordingly.

Let $\{\boldsymbol{Z}_1, \ldots, \boldsymbol{Z}_N\}$ be a sample from $g(\mathbf{z})$. Then an unbiased estimator of $\ell(\boldsymbol{v})$ is

$$\bar{\ell}_N(\boldsymbol{v}) = \frac{1}{N} \sum_{i=1}^{N} L(\boldsymbol{Z}_i) W(\boldsymbol{Z}_i, \boldsymbol{v}). \tag{5.2.3}$$

The estimator $\bar{\ell}_N(\boldsymbol{v})$ will be referred to as the *likelihood ratio* (LR) estimator of $\ell(\boldsymbol{v})$. In the special case where $g(\boldsymbol{y}) = f(\boldsymbol{y}, \boldsymbol{v})$, the LR estimator reduces to

$$\tilde{\ell}_N(\boldsymbol{v}) = \frac{1}{N} \sum_{i=1}^{N} L(\boldsymbol{Y}_i), \tag{5.2.4}$$

to be referred to as the *crude Monte Carlo* (CMC) estimator of $\ell(\boldsymbol{v})$. Here $\{\boldsymbol{Y}_1, \ldots, \boldsymbol{Y}_N\}$ is a sample from the original pdf $f(\boldsymbol{y}, \boldsymbol{v})$.

It is not difficult to see that once the dominating pdf, $g(\boldsymbol{y})$, is chosen, the LR estimator $\bar{\ell}_N(\boldsymbol{v})$ may be used to estimate the *entire surface* $\ell(\boldsymbol{v})$ (response surface) from a *single simulation run*, while the CMC estimator $\tilde{\ell}_N(\boldsymbol{v})$ requires *separate simulation runs* for *each* value of the parameter vector $\boldsymbol{v}$.

The following simple example illustrates this important issue in greater detail.

Example 5.2.1. Let $L(Y) = Y$, where $Y \sim \text{Ber}(p)$. Suppose we wish to estimate the parameter $p = \mathbb{E}_p\{Y\}$ by simulation. Let us choose

$$g(y) = p_0^y (1 - p_0)^{1-y}, \qquad y = 0, 1$$

to be the dominating pdf.

In this case (using likelihood ratios) we can write $\ell(p)$ as

$$\ell(p) = \mathbb{E}_g \left\{ Z \frac{p^Z (1-p)^{1-Z}}{p_0^Z (1-p_0)^{1-Z}} \right\},$$

where $Z \sim \text{Ber}(p_0)$. The LR estimator is then

$$\bar{\ell}_N(p) = \frac{1}{N} \sum_{i=1}^{N} Z_i \frac{p^{Z_i} (1-p)^{1-Z_i}}{p_0^{Z_i} (1-p_0)^{1-Z_i}}, \tag{5.2.5}$$

where $\{Z_i : i = 1, \ldots, N\}$ is a random sample from $\text{Ber}(p_0)$. In the special case where $g(y) = f(y, p)$, i.e., $p_0 = p$, the estimator $\bar{\ell}_N(p)$ reduces to the CMC estimator

$$\tilde{\ell}_N(p) = \frac{1}{N} \sum_{i=1}^{N} Y_i,$$

where $\{Y_i : i = 1, \ldots, N\}$ is a random sample from $\text{Ber}(p)$.

As a simple illustration, suppose we observed the random sample

$$\{Z_1, \ldots, Z_{20}\} = \{0, 1, 0, 0, 1, 0, 0, 1, 1, 1, 0, 1, 0, 1, 1, 0, 1, 0, 1, 1\}$$

of size $N = 20$ from $g(y) \sim \text{Ber}(p_0 = \frac{1}{2})$, and we wish to estimate from it the quantities $\ell(p) \sim \mathbb{E}_p\{Y\}$, *simultaneously* for $p = \frac{1}{4}$ and $p = \frac{1}{10}$. From (5.2.5), we then have

$$\begin{aligned}\bar{\ell}_N\left(p = \tfrac{1}{4}\right) &= \frac{1}{N} \sum_{i=1}^{N} Z_i \frac{(\frac{1}{4})^{Z_i} (1 - \frac{1}{4})^{1-Z_i}}{(\frac{1}{2})^{Z_i} (1 - \frac{1}{2})^{1-Z_i}} \\ &= \tfrac{1}{20}(0 + \tfrac{1}{2} + 0 + 0 + \tfrac{1}{2} + 0 + 0 + \tfrac{1}{2} + \tfrac{1}{2} + \tfrac{1}{2} + 0 + \tfrac{1}{2} + 0 + \tfrac{1}{2} + \tfrac{1}{2} + 0 \\ &\quad + \tfrac{1}{2} + 0 + \tfrac{1}{2} + \tfrac{1}{2}) = \tfrac{11}{40},\end{aligned}$$

and

$$\bar{\ell}_N(p = \tfrac{1}{10}) = \frac{1}{N}\sum_{i=1}^{N} Z_i \frac{(\frac{1}{10})^{Z_i}(1-\frac{1}{10})^{1-Z_i}}{(\frac{1}{2})^{Z_i}(1-\frac{1}{2})^{1-Z_i}}$$
$$= \tfrac{1}{20}(0 + \tfrac{1}{5} + 0 + 0 + \tfrac{1}{5} + 0 + 0 + \tfrac{1}{5} + \tfrac{1}{5} + \tfrac{1}{5} + 0 + \tfrac{1}{5} + 0 + \tfrac{1}{5} + \tfrac{1}{5} + 0 + \tfrac{1}{5} + 0 + \tfrac{1}{5} + \tfrac{1}{5}) = 0.11.$$

It is readily seen that the corresponding estimation using the CMC estimator, $\tilde{\ell}_N(p)$, would require two random samples of size $N = 20$ each: One from Ber($\frac{1}{4}$) and another from Ber($\frac{1}{10}$).

Example 5.2.2. Let $L(Y) = Y$, where $Y \sim \text{Exp}(\lambda)$, and select $g(y) = f(y, \lambda_0)$, that is,

$$g(y) = \lambda_0 e^{-\lambda_0 y}, \qquad \lambda_0 > 0, \quad y > 0.$$

Using the likelihood ratio we can express $\ell(\lambda) = \mathbb{E}_\lambda\{Y\} = 1/\lambda$ as

$$\ell(\lambda) = \mathbb{E}_g\left\{Z \frac{\lambda e^{-\lambda Z}}{\lambda_0 e^{-\lambda_0 Z}}\right\},$$

where $Z \sim \text{Exp}(\lambda_0)$. The LR estimator of $\ell(\lambda)$ is

$$\bar{\ell}_N(\lambda) = \frac{1}{N}\sum_{i=1}^{N} Z_i \frac{\lambda e^{-\lambda Z_i}}{\lambda_0 e^{-\lambda_0 Z_i}},$$

where $\{Z_i: i = 1, \ldots, N\}$ is a random sample from $\text{Exp}(\lambda_0)$.

We now derive an expression for the performance gradient, $\nabla\ell(\boldsymbol{v})$. Assume first that the parameter v is a scalar and the parameter set V is an open interval on the real line. Suppose that for all $\mathbf{y}$, the function $f(\mathbf{y}, v)$ is continuously differentiable in v and that there exists a Lebesgue-integrable function $h(\mathbf{y})$, such that

$$\left| L(\mathbf{y}) \frac{\partial W(\mathbf{y}, v)}{\partial v} \right| \leq h(\mathbf{y}) \tag{5.2.6}$$

for all $v \in V$. Then by the Lebesgue dominated convergence theorem, the differentiation and expectation (integration) operators are interchangeable, so that differentiation of (5.2.2) yields

$$\frac{d\ell(v)}{dv} = \frac{d}{dv}\int L(\mathbf{y}) f(\mathbf{y}, v)\, d\mathbf{y} = \int L(\mathbf{y}) \frac{\partial f(\mathbf{y}, v)}{\partial v}\, d(\mathbf{y})$$
$$= \int L(\mathbf{y}) \frac{\partial W(\mathbf{y}, v)}{\partial v}\, dG(\mathbf{y}) = \mathbb{E}_g\left\{L(\mathbf{Z}) \frac{\partial W(\mathbf{Z}, v)}{\partial v}\right\}.$$

Consider next the multidimensional case. Similar arguments allow us to represent the gradient of $\ell(\boldsymbol{v})$ in the form

$$\begin{aligned}\nabla \ell(\boldsymbol{v}) &= \int L(\boldsymbol{y}) \nabla f(\boldsymbol{y}, \boldsymbol{v})\, d\boldsymbol{y} = \int L(\boldsymbol{y}) \nabla W(\boldsymbol{y}, \boldsymbol{v})\, dG(\boldsymbol{y}) \\ &= \mathbb{E}_g\{L(\boldsymbol{Z}) \nabla W(\boldsymbol{Z}, \boldsymbol{v})\}. \end{aligned} \tag{5.2.7}$$

Differentiating $\nabla \ell(\boldsymbol{v})$ repeatedly and assuming that the differentiation and expectation operators are interchangeable, we obtain

$$\nabla^k \ell(\boldsymbol{v}) = \mathbb{E}_g\{L(\boldsymbol{Z}) \nabla^k W(\boldsymbol{Z}, \boldsymbol{v})\}, \qquad k = 1, 2, \ldots. \tag{5.2.8}$$

Formula (5.2.8) is also valid for $k = 0$ if we adopt the conventions $\nabla^0 \ell(\boldsymbol{v}) \equiv \ell(\boldsymbol{v})$ and $\nabla^0 W(\boldsymbol{y}, \boldsymbol{v}) \equiv W(\boldsymbol{y}, \boldsymbol{v})$. Under similar assumptions, it is also valid for discrete distributions; one only needs to replace the integrals by appropriate sums, with all other data remaining unchanged.

In general, the quantities $\nabla^k \ell(\boldsymbol{v})$, $k = 0, 1, \ldots$, are not available analytically, since the performance $\ell(\boldsymbol{v})$ is not. They can be evaluated, however, either by conventional deterministic numerical methods (see, e.g., Szidarovszky and Yakowitz, 1978) or via simulation. Both methods are applicable here, since formula (5.2.8) presents *closed-form expressions* for $\nabla^k \ell(\boldsymbol{v})$, $k = 0, 1, \ldots$. Simulation is particularly convenient, as the performance, $\ell(\boldsymbol{v})$, and *all* the sensitivities, $\nabla^k \ell(\boldsymbol{v})$, $k = 1, 2, \ldots$, are expressed as expectations with respect to the same dominating pdf, $g(\boldsymbol{y})$.

Let $\{\boldsymbol{Z}_1, \ldots, \boldsymbol{Z}_N\}$ be a sample from $g(\boldsymbol{y})$. Then all the quantities $\nabla^k \ell(\boldsymbol{v})$ can be estimated *simultaneously* from a *single simulation run* by estimators of the form

$$\bar{\nabla}^k \ell_N(\boldsymbol{v}) = \frac{1}{N} \sum_{i=1}^{N} L(\boldsymbol{Z}_i) \nabla^k W(\boldsymbol{Z}_i, \boldsymbol{v}). \tag{5.2.9}$$

It is not difficult to deduce from (5.2.3), (5.2.8) and (5.2.9) that

$$\bar{\nabla}^k \ell_N(\boldsymbol{v}) = \nabla^k \bar{\ell}_N(\boldsymbol{v}), \qquad k = 1, 2, \ldots. \tag{5.2.10}$$

It is important to note that the estimators $\nabla^k \bar{\ell}_N(\boldsymbol{v})$, $k = 0, 1, \ldots$, allow us to estimate the corresponding $\nabla^k \ell(\boldsymbol{v})$ at virtually any point $\boldsymbol{v} \in V$, provided the interchangeability of integration and differentiation is valid. As will be seen below, this fact renders the above estimators particularly suitable for solving the optimization problem (P_0) introduced earlier.

The following two simple examples illustrate how to estimate $\nabla \ell(\boldsymbol{v})$ from a single simulation run.

Example 5.2.3 (Example 5.2.1 continued). Since $\ell(p) = p$, one has $\nabla\ell(p) \equiv 1$. Using likelihood ratios, $\nabla\ell(p)$ can be written as

$$\nabla\ell(p) = \mathbb{E}_g\left\{Z\frac{\nabla(p^Z(1-p)^{1-Z})}{p_0^Z(1-p_0)^{1-Z}}\right\} = \mathbb{E}_g\left\{Z\frac{p^Z(1-p)^{1-Z}}{p_0^Z(1-p_0)^{1-Z}} \times \frac{Z-p}{p(1-p)}\right\} = 1,$$

and estimated by

$$\nabla\bar{\ell}_N(p) = \frac{1}{N}\sum_{i=1}^{N} Z_i \frac{p^{Z_i}(1-p)^{1-Z_i}}{p_0^{Z_i}(1-p_0)^{1-Z_i}} \times \frac{Z_i - p}{p(1-p)}, \tag{5.2.11}$$

where as before, $\{Z_i : i = 1, \dots, N\}$ is a random sample from Ber(p_0). Suppose again that we observed the sample

$$\{Z_1, \dots, Z_{20}\} = \{0, 1, 0, 0, 1, 0, 0, 1, 1, 1, 0, 1, 0, 1, 1, 0, 1, 0, 1, 1\}$$

of size $N = 20$ from Ber($p_0 = \frac{1}{2}$), and we wish to estimate $\nabla\ell(p) \equiv 1$ for $p = \frac{1}{4}$ and $p = \frac{1}{10}$, *simultaneously*. From (5.2.11), we have

$$\begin{aligned}\nabla\bar{\ell}_N(p = \tfrac{1}{4}) &= \frac{1}{N}\sum_{i=1}^{N} Z_i \frac{(\frac{1}{4})^{Z_i}(1-\frac{1}{4})^{1-Z_i}}{(\frac{1}{2})^{Z_i}(1-\frac{1}{2})^{1-Z_i}} \times \frac{Z_i - \frac{1}{4}}{(\frac{1}{4})(\frac{3}{4})}\\ &= \tfrac{1}{20}(0 + \tfrac{1}{2}\cdot 4 + 0 + 0 + \tfrac{1}{2}\cdot 4 + 0 + 0 + \tfrac{1}{2}\cdot 4 + \tfrac{1}{2}\cdot 4 + \tfrac{1}{2}\cdot 4 + 0\\ &\quad + \tfrac{1}{2}\cdot 4 + 0 + \tfrac{1}{2}\cdot 4 + \tfrac{1}{2}\cdot 4 + 0 + \tfrac{1}{2}\cdot 4 + 0 + \tfrac{1}{2}\cdot 4 + \tfrac{1}{2}\cdot 4) = 1.1\end{aligned}$$

and

$$\begin{aligned}\nabla\bar{\ell}_N(p = \tfrac{1}{10}) &= \frac{1}{N}\sum_{i=1}^{N} Z_i \frac{(\frac{1}{10})^{Z_i}(1-\frac{1}{10})^{1-Z_i}}{(\frac{1}{2})^{Z_i}(1-\frac{1}{2})^{1-Z_i}} \times \frac{Z_i - \frac{1}{10}}{(\frac{1}{10})\cdot(\frac{9}{10})}\\ &= \tfrac{1}{20}(0 + \tfrac{1}{5}\cdot 10 + 0 + 0 + \tfrac{1}{5}\cdot 10 + 0 + 0 + \tfrac{1}{5}\cdot 10 + \tfrac{1}{5}\cdot 10 + \tfrac{1}{5}\cdot 10\\ &\quad + 0 + \tfrac{1}{5}\cdot 10 + 0 + \tfrac{1}{5}\cdot 10 + \tfrac{1}{5}\cdot 10 + 0 + \tfrac{1}{5}\cdot 10 + 0 + \tfrac{1}{5}\cdot 10 + \tfrac{1}{5}\cdot 10)\\ &= 1.1,\end{aligned}$$

respectively. Notice that the estimates $\nabla\bar{\ell}_N(p = \frac{1}{4}) = 1.1$ and $\nabla\bar{\ell}_N(p = \frac{1}{10}) = 1.1$ coincide.

Example 5.2.4 (Example 5.2.2 continued). In this case,

$$\nabla\ell(\lambda) = \mathbb{E}_g\{Z\nabla W(Z, \lambda)\} = \mathbb{E}_g\left\{Z\frac{\nabla\lambda e^{-\lambda Z}}{\lambda_0 e^{-\lambda_0 Z}}\right\} = -1/\lambda^2,$$

where $Z \sim \text{Exp}(\lambda_0)$ and $\nabla\lambda e^{-\lambda Z} \equiv \partial\lambda e^{-\lambda Z}/\partial\lambda$. (The calculation of higher-order derivatives is similar.) An unbiased estimator of $\nabla\ell(\lambda)$ is

$$\nabla\bar{\ell}_N(\lambda) = \frac{1}{N}\sum_{i=1}^{N} Z_i \frac{\nabla\lambda e^{-\lambda Z_i}}{\lambda_0 e^{-\lambda_0 Z_i}} = \frac{1}{N}\sum_{i=1}^{N} Z_i \frac{\lambda e^{-\lambda Z_i}(\lambda^{-1} - Z_i)}{\lambda_0 e^{-\lambda_0 Z_i}}, \tag{5.2.12}$$

where $\{Z_i: i = 1, \ldots, N\}$ is a random sample from $\text{Exp}(\lambda_0)$. For the special case of $g(y) = f(y, \lambda)$, (5.2.12) reduces to

$$\nabla\bar{\ell}_N(\lambda) = \frac{1}{N}\sum_{i=1}^{N} Y_i(\lambda^{-1} - Y_i), \tag{5.2.13}$$

where $\{Y_1, \ldots, Y_N\}$ is a random sample from $\text{Exp}(\lambda)$.

For a given dominating pdf $g(\mathbf{z})$, the algorithm for estimating the sensitivities $\nabla^k\ell(\boldsymbol{v})$, $k = 0, 1, \ldots$, for *multiple values* $\boldsymbol{v}$ *from a single simulation run*, is sketched below.

Algorithm 5.2.1

1. *Generate a sample* $\{\boldsymbol{Z}_1, \ldots, \boldsymbol{Z}_N\}$ *from the dominating pdf,* $g(\mathbf{z})$.
2. *Calculate the sample performance* $L(\boldsymbol{Z}_i)$ *and the sensitivities* $\nabla^k W(\boldsymbol{Z}_i, \boldsymbol{v})$, $i = 1, \ldots, N$, *for the desired parameter vectors* $\boldsymbol{v}_1, \ldots, \boldsymbol{v}_k$.
3. *Calculate* $\nabla^k\bar{\ell}_N(\boldsymbol{v})$ *from (5.2.9).*

It is important to note that in order to estimate the sensitivities $\nabla^k\ell(\boldsymbol{v})$, $k = 1, 2, \ldots$, there is no need to differentiate $L(\boldsymbol{Y})$, which in most practical applications is not a smooth (differentiable) function. All we need is a sample $\{\boldsymbol{Z}_i: i = 1, \ldots, N\}$, from the dominating pdf $g(\boldsymbol{y})$, from which the sample performance, $L(\boldsymbol{Z}_i)$, and the derivatives $\nabla^k W(\boldsymbol{Z}_i, \boldsymbol{v})$, may be computed.

Confidence regions for $\nabla^k\ell(\boldsymbol{v})$ can be obtained by standard techniques. In particular, $N^{1/2}[\nabla\bar{\ell}_N(\boldsymbol{v}) - \nabla\ell(\boldsymbol{v})]$ converges in distribution to a multivariate normal with mean zero and covariance matrix

$$\text{cov}\{L\nabla W\} = \mathbb{E}_g\{L^2(\boldsymbol{Z})\nabla W(\boldsymbol{Z}, \boldsymbol{v})\nabla W(\boldsymbol{Z}, \boldsymbol{v})'\} - [\nabla\ell(\boldsymbol{v})][\nabla\ell(\boldsymbol{v})]' \tag{5.2.14}$$

(see Rubinstein and Shapiro, 1993).

We shall next derive expressions for the sensitivities $\nabla^k\ell(\boldsymbol{v})$ for the special case $g(\boldsymbol{y}) = f(\boldsymbol{y}, \boldsymbol{v})$. From (5.2.8), for $k = 1$,

$$\nabla\ell(\boldsymbol{v}) = \mathbb{E}_{\boldsymbol{v}}\{L(\boldsymbol{Y})\boldsymbol{S}^{(1)}(\boldsymbol{Y}, \boldsymbol{v})\} = \mathbb{E}_{\boldsymbol{v}}\{L(\boldsymbol{Y})\nabla \log f(\boldsymbol{Y}, \boldsymbol{v})\}, \tag{5.2.15}$$

where

$$\boldsymbol{S}^{(1)}(y, \boldsymbol{v}) = \frac{\nabla f(y, \boldsymbol{v})}{f(y, \boldsymbol{v})} = \nabla \log f(y, \boldsymbol{v}) \tag{5.2.16}$$

is called the *score function* (SF). The latter plays an important role in sensitivity analysis and optimization of DES.

Proceeding with $\nabla \ell(\boldsymbol{v})$ and assuming that the differentiation and expectation operators are interchangeable, we obtain

$$\nabla^k \ell(\boldsymbol{v}) = \mathbb{E}_{\boldsymbol{v}}\{L(\boldsymbol{Y})\boldsymbol{S}^{(k)}(\boldsymbol{Y}, \boldsymbol{v})\}, \tag{5.2.17}$$

where

$$\boldsymbol{S}^{(k)}(y, \boldsymbol{v}) = \frac{\nabla^k f(y, \boldsymbol{v})}{f(y, \boldsymbol{v})}. \tag{5.2.18}$$

In particular, the second-order partial derivatives

$$\frac{\partial^2 \ell(\boldsymbol{v})}{\partial v_i \partial v_j}, \qquad i, j = 1, \ldots, n,$$

which are the components of the Hessian matrix $\nabla^2 \ell(\boldsymbol{v}) = \mathbb{E}_{\boldsymbol{v}}\{L(\boldsymbol{Y})\boldsymbol{S}^{(2)}(\boldsymbol{Y}, \boldsymbol{v})\}$, can be written as

$$\begin{aligned}
\frac{\partial^2 \ell(\boldsymbol{v})}{\partial v_i \partial v_j} &= \int L(y) \frac{\partial^2 f(y, \boldsymbol{v})}{\partial v_i \partial v_j} dy = \int L(y) \frac{\partial}{\partial v_i}\left[\frac{\partial \log f(y, \boldsymbol{v})}{\partial v_j} f(y, \boldsymbol{v})\right] dy \\
&= \int L(y)\left[\frac{\partial^2 \log f(y, \boldsymbol{v})}{\partial v_i \partial v_j} + \frac{\partial \log f(y, \boldsymbol{v})}{\partial v_i} \times \frac{\partial \log f(y, \boldsymbol{v})}{\partial v_j}\right] f(y, \boldsymbol{v})\, dy \\
&= \mathbb{E}_{\boldsymbol{v}}\left\{L(\boldsymbol{Y})\left[\frac{\partial^2 \log f(\boldsymbol{Y}, \boldsymbol{v})}{\partial v_i \partial v_j} + \frac{\partial \log f(\boldsymbol{Y}, \boldsymbol{v})}{\partial v_i} \times \frac{\partial \log f(\boldsymbol{Y}, \boldsymbol{v})}{\partial v_j}\right]\right\},
\end{aligned}$$

where $\boldsymbol{S}^{(2)}(y, \boldsymbol{v})$ can be represented as

$$\begin{aligned}
\boldsymbol{S}^{(2)}(y, \boldsymbol{v}) &= \nabla \boldsymbol{S}^{(1)}(y, \boldsymbol{v}) + \boldsymbol{S}^{(1)}(y, \boldsymbol{v})\boldsymbol{S}^{(1)}(y, \boldsymbol{v})' \\
&= \nabla^2 \log f(y, \boldsymbol{v}) + \nabla \log f(y, \boldsymbol{v}) \nabla \log f(y, \boldsymbol{v})'.
\end{aligned} \tag{5.2.19}$$

The prime notation above denotes the transpose operator applied to a vector or a matrix, and all partial derivatives are taken with respect to the components of the parameter vector $\boldsymbol{v}$.

We next calculate the score function $\boldsymbol{S}^{(1)}(y, \boldsymbol{v})$ for several commonly used pdf's.

Example 5.2.5 Bernoulli Distribution. Let $Y_k \sim \text{Ber}(p_k)$, $k = 1, \ldots, m$, be independent Bernoulli random variables, so that the corresponding pdf's have the form

$$f_k(y_k, p_k) = p_k^{y_k}(1 - p_k)^{1-y_k}, \qquad y_k = 0, 1.$$

Then the kth component of the vector $\boldsymbol{S}^{(1)}(\boldsymbol{y}, \boldsymbol{v})$ is

$$[\boldsymbol{S}^{(1)}(\boldsymbol{y}, \boldsymbol{p})]_k = \frac{y_k - p_k}{p_k(1 - p_k)}, \qquad k = 1, \ldots, m. \tag{5.2.20}$$

Example 5.2.6 Geometric Distribution. Let $Y_k \sim \text{Ge}(p_k)$, $k = 1, \ldots, m$, be independent geometric random variables, so that the corresponding pdf's have the form

$$f_k(y_k, p_k) = p_k(1 - p_k)^{y_k - 1}, \qquad y_k = 1, 2, \ldots. \tag{5.2.21}$$

Then

$$[\boldsymbol{S}^{(1)}(\boldsymbol{y}, \boldsymbol{p})]_k = \frac{1 - p_k y_k}{p_k(1 - p_k)}.$$

Example 5.2.7 Gamma Distribution. Let $Y_k \sim \text{G}(\lambda_k, \beta_k)$, $k = 1, \ldots, m$, be independent gamma random variables, so that the corresponding pdf's have the form

$$f_k(y_k, \lambda_k, \beta_k) = \frac{\lambda_k e^{-\lambda_k y_k}(\lambda_k y_k)^{\beta_k - 1}}{\Gamma(\beta_k)}, \qquad y_k > 0, \quad \lambda_k > 0, \quad \beta_k > 0, \tag{5.2.22}$$

and

$$f(\boldsymbol{y}, \boldsymbol{\lambda}, \boldsymbol{\beta}) = \prod_{k=1}^{m} f_k(y_k, \lambda_k, \beta_k).$$

Suppose we are interested in the sensitivities with respect to $\boldsymbol{\lambda} = (\lambda_1, \ldots, \lambda_m)'$ only. Then,

$$\boldsymbol{S}^{(1)}(\boldsymbol{y}, \boldsymbol{\lambda}) = \boldsymbol{\beta}\boldsymbol{\lambda}^{-1} - \boldsymbol{y}, \tag{5.2.23}$$

where $\boldsymbol{\beta}\boldsymbol{\lambda}^{-1}$ is the m-dimensional vector with components $\beta_1\lambda_1^{-1}, \ldots, \beta_m\lambda_m^{-1}$.

The above examples constitute special cases of the following exponential family.

Example 5.2.8 Exponential Family of Distributions. Let Y be a random variable from an m-parameter exponential family of distributions, given in the following canonical form:

$$f(y, \boldsymbol{v}) = c(\boldsymbol{v}) \exp\left(\sum_{k=1}^{m} v_k t_k(y) \right) h(y), \tag{5.2.24}$$

where $c(\boldsymbol{v}) > 0$ is a real-valued function of the parameter vector $\boldsymbol{v}$, and $t_k(y)$ and $h(\boldsymbol{y})$ are real-valued functions of $\boldsymbol{y}$. Notice that the density $f(y, \boldsymbol{v})$ is parameterized by the vector $\boldsymbol{v} = (v_1, \ldots, v_m)' \in V$, and that the function $c(\boldsymbol{v})$ in (5.2.24) is determined by the functions $t_1(y), \ldots, t_m(y)$ and $h(y)$ through the identity

$$c(\boldsymbol{v}) \int \exp\left(\sum_{k=1}^{m} v_k t_k(y) \right) h(y)\, dy = 1.$$

[For more details see Rubinstein and Shapiro (1993)].

The score function is

$$\boldsymbol{S}^{(1)}(y, \boldsymbol{v}) = \frac{\nabla c(\boldsymbol{v})}{c(\boldsymbol{v})} + \sum_{k=1}^{m} t_k(y) \nabla \boldsymbol{v}. \tag{5.2.25}$$

Table 5.2.1 displays the functions $c(\boldsymbol{v})$, $t_k(y)$, and $h(y)$ for several commonly used pdf's (a dash means that the corresponding value is not used).

Table 5.2.2 displays the corresponding score functions calculated from (5.2.25) for the same commonly used pdf's.

Example 5.2.9 Stochastic Shortest-Path System. Consider the sensitivities $\nabla \ell(\boldsymbol{v})$ and their corresponding estimators $\bar{\nabla} \ell_N(\boldsymbol{v})$, for the stochactic shortest-path system (4.2.7). We have

$$\nabla \ell(\boldsymbol{v}) = \mathbb{E}_{\boldsymbol{v}} \left\{ \min_{j=1,\ldots,p} \left(\sum_{i \in \mathcal{L}_j} Y_i \right) \boldsymbol{S}^{(1)}(\boldsymbol{Y}, \boldsymbol{v}) \right\}.$$

Assume that all Y_k are mutually independent, gamma random variables. In view of (5.2.23), we can estimate $\nabla \ell(\boldsymbol{\lambda})$ by

$$\bar{\nabla} \ell_N(\boldsymbol{\lambda}) = \frac{1}{N} \sum_{t=1}^{N} \min_{j=1,\ldots,p} \left(\sum_{i \in \mathcal{L}_j} Y_{it} \right) (\boldsymbol{\beta} \boldsymbol{\lambda}^{-1} - \boldsymbol{Y}_t). \tag{5.2.26}$$

It is important to note that without change of the measure, the estimators $\nabla^k \bar{\ell}_N(\boldsymbol{v})$, $k = 0, 1, \ldots,$ allow us to evaluate the performance $\ell(\boldsymbol{v})$ and the sensitivities $\nabla^k \ell(\boldsymbol{v})$, $k = 0, 1, \ldots,$ *only at a fixed parameter value* $\boldsymbol{v}$, whereas the

Table 5.2.1 Functions $c(\boldsymbol{v})$, $t_k(y)$, and $h(y)$ for Commonly Used pdf's

pdf	$f(y, \boldsymbol{v})$	$\mathbb{E}(Y)$	$t_1(y),\ t_2(y)$	$c(\boldsymbol{v})$	$v_1,\ v_2$	$h(y)$
Gamma	$\frac{\lambda^\beta y^{\beta-1}}{\Gamma(\beta)} \exp(-\lambda y)$	$\frac{\beta}{\lambda}$	$\boldsymbol{y},\quad \ln \boldsymbol{y}$	$\frac{\lambda^\beta}{\Gamma(\beta)}$	$-\lambda,\quad \beta-1$	1
Normal	$\frac{1}{\sigma(2\pi)^{1/2}} \exp\left(\frac{-(y-\mu)^2}{2\sigma^2}\right)$	μ	$y,\quad y^2$	$\frac{\exp\left(\frac{-\mu^2}{2\sigma^2}\right)}{(2\pi)^{1/2}\sigma}$	$\frac{\mu}{\sigma^2},\quad -\frac{1}{2\sigma^2}$	1
Weibull	$\frac{\alpha}{\beta} \boldsymbol{y}^{\alpha-1} \exp(-\beta^{-1} \boldsymbol{y}^\alpha)$	$\beta^{1/\alpha}\Gamma\left(1+\frac{1}{\alpha}\right)$	$\ln \boldsymbol{y},\quad -\boldsymbol{y}^\alpha$	$\alpha\beta^{-1}$	$\alpha - 1,\quad \beta^{-1}$	1
Binomial	$\binom{n}{y} p^y (1-p)^{n-y}$	np	$y,\quad$ —	$(1-p)^n$	$\ln \frac{p}{1-p},\quad$ —	$\binom{n}{y}$
Poisson	$\frac{\lambda^y e^{-\lambda}}{y!}$	λ	$y,\quad$ —	$e^{-\lambda}$	$\ln \lambda,\quad$ —	$\frac{1}{y!}$
Geometric	$p(1-p)^y$	$\frac{1-p}{p}$	$y,\quad$ —	p	$\ln(1-p),\quad$ —	1

Table 5.2.2 Score Functions for Commonly Used pdf's

pdf	$f(y, \boldsymbol{v})$	$\boldsymbol{S}^{(1)}$
Gamma	$\frac{\lambda^\beta y^{\beta-1}}{\Gamma(\beta)}\exp(-\lambda y)$	$\beta\lambda^{-1} - y;\ \log\lambda + \log y - \frac{\Gamma'(\beta)}{\Gamma(\beta)}$
Normal	$\frac{1}{(2\pi)^{1/2}\sigma}\exp\left(\frac{-(y-\mu)^2}{2\sigma^2}\right)$	$\sigma^{-2}(y-\mu);\ \sigma^{-1} + \sigma^{-3}(y-\mu)^2$
Weibull	$\frac{\alpha}{\beta}y^{\alpha-1}\exp(-\beta^{-1}y^\alpha)$	$\alpha^{-1} + \log\, y - \frac{y^\alpha}{\beta}\log\, y;\ -\beta^{-1} + \frac{y^\alpha}{\beta^2}$
Binomial	$\binom{n}{y} p^y(1-p)^{n-y}$	$\frac{y - np}{p(1-p)}$
Poisson	$\frac{\lambda^y e^{-\lambda}}{y!}$	$\frac{y}{\lambda} - 1$
Geometric	$p(1-p)^{y-1}$	$\frac{1 - py}{p(1-p)}$

estimators (5.2.8), based on likelihood ratios, allow us to estimate $\ell(\boldsymbol{v})$ and $\nabla^k\ell(\boldsymbol{v}),\ k = 1, 2, \ldots$, essentially at any parameter value $\boldsymbol{v} \in V$.

Remark 5.2.1. Take $L(\boldsymbol{y}) = C$, for some constant C. Then $\ell(\boldsymbol{v}) = \mathbb{E}_{\boldsymbol{v}}\{L(\boldsymbol{Y})\} = C$ for all $\boldsymbol{v}$. Clearly, for all partial derivatives, $\nabla^k\ell(\boldsymbol{v}) = \nabla^k C = 0$. Take now $L(\boldsymbol{y}) = 1$ and consider (5.2.17) and (5.2.18). Then, $\nabla^k\ell(\boldsymbol{v}) = \mathbb{E}_{\boldsymbol{v}}\{1 \cdot \boldsymbol{S}^{(k)}(\boldsymbol{Y}, \boldsymbol{v})\} = \mathbb{E}_{\boldsymbol{v}}\{\boldsymbol{S}^{(k)}(\boldsymbol{Y}, \boldsymbol{v})\} = \mathbf{0},\ k = 1, 2, \ldots$. Consequently we can write $\nabla^k\ell(\boldsymbol{v})$ in (5.2.17) as

$$\begin{aligned}\nabla^k\ell(\boldsymbol{v}) &= \mathbb{E}_{\boldsymbol{v}}\{L\boldsymbol{S}^{(k)}\} = \mathrm{cov}_{\boldsymbol{v}}\{L, \boldsymbol{S}^{(k)}\} \\ &\quad + \mathbb{E}_{\boldsymbol{v}}\{L\}\mathbb{E}_{\boldsymbol{v}}\{\boldsymbol{S}^{(k)}\} = \mathrm{cov}_{\boldsymbol{v}}\{L, \boldsymbol{S}^{(k)}\}, \qquad k = 1, 2, \ldots. \end{aligned} \tag{5.2.27}$$

Thus, the sensitivities $\nabla^k\ell(\boldsymbol{v}),\ k = 1, 2, \ldots$, can be represented as the *covariances* of the sample performance, L, and the associated score functions, $\boldsymbol{S}^{(k)}$.

Consider now a change of measure with some dominating pdf $g(\boldsymbol{y})$. From the definition of W and $\boldsymbol{S}^{(k)}$ it readily follows that

$$\nabla^k W(\boldsymbol{Z}, \boldsymbol{v}) = \frac{\nabla^k f(\boldsymbol{Z}, \boldsymbol{v})}{f(\boldsymbol{Z}, \boldsymbol{v})}\frac{f(\boldsymbol{Z}, \boldsymbol{v})}{g(\boldsymbol{y})} = \boldsymbol{S}^{(k)}(\boldsymbol{Z}, \boldsymbol{v})W(\boldsymbol{Z}, \boldsymbol{v}). \tag{5.5.28}$$

Thus, the previously derived sensitivities, $\nabla^k\ell(\boldsymbol{v}) = \mathbb{E}_{\boldsymbol{v}}\{L\boldsymbol{S}^{(k)}\}$ represent a special case of (5.2.8) with $g(\boldsymbol{y}) = f(\boldsymbol{y}, \boldsymbol{v})$ and $W(\boldsymbol{y}, \boldsymbol{v}) = 1$.

Since the sensitivity estimators $\nabla^k\bar{\ell}_N(\boldsymbol{v}),\ k = 1, 2, \ldots$ are based on the *efficient score*, $\boldsymbol{S}^{(1)}(\boldsymbol{Y}, \boldsymbol{v}) = \nabla\log f(\boldsymbol{Y}, \boldsymbol{v})$, we shall refer to this approach as the SF method, and to the associated estimator (5.2.8) as the SF estimator. In a similar vein, we shall refer to $\nabla^k W(\boldsymbol{Z}, \boldsymbol{v})$ and $\boldsymbol{S}^{(k)}(\boldsymbol{Y}, \boldsymbol{v}),\ k = 1, 2, \ldots,$ as the *generalized scores* and the *k*th order SFs, respectively.

Noting that $\mathbb{E}_g\{W(\boldsymbol{Z}, \boldsymbol{v})\} = 1$ for all $\boldsymbol{v}$, we have (see also Remark 5.2.1)

$$\mathbb{E}_g\{\nabla^k W(\boldsymbol{Z}, \boldsymbol{v})\} = \mathbb{E}_g\{W(\boldsymbol{Z}, \boldsymbol{v})\boldsymbol{S}^{(k)}(\boldsymbol{Z}, \boldsymbol{v})\} = \boldsymbol{0}, \qquad k = 1, 2, \ldots,$$

and

$$\nabla^k \ell(\boldsymbol{v}) = \mathrm{cov}_g\{L(\boldsymbol{Z}), \nabla^k W(\boldsymbol{Z}, \boldsymbol{v})\}, \qquad k = 1, 2, \ldots. \tag{5.5.29}$$

Thus, the sensitivities $\nabla^k \ell(\boldsymbol{v})$ can be represented as the covariances of the sample performance, L, and the generalized scores, $\nabla^k W$.

5.3 WHAT-IF PROBLEM AND RESPONSE-SURFACE METHODOLOGY

The *what-if* problem is a generic name for the following problem: "What will be the values of the performance measures $\nabla^k \ell(\boldsymbol{v}),\ k = 0, 1, \ldots,$ if we perturb some (all) of the components of the parameter vector $\boldsymbol{v}$?" More precisely, it is the problem of estimating the performance measure, $\ell(\boldsymbol{v})$, and the associated sensitivities, $\nabla^k \ell(\boldsymbol{v}),\ k = 1, 2 \ldots,$ for multiple values of $\boldsymbol{v}$. The proposed solution will utilize the estimators $\nabla^k \bar{\ell}_N(\boldsymbol{v}),\ k = 0, 1, \ldots,$ in (5.2.9).

The what-if problem is directly related to the *response-surface methodology*, which seeks to estimate the response functions $\nabla^k \ell(\boldsymbol{v}),\ k = 0, 1, \ldots$ for arbitrary parameter vectors $\boldsymbol{v}_1, \ldots, \boldsymbol{v}_k$. This in turn can be achieved by using Algorithm 5.2.1.

The rest of this section deals with the accuracy and efficiency of the SF estimators $\nabla^k \bar{\ell}_N(\boldsymbol{v}),\ k = 0, 1, \ldots.$ We assume that the dominating density function, $g(z)$, is of the form $g(z) = f(z, \boldsymbol{v}_0)$, where $\boldsymbol{v}_0$ is a fixed vector, called the *reference parameter vector* or simply the *reference parameter*. Whenever convenient, we shall use interchangeably the notations $\nabla^k \bar{\ell}_N(\boldsymbol{v}, \boldsymbol{v}_0)$ and $\nabla^k \bar{\ell}_N(\boldsymbol{v})$.

We distinguish between the following two what-if cases:

A. $\boldsymbol{v}$ is *fixed*, while $\boldsymbol{v}_0$ *varies*.

B. $\boldsymbol{v}_0$ is *fixed*, while $\boldsymbol{v}$ *varies*.

Case A. Let

$$\varepsilon^k(\boldsymbol{v}, \boldsymbol{v}_0) = \frac{\mathrm{var}_{\boldsymbol{v}_0}\{\nabla^k \bar{\ell}_N(\boldsymbol{v}, \boldsymbol{v}_0) t(\boldsymbol{v}_0)\}}{\mathrm{var}_{\boldsymbol{v}}\{\nabla^k \bar{\ell}_N(\boldsymbol{v}, \boldsymbol{v}) t(\boldsymbol{v})\}}, \qquad k = 0, 1, \tag{5.3.1}$$

be the efficiency of the what-if estimators, $\nabla^k \bar{\ell}_N(\boldsymbol{v}_0) = \nabla^k \bar{\ell}_N(\boldsymbol{v}, \boldsymbol{v}_0)$, relative to their standard counterpart, $\nabla^k \bar{\ell}_N(\boldsymbol{v}) = \nabla^k \bar{\ell}_N(\boldsymbol{v}, \boldsymbol{v})$ (corresponding to no change of measure); for $k=1$, we define $\text{var}\{\nabla \bar{\ell}_N\}$ as the trace of the covariance matrix of $\nabla \bar{\ell}_N$. Recall that the subscript $\boldsymbol{v}_0$ signifies that all the expectations involved in the calculation of $\text{var}_{\boldsymbol{v}_0}(\cdot)$ are taken with respect to the density $f(\boldsymbol{y}, \boldsymbol{v}_0)$. Here, $t(\boldsymbol{v}_0)$ and $t(\boldsymbol{v})$ denote the CPU times needed to compute the what-if and the standard estimators, respectively. Although, the CPU time of the what-if estimators, $\nabla^k \bar{\ell}_N(\boldsymbol{v}, \boldsymbol{v}_0)$, $k = 0, 1$, is typically a bit larger than that of its standard counterpart, $\nabla^k \bar{\ell}_N(\boldsymbol{v}, \boldsymbol{v})$, we assume below that $t(\boldsymbol{v}_0) = t(\boldsymbol{v})$, for simplicity. Therefore (5.3.1) becomes

$$\varepsilon^k(\boldsymbol{v}, \boldsymbol{v}_0) = \frac{\text{var}_{\boldsymbol{v}_0}\{\nabla^k \bar{\ell}_N(\boldsymbol{v}, \boldsymbol{v}_0)\}}{\text{var}_{\boldsymbol{v}}\{\nabla^k \bar{\ell}_N(\boldsymbol{v}, \boldsymbol{v})\}}, \qquad k = 0, 1. \tag{5.3.2}$$

We say that the LR estimator is the *more efficient* of the two, or that a *variance reduction* is attained, if

$$\varepsilon^k(\boldsymbol{v}, \boldsymbol{v}_0) < 1, \qquad k = 0, 1.$$

Our main goal in Case A is to find a nonempty set of reference parameters, $\boldsymbol{v}_0$, such that the LR estimators $\nabla^k \bar{\ell}_N(\boldsymbol{v}, \boldsymbol{v}_0)$, $k = 0, 1$, are the more efficient ones. To this end we need to calculate the corresponding variances

$$\text{var}_{\boldsymbol{v}_0}\{\nabla^k \bar{\ell}_N(\boldsymbol{v}, \boldsymbol{v}_0)\} = \frac{1}{N}\text{var}_{\boldsymbol{v}_0}\{L(\boldsymbol{Z})\nabla^k W(\boldsymbol{Z}, \boldsymbol{v}, \boldsymbol{v}_0)\}, \qquad k = 0, 1, \tag{5.3.3}$$

for various values $\boldsymbol{v}_0$, where

$$W(\boldsymbol{Z}, \boldsymbol{v}, \boldsymbol{v}_0) = \frac{f(\boldsymbol{Z}, \boldsymbol{v})}{f(\boldsymbol{Z}, \boldsymbol{v}_0)},$$

and $\boldsymbol{v}$ is fixed.

To glean a better insight into the above, we shall calculate the variance of the estimators (i) $\bar{\ell}_N(\boldsymbol{v}, \boldsymbol{v}_0)$ and (ii) $\nabla \bar{\ell}_N(\boldsymbol{v}, \boldsymbol{v}_0)$, for the case where the components $Y_1, \ldots, Y_m$, of the random vector $\boldsymbol{Y}$ are drawn independently from the gamma pdf given in (5.2.22); subsequently, the results will be extended to more general distributions. Suppose that only the reference parameter vector $\boldsymbol{\lambda}_0 = (\lambda_{01}, \ldots, \lambda_{0m})'$ varies in the gamma pdf, while the parameter vector $\boldsymbol{\beta} = (\beta_1, \ldots, \beta_m)'$ is fixed and known.

(i) Consider first the case $m = 1$. We shall prove that

$$\mathbb{E}_{\lambda_0}\{(LW)^2\} = \mathbb{E}_{\lambda_0}\{W^2\}\mathbb{E}_{2\lambda-\lambda_0}\{L^2\}, \tag{5.3.4}$$

where

$$\mathbb{E}_{\lambda_0}\{W^2\} = \left(\frac{1}{1-\alpha^2}\right)^{\beta}, \tag{5.3.5}$$

and

$$\alpha = \Delta\lambda/\lambda \quad \text{and} \quad \Delta\lambda = \lambda_0 - \lambda. \tag{5.3.6}$$

Indeed, by assumption,

$$W(z, \lambda, \beta) = \frac{f(z, \lambda, \beta)}{f(z, \lambda_0, \beta)} = \left(\frac{\lambda}{\lambda_0}\right)^{\beta} \exp(-(\lambda - \lambda_0)z),$$

and

$$\begin{aligned}
\mathbb{E}_{\lambda_0}\{(LW)^2\} &= \int L^2(y)\left(\frac{\lambda}{\lambda_0}\right)^{2\beta} e^{-2(\lambda-\lambda_0)y}\frac{\lambda_0 e^{-\lambda_0 y}(\lambda_0 y)^{\beta-1}}{\Gamma(\beta)}\,dy \\
&= \left(\frac{\lambda}{\lambda_0}\right)^{2\beta}\int L^2(y)e^{[-2(\lambda-\lambda_0)-\lambda_0]y}\lambda_0^{\beta}\frac{y^{\beta-1}}{\Gamma(\beta)}\,dy \\
&= \left(\frac{\lambda}{\lambda_0}\right)^{2\beta}\lambda_0^{\beta}\int L^2(y)e^{-(2\lambda-\lambda_0)y}\left[\frac{2\lambda-\lambda_0}{2\lambda-\lambda_0}\right]^{\beta-1}\frac{y^{\beta-1}}{\Gamma(\beta)}\,dy \\
&= \left[\frac{\lambda^2}{\lambda_0(2\lambda-\lambda_0)}\right]^{\beta}\mathbb{E}_{2\lambda-\lambda_0}\{L^2(Z)\}.
\end{aligned}$$

It remains to note that

$$\left[\frac{\lambda^2}{\lambda_0(2\lambda-\lambda_0)}\right]^{\beta} = \left[\frac{\lambda^2}{(\lambda-\Delta\lambda)(\lambda+\Delta\lambda)}\right]^{\beta} = \left[\frac{1}{1-\alpha^2}\right]^{\beta}.$$

It is important to realize that the formulas (5.3.4) and (5.3.5) must satisfy the condition

$$2\lambda - \lambda_0 > 0. \tag{5.3.7}$$

Example 5.3.1 [Example 5.2.2 continued (λ fixed)]. For $L(Y) = Y,\ Y \sim \text{Exp}(\lambda)$, one has

$$\ell(\lambda) = \frac{1}{\lambda}, \qquad \mathbb{E}_{2\lambda-\lambda_0}\{L^2\} = \frac{2}{(2\lambda-\lambda_0)^2}, \qquad \mathbb{E}_{\lambda_0}\{W^2\} = \frac{1}{1-\alpha^2}$$

and

$$\mathbb{E}_{\lambda_0}\{(LW)^2\} = \mathbb{E}_{\lambda_0}\{W^2\}\mathbb{E}_{2\lambda-\lambda_0}\{L^2(Z)\} = \frac{1}{1-\alpha^2} \times \frac{2}{(2\lambda-\lambda_0)^2},$$

whence

$$\varepsilon^0 = \frac{\operatorname{var}_{\lambda_0}\{LW\}}{\operatorname{var}_{\lambda}\{L\}} = \frac{2}{(1+\alpha)(1-\alpha)^3} - 1.$$

It is readily seen that variance reduction ($\varepsilon^0 < 1$) is attained when λ_0 belongs to the interval $(\hat{\lambda}, \lambda)$, where $\hat{\lambda}$ satisfies the equation

$$\operatorname{var}_{\lambda}\{L\} = \operatorname{var}_{\hat{\lambda}}\{LW\}.$$

Some simple algebra yields $\hat{\lambda} \approx 0.16\lambda$ and $\hat{\alpha} \approx -0.84$.

Example 5.3.2. Let $\ell(x) = P\{L > x\}$, where $L(Y) = Y$ and $Y \sim \text{Exp}(\lambda)$. Then,

$$\ell(x) = P_{\lambda}\{Y > x\} = \mathbb{E}_{\lambda}\{I_{\{Y>x\}}\} = e^{-\lambda x} \quad \text{and} \quad \mathbb{E}_{\lambda_0}\{W^2\} = \frac{1}{1-\alpha^2}.$$

Hence,

$$\mathbb{E}_{\lambda_0}\{(LW)^2\} = \mathbb{E}_{\lambda_0}\{W^2\}\mathbb{E}_{2\lambda-\lambda_0}\{I_{\{Z>x\}}\} = \frac{\lambda^2 e^{-(2\lambda-\lambda_0)x}}{\lambda_0(2\lambda-\lambda_0)}, \tag{5.3.8}$$

where $Z \sim \text{Exp}(\lambda)$. The optimal value of the reference parameter $\lambda_0^* = \lambda_0^*(x)$, which minimizes $\mathbb{E}_{\lambda_0}\{(LW)^2\}$ on the interval $(0, 2\lambda)$, equals

$$\lambda_0^*(x) = \lambda + x^{-1} - (\lambda^2 + x^{-2})^{1/2}. \tag{5.3.9}$$

Let $x \gg \lambda$ and assume that λ is fixed, say, $\lambda = 1$. Then $\{L > x\}$ is a rare event, and $\lambda_0^*(x)$ can be approximated from (5.3.9) as $\lambda_0^*(x) \approx x^{-1}$.

For $\lambda_0 = \lambda_0^*$, the relative efficiency, $\varepsilon^0(\lambda, \lambda_0^*, x)$, is

$$\varepsilon^0(\lambda, \lambda_0^*, x) \approx 0.5x\lambda e^{1-\lambda x}.$$

For $\lambda = 1$ and $x = 12$, one has

$$\mathbb{E}_{\lambda=1}\{I_{\{Y>12\}}\} = P_{\lambda=1}\{Y > 12\} = e^{-12} \approx 10^{-6},$$

$\lambda_0^*(x) \approx \frac{1}{12}$ and $\varepsilon^0(1, \lambda_0^*, 12) \approx 10^{-4}$. Thus, using the optimal value, $\lambda_0^* \approx \frac{1}{12}$, one can attain a variance reduction on the order of 10^4. Further, the equality

$$\varepsilon^0(\lambda, \lambda) = \varepsilon^0(\lambda, \hat{\lambda}) = 1$$

implies that for large x,

$$\hat{\lambda} \approx 0.5\lambda e^{-\lambda x}.$$

Again for $\lambda = 1$ and $x = 12$, one has $\hat{\lambda} = 0.5e^{-12} \approx 3 \times 10^{-6}$ and variance reduction is effected when λ_0 belongs to the interval $(\hat{\lambda}, \lambda) \approx (3 \times 10^{-6}, 1)$.

Consider now the case $m > 1$. The reader is asked to prove that for $m > 1$, (5.3.4) and (5.3.5) can be generalized to

$$\mathrm{var}_{\boldsymbol{\lambda}_0}\{LW\} = \mathbb{E}_{\boldsymbol{\lambda}_0}\{W^2\}\mathbb{E}_{2\boldsymbol{\lambda}-\boldsymbol{\lambda}_0}\{L^2(\boldsymbol{Z})\} - \ell^2(\boldsymbol{v}), \tag{5.3.10}$$

and

$$\mathbb{E}_{\boldsymbol{\lambda}_0}\{W^2\} = \prod_{k=1}^{m}\left(\frac{1}{1-\alpha_k^2}\right)^{\beta_k}, \tag{5.3.11}$$

respectively. Here, $\boldsymbol{\lambda}_0 = (\lambda_{01}, \ldots, \lambda_{0m})'$, $\Delta\boldsymbol{\lambda} = \boldsymbol{\lambda}_0 - \boldsymbol{\lambda}$, and $\alpha_k = \Delta\lambda_k/\lambda_k$, $k = 1, \ldots, m$. Recall that the subscripts $\boldsymbol{\lambda}_0$ and $2\boldsymbol{\lambda} - \boldsymbol{\lambda}_0$ in (5.3.10) stand for vector values of the parameter vector $\boldsymbol{\lambda}$ with respect to which the expectations are taken. The quantities $\Delta\lambda_k$ and α_k will be referred to as the *absolute* perturbation and *relative* perturbation, respectively. It is interesting to note that the second moments of LW and $L\nabla W$ can be represented in terms of the product of $\mathbb{E}_{\boldsymbol{\lambda}_0}\{W^2\}$ with $\mathbb{E}_{2\boldsymbol{\lambda}-\boldsymbol{\lambda}_0}\{L(\boldsymbol{Z})^2\}$; for large m, the term $\mathbb{E}_{\boldsymbol{\lambda}_0}\{W^2\}$ is typically dominant. Similar results (see below) hold for the exponential family of distributions in the canonical form.

It is important to note that the formulas (5.3.10) and (5.3.11) are derived under the assumption [see (5.3.7)] that

$$2\boldsymbol{\lambda} - \boldsymbol{\lambda}_0 > \mathbf{0},$$

or equivalently,

$$\lambda_k > \tfrac{1}{2}\lambda_{0k}, \qquad k = 1, \ldots, m. \tag{5.3.12}$$

For practical purposes it is desirable to choose

$$\lambda_k \geq \lambda_{0k}, \qquad k = 1, \ldots, m.$$

Example 5.3.3. Let $L(Y) = \min(Y_1, \ldots, Y_m)$, where the random variables Y_i, $i = 1, \ldots, m$, are iid from $\mathrm{Exp}(\lambda)$. In this case, $\ell(\lambda)$, $\mathbb{E}_\lambda\{L^2\}$, and ε^0 become

$$\ell(\lambda) = \frac{1}{m\lambda},$$

$$\mathbb{E}_\lambda\{L^2\} = \frac{2}{(m\lambda)^2}, \tag{5.3.13}$$

and

$$\varepsilon^0 = \frac{\mathrm{var}_{\lambda_0}\{LW\}}{\mathrm{var}_\lambda\{L\}} = \frac{2}{(1-\alpha^2)^m(1-\alpha)^2} - 1, \tag{5.3.14}$$

respectively, where

$$\mathrm{var}_{\lambda_0}\{LW\} = \frac{2}{(m(2\lambda - \lambda_0))^2}\left(\frac{\lambda^2}{\lambda_0(2\lambda - \lambda_0)}\right)^m - \frac{1}{m^2\lambda^2}. \tag{5.3.15}$$

Table 5.3.1 displays the efficiency ε^0 [see (5.3.2)] of the LR estimator, $\bar{\ell}_N$, relative to its CMC counterpart, as a function of α and m, for fixed λ and varying λ_0. Here $\mathcal{M}$ denotes a large number (say, $\mathcal{M} > 10^3$), while $\Delta\lambda = \lambda_0 - \lambda$ and $\alpha = \Delta\lambda/\lambda$, as before.

Note that some of the entries in Table 5.3.1 and elsewhere in the book are marked with an asterisk; these correspond to the maximal efficiency (minimal variance) in the corresponding column. Table 5.3.1 shows that that for $m \le 100$, variance reduction ($\varepsilon^0 < 1$) is attained for some specific values of α. For example, the efficiency corresponding to $m = 10$ and $\alpha = -9\%$ is $\varepsilon^0 = 0.825$. On the other

Table 5.3.1 Efficiency ε^0 of the LR Estimator $\bar{\ell}_N$ as Function of m and α, for $L(Y) = \min(Y_1, \ldots, Y_m)$

α	$m=1$	$m=10$	$m=10^2$	$m=10^3$	$m=10^4$	$m=10^5$
0.96	1.7E+5	1.8E+14	$\mathcal{M}$	$\mathcal{M}$	$\mathcal{M}$	$\mathcal{M}$
0.67	31.40	6425	$\mathcal{M}$	$\mathcal{M}$	$\mathcal{M}$	$\mathcal{M}$
0.25	2.792	5.779	22.57	3.8E+22	$\mathcal{M}$	$\mathcal{M}$
0.11	1.562	1.866	7.766	6.2E+06	$\mathcal{M}$	$\mathcal{M}$
0.000	1.000	1.000	1.000	1.000*	1.000*	1.000*
−0.01	0.961	0.962	0.980*	1.162	4.226	3.5E+05
−0.03	0.889	0.904	1.055	3.412	9159	1.3E+27
−0.05	0.826	0.864	1.286	16.64	1.3E+10	$\mathcal{M}$
−0.09	0.694	0.825*	2.853	6753	$\mathcal{M}$	$\mathcal{M}$
−0.17	0.511	0.947	23.58	2.5E+12	$\mathcal{M}$	$\mathcal{M}$
−0.33	0.265	2.653	14E+06	$\mathcal{M}$	$\mathcal{M}$	$\mathcal{M}$
−0.50	0.1851*	14.78	2.7E+12	$\mathcal{M}$	$\mathcal{M}$	$\mathcal{M}$

hand, for larger values of α (say, $\alpha = -50\%$) and the same $m = 10$, we have $\varepsilon^0 = 14.78$, to wit, a rather substantial increase in the variance of LW.

(ii) Consider the variance (trace of the covariance matrix) of $\nabla \bar{\ell}_N(\boldsymbol{v}, \boldsymbol{v}_0)$, for the case of gamma distribution. Starting with $m = 1$, and in view of the relation $\nabla W = WS^{(1)}$ [see (5.2.28)], we have analogously (5.3.4),

$$\mathbb{E}_{\lambda_0}\{(L\nabla W)^2\} = \mathbb{E}_{\lambda_0}\{W^2\}\mathbb{E}_{2\lambda-\lambda_0}\{L^2[S^{(1)}]^2\}, \tag{5.3.16}$$

where $\mathbb{E}_{\lambda_0}\{W^2\}$ is given in (5.3.5).

Example 5.3.4 [Example 5.3.1 continued (λ fixed)]. For $L(Y) = Y$, $Y \sim \text{Exp}(\lambda)$, one has

$$\begin{aligned} \text{var}_{\lambda_0}\left\{L\frac{\partial W(Z, \lambda)}{\partial \lambda}\right\} &= \mathbb{E}_{\lambda_0}\{W^2\}\mathbb{E}_{2\lambda-\lambda_0}\{Z^2(\lambda^{-1} - Z)^2\} - \left(\frac{\partial \ell(\lambda)}{\partial \lambda}\right)^2 \\ &= \frac{\lambda^2}{\lambda_0(2\lambda - \lambda_0)^3}\left(\frac{2}{\lambda^2} - \frac{12}{\lambda_0(2\lambda - \lambda_0)} + \frac{24}{\lambda_0(2\lambda - \lambda_0)^2}\right) - \frac{1}{\lambda^4}. \end{aligned} \tag{5.3.17}$$

Thus, variance reduction

$$\varepsilon^0 = \frac{\text{var}_{\lambda_0}\left\{L\dfrac{\partial W(\lambda)}{\partial \lambda}\right\}}{\text{var}_\lambda\{LS^{(1)}\}} < 1 \tag{5.3.18}$$

is attained for $\lambda_0 \in (\hat{\lambda}, \lambda)$, where $\hat{\lambda}$ satisfies the equation

$$\text{var}_\lambda\{LS^{(1)}\} = \text{var}_{\hat{\lambda}}\left\{L\frac{\partial W(\lambda)}{\partial \lambda}\right\}.$$

Here, $\text{var}_\lambda\{LS^{(1)}\}$ is a special case of $\text{var}_{\lambda_0}\{L[\partial W(\lambda)/\partial \lambda]\}$ with $W = 1$ and $\lambda_0 = \lambda$. Some simple algebra yields $\hat{\lambda} \approx 0.19\lambda$ and $\hat{\alpha} \approx -0.78$.

Consider now the case $m > 1$ for which the variance of $L\nabla W$ is the trace of the covariance matrix $\text{cov}_{\lambda_0}\{L\nabla W\}$. For the case of gamma distribution, one has

$$\text{cov}_{\lambda_0}\{L\nabla W\} = \mathbb{E}_{\lambda_0}\{L^2(\nabla W)(\nabla W)'\} - (\nabla \ell)(\nabla \ell)'.$$

Arguing similarly to (5.3.10) and in view of

$$W(\boldsymbol{z}, \boldsymbol{\lambda}, \boldsymbol{\beta}) = \prod_{k=1}^{m}\left(\frac{\lambda_k}{\lambda_{0k}}\right)^{\beta_k} \exp\{-(\lambda_k - \lambda_{0k})z_k\}$$

one has

$$\frac{\partial W(\mathbf{z}, \boldsymbol{\lambda}, \boldsymbol{\beta})}{\partial \lambda_k} = (\beta_k \lambda_k^{-1} - z_k) \frac{f(\mathbf{z}, \boldsymbol{\lambda}, \boldsymbol{\beta})}{f(\mathbf{z}, \boldsymbol{\lambda}_0, \boldsymbol{\beta})}, \tag{5.3.19}$$

for $k = 1, \ldots, m$. It follows that

$$\mathbb{E}_{\boldsymbol{\lambda}_0}\{L^2(\nabla W)(\nabla W)'\} = \mathbb{E}_{\boldsymbol{\lambda}_0}\{W^2\}\mathbb{E}_{2\boldsymbol{\lambda}-\boldsymbol{\lambda}_0}\{L^2(\boldsymbol{Z})(\boldsymbol{\beta}\boldsymbol{\lambda}^{-1} - \boldsymbol{Z})(\boldsymbol{\beta}\boldsymbol{\lambda}^{-1} - \boldsymbol{Z})'\}, \tag{5.3.20}$$

and hence

$$\mathrm{cov}_{\boldsymbol{\lambda}_0}\{L\nabla W\} = \mathbb{E}_{\boldsymbol{\lambda}_0}\{W^2\}\mathbb{E}_{2\boldsymbol{\lambda}-\boldsymbol{\lambda}_0}\{L^2(\boldsymbol{Z})(\boldsymbol{\beta}\boldsymbol{\lambda}^{-1} - \boldsymbol{Z})(\boldsymbol{\beta}\boldsymbol{\lambda}^{-1} - \boldsymbol{Z})'\} - (\nabla\ell)(\nabla\ell)', \tag{5.3.21}$$

where $\mathbb{E}_{\boldsymbol{\lambda}_0}\{W^2\}$ is given in (5.3.11). [Recall that $\boldsymbol{\lambda}^{-1} = (\lambda_1^{-1}, \ldots, \lambda_m^{-1})$].

Case B. Consider next the case where the parameter vector $\boldsymbol{v}$ *varies*, while the reference parameter $\boldsymbol{v}_0$ is *fixed*. As mentioned earlier, this corresponds to estimating the response function (surface) $\nabla^k \ell(\boldsymbol{v}) = \nabla^k \ell(\boldsymbol{v}|\boldsymbol{v}_0)$, $k = 0, 1$, *simultaneously for multiple values* $\boldsymbol{v}$, *from a single simulation run*.

For a given set of values of $\boldsymbol{v}$ (say, $\{\boldsymbol{v}_1, \ldots, \boldsymbol{v}_k\}$), our goal is to find a reference parameter $\boldsymbol{v}_0$, for which variance reduction ($\varepsilon < 1$) is achieved; for more details see also Chapter 8.

To glean a better insight, we shall again calculate (i) the variance of the estimator $\bar{\ell}_N(\boldsymbol{v})$ and (ii) the covariance matrix of the estimator $\nabla\bar{\ell}_N(\boldsymbol{v})$, assuming that the components $Y_1, \ldots, Y_m$, of the random vector $\boldsymbol{Y}$ are independent and gamma distributed.

(i) First let $m = 1$, and denote

$$\alpha_0 = \Delta\lambda_0/\lambda_0 \quad \text{and} \quad \Delta\lambda_0 = \lambda - \lambda_0. \tag{5.3.22}$$

Comparing (5.3.22) with (5.3.6) then yields

$$\alpha_0 = -\frac{\alpha}{\alpha + 1}. \tag{5.3.23}$$

Next, consider again the formulas (5.3.4) and (5.3.5). In view of (5.3.23), it is not difficult to see that in this case (5.3.4) remains valid, while (5.3.5) results in

$$\mathbb{E}_{\lambda_0}\{W^2\} = \left[1 + \frac{\alpha_0^2}{1 + 2\alpha_0}\right]^{\beta}. \tag{5.3.24}$$

Example 5.3.5 [Example 5.3.1 continued (λ_0 fixed)]. One has

$$\ell(\lambda) = \frac{1}{\lambda}, \qquad \mathbb{E}_{2\lambda-\lambda_0}\{L^2\} = \frac{2}{(2\lambda - \lambda_0)^2}, \qquad \mathbb{E}_{\lambda_0}\{W^2\} = 1 + \frac{\alpha_0^2}{1 + 2\alpha_0},$$

and

$$\varepsilon^0 = \frac{\text{var}_{\lambda_0}\{LW\}}{\text{var}_\lambda\{L\}} = 2\frac{(1+\alpha_0)^4}{(1+2\alpha_0)^3} - 1.$$

It is readily seen that variance reduction ($\varepsilon^0 < 1$) is attained if λ belongs to the interval $(\lambda_0, \tilde{\lambda}_0)$, where $\tilde{\lambda}_0$ satisfies

$$\text{var}_\lambda\{L\} = \text{var}_{\tilde{\lambda}_0}\{LW\}.$$

This corresponds to $\tilde{\lambda}_0 \approx 6.22\lambda_0$ and $\alpha_0 \approx 5.22$. Thus, the variance reduction interval is rather wide.

A simple practical suggestion for choosing a "good" λ_0, when estimating the response curve $\ell(\lambda)$ for a reasonable interval $\lambda \in (\lambda_1, \lambda_2)$, is simply to take $\lambda_0 = \lambda_1$ (why not $\lambda_0 = \lambda_2$?). The selection of other parameters of the exponential family of distributions is similar.

Consider next $m > 1$. It is readily seen that in this case, formula (5.3.10) remains valid, while formula (5.3.11) becomes

$$\mathbb{E}_{\boldsymbol{\lambda}_0}\{W^2\} = \prod_{k=1}^{m}\left(1 + \frac{\alpha_{0k}^2}{1 + 2\alpha_{0k}}\right)^{\beta_k}. \tag{5.3.25}$$

Here, $\Delta\boldsymbol{\lambda}_0 = \boldsymbol{\lambda} - \boldsymbol{\lambda}_0$ and $\alpha_{0k} = \Delta\lambda_{0k}/\lambda_{0k}, \; k = 1, \ldots, m$.

For the special case where α_{0k} and β_k do not depend on k, say, $\alpha_{0k} = \alpha_0$ and $\beta_k = \beta, \; k = 1, \ldots, m$, formula (5.3.10) reduces to

$$\text{var}_{\boldsymbol{\lambda}_0}\{LW\} = \left(1 + \frac{\alpha_0^2}{1 + 2\alpha_0}\right)^{m\beta} \mathbb{E}_{2\boldsymbol{\lambda}-\boldsymbol{\lambda}_0}\{L^2\} - \ell^2(\boldsymbol{\lambda}). \tag{5.3.26}$$

It follows from (5.3.26) that

$$1 + 2\alpha_0 > 0.$$

We point out that for fixed α_0 and β (even with $1 + 2\alpha_0 > 0$), the variance of LW increases exponentially in m. For small values of α_0, the first term on the right-hand side of (5.3.26) can be approximated by

$$\left(1 + \frac{\alpha_0^2}{1 + 2\alpha_0}\right)^{m\beta} = \exp\left\{m\beta \log\left(1 + \frac{\alpha_0^2}{1 + 2\alpha_0}\right)\right\} \cong \exp\left\{\frac{m\beta\alpha_0^2}{1 + 2\alpha_0}\right\},$$

using the fact that for small x, $\log(1+x) \cong x$. This shows that in order for the variance of LW to be manageably small, the value of $m\beta\alpha_0^2/(1+2\alpha_0)$ must not be "too large." That is, as m increases, $\alpha_0^2\beta$ should satisfy

$$\alpha_0^2\beta = O(m^{-1}). \tag{5.3.27}$$

It will be shown later on that an assumption similar to (5.3.27) must hold for rather general distributions, and in particular, for the exponential family.

Example 5.3.6 [Example 5.3.3 continued (λ_0 fixed)]. Let again $L(Y) = \min(Y_1, \ldots, Y_m)$, where the random variables Y_i, $i = 1, \ldots, m$, are iid from Exp(λ).

Tables 5.3.2 and 5.3.3 display $\mathrm{var}_{\lambda_0}\{LW\}$ and ε^0, respectively, as functions of α_0 and m, for $\lambda_0 = 1$ [see (5.3.15) and (5.3.14)].

Note that the data in Table 5.3.2 coincide with those of Table 5.3.3, provided α is replaced by $\alpha_0 = \alpha/(\alpha+1)$. In particular, it can be seen that the results of Tables 5.3.2 and 5.3.3 conform well to requirement (5.3.27), in the sense that in order for the efficiency to satisfy $\varepsilon^0 = O(1)$, requirement (5.3.27) must hold. Formula (5.3.27) is associated with the so-called *trust region*, that is, the region where the likelihood ratio estimator $\nabla^k \bar{\ell}_N(\boldsymbol{v})$ can be trusted to give a resonably good approximation of $\nabla^k \ell(\boldsymbol{v})$. As an illustration, consider $m = 100$. It can be seen that the what-if estimator, $\bar{\ell}_N(\lambda)$, performs reasonably well for α_0 in the interval (trust region corresponding to) $(-0.10, 0.10)$, that is, when the relative perturbation in λ_0 is from -10% to 10%. For larger relative perturbations, the term $\mathbb{E}_{\lambda_0}\{W^2\}$ "blows up" the variance of the estimators. Similar results are obtained for other m as well as for different distributions.

With $\mathrm{var}_{\lambda_0}\{L\nabla^k W\}$ at hand, we can derive various confidence intervals for an

Table 5.3.2 $\mathrm{var}_{\lambda_0}\{LW(\lambda|\lambda_0)\}$ **as Function of** m **and** α_0**, for** $L(Y) = \min(Y_1, \ldots, Y_m)$

α_0	$m=1$	$m=10$	$m=10^2$	$m=10^3$	$m=10^4$	$m=10^5$
−0.49	1.7E+5	1.8E+12	$\mathcal{M}$	$\mathcal{M}$	$\mathcal{M}$	$\mathcal{M}$
−0.40	87.22	178.5	$\mathcal{M}$	$\mathcal{M}$	$\mathcal{M}$	$\mathcal{M}$
−0.20	4.36	0.090	0.35	5.9E+22	$\mathcal{M}$	$\mathcal{M}$
−0.10	1.929	0.023	0.0009	0.776	$\mathcal{M}$	$\mathcal{M}$
0.000	1.000	0.010	1E-04	1E-06*	1E-08*	1E-10*
0.01	0.942	0.009	9.6E-05*	1.1E-06	4.2E-08	3.5E-06
0.03	0.839	0.009	1E-04	3.3E-06	8.6E-05	1.3E+27
0.05	0.749	0.008	1.2E-04	1.6E-05	119.5	$\mathcal{M}$
0.10	0.574	0.006*	2E-04	0.006	$\mathcal{M}$	$\mathcal{M}$
0.20	0.355	0.007	0.002	1.7E+7	$\mathcal{M}$	$\mathcal{M}$
0.50	0.18	0.012	6.51	$\mathcal{M}$	$\mathcal{M}$	$\mathcal{M}$
1.00	0.046*	0.03	6.9E+08	$\mathcal{M}$	$\mathcal{M}$	$\mathcal{M}$

Table 5.3.3 Efficiency ε^0 as Function of m and α_0, for $L(Y) = \min(Y_1, \ldots, Y_m)$

α_0	$m=1$	$m=10$	$m=10^2$	$m=10^3$	$m=10^4$	$m=10^5$
−0.49	1.7E+5	1.8E+14	$\mathcal{M}$	$\mathcal{M}$	$\mathcal{M}$	$\mathcal{M}$
−0.40	31.40	6425	$\mathcal{M}$	$\mathcal{M}$	$\mathcal{M}$	$\mathcal{M}$
−0.20	2.792	5.779	22.57	3.8E+22	$\mathcal{M}$	$\mathcal{M}$
−0.10	1.562	1.866	7.766	6.2E+06	$\mathcal{M}$	$\mathcal{M}$
0.000	1.000	1.000	1.000	1.000*	1.000*	1.000*
0.01	0.961	0.962	0.980*	1.162	4.226	3.5E+05
0.03	0.889	0.904	1.055	3.412	9159	1.3E+27
0.05	0.826	0.864	1.286	16.64	1.3E+10	$\mathcal{M}$
0.10	0.694	0.825*	2.853	6753	$\mathcal{M}$	$\mathcal{M}$
0.20	0.511	0.947	23.58	2.5E+12	$\mathcal{M}$	$\mathcal{M}$
0.50	0.265	2.653	14E+06	$\mathcal{M}$	$\mathcal{M}$	$\mathcal{M}$
1.00	0.1851*	14.78	2.7E+12	$\mathcal{M}$	$\mathcal{M}$	$\mathcal{M}$

entire response surface (function), $\nabla^k \ell(\boldsymbol{v})$, with fixed λ_0. We can then see how well the SF estimators perform for different values of λ.

Let $T(\lambda|\lambda_0)$ denote a statistic, estimated for parameter λ, but computed from a simulation run corresponding to λ_0. Figure 5.1 depicts the estimated curve $\bar{\ell}_N(\lambda) = \bar{\ell}_N(\lambda|\lambda_0)$ (denoted by $\bar{\ell}_N$) along with the curves

$$J_1 = \{\bar{\ell}_N(\lambda|\lambda_0) - w_r\}, \qquad J_2 = \{\bar{\ell}_N(\lambda|\lambda_0) + w_r\} \tag{5.3.28}$$

[denoted by 95% confidence interval (CI)], as functions of λ, for $m = 10$. The estimators were computed from a sample size $N = 10^4$,

$$w_r = \frac{1.96S(\lambda|\lambda_0)}{N^{1/2}|\bar{\ell}_N(\lambda|\lambda_0)|}$$

represents the half width of the 95% (relative) confidence interval, and S^2 is the sample variance of $\mathrm{var}_{\lambda_0}\{LW(\lambda|\lambda_0)\}$. Note that $\bar{\ell}_N(\lambda|\lambda_0)$ and the dimensionless quantity w_r (in J_1 and J_2) are given in different units.

(ii) Consider next the covariance matrix $\mathrm{cov}_{\lambda_0}\{L\nabla W\}$ for the case of gamma distributions. Here, the basic formula (5.3.21) remains valid, where $\mathbb{E}_{\lambda_0}\{W^2\}$ is given in (5.3.25).

Example 5.3.7 [Example 5.3.4 continued (λ_0 fixed)]. This example may be considered as a dual to Example 5.3.4, in the sense that λ and λ_0 interchange their

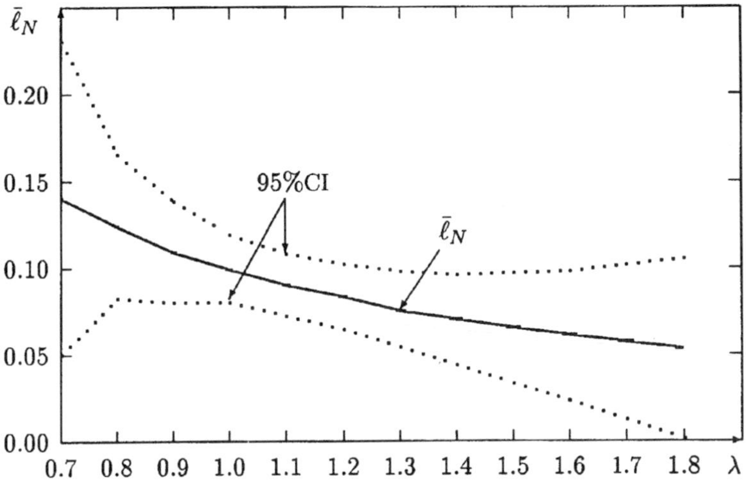

Figure 5.1. Performance of the LR estimates $\bar{\ell}_N(\lambda|\lambda_0)$ for $m=10$, $N=10^4$, and $\lambda_0=1$.

roles and $\alpha_0 = \alpha/(\alpha+1)$. Clearly, formula (5.3.17) for $\mathrm{var}_{\lambda_0}\{L(\partial W(Z,\lambda)]/\partial\lambda\}$ remains valid, while formula (5.3.18) for ε^1 can be written as

$$\varepsilon^1 = \frac{\mathrm{var}_{\lambda_0}\left\{L\dfrac{\partial W(\lambda)}{\partial\lambda}\right\}}{\mathrm{var}_\lambda\{LS^{(1)}\}} = \frac{\lambda_0(1+\alpha_0)^6}{13(1+2\alpha_0)^3} \times \left(\frac{2}{(1+\alpha_0)^2} - \frac{12}{(1+\alpha_0)(1+2\alpha_0)} + \frac{24}{(1+2\alpha_0)^2}\right) - \frac{1}{13}. \tag{5.3.29}$$

Variance reduction ($\varepsilon^1 < 1$) is attained in the interval $(\lambda_0, \tilde{\lambda}_0)$, where $\tilde{\lambda}_0$ satisfies

$$\mathrm{var}_{\tilde{\lambda}}\left\{L\frac{\partial W(\lambda)}{\partial\lambda}\right\} = \mathrm{var}_\lambda\{LS^{(1)}\}.$$

This corresponds to $\tilde{\lambda}_0 \approx 4.62\lambda$ and $\alpha_0 = \Delta\lambda_0/\lambda_0 \approx 3.62$.

Again, a simple practical suggestion for choosing a "good" λ_0, when estimating the response surface $\nabla\ell(\lambda|\lambda_0)$ in the interval $\lambda \in (\lambda_1, \lambda_2)$, is to take $\lambda_0 = \lambda_1$.

Figure 5.2 depicts data similar to that of Figure 5.1 for the derivative $\nabla\ell(v)$ with respect to λ; more specifically, it depicts its estimated curve $\nabla\bar{\ell}_N(\lambda) = \nabla\bar{\ell}_N(\lambda|\lambda_0)$ (denoted by $\nabla\bar{\ell}_N$), along with the curves

$$J_1 = \{\nabla\bar{\ell}_N(\lambda|\lambda_0) - w_r\} \quad \text{and} \quad J_2 = \{\nabla\bar{\ell}_N(\lambda|\lambda_0) + w_r\} \tag{5.3.30}$$

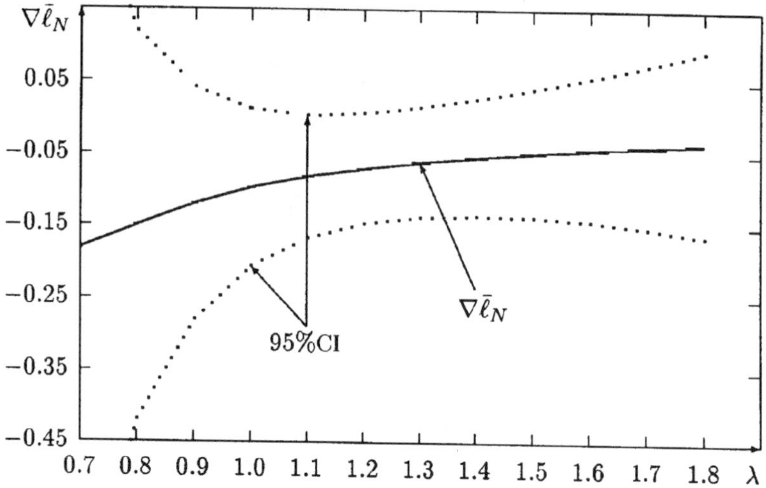

Figure 5.2. Performance of the LR estimators $\nabla\bar{\ell}_N(\lambda|\lambda_0)$, for $m=10$, $N=10^4$, and $\lambda_0=1$.

(denoted by 95% CI), as a function of λ, for $m=10$. The estimators were calculated from a sample size $N=10^4$. Here,

$$w_r = \frac{1.96S(\lambda|\lambda_0)}{N^{1/2}|\nabla\bar{\ell}_N(\lambda|\lambda_0)|}$$

represents the half width of the 95% (relative) confidence interval and S^2 is the sample variance of

$$\mathrm{var}_{\lambda_0}\{\nabla LW(\lambda|\lambda_0)\} = \mathbb{E}_{\lambda_0}\{W^2(\lambda|\lambda_0)\}\mathbb{E}_{2\lambda-\lambda_0}\left\{L^2(m\lambda^{-1} - \sum_{i=1}^{m} Z^2\right\} + (\nabla\lambda)^2. \quad (5.3.31)$$

Note again that the results of the tables and figures above are in agreement with the basic formula (5.3.27), requiring that α_0^2 satisfy $\alpha_0^2 = O(m^{-1})$.

One can extend formulas (5.3.10) and (5.3.21) to the case of an m-parameter exponential family, given in the canonical form (5.2.24), in which case

$$W(y, v) = \frac{f(y, v)}{f(y, v_0)} = \frac{c(v)}{c(v_0)}\exp\left\{\sum_{k=1}^{m}(v_k - v_{0k})t_k(y)\right\}. \quad (5.3.32)$$

Similarly to (5.3.10), it can be shown (see, e.g., Rubinstein and Shapiro, 1993) that in this case

$$\mathrm{var}_{v_0}\{LW\} = \mathbb{E}_{v_0}\{W^2\}\mathbb{E}_{2v-v_0}\{L^2\} - \ell^2(v), \quad (5.3.33)$$

where

$$\mathbb{E}_{v_0}\{W^2\} = \frac{c^2(\boldsymbol{v})}{c(\boldsymbol{v}_0)c(2\boldsymbol{v} - \boldsymbol{v}_0)}. \tag{5.3.34}$$

Notice that (5.3.33) holds under the condition that the vector $2\boldsymbol{v} - \boldsymbol{v}_0$ belongs to the admissible parameter space V. For gamma distributions, this reduces to the condition (5.3.12).

Table 5.3.34 displays $\mathbb{E}_{\boldsymbol{v}_0}\{W^2\}$, calculated from (5.3.34), for commonly used pdf's. Note that in calculating $\mathbb{E}_{\boldsymbol{v}_0}\{W^2\}$ we made use of the particular representations of $c(\boldsymbol{v})$ from Table 5.2.1. Note also that for the two-parametric distributions (gamma, normal, beta and binomial) the quantity $\mathbb{E}_{\boldsymbol{v}_0}\{W^2\}$ is calculated with respect to the first parameter only. The calculation of $\mathbb{E}_{\boldsymbol{v}_0}\{W^2\}$ with respect to the second parameter is left as an exercise to the reader.

Arguing similarly to (5.3.21), it is not difficult to show that for the exponential family (5.2.24), the following expression holds for the covariance matrix:

$$\text{cov}_{\boldsymbol{v}_0}\{L\nabla W\} = \mathbb{E}_{\boldsymbol{v}_0}\{W^2\}\mathbb{E}_{2\boldsymbol{v}-\boldsymbol{v}_0}\{L^2(\boldsymbol{Z})\boldsymbol{S}^{(1)}(\boldsymbol{Z}, \boldsymbol{v})\boldsymbol{S}^{(1)}(\boldsymbol{Z}, \boldsymbol{v})'\} - (\nabla\ell)(\nabla\ell)', \tag{5.3.35}$$

where

$$\boldsymbol{S}^{(1)}(y, \boldsymbol{v}) = \frac{\nabla c(\boldsymbol{v})}{c(\boldsymbol{v})} + \boldsymbol{t}(y) \tag{5.3.36}$$

and $\boldsymbol{t}(y) = (t_1(y), \ldots, t_m(y))'$.

Table 5.3.4 $\mathbb{E}_{v_0}\{W^2\}$ **for Commonly Used pdf's**

pdf	$f(y, \boldsymbol{v})$	$\mathbb{E}_{v_0}\{W^2\}$
Gamma	$\dfrac{\lambda^\beta y^{\beta-1}}{\Gamma(\beta)}\exp(-\lambda y)$	$\left[\dfrac{\lambda^2}{\lambda_0(2\lambda - \lambda_0)}\right]^\beta$
Normal	$\dfrac{1}{\sigma(2\pi)^{1/2}}\exp\dfrac{-(y-\mu)^2}{2\sigma^2}$	$\exp\left(\dfrac{(\mu - \mu_0)^2}{\sigma^2}\right)$
Beta	$\dfrac{y^{\alpha-1}(1-y)^{\beta-1}}{B(\alpha, \beta)}$	$\dfrac{B(\alpha_0, \beta)B(2\alpha - \alpha_0, \beta)}{B(\alpha, \beta)}$
Binomial	$\binom{n}{y}p^y(1-p)^{n-y}$	$\left[\dfrac{p_0 - 2pp_0 + p^2}{(1-p)p_0}\right]^n$
Poisson	$\dfrac{\lambda^y e^{-\lambda}}{y!}$	$\exp\left(\dfrac{(\lambda - \lambda_0)^2}{\lambda_0}\right)$
Geometric	$p(1-p)^{y-1}$	$\dfrac{p^2}{p_0(2p - p_0)}$

As in the case of the gamma pdf, the second moments of LW and $L\nabla W$ [see (5.3.33) and (5.3.35)] are represented as the products of $\mathbb{E}_{v_0}\{W^2\}$ with $\mathbb{E}_{2v-v_0}\{L^2\}$ and with $\mathbb{E}_{2v-v_0}\{L^2 \boldsymbol{S}^{(1)} \boldsymbol{S}^{(1)\prime}\}$, respectively. Finally, for large m, the term $\mathbb{E}_{v_0}\{W^2\}$ in (5.3.33) and (5.3.35) is typically dominant.

5.4 OPTIMIZATION OF DESS

Consider again the mathematical programming problem (II.1),

$$(P_0) = \begin{cases} \text{minimize} & \ell_0(\boldsymbol{v}), & \boldsymbol{v} \in V, \\ \text{subject to} & \ell_j(\boldsymbol{v}) \leq 0, & j = 1, \ldots, k, \\ & \ell_j(\boldsymbol{v}) = 0, & j = k+1, \ldots, M, \end{cases} \tag{5.4.1}$$

where

$$\ell_j(\boldsymbol{v}) = \mathbb{E}_{\boldsymbol{v}}\{L_j(\boldsymbol{Y})\} = \int L_j(\boldsymbol{y}) f(\boldsymbol{y}, \boldsymbol{v})\, d\boldsymbol{y}, \qquad j = 0, 1, \ldots, M. \tag{5.4.2}$$

Suppose the objective function $\ell_0(\boldsymbol{v})$ and some of the constraint functions $\ell_j(\boldsymbol{v})$ are not available in an analytical form, so that in order to solve (P_0) we must resort to Monte Carlo simulation. This section shows how the SF method can be used to estimate an optimal solution of the program (P_0) from a *single simulation* experiment. The underlying idea (see, e.g., Rubinstein, 1986) is to replace the deterministic program (P_0) by a *stochastic counterpart* (see below), which then can be solved by standard mathematical programming techniques. The resultant optimal solution provides an estimator of the corresponding "true" optimal solution for the (original) program (P_0).

To fix the ideas, consider first the following unconstrained case:

$$(P_0) \qquad \text{minimize} \quad \ell(\boldsymbol{v}) = \mathbb{E}_{\boldsymbol{v}}\{L(\boldsymbol{Y})\}, \qquad \boldsymbol{v} \in V. \tag{5.4.3}$$

Assume that V is an open set and that $\ell(\boldsymbol{v})$ is continuously differentiable on V. Then, by the first-order necessary conditions the gradient of $\ell(\boldsymbol{v})$ at an optimal solution point, $\boldsymbol{v}^*$, must vanish. Consequently, the optimal solution, $\boldsymbol{v}^*$, can be found by solving the equation system:

$$\nabla \ell(\boldsymbol{v}) = \boldsymbol{0}. \tag{5.4.4}$$

It is natural to consider the program

$$(\bar{P}_N) \qquad \text{minimize} \quad \bar{\ell}_N(\boldsymbol{v}), \qquad \boldsymbol{v} \in V, \tag{5.4.5}$$

as a stochastic counterpart of (5.4.3), (see also Rubinstein, 1986), where

$$\bar{\ell}_N(\boldsymbol{v}) = \frac{1}{N}\sum_{i=1}^{N} L(\boldsymbol{Z}_i)W(\boldsymbol{Z}_i, \boldsymbol{v})$$

is *viewed as a function of* $\boldsymbol{v}$, *rather than as an estimator for fixed* $\boldsymbol{v}$, $\{\boldsymbol{Z}_1, \ldots, \boldsymbol{Z}_N\}$ is a random sample from the dominating pdf $g(\boldsymbol{z})$, and

$$W(\boldsymbol{z}, \boldsymbol{v}) = \frac{f(\boldsymbol{z}, \boldsymbol{v})}{g(\boldsymbol{z})}.$$

Assuming further that $\bar{\ell}_N(\boldsymbol{v})$ is continuously differentiable on V, the optimal solution of (5.4.3) can be estimated by solving the equation system

$$\nabla\bar{\ell}_N(\boldsymbol{v}) = \mathbf{0}, \qquad \boldsymbol{v} \in V, \tag{5.4.6}$$

which itself may be viewed as a stochastic counterpart of the deterministic system (5.4.4).

Example 5.4.1. Consider the unconstrained minimization program

$$\min_p \ell(p) = \min_p \left\{ \mathbb{E}_p\{Y\} + \frac{b}{p} \right\}, \qquad p \in (p_1, p_2), \tag{5.4.7}$$

where $Y \sim \text{Ber}(p)$. In this case $\ell(p) = p + b/p$, and the stochastic counterpart of

$$\nabla\ell(p) = 1 - \frac{b}{p^2} = 0$$

can be written as

$$\nabla\bar{\ell}_N(p) = \frac{1}{N}\sum_{i=1}^{N} Z_i \frac{p^{Z_i}(1-p)^{1-Z_i}}{p_0^{Z_i}(1-p_0)^{1-Z_i}} \times \frac{Z_i - p}{p(1-p)} - \frac{b}{p^2} = 0, \tag{5.4.8}$$

where $\{Z_i\colon i = 1, \ldots, N\}$ is a random sample from $\text{Ber}(p_0)$.

Assume as before that we generated a random sample

$$\{Z_1, \ldots, Z_{20}\} = \{0, 1, 0, 0, 1, 0, 0, 1, 1, 1, 0, 1, 0, 1, 1, 0, 1, 0, 1, 1\}$$

of size $N = 20$ from Ber($p_0 = \frac{1}{2}$) in order to estimate the optimal solution p^*, which is $p^* = b^{1/2}$. From (5.4.8) we have

$$\nabla \bar{\ell}_N(p) = \tfrac{1}{20}(0+2+0+0+2+0+0+2+2+2+0+2+0+2+2+0$$
$$+2+0+2+2) - \frac{b}{p^2} = 1.1 - \frac{b}{p^2} = 0,$$

yielding

$$\bar{p}_N^* = \left(\frac{b}{1.1}\right)^{1/2}.$$

To illustrate, for $b = \frac{1}{16}$, the exact optimal value is $p^* = \frac{1}{4}$, while its estimate is $\bar{p}_N^* = ((1/16)/1.1)^{1/2} \approx 0.23$.

Example 5.4.2. Consider the unconstrained minimization program

$$\min_\lambda \ell(\lambda) = \min_\lambda \{\mathbb{E}_\lambda\{Y\} + b\lambda\}, \qquad \lambda \in (\lambda_1, \lambda_2) = (0.5, 2.0), \tag{5.4.9}$$

where $Y \sim \text{Exp}(\lambda)$. We have $\ell(\lambda) = 1/\lambda + b\lambda$, and the stochastic counterpart of $\nabla\ell(\lambda) = -1/\lambda^2 + b = 0$ can be written as

$$\nabla \bar{\ell}_N(\lambda) = \frac{1}{N}\sum_{i=1}^{N} Z_i \frac{\nabla \lambda e^{-\lambda Z_i}}{\lambda_0 e^{-\lambda_0 Z_i}} + b = \frac{1}{N}\sum_{i=1}^{N} Z_i \frac{\lambda e^{-\lambda Z_i}(\lambda^{-1} - Z_i)}{\lambda_0 e^{-\lambda_0 Z_i}} + b = 0, \tag{5.4.10}$$

where $\{Z_i : i = 1, \ldots, N\}$ is a random sample from Exp(λ_0).

Figure 5.3 depicts the theoretical curve $\ell(\lambda)$ and its stochastic counterpart $\bar{\ell}_N(\lambda)$, as functions of λ, for various sample sizes N. Figure 5.4 depicts similar (derivative) curves $\nabla\ell(\lambda)$ and $\nabla\bar{\ell}_N(\lambda)$. The parameters chosen were $b = 1$ (this corresponds to $\lambda^* = 1$), and the reference parameter was set at $\lambda = 0.5$. The figures indicate that the estimated curves, $\bar{\ell}_N(\lambda)$ and $\nabla\bar{\ell}_N(\lambda)$, converge to their theoretical counterparts, $\ell(\lambda)$ and $\nabla\ell(\lambda)$, respectively.

Figure 5.5 depicts the theoretical values of λ^*, the point estimators $\bar{\lambda}_N$, and the 95% relative confidence interval as functions of b for $N = 500$. We stress that *all* the estimators $\bar{\lambda}_N = \bar{\lambda}_N(b)$ were obtained *simultaneously, from a single simulation run* (with $\lambda_0 = 0.5$), by solving the system of equations (5.4.10) for various values of b. Figure 5.5 indicates that the estimator $\bar{\lambda}_N$ performs well.

It is important to note that while solving the stochastic counterpart (5.4.6) based on likelihood ratios, two requirements must hold: The reference parameter vector $\boldsymbol{v}_0$ must be carefuly selected and the parameter set V (trust region) must not be too wide [see Example 5.3.6 and formula (5.3.27)]. If the region V is too wide, then alternative methods might be used. One such alternative is to use a Newton type iterative procedure based on score function derivatives [see (5.2.17)],

$$\nabla^k \ell(\boldsymbol{v}) = \mathbb{E}_{\boldsymbol{v}}\{L(\boldsymbol{Y})\boldsymbol{S}^{(k)}(\boldsymbol{Y}, \boldsymbol{v})\}, \qquad k = 1, 2.$$

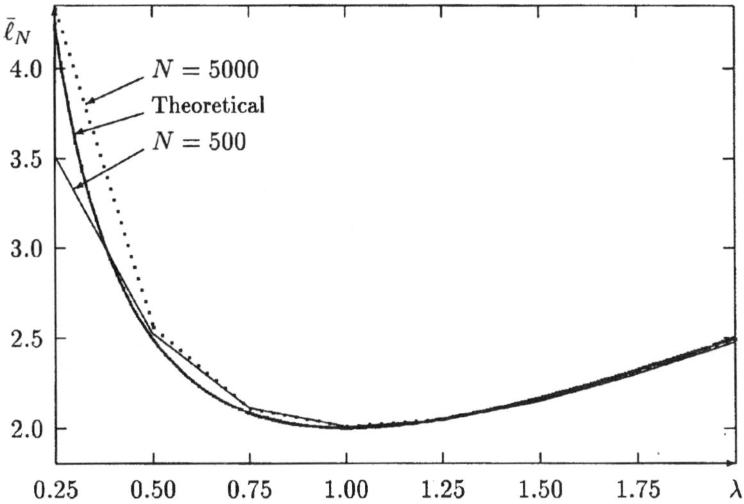

Figure 5.3. Theoretical curve $\ell(\lambda)$ and the estimated curves $\bar{\ell}_N(\lambda)$ as functions of λ, for $b=1$ and $\lambda_0=0.5$.

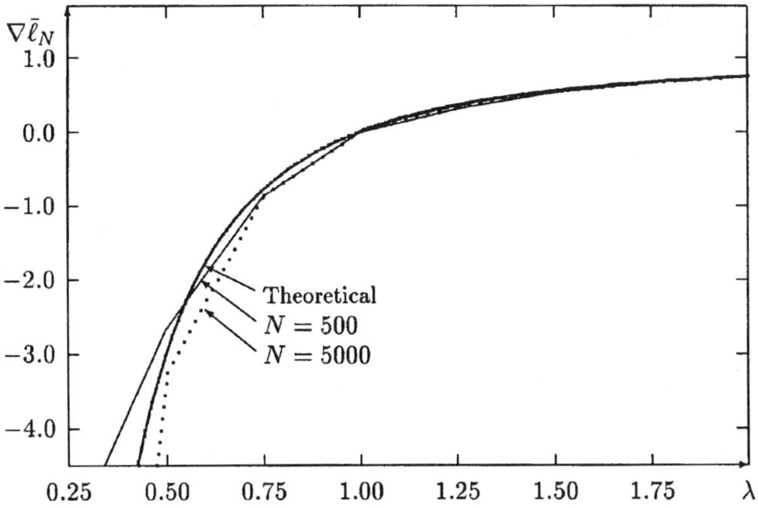

Figure 5.4. Theoretical curve $\nabla\ell(\lambda)$ and the estimated curves $\nabla\bar{\ell}_N(\lambda)$ as functions of λ, for $b=1$ and $\lambda_0=0.5$.

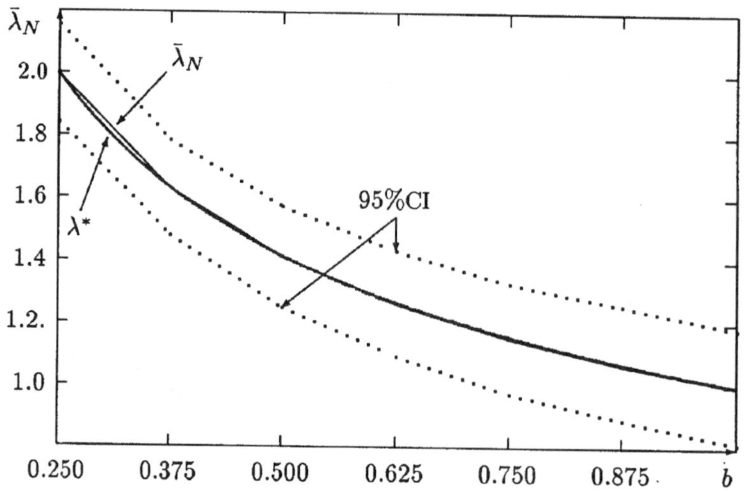

Figure 5.5. Theoretical values of λ^*, the point estimators $\bar{\lambda}_N$, and the 95% relative confidence intervals of λ^* as functions of b, for $N=500$ and $\lambda_0=0.5$.

Optimal solutions of the program (5.4.5) correspond to the so-called M estimators (Huber, 1964, 1981). Technically, M estimators are extensions of the conventional maximum likelihood estimators; their statistical properties have been studied in Huber (1981). It should be pointed out that the goal of traditional M estimators, as used in statistics, differs from the estimators $\bar{\boldsymbol{v}}_N$ considered here. In the former context, M estimators are employed to estimate unknown parameters, whereas in the latter context, the estimators $\bar{\boldsymbol{v}}_N$ are used to evaluate the unknown optimal solution, $\boldsymbol{v}^*$, of the program (P_0).

Returning to the general program (P_0) in (5.4.1), it follows from the foregoing discussion that the stochastic counterpart of (P_0) can be written as

$$(P_N) = \begin{cases} \text{minimize} & \bar{\ell}_{0N}(\boldsymbol{v}), & \boldsymbol{v} \in V, \\ \text{subject to} & \bar{\ell}_{jN}(\boldsymbol{v}) \leq 0, & j = 1, \ldots, k, \\ & \bar{\ell}_{jN}(\boldsymbol{v}) = 0, & j = k+1, \ldots, M, \end{cases} \tag{5.4.11}$$

where

$$\bar{\ell}_{jN}(\boldsymbol{v}) = \frac{1}{N} \sum_{i=1}^{N} L_j(\boldsymbol{Z}_i) W(\boldsymbol{Z}_i, \boldsymbol{v}), \qquad j = 0, 1, \ldots, M, \tag{5.4.12}$$

$\{\boldsymbol{Z}_1, \ldots, \boldsymbol{Z}_N\}$ is a random sample from the dominating pdf $g(\boldsymbol{z})$, and the $\bar{\ell}_{jN}(\boldsymbol{v})$ are viewed as functions of $\boldsymbol{v}$ rather than as estimators for a fixed $\boldsymbol{v}$.

Note that once the sample $\{\boldsymbol{Z}_1, \ldots, \boldsymbol{Z}_N\}$ is generated, the functions $\bar{\ell}_{jN}(\boldsymbol{v})$, $j = 0, \ldots, M$, become *explicitly* determined via the known density functions $f(\boldsymbol{Z}_i, \boldsymbol{v})$. The corresponding gradients, $\nabla \bar{\ell}_{jN}(\boldsymbol{v})$, and Hessian matrices,

$\nabla^2 \bar{\ell}_{jN}(\boldsymbol{v})$, can be calculated, in principle, via the SF method (5.2.9) for any $\boldsymbol{v}$, from a single simulation run. Consequently, the optimization problem $(\bar{P}_N)$ can be solved by standard methods of mathematical programming. The resultant optimal value, $\bar{\varphi}_N$, and optimal solution, $\bar{\boldsymbol{v}}_N$, of the program $(\bar{P}_N)$ provide estimators of the optimal value, φ^*, and the optimal solution, $\boldsymbol{v}^*$, respectively, of the original program (P_0). It is important to realize that what makes this approach feasible is the fact that once the sample $\{\boldsymbol{Z}_1, \ldots, \boldsymbol{Z}_N\}$ is generated, the functions $\bar{\ell}_{jN}(\boldsymbol{v})$, $j = 0, \ldots, M$, become known explicitly, since the sample functions $L_j(\boldsymbol{y})$ *do not depend* on $\boldsymbol{v}$. Other optimization problems will be treated in the sequel.

Example 5.4.3. Consider the optimization problem (5.1.2). Its stochastic counterpart can be written as

$$(\bar{P}_N) = \begin{cases} \text{maximize} & \bar{\ell}_{0N}(\boldsymbol{v}), \qquad \boldsymbol{v} \in V, \\ \text{subject to} & \bar{\ell}_{1N} \leq 0, \\ & \ell_2(\boldsymbol{v}) \leq 0, \end{cases} \tag{5.4.13}$$

where

$$\bar{\ell}_{0N}(\boldsymbol{v}) = \frac{1}{N} \sum_{i=1}^{N} L(\boldsymbol{Z}_i) W(\boldsymbol{Z}_i, \boldsymbol{v}),$$

$$\bar{\ell}_{1N}(\boldsymbol{v}) = \frac{1}{N} \sum_{i=1}^{N} I_{L(\boldsymbol{Z}_i > x)} W(\boldsymbol{Z}_i, \boldsymbol{v}) - b_1,$$

$$L(\boldsymbol{Z}_i) = \min_{j=1,\ldots,p} \sum_{k \in \mathcal{L}_j} Z_{ki},$$

and the function $\ell_2(\boldsymbol{v})$ is identical to the one in Example 5.1.1. The only difference between (5.4.13) and (5.1.2) is that $\ell_0(\boldsymbol{v})$ and $\ell_1(\boldsymbol{v})$ in (5.1.2) are replaced here by $\bar{\ell}_{0N}(\boldsymbol{v})$ and $\bar{\ell}_{1N}(\boldsymbol{v})$, while the deterministic constraint $\ell_2(\boldsymbol{v}) \leq 0$ remains the same.

The algorithm for estimating the optimal solution, $\boldsymbol{v}^*$, of the program (P_0) via the stochastic counterpart $(\bar{P}_N)$, can be written as follows:

Algorithm 5.4.1 Estimation of $\boldsymbol{v}^*$

1. *Generate a random sample* $\{\boldsymbol{Z}_1, \ldots, \boldsymbol{Z}_N\}$ *from* $g(\boldsymbol{z})$.
2. *Calculate* $L_j(\boldsymbol{Z}_i)$, $j = 0, \ldots, M$, $i = 1, \ldots, N$ *via simulation.*
3. *Solve the program* $(\bar{P}_N)$ *by standard mathematical programming methods.*
4. *Return the resultant optimal solution,* $\bar{\boldsymbol{v}}_N$ *of* $(\bar{P}_N)$, *as an estimate of* $\boldsymbol{v}^*$.

The third step of Algorithm 5.4.1 would typically call for iterative numerical procedures, which may require, in turn, calculation of the functions $\bar{\ell}_{jN}(\boldsymbol{v})$, $j = 0, \ldots, M$ (and possibly their gradients and Hessians), for multiple values of the parameter vector $\boldsymbol{v}$. Clearly, the CPU time for the program $(\bar{P}_N)$ is an increasing function in the sample size N. Extensive simulation studies with typical

DES show that an approximate optimal solution $\bar{\boldsymbol{v}}_N$ of $(\bar{P}_N)$ constitutes a reliable estimator of the "exact" optimal solution, $\boldsymbol{v}^*$, provided the sample size, N, is on the order of 1000 or more, and the program $(\bar{P}_N)$ is convex. For nonconvex programs, one can use methods of global Monte Carlo optimization (e.g., Katkovnik, 1976; Rubinstein, 1986).

The following theorem summarizes the basic statistical properties of $\bar{\boldsymbol{v}}_N$ (consistency and asymptotic normality) for the unconstrained program formulation. Additional discussion, including proofs for both the unconstrained and constrained programs, may be found in Rubinstein and Shapiro (1993).

Theorem 5.4.1. *Let $\boldsymbol{v}^*$ be a unique minimizer of $\ell(\boldsymbol{v})$ over V.*

(a) Suppose that

(i) The set V is compact.
(ii) For almost every $\boldsymbol{y}$, the function $f(\boldsymbol{y}, \cdot)$ is continuous on V.
(iii) The family of functions $\{|L(\boldsymbol{y}) f(\boldsymbol{y}, \boldsymbol{v})| : \boldsymbol{v} \in V\}$ is dominated by a Lebesgue-integrable function $h(\boldsymbol{y})$, that is,

$$|L(\boldsymbol{y}) f(\boldsymbol{y}, \boldsymbol{v})| \leq h(\boldsymbol{y}) \qquad \text{for all} \quad \boldsymbol{v} \in V.$$

Then an optimal solution $\bar{\boldsymbol{v}}_N$ of (5.4.5) converges to $\boldsymbol{v}^*$, as $N \to \infty$, with probability one.

(b) Suppose further that

(i) $\boldsymbol{v}^*$ is an interior point of V.
(ii) For almost every $\boldsymbol{y}, f(\boldsymbol{y}, \cdot)$ is twice continuously differentiable in a neighborhood U of $\boldsymbol{v}^*$, and the families of functions

$$\{\|L(\boldsymbol{y})\nabla^s f(\boldsymbol{y}, \boldsymbol{v})\| : \boldsymbol{v} \in U, \; s = 1, 2\}, \quad \text{where} \quad \|\boldsymbol{x}\| = (x_1^2 + \cdots + x_n^2)^{1/2},$$

are dominated by a Lebesgue-integrable function.
(iii) The matrix

$$\boldsymbol{B} = \mathbb{E}_g\{L(\boldsymbol{Z})\nabla^2 W(\boldsymbol{Z}, \boldsymbol{v}^*)\} \tag{5.4.14}$$

is nonsingular.
(iv) The covariance matrix of the vector $L(\boldsymbol{Z})\nabla W(\boldsymbol{Z}, \boldsymbol{v}^*)$, given by

$$\begin{aligned} \text{cov}\{L(\boldsymbol{Z})\nabla W(\boldsymbol{Z}, \boldsymbol{v}^*)\} = \mathbb{E}_g\{L^2(\boldsymbol{Z})\nabla W(\boldsymbol{Z}, \boldsymbol{v}^*)\nabla' W(\boldsymbol{Z}, \boldsymbol{v}^*)\} \\ - \nabla\ell(\boldsymbol{v}^*)\nabla'\ell(\boldsymbol{v}^*), \end{aligned}$$

exists.

Then the random vector $N^{1/2}(\bar{\boldsymbol{v}}_N - \boldsymbol{v}^*)$ converges in distribution to a normal random vector with zero mean and covariance matrix

$$\boldsymbol{B}^{-1}\text{cov}\{L(\boldsymbol{Z})\nabla W(\boldsymbol{Z}, \boldsymbol{v}^*)\}\boldsymbol{B}^{-1}. \tag{5.4.15}$$

The asymptotic efficiency of the estimator $N^{1/2}(\bar{\boldsymbol{v}}_N - \boldsymbol{v}^*)$ is controlled by the covariance matrix given in (5.4.15). Under the assumptions of Theorem 5.4.1, this covariance matrix can be consistently estimated by $\boldsymbol{B}_N^{-1}\boldsymbol{\Sigma}_N\boldsymbol{B}_N^{-1}$, where

$$\boldsymbol{B}_N = \frac{1}{N}\sum_{i=1}^{N} L(\boldsymbol{Z}_i)\nabla^2 W(\boldsymbol{Z}_i, \bar{\boldsymbol{v}}_N) \tag{5.4.16}$$

and

$$\boldsymbol{\Sigma}_N = \frac{1}{N}\sum_{i=1}^{N} L(\boldsymbol{Z}_i)^2 \nabla W(\boldsymbol{Z}_i, \bar{\boldsymbol{v}}_N)\nabla' W(\boldsymbol{Z}_i, \bar{\boldsymbol{v}}_N) - \nabla\bar{\ell}_N(\bar{\boldsymbol{v}}_N)\nabla'\bar{\ell}_N(\bar{\boldsymbol{v}}_N), \tag{5.4.17}$$

are consistent estimators of the matrices $\boldsymbol{B}$ and $\text{cov}\{L(\boldsymbol{Z})\nabla W(\boldsymbol{Z}, \boldsymbol{v}^*)\}$, respectively. Observe that these matrices can be estimated from the same sample $\{\boldsymbol{Z}_1, \ldots, \boldsymbol{Z}_N\}$ simultaneously with the estimator $\bar{\boldsymbol{v}}_N$. Observe also that the matrix $\boldsymbol{B}$ coincides with the Hessian matrix $\nabla^2\ell(\boldsymbol{v}^*)$ and is, therefore, independent of the choice of a dominating pdf $g(z)$.

Example 5.4.4 (Example 5.4.2 continued). In this case, λ is a scalar and formulas (5.4.14) and (5.4.15) reduce to

$$\boldsymbol{B} = \mathbb{E}_{\lambda_0}\{Z\nabla^2 W(Z, \lambda^*)\} = \frac{2}{\lambda^{3*}},$$

$$\begin{aligned}\text{var}\{Z\nabla W(Z, \lambda^*)\} &= \mathbb{E}_{\lambda_0}\{Z^2\nabla W^2(Z, \lambda^*)\} - \{\mathbb{E}_{\lambda_0}\{Z\nabla W(Z, \lambda^*)\}\}^2 \\ &= \frac{\lambda^{2*}}{\lambda_0(2\lambda^* - \lambda_0)^3}\left(\frac{2}{\lambda^{*2}} - \frac{12}{\lambda_0(2\lambda^* - \lambda_0)} + \frac{24}{\lambda_0(2\lambda^* - \lambda_0)^2}\right) - \frac{1}{\lambda^{*4}},\end{aligned}$$

(see 5.3.18), and

$$\begin{aligned}\text{var}\{N^{1/2}(\bar{\lambda}_N - \lambda^*)\} \approx & \frac{\lambda^{*8}}{4\lambda_0(2\lambda^* - \lambda_0)^3} \\ & \times\left(\frac{2}{\lambda^{*2}} - \frac{12}{\lambda_0(2\lambda^* - \lambda_0)} + \frac{24}{\lambda_0(2\lambda^* - \lambda_0)^2}\right) - \frac{\lambda^{*2}}{\lambda^{*4}},\end{aligned}$$

respectively, where

$$\nabla W(Z, \lambda^*) = \frac{\partial W(Z, \lambda^*)}{\partial \lambda^*}.$$

As an example, assume that $\lambda \in (\lambda_1, \lambda_2) = (0.5, 2)$. Let $\lambda_0 = 1$ and consider the optimal values $\lambda^* = 0.5$, 1, 2. [Observe that the solution of the program (5.4.9) is $\lambda^* = b^{-1/2}$.] Then, $\text{var}\{N^{1/2}(\bar{\lambda}_N - \lambda^*)\} \approx \infty$, 3.25, 1.765, respectively. Notice that λ_0 and λ^* play a role, similar to that of λ_0 and λ in Example 5.3.7, in estimating the response curve $\partial\ell(\lambda)/\partial\lambda$. Furthermore, since we require that $\lambda \in (\lambda_1, \lambda_2)$, then by analogy to Example 5.3.7, a simple practical suggestion for choosing a good λ_0 is to take $\lambda_0 = \lambda_1$. Finally, observe that the infinite variance in the case $\lambda^* = 0.5$ is a consequence of choosing a wrong λ_0 (namely, $\lambda_0 = 10$, rather than $\lambda_0 = \lambda_1 = 0.5$). The reader is asked to check that in the latter case, $\text{var}\{N^{1/2}(\bar{\lambda}_N - \lambda^*)\}$ is finite for the above three values $\lambda^* = 0.5$, 1, 2, and $N \geq 1000$, the relative width of the 95% confidence interval for λ^* (for all three values of λ^*) is less than 5%.

5.5 EXERCISES

1. Let Y_k, $k = 1, \ldots, m$, be independent random variables, and let $\boldsymbol{Y} = (Y_1, \ldots, Y_m)'$ be the corresponding random vector.

(i) Show that if each Y_k is Weibull distributed, that is

$$f_k(y_k, \alpha_{0k}) = \alpha_{0k} y_k^{\alpha_{0k}-1} \exp(-y_k^{\alpha_{0k}}), \tag{5.6.1}$$

then for $\alpha_0 = (\alpha_{01}, \ldots, \alpha_{0m})'$, the efficient scores are given by

$$[\boldsymbol{S}^{(1)}(\boldsymbol{y}, \alpha_0)]_k = \alpha_{0k}^{-1} + \log y_k - y_k^{\alpha_{0k}} \log y_k, \qquad k = 1, \ldots, m,$$

and

$$[\boldsymbol{S}^{(2)}(\boldsymbol{y}, \alpha_0)]_{jk} = \delta_{jk}[-\alpha_{0k}^{-2} - y_k^{\alpha_{0k}}(\log y_k)^2] \\ + (\alpha_{0j}^{-1} + \log y_j - y_j^{\alpha_{0j}} \log y_j)(\alpha_{0k}^{-1} + \log y_k - y_k^{\alpha_{0k}} \log y_k),$$

for j, $k = 1, \ldots, m$, with $\delta_{jk} = 1$ if $j = k$ and $\delta_{jk} = 0$ if $j \neq k$.

(ii) Show that if each Y_k is Poisson distributed, that is,

$$P_k(y_k, \lambda_k) = \frac{\lambda_k^{y_k}}{y_k!} e^{-\lambda_k}, \qquad y_k = 0, 1, \ldots,$$

then

$$[\boldsymbol{S}^{(1)}(\boldsymbol{y}, \lambda)]_k = \lambda_k^{-1} y_k - 1, \quad k = 1, \ldots, m,$$

and

$$[\boldsymbol{S}^{(2)}(\boldsymbol{y}, \boldsymbol{\lambda})]_{jk} = -\delta_{jk}\lambda_k^{-2}y_k + (\lambda_j^{-1}y_j - 1)(\lambda_k^{-1}y_k - 1),$$

for $j,\ k = 1, \ldots, m$.

2. Consider a function $W(\boldsymbol{y}, \boldsymbol{v})$ of the form (5.3.32). Show that

$$\nabla^2 W(\boldsymbol{y}, \boldsymbol{v}) = W(\boldsymbol{y}, \boldsymbol{v})\boldsymbol{S}^{(2)}(\boldsymbol{y}, \boldsymbol{v}),$$

where $\boldsymbol{S}^{(2)}(\mathbf{y}, \boldsymbol{v})$ is an $m \times m$ matrix given by

$$\begin{aligned}\boldsymbol{S}^{(2)}(\boldsymbol{y}, \boldsymbol{v}) &= (1/c(\boldsymbol{v}))\nabla^2 c(\boldsymbol{v}) - (1/c^2(\boldsymbol{v}))\nabla c(\boldsymbol{v})\nabla c(\boldsymbol{v})' \\ &\quad + [(1/c(\boldsymbol{v}))\nabla c(\boldsymbol{v}) + \boldsymbol{t}(\boldsymbol{y})][(1/c(\boldsymbol{v}))\nabla c(\boldsymbol{v}) + \boldsymbol{t}(\boldsymbol{y})]'.\end{aligned}$$

3. Let $Y_k,\ k = 1, \ldots, m$, be independent random variables, where Y_k is normally distributed with mean μ_k and variance σ_k^2, that is,

$$f_k(y_k, \mu_k, \sigma_k) = \frac{1}{(2\pi)^{1/2}\sigma_k} \exp\left(\frac{-(y_k - \mu_k)^2}{2\sigma_k^2}\right).$$

Suppose we are interested in sensitivities with respect to $\boldsymbol{\mu} = (\mu_1, \ldots, \mu_m)'$ only. Show that for $k = 1, \ldots, m$,

$$\mathbb{E}_{\boldsymbol{\mu}_0}\{W^2\} = \exp\left(\sum_{k=1}^{m} \frac{(\Delta\mu_k)^2}{\sigma_k^2}\right)$$

where $\Delta\mu = \mu - \mu_0$, and

$$[\boldsymbol{S}^{(1)}(\boldsymbol{y}, \boldsymbol{\mu})]_k = \sigma_k^{-2}(y_k - \mu_k).$$

4. Present a table similar to Table 5.3.4 for

$$\mathrm{cov}_{\boldsymbol{v}_0}\{\nabla W\} = \mathbb{E}_{\boldsymbol{v}_0}\{W^2\}\mathbb{E}_{2\boldsymbol{v}-\boldsymbol{v}_0}\{\boldsymbol{S}^{(1)}(\boldsymbol{Z}, \boldsymbol{v})\boldsymbol{S}^{(1)\prime}(\boldsymbol{Z}, \boldsymbol{v})\},$$

[see (5.3.35)].

***5.** Let the components $Y_k,\ k = 1, \ldots, m$, of the random vector $\boldsymbol{Y}$, be independent, each distributed according to the pdf

$$f_k(y_k, v_k) = a_k(v_k)\exp(b_k(v_k)t_k(y_k))h_k(y_k),$$

where $a_k(\boldsymbol{v}_k)$ is a positive-valued function of $\boldsymbol{v}_k$, and $b_k(\boldsymbol{v}_k)$, $t_k(y_k)$, and $h_k(y_k)$ are real-valued functions. Consider the corresponding pdf of $\boldsymbol{Y}$, given by

$$f(\boldsymbol{y}, \boldsymbol{v}) = a(\boldsymbol{v}) \exp\left(\sum_{k=1}^{m} b_k(\boldsymbol{v}_k) t_k(y_k) \right) h(\boldsymbol{y}),$$

where $\boldsymbol{v} = (\boldsymbol{v}_1', \ldots, \boldsymbol{v}_m')'$, $a(\boldsymbol{v}) = \prod_{k=1}^{m} a_k(\boldsymbol{v}_k)$, and $h(\boldsymbol{y}) = \prod_{k=1}^{m} h_k(y_k)$. Show that

(i)

$$\mathrm{var}_{\boldsymbol{v}_0}\{LW\} = \mathbb{E}_{\boldsymbol{v}_0}\{W^2\} \mathbb{E}_{(a^*, b^*)}\{L^2\} - \ell^2(\boldsymbol{v}),$$

where the subscript (a^*, b^*) signifies that the expectation is taken with respect to the exponential family, determined by the functions

$$b_k^*(\boldsymbol{v}_k, \boldsymbol{v}_{0k}) = 2b_k(\boldsymbol{v}_k) - b_k(\boldsymbol{v}_{0k}), \qquad k = 1, \ldots, m,$$

and by the function $a^*(\boldsymbol{v}, \boldsymbol{v}_0)$ which satisfies the condition

$$a^*(\boldsymbol{v}, \boldsymbol{v}_0) \int \exp\left(\sum_{k=1}^{m} b_k^*(\boldsymbol{v}_k, \boldsymbol{v}_{0k}) t_k(y_k) \right) h(\boldsymbol{y})\, d\boldsymbol{y} = 1.$$

(ii)

$$\mathbb{E}_{\boldsymbol{v}_0}\{W^2\} = \frac{a^2(\boldsymbol{v})}{a(\boldsymbol{v}_0) a^*(\boldsymbol{v}, \boldsymbol{v}_0)}.$$

(iii)

$$a^*(\boldsymbol{v}, \boldsymbol{v}_0) = \prod_{k=1}^{m} a_k^*(\boldsymbol{v}_k, \boldsymbol{v}_{0k}),$$

where $a_k^*(\boldsymbol{v}_k, \boldsymbol{v}_{0k})$ are defined by the equations

$$a_k^*(\boldsymbol{v}_k, \boldsymbol{v}_{0k}) \int \exp\{b_k^*(\boldsymbol{v}_k, \boldsymbol{v}_{0k}) t_k(y_k)\} h_k(y_k)\, dy_k = 1.$$

(iv)

$$\mathrm{cov}_{\boldsymbol{v}_0}\{L \nabla W\} = \mathbb{E}_{\boldsymbol{v}_0}\{W^2\} \mathbb{E}_{(a^*, b^*)})\{L^2 \boldsymbol{S}^{(1)} \boldsymbol{S}^{(1)\prime}\} - (\nabla \ell)(\nabla \ell)',$$

where $\boldsymbol{S}^{(1)} = (S_1^{(1)}(Z_1, \boldsymbol{v}_1)', \ldots, S_m^{(1)}(Z_m, \boldsymbol{v}_m)')'$ with

$$\boldsymbol{S}_k^{(1)}(z_k, \boldsymbol{v}_k) = \frac{\nabla a_k(\boldsymbol{v}_k)}{a_k(\boldsymbol{v}_k)} + t_k(z_k) \nabla b_k(\boldsymbol{v}_k), \qquad k = 1, \ldots, m.$$

(The gradients are taken with respect to $\boldsymbol{v}_k$.)

***6.** Consider the exponential pdf $f(y, v) = v \exp(-vy)$. Show that if $L(y)$ is a monotonically increasing function, then the expected performance $\ell(v) = \mathbb{E}_v\{L(Y)\}$ is a monotonically decreasing, convex function of $v \in (0, \infty)$.

7. Prove that

$$\mathrm{cov}_g\{L(\boldsymbol{Z})W(\boldsymbol{Z}, \boldsymbol{v})\} = \mathbf{0}.$$

Hint. Use the formula

$$\ell(\boldsymbol{v}) = \mathbb{E}_g\{L(\boldsymbol{Z})W(\boldsymbol{Z}, \boldsymbol{v})\}.$$

8. Let $L = \min(Y_1, \ldots, Y_m)$. Run a simulation program with Y_i from the Bernoulli distribution, Ber(p), and present figures similar to those in Figures 5.1 and 5.2 for $p \in (\frac{1}{4}, \frac{3}{4})$, $p_0 = \frac{1}{2}$, and $m = 10$, $N = 10^4$. Repeat the simulation for the Weibull distribution (5.6.1) with $\alpha_0 = \frac{1}{2}$ and $\alpha_0 = 2$, for $\alpha \in (\frac{1}{2}, 2)$.

9. Compare the numerical results of Figure 5.1 with their theoretical counterparts, and plot the response curve $\ell(\lambda)$ along with the (theoretical) curves

$$\hat{J}_1 = \{\ell(\lambda) - \hat{w}_r\}, \qquad \hat{J}_2 = \{\ell(\lambda) + \hat{w}_r\}, \tag{5.5.2}$$

as functions of λ, for $m = 10$. Here,

$$\hat{w}_r = \frac{1.96\sigma}{N^{1/2}\ell(\lambda)}$$

represents the (theoretical) relative half width of the 95% confidence interval, and $\sigma^2 = \mathrm{var}_{\lambda_0}\{LW(\lambda|\lambda_0)\}$ is given in Table 5.3.2. Are the theoretical values in agreement with the estimated values of Figure 5.1?

10. Consider the optimization problem (5.4.9) in Example 5.4.2. Find a sequence of appropriate values b in (5.4.9), such that the optimal values of λ^* are $\lambda^* = 1, 2, 5, 10$. Run a simulation program with a sample size $N = 10^4$ at $\lambda_0 = 1$ and estimate the optimal $\lambda^* = 1, 2, 5, 10$, using the stochastic counterpart $\nabla\bar{\ell}_N(\lambda) = 0$. For $\lambda^* = 1, 2, 5, 10$, estimate

$$\mathrm{var}\{N^{1/2}(\bar{\lambda}_N - \lambda^*)\} \approx (\mathbb{E}_{\lambda_0}\{Z\nabla^2 W(Z, \lambda^*)\})^{-2}\mathbb{E}_{\lambda_0}\{Z^2(\nabla W(Z, \lambda^*))^2\}.$$

REFERENCES

Huber, P. J. (1964). "Robust estimation of a location parameter," *Ann. Math. Statist.*, **35**, 73–101.

Huber, P. J. (1981). *Robust Statistics*, Wiley, New York.

Katkovnik, V. Y. (1976). *Linear Estimation and Stochastic Optimization Problems (Parametric Smoothing Operators Method)*, Nauka, Moscow (in Russian).

Rubinstein, R. Y. (1986). *Monte Carlo Optimization Simulation and Sensitivity of Queueing Network*, Wiley, New York.

Rubinstein, R. Y. and Shapiro, A. (1993). *Discrete Event Systems: Sensitivity Analysis and Stochastic Optimization via the Score Function Method*, Wiley, New York.

CHAPTER 6

Sensitivity Analysis and Optimization of Discrete-Event Dynamic Systems: Distributional Parameters

6.1 INTRODUCTION

This chapter deals with sensitivity analysis and optimization of discrete-event dynamic systems (DEDS) and, in particular, with non-product-form queueing networks. As we shall see later, the results of this section extend those of Chapter 5 to discrete-event static systems (DESS). For example, in regenerative queues, the latter can be viewed as a particular case of DEDS with regenerative cycles of length one.

Assume that the DEDS under consideration is driven by an independent identically distributed (iid) sequence of random vectors $\boldsymbol{Y}_t$ generated from the joint probability density function (pdf) $f(\boldsymbol{y}, \boldsymbol{v})$, $\boldsymbol{v}$ being a vector of parameters (e.g., interarrival rates, service rates, and routing probabilities), called the *distributional parameter vector*. Assume further that the output process $\{L_t\}$ (e.g., sojourn times in a queueing model, the number of customers in the system, throughput, and utilization processes), which is called the *sample performance process*, is ergodic and settles into steady state. In practice, a DEDS is started from some initial state and is allowed to run until "reaching stationarity". Thereafter, T consecutive observations, $L_1, L_2, \ldots, L_T$, are taken to evaluate (estimate) the expected performance, $\ell(\boldsymbol{v})$, and the associated sensitivities or to optimize an entire queueing network, that is, to solve an optimization problem of the form (P_0) [see, e.g., the program (5.4.1) in Section 5.4]. In particular, we might be interested in evaluating the expected sojourn time in an open queueing network and the associated sensitivities, say, with respect to service rate parameters and routing probabilities. In addition, we might be interested in minimizing the expected sojourn time with respect to the service rate vector, subject to some cost constraints.

For Markovian product-form queueing networks, the program (P_0) can be solved by conventional methods of mathematical programming, since in this case both the objective function and the constraints are available analytically. Optimization of such Markovian networks is treated in Bertsekas and Gallager (1987) and

Kleinrock (1975). Here, however, the discussion is not restricted to product-form queueing networks; rather, it admits fairly general queueing networks with arbitrary distributions of interarrival and service times, finite buffer sizes, multiple servers, priorities, and so forth. The price paid is that the functions $\ell_j(\boldsymbol{v}) = \mathbb{E}_{\boldsymbol{v}}\{L_j\}$, $j = 0, 1, \ldots, M$ are no longer available analytically and one has to resort to simulation.

In this chapter we consider sensitivity analysis and optimization with respect to the distributional parameter vector $\boldsymbol{v}$, that is, we assume that the sample performance functions L_j [e.g., the service rate in a $GI/D/1$ queue, the (s, S) parameter vector in an inventory, and the (s, S) policy model] do not depend on a parameter vector, called the *structural parameter vector*. Chapter 7 extends the treatment of sensitivity analysis and stochastic optimization to structural parameters.

As in Chapter 3, a distinction is made between finite-horizon and steady-state simulations. We first show that sensitivity analysis and optimization of DEDS over a finite-horizon is similar to that of the static models discussed in Chapter 5.

Example 6.1.1 *GI/G/1* Queue. Consider estimation of performance measures of the form $\nabla^k \ell(\boldsymbol{v})$, $k = 0, 1, \ldots$, simultaneously, for different values of $\boldsymbol{v}$, where, say

$$\ell(\boldsymbol{v}) = \mathbb{E}_{\boldsymbol{v}}\left\{\frac{1}{q} \sum_{t=r+1}^{r+q} L_t(\underline{\boldsymbol{Y}}_t)\right\}, \tag{6.1.1}$$

for some given initial state of the queue, $\mathscr{L}(0)$ (say, the initial number of customers at time $t=0$). Here, $\underline{\boldsymbol{Y}}_t = (\boldsymbol{Y}_1, \ldots, \boldsymbol{Y}_t)$; $t = r+1, \ldots, r+q$, is a sequence of iid random vectors from the pdf $f(\boldsymbol{y}, \boldsymbol{v})$, q and r being fixed. If, for example, L_t is the sojourn time in the $GI/G/1$ queue, then (6.1.1) represents the average sojourn time of the k customers indexed contiguously by $r+1$ through $r+q$.

It is straightforward to see that in this case, the what-if estimator of $\nabla^k \ell(\boldsymbol{v})$ can be written, by analogy to (5.3.4), as

$$\bar{\nabla}^k \ell_N(\boldsymbol{v}) = \frac{1}{N} \sum_{i=1}^{N} \frac{1}{q} \sum_{t=r+1}^{r+q} L_{ti}(\underline{\boldsymbol{Z}}_{ti}) \nabla^k \tilde{W}_{ti}(\underline{\boldsymbol{Z}}_{ti}, \boldsymbol{v}), \tag{6.1.2}$$

where for $t = 1, \ldots, q$,

$$\tilde{W}_{ti}(\underline{\boldsymbol{Z}}_{ti}, \boldsymbol{v}) = \frac{f_t(\underline{\boldsymbol{Z}}_{ti}, \boldsymbol{v})}{g_t(\underline{\boldsymbol{Z}}_{ti})},$$

$$f_t(\underline{\mathbf{z}}_t, \boldsymbol{v}) = \prod_{j=r+1}^{r+t} f(\mathbf{z}_j, \boldsymbol{v}), \qquad g_t(\underline{\mathbf{z}}_t) = \prod_{j=r+1}^{r+t} g(\mathbf{z}_j),$$

and $\boldsymbol{Z}_{ti} \sim g(\mathbf{z})$.

The algorithm for estimating the performance measures, $\nabla^k \ell(\boldsymbol{v})$, simultaneously for multiple values $\boldsymbol{v}$, is similar to Algorithm 5.2.1 for DESS.

Algorithm 6.1.1 Finite-Horizon DEDS Simulation

1. *Generate a sample* $\{\mathbf{Z}_{1i}, \ldots, \mathbf{Z}_{ki}\}$, $i = 1, \ldots, N$, *from the dominating pdf,* $g(\mathbf{z})$.
2. *Generate the output processes* L_{ti} *and* $\nabla^k \tilde{W}_{ti}$, *starting each replication (simulation run) i,* $i = 1, \ldots, N$, *from the same initial state* $\mathscr{L}(0)$.
3. *Perform N independent replications and estimate* $\nabla^k \ell(\boldsymbol{v})$ *from (6.1.2).*

Consider another example.

Example 6.1.2 Insurance Risk Model. Assume that claims arrive according to a Poisson process $\{N_t : t \geq 0\}$ with rate $v_1 > 0$. Assume further that the claim sizes are iid non-negative random variables U_i, $i = 1, 2, \ldots$, with a common cumulative distribution function (cdf) $F(y, \boldsymbol{v}_2)$. Finally, assume that the sequences $\{U_i\}$ and $\{N_t\}$ are independent. Consider the following (compound Poisson) risk process with state-dependent premium,

$$R_t = u - A_t + \int_0^t p(R(s))\, ds, \tag{6.1.3}$$

where

$$A_t = \sum_{i=1}^{N_t} U_i$$

is the cumulative claim, u is the initial reserve, N_t is the number of claims in the interval (0, t), and p is the premium flow per unit time. A standard performance measure is the *ruin probability*, defined as

$$\ell(\boldsymbol{v}) = P_{\boldsymbol{v}}\left\{\inf_{t \geq 0} R(t) < 0\right\}, \tag{6.1.4}$$

where $\boldsymbol{v} = (v_1, \boldsymbol{v}_2)$. An estimator for $\nabla \ell(\boldsymbol{v})$, similar to (6.1.2), and an algorithm similar to Algorithm 6.1.1 are easily derived. For good sources on insurance risk models, see Asmussen (1985, 1996).

Consider now the program (P_0) [see, e.g., (5.4.1)] for DEDS in the finite-horizon regime. It is readily seen that in order to construct a stochastic counterpart $(\bar{P}_N)$, similar to that given in (5.4.4) for static models, we have to replace $\bar{\ell}_{jN}(\boldsymbol{v})$ in (5.4.3) by the corresponding estimators (6.1.2), with all other data unchanged. Note that Theorem 5.4.1 for DESS, concerning the convergence of the optimal solution, $\bar{\boldsymbol{v}}_N^*$ of the program $(\bar{P}_N)$, to the optimal solution, $\boldsymbol{v}^*$ of the program (P_0), can be readily modified for the DEDS case in the finite-horizon regime.

The organization of this chapter mirrors that of Chapter 5. Section 6.2 deals with sensitivity analysis of system performance, to wit, with estimating the performance $\ell(\boldsymbol{v})$ and the associated sensitivities $\nabla^k \ell(\boldsymbol{v})$, $k \geq 1$, simultaneously, for multiple values of $\boldsymbol{v}$. It is shown that if $\{L_t\}$ is a regenerative process, then $\nabla^k \ell(\boldsymbol{v})$ can be represented as the cross-covariance of two processes: the standard output process, $\{L_t\}$, and the so-called *kth order score process*. As its name suggests, the latter is based on the score function. A convenient recursive formula is presented for computing higher-order sensitivities via lower-order ones (for multiple values of the parameter vector $\boldsymbol{v}$) from a *single simulation run*.

Section 6.3 introduces the so-called *decomposable and truncated estimators*, but only treats the former. It is shown that although, in general, these estimators introduce some bias, the resulting variance reduction is still substantial as compared to the regenerative score function (SF) estimators of Section 6.2. An additional advantage of decomposable and truncated estimators (as compared to regenerative score function estimators) is that they can be applied in both regenerative and nonregenerative (e.g., batch-means) settings.

Section 6.4 shows how to optimize the program (P_0) for non-Markovian queueing models, again from a *single* simulation experiment. A distinction is made here between so-called *on-line* and *off-line* models. For on-line models, we replace the deterministic program (P_0) by its stochastic counterpart $(\bar{P}_N)$ (see below), similarly to the treatment of static models in Section 5.4. Solving $(\bar{P}_N)$ by standard deterministic procedures allows one to estimate the optimal solution, $\boldsymbol{v}^*$, of the program (P_0). For on-line models, we use adaptive procedures of stochastic approximation type.

Finally, Section 6.5 deals with network topological design or the so-called network *what-if* design problem, which can be stated as follows: What would the expected performance and the associated sensitivities of the network be if we perturbed the topology of the system, that is, if we add (eliminate) a node (several nodes) to (from) the network.

6.2 SENSITIVITY ANALYSIS OF SYSTEM PERFORMANCE

Let $\boldsymbol{Y}_1, \boldsymbol{Y}_2, \ldots,$ be an input sequence of iid m-dimensional random vectors, generated from a joint pdf $f(\boldsymbol{y}, \boldsymbol{v})$ that depends on some n-dimensional parameter vector $\boldsymbol{v}$.

Consider an output process, $\{L_t : t > 0\}$ driven by the input sequence $\{\underline{\boldsymbol{Y}}_t\}$. That is, $L_t = L_t(\underline{\boldsymbol{Y}}_t)$, where the vector $\underline{\boldsymbol{Y}}_t = (\boldsymbol{Y}_1, \boldsymbol{Y}_2, \ldots, \boldsymbol{Y}_t)$ represents a history of the input process up to time t, and $\{L_t(\cdot)\}$ is a sequence of real-valued functions. Assume that $\{L_t\}$ is a regenerative process with a regenerative cycle of length τ, a typical example being the steady-state waiting time process in the $GI/G/1$ queue with first-in/first-out (FIFO) discipline. In this case (see Section 3.7), the expected steady-state performance, $\ell(\boldsymbol{v})$, can be written as

$$\ell(\boldsymbol{v}) = \frac{\mathbb{E}_{\boldsymbol{v}}\{X\}}{\mathbb{E}_{\boldsymbol{v}}\{\tau\}}, \qquad X = \sum_{t=1}^{\tau} L_t. \tag{6.2.1}$$

It will now be shown how to estimate the performance $\ell(\boldsymbol{v})$, and the derivatives $\nabla^k \ell(\boldsymbol{v})$, $k = 1, 2, \ldots$, from a *single simulation run essentially at any point* $\boldsymbol{v}$, subject to certain regularity conditions. To this end, choose a cdf G on $\mathbb{R}^m$ with a corresponding pdf, $g(\mathbf{z})$. Assume that $g(\mathbf{z})$ dominates the densities $f(\mathbf{z}, \boldsymbol{v})$ in the sense that

$$\text{supp}\{f(\mathbf{z}, \boldsymbol{v})\} \subset \text{supp}\{g(\mathbf{z})\}, \qquad \boldsymbol{v} \in V,$$

where supp$\{f\}$ is the support of the function f.

Consider first $\ell_1(\boldsymbol{v}) = \mathbb{E}_{\boldsymbol{v}}\{X\}$. It will be shown that $\ell_1(\boldsymbol{v})$ can be represented as

$$\ell_1(\boldsymbol{v}) = \mathbb{E}_g\left\{\sum_{t=1}^{\tau} L_t(\underline{\boldsymbol{Z}}_t)\tilde{W}_t(\underline{\boldsymbol{Z}}_t, \boldsymbol{v})\right\}, \tag{6.2.2}$$

where

$$\underline{\boldsymbol{Z}}_t = (\boldsymbol{Z}_1, \ldots, \boldsymbol{Z}_t) \sim g_t(\underline{\mathbf{z}}_t) \quad \text{and} \quad \tilde{W}_t(\underline{\boldsymbol{Z}}_t, \boldsymbol{v}) = \prod_{j=1}^{t} f(\boldsymbol{Z}_j, \boldsymbol{v})/g(\boldsymbol{Z}_j).$$

To this end first write

$$\sum_{t=1}^{\tau} L_t = \sum_{t=1}^{\infty} L_t I_{\{\tau \ge t\}}. \tag{6.2.3}$$

The reader should note that the notation in (6.2.3) is abbreviated. More specifically, from here and on, and with a slight abuse of notation, we shall routinely abbreviate $L_t(\underline{\boldsymbol{Z}}_t)$ as L_t and $\tilde{W}_t(\underline{\boldsymbol{Z}}_t, \boldsymbol{v})$ as $\tilde{W}_t$. Furthermore, since $\tau = \tau(\underline{\boldsymbol{Z}}_t)$ is completely determined by $\underline{\boldsymbol{Z}}_t$, we shall use the notation $I_{\{\tau \ge t\}}(\underline{\mathbf{z}}_t)$, where $\underline{\mathbf{z}}_t$ is a realization of $\underline{\boldsymbol{Z}}_t$; this again constitutes a slight abuse of notation, since formally the ordinary $I_{\{\tau \ge t\}}(\omega)$ is a function of the sample point ω, rather than of $\underline{\mathbf{z}}_t$. Accordingly, the expectation of $L_t I_{\{\tau \ge t\}}$ is

$$\begin{aligned}
\mathbb{E}_{\boldsymbol{v}}\{L_t I_{\{\tau \ge t\}}\} &= \int L_t(\underline{\mathbf{z}}_t) I_{\{\tau \ge t\}}(\underline{\mathbf{z}}_t) f_t(\underline{\mathbf{z}}_t, \boldsymbol{v})\, d\underline{\mathbf{z}}_t \\
&= \int L_t(\underline{\mathbf{z}}_t) I_{\{\tau \ge t\}}(\underline{\mathbf{z}}_t) \tilde{W}_t(\underline{\mathbf{z}}_t, \boldsymbol{v}) g_t(\underline{\mathbf{z}}_t)\, d\underline{\mathbf{z}}_t \\
&= \mathbb{E}_g\{L_t(\underline{\boldsymbol{Z}}_t) I_{\{\tau \ge t\}}(\underline{\boldsymbol{Z}}_t) \tilde{W}_t(\underline{\boldsymbol{Z}}_t, \boldsymbol{v})\},
\end{aligned} \tag{6.2.4}$$

where

$$\tilde{W}_t(\underline{\mathbf{z}}_t, \boldsymbol{v}) = \frac{f_t(\underline{\mathbf{z}}_t, \boldsymbol{v})}{g_t(\underline{\mathbf{z}}_t)}$$

and

$$f_t(\underline{\mathbf{z}}_t, \boldsymbol{v}) = \prod_{i=1}^{t} f(\mathbf{z}_i, \boldsymbol{v}), \qquad g_t(\underline{\mathbf{z}}_t) = \prod_{i=1}^{t} g(\mathbf{z}_i).$$

The result follows by combining (6.2.3) and (6.2.4).

For the special case where $L_t \equiv 1$, (6.2.2) reduces to

$$\ell_2(\boldsymbol{v}) = \mathbb{E}_{\boldsymbol{v}}\{\tau\} = \mathbb{E}_g\left\{\sum_{t=1}^{\tau} \tilde{W}_t(\underline{\boldsymbol{Z}}_t, \boldsymbol{v})\right\}.$$

Under standard regularity conditions ensuring the interchangeability of the differentiation and expectation operators, one can write

$$\nabla^k \ell_1(\boldsymbol{v}) = \mathbb{E}_g\left\{\sum_{t=1}^{\tau} L_t(\underline{\boldsymbol{Z}}_t)\nabla^k \tilde{W}_t(\underline{\boldsymbol{Z}}_t, \boldsymbol{v})\right\}. \tag{6.2.5}$$

For $k = 1, 2, \ldots, \boldsymbol{G}_t^k = L_t(\underline{\boldsymbol{Z}}_t)\nabla^k \tilde{W}_t(\underline{\boldsymbol{Z}}_t, \boldsymbol{v})$ and $\nabla^k \tilde{W}_t$ are called the *sensitivity process* and the *generalized score function process of order k*, respectively, while for $k=0$, $\boldsymbol{G}_{\underline{t}}^0 = L_t \nabla^0 \tilde{W}_t = L_t \tilde{W}_t$ is called *the sample performance process* and $\nabla^0 \tilde{W}_t = \tilde{W}_t$ is called the *likelihood ratio process*.

Let now $\{\boldsymbol{Z}_{11}, \ldots, \boldsymbol{Z}_{\tau_1 1}, \ldots, \boldsymbol{Z}_{1N}, \ldots, \boldsymbol{Z}_{\tau_N N}\}$ be a sample of N-regenerative cycles from the pdf $g(\mathbf{z})$. Then by virtue of (6.2.5), we can estimate $\nabla^k \ell_1(\boldsymbol{v})$, $k = 0, 1, \ldots,$ from a *single simulation run* by

$$\bar{\nabla}^k \ell_{1N}(\boldsymbol{v}) = \frac{1}{N}\sum_{i=1}^{N}\sum_{t=1}^{\tau_i} L_{ti}(\underline{\boldsymbol{Z}}_{ti})\nabla^k \tilde{W}_{ti}(\underline{\boldsymbol{Z}}_{ti}, \boldsymbol{v}), \tag{6.2.6}$$

where

$$\tilde{W}_{ti} = \prod_{j=1}^{t} W_{ji}, \qquad W_{ji}(Z_{ji}, \boldsymbol{v}) = \frac{f(Z_{ji}, \boldsymbol{v})}{g(Z_{ji})}$$

and $Z_{ji} \sim g(\underline{z})$. Notice that here $\bar{\nabla}^k \ell_{1N}(\boldsymbol{v}) = \nabla^k \bar{\ell}_{1N}(\boldsymbol{v})$ with the convention $\bar{\nabla}^0 \ell_{1N}(\boldsymbol{v}) \equiv \bar{\ell}_{1N}(\boldsymbol{v})$.

Consider now the estimation of $\nabla^k \ell_1(\boldsymbol{v})$ for the special case where $g(y) = f(y, \boldsymbol{v})$, that is, when using the original probability measure, $f(y, \boldsymbol{v})$. One has

$$\nabla^k \ell_1(\boldsymbol{v}) = \mathbb{E}_{\boldsymbol{v}}\left\{\sum_{t=1}^{\tau} L_t \tilde{\boldsymbol{S}}_t^{(k)}\right\}, \tag{6.2.7}$$

where $\tilde{\boldsymbol{S}}_t^{(k)} = \nabla^k f_t(\underline{\boldsymbol{Y}}_t, \boldsymbol{v}) / f_t(\underline{\boldsymbol{Y}}_t, \boldsymbol{v})$ is called *the score function process of order k*. For $k = 1$, the score function process reduces to

$$\tilde{\boldsymbol{S}}_t^{(1)} = \sum_{j=1}^{t} \nabla \log f(\boldsymbol{Y}_j, \boldsymbol{v}) \tag{6.2.8}$$

and is then called simply *the score function process*. Note that for $k = 1, 2, \ldots,$ we have

$$\nabla^k \tilde{W}_t = \tilde{W}_t \tilde{\boldsymbol{S}}_t^{(k)}.$$

Example 6.2.1. Let $\underline{Y}_t = (Y_1, \ldots, Y_t)$ be a sequence of iid variates, each distributed geometrically with common pdf

$$P(y, p) = p(1-p)^{y-1}, \qquad y = 1, 2, \ldots$$

Then,

$$\tilde{S}_t^{(1)}(\underline{Y}_t, p) = \frac{\partial}{p} \log f_t(\underline{Y}_t, p) = \frac{t - p\sum_{j=1}^{t} Y_j}{p(1-p)}.$$

Example 6.2.2. Let $\underline{Y}_t = (Y_1, \ldots, Y_t)$ be a sequence of iid variates, each distributed gamma with pdf

$$f(y, \lambda, \beta) = \frac{\lambda^{\beta} y^{\beta-1} e^{-\lambda y}}{\Gamma(\beta)}, \qquad y > 0.$$

Suppose we are interested in the sensitivities with respect to λ. Then,

$$\tilde{S}_t^{(1)}(\underline{Y}_t, \lambda) = \frac{\partial}{\lambda} \log f_t(\underline{Y}_t, \lambda, \beta) = \frac{t\beta}{\lambda} - \sum_{i=1}^{t} Y_i.$$

Note that for $\tau = 1$, (6.2.7) corresponds to the kth derivative of the expected system performance for *static queueing models*. As examples of such static models, consider the $GI/G/1/0$ queue (without waiting room), the $GI/G/\infty$ queue (the infinite-server queue), or a closed queueing network with a single customer circulating in it.

Let us turn now to $\ell(\boldsymbol{v}) = \mathbb{E}_{\boldsymbol{v}}\{X\}/\mathbb{E}_{\boldsymbol{v}}\{\tau\}$. In view of (6.2.5) and the fact that $\tau_i = \sum_{t=1}^{\tau_i}$ is a special case of $X = \sum_{t=1}^{\tau} L_t$ with $L_t \equiv 1$, one can write $\ell(\boldsymbol{v})$ and $\nabla\ell(\boldsymbol{v})$ as

$$\ell(\boldsymbol{v}) = \frac{\mathbb{E}_g\{\sum_{t=1}^{\tau} L_t \tilde{W}_t\}}{\mathbb{E}_g\{\sum_{t=1}^{\tau} \tilde{W}_t\}} \tag{6.2.9}$$

and

$$\nabla\ell(\boldsymbol{v}) = \frac{\mathbb{E}_g\{\sum_{t=1}^{\tau} L_t \nabla\tilde{W}_t\}}{\mathbb{E}_g\{\sum_{t=1}^{\tau} W_t\}} - \frac{\mathbb{E}_g\{\sum_{t=1}^{\tau} L_t \tilde{W}_t\}}{\mathbb{E}_g\{\sum_{t=1}^{\tau} \tilde{W}_t\}} \times \frac{\mathbb{E}_g\{\sum_{t=1}^{\tau} \nabla\tilde{W}_t\}}{\mathbb{E}_g\{\sum_{t=1}^{\tau} \tilde{W}_t\}}, \tag{6.2.10}$$

respectively [observe that $\tilde{W} = \tilde{W}_t(\underline{\boldsymbol{Z}}_t, \boldsymbol{v})$ is a function of $\boldsymbol{v}$, but $L_t = L_t(\underline{\boldsymbol{Z}}_t)$ is not]. Higher-order partial derivatives with respect to parameters of interest can then be obtained from (6.2.10).

Utilizing (6.2.9) and (6.2.10), one can estimate $\ell(\boldsymbol{v})$ and $\nabla\ell(\boldsymbol{v})$, for all $\boldsymbol{v}$, as

$$\ell_N(\boldsymbol{v}) = \frac{\sum_{i=1}^{N} \sum_{t=1}^{\tau_i} L_{ti} \tilde{W}_{ti}}{\sum_{i=1}^{N} \sum_{t=1}^{\tau_i} \tilde{W}_{ti}} \tag{6.2.11}$$

and

$$\bar{\nabla}\ell_N(\boldsymbol{v}) = \frac{\sum_{i=1}^{N} \sum_{t=1}^{\tau_i} L_{ti} \nabla\tilde{W}_{ti}}{\sum_{i=1}^{N} \sum_{t=1}^{\tau_i} \tilde{W}_{ti}} - \frac{\sum_{i=1}^{N} \sum_{t=1}^{\tau_i} L_{ti} \tilde{W}_{ti}}{\sum_{i=1}^{N} \sum_{t=1}^{\tau_i} \tilde{W}_{ti}} \times \frac{\sum_{i=1}^{N} \sum_{t=1}^{\tau_i} \nabla\tilde{W}_{ti}}{\sum_{i=1}^{N} \sum_{t=1}^{\tau_i} \tilde{W}_{ti}}, \tag{6.2.12}$$

respectively, and similarly for higher-order derivatives. Notice again that in this case, $\bar{\nabla}\ell_N(\boldsymbol{v}) = \nabla\bar{\ell}_N(\boldsymbol{v})$, where $\nabla\bar{\ell}_N(\boldsymbol{v})$ denotes the gradient of the estimator $\bar{\ell}_N(\boldsymbol{v})$.

The algorithm for estimating the gradient $\nabla\ell(\boldsymbol{v})$ by means of the sensitivity estimator (6.2.12), for all $\boldsymbol{v}$, can be written as follows.

Algorithm 6.2.1 $\nabla\ell(\boldsymbol{v})$ Estimation

1. *Generate a random sample* $\{\boldsymbol{Z}_1, \ldots, \boldsymbol{Z}_T\}$, $T = \sum_{i=1}^{N} \tau_i$, *from* $g(\mathbf{z})$.
2. *Generate the output processes* $\{L_t\}$ *and* $\{\nabla\tilde{W}_t\} = \{\tilde{W}_t \tilde{\boldsymbol{S}}_t^{(1)}\}$.
3. *Calculate* $\nabla\bar{\ell}_N(\boldsymbol{v}) = \bar{\nabla}\ell_N(\boldsymbol{v})$ *from (6.2.12).*

Confidence intervals (regions) for the sensitivities $\nabla^k\ell(\boldsymbol{v})$, $k = 0, 1$, utilizing the SF estimators $\nabla^k\bar{\ell}_N(\boldsymbol{v})$, $k = 0, 1$, can be derived analogously to those for the standard regenerative estimator (3.7.2) of Chapter 3, and are left as an exercise. For the special case where $g(\boldsymbol{y}) = f(\boldsymbol{y}, \boldsymbol{v})$, (6.2.10) reduces to

$$\begin{aligned}\nabla\ell(\boldsymbol{v}) &= \frac{\nabla\mathbb{E}_{\boldsymbol{v}}\{X\}}{\mathbb{E}_{\boldsymbol{v}}\{\tau\}} - \frac{\mathbb{E}_{\boldsymbol{v}}\{X\}}{\mathbb{E}_{\boldsymbol{v}}\{\tau\}} \times \frac{\nabla\mathbb{E}_{\boldsymbol{v}}\{\tau\}}{\mathbb{E}_{\boldsymbol{v}}\{\tau\}} \\ &= \frac{\mathbb{E}_{\boldsymbol{v}}\left\{\sum_{t=1}^{\tau} L_t \tilde{\boldsymbol{S}}_t^{(1)}\right\}}{\mathbb{E}_{\boldsymbol{v}}\{\tau\}} - \frac{\mathbb{E}_{\boldsymbol{v}}\{\sum_{t=1}^{\tau} L_t\}}{\mathbb{E}_{\boldsymbol{v}}\{\tau\}} \times \frac{\mathbb{E}_{\boldsymbol{v}}\left\{\sum_{t=1}^{\tau} \tilde{\boldsymbol{S}}_t^{(1)}\right\}}{\mathbb{E}_{\boldsymbol{v}}\{\tau\}}.\end{aligned} \tag{6.2.13}$$

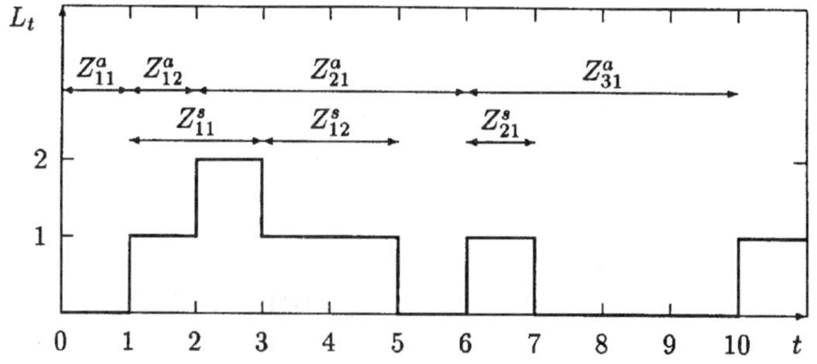

Figure 6.1. Number of customers process, $\{L_t\}$, in a $GI/G/1$ queue.

Example 6.2.3. Figure 6.1 depicts two complete cycles, $(X_1, \tau_1) = (5, 5)$ and $(X_2, \tau_2) = (1, 4)$, for the number of customers, L_t, in a $GI/G/1$ queue over the period $0 \le t \le 10$. The total number of customers that arrived and were served during the time interval [0, 10] is $N_a = 4$ and $N_s = 3$, respectively. Assume that the sojourn times, corresponding to those three customers (in the two cycles) are $\{L_1, L_2, L_3\} \equiv \{L_{11}, L_{12}, L_{21}\} = \{2, 3, 1\}$. As a simple illustration, suppose that the interarrival and service times are $4 + Y^a$, $Y^a \sim \text{Ber}(p_a)$ and $1 + Y^s$, $Y^s \sim \text{Ber}(p_s)$, with the dominating pdf's $g_a(z)$ and $g_s(z)$ corresponding to the following interarrival and service times random variables: $4 + Z^a$, $Z^a \sim \text{Ber}(p_{a0} = 0.4)$ and $1 + Z^s$, $Z^s \sim \text{Ber}(p_{s0} = 0.2)$. Assume next that the sequence of random variables generated from $g_a(z)$ and $g_s(z)$ for those three customers are $\{Z^a_1, Z^a_2, Z^a_3, Z^a_4\} \equiv \{Z^a_{11}, Z^a_{12}, Z^a_{21}, Z^a_{31}\} = \{0, 0, 1, 1\}$ and $\{Z^s_1, Z^s_2, Z^s_3\} \equiv \{Z^s_{11}, Z^s_{12}, Z^s_{21}\} = \{1, 1, 0\}$, respectively. Our goal is to estimate by simulation: (i) the expected steady-state sojourn time, and (ii) the expected steady-state number of customers in the system, and the associated derivatives, *simultaneously* for $p_a = 0.5$, $p_s = 0.1$ and $p_a = 0.55$, $p_s = 0.15$. We present the calculations for $p_a = 0.5$, $p_s = 0.1$ only. The reader can readily repeat them for $p_a = 0.55$, $p_s = 0.15$.

6.2.1 Estimation of Expected Sojourn Time

It is not difficult to see that in this case

$$\tilde{W}_{it} = \begin{cases} \prod_{j=1}^{t} W^s_{ij} \prod_{j=2}^{t} W^a_{ij} & \text{if } t > 1 \\ W^s_{i1} & \text{if } t = 1, \end{cases}$$

$$W^a_{ij} = \frac{p_a^{Z^a_{ij}} (1 - p_a)^{1 - Z^a_{ij}}}{p_{a0}^{Z^a_{ij}} (1 - p_{a0})^{1 - Z^a_{ij}}}, \qquad W^s_{ij} = \frac{p_s^{Z^s_{ij}} (1 - p_s)^{1 - Z^s_{ij}}}{p_{s0}^{Z^s_{ij}} (1 - p_{s0})^{1 - Z^s_{ij}}},$$

$$\nabla_{p_a} \tilde{W}_{it} = \begin{cases} \tilde{W}_{it} \dfrac{\sum_{j=2}^{t} Z^a_{ij} - (t - 1) p_a}{p_a (1 - p_a)} & \text{if } t > 1 \\ 0 & \text{if } t = 1 \end{cases}$$

and

$$\nabla_{p_s}\tilde{W}_{it} = \tilde{W}_{it}\frac{\sum_{j=1}^{t} Z_{ij}^s - tp_s}{p_s(1-p_s)},$$

where $Z_{ij}^a \sim \text{Ber}(p_{a0} = 0.4)$ and $Z_{ij}^s \sim \text{Ber}(p_{s0} = 0.2)$. Here the symbols ∇_{p_a} and ∇_{p_s} stand for the derivatives with respect to p_a and p_s, respectively. Note that $W_{i1}^a \equiv 1$.

For the first cycle, one has $\{Z_{11}^a, Z_{12}^a\} = \{0, 0\}$ and $\{Z_{11}^s, Z_{12}^s\} = \{1, 1\}$. The corresponding likelihood ratios are

$$\begin{aligned}
\tilde{W}_{11} &= \frac{0.1^{Z_{11}^s}(1-0.1)^{1-Z_{11}^s}}{0.2^{Z_{11}^s}(1-0.2)^{1-Z_{11}^s}} = 0.5,\\
\tilde{W}_{12} &= W_{11}^s W_{12}^s W_{12}^a\\
&= 0.5\frac{0.1^{Z_{12}^s}(1-0.1)^{1-Z_{12}^s}}{0.2^{Z_{12}^s}(1-0.2)^{1-Z_{12}^s}} \times \frac{0.5^{Z_{12}^a}(1-0.5)^{1-Z_{12}^a}}{0.4^{Z_{12}^a}(1-0.4)^{1-Z_{12}^a}}\\
&= (0.5)(0.5)(\tfrac{5}{6}) \approx 0.21
\end{aligned}$$

and

$$\begin{aligned}
\nabla_{p_s}\tilde{W}_{11} &= \tilde{W}_{11}\frac{1-0.1}{0.1(1-0.1)} = (0.5)10 = 5,\\
\nabla_{p_a}\tilde{W}_{12} &= \tilde{W}_{12}\frac{0-0.5}{0.5(1-0.5)} \approx 0.21(-2) = -0.42,\\
\nabla_{p_s}\tilde{W}_{12} &= \tilde{W}_{12}\frac{\sum_{j=1}^{2} Z_{ij}^s - 2p_s}{p_s(1-p_s)} \approx 0.21\frac{2-0.1(2)}{0.1(1-0.1)} = 0.21(20) = 4.2.
\end{aligned}$$

Similarly, for the second cycle, one has

$$\tilde{W}_{21} = 1.125, \qquad \nabla_{p_s}\tilde{W}_{21} = -1.25.$$

To calculate the resulting estimators [see (6.2.11) and (6.2.12)]

$$\begin{aligned}
\bar{\ell}_N(p) &= \frac{\bar{\ell}_{1N}(p)}{\bar{\ell}_{2N}(p)},\\
\nabla\bar{\ell}_N(p) &= \frac{\nabla\bar{\ell}_{1N}(p)\bar{\ell}_{2N}(p) - \nabla\bar{\ell}_{2N}(p) \times \bar{\ell}_{1N}(p)}{[\bar{\ell}_{2N}(p)]^2}
\end{aligned}$$

we need to calculate separately $\bar{\ell}_{1N}(p)$, $\bar{\ell}_{2N}(p)$, $\nabla\bar{\ell}_{1N}(p)$, and $\nabla\bar{\ell}_{2N}(p)$. Those are

$$\begin{aligned}
\bar{\ell}_{1N}(p) &= \frac{1}{2}\sum_{i=1}^{2}\sum_{j=1}^{\tau_i} L_{ij}\tilde{W}_{ij} \\
&\approx \tfrac{1}{2}[(2 \times 0.5 + 2 \times 0.21) + (1 \times 1.125)] = 1.27, \\
\nabla_{p_a}\bar{\ell}_{1N}(p) &= \frac{1}{2}\sum_{i=1}^{2}\sum_{j=1}^{\tau_i} L_{ij} \times \nabla_{p_a}\tilde{W}_{ij} \\
&\approx \tfrac{1}{2}[(2 \times 0 + 2 \times (-0.42)) + (1 \times 0)] = -0.42, \\
\nabla_{p_s}\bar{\ell}_{1N}(p) &= \frac{1}{2}\sum_{i=1}^{2}\sum_{j=1}^{\tau_i} L_{ij} \times \nabla_{p_s}\tilde{W}_{ij} \\
&\approx \tfrac{1}{2}[(2 \times 5 + 2 \times 4.2) + (1 \times (-1.25))] = 8.575, \\
\bar{\ell}_{2N}(p) &= \frac{1}{2}\sum_{i=1}^{2}\sum_{j=1}^{\tau_i} \tilde{W}_{ij} \\
&\approx \tfrac{1}{2}[(0.5 + 0.21) + (1.125)] = 0.918, \\
\nabla_{p_a}\bar{\ell}_{2N}(p) &= \frac{1}{2}\sum_{i=1}^{2}\sum_{j=1}^{\tau_i} \nabla_{p_a}\tilde{W}_{ij} \\
&\approx \tfrac{1}{2}[0 - 0.42 + 0] = -0.21, \\
\nabla_{p_s}\bar{\ell}_{2N}(p) &= \frac{1}{2}\sum_{i=1}^{2}\sum_{j=1}^{\tau_i} \nabla_{p_s}\tilde{W}_{ij} \\
&\approx \tfrac{1}{2}[5 + 4.2 - 1.25] = 3.975.
\end{aligned}$$

Finally,

$$\begin{aligned}
\bar{\ell}_N(p) &\approx \frac{1.27}{0.918} \approx 1.38 \\
\nabla_{p_a}\bar{\ell}_N(p) &\approx \frac{-0.42 \times 0.918 + 0.21 \times 1.27}{0.918^2} \approx -0.14, \\
\nabla_{p_s}\bar{\ell}_N(p) &\approx \frac{8.575 \times 0.918 - 3.975 \times 1.27}{0.918^2} \approx 3.35.
\end{aligned}$$

6.2.2 Estimation of Expected Number of Customers in System

In this case the processes, $\tilde{W}_t$ and $\nabla\tilde{W}_t$, are continuous (see Figure 6.2.). Let $t_{i1}, \ldots, t_{iT_i}$ be times of consecutive events in cycle i. By an event we mean an arrival of a customer in the system or departure of any customer from the system. Here t_{i1} corresponds to the time when cycle i begins (arrival of a customer at an

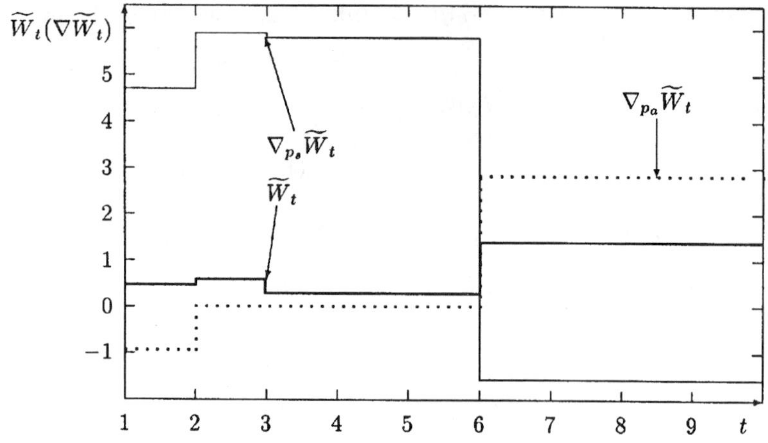

Figure 6.2. LR and SF processes, $\tilde{W}_t$ and $\nabla\tilde{W}_t$, in $GI/G/1$ queue.

empty system) and t_{iT_i} corresponds to the time when cycle i ends. Clearly, $t_{iT_i} = t_{(i+1)1}$. In our case,

$$T_1 = 5, \qquad T_2 = 3,$$

$$t_{11} = 1, \qquad t_{12} = 2, \qquad t_{13} = 3, \qquad t_{14} = 5, \qquad t_{15} = 6,$$

$$t_{21} = 6, \qquad t_{22} = 7, \qquad t_{23} = 10.$$

Then L_t, $\tilde{W}_t$, $\nabla_{p_a}\tilde{W}_t$, and $\nabla_{p_s}\tilde{W}_t$ remain constant over each interval

$$(t_{ij}, t_{i(j+1)}), \qquad i = 1, 2; \qquad j = 1, \ldots, T_i - 1.$$

[Note also that $\tilde{W}_t$, $\nabla_{p_a}\tilde{W}_t$ and $\nabla_{p_s}\tilde{W}_t$ do not change over intervals $(t_{i(T_i-2)}, t_{i(T_i-1)})$ and $(t_{i(T_i-1)}, t_{iT_i})$, $i = 1, 2$.]

Denote by L_{ij}, $\tilde{W}_{ij}$, $\nabla_{p_a}\tilde{W}_{ij}$, and $\nabla_{p_s}\tilde{W}_{ij}$ the corresponding values of L_t, $\tilde{W}_t$, $\nabla_{p_a}\tilde{W}_t$, and $\nabla_{p_s}\tilde{W}_t$ on $(t_{ij},\ t_{i(j+1)})$. Then,

$$\begin{aligned}
\tilde{W}_{11} &= W^a_{12}W^s_{11} = \tfrac{5}{6}0.5 \approx 0.47, \\
\tilde{W}_{12} &= \tilde{W}_{11}W^a_{21} \approx 0.47 \times 1.25 \approx 0.59, \\
\tilde{W}_{13} &= \tilde{W}_{12}W^s_{12} \approx 0.59 \times 0.5 \approx 0.29,
\end{aligned}$$

$$\nabla_{p_a}\tilde{W}_{11} = \tilde{W}_{11}\frac{Z_{12}^a - p_a}{p_a(1-p_a)} \approx 0.47(-2) \approx -0.94,$$
$$\nabla_{p_s}\tilde{W}_{11} = \tilde{W}_{11}\frac{Z_{11}^s - p_s}{p_s(1-p_s)} \approx 0.47 \times 10 \approx 4.7,$$
$$\nabla_{p_a}\tilde{W}_{12} = \tilde{W}_{12}\frac{Z_{12}^a + Z_{21}^a - 2p_a}{p_a(1-p_a)} = 0.59 \times 0 = 0,$$
$$\nabla_{p_s}\tilde{W}_{12} = \tilde{W}_{12}\frac{Z_{11}^s - p_s}{p_s(1-p_s)} = 0.59 \times 10 = 5.9,$$
$$\nabla_{p_a}\tilde{W}_{13} = \tilde{W}_{13}\frac{Z_{12}^a + Z_{21}^a - 2p_a}{p_a(1-p_a)} = 0.29 \times 0 = 0,$$
$$\nabla_{p_s}\tilde{W}_{13} = \tilde{W}_{13}\frac{Z_{11}^s + Z_{22}^s - 2p_s}{p_s(1-p_s)} = 0.29 \times 20 = 5.8,$$
$$\tilde{W}_{21} = W_{31}^a W_{21}^s = 1.25 \times 1.125 \approx 1.41,$$
$$\nabla_{p_a}\tilde{W}_{21} = \tilde{W}_{21}\frac{Z_{31}^a - p_a}{p_a(1-p_a)} \approx 1.41 \times 2 \approx 2.82,$$
$$\nabla_{p_s}\tilde{W}_{21} = \tilde{W}_{21}\frac{Z_{21}^s - p_s}{p_s(1-p_s)} \approx 1.41(-1.11) \approx -1.57.$$

The resulting quantities $\bar{\ell}_{1N}(p)$, $\bar{\ell}_{2N}(p)$, $\nabla\bar{\ell}_{1N}(p)$, $\nabla\bar{\ell}_{2N}(p)$, $\bar{\ell}(p)$, and $\nabla\bar{\ell}(p)$ are

$$\begin{aligned}
\bar{\ell}_{1N}(p) &= \frac{1}{2}\sum_{i=1}^{2}\int_{t_{i1}}^{t_{iT_i}} L_t\tilde{W}_t\,dt = \frac{1}{2}\sum_{i=1}^{2}\sum_{j=1}^{T_i-1} L_{ij}\tilde{W}_{ij}(t_{i(j+1)} - t_{ij}) \\
&\approx \tfrac{1}{2}[(1 \times 0.47 \times 1 + 2 \times 0.59 \times 1 + 1 \times 0.29 \times 2 + 0 \times 0.29 \times 1) \\
&\quad + (1 \times 1.41 \times 1 + 0 \times 1.41 \times 3)] = 1.82,
\end{aligned}$$

$$\begin{aligned}
\bar{\ell}_{2N}(p) &= \frac{1}{2}\sum_{i=1}^{2}\int_{t_{i1}}^{t_{iT_i}} \tilde{W}_t\,dt = \frac{1}{2}\sum_{i=1}^{2}\sum_{j=1}^{T_i-1} \tilde{W}_{ij}(t_{i(j+1)} - t_{ij}) \\
&\approx \tfrac{1}{2}[(0.47 \times 1 + 0.59 \times 1 + 0.29 \times 3) + (1.41 \times 4)] = 7.57,
\end{aligned}$$

$$\begin{aligned}
\nabla_{p_a}\bar{\ell}_{1N}(p) &= \frac{1}{2}\sum_{i=1}^{2}\int_{t_{i1}}^{t_{iT_i}} L_t\nabla_{p_a}\tilde{W}_t\,dt = \frac{1}{2}\sum_{i=1}^{2}\sum_{j=1}^{T_i-1} L_{ij}\nabla\tilde{W}_{ij}(t_{i(j+1)} - t_{ij}) \\
&\approx \tfrac{1}{2}[(1(-0.94)1 + 2 \times 0 \times 1 + 1 \times 0 \times 2 + 0 \times 0 \times 1) \\
&\quad + (1 \times 2.82 \times 1 + 0 \times 2.82 \times 3)] = 0.94,
\end{aligned}$$

$$\begin{aligned}
\nabla_{p_s}\bar{\ell}_{1N}(p) &= \frac{1}{2}\sum_{i=1}^{2}\int_{t_{i1}}^{t_{iT_i}} L_t\nabla_{p_s}\tilde{W}_t\,dt = \frac{1}{2}\sum_{i=1}^{2}\sum_{j=1}^{T_i-1} L_{ij}\nabla\tilde{W}_{ij}(t_{i(j+1)} - t_{ij}) \\
&\approx \tfrac{1}{2}[(1 \times 4.7 \times 1 + 2 \times 5.9 \times 1 + 1 \times 5.8 \times 2 + 0 \times 5.8 \times 1) \\
&\quad + (1(-1.57)1 + 0(-1.57)3)] = 13.27,
\end{aligned}$$

$$\begin{aligned}
\nabla_{p_a}\bar{\ell}_{2N}(p) &= \frac{1}{2}\sum_{i=1}^{2}\int_{t_{i1}}^{t_{iT_i}} \nabla_{p_a}\tilde{W}_t = \frac{1}{2}\sum_{i=1}^{2}\sum_{j=1}^{T_i-1} \nabla\tilde{W}_{ij}(t_{i(j+1)} - t_{ij}) \\
&\approx \tfrac{1}{2}[-0.94 + 2.82] = 0.94,
\end{aligned}$$

$$\nabla_{p_s}\bar{\ell}_{2N}(p) = \frac{1}{2}\sum_{i=1}^{2}\int_{t_{i1}}^{t_{iT_i}} \nabla_{p_s}\tilde{W}_t = \frac{1}{2}\sum_{i=1}^{2}\sum_{j=1}^{T_i-1} \nabla\tilde{W}_{ij}(t_{i(j+1)} - t_{ij})$$
$$\approx \tfrac{1}{2}[4.7 \times 1 + 5.9 \times 15.8 \times 3 - 1.57 \times 4] = 10.86,$$

so that

$$\bar{\ell}_N(p) = \frac{\bar{\ell}_{1N}(p)}{\bar{\ell}_{2N}(p)} \approx \frac{1.82}{7.57} \approx 0.24,$$

and

$$\nabla_{p_a}\bar{\ell}_N(p) = \frac{\nabla_{p_a}\bar{\ell}_{1N}(p) \times \bar{\ell}_{2N}(p) - \nabla_{p_a}\bar{\ell}_{2N}(p) \times \bar{\ell}_{1N}(p)}{[\bar{\ell}_{2N}(p)]^2} \approx 0.09,$$
$$\nabla_{p_s}\bar{\ell}_N(p) = \frac{\nabla_{p_s}\bar{\ell}_{1N}(p) \times \bar{\ell}_{2N}(p) - \nabla_{p_s}\bar{\ell}_{2N}(p) \times \bar{\ell}_{1N}(p)}{[\bar{\ell}_{2N}(p)]^2} \approx 1.41.$$

We now proceed to derive a more compact representation of formula (6.2.13). Denote by $\tilde{\boldsymbol{s}}^{(k)} = \tilde{\boldsymbol{s}}^{(k)}(\boldsymbol{v})$ the expected steady-state performance of the process $\tilde{\boldsymbol{S}}_t^{(k)}$, and observe that the processes $\tilde{\boldsymbol{S}}_t^{(k)}$ and $L_t\tilde{\boldsymbol{S}}_t^{(k)}$ are regenerative [for details see Asmussen and Rubinstein (1992)]. Renewal theory ensures that $\nabla\ell(\boldsymbol{v})$ can be represented as the expected steady-state performance of the process

$$\boldsymbol{Q}_t^{(1)} = (L_t - \ell(\boldsymbol{v}))\left(\tilde{\boldsymbol{S}}_t^{(1)} - \tilde{\boldsymbol{s}}^{(1)}\right), \tag{6.2.14}$$

where

$$\tilde{\boldsymbol{s}}^{(1)} = \frac{\mathbb{E}_{\boldsymbol{v}}\{\sum_{t=1}^{\tau} \tilde{\boldsymbol{S}}_t^{(1)}\}}{\mathbb{E}_{\boldsymbol{v}}\{\tau\}}.$$

Hence, we can rewrite (6.2.13) as

$$\nabla\ell(\boldsymbol{v}) = \mathbb{E}_{\boldsymbol{v}}\{\boldsymbol{Q}_t^{(1)}\} = \text{cov}_{\boldsymbol{v}}\{L_t, \tilde{\boldsymbol{S}}_t^{(1)}\} = \frac{\mathbb{E}_{\boldsymbol{v}}\{\sum_{t=1}^{\tau} \boldsymbol{Q}_t^{(1)}\}}{\mathbb{E}_{\boldsymbol{v}}\{\tau\}}. \tag{6.2.15}$$

Thus, the gradient, $\nabla\ell(\boldsymbol{v})$, can be expressed as the *covariance* of the steady-state sample performance process, $\{L_t\}$, and the steady-state score function process, $\{\boldsymbol{S}_t^{(1)}\}$, the latter being a function of the score function, $\nabla \log f(\boldsymbol{Y}, \boldsymbol{v})$. It is also important to note that the only difference between (6.2.1) and (6.2.15) is that L_t is replaced by

$$\boldsymbol{Q}_t^{(1)} = (L_t - \ell(\boldsymbol{v}))(\tilde{\boldsymbol{S}}_t^{(1)} - \tilde{\boldsymbol{s}}^{(1)}).$$

In the latter case, we say that the gradient $\nabla\ell(\boldsymbol{v})$ is *embedded* in the regenerative framework, in the sense that it is expressed as the ratio of the expected reward, $\mathbb{E}_{\boldsymbol{v}}\{\sum_{t=1}^{\tau} \boldsymbol{Q}_t^{(1)}\}$, obtained during a cycle of length τ, and the expected length of the cycle, $\mathbb{E}_v\{\tau\}$. Notice that the process $\{\boldsymbol{Q}_t^{(1)}\}$ is regenerative, being the product of two regenerative processes. Let $\tau = 1$ (as is the case, e.g., for the $GI/G/1/0$ queue). Then, $\tilde{\boldsymbol{S}}_t^{(1)} = \boldsymbol{S}_t^{(1)}$, and (6.2.13) reduces to

$$\nabla \ell(\boldsymbol{v}) = \mathbb{E}_v\{L\boldsymbol{S}^{(1)}\} = \mathrm{cov}_v\{L, \boldsymbol{S}^{(1)}\},$$

which coincides with the representation of $\nabla \ell(\boldsymbol{v})$ for *static models*.

Proceeding recursively with (6.2.15) and in view of (6.2.7), one obtains

$$\nabla^k \ell(\boldsymbol{v}) = \mathbb{E}_v\{\boldsymbol{Q}^{(k)}\} = \frac{\mathbb{E}_v\left\{\sum_{t=1}^{\tau} \boldsymbol{Q}_t^{(k)}\right\}}{\mathbb{E}_v\{\tau\}}, \qquad k = 0, 1, 2, \ldots, \tag{6.2.16}$$

where, by convention, $\nabla^0 \ell(\boldsymbol{v}) \equiv \ell(\boldsymbol{v})$ and $\boldsymbol{Q}_t^{(0)} \equiv L_t$.

Let $\{\boldsymbol{Y}_{11}, \ldots, \boldsymbol{Y}_{\tau_1 1}, \ldots, \boldsymbol{Y}_{1N}, \ldots, \boldsymbol{Y}_{\tau_N N}\}$ be a sample of N-regenerative cycles, generated from the pdf $f(\boldsymbol{y}, \boldsymbol{v})$. In view of (6.2.16), we can estimate all the quantities $\nabla^k \ell(\boldsymbol{v}),\ k = 0, 1, \ldots,$ from a *single* simulation run, by

$$\bar{\nabla}^k \ell_N(\boldsymbol{v}) = \frac{\sum_{i=1}^{N} \sum_{t=1}^{\tau_i} \tilde{\boldsymbol{Q}}_{ti}^{(k)}}{\sum_{i=1}^{N} \sum_{t=1}^{\tau_i} 1}, \qquad k = 0, 1, \tag{6.2.17}$$

where

$$\tilde{\boldsymbol{Q}}_{ti}^{(0)} = \boldsymbol{Q}_{ti}^{(0)} = L_{ti}, \qquad \tilde{\boldsymbol{Q}}_{ti}^{(k)} = (L_{ti} - \bar{\ell}_N)(\tilde{\boldsymbol{S}}_{ti}^{(k)} - \bar{\boldsymbol{s}}_N^{(k)}), \qquad k = 1, 2, \ldots. \tag{6.2.18}$$

Here, $\bar{\ell}_N$ and $\bar{\boldsymbol{s}}_N^{(1)}$ are the sample estimators of $\ell = \mathbb{E}_v\{\sum_{t=1}^{\tau} L_t\}/\mathbb{E}_v\{\tau\}$ and $\tilde{\boldsymbol{s}}^{(1)} = \mathbb{E}_v\{\sum_{t=1}^{\tau} \tilde{\boldsymbol{S}}_t^{(1)}\}/\mathbb{E}_v\{\tau\}$, respectively.

Example 6.2.4 The *GI/G/1* Queue. Consider Lindley's equation

$$L_{t+1} = \max\{0, L_t + U_t\}, \qquad L_0 = 0$$

for the waiting time of the tth customer in a $GI/G/1$ queue. Here $U_j = Y_{1j} - Y_{2(j+1)}$; Y_{1j} and Y_{2j} are the service and the interarrival times of the jth customer, respectively; $Y_{2j} = 0$ for $j = 1$ and $Y_{2j} = A_j - A_{j-1}$ for $j \geq 2$; and A_j is the arrival time of the jth customer. The steady-state expected waiting time, $\ell(\boldsymbol{v}) = \mathbb{E}_v\{L_t\}$, can be written as

$$\ell(\boldsymbol{v}) = \frac{\mathbb{E}_v\{\sum_{t=1}^{\tau} L_t\}}{\mathbb{E}_v\{\tau\}} = \frac{\mathbb{E}_v\left\{\sum_{t=1}^{\tau} \sum_{j=1}^{t-1} U_j\right\}}{\mathbb{E}_v\{\tau\}}, \tag{6.2.19}$$

where $\tau = \min\{t: \sum_{j=1}^{t} U_j \leq 0\}$ is the number of customers served during the busy period. We have,

$$\nabla \ell(\boldsymbol{v}) = \frac{\mathbb{E}_{\boldsymbol{v}}\left\{\sum_{t=1}^{\tau} \boldsymbol{Q}_t^{(1)}\right\}}{\mathbb{E}_{\boldsymbol{v}}\{\tau\}}, \tag{6.6.20}$$

where

$$\begin{aligned} \boldsymbol{Q}_t^{(1)} &= \left(\sum_{j=1}^{t-1} U_j - \ell\right)(\tilde{\boldsymbol{S}}_t^{(1)} - \tilde{\boldsymbol{s}}^{(1)}), \\ \tilde{\boldsymbol{S}}_t^{(1)} &= \sum_{j=1}^{t} \nabla \log f(\boldsymbol{Y}_j, \boldsymbol{v}), \qquad f(\boldsymbol{y}, \boldsymbol{v}) = f_1(y_1, v_1) f_2(y_2, v_2), \\ \boldsymbol{Y} &= (Y_1, Y_2), \qquad Y_1 \sim f_1(y_1, v_1), \qquad Y_2 \sim f_2(y_2, v_2). \end{aligned} \tag{6.2.21}$$

Note that $\tilde{\boldsymbol{Q}}_{ti}^{(1)}$ in (6.2.18) reduces to

$$\tilde{\boldsymbol{Q}}_{ti}^{(1)} = \left(\sum_{j=1}^{t-1} U_{ji} - \bar{\ell}_N\right)(\boldsymbol{S}_{ti}^{(1)} - \bar{\boldsymbol{s}}_N^{(1)}). \tag{6.2.22}$$

It is not difficult to see that (6.2.19) and (6.2.20) hold again with U_j defined as $U_j = Y_{1j} - Y_{2j}$, where $Y_{20} = 0$ and $Y_{2j} = A_{j+1} - A_j$ for $j \geq 1$.

Example 6.2.5 Production Inventory Model. Consider a single-commodity production-inventory model (PIM) that operates under the following (base-stock) policy: In each *fixed-length* period, production is set to a target level S (not exceeding the capacity s of the production facility), and so as to satisfy the demand in that period. It is assumed that the demand is either filled immediately or back-ordered. Let N_t denote the net inventory (on-hand inventory minus back orders) at the start of period t, and let D_t denote the demand in period t. Then, the production in period t is set to $\min\{s, S - N_t + D_t\}$ and the inventory at the end of period t is

$$N_{t+1} = N_t - D_t - \min\{s, S - N_t + D_t\} = \min\{N_t + s - D_t, S\}.$$

If demands are iid with $\mathbb{E}\{D_t\} < s$, then the net inventory process, $\{L_t\}$, converges in distribution to a random variable N satisfying

$$N_t \Longrightarrow \min\{N + s - D, S\}.$$

Recall that $\Longrightarrow$ denotes weak convergance and that the subscript ∞ was omitted in N and D. The process $\{N_t\}$ is regenerative, with regeneration points

$$\tau = \min\{t: N_t \geq S\}.$$

Let $f(y, \boldsymbol{v})$ be the common pdf of demands. The gradient $\nabla \ell(\boldsymbol{v})$ of the expected steady-state net inventory can be estimated via (6.2.17) and (6.2.18).

If, instead of net inventory, N_t, we consider $L_t = S - N_t$ (the amount by which the target inventory exceeds the net inventory, called *shortfall*), then

$$L_t = S - \min\{L_t + s - D_t, S\} = \max\{0, L_t + D_t - s\}. \tag{6.2.23}$$

Thus, the shortfall process satisfies a Lindley equation coinciding with the waiting time process in a $D/G/1$ queue with fixed interarrival time s and service time D_t. Clearly, in this case, the sensitivity process $\{Q_t^{(1)}\}$ coincides with (6.2.21), with $U_j = D_j - s$.

The accuracy (variance) of the estimator (6.2.17), for $k=1$, was studied in Asmussen and Rubinstein (1992). In particular, it was shown there that in heavy traffic ($\rho \to 1$), the score function process, $\tilde{\boldsymbol{S}}_t^{(1)}$, properly normalized, has a heavy traffic limit involving a certain variant of two-dimensional Brownian motion (see also below). A diffusion-based approximation is also derived for the variance of $\bar{\nabla}^k \ell_N(\boldsymbol{v})$.

The following example provides insight into the performance of the what-if estimators $\nabla^k \ell_N(\boldsymbol{v})$, $k = 0, 1$, in (6.2.11) and (6.2.12) and the response surface methodology (see Chapter 8).

Example 6.2.6. Let $\ell(\boldsymbol{v})$ be the steady-state expected waiting time in a $M/G/1$ queue, with the following two-point service time pdf

$$f(y, v) = \begin{cases} v & \text{if } y = 1.2 \\ 1 - v & \text{if } y = 0.2 \end{cases} \tag{6.2.24}$$

where $0 < v < 1$.

Table 6.2.1 displays how the SF estimator, $\nabla^k \bar{\ell}_N(\rho)$, performs for an $M/G/1$ queue, for various values of ρ, by exhibiting estimated values of the response surfaces $\nabla^k \ell(\rho)$, $k = 0, 1$ for different values of ρ, obtained from a simulation run with reference parameter $\rho_0 = 0.8$. Table 6.2.1 also displays the sample relative efficiency $\bar{\varepsilon}^k(\rho|\rho_0)$, $k = 0, 1$, defined analogously to $\bar{\varepsilon}^k(\rho, \rho_0)$ as

$$\bar{\varepsilon}^k(\rho|\rho_0) = \frac{\hat{\sigma}^2(\rho, k|\rho_0)}{\hat{\sigma}^2(\rho_0, k|\rho_0)}, \tag{6.2.25}$$

along with the sample performance, $\nabla^k \bar{\ell}_N(\rho|\rho_0)$, the sample variance $\hat{\sigma}^2(\rho, k|\rho_0)$ of $\text{var}_{\rho_0}\{\bar{\ell}_N(\rho|\rho_0)\}$, and the half width

$$w_r(\rho|\rho_0) = \frac{1.96\hat{\sigma}(\rho|\rho_0)}{|\bar{\ell}_N(\rho|\rho_0)|}$$

Table 6.2.1 $\nabla^k \bar{\ell}_N(\rho|\rho_0)$, $\hat{\sigma}^2(\rho, k|\rho_0)$, $\bar{\varepsilon}^k(\rho|\rho_0)$ **and** $w_r(\rho|\rho_0)$ **as Functions of** ρ**, for the** $M/G/1$ **Queue with** $\rho_0 = 0.8$

ρ	$\Delta\delta$	$\bar{\ell}_N(\rho\|\rho_0)$ $\hat{\sigma}^2(\rho, 0\|\rho_0)$ $\bar{\varepsilon}^0(\rho\|\rho_0)$	$\nabla\bar{\ell}_N(\rho\|\rho_0)$ $\hat{\sigma}^2(\rho, 1\|\rho_0)$ $\bar{\varepsilon}^1(\rho\|\rho_0)$	$w_r(\rho\|\rho_0)$ for $\ell(v)$	$w_r(\rho\|\rho_0)$ for $\nabla\ell(v)$
0.40	−0.50	2.66E-01 4.42E-06 3.39	1.62E+00 4.31E-04 1.37	0.015	0.025
0.45	−0.44	3.54E-01 8.61E-06 3.05	1.92E+00 3.88E-04 0.84	0.016	0.020
0.50	−0.37	4.59E-01 1.25E-05 2.82	2.31E+00 2.87E-04 0.34	0.015	0.014
0.55	−0.31	5.87E-01 1.42E-05 1.74	2.84E+00 6.11E-04 0.32	0.012	0.017
0.60	−0.25	7.47E-01 1.63E-05 0.99	3.60E+00 1.23E-03 0.28	0.011	0.019
0.65	−0.18	9.53E-01 2.39E-05 0.81	4.72E+00 2.18E-03 0.23	0.010	0.019
0.70	−0.12	1.23E+00 4.27E-05 0.71	6.41E+00 5.00E-03 0.18*	0.010	0.021
0.75	−0.06	1.61E+00 9.16E-05 0.61*	9.15E+00 1.94E-02 0.24	0.012	0.030
0.80	0.00	2.18E+00 3.64E-04 1.00	1.41E+01 2.44E-01 1.00	0.017	0.068
0.85	0.06	3.07E+00 5.03E-03 3.87	2.14E+01 5.60E+00 4.24	0.045	0.216

of the 95% (relative) confidence interval (CI) as functions of ρ. Note that the relative perturbation in ρ_0 is defined here as $\Delta\delta = (\rho - \rho_0)/\rho_0$. The interarrival rate was set to 1, and 10^6 customers were simulated at reference parameter $v_0 = 0.6$ ($\rho_0 = 0.8$).

In analogy to Figure 5.1 in Section 5.3, Figure 6.3 depicts the estimated response curve $\bar{\ell}_N(\rho|\rho_0)$ (denoted by $\bar{\ell}_N$ in the figure) along with the curves

$$\begin{aligned} J_1(\rho|\rho_0) &= \{\bar{\ell}_N(\rho|\rho_0) - w_r(\rho|\rho_0)\}, \\ J_2(\rho|\rho_0) &= \{\bar{\ell}_N(\rho|\rho_0) + w_r(\rho|\rho_0)\} \end{aligned} \tag{6.2.26}$$

(denoted by 95% CI in the figure), as functions of ρ ($\rho = 0.2, 0.3, \ldots, 0.8$). Note again that $\bar{\ell}_N(\rho|\rho_0)$ and $w_r(\rho|\rho_0)$ are plotted using different scales (w_r is dimensionless).

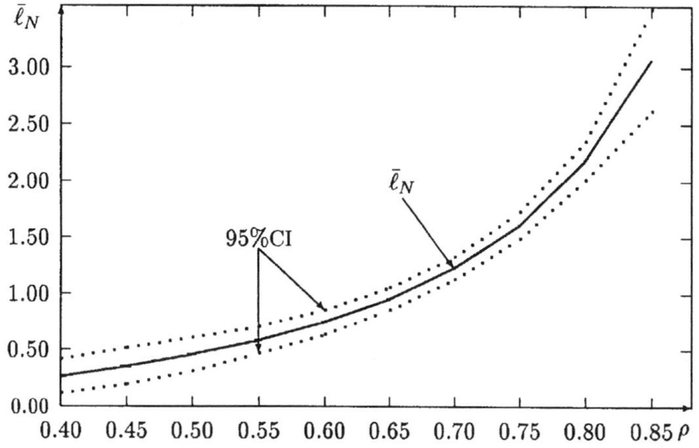

Figure 6.3. Performance of the SF estimator $\bar{\ell}_N(\rho|\rho_0)$ for various values of ρ in the $M/G/1$ queue with reference traffic intensity $\rho_0 = 0.8$.

Figure 6.4 depicts similar data for the derivative of the expected waiting time in the $M/G/1$ queue with respect to v. We point out that the performance, $\ell(\rho)$, and its derivatives, $\nabla\ell(\rho)$, were estimated *simultaneously*, *from a single simulation run*, based on $N = 10{,}000$ customers; the notation $\bar{\ell}_N(\rho|\rho_0)$, $w_r(\rho|\rho_0)$, and so forth is designed to reflect this fact.

It is readily seen that the the SF estimators, $\bar{\ell}_N(\rho|\rho_0)$ and $\nabla\bar{\ell}_N(\rho|\rho_0)$, perform reasonably well in the range $\rho \in (0.4, 0.8)$, that is, when the relative perturbation in ρ does not exceed 200%. For large relative perturbations ($\geq$200%), the SF process $\nabla^k \tilde{W}$ causes the variance of the estimators $\bar{\ell}_N(\rho|\rho_0)$ and $\nabla\bar{\ell}_N(\rho|\rho_0)$ to "blow-up." More details on the response surface methodology for queueing models will be provided in Chapter 8.

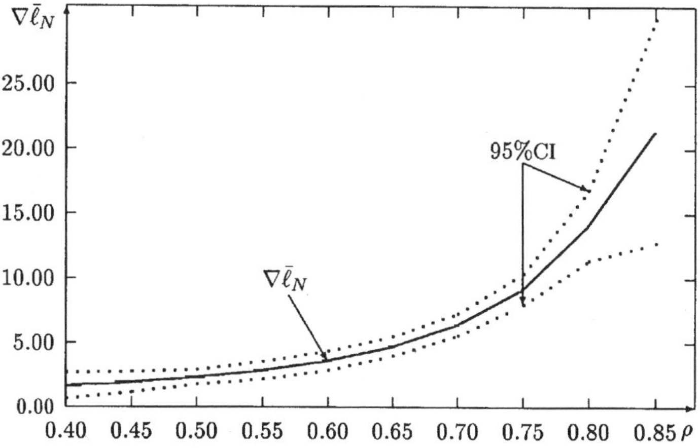

Figure 6.4. Performance of the SF estimator $\nabla\bar{\ell}_N(\rho|\rho_0)$ for various values of ρ in the $M/G/1$ queue with reference traffic intensity $\rho_0 = 0.8$.

Consider next the SF estimators of $\nabla^k \ell(\boldsymbol{v})$ for more complex queueing models. It is not difficult to see that the basic formulas (6.2.9)–(6.2.12), developed for the $GI/G/1$ queue with the FIFO discipline, can be extended to rather general queueing models; formulas (6.2.9)–(6.2.12) will still hold, provided the *indexing* in the likelihood ratio process, $\{\tilde{W}_t\}$, is properly modified. This will now be exemplified for the following models:

1. $GI/G/1/b$ queues, where b is the buffer size
2. General queueing networks consisting of r nodes

In case 1 we vary (perturb) the parameter vector $\boldsymbol{v}$ of the interarrival time pdf, while in case 2 we vary the parameter vector $\boldsymbol{v} = (v_1, \ldots, v_r)$ of the service time pdf's.

1. In this case, the index $t = 1, \ldots, \tau$ in L_t and $\tilde{W}_t$ must correspond to the order in which customers *arrive* at the system and in which they *enter service*; the interarrival time random variables, associated with rejected (blocked) customers, must also be taken into account in both L_t and $\tilde{W}_t$. As an example, consider the $GI/G/1/2$ queue. Assume that six customers attempted to enter the system during a given regenerative cycle. Assume further that customers 4 and 5 were prevented from entering the system (due to a full buffer). Then the corresponding cycle has length $\tau = 4$, and the associated likelihood ratios are

$$\tilde{W}_t = \prod_{i=1}^{t} \tilde{W}_i, \quad t = 1, 2, 3; \qquad \tilde{W}_4 = \prod_{i=1}^{6} \tilde{W}_i.$$

2. Consider the sojourn time in a general open queueing network of r nodes, and suppose that only the service time densities $f_i(\cdot, \boldsymbol{v}_i)$, $i = 1, \ldots, r$, are varied. The indices t correspond here to events in the underlying discrete-event simulation; these events should include at least all instances of a customer entering service and departing from the system. If t corresponds to an event whereby a customer enters service at node i, then a service time is generated from $f_i(\cdot, \boldsymbol{v})$. Note that it may not always be necessary to generate a random vector $\underline{\boldsymbol{Y}}_t$, but formally, we may always assume that this is the case (certain $\underline{\boldsymbol{Y}}_t$ may be dummy). If t corresponds to a departure event, we let L_t be the sojourn time of the departing customer; in all other cases, we set $L_t = 0$.

More formally, let

$$\underline{\boldsymbol{Y}}_t = \{\underline{\boldsymbol{Y}}_t^1, \underline{\boldsymbol{Y}}_t^2, \ldots, \underline{\boldsymbol{Y}}_t^r\}$$

be an iid sequence, where each $\underline{\boldsymbol{Y}}_t^i$ is itself a sequence of independent random vectors generated from the same pdf, $f_i(\boldsymbol{y}, \boldsymbol{v}_i)$, where each $\boldsymbol{v}_i$, $i = 1, \ldots, r$, is a vector of parameters. Further, let C_t^i be the set of indices of the random variables comprising

$\underline{Y}^i_t$, such that $\underline{Y}^i_t = \{Y_j : j \in C^i_t\}$. (Observe that each Y_j represents a random variable rather than a random vector.) With these definitions, we may now represent $\tilde{W}_t$ as

$$\tilde{W}_t = \prod_{i=1}^{r} \prod_{j \in C^i_t} \tilde{W}j, \tag{6.2.27}$$

and similarly for $\nabla^k \tilde{W}_t$. We mention that for a tandem topology with the FIFO discipline, (6.2.27) reduces to $\tilde{W}_t = \prod_{i=1}^{r} \prod_{j=1}^{t} \tilde{W}_{ji}$. More details concerning the issue of indexing the likelihood ratio process, $\{\tilde{W}_t\}$, for open queueing networks with various topologies and their implementation in the queueing network stabilizer and optimizer (QNSO) simulation package are given in Perez-Luna (1990).

Similar methods may also be applied to closed queueing networks, provided certain conditions are imposed on the service time distributions (see Kaspi and Mandelbaum, 1991, and Sigman, 1990). Although these conditions vary in the above references, all of them require that the regeneration scheme be positive recurrent, and that regeneration points occur when all the customers reside in one specific server and the first one is about to depart.

Two notable alternative methods for sensitivity analysis and stochastic optimization of DEDS should be mentioned at this juncture: *infinitesimal perturbation analysis* (IPA) and *weak derivatives*. The IPA method was introduced by Ho et al. (1979) and will be discussed at some length in Chapter 7. Weak derivatives were introduced by Pflug; see Pflug (1988, 1989, 1990, 1992, 1996). These studies give convergence results for weak derivatives and present interesting examples where weak derivatives outperform the SF method under certain criteria.

To demonstrate Pflug's weak derivative method, consider

$$\nabla \ell(v) = \int L(y) \nabla f(y, v)\, dy,$$

and represent $\nabla f(y, v) \equiv \nabla_v f(y, v)$ as

$$\nabla_v f(y, v) = c_v[\dot{f}(y, v) - \ddot{f}(y, v)], \tag{6.2.28}$$

where $\dot{f}(y, v)$ and $\ddot{f}(y, v)$ are *densities* and c_v is some constant. The simplest way of decomposing $\nabla_v f(y, v)$ in (6.2.28) is to set

$$\dot{f}(y, v) = c_v^{-1}[\nabla_v f(y, v)]^+,$$
$$\ddot{f}(y, v) = c_v^{-1}[\nabla_v f(y, v)]^-$$

and

$$c_v = \int [\nabla_v f(v, y)]^+\, dy$$
$$= \int [\nabla_v f(v, y)]^-\, dy,$$

where $[x]^+ = \max(x, 0)$ is the positive part of x, and $[x]^- = -\min(x, 0)$ is the negative part of x. An unbiased estimator of $\nabla\ell(v)$ is

$$\nabla\hat{\ell}_N(v) = \frac{c_v}{N}\sum_{i=1}^{N} [L(\dot{Z}_i) - L(\ddot{Z}_i)], \tag{6.2.29}$$

where $\dot{Z}_i$ and $\ddot{Z}_i$ have densities $\dot{f}(y, v)$ and $\ddot{f}(y, v)$, respectively. The estimator $\nabla\hat{\ell}_N(v)$ is called the *weak derivative estimator* of $\nabla\ell(v)$.

Note that to estimate $\nabla_v\ell(v)$ one has to generate *two sequences* of variates $\{\dot{Z}_i, \ddot{Z}_i\}$ and two associated sequences of sample performances, $\{L(\dot{Z}_i), L(\ddot{Z}_i)\}$.

Example 6.2.7. Consider the normal density

$$f(y, v) = \frac{1}{\sigma\sqrt{2\pi}}\exp\left(-\frac{1}{2}\left(\frac{y-v}{\sigma}\right)^2\right)$$

with mean v and variance σ^2. The derivative with respect to (w.r.t) v is

$$\begin{aligned}\nabla_v f(y, v) &= \frac{1}{\sqrt{2\pi}}\frac{(y-v)}{\sigma^2}\exp\left(-\frac{1}{2}\left(\frac{y-v}{\sigma}\right)^2\right)\\ &= \frac{1}{\sqrt{2\pi}}\frac{(y-v)}{\sigma^2}\exp\left(-\frac{1}{2}\left(\frac{y-v}{\sigma}\right)^2\right)1_{\{y\geq v\}}\\ &\quad - \frac{1}{\sqrt{2\pi}}\frac{(v-y)}{\sigma^2}\exp\left(-\frac{1}{2}\left(\frac{y-v}{\sigma}\right)^2\right)1_{\{y< v\}}.\end{aligned}$$

The constant c_v is readily identified as $1/\sqrt{2\pi}$, while the positive and the negative parts, $\dot{f}(y, v)$ and $\ddot{f}(y, v)$, are densities of $v + Z$ and $v - Z$, respectively, where

$$Z \sim \text{Weibull}\left(2, \frac{1}{2\sigma^2}\right).$$

The resulting weak derivative estimator is

$$\nabla\hat{\ell}_N(v) = \frac{1}{\sqrt{2\pi}N}\sum_{i=1}^{N} [L(v + Z_i) - L(v - Z_i)], \tag{6.2.30}$$

where

$$Z_i \sim \text{Weibull}\left(2, \frac{1}{2\sigma^2}\right).$$

Note that according to (6.2.30), only *one sequence* of variates $\{Z_i\}$ is needed.

Example 6.2.8. Consider the gamma density

$$f(y, v) = \frac{v^\alpha}{\Gamma(\alpha)} y^{\alpha-1} \exp(-yv) \sim G(\alpha, v).$$

The derivative of $f(y, v)$ with respect to v is

$$\begin{aligned}\nabla_v f(y, v) &= \frac{\alpha v^{\alpha-1}}{\Gamma(\alpha)} y^{\alpha-1} \exp(-yv) - \frac{v^\alpha}{\Gamma(\alpha)} y^\alpha \exp(-yv) \\ &= \frac{\alpha}{v}\left[\frac{x^\alpha}{\Gamma(\alpha)} y^{\alpha-1} \exp(-yv) - \frac{y^{\alpha+1}}{\Gamma(\alpha+1)} y^\alpha \exp(-yv)\right].\end{aligned}$$

One might use the following decomposition:

$$\nabla_v f(y, v) = \frac{\alpha}{v}[\dot{f}(y, v) - \ddot{f}(y, v)],$$

where

$$\dot{f}(y, v) = \frac{y^\alpha}{\Gamma(\alpha)} y^{\alpha-1} \exp(-yv)$$

and

$$\ddot{f}(y, v) = \frac{y^{\alpha+1}}{\Gamma(\alpha+1)} y^\alpha \exp(-yv)$$

are $G(\alpha, v)$ and $G(\alpha + 1, v)$ pdf's, respectively. Thus, in this case, the constant c_v is α/v, while the positive and the negative parts, $\dot{f}(y, v)$ and $\ddot{f}(y, v)$, are $G(\alpha, v)$ and $G(\alpha + 1, v)$ densities, respectively.

The statistical properties of the estimators of weak derivatives, and in particular, the use of common random numbers in $\dot{Z}_i$ and $\ddot{Z}_i$ for variance reduction [see (6.2.29)] are treated in Pflug (1996).

*6.3 DECOMPOSABLE SCORE FUNCTION ESTIMATORS

It is readily seen from the results of Section 6.2 that the SF estimators $\nabla^k \ell_N(\boldsymbol{v})$, $k = 0, 1, \ldots$, perform well (have a reasonably small variance) if the regenerative cycles τ_i, $i = 1, \ldots, N$, are not too large.

Consider, for example, r $GI/G/1$ queues in tandem, and let $\{L_t\}$ be the sojourn time process. For large r, the SF estimators $\nabla^k \ell_N(\boldsymbol{v})$, $k = 0, 1, \ldots$, become useless (have excessive variance), since the probability of regeneration (the event of an empty system) is very small. This typically gives rise to long regenerative cycles, τ_i, and subsequent large variance of the random variables

$$\sum_{t=1}^{\tau} L_t \tilde{W}_t, \qquad \sum_{t=1}^{\tau} \tilde{W}_t, \qquad \sum_{t=1}^{\tau} L_t \nabla \tilde{W}_t, \quad \text{and} \quad \sum_{t=1}^{\tau} \nabla \tilde{W}_t,$$

used to form the estimators $\nabla^k \ell_N(\boldsymbol{v}),\ k = 0, 1, \ldots$.

To overcome this difficulty one can use the so-called *decomposable score function* (DSF) and *truncated score function* (TSF) estimators, which are useful for both regenerative and nonregenerative (though stationary and ergodic) queueing systems. DSF estimators (see Lirov and Melamed, 1992, and Rubinstein, 1992) are based on *local* regenerative cycles, τ_i, at each individual queue, rather than on the *global* one, τ, of the entire system. TSF estimators are based on truncation of the generalized SF process.

Before applying DSF and TSF estimators, one has to take into consideration the following facts:

1. Although for general networks, DSF and TSF estimators contain some *bias*, they nevertheless lead to *dramatic variance reduction* as compared to the SF estimators of Section 6.2, which we call below the *regenerative score function* (RSF) estimators. This fact provides some insight into why the bias is typically not overly large.
2. Unlike RSF estimators, DSF and TSF estimators can be used not only for *regenerative processes*, $\{L_t\}$, but also for *stationary and ergodic processes* as well. Thus, DSF and TSF estimators are suitable for both the regenerative method and the batch-means method.
3. The relative efficiencies of the DSF and TSF estimators, as compared to their RSF counterparts, increase with the size of the network.

RSF, DSF, and TSF estimators have been implemented in the simulation package QNSO, described in Perez-Luna (1990). This package is suitable for performance evaluation, sensitivity analysis, and optimization of general open non-Markovian queueing networks, with respect to the parameter vector, $\boldsymbol{v}$, of an exponential family of distributions.

We consider here DSF estimators only; TSF and RSF estimators are discussed in some detail in Rubinstein and Shapiro (1993). The idea underlying the DSF estimator concept is to first decompose the queueing network into smaller units, called *modules*, each containing several connected queues, and then to approximate (estimate) the unknown quantities $\nabla^k \ell(\boldsymbol{v}),\ k = 1, 2, \ldots,$ by treating these modules *as if* they were completely independent. In other words, we want to use frequently occurring *local* regenerative cycles at each *individual module* instead of *true* but seldom-occurring *global* ones of the *entire system*. Although the local cycles at each module interact with their neighbors, numerical studies show that, when properly chosen, neighbor contributions are comparably small. Thus, DSF estimators would estimators would approximate the unknown quantities, $\nabla^k \ell(\boldsymbol{v})$, rather well, in the sense that their variance is manageable and their bias is not excessively large. The variance reduction obtained via DSF estimators is typically dramatic as compared to their RSF counterparts.

Some insight into DSF estimators can be gained from considering r static $GI/G/\infty$ queues in tandem. Letting $L(\boldsymbol{Y})$ be the sojourn time of a customer, one has

$$L(\boldsymbol{y}) = \sum_{k=1}^{r} L_k(y_k), \tag{6.3.1}$$

where L_k is the sojourn time at node k. Hence,

$$\ell(\boldsymbol{v}) = \sum_{k=1}^{r} \ell_k(v_k),$$

where $\ell_k(v_k) = \mathbb{E}_{v_k}\{L_k(Y_k)\}$ and $Y_k \sim f_k(y_k, v_k)$, $k = 1, \ldots, r$. Assume further that the components $Y_1, \ldots, Y_r$ of the random vector $\boldsymbol{Y}$ are independent random variables representing the service times at the constituent $GI/G/\infty$ queues. Each $\ell_k(v_k)$, $k = 1, \ldots, r$ can be written as

$$\ell_k(v_k) = \mathbb{E}_{\boldsymbol{v}_{0k}}\{L_k(Z_k) W_k(Z_k, v_k)\}, \tag{6.3.2}$$

where

$$W_k(z_k, v_k) = f(z_k, v_k)/f(z_k, v_{0k}),$$

and $Z_k \sim f_k(z_k, v_{0k})$; each $\ell_k(v_k)$ can be estimated by the corresponding what-if estimator, $\bar{\ell}_{kN}(v_k)$. Consequently, $\ell(\boldsymbol{v})$ may be estimated by

$$\bar{\ell}_N^d(\boldsymbol{v}) = \sum_{k=1}^{r} \bar{\ell}_{kN}(v_k) = \frac{1}{N} \sum_{k=1}^{r} \sum_{i=1}^{N} L_{ki} \tilde{W}_{ki}. \tag{6.3.3}$$

Notice that the variance

$$\mathrm{var}_{\boldsymbol{v}_0}\left\{\sum_{k=1}^{r} \bar{\ell}_{kN}(v_k)\right\} = \sum_{k=1}^{r} \mathrm{var}_{\boldsymbol{v}_{0k}}\{\bar{\ell}_{kN}(v_k)\}$$

of the "decomposable" estimator, $\bar{\ell}_N^d(\boldsymbol{v})$, is typically much smaller than the corresponding variance of the naive estimator,

$$\bar{\ell}_N(\boldsymbol{v}) = \frac{1}{N} \sum_{k=1}^{r} \sum_{i=1}^{N} L_{ki} \tilde{W}_i,$$

where

$$W_i = \prod_{k=1}^{r} \tilde{W}_{ki}.$$

Consider next a general non-product queueing network consisting of r nodes. Let $\{L_t : t > 0\}$ be a regenerative process (say, of sojourn times), and let τ be the length of a regenerative cycle of the queueing network. Assume that all interarrival times, service times, and routing decisions are mutually independent. Then L_t can be written as

$$L_t = \sum_{q=1}^{r} L_{qt} = \sum_{q=1}^{r} \sum_{k=1}^{P_q} L_{qkt}, \tag{6.3.4}$$

where $\{L_{qkt} : t > 0\}$ is the partial sojourn time of the tth customer on its kth visit at node q, and P_q is the (random) total number of visits to node q. Observe that for r queues in tandem, $P_q = 1$. Taking advantage of (6.3.4) and assuming that $\{L_t\}$ is a regenerative process, rewrite $\ell(\boldsymbol{v})$ in (6.2.9) as

$$\ell(\boldsymbol{v}) = \sum_{q=1}^{r} \frac{\mathbb{E}_g\{\sum_{t=1}^{\tau} L_{qt} \tilde{W}_t\}}{\mathbb{E}_g\{\sum_{t=1}^{\tau} \tilde{W}_t\}}, \tag{6.3.5}$$

where

$$L_{qt} = L_{qt}(\underline{\boldsymbol{Z}}_t), \qquad \tilde{W}_t = \prod_{q=1}^{r} \prod_{j \in C_t^q} \tilde{W}_j, \qquad W_j = W(\boldsymbol{Z}_j, \boldsymbol{v}) = \frac{f(\boldsymbol{Z}_j, \boldsymbol{v})}{g(\boldsymbol{Z}_j)},$$

and C_t^q, $q = 1, \ldots, r$, is the set of indices defined in (6.2.27). The estimation of $\nabla^k \ell(\boldsymbol{v})$, $k = 1, 2, \ldots$, is similar.

Observe that for r $GI/G/1$ queues in tandem with the FIFO discipline, $\tilde{W}_t$ reduces to $\tilde{W}_t = \prod_{k=1}^{r} \prod_{j=1}^{t} \tilde{W}_{jk}$, where $\tilde{W}_{jk}$ has the form

$$\tilde{W}_{jk} = W(\boldsymbol{Z}_{jk}, \boldsymbol{v}) = \frac{f(\boldsymbol{Z}_{jk}, \boldsymbol{v})}{g(\boldsymbol{Z}_{jk})}.$$

Observe further that for r $GI/G/\infty$ (static) queues in tandem, $\tilde{W}_t$ and $\ell(\boldsymbol{v})$ reduce to $\tilde{W}_t = \prod_{k=1}^{r} \tilde{W}_k$ and

$$\ell(\boldsymbol{v}) = \sum_{q=1}^{r} \frac{\mathbb{E}_g\{L_q \tilde{W}\}}{\mathbb{E}_g\{\tilde{W}\}}, \tag{6.3.6}$$

respectively. Finally, in view of the equalities

$$\mathbb{E}_g\{\tilde{W}\} = \mathbb{E}_g\left\{\prod_{k=1}^{r} \tilde{W}_k\right\} = \prod_{k=1}^{r} \mathbb{E}_g\{\tilde{W}_k\} = 1,$$

one can represent $\ell(v)$ as

$$\ell(\boldsymbol{v}) = \sum_{q=1}^{r} \mathbb{E}_g\{L_q \tilde{W}\} = \sum_{q=1}^{r} \mathbb{E}_g\{L_q W_k\}. \tag{6.3.7}$$

In fact, the estimator of (6.3.7) coincides with that of (6.3.3).

Consider first the case where each module consists of a *single queue*. In this case, $\ell(\boldsymbol{v})$ in (6.3.5) can be approximated by

$$\ell^d(\boldsymbol{v}) = \sum_{q=1}^{r} \frac{\mathbb{E}_g\left\{\sum_{t=1}^{\tau_q} L_{qt} \tilde{W}_{qt}^d\right\}}{\mathbb{E}_g\left\{\sum_{t=1}^{\tau_q} \tilde{W}_{qt}^d\right\}}. \tag{6.3.8}$$

Here, $\tau_q,\ q = 1, \ldots, r$, is the length of the regenerative (*local*) cycle at the qth queue,

$$\tilde{W}_{qt}^d = \prod_{j \in \tilde{C}_t^q} \tilde{W}_{qj}, \qquad \tilde{W}_{qj} = W(\boldsymbol{Z}_{qj}, \boldsymbol{v}) = \frac{f(\boldsymbol{Z}_{qj}, \boldsymbol{v})}{g(\boldsymbol{Z}_{qj})},$$

and $\tilde{C}_t^q,\ q = 1, \ldots, r$, is the set of indices, defined analogously to (6.2.27). Observe that for r tandem $GI/G/1$ queues with the FIFO discipline, $\tilde{W}_{qt}^d$ reduces to $\tilde{W}_{qt}^d = \prod_{j=1}^{t} \tilde{W}_{qj}$, which can be equivalently written as

$$\tilde{W}_{qt}^d = \tilde{W}_{qt}^d(\underline{\boldsymbol{Y}}_{qt}, \boldsymbol{v}_q) = f_{qt}(\underline{\boldsymbol{Y}}_{qt}, \boldsymbol{v}_q)/g_{qt}(\underline{\boldsymbol{Y}}_{qt}),$$

$$f_{qt}(\underline{\boldsymbol{Y}}_{qt}, \boldsymbol{v}_q) = \prod_{i=1}^{t} f_q(\boldsymbol{Y}_{qi}, \boldsymbol{v}_q),$$

$$g_{qt}(\underline{\boldsymbol{Y}}_{qt}) = \prod_{i=1}^{t} g_q(\boldsymbol{Y}_{qi}), \qquad \underline{\boldsymbol{Y}}_{qt} = (\boldsymbol{Y}_{q1}, \ldots, \boldsymbol{Y}_{qt}), \qquad t = 1, \ldots, \tau_q.$$

Next, note that (6.3.8) is obtained from (6.3.5) by replacing the *global* cycle, τ, with the local ones, $\tau_q,\ q = 1, \ldots, r$. In view of (6.3.8), one can estimate $\ell^d(\boldsymbol{v})$ by

$$\bar{\ell}_N^d(\boldsymbol{v}) = \sum_{q=1}^{r} \frac{\sum_{i=1}^{N} \sum_{t=1}^{\tau_{qi}} L_{qti} \tilde{W}_{qti}^d}{\sum_{i=1}^{N} \sum_{t=1}^{\tau_{qi}} \tilde{W}_{qti}^d}. \tag{6.3.9}$$

Differentiating $\bar{\ell}_N^d(\boldsymbol{v})$ above with respect to $\boldsymbol{v}$ yields estimators for $\nabla^k \ell(\boldsymbol{v})$. In particular, the decomposable estimator of $\nabla \ell(\boldsymbol{v})$ can be written as

$$\nabla \bar{\ell}_N^d(\boldsymbol{v}) = \sum_{q=1}^{r} \left(\frac{\sum_{i=1}^{N} \sum_{t=1}^{\tau_{qi}} L_{qti} \nabla \tilde{W}_{qti}^d}{\sum_{i=1}^{N} \sum_{t=1}^{\tau_{qi}} \tilde{W}_{qti}^d} - \frac{\sum_{i=1}^{N} \sum_{t=1}^{\tau_{qi}} L_{qti} \tilde{W}_{qti}^d}{\sum_{i=1}^{N} \sum_{t=1}^{\tau_{qi}} \tilde{W}_{qti}^d} \times \frac{\sum_{i=1}^{N} \sum_{t=1}^{\tau_{qi}} \nabla \tilde{W}_{qti}^d}{\sum_{i=1}^{N} \sum_{t=1}^{\tau_{qi}} \tilde{W}_{qti}^d} \right). \tag{6.3.10}$$

Since $\tau_q < \tau$, the estimators $\nabla^k \bar{\ell}^d_N(\boldsymbol{v}),\ k = 0, 1, \ldots,$ represent *truncated* versions of the RSF estimators $\nabla^k \bar{\ell}_N(\boldsymbol{v}),\ k = 0, 1, \ldots$ [see (6.2.11) and (6.2.12)], and thus will typically be biased. It is important to note, however, that $\nabla^k \bar{\ell}^d_N(\boldsymbol{v}),\ k = 0, 1, \ldots,$ yield consistent estimators of $\nabla^k \ell(\boldsymbol{v})$ for $M/M/1$ queues in tandem and for static queueing networks consisting of $G/G/\infty$ and $GI/G/1/0$ queues. For $r=1$, $\nabla^k \bar{\ell}^d_N(\boldsymbol{v})$ reduces to $\nabla^k \bar{\ell}_N(\boldsymbol{v})$ for all $k = 0, 1, \ldots,$ that is, DSF estimators coincide with RSF estimators.

In order to reduce the bias of the DSF estimators, (6.3.9) and (6.3.10), one may use larger modules rather than single queues, and then treat the data from different modules *as if* they were completely independent. In other words, one first decomposes the underlying network into a partition of r modules, and then produces DSF estimators by combining several local cycles of the individual queues in a given module. To this end, let γ denote the number of nodes in a module, and define $\Delta_\gamma = (\delta_1, \ldots, \delta_\gamma)$, where $\delta_m,\ m = 1, \ldots, r$, denotes the number of local regenerative cycles at node m. We next represent $\tilde{W}^d_{qt}$ as a function of γ and Δ_γ, that is, $\tilde{W}^d_{qt} = \tilde{W}^d_{qt}(\gamma, \Delta_\gamma),\ \gamma = 1, \ldots, r$. Analogously to (6.3.9) and (6.3.10), define for $k=0$ and $k=1$ the following decomposable estimators:

$$\bar{\ell}^d_N(\boldsymbol{v}, \gamma, \Delta_\gamma) = \sum_{q=1}^{r} \frac{\sum_{i=1}^{N} \sum_{t=1}^{\tau_i} L_{qti} \tilde{W}^d_{qt}(\gamma, \Delta_\gamma)}{\sum_{i=1}^{N} \sum_{t=1}^{\tau_i} \tilde{W}^d_{qti}(\gamma, \Delta_\gamma)} \tag{6.3.11}$$

and

$$\nabla \bar{\ell}^d_N(\boldsymbol{v}, \gamma, \Delta_\gamma) = \sum_{q=1}^{r} \left(\frac{\sum_{i=1}^{N} \sum_{t=1}^{\tau_i} L_{qti} \nabla \tilde{W}^d_{qti}(\gamma, \Delta_\gamma)}{\sum_{i=1}^{N} \sum_{t=1}^{\tau_i} \tilde{W}^d_{qti}(\gamma, \Delta_\gamma)} - \frac{\sum_{i=1}^{N} \sum_{t=1}^{\tau_i} L_{qti} \tilde{W}^d_{qti}(\gamma, \Delta_\gamma)}{\sum_{i=1}^{N} \sum_{t=1}^{\tau_i} \tilde{W}^d_{qti}(\gamma, \Delta_\gamma)} \times \frac{\sum_{i=1}^{N} \sum_{t=1}^{\tau_i} \nabla \tilde{W}^d_{qti}(\gamma, \Delta_\gamma)}{\sum_{i=1}^{N} \sum_{t=1}^{\tau_i} \tilde{W}^d_{qti}(\gamma, \Delta_\gamma)} \right). \tag{6.3.12}$$

Decomposable estimators for $\nabla^k \bar{\ell}^d_N(\boldsymbol{v}, \gamma, \Delta_\gamma),\ k > 1$, may be similarly defined.

From (6.3.11) and (6.3.12) we can identify the following two extreme cases. At one extreme we have $\gamma=1$ and $\delta_1=1$, with the corresponding DSF estimators, (6.3.9) and (6.3.10), based on the local cycles of a single node. At the other extreme we have $\gamma=r$, and δ_m equals the total number of local cycles at node $m,\ m = 1, \ldots, r$. That is, each δ_m is associated with the global cycle length τ through the relationship $\tau = \sum_{k=1}^{\delta_m} \tau_{mk}$, where τ_{mk} is the length of the kth local cycle at node m. Clearly, in this case, the DSF estimators (6.3.9) and (6.3.10) coincide with the RSF estimators (6.2.11) and (6.2.12), respectively. In between these two extremes of γ and $\Delta_\gamma = (\delta_1, \ldots, \delta_\gamma)$, we have a finite set of estimators of $\nabla^k \ell(\boldsymbol{v})$. Note again that for the two extreme cases above, the estimators $\nabla^k \bar{\ell}^d_N(\boldsymbol{v}, \gamma, \Delta_\gamma)$ reduce to $\nabla^k \bar{\ell}^d_N(\boldsymbol{v}) = \nabla^k \bar{\ell}^d_N(\boldsymbol{v}, 1, 1)$ and $\nabla^k \bar{\ell}_N(\boldsymbol{v})$, respectively, while for $r=1$,

$$\nabla^k \bar{\ell}^d_N(\boldsymbol{v}, \gamma, \Delta_\gamma) = \nabla^k \bar{\ell}^d_N(\boldsymbol{v}) = \nabla^k \bar{\ell}_N(\boldsymbol{v}).$$

Since the estimators $\nabla^k \bar{\ell}_N^d(\boldsymbol{v}, \gamma, \Delta_\gamma)$, $k = 0, 1, \ldots,$ are based on a network decomposition into modules using *local* cycles, they are termed *DSF estimators of order* (γ, Δ_γ), or simply *DSF estimators*. Note that the $\nabla^k \bar{\ell}_N^d(\boldsymbol{v})$ represent DSF estimators of order (1, 1), to be referred to as *totally decomposable score function* (TDSF) estimators.

Some guidance will next be offered on how to choose a "good" set of parameters (γ, Δ_γ), in the sense that the variance of the DSF estimator $\bar{\ell}_N^d(\boldsymbol{v}, \gamma, \Delta_\gamma)$ is still manageable, and the relative bias of the system performance, defined as

$$b = \left| \frac{\mathbb{E}\{\bar{\ell}_N^d(\boldsymbol{v}, \gamma, \Delta_\gamma)\} - \ell(\boldsymbol{v})}{\ell(\boldsymbol{v})} \right|,$$

is relatively small, say, no larger than 0.1. Formally, we would like to minimize the mean-squared error of $\nabla^k \bar{\ell}_N^d(\boldsymbol{v}, \gamma, \Delta_\gamma)$ with respect to (γ, Δ_γ). Since an analytical solution of such a minimization problem is not feasible (even for simple queueing networks), one must resort to simulation. Extensive simulation studies have shown that the bias of the estimators $\nabla^k \bar{\ell}_N^d(\boldsymbol{v}, \gamma, \Delta_\gamma)$ is typically a nonincreasing function in each component of the vector (γ, Δ_γ); moreover, in order for the relative bias, b, to be less than 0.05, one may use the following choices for (γ, Δ_γ):

$$(\gamma = 2, \delta = 2), \qquad (\gamma = 2, \delta = 5), \qquad (\gamma = 3, \ \delta = 10), \tag{6.3.13}$$

for light, moderate and heavy traffic, respectively. (Here $\delta = \delta_\gamma, \ \gamma = 1, \ldots, r$.) For example, it was found empirically that when the queueing network operates in moderate traffic and one wishes to keep the variance of the DSF estimators reasonably small and the relative bias less than 0.1, then one should decompose the network into modules (in quite an arbitrary fashion), such that the size (number of queues) of each module is $\gamma = 2$, and the number of local regenerative cycles at each queue in a fixed module is $\delta = 4$.

Since the DSF estimators use local, rather than global, regenerative cycles, they can be used for both the regenerative and the standard batch-means methods (see Section 1.2.2); variance reduction obtained via DSF estimators versus their RSF counterparts will typically be very substantial. Note, finally, that confidence intervals (regions) for $\bar{\ell}_N^d(\boldsymbol{v}, \gamma, \Delta_\gamma)$ and $\nabla \bar{\ell}_N^d(\boldsymbol{v}, \gamma, \Delta_\gamma)$ can be derived by using standard statistical methods, similar to those given in Section 3.3.2.

6.3.1 Numerical Results

This subsection presents numerical results for the DSF estimators, $\nabla^k \bar{\ell}_N^d(\boldsymbol{v}, \gamma, \Delta_\gamma)$, and the RSF estimators, $\nabla^k \bar{\ell}_N(\boldsymbol{v})$, for the queueing network of Figure 6.5 in (a) Markovian setting and (b) non-Markovian setting. The steady-state mean waiting time, $\ell(\boldsymbol{v})$, as well as the associated derivatives, $\nabla \ell(\boldsymbol{v})$, were estimated, in both cases, with respect to the parameters of the service time distribution. All estimates were based on simulation runs of $N = 1.5 \times 10^6$ customers. In Tables 6.3.1 and

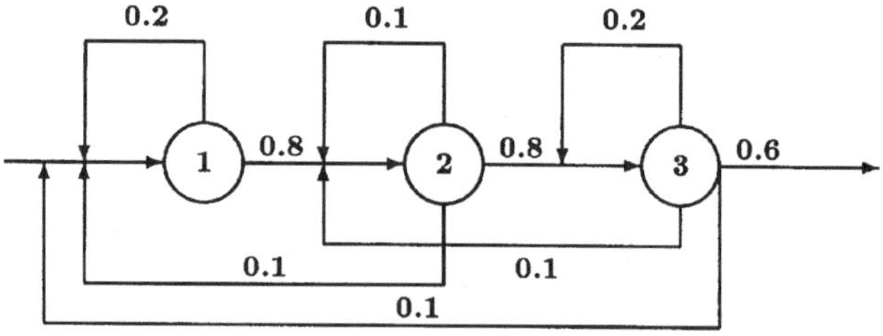

Figure 6.5. Queueing network consisting of three queues with feedbacks.

6.3.2, the column headings Estim, $\hat{\sigma}^2$, and w_r, respectively, refer to point estimators, sample variances [sample mean-squared error (MSE)], and the relative width of the confidence intervals, defined as

$$w_r = \frac{2\hat{\sigma} z_{1-\alpha_0/2}}{|\nabla^k \bar{\ell}_N(\boldsymbol{v})|}, \qquad k = 0, 1.$$

In the same tables, the DSF estimators correspond to $\gamma = 1$ and $\delta_1 = 1$, and are based on the TDSF estimators, (6.3.9) and (6.3.10).

Table 6.3.1 Performance of the RSF and DSF Estimators for Mean Waiting Times in the Jackson Network of Figure 6.5

i	Performance $\ell(\boldsymbol{v})$						
	Regenerative			Decomposition			
	Estim	$\hat{\sigma}^2$	w_r	Estim	$\hat{\sigma}^2$	w_r	$\hat{\varepsilon}^0(\boldsymbol{v})$
1	1.334	1.6E-05	0.012	1.334	1.6E-05	0.012	1
2	0.485	1.3E-04	0.091	0.488	1.2E-06	0.009	108
3	0.223	4.1E-05	0.110	0.253	3.6E-07	0.009	114
	Derivative $\nabla_{v_1}\ell(\boldsymbol{v}) = \partial\ell(\boldsymbol{v})/\partial v_1$						$\hat{\varepsilon}^1(v_1)$
1	−0.396	1.0E-04	0.103	−0.397	2.3E-05	0.046	4.5
2	−0.091	2.5E-04	0.680	−0.085	6.5E-08	0.018	3800
3	−0.033	2.1E-05	0.537	−0.031	8.5E-09	0.011	2500
	Derivative $\nabla_{v_2}\ell(\boldsymbol{v}) = \partial\ell(\boldsymbol{v})/\partial v_2$						$\hat{\varepsilon}^1(v_2)$
1	−0.418	9.8E-05	0.093	−0.398	2.4E-05	0.048	4.08
2	−0.085	5.8E-05	0.350	−0.085	6.2E-08	0.011	940
3	−0.028	1.5E-05	0.540	−0.031	6.8E-09	0.010	2200
	Derivative $\nabla_{v_3}\ell(\boldsymbol{v}) = \partial\ell(\boldsymbol{v})/\partial v_3$						$\hat{\varepsilon}^1(v_3)$
1	−0.432	1.2E-04	0.101	−0.401	2.7E-05	0.045	4.4
2	−0.090	2.9E-04	0.793	−0.084	6.7E-08	0.012	4300
3	−0.037	3.8E-05	0.650	−0.031	7.7E-09	0.011	4900

Table 6.3.2 Performance of the RSF and DSF Estimators for Mean Waiting Times in the Queueing Network of Figure 6.5 with Gamma Interarrival Time and Service Time pdf's

	Performance $\ell(\boldsymbol{v})$						
i	Regenerative			Decomposition			
	Estim	$\hat{\sigma}^2$	w_r	Estim	$\hat{\sigma}^2$	w_r	$\hat{\varepsilon}^0(\boldsymbol{v})$
1	1.267	1.6E-05	0.013	1.267	1.6E-05	0.013	1
2	0.766	7.2E-06	0.014	0.764	3.5E-06	0.009	2
3	0.551	1.8E-05	0.031	0.544	1.5E-06	0.009	12
	Derivative $\nabla_{v_1}\ell(\boldsymbol{v}) = \partial\ell(\boldsymbol{v})/\partial v_1$						$\hat{\varepsilon}^1(v_1)$
1	−0.296	6.9E-05	0.110	−0.330	1.4E-05	0.043	4.9
2	−0.137	1.5E-05	0.110	−0.154	5.6E-07	0.019	27
3	−0.075	3.4E-05	0.314	−0.083	8.8E-09	0.013	390
	Derivative $\nabla_{v_2}\ell(\boldsymbol{v}) = \partial\ell(\boldsymbol{v})/\partial v_2$						$\hat{\varepsilon}^1(v_2)$
1	−0.444	1.0E-04	0.091	−0.434	2.7E-05	0.047	3.7
2	−0.207	1.8E-05	0.081	−0.203	1.0E-06	0.020	18
3	−0.116	2.5E-05	0.172	−0.113	1.6E-07	0.014	156
	Derivative $\nabla_{v_3}\ell(\boldsymbol{v}) = \partial\ell(\boldsymbol{v})/\partial v_3$						$\hat{\varepsilon}^1(v_3)$
1	−0.912	2.4E-04	0.067	−0.868	1.0E-04	0.046	2.4
2	−0.370	2.7E-05	0.055	−0.359	3.7E-06	0.021	7.3
3	−0.239	2.5E-05	0.083	−0.226	8.5E-07	0.016	29

Markovian Setting

Table 6.3.1 displays the point estimators $\bar{\ell}_N(\boldsymbol{v})$, $\nabla\bar{\ell}_N(\boldsymbol{v})$, $\bar{\ell}_N^d(\boldsymbol{v})$, and $\nabla\bar{\ell}_N^d(\boldsymbol{v})$ (in the columns marked Estim) of the RSF and DSF estimators, respectively, their associated sample variances (in the columns marked $\hat{\sigma}^2$), the relative width of the confidence intervals (in the columns marked w_r) and the sample efficiency $\hat{\varepsilon}^k(\boldsymbol{v})$, $k = 0, 1$. The rows correspond to various values of the relative parameter perturbation, δ_i, defined as $\delta_i = |(v_i - v_{0i})/v_{0i}|$, $i = 1, 2$, where $\boldsymbol{v} = (v_1, v_2) = (\mu_1, \mu_2)$ is the vector of service rates. The arrival rate was set to $\lambda = 0.8$, and the reference service rates were selected as $\mu_{0i} = 3$, $i = 1, 2, 3$. The waiting time performance and its gradient were estimated for $\mu_i = 3.0, 4.5, 6.0$, $i = 1, 2, 3$, for $\delta = 0, 0.5, 1$, respectively. For $k = 0$, the efficiency is defined by

$$\hat{\varepsilon}^0(\boldsymbol{v}) = \frac{\text{sample variance of } \bar{\ell}_N(\boldsymbol{v})}{\text{sample MSE of } \bar{\ell}_N^d(\boldsymbol{v})},$$

and similarly for $k = 1$. Here MSE is defined as

$$\text{MSE}\{\bar{\ell}_N^d(\boldsymbol{v})\} = \mathbb{E}_{\boldsymbol{v}}\{(\bar{\ell}_N^d(\boldsymbol{v}) - \bar{\ell}_N(\boldsymbol{v}))^2\}.$$

Note that the MSE is used in the denumerator in lieu of the variance, since in general, $\bar{\ell}_N^d(\boldsymbol{v})$ is a biased (inconsistent) estimator of $\ell(\boldsymbol{v})$, while its counterpart, $\bar{\ell}_N(\boldsymbol{v})$, is a consistent one. The same holds true for $\nabla\bar{\ell}_N^d(\boldsymbol{v})$ versus $\nabla\bar{\ell}_N(\boldsymbol{v})$. Table 6.3.1 indicates that the TDSF estimators perform "well" in the sense that their bias is bounded by 0.1, and the efficiency, $\hat{\varepsilon}^k(\boldsymbol{v})$, increases with the number of nodes (stations) and the relative perturbation δ, as well as with k ($k = 0, 1$).

Non-Markovian Setting

Table 6.3.2 displays data similar to those of Table 6.3.1. Here, the interarrival time and the three service time (dominating) pdf's are gamma distributed $G_i(\beta_i, \lambda_i)$, $i = 1, 2, 3, 4$, to wit,

$$G_i(\beta_i, \lambda_i) = \frac{\lambda_i^{\beta_i} y^{\beta_i - 1} e^{-\lambda_i y}}{\Gamma(\beta_i)}, \qquad y \geq 0.$$

More specifically, the parameters of the interarrival time pdf were set to $\boldsymbol{v}_1 = (\beta_1, \lambda_1) = (0.5, 1)$, and the parameters of the service time pdf's were set to $\boldsymbol{v}_2 = (0.5, 2.5)$, $\boldsymbol{v}_3 = (0.4, 2.0)$, and $\boldsymbol{v}_4 = (0.25, 1.25)$. The performance $\ell(\boldsymbol{v})$ (the mean waiting time in the system), and the gradient

$$\nabla\ell(\boldsymbol{v}) = \left(\frac{\partial\ell(\boldsymbol{v})}{\partial v_1}, \frac{\partial\ell(\boldsymbol{v})}{\partial v_2}, \frac{\partial\ell(\boldsymbol{v})}{\partial v_3}\right)$$

were estimated for various values of the scale parameters $(\lambda_2, \lambda_3, \lambda_4)$, while holding the shape parameters at $(\beta_2, \beta_3, \beta_4) = (0.5, 0.4, 0.25)$. Those values were: $(\lambda_2, \lambda_3, \lambda_4) = (2.5, 2.0, 1.25)$, $(3.125, 2.5, 1.625)$, $(3.75, 3.0, 1.875)$, which correspond (componentwise) to $(\delta_2, \delta_3, \delta_4) = (0, 0, 0), (0.25, 0.25, 3.0), (0.5, 0.5, 0.5)$.

The results of Table 6.3.2 (for non-Markovian networks) are similar to those of Table 6.3.1 (for Markovian networks). The relative bias of the DSF estimators is still less than 0.15. Extensive simulation studies were also performed for the TDSF estimators, in more complex queueing networks. The simulation results agreed with the theoretical ones and were numerically similar to those given in Tables 6.3.1 and 6.3.2. In particular, it was found that the RSF estimators become useless (have excessive variance) for $r > 3$ and $\rho_i \geq 0.5$, $i = 1, \ldots, r$, where r is the number of nodes in the network. The reason for the estimator deterioration is that the asymptotic variance of the regenerative estimators increases dramatically in r, while the asymptotic variance of the TDSF estimators increases rather moderately in r.

Using the TDSF estimators with $(\gamma, \Delta_\gamma) = (1, 1)$, the bias was often observed, however, to exceeds 0.1. To reduce the bias, the DSF estimators $\nabla^k\bar{\ell}_N^d(\boldsymbol{v}, \gamma, \Delta_\gamma)$, $k = 0, 1$, were utilized as well as larger modules; that is, the corresponding parameters, (γ, Δ_γ), were used following the suggestion in (6.3.13), and similarly for the TSF estimators. The resulting relative bias was subsequently reduced to less than 5%.

6.4 OPTIMIZATION OF DEDS

This section treats optimization programs for non-Markovian queueing networks, given in the following form [see also (5.4.1)]:

$$(P_0) = \begin{cases} \text{minimize} & \ell_0(\boldsymbol{v}), & \boldsymbol{v} \in V, \\ \text{subject to} & \ell_j(\boldsymbol{v}) \leq 0, & j = 1, \ldots, k, \\ & \ell_j(\boldsymbol{v}) = 0, & j = k+1, \ldots, M, \end{cases} \tag{6.4.1}$$

where

$$\ell_j(\boldsymbol{v}) = \mathbb{E}_{\boldsymbol{v}}\{L_j\} = \frac{\mathbb{E}_{\boldsymbol{v}}\left\{\sum_{t=1}^{\tau} L_{jt}\right\}}{\mathbb{E}_{\boldsymbol{v}}\{\tau\}}, \qquad j = 0, 1, \ldots, M, \tag{6.4.1}$$

are the steady-state expected performances associated with a regenerative output process, $\{L_{jt}\}$. As an illustration, consider a manufacturing system in which:

1. The objective function, $\ell_0(\boldsymbol{v})$, might be the average make-span (sojourn time) of items to be processed at several workstations, according to a given schedule and route.
2. The decision vector, $\boldsymbol{v}$, might be the average rate at which the workstations process items.
3. The constraints might be stipulated as the probability that the buffer overflow at the jth workstation (queue) not exceed a fixed (small) value α_j, namely,

$$P_{\boldsymbol{v}}\{L_{jt} > b_j\} \leq \alpha_j, \qquad j = 1, \ldots, r,$$

where L_{jt} denotes the total number of items present at the jth workstation at time t.

Suppose, as in Section 5.4, that some of the functions ℓ_j are not available analytically, so that one must resort to Monte Carlo optimization methods. In this case, we replace the original program, P_0, by the following approximation, called *the stochastic counterpart*:

$$(\bar{P}_N) = \begin{cases} \text{minimize} & \bar{\ell}_{0N}(\boldsymbol{v}), & \boldsymbol{v} \in V, \\ \text{subject to} & \bar{\ell}_{jN}(\boldsymbol{v}) \leq 0, & j = 1, \ldots, k, \\ & \bar{\ell}_{jN}(\boldsymbol{v}) = 0, & j = k+1, \ldots, M, \end{cases} \tag{6.4.3}$$

where, as in Section 5.4, N is a positive integer and $\{\bar{\ell}_{jN}(\boldsymbol{v})\}$ is a sequence of random variables that converges asymptotically in N to the corresponding original functions, $\ell_j(\boldsymbol{v})$.

Assume that we are given a sequence $\{\boldsymbol{Z}_1, \boldsymbol{Z}_2, \ldots, \boldsymbol{Z}_N\}$ of iid random vectors from the dominating pdf, $g(\mathbf{z})$. Computational considerations lead us to distinguish between two cases.

Case A. The following hold true:

1. It is easy to precompute and store the whole sample $\{\boldsymbol{Z}_1, \boldsymbol{Z}_2, \ldots, \boldsymbol{Z}_N\}$, prior to running the simulation.
2. Given a sample $\{\boldsymbol{Z}_1, \boldsymbol{Z}_2, \ldots, \boldsymbol{Z}_N\}$, it is easy to compute the sample performance $\bar{\ell}_{jN}(\boldsymbol{v})$, for any desired value $\boldsymbol{v}$.

Case B. Either of the following holds true:

1. It is too expensive to store long samples $\{\boldsymbol{Z}_1, \boldsymbol{Z}_2, \ldots, \boldsymbol{Z}_N\}$, and the associated sequences, $\{\bar{\ell}_{jN}(\boldsymbol{v})\}$.
2. Optimization cannot be deferred (e.g., it must be performed after each regenerative cycle).
3. The sample performance $\bar{\ell}_{jN}(\boldsymbol{v})$ cannot be computed simultaneously for different values of $\boldsymbol{v}$. However, we are allowed to set the control vector, $\boldsymbol{v}$, at any desired value $\boldsymbol{v}_t$, and then observe (compute) the random variables $\bar{\ell}_{jN}(\boldsymbol{v}_t)$, and optionally, the associated derivatives (gradients) $\nabla\bar{\ell}_{jN}(\boldsymbol{v})$ at $\boldsymbol{v} = \boldsymbol{v}_t$.

From an application-oriented viewpoint the main difference between case A and case B is that the former is associated with *off-line optimization (control)*, and the latter with *on-line (real-time) optimization (control)*. Case A is treated in Rubinstein (1986) and Rubinstein and Shapiro (1993), which establishes the asymptotic normality of the estimators $\bar{\ell}_N^*$ and $\bar{\boldsymbol{v}}_N^*$ of the program $(\bar{P}_N)$ and conditions of convergence to their optimal counterparts, φ^* and $\boldsymbol{v}^*$, in the original program (P_0). Case B is based on algorithms of the *stochastic approximation* (SA) type and is treated in many references, including Ermoliev (1969), Ermoliev and Gaivoronski (1992), L'Ecuyer (1992), L'Ecuyer and Glynn (1994), L'Ecuyer et al. (1994), Pflug (1992), and Rubinstein (1986).

Remark 6.4.1. It is important to note that the on-line approach can be applied to both case A and case B, while the *off-line approach can only be applied to* Case B. This advantage of the on-line approach should be kept in mind, since it is difficult, in general, to determine in advance which approach is better (off-line or on-line) for case A.

The following two subsections deal separately with the off-line and on-line optimization approaches.

6.4.1 Off-line Optimization

When the output process $\{L_{jt}\}$ is regenerative, one may approximate each function

$$\ell_j(\boldsymbol{v}) = \mathbb{E}_{\boldsymbol{v}}\{L_j(\boldsymbol{Y})\}, \qquad j = 0, 1, \ldots, M,$$

by its sample counterpart,

$$\bar{\ell}_{jN}(\boldsymbol{v}) = \frac{\sum_{i=1}^{N}\sum_{t=1}^{\tau_i} L_{jti}\tilde{W}_{ti}}{\sum_{i=1}^{N}\sum_{t=1}^{\tau_i} \tilde{W}_{ti}}, \tag{6.4.4}$$

where

$$L_{jti} = L_{jt}(\underline{\boldsymbol{Z}}_{ti}), \qquad \tilde{W}_{ti} = \prod_{k=1}^{r}\prod_{p\in C_t^k} \tilde{W}_{pki},$$

and the C_t^k, $k = 1, \ldots, r$, are the sets of indices defined in (6.2.27).

Viewing the $\bar{\ell}_{jN}(\boldsymbol{v})$ as *functions of* $\boldsymbol{v}$ *rather than estimators for fixed* $\boldsymbol{v}$, one may estimate the optimal solution, $\boldsymbol{v}^*$, by solving the stochastic program $(\bar{P}_N)$. Observe that as soon as the sample $\{\boldsymbol{Z}_{11}, \ldots, \boldsymbol{Z}_{\tau_1 1}, \ldots, \boldsymbol{Z}_{1N}, \ldots, \boldsymbol{Z}_{\tau_N N}\}$ is generated from $g(\mathbf{z})$, the functions $\bar{\ell}_{jN}(\boldsymbol{v})$, $j = 0, \ldots, m$, become available for all $\boldsymbol{v} \in V$, and the corresponding gradients, $\nabla\bar{\ell}_{jN}(\boldsymbol{v})$, and Hessian matrices, $\nabla^2\bar{\ell}_{jN}(\boldsymbol{v})$, can be calculated from the same *single simulation* run. Consequently, the optimization problem $(\bar{P}_N)$ can be solved, in principle, by *standard methods of mathematical programming.* The resultant optimal value, $\bar{\varphi}_N^*$, and optimal solution, $\bar{\boldsymbol{v}}_N^*$, of the program $(\bar{P}_N)$ provide estimators for the optimal value, φ^*, and the optimal solution, $\boldsymbol{v}^*$, of the original program (P_0), respectively. Following is the algorithm for solving the stochastic counterpart for $\bar{\boldsymbol{v}}_N^*$.

Algorithm 6.4.1 Off-line Optimization

1. *Generate a random sample* $\{\boldsymbol{Z}_{11}, \ldots, \boldsymbol{Z}_{\tau_1 1}, \ldots, \boldsymbol{Z}_{1N}, \ldots, \boldsymbol{Z}_{\tau_N N}\}$, *from the dominating pdf,* $g(\mathbf{z})$.
2. *Generate the output (sample performance) processes* $\{L_{jti} : j = 0, \ldots, M\}$, *and the likelihood ratio (weight) process* $\{\tilde{W}_{ti}(\boldsymbol{v}) : t = 1, \ldots, \tau_i;\ i = 1, \ldots, N\}$.
3. *Solve the program* $(\bar{P}_N)$ *by mathematical programming methods.*
4. *Return the solution,* $\bar{\boldsymbol{v}}_N$ of $(\bar{P}_N)$, *as an estimator of* $\boldsymbol{v}^*$.

Convergence conditions for Algorithm 6.4.1 and asymptotic confidence regions for the unknown parameter vector, $\boldsymbol{v}^*$, may be found in Rubinstein and Shapiro (1993).

Assume further that the parameter set V is of the form

$$V = \{\boldsymbol{v} : \mathbf{0} \le \boldsymbol{\rho}^- \le \boldsymbol{\rho}(\boldsymbol{v}) \le \boldsymbol{\rho}^+ < \mathbf{1},\ \ \boldsymbol{\rho} = (\rho_1, \ldots, \rho_r)\}, \tag{6.4.5}$$

where $\rho_k = \rho_k(\boldsymbol{v}),\ k = 1, \ldots, r$, is the traffic intensity at the kth queue, $\boldsymbol{\rho}^-$ and $\boldsymbol{\rho}^+$ are fixed, r is the number of nodes in the network, $\mathbf{0}$ and $\mathbf{1}$ are vectors of size r of 0's and 1's, respectively, and inequalities among vectors are componentwise.

Remark 6.4.2. Rubinstein and Shapiro (1993) (see also Chapter 8) show that in solving the program (6.4.1) with the stochastic counterpart (6.4.3) and (6.4.4), a "good" vector of reference parameters for $\boldsymbol{\rho}_0 = \boldsymbol{\rho}_0(\boldsymbol{v}_0)$ must be chosen greater or equal to $\boldsymbol{\rho}^+$. We assume below $\boldsymbol{\rho}_0 = \boldsymbol{\rho}^+$. This means that a good reference parameter $\boldsymbol{v}_0$ should correspond to the *highest traffic intensity vector*, $\boldsymbol{\rho}^+$, among all traffic intensities defined in the constraint (6.4.5). Since $\boldsymbol{\rho}^+$ represents the "noisiest" (most variable) system configuration (regime), this in turn means that in choosing a good $\boldsymbol{v}_0$, we *ignore* all configurations except for the latter. It turns out that this noisiest configuration is the "most informative" one in the sense that through it we can exercise good control over the entire system with the decision vector set $\boldsymbol{v} \in V$ given in (6.4.5).

Remark 6.4.2 can be readily reformulated in terms of the original parameter vector $\boldsymbol{v}_0$, instead of $\boldsymbol{\rho}_0$. If, for example, $\boldsymbol{v}$ is the parameter vector of the interarrivals rates in a queueing network, then the noisiest configuration corresponds to the one with $\boldsymbol{\rho}_0$, $\boldsymbol{\rho}^-$, and $\boldsymbol{\rho}^+$, in (6.4.5) replaced by $\boldsymbol{v}_0$, $\boldsymbol{v}^-$, and $\boldsymbol{v}^+$, respectively. If, however, $\boldsymbol{v}$ is the parameter vector of the service rates, then one has to replace $\boldsymbol{\rho}^+$ by $\boldsymbol{v}^-$ and $\boldsymbol{\rho}^-$ by $\boldsymbol{v}^+$.

Remark 6.4.3. It is shown in Rubinstein and Shapiro (1993) (see also Chapter 8) that for typical performance measures, there exists $\hat{\boldsymbol{\rho}} = \hat{\boldsymbol{\rho}}(\hat{v}, k)$, such that if the reference traffic intensity vector, $\boldsymbol{\rho}_0$, satisfies

$$\boldsymbol{\rho} \le \boldsymbol{\rho}_0 \le \hat{\boldsymbol{\rho}}, \tag{6.4.6}$$

then variance reduction is obtained in the sense of

$$\operatorname{var}_{\boldsymbol{\rho}_0}\{\nabla^k \bar{\ell}_N(\boldsymbol{\rho}, \boldsymbol{\rho}_0)\} \le \operatorname{var}_{\boldsymbol{\rho}}\{\nabla^k \bar{\ell}_N(\boldsymbol{\rho}, \boldsymbol{\rho})\}, \qquad k = 0, 1, \tag{6.4.7}$$

with strict equality holding for $\boldsymbol{\rho} = \hat{\boldsymbol{\rho}}$. Here $\bar{\ell}_N(\boldsymbol{\rho}, \boldsymbol{\rho})$ corresponds to the crude Monte Carlo (CMC) estimator. In other words, for the score function (SF) estimator $\nabla^k \bar{\ell}_N(\boldsymbol{\rho}_0)$ to perform *well* [in the sense of (6.4.7)], the reference traffic intensity, $\boldsymbol{\rho}_0$, must belong to the set $(\boldsymbol{\rho}, \hat{\boldsymbol{\rho}})$. Of course, it may happen that

$$\boldsymbol{\rho}^- = \min_{v \in V} \boldsymbol{\rho}(\boldsymbol{v}_k) < \hat{\boldsymbol{\rho}}.$$

In this case, the variance is reduced in the sense of (6.4.7) whenever $\boldsymbol{\rho} = \boldsymbol{\rho}(\boldsymbol{v})$ is in $[\hat{\boldsymbol{\rho}}, \boldsymbol{\rho}^+]$; otherwise, the variance is increased. In the latter case, the variance increase is typically moderate when $\hat{\boldsymbol{\rho}} - \boldsymbol{\rho}$ is not too large and the cycle length, τ, tends to be small [see Rubinstein and Shapiro (1993) for more details]. If such is not the case, then the set V should be partitioned into smaller subsets, and a different $\boldsymbol{v}_0$ should be chosen over each subset, or a recursive algorithm of the stochastic approximation type can be used.

Keeping the foregoing remarks in mind, consider first the following unconstrained program (see also Section 5.4):

$$(P_0) \qquad \text{minimize} \quad \ell(\boldsymbol{v}), \quad \boldsymbol{v} \in V. \tag{6.4.8}$$

The stochastic counterparts of the program (6.4.8) can be written as

$$(\bar{P}_N) \qquad \text{minimize} \quad \bar{\ell}_N(\boldsymbol{v}), \quad \boldsymbol{v} \in V. \tag{6.4.9}$$

Assume for simplicity that the expected performance, $\ell(\boldsymbol{v})$, has the representation

$$\ell(\boldsymbol{v}) = c\mathbb{E}_{\boldsymbol{v}}\{L_t\} + \sum_{k=1}^{r} b_k v_k, \tag{6.4.10}$$

where $\{L_t\}$ is the sojourn time process, c is the cost incurred by a waiting customer, $\boldsymbol{v} = (v_1, \ldots, v_r)$, and b_k is the unit cost of v_k. We remark that under some mild regularity conditions (see, e.g., Rubinstein and Shapiro, 1993), $\mathbb{E}_{\boldsymbol{v}}\{L_t\}$ is a strictly convex differentiable function with respect to $\boldsymbol{v}$. Thus, $\boldsymbol{v}^*$ is a unique minimizer of (P_0) over the convex region V.

With these assumptions, we may solve the stochastic counterpart (6.4.9), using the following nonlinear system of equations [see also (5.4.6)]:

$$\nabla\bar{\ell}_N(\boldsymbol{v}) = \mathbf{0}, \qquad \boldsymbol{v} \in V.$$

In particular, for a single-node queueing model with FIFO discipline, the above reduces to

$$\begin{aligned} \frac{\nabla\bar{\ell}_N(\boldsymbol{v})}{c} = \frac{\sum_{i=1}^{N}\sum_{t=1}^{\tau_i} L_{ti}\nabla\tilde{W}_{ti}}{\sum_{i=1}^{N}\sum_{t=1}^{\tau_i}\tilde{W}_{ti}} - \frac{\sum_{i=1}^{N}\sum_{t=1}^{\tau_i} L_{ti}\tilde{W}_{ti}}{\sum_{i=1}^{N}\sum_{t=1}^{\tau_i}\tilde{W}_{ti}} \\ \times \frac{\sum_{i=1}^{N}\sum_{t=1}^{\tau_i}\nabla\tilde{W}_{ti}}{\sum_{i=1}^{N}\sum_{t=1}^{\tau_i}\tilde{W}_{ti}} + \frac{\boldsymbol{b}}{c} = \mathbf{0}, \qquad \boldsymbol{v} \in V, \end{aligned} \tag{6.4.11}$$

where

$$\tilde{W}_{ti} = \prod_{k=1}^{t} W(\boldsymbol{Z}_{ki}, \boldsymbol{v})$$

and

$$W(\boldsymbol{Z}, \boldsymbol{v}) = \frac{f(\boldsymbol{Z}, \boldsymbol{v})}{g(\boldsymbol{Z})},$$

and $\boldsymbol{b} = (b_1, \ldots, b_r)$ is the vector of unit costs.

We next present numerical results, illustrating the solution of the stochastic counterpart (6.4.11) for several queueing decision models. It will be assumed below that $g(\mathbf{y}) = f(\mathbf{y}, \mathbf{v}_0)$, where $\mathbf{v}_0$ is the reference parameter.

Example 6.4.1 The *M/M/1* Queue. Let λ and v be the interarrival and service rates, respectively, and let v be the decision parameter. Since $\mathbb{E}_v\{L\} = 1/(v - \lambda)$, it is readily seen that the optimal v^* that minimizes the performance measure given in (6.4.10) is $v^* = \lambda + (c/b)^{1/2}$.

Table 6.4.1 displays the theoretical curves, $\ell(\rho)$ and $\nabla\ell(\rho)$, and their stochastic analogs $\bar{\ell}_N(\rho)$ and $\nabla\bar{\ell}_N(\rho)$ as functions of ρ $(\rho = \lambda/v)$. The parameters chosen were $\lambda = 1$, $b = 1$, $c = 1$ (these correspond to $v^* = 2$ and $\rho^* = \rho(v^*) = 0.5$), and the reference traffic intensity was set at $\rho_0 = 0.8$. The $M/M/1$ queue was then simulated for various sample sizes, N. Figures 6.6 and 6.7 compare the curves $\bar{\ell}_N(\rho)$ and $\nabla\bar{\ell}_N(\rho)$, respectively, in the vicinity of the optimal solution, $\rho^* = 0.5$, to their respective theoretical counterparts.

Table 6.4.1 and Figures 6.5 and 6.6 indicate that the estimated curves, $\bar{\ell}_N(\rho)$ and $\nabla\bar{\ell}_N(\rho)$, *converge* to their respective theoretical counterparts, $\ell(\rho)$ and $\nabla\ell(\rho)$, for $\rho \le \rho_0 = 0.8$, and *diverge* from $\ell(\rho)$ and $\nabla\ell(\rho)$, respectively, for $\rho > \rho_0 = 0.8$. These phenomena are in agreement with the theoretical results.

Table 6.4.2 displays the point estimators, $\bar{v}_N = \bar{v}_N(b)$, the corresponding theoretical values $\boldsymbol{v}^*$, and the 95% confidence intervals of v^* (denoted by 95% CI) as functions of b, $\rho^* = \lambda/v^*$ and $\alpha = [\rho_0 - \rho^*]/\rho^*$. The parameters were again chosen as $\lambda = 1$, $c = 1$, and the reference traffic intensity was set at $\rho_0 = 0.8$. The $M/M/1$ queue was then simulated for $N = 10{,}000$ cycles (approximately 50,000 customers). Note that *all* estimators $\bar{v}_N(b)$ were obtained *simultaneously from a single simulation run* (with $\rho_0 = 0.8$), by solving the system of Eqs. (6.4.11) for different values of b.

Figure 6.8 visualizes the theoretical values of v^*, the point estimators $\bar{v}_N$, and the relative 95% confidence intervals for v^*, as functions of ρ, all based on the values of Table 6.4.2. It is readily seen that the estimator $\bar{v}_N$ performs reasonably well for $\rho \in (0.3, 0.8)$. Poor performance of the estimator $\bar{v}_N$ may be noted for the following cases:

1. For $\rho > 0.8$, this is a consequence of violating the requirement of Remark 6.4.2, which calls for ρ in the range $\rho \le \rho_0 = 0.8$; in fact, we have $0.8 = \rho_0 \le \rho \le 0.88$.
2. For $\rho < 0.2$ $(\alpha = [\rho_0 - \rho^*]/\rho^* > 3)$, this is a consequence of the fact that ρ is perturbed from ρ^* by more than 300%.

Example 6.4.2 The *M/G/1* Queue. Assume that the service time random variable, Y, has the following two-point service time pdf [see (6.2.24)]:

$$f(y, p) = \begin{cases} p & \text{if } y = a_1 \\ 1 - p & \text{if } y = a_2, \end{cases}$$

where $a_1 > 0$, $a_2 > 0$, and $0 < p < 1$.

Table 6.4.1 **Estimated Values, $\bar{\ell}_N(\rho)$, $\nabla\bar{\ell}_N(\rho)$, and the Corresponding Theoretical Values, $\ell(\rho)$ and $\nabla\ell(\rho)$, as Functions of ρ, for $b=1$ and $c=1$, $\lambda=1$, and Reference Traffic Intensity $\rho_0=0.8$**

	N = 500		N = 5,000		N = 50,000		N = 500,000		Theoretical Value	
ρ	$\bar{\ell}_N$	$\nabla\bar{\ell}_N$	$\bar{\ell}_N$	$\nabla\bar{\ell}_N$	$\bar{\ell}_N$	$\nabla\bar{\ell}_N$	$\bar{\ell}_N$	$\nabla\bar{\ell}_N$	ℓ	$\nabla\ell$
0.1	10.1	0.99	10.1	0.99	10.1	0.99	10.1	0.99	10.1	0.99
0.2	5.22	0.94	5.24	0.94	5.24	0.94	5.25	0.94	5.25	0.94
0.3	3.72	0.82	3.74	0.82	3.75	0.82	3.76	0.82	3.76	0.82
0.4	3.13	0.56	3.14	0.59	3.15	0.57	3.16	0.56	3.17	0.56
0.5	2.94	0.08	2.94	0.11	2.97	0.06	2.99	0.00	3.00	0.00
0.6	3.07	−1.03	3.07	−1.21	3.11	−1.16	3.16	−1.25	3.17	−1.25
0.7	3.46	−2.08	3.66	−4.26	3.70	−4.58	3.77	−4.59	3.76	−4.44
0.8	3.84	−3.01	4.97	−12.0	5.30	−16.9	5.31	−15.2	5.25	−15.0
0.85	5.12	−4.33	6.15	−28.6	6.44	−30.1	6.76	−31.6	6.84	−31.2
0.88	9.87	−8.01	7.87	−37.5	8.06	−48.7	8.24	−55.1	8.33	−53.2

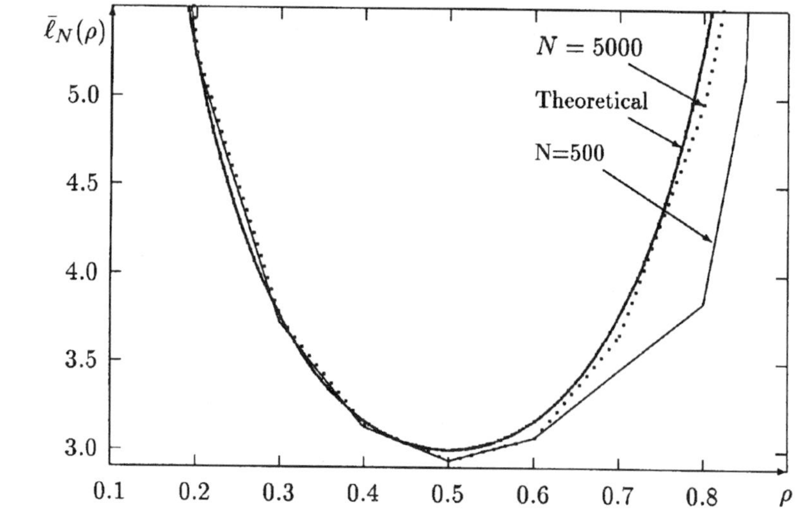

Figure 6.6. Behavior of $\ell(\rho)$ and $\bar{\ell}_N(\rho)$ in the vicinity of ρ^*, for $b=1$, $c=1$, $\lambda=2$.

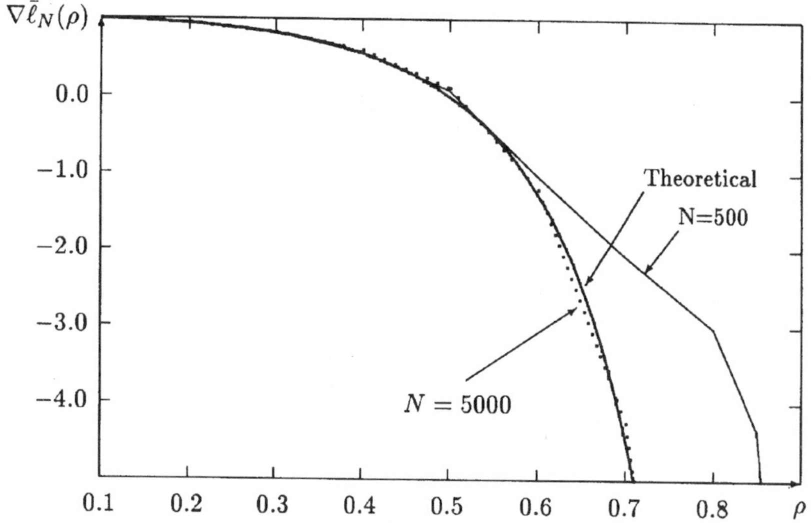

Figure 6.7. Behavior of $\nabla\ell(\rho)$ and $\nabla\bar{\ell}_N(\rho)$ in the vicinity of ρ^*, for $b=1$, $c=1$, $\lambda=2$.

Table 6.4.2 Performance of Stochastic Counterpart (6.4.11), for the $M/M/1$ Queue with Reference Traffic Intensity $\rho_0 = 0.8$

ρ^*	b	α	$\bar{v}_N(b)$	v^*	95% CI
0.88	53.77	−0.091	0.675	1.136	0.07, 1.27
0.85	32.11	−0.058	0.978	1.176	0.78, 1.18
0.8	16.00	0.000	1.261	1.255	1.15, 1.37
0.7	5.444	0.143	1.433	1.429	1.33, 1.54
0.6	2.225	0.333	1.654	1.667	1.58, 1.73
0.5	1.000	0.600	1.971	2.000	1.91, 2.03
0.4	0.444	1.000	2.467	2.500	2.42, 2.51
0.3	0.184	1.667	3.324	3.333	3.25, 3.39
0.2	0.063	3.000	4.947	5.000	4.74, 5.15
0.1	0.012	7.000	9.741	10.00	9.42, 10.06

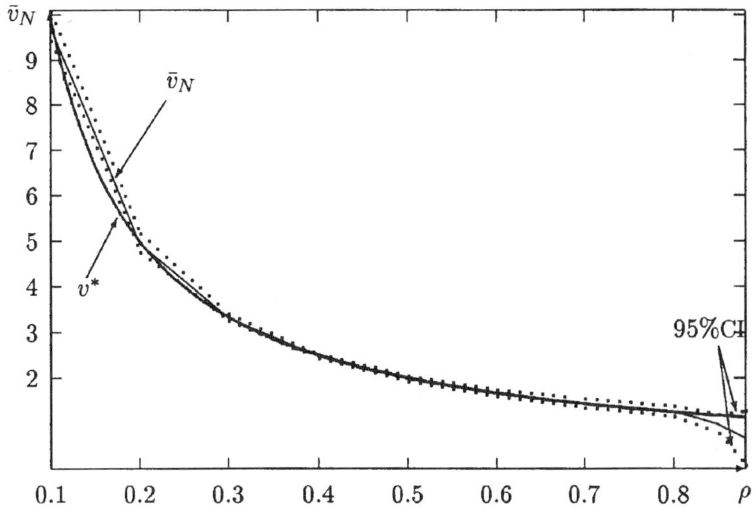

Figure 6.8. Performance of the estimator $\bar{v}_N$ for the $M/M/1$ queue.

By the Pollaczek-Khinchin formula (see, e.g., Gross and Harris, 1985), the steady-state expected sojourn time can be written as

$$\ell(p) = \mathbb{E}_p\{L\} = \beta_1 + \frac{\lambda\beta_2}{2(1-\lambda\beta_1)},$$

where λ is the arrival rate, and

$$\beta_1 = \mathbb{E}\{Y\} = pa_1 + (1-p)a_2, \ \beta_2 = \mathbb{E}\{Y^2\} = pa_1^2 + (1-p)a_2^2.$$

In this case, the likelihood ratio, W in (6.4.11), reduces to

$$W(Z,p) = \left(\frac{p}{p_0}\right)^{(Z-a_2)/(a_2-a_1)} \left(\frac{1-p}{1-p_0}\right)^{(Z-a_1)/(a_2-a_1)}$$

Table 6.4.3 displays data similar to those of Table 6.4.2. Here $\lambda = 0.6$, $p_0 = 0.5$, $a_1 = 0.5$, $a_2 = 1.2$ ($\rho_0 = 0.85$), $c = 1$, and $\rho^+ = 0.8$. A simulation of the $M/G/1$ queue was run for $N = 15{,}000$ cycles (approximately 100,000 customers). In view of Remark 6.4.2 and the selected value $\rho^+ = 0.8$, the reference traffic intensity $\rho_0 = 0.85$ was chosen moderately larger than ρ^+.

Table 6.4.4 displays data similar to those in Table 6.4.2, but for two $M/M/1$ queues in tandem, corresponding to the stochastic counterpart ($\bar{P}_N^d$). Specifically, it displays the theoretical values $\ell(\boldsymbol{v}^*)$, the point estimators $\ell(\bar{\boldsymbol{v}}_N)$, the 95% confidence intervals for $\ell(\boldsymbol{v}^*)$ (denoted 95% CI) along with the values $\gamma = (\|\boldsymbol{v}^* - \bar{\boldsymbol{v}}_N\|)/\|\boldsymbol{v}^*\|$ ($\|\boldsymbol{v}\| = (v_1^2 + v_2^2)^{1/2}$), as functions of (b_1, b_2), $(\rho_1^*, \rho_2^*) = (\lambda/v_1^*, \lambda/v_2^*)$, and $(\alpha_1, \alpha_2) = ([\rho_{01} - \rho_1^*]/\rho_1^*, [\rho_{02} - \rho_2^*]/\rho_2^*)$. Selecting again $\lambda = 1$, $c = 1$, the reference traffic intensities were set at $\rho_{01} = 0.8$, $\rho_{02} = 0.6$, and a simulation was run for

Table 6.4.3 Performance of Stochastic Counterpart (6.4.11), for the $M/G/1$ Queue with Reference Traffic Intensity $\rho_0 = 0.85$

ρ^*	b	α	p^*	$\bar{p}_N$	95% CI
0.80	10.32	0.062	0.571	0.559	0.34, 0.78
0.70	4.978	0.214	0.714	0.716	0.65, 0.73
0.65	3.843	0.307	0.786	0.804	0.77, 0.83
0.60	3.106	0.417	0.857	0.868	0.82, 0.91
0.55	2.601	0.545	0.928	0.951	0.91, 0.99
0.50	2.240	0.700	1.000	1.000	0.95, 1.05

Table 6.4.4 Performance of Stochastic Counterpart (6.4.11), for Two $M/M/1$ Queues in Tandem

ρ_1^*	ρ_2^*	b_1	b_2	α_1	α_2	$\ell(\boldsymbol{v}^*)$	$\ell(\bar{\boldsymbol{v}}_N)$	95% CI	γ
0.60	0.88	2.25	53.8	0.33	−0.09	73.7	76.0	72.3, 79.7	0.200
0.60	0.80	2.25	16.00	0.33	0.00	29.3	29.3	28.0, 30.6	0.010
0.60	0.60	2.25	2.25	0.33	0.143	10.50	10.48	10.3, 10.6	0.004
0.60	0.40	2.25	0.44	0.33	1.00	7.03	7.02	6.92, 7.12	0.004
0.60	0.20	2.25	0.06	0.33	3.00	5.81	5.81	5.72, 5.90	0.002
0.60	0.10	2.25	0.01	0.33	7.00	5.49	5.48	5.38, 5.58	0.003
0.88	0.60	53.8	2.25	−0.09	0.33	73.7	75.7	69.3, 82.0	0.154
0.80	0.60	16.0	2.25	0.00	0.33	29.25	29.30	28.1, 30.5	0.008
0.40	0.60	0.44	2.25	1.00	0.33	7.03	7.02	6.93, 7.12	0.002
0.20	0.60	0.06	2.25	3.00	0.33	5.81	5.80	5.71, 5.90	0.013
0.10	0.60	0.01	2.25	7.00	0.33	5.48	5.47	5.38, 5.57	0.007

$N = 10{,}000$ cycles (approximately 50,000 customers). The results of Tables 6.4.3 and 6.4.4 indicate good performance of the SF method.

Consider now the constrained program (6.4.1). Suppose that $\ell_0(\boldsymbol{v}) = \mathbb{E}_{\boldsymbol{v}}\{L_0\}$ represents the steady-state sojourn time performance in a general queueing network consisting of r nodes, and consider the following constraints:

$$\ell_1(\boldsymbol{v}) = \sum_{k=1}^{r} b_k v_k \leq c_1, \tag{6.4.12}$$

$$\ell_2(\boldsymbol{v}) = \mathbb{E}_{\boldsymbol{v}}\{L_1\} \leq c_2, \tag{6.4.13}$$

$$V = \{\boldsymbol{v}\colon \mathbf{0} \leq \boldsymbol{\rho}^- \leq \boldsymbol{\rho}(\boldsymbol{v}) \leq \boldsymbol{\rho}^+ < \mathbf{1}, \boldsymbol{\rho} = (\rho_1, \ldots, \rho_r)\}. \tag{6.4.14}$$

Here, b_k is the unit cost of v_k, L_1 is the number of customers in the network, and c_1 and c_2 are fixed numbers. Observe that the functions $\ell_0(\boldsymbol{v})$ in the stochastic counterparts, $(\bar{P}_N)$ and $(\bar{P}_N^d)$, must be replaced by their stochastic equivalents, $\bar{\ell}_{0N}(\boldsymbol{v})$ [see (6.4.4)] and $\bar{\ell}_{0N}^d(\boldsymbol{v})$ [see (6.3.9)], and similarly for the function $\ell_2(\boldsymbol{v})$. However, the "deterministic" function $\ell_1(\boldsymbol{v})$ remains the same in all three programs: (P_0), $(\bar{P}_N)$, and $(\bar{P}_N^d)$. Notice also that our approach to the constrained programs (stochastic counterparts) $(\bar{P}_N)$ and $(\bar{P}_N^d)$ is to first reduce them to unconstrained ones via the *augmented Lagrangian* method (see Avriel, 1976), and then to solve them by the *quasi-Newton* algorithm.

Table 6.4.5 displays data similar to those of Table 6.4.2, for the $M/M/1$ queue, obtained by considering the program

$$(P_0) = \begin{cases} \text{minimize} & \ell_0(v) = \mathbb{E}_v\{L_0\} + bv, \quad v \in V, \\ \text{subject to} & \ell_1(v) = v - 3 \leq 0, \\ & \ell_2(v) = \mathbb{E}_v\{L_1\} - 2 \leq 0, \\ & V = \{v : \rho^- = 0.2 \leq \rho \leq \rho^+ = 0.85\}, \end{cases} \tag{6.4.15}$$

and simulating $N = 10{,}000$ cycles (approximately 50,000 customers). Here L_0 and L_1 are the sojourn time and the number of customers in the $M/M/1$ system, respectively.

Table 6.4.5 Performance of Stochastic Counterpart $(\bar{P}_N)$ for Program (6.4.15)

ρ^*	b	α	$\bar{v}_N(b)$	v^*	95% CI
0.85	32.11	−0.058	1.477	1.500	1.43, 1.53
0.8	16.00	0.000	1.477	1.500	1.44, 1.52
0.7	5.444	0.143	1.477	1.500	1.44, 1.52
0.6	2.225	0.333	1.654	1.667	1.63, 1.68
0.5	1.000	0.600	1.971	2.000	1.93, 2.00
0.4	0.444	1.000	2.467	2.500	2.42, 2.51
0.3	0.184	1.667	3.000	3.000	2.98, 3.02
0.2	0.063	3.000	3.000	3.000	2.98, 3.03
0.1	0.012	7.000	3.000	3.000	2.96, 3.04

It is readily seen that the data of Table 6.4.5 are in agreement with the theoretical results. At this juncture, it is also worth pointing out that extensive supporting numerical results are given in Itzhaki (1994) for both the unconstrained and constrained programs (P_0), for various network topologies and for various dimensions n ($n \leq 1000$) of the decision vector $\boldsymbol{v}$ (in interarrival time and service time distributions, and in routing probabilities), which handles problems with hundreds of parameters. These were obtained by means of the research QNSO (queueing network optimizer and stabilizer). For $r \geq 3$ nodes, Itzhaki (1994) utilized the stochastic counterpart $(\bar{P}_N^d)$ rather than $(\bar{P}_N)$. As discussed earlier, the choice of $(\bar{P}_N^d)$ reduces dramatically the variance of the estimator of $\boldsymbol{v}^*$, at the cost of introducing a modest bias.

It is important to note, however, that if the constrained region is too wide to be trusted, [say the interval V, defined in (6.4.15) as

$$V = \{v: \rho^- = 0.2 \leq \rho \leq \rho^+ = 0.85\}$$

is too wide], then alternatives to stochastic counterpart methods (e.g., Shapiro, 1996) might be employed. One of the alternatives is to use stochastic approximation-type algorithms, as below.

6.4.2 On-line Optimization

In on-line optimization setting, one cannot construct directly a stochastic counterpart of the type $(\bar{P}_N)$; it is then the duty of the mathematical programmer to come up with specific optimization techniques. Two algorithms are presented below for on-line optimization: The first is called the *classical stochastic approximation* (CSA) algorithm, and the second is called the *robust stochastic approximation* (RSA) algorithm.

Classical Stochastic Approximation

The classical stochastic approximation originated with the seminal studies of Robbins-Monro and Kiefer-Wolfowitz. In its original form, CSA deals with *smooth unconstrained* problems of the form

$$\min_{\boldsymbol{v}} \ell(\boldsymbol{v}) = \min_{\boldsymbol{v}} \mathbb{E}\{h(\boldsymbol{Y}, \boldsymbol{v})\}, \tag{6.4.16}$$

where $\boldsymbol{v}$ belongs to an n-dimensional Euclidean space, $\boldsymbol{Y}$ is a random vector with unknown distribution, ℓ is a smooth and convex function that attains its minimum at a unique point, and the function is nondegenerate at this point, that is, its second-order derivative is positive definite. It is further assumed that h is differentiable in $\boldsymbol{v}$ and sufficiently regular to permit the interchange of the expectation and differentiation operators, namely,

$$\nabla \ell(\boldsymbol{v}) = \mathbb{E}\{\nabla h(\boldsymbol{Y}, \boldsymbol{v})\}. \tag{6.4.17}$$

The basic assumption of CSA is that a solution of the program (6.4.16) can be obtained by estimating $\nabla\ell(\boldsymbol{v})$. We assume that at each step t, one can compute the random vector $\nabla(\boldsymbol{Y}_t, \boldsymbol{v}_t) \equiv G(\boldsymbol{Y}_t, \boldsymbol{v}_t)$, where $\boldsymbol{Y}_t$ is the input sequence and $\boldsymbol{v}_t$ is updated on the basis of information obtained at the previous steps. We further assume that for fixed $\boldsymbol{Y}_1, \ldots, \boldsymbol{Y}_{t-1}$, the conditional expectation of G is precisely the gradient of h at $\boldsymbol{v}_t$, which is equivalent to

$$G(\boldsymbol{Y}_t, \boldsymbol{v}_t) = \nabla\ell(\boldsymbol{v}_t) + \boldsymbol{\delta}(\boldsymbol{Y}_t, \boldsymbol{v}_t), \tag{6.4.18}$$

for some vector function $\boldsymbol{\delta}$ such that

$$\mathbb{E}\{\boldsymbol{\delta}(\boldsymbol{Y}_t, \boldsymbol{v}_t) | \boldsymbol{Y}_1, \ldots, \boldsymbol{Y}_{t-1}\} = \mathbf{0}.$$

This basic relation [which follows immediately from (6.4.17) and the assumption that the random vectors $\boldsymbol{Y}_t$ are independent] means that we observe unbiased estimates of the gradient of our objective function $\nabla\ell(\boldsymbol{v})$. Under these assumptions, one may use the following iterative procedure:

$$\boldsymbol{v}_{t+1} = \Pi_X\{\boldsymbol{v}_t - \gamma_t G(\boldsymbol{Y}_t, \boldsymbol{v}_t)\}, \tag{6.4.19}$$

where $\{\gamma_t\}$ is a positive sequence, the initial value (starting point) $\boldsymbol{v}_1$ is fixed, and $\Pi_X(z)$ denotes the projection operator, which maps the point z onto the set X [i.e., for every $x \in \mathbb{R}^n$ we have that $\Pi(x) \in X$, and $\|x - \Pi(x)\| = \min_{y \in X} \|x - y\|$]. It is readily seen that (6.4.19) represents a *gradient descent procedure* in which the exact gradients are replaced by their estimates. Under certain reasonable assumptions (e.g., Kushner and Clark, 1978), the sequence $\{\boldsymbol{v}_t\}$ converges almost surely and in squared mean to the minimizer of the objective function ℓ. Intuitively, convergence can be explained by the following heuristic argument. Each step in (6.4.19) consists of two components: a deterministic component, that is, the gradient of the objective function that moves $\boldsymbol{v}_t$ toward the minimum, and a random component, δ, which moves $\boldsymbol{v}_t$ in an unknown direction. However, if the step sizes are small enough and the vector field formed by the gradients is smooth, then the random components in (6.4.19) will approximately cancel each other after a reasonable number of steps, while the deterministic ones will jointly move the system toward the optimal point. There are many theorems on convergence and the corresponding convergence rate for SA (e.g., Kushner and Clark, 1978). One of the simplest is presented next.

Theorem 6.4.1. *Assume that ℓ is smooth and strictly convex, namely,*

$$\ell(\boldsymbol{v} + \Delta\boldsymbol{v}) \geq \ell(\boldsymbol{v}) + (\nabla_{\boldsymbol{v}}\ell(\boldsymbol{v}))'\Delta\boldsymbol{v} + \frac{\alpha}{2}(\Delta\boldsymbol{v})'\Delta\boldsymbol{v}, \qquad \alpha > 0, \tag{6.4.20}$$

where prime denotes the transpose operator. Assume further that the observation errors in the derivative vector $G(x, Y)$ possess a bounded second moment, that is,

$$\mathbb{E}\{\|G(x, Y) - \nabla\ell(\boldsymbol{v})\|^2\} \leq L^2 < \infty.$$

Then for an arbitrary (deterministic) positive sequence $\{\gamma_t\}$, *such that*

$$\sum_{t=1}^{\infty} \gamma_t = \infty, \qquad \sum_{t=1}^{\infty} \gamma_t^2 < \infty,$$

the vector $\boldsymbol{v}_t$ *converges asymptotically to* $\boldsymbol{v}^*$ *(the minimizer of* ℓ*) in the sense of mean square. If, moreover,*

$$\gamma_t = c/t$$

with an appropriate constant c [*whether a given c is appropriate depends only on* α *in* (6.4.20)], *then for all t we have the following bounds:*

$$\mathbb{E}\{\|\boldsymbol{v}_t - \boldsymbol{v}^*\|^2\} \leq \frac{A(\alpha, c)}{t} \|v_0 - v\|^2$$

and

$$\mathbb{E}\{\ell(\boldsymbol{v}_t) - \ell(\boldsymbol{v}^*)\} \leq O(1/t),$$

where $A(\alpha, c)$ *is some constant depending on* α *and c.*

The assumptions for asymptotic normality of the random vector

$$t^{1/2}(\boldsymbol{v}_t - \boldsymbol{v}^*)$$

and its exact asymptotic covariance matrix may be found, for example, in Kushner and Clark (1978), and Rubinstein (1986).

We next look further into asymptotics of SA. First, it turns out that the convergence rate $O(1/t)$ is *unimprovable* in a very strong sense. Therefore, the classical SA procedure is, in fact, optimal in the order of convergence. It is even possible to define asymptotically efficient procedures with unimprovable first-order term of the asymptote. Note, however, that "small" step sizes,

$$\gamma_t = O(1/t),$$

which are optimal from the viewpoint of *asymptotic behavior*, may result in very slow convergence. In this sense, CSA is known as a *nonrobust procedure* [see Nemirovskii and Rubinstein (1996) for more details].

In order to obtain a reasonable rate of convergence, one needs to adjust the step sizes to the actual values of the smoothness/nondegeneracy parameters, as even moderate misestimation of these parameters can result in a dramatic decrease in the convergence rate. To overcome this difficulty, many on-line adjusting strategies for step sizes have been proposed. Even with these strategies, the rate of convergence of SA remains sensitive to smoothness and nondegeneracy of the objective function. Moreover, if even one of these assumptions does not hold, then almost nothing may be concluded as regards the convergence rate of SA.

Fortunately, there exists a *robust* version of the classical SA procedure. Note that "robustness" is meant here with respect to the *a priori assumptions on smoothness and nondegeneracy of the objective function*, and not in the classical statistical sense (see, e.g., Huber, 1981). Here, we shall consider only robust SA (RSA) for solving the simplest unconstrained program, with $\boldsymbol{Y}_1, \boldsymbol{Y}_2, \ldots,$ being an iid sequence. A more general setting is considered in Nemirovskii and Rubinstein (1996).

Robust SA

Consider the following SA algorithm:

$$\boldsymbol{v}_{t+1} = \Pi\{\boldsymbol{v}_t - \gamma_t G(\boldsymbol{Z}_t, \boldsymbol{v}_t)\}, \qquad \boldsymbol{v} \in V. \tag{6.4.21}$$

Here Π is the projection operator,

$$\gamma_t = \frac{D}{M\sqrt{t}} \tag{6.4.22}$$

is the step size, D is the diameter of V, and M satisfies

$$\mathbb{E}\{\|G(\boldsymbol{Z}_t, \boldsymbol{v})\|^2\} \le M^2. \tag{6.4.23}$$

As approximate solutions of the program (6.4.16), we take

$$\boldsymbol{v}^t = \frac{1}{\lfloor t/2 \rfloor} \sum_{\lfloor t/2 \rfloor + 1 \le r \le t} \boldsymbol{v}_r, \tag{6.4.24}$$

rather than the $\boldsymbol{v}_t$ of (6.4.21). Notice that the sequence $\{\boldsymbol{v}^t\}$ can be viewed as *moving (Cesaro) averages* of the sequence $\{\boldsymbol{v}_t\}$. As already stated, procedure (6.4.21)–(6.4.24) is called the robust stochastic approximation (RSA) procedure. It readily follows that RSA differs from classical SA in that it uses:

- A larger step size, namely, $\gamma_t = O(t^{-1/2})$ instead of $\gamma_t = O(t^{-1})$.
- Moving averages $\boldsymbol{v}^t$ [see (6.4.24)] instead of $\boldsymbol{v}_t$ [see (6.4.21)].

The convergence proof and supporting numerical results pertaining to the RSA procedure (6.4.21)–(6.4.24) for simple queueing decision models are given in Nemirovskii and Rubinstein (1996).

It is shown in Shapiro (1996) that for smooth unconstrained problems, the SA estimators based on an optimal choice of the step sizes are asymptotically equivalent to the stochastic counterpart estimators. Their asymptotic properties for nonsmooth and constrained problems is still an open problem.

6.5 TOPOLOGICAL DESIGN OF NETWORKS

This section deals with topological design of networks, namely, with the network version of the so-called what-if design problem. The problem may be stated as follows: How would the expected performance, $\ell(\boldsymbol{v})$, and the associated sensitivities, $\nabla^k \ell(\boldsymbol{v})$, change as the topology of the network is perturbed? Permissible perturbations consist of adding (deleting) a node (several nodes) to (from) the network.

It can be shown that the DSF estimators [see Section 6.3], $\nabla^k \bar{\ell}_N^d(\boldsymbol{v}) = \nabla^k \bar{\ell}_N^d(\boldsymbol{v}, \gamma = 1, \delta_1 = 1)$, can successfully handle such what-if problems. The treatment of the TSF estimators turns out to be similar.

Consider first a tandem queueing network consisting of r nodes. Suppose we wish to estimate *simultaneously, from a single simulation run*, the performance measures $\nabla^k \ell(\boldsymbol{v})$, in both the original r-node network and the perturbed one; the latter is obtained from the original network, by *eliminating*, say, the last node (the one with the highest index). To this end, we argue as follows. First, note that for the perturbed $(r-1)$-node network, the output process, $\{L_t\}$ in (6.3.4), can be written as

$$L_t = \sum_{q=1}^{r-1} L_{qt} = \sum_{q=1}^{r} L_{qt},$$

where $L_{rt} = 0$. This clearly suggests that one may use the original network as the reference case (the case to be simulated), and then consider the perturbed network as a special case, namely, with the traffic intensity at the last node set to zero ($\rho_r = 0$), but with all other data unchanged.

Modeling such a special case can be done in several ways. One alternative is to model the rth queue in the perturbed network as an auxiliary $GI/G/1$ queue with an infinite (very large) service rate. It will now be shown that in this case, the rth term in the TDSF $\bar{\ell}_N^d(\boldsymbol{v})$ [see (6.3.9)] vanishes. More accurately, one gets

$$\frac{\sum_{i=1}^{N} \sum_{t=1}^{\tau_{ri}} L_{rti} \tilde{W}_{rti}^d}{\sum_{i=1}^{N} \sum_{t=1}^{\tau_{ri}} \tilde{W}_{rti}^d} = \frac{0}{0}, \tag{6.5.1}$$

where we use the convention $0/0 = 0$, and similarly for the rth terms in $\nabla^k \bar{\ell}_N^d(\boldsymbol{v})$, $k > 0$. To justify (6.5.1), observe that since the service rate in the auxiliary rth queue equals infinity (equivalently, $\rho_r = 0$), the values of the output processes $\{L_{rti}\}$, and likelihood ratios $\{\tilde{W}_{rti}^d\}$, will both vanish with probability 1. The validity of (6.5.1) now follows, in view of the convention $0/0 = 0$. In practice, one chooses the service rate at the rth queue to be very large (yielding very small ρ_r) rather than infinity ($\rho_r = 0$). Note also that if we need to *add*, rather than *delete*, a node to the network, we may select a network consisting of $r + 1$ nodes as the reference network, and then consider the original r-node network as a special case of the reference network with the $(r + 1)$st queue having an infinite (very large) service rate. Again, the resultant what-if estimators will be biased, since so are the TDSF estimators, $\bar{\ell}_N^d(\boldsymbol{v})$.

The above ideas can be adapted to rather general queueing networks. More specifically, assume that we have several "related" topological designs, and we wish to evaluate their performance measures, $\nabla^k \ell(\boldsymbol{v})$, *simultaneously from a single simulation run*. The term "related" means here that all network topologies under consideration can be obtained from each other only by addition and deletion operations of a few nodes.

The algorithm for the topological what-if network design is given below.

Algorithm 6.5.1 Topological Design

1. *Construct a reference network by combining all topologically related networks under study, such that any network under consideration is a subnetwork of the reference network.*
2. *Expand each such subnetwork to the reference network by adding one more dummy node with very small traffic intensities.*
3. *Run a simulation of the reference network and estimate (from a single simulation run) the performance measures,* $\nabla^k \ell(\boldsymbol{v})$, $k = 0, 1,$ *for all networks under consideration, using the TDSF estimators* $\nabla^k \bar{\ell}_N^d(\boldsymbol{v}) = \nabla^k \bar{\ell}_N^d(\boldsymbol{v}, \gamma = 1, \delta_1 = 1), k = 0, 1$ *in* (6.3.9).

6.6 EXERCISES

1. Consider the $M/G/1$ queue. Let $\ell(\boldsymbol{v}) = \mathbb{E}_{\boldsymbol{v}}\{L_t\}$, $\boldsymbol{v} = (\lambda, \mu)$, where λ and μ are the interarrival and service rates, respectively. Find analytical expressions for $\nabla \ell(\boldsymbol{v}) = \nabla \mathbb{E}_{\boldsymbol{v}}\{L_t\}$, where $\nabla \ell(\boldsymbol{v}) = (\partial \ell(\boldsymbol{v})/\partial \lambda, \ \partial \ell(\boldsymbol{v})/\partial \mu)$, assuming that the process $\{L_t\}$ is

(i) the waiting times in the queue
(ii) the sojourn times through the queue
(iii) the numbers of customers in the queue
Hint. Use the Pollaczek-Khinchin formula. For example, for sojourn times,

$$\ell(\boldsymbol{v}) = \mathbb{E}\{S\} + \frac{\lambda \mathbb{E}\{S^2\}}{2(1 - \lambda \mathbb{E}\{S\})},$$

where S denotes the service time random variable.

2. Consider a cyclic queueing network consisting of m $GI/G/1$ queues with ring topology. Assuming that only a *single customer* circulates in the network, find:

(i) $\ell(\boldsymbol{v}) = \mathbb{E}_{\boldsymbol{v}}\{L\}$, where $L = \sum_{j=1}^{m} L_j$, and L_j is the sojourn time at the jth queue. In this case, $L_j = Y_j$, where Y_j is the service time at the jth queue.

(ii) $\nabla\ell(\boldsymbol{v}) = \nabla\mathbb{E}_{\boldsymbol{v}}\{L\} = \mathbb{E}_{\boldsymbol{v}_0}\{L\nabla W\}$.

(iii) $\text{var}_{\boldsymbol{v}_0}\{L\nabla W\}$, where $W = f(\boldsymbol{Z}, \boldsymbol{v})/f(\boldsymbol{Z}, \boldsymbol{v}_0)$, $\boldsymbol{Z} \sim f(\mathbf{z}, \boldsymbol{v}_0)$, $f(\mathbf{z}, \boldsymbol{v}) = \prod_{j=1}^{m} f_j(z_j, \boldsymbol{v}_j)$, and the distributions of the components of the random vector $\boldsymbol{Z} = (Z_1, \ldots, Z_m)$ belong to the canonical exponential family of distributions, given in (5.2.24).

3. Consider the $GI/\mathrm{G}/1$ queue. Write expressions for the score function process, $\{\tilde{S}_t^{(1)}(\boldsymbol{v})\}$, utilizing the pdf's given in Examples 5.2.5–5.2.8 of Chapter 5.

4. Consider m $GI/G/\infty$ queues in tandem. Assuming that each queue is served by an *infinite number* of servers and that the service time distributions are $G(\lambda_0, \beta)$, where G denotes the gamma distribution, show that

(i) $\text{var}_{\boldsymbol{v}_0}\{L\tilde{W}\} = (1 + \alpha_0^2/(1 + 2\alpha_0))^{\beta m}\mathbb{E}_{2\boldsymbol{v}-\boldsymbol{v}_0}\{L^2(Z_1, \ldots, Z_m)\} - \ell^2(\boldsymbol{v})$, for $\boldsymbol{v}_0 = (\lambda_0, \beta)$ (i.e., only λ is perturbed), $\alpha_0 = (\lambda - \lambda_0)/\lambda_0$, $\tilde{W} = \prod_{j=1}^{m} W_j$, and $W_j = f_j(Z_j, \boldsymbol{v})/f_j(Z_j, \boldsymbol{v}_0)$.

(ii) Find $\text{var}_{\boldsymbol{v}_0}\{L\nabla\tilde{W}\}$, for $\boldsymbol{v}_0 = (\lambda_0, \beta)$.

(iii) Let $L(Z_1, \ldots, Z_m) = \sum_{j=1}^{m} Z_j$. Show that for any given λ, there exists λ_0, such that if $\lambda_0 < \lambda$, then $\text{var}_{\lambda_0}\{L\tilde{W}\} \le \text{var}_\lambda\{L\}$. In other words, the what-if estimator is more accurate than the ordinary one.

(iv) Similarly to (iii), show that there exists $\tilde{\lambda}_0$, such that if $\tilde{\lambda}_0 < \lambda$, then $\text{var}_{\tilde{\lambda}_0}\{L\nabla\tilde{W}\} \le \text{var}_\lambda\{L\tilde{S}\}$, where $\tilde{S} = \sum_{j=1}^{m} \nabla \log f(Y_j, \lambda)$ and $Y_j \sim G(\lambda, \beta)$.

(v) Let $m = 1$, $\beta = 1$, and $L(Y) = I_{(-\infty, a)}(Y)$. Assuming that $Y \sim \text{Exp}(v)$ and a is given, find the solution of the following optimization problem:

$$\min_{v_0} \text{var}_{v_0}\{LW(v)\}.$$

(vi) Repeat (v), replacing LW by $L\nabla W$.

5. Consider the $GI/G/1/b$ queue (b is the buffer size). Devise an algorithm for computing the likelihood process, $\{\tilde{W}_t\}$, while estimating the performance measures

$$\nabla^k\ell(\boldsymbol{v}) = \mathbb{E}_{\boldsymbol{v}_0}\left\{\sum_{t=1}^{\tau} L_t \nabla^k W_t(\boldsymbol{v})\right\}, \qquad k = 0, 1, \ldots,$$

under the FIFO and last-in/first-out (LIFO) disciplines, for $\boldsymbol{v} = (\lambda, \mu)$.
Hint: Note that for the $GI/G/1$ queue,

$$W_t = \prod_{j=1}^{t} \tilde{W}_j, \qquad \tilde{W}_j = \tilde{W}_{j1}\tilde{W}_{j2}, \qquad \tilde{W}_{ji} = \frac{f_i(Z_{ji}, \boldsymbol{v}_i)}{f_i(Z_{ji}, \boldsymbol{v}_{0i})},$$

$$Z_{ji} \sim f_i(z, \boldsymbol{v}_i), \qquad i = 1, 2.$$

6. Consider the $GI/G/1$ queue with a feedback loop, and repeat Exercise 5.

7. Consider an open (closed) queueing network with a routing-probability matrix $P(v) = \{p_{ij}(v)\}$ that depends on the parameter v. Find an expression for the SF estimator of

$$\nabla^k \ell(\boldsymbol{v}) = \mathbb{E}_{\boldsymbol{v}_0}\left\{\sum_{t=1}^{\tau} L_t \nabla^k W_t(\boldsymbol{v})\right\}, \qquad k = 0, 1, \ldots$$

8. Run a simulation of the $M/M/1/m$ queue and the $M/M/1$ queue with a feedback loop, and estimate

$$\nabla^k \ell(\boldsymbol{v}) = \mathbb{E}_{\boldsymbol{v}_0}\left\{\sum_{t=1}^{\tau} L_t \nabla^k W_t(\boldsymbol{v})\right\}, \qquad k = 0, 1$$

using the SF method.

9. Find confidence intervals (regions) for the DSF estimators (6.3.11) and (6.3.12).

10. Derive confidence intervals for $\nabla^k \ell(\boldsymbol{v})$, using the SF estimators (6.2.11) and (6.2.12).

11. Simulate the queueing network of Figure 6.5, and repeat the numerical results in Tables 6.3.1 and 6.3.2.

12. Simulate an $M/M/1$ queue with the reference traffic intensity set at $\rho_0 = \lambda/\mu_0 = 0.8$. Set $\mu_0 = 1$, and sample size $N = 10^4$, and estimate the steady-state response surfaces $\nabla^k \ell(\rho)$, $k = 0, 1$, for $0.2 \leq \rho \leq 0.85$, where $\ell(\rho)$ is the expected number of customers in the queue. Generate figures similar to Figures 6.3 and 6.4 and discuss your results.

13. Simulate the $M/M/1$ and the $M/G/1$ queues in Examples 6.4.1 and 6.4.2, respectively, and repeat the results.

14. Is it possible to represent $\nabla \ell(\boldsymbol{v})$ in (6.2.12), analogously to (6.2.15), as

$$\nabla \ell(\boldsymbol{v}) = \mathbb{E}_g\{\boldsymbol{G}_t\} = cov_{\boldsymbol{v}}\{L_t, \nabla \tilde{W}_t\} = \frac{\mathbb{E}_g\{\sum_{t=1}^{\tau} \boldsymbol{G}_t\}}{\mathbb{E}_g\{\sum_{t=1}^{\tau} \tilde{W}_t\}},$$

where

$$\boldsymbol{G}_t = (L_t - \ell(\boldsymbol{v}))(\nabla \tilde{W}_t - \nabla \tilde{w})$$

and $\nabla \tilde{w} = \mathbb{E}_g\{\nabla \tilde{W}_t\}$?

REFERENCES

Avriel, M. (1976). *Nonlinear Programming Analysis and Methods*, Prentice-Hall, Englewood Cliffs, NJ.

Asmussen, S. (1985). "Conjugate processes and the simulation of ruin problems," *Stoch. Proc. Appl.*, **20**, 213–229.

Asmussen, S. (1996). *Ruin Probabilities*, World Scientific, Singapore.

Asmussen, S. and Rubinstein, R. Y. (1992). "The efficiency and heavy traffic properties of the score function method in sensitivity analysis of queueing models," *Adv. Appl. Prob.*, **24**(1), 172–201.

Bertsekas, A. and Gallager, R. (1987). *Data Networks*, Prentice-Hall, Englewood Cliffs, NJ.

Ermoliev, M. (1969). "On the method of generalized stochastic gradients and quasi-Fejer sequences," *Cybernetics*, **5**(2), 208–220.

Ermoliev, Y. M. and Gaivoronski, A. A. (1992). "Stochastic programming techniques for optimization of discrete event systems," *Ann. Oper. Res.*, **39**, 1–41.

Gross, D. and Harris, C. (1985). *Fundamentals of Queueing Theory*, Wiley, New York.

Ho, Y. C., Eyler, M. A., and Chien, T. T. (1979). "A gradient technique for general buffer strorage design in a serial production line," *Int. J. Product. Res.*, **17**(6), 557–580.

Huber, P. J. (1981). *Robust Statistics*, Wiley, New York.

Itzhaki, Ya. (1994). "Stochastic optimization of open queueing networks by the score function method," Master's Thesis, Technion, Haifa, Israel.

Kaspi, H. and Mandelbaum, A. (1991). "Regenerative closed queueing networks," Manuscript, Technion, Haifa, Israel.

Kleinrock, L. (1975). *Queueing Systems*, Vols. I and II, Wiley, New York.

Kushner, H. J. and Clark, D. S. (1978). *Stochastic Approximation Methods for Constrained and Unconstrained Systems*, Springer-Verlag, Applied Math. Sciences, Vol. 26. New York.

L'Ecuyer, P. L. (1992). "Convergence rates for steady-state derivative estimators," *Ann. Oper. Res.*, **39**, 121–137.

L'Ecuyer, P. and Glynn, P. W. (1994). "Stochastic optimization by simulation: Convergence proofs for the GI/G/1 queue in steady-state," *Manage. Sci.*, **40**, 1562–1578.

L'Ecuyer, P., Giroux, N., and Glynn, P. W. (1994). "Stochastic optimization by simulation: Numerical experiments for the M/M/1 queue in steady-state," *Manage. Sci.*, **40**, 1245–1261.

Lirov, Y. and Melamed, B. (1992). "Distributed expert systems for queueing networks capacity planning," *Ann. Oper. Res.*, **39**, 137–157.

Nemirovskii, A. and Rubinstein, R. Y. (1996). "An efficient stochastic approximation algorithm for stochastic saddle point problems," Manuscript, Technion, Haifa, Israel.

Perez-Luna, A. (1990). "Sensitivity analysis and optimization of queueing networks by the score function method," Master's Thesis, Technion, Haifa, Israel.

Pflug, G.Ch. (1988). "Sensitivity analysis of semi-Markovian processes," Manuscript University of Giessen, Germany.

Pflug, G.Ch. (1989). "Sampling derivatives of probabilities," *Computing*, **42**, 315–328.

Pflug, G.Ch. (1990). "On-line optimization of simulated Markovian processes," *Math. Oper. Res.*, **15**(3), 381–395.

Pflug, G.Ch. (1992). "Optimization of simulated discrete events processes," *Ann. Oper. Res.*, **39**, 173–195.

Pflug, G.Ch. (1996). *Optimization of Stochastic Models: The Interface Between Simulation and Optimization*, Kluwer Academic.

Rubinstein, R. Y. (1986). *Monte Carlo Optimization Simulation and Sensitivity of Queueing Network*, Wiley, New York.

Rubinstein, R. Y. (1992). "Decomposable score function estimators for sensitivity analysis and optimization of queueing networks," *Ann. Oper. Res.*, **39**, 195–229.

Rubinstein, R. Y. and Shapiro, A. (1993). *Discrete Event Systems: Sensitivity Analysis and Stochastic Optimization via the Score Function Method*, Wiley, New York.

Shapiro A. (1996). "Simulation based optimization-convergence analysis and statistical inference," *Stoch. Models*, **12**(3), 425–454.

Sigman, K. (1990). "The stability of open queueing networks," *Stoch. Proc. Appl.*, **35**, 11–25.

CHAPTER 7

Sensitivity Analysis of Discrete-Event Dynamic Systems: Structural Parameters

7.1 INTRODUCTION

This chapter treats extensions of the score function (SF) method for sensitivity analysis and stochastic optimization to the following discrete-event system (DES) model:

$$\ell(\boldsymbol{v}) = \mathbb{E}_{\boldsymbol{v}_1}\{L(\underline{\boldsymbol{Y}}_t, \boldsymbol{v}_2)\}, \tag{7.1.1}$$

where $L(\underline{\boldsymbol{Y}}_t, \boldsymbol{v}_2)$ is the sample performance depending on the parameter vector $\boldsymbol{v}_2$ and driven by an input sequence $\underline{\boldsymbol{Y}}_t = \{\boldsymbol{Y}_1, \ldots, \boldsymbol{Y}_t\}$ of independent identically distributed (iid) random vectors with common probability density function (pdf) $f(\boldsymbol{y}, \boldsymbol{v}_1)$. The subscript $\boldsymbol{v}_1$ in $\mathbb{E}_{\boldsymbol{v}_1}\{L\}$ indicates that the expectation is taken with respect to the pdf $f(\boldsymbol{y}, \boldsymbol{v}_1)$, and the combined vector of parameters is given here by $\boldsymbol{v} = (\boldsymbol{v}_1, \boldsymbol{v}_2)$. We assume that f depends on the parameter vector $\boldsymbol{v}_1$ but *not* on $\boldsymbol{v}_2$, and that L depends on $\boldsymbol{v}_2$, but *not* on $\boldsymbol{v}_1$. Note that in the terminology of Pflug (1996), the parameter vectors $\boldsymbol{v}_1$ and $\boldsymbol{v}_2$ are called the parameters of the *probability measure* and the *random process*, respectively. We shall call $\boldsymbol{v}_1$ and $\boldsymbol{v}_2$, however, the *distributional parameter vector* and the *structural parameter vector*, respectively. Note that the previous model, $\ell(\boldsymbol{v}) = \mathbb{E}_{\boldsymbol{v}}\{L(\underline{\boldsymbol{Y}}_t)\}$, can be considered as a special case of the model (7.1.1) in which L does not depend on $\boldsymbol{v}_2$, and $\boldsymbol{v} = \boldsymbol{v}_1$.

As before, suppose that $\ell(\boldsymbol{v})$ is not available analytically, and we wish to estimate via simulation both $\ell(\boldsymbol{v})$ and the associated sensitivities, $\nabla^k \ell(\boldsymbol{v})$, $k = 1, 2, \ldots$, for multiple values of $\boldsymbol{v} = (\boldsymbol{v}_1, \boldsymbol{v}_2)$.

Consider the following motivating examples.

1. *Stochastic Shortest Path Networks.* Consider the stochastic shortest path network (4.2.7) with the sample performance

$$L(\boldsymbol{Y}) = \min_{j=1\ldots,p} \left\{ \sum_{i \in \mathscr{L}_j} Y_i \right\}, \tag{7.1.2}$$

where $\mathscr{L}_j$ is the jth complete path from a source to a sink in the system, $\boldsymbol{Y} = (Y_1, \ldots, Y_m)$ is the vector of component durations, and p is the number of complete paths in the system. Suppose that the vector $\boldsymbol{Y}$ can be decomposed into two parts: stochastic and deterministic. Specifically, $\boldsymbol{Y} = (\boldsymbol{Y}_1, \boldsymbol{Y}_2)$, where $\boldsymbol{Y}_1 \sim f_1(\boldsymbol{y}_1, \boldsymbol{v}_1)$, and $\boldsymbol{Y}_2$ is the deterministic part, say, $\boldsymbol{Y}_2 = \boldsymbol{v}_2$. Then $L(\boldsymbol{Y}) = L(\boldsymbol{Y}_1, \boldsymbol{v}_2)$, and the corresponding expected performance takes on the form (7.1.1).

2. *GI/G/1 Queues.* Suppose it is desired to estimate the cumulative distribution function (cdf)

$$H_L(x, \boldsymbol{v}) = P_{\boldsymbol{v}}\{L \leq x\} \tag{7.1.3}$$

of the sample performance, L, in steady state, and the associated derivative, $\partial H_L(x, \boldsymbol{v})/\partial x$, *simultaneously* for multiple values of $\boldsymbol{v}$ and x. In this case, we can represent $H_L(x, \boldsymbol{v})$ as

$$H_L(x, \boldsymbol{v}) = \mathbb{E}_{\boldsymbol{v}_1}\{I_{(-\infty,0]}(L - x)\}, \tag{7.1.4}$$

where $I_{(-\infty,0]}(\cdot)$ is the indicator function of the interval $(-\infty, 0]$.

3. *GI/D/1 and D/G/1 Queues.* Suppose it is desired to estimate the performance of a stable $GI/D/1$ or $D/G/1$ queue in steady state, say, the expected waiting time $\ell(\boldsymbol{v}) = \mathbb{E}_{\boldsymbol{v}_1}\{L(\boldsymbol{Y}, \boldsymbol{v}_2)\}$ and the associated derivatives, for multiple values of $\boldsymbol{v}_1$ and $\boldsymbol{v}_2$. For the $GI/D/1$ queue, $Y \sim f_1(y_1, \boldsymbol{v}_1)$ represents the interarrival time random variable, and v_2 is the (deterministic) service time. The formulation for the $D/G/1$ queue is analogous.

4. *GI/G/1/m Queues.* Suppose it is desired to estimate the steady-state expected waiting time $\ell(\boldsymbol{v}) = \mathbb{E}_{\boldsymbol{v}_1}\{L(\boldsymbol{Y}, v_2)\}$ in a stable $GI/G/1/m$ queue, for multiple values of $v_2 = m$, where m is the buffer size. [This problem is treated rigorously in Kriman (1995)].

5. *$s - S$ Policy Inventory Model.* Suppose it is desired to estimate the performance of an $s - S$ policy inventory model in steady state. For example, we may wish to estimate the on-hand inventory $\ell(\boldsymbol{v}) = \mathbb{E}_{\boldsymbol{v}_1}\{L(\boldsymbol{Y}, \boldsymbol{v}_2)\}$, and the associated derivatives, for multiple values of $\boldsymbol{v}_1$ and $\boldsymbol{v}_2$. In this case, $Y \sim f_1(y_1, \boldsymbol{v}_1)$ represents the interdemand time random variable and $\boldsymbol{v}_2 = (s, S)$.

6. *Insurance Risk Model.* Consider the ruin probability

$$\ell(\boldsymbol{v}) = P_{\boldsymbol{v}}\left\{\inf_{t \geq 0} R(t) < 0\right\},$$

of the insurance risk model with the risk process

$$R_t = u - A_t + \int_0^t p(R(s))\, ds,$$

of (6.1.3), where

$$A_t = \sum_{i=1}^{N_t} U_i,$$

u is the initial reserve, N_t is the number of claims in the interval $(0, t)$, A_t is the cumulative (total) claims at time t, and p is the premium flow per unit time. We might wish to estimate the ruin probability $\ell(\boldsymbol{v})$, $\boldsymbol{v} = (\boldsymbol{v}_1, \boldsymbol{v}_2)$ and the associated derivatives, for multiple values of $\boldsymbol{v}_1$ and $\boldsymbol{v}_2$. In this case, $\boldsymbol{v}_1$ might represent the parameter vector in the joint pdf of claim size and claim interarrival time, and $\boldsymbol{v}_2 = (u, a)$, where u is the initial reserve and a is a parameter in the following two-step premium rule function:

$$p(x) = \begin{cases} p_1 & 0 < x \leq a \\ p_2 & a < x < \infty. \end{cases} \tag{7.1.5}$$

7. *DES with Autocorrelated Inputs.* Consider an output process, $\{L_t : t > 0\}$, driven by an autocorrelated input sequence, $\{\boldsymbol{X}_t : t > 0\}$. Thus, $L_t(\cdot) = L_t(\underline{\boldsymbol{X}}_t)$, where $\underline{\boldsymbol{X}}_t = (\boldsymbol{X}_1, \boldsymbol{X}_2, \ldots, \boldsymbol{X}_t)$. As an example, consider autocorrelated interarrival time and service time processes in a $G/G/1$ queue, with $\{L_t : t > 0\}$ being the waiting time process. Assume, for simplicity, that $\{X_t : t > 0\}$ is a scalar-valued process that obeys the recursive relation

$$X_t(\boldsymbol{v}_2) = h(X_{t-1}, Y_t, \boldsymbol{v}_2), \qquad t > 0, \tag{7.1.6}$$

where $h(x, y, \boldsymbol{v}_2)$ is a real-valued function, and $\{Y_t : t > 0\}$ is an iid random sequence. Thus, $\{X_t : t > 0\}$ is a function of $\{Y_t : t > 0\}$, namely, $X_t = X_t(\boldsymbol{v}_2) = x_t(\underline{Y}_t, \boldsymbol{v}_2)$, where $x_t(\cdot)$ is a real-valued function calculated from $h(x, y, \boldsymbol{v}_2)$, and $\boldsymbol{v}_2$ is the parameter vector associated with the autocorrelation of the variates $X_1, X_2, \ldots, X_t$. In this context, $\boldsymbol{v}_2$ will be called the *autocorrelation parameter vector*. Assume that both processes, $\{X_t : t > 0\}$ and $\{L_t : t > 0\}$, are ergodic and that they approach stationarity as t approaches infinity.

As examples of $\{X_t : t > 0\}$ consider:

- **First-Order Autoregressive [AR(1)] Processes**

$$X_t(v_2) = \begin{cases} X_1 & \text{if } t = 1, \\ v_2 X_{t-1} + Y_t & \text{if } t > 1, \end{cases} \tag{7.1.7}$$

  where $-1 < v_2 < 1$.

- **Transform-Expand-Sample (TES) Processes** (see also Section 2.8)

$$X_t(\boldsymbol{v}_2) = F^{-1}[U_t^+(\boldsymbol{v}_2)], \tag{7.1.8}$$

where F^{-1} is the inverse of a cdf $F(y, \boldsymbol{v})$ and U_t^+ is defined in (2.8.3). For example, if $Y \sim \text{Exp}(v)$, then $X_t = (-1/v)\ln(1 - U_t^+)$, and $U_t^+(\boldsymbol{v}_2)$ is given recursively as a basic background TES$^+$ process (see Melamed, 1991; Jagerman and Melamed, 1992a,b):

$$U_t^+(\boldsymbol{v}_2) = \begin{cases} U_1 & t = 1, \\ \langle U_{t-1}^+ + \theta_1 + (\theta_2 - \theta_1)Y_t \rangle & t > 1. \end{cases} \tag{7.1.9}$$

Here, $\boldsymbol{v}_2 = (\theta_1, \theta_2)$, $-\frac{1}{2} \le \theta_1 < \theta_2 < \frac{1}{2}$, $\langle x \rangle = x - \lfloor x \rfloor$ is the fractional part of x, $\lfloor x \rfloor = \max\{n \text{ integer} : n \le x\}$ is the integral part of x, and $\{Y_t\}$ is an iid $\mathcal{U}(0, 1)$ sequence of random variables. It can be shown (Melamed, 1991) that the sequence $\{U_t^+\}$ is marginally uniform on [0, 1) and covers all positive lag-1 autocorrelations in the range [0, 1) as $(\theta_1, \theta_2) = \boldsymbol{v}_2$ varies in its domain. Recall from Section 2.8 that to produce negative autocorrelations, one can use the TES$^-$ background process (see Melamed, 1991):

$$U_t^- = \begin{cases} U_t^+ & t \text{ even}, \\ 1 - U_t^+ & t \text{ odd}. \end{cases} \tag{7.1.10}$$

Observe that when $\theta_2 - \theta_1 = 1$, the autocorrelated sequence $\{X_t\}$ reduces to an iid sequence. Recall from Section 2.8 that for all $t \ge 1$, the marginals of both random sequences, $\{U_t^+\}$ and $\{U_t^-\}$, are uniformly distributed on the interval [0, 1). For more details on TES sequences see Melamed (1991) and Jagerman and Melamed (1992a,b, 1994).

This chapter is organized as follows. Section 7.2 deals with sensitivity analysis of the model (7.1.1). It presents two transformation techniques, called *push-out* and *push-in*, respectively. These terms derive from the fact (see below) that in the former case we "push out" the parameter vector $\boldsymbol{v}_2$ from the original sample performance, $L(\boldsymbol{Y}, \boldsymbol{v}_2)$, into an auxiliary pdf via a suitable transformation, and then apply the standard SF method to perform sensitivity analysis and optimization; in the latter case, we operate the other way around, namely, we first "push in" (via a suitable transformation) the parameter vector $\boldsymbol{v}_1$ into the sample performance, $L(\boldsymbol{Y}, \boldsymbol{v}_2)$, and then differentiate the resulting (auxiliary) sample performance with respect to $\boldsymbol{v} = (\boldsymbol{v}_1, \boldsymbol{v}_2)$. Conditions will be discussed under which such transformations are useful, in the sense that they either generate *smooth sample performances* or lead to *variance reduction*. Although the push-out technique is model dependent, the reader will learn in Sections 7.3–7.5 how to use it for many interesting applications including static, queueing, and inventory models. The discussion of the push-out method follows Rubinstein (1992). [For relevant references see also Marti (1990) and Uryas'ev (1994, 1997)]. It will also be shown that the *infinitesimal perturbation analysis* (IPA) method, introduced by Ho and his co-workers, corresponds to the push-in technique; the latter can be viewed as a dual of the push-out technique. Section 7.6 deals with autocorrelated input sequences, extending earlier results on sensitivity analysis of DES from independent input sequences to autocorrelated ones. Finally, Section 7.7 shows how to combine the SF method with the crude Monte Carlo method, so as to efficiently estimate the expected performance (response surface), $\ell(\boldsymbol{v}) = \ell(\boldsymbol{v}_1, \boldsymbol{v}_2)$, *simultaneously* for different scenarios (combinations) of $(\boldsymbol{v}_1, \boldsymbol{v}_2)$. Numerical results supporting this theory will also be presented.

7.2 DIRECT, PUSH-OUT, AND PUSH-IN ESTIMATORS

7.2.1 Direct Estimators

This section deals with direct estimators for the derivatives $\nabla^k \ell(\boldsymbol{v})$ of the model (7.1.1). Discrete-event static systems (DESS) and discrete-event dynamic systems (DEDS) models will be considered separately.

DESS Models

Let G be a probability measure with pdf $g(\mathbf{z})$ so that $dG(\mathbf{z}) = g(\mathbf{z})d\mathbf{z}$. Suppose that for every permissible value of the parameter vector $\boldsymbol{v}_1$, one has

$$\text{supp}\{f(\mathbf{z}, \boldsymbol{v}_1)\} \subset \text{supp}\{g(\mathbf{z})\}.$$

Then $\ell(\boldsymbol{v})$ can be represented as

$$\ell(\boldsymbol{v}) = \mathbb{E}_g\{L(\boldsymbol{Z}, \boldsymbol{v}_2)W(\boldsymbol{Z}, \boldsymbol{v}_1)\}, \tag{7.2.1}$$

where

$$W(\mathbf{z}, \boldsymbol{v}_1) = f(\mathbf{z}, \boldsymbol{v}_1)/g(\mathbf{z}) \tag{7.2.2}$$

is the likelihood ratio, $\boldsymbol{Z} \sim g(\mathbf{z})$, and the subscript g indicates that the expectation is taken with respect to the dominating pdf, $g(\mathbf{z})$. Under the standard regularity conditions admitting the interchangeability of the expectation and differentiation operators, we have

$$\nabla^k_{\boldsymbol{v}_1} \ell(\boldsymbol{v}) = \mathbb{E}_g\{L(\boldsymbol{Z}, \boldsymbol{v}_2)\nabla^k_{\boldsymbol{v}_1} W(\boldsymbol{Z}, \boldsymbol{v}_1)\}, \qquad k = 1, 2, \ldots, \tag{7.2.3}$$

and

$$\nabla^k_{\boldsymbol{v}_2} \ell(\boldsymbol{v}) = \mathbb{E}_g\{\nabla^k_{\boldsymbol{v}_2} L(\boldsymbol{Z}, \boldsymbol{v}_2)W(\boldsymbol{Z}, \boldsymbol{v}_1)\}, \qquad k = 1, 2, \ldots. \tag{7.2.4}$$

The corresponding estimators, $\nabla^k_{\boldsymbol{v}_1} \bar{\ell}_N(\boldsymbol{v})$ and $\nabla^k_{\boldsymbol{v}_2} \bar{\ell}_N(\boldsymbol{v})$, of $\nabla^k_{\boldsymbol{v}_1} \ell(\boldsymbol{v})$ and $\nabla^k_{\boldsymbol{v}_2} \ell(\boldsymbol{v})$, are straightforward. They are left as an exercise to the reader.

Example 7.2.1. Let $L(\boldsymbol{Y}, v_1, v_4) = \max\{v_1 + Y_2, Y_3 + v_4\}$, where $\boldsymbol{Y} = (Y_2, Y_3)$ is a two-dimensional vector with independent components, $\boldsymbol{v} = (v_1, v_2, v_3, v_4)$, and $Y_i \sim F_i(y, v_i),\ i = 2, 3$. Then,

$$\ell(\boldsymbol{v}) = \mathbb{E}_g\{L(\boldsymbol{Z}, v_1, v_4)W(\boldsymbol{Z}, \boldsymbol{v}_1)\},$$

$$\frac{\partial \ell(\boldsymbol{v})}{\partial v_1} = \mathbb{E}_g\left\{\frac{\partial L(\boldsymbol{Z}, v_1, v_4)}{\partial v_1} W(\boldsymbol{Z}, \boldsymbol{v}_1)\right\},$$

$$\frac{\partial \ell(\boldsymbol{v})}{\partial v_2} = \mathbb{E}_g\left\{L(\boldsymbol{Z}, v_1, v_4)\frac{\partial W(\boldsymbol{Z}, \boldsymbol{v}_1)}{\partial v_2}\right\},$$

and similarly for $\partial \ell(\boldsymbol{v})/\partial v_3$ and $\partial \ell(\boldsymbol{v})/\partial v_4$. Here,

$$\begin{aligned} L(\boldsymbol{Z}, v_1, v_4) &= \max\{v_1 + Z_2, Z_3 + v_4\}, \\ W(\boldsymbol{Z}, \boldsymbol{v}_1) &= \frac{f_2(Z_2, v_2)}{g_2(Z_2)} \times \frac{f_3(Z_3, v_3)}{g_3(Z_3)}, \qquad Z_i \sim g_i(z_i),\ i = 2, 3, \\ \frac{\partial L(\boldsymbol{Z}, v_1, v_4)}{\partial v_1} &= \begin{cases} 1 & \text{if } v_1 + Z_2 > Z_3 + v_4, \\ 0 & \text{otherwise} \end{cases} \end{aligned} \tag{7.2.5}$$

and similarly for $\partial L(\boldsymbol{Z}, v_1, v_4)/\partial v_4$.

DEDS Models

Consider again (6.2.9). Differentiating $\ell(\boldsymbol{v})$ with respect to $\boldsymbol{v}_1$ and $\boldsymbol{v}_2$ we obtain

$$\begin{aligned} \nabla_{\boldsymbol{v}_1}\ell(\boldsymbol{v}) &= \frac{\mathbb{E}_g\{\sum_{t=1}^{\tau} L_t \nabla \tilde{W}_t(\boldsymbol{v}_1)\}}{\mathbb{E}_g\{\sum_{t=1}^{\tau} \tilde{W}_t(\boldsymbol{v}_1)\}} \\ &\quad - \frac{\mathbb{E}_g\{\sum_{t=1}^{\tau} L_t \tilde{W}_t(\boldsymbol{v}_1)\}}{\mathbb{E}_g\{\sum_{t=1}^{\tau} \tilde{W}_t(\boldsymbol{v}_1)\}} \times \frac{\mathbb{E}_g\{\sum_{t=1}^{\tau} \nabla \tilde{W}_t(\boldsymbol{v}_1)\}}{\mathbb{E}_g\{\sum_{t=1}^{\tau} \tilde{W}_t(\boldsymbol{v}_1)\}}, \end{aligned} \tag{7.2.6}$$

and

$$\nabla_{\boldsymbol{v}_2}\ell(\boldsymbol{v}) = \frac{\mathbb{E}_g\{\sum_{t=1}^{\tau} \nabla L_t(\boldsymbol{v}_2)\tilde{W}_t(\boldsymbol{v}_1)\}}{\mathbb{E}_g\{\sum_{t=1}^{\tau} \tilde{W}_t(\boldsymbol{v}_1)\}}, \tag{7.2.7}$$

provided the usual interchangeability conditions hold for expectation and differentiation. With (7.2.6) and (7.2.7) in hand, one can derive statistical estimators of $\nabla_{\boldsymbol{v}_1}\ell(\boldsymbol{v})$ and $\nabla_{\boldsymbol{v}_2}\ell(\boldsymbol{v})$ and their higher-order partial and mixed derivatives.

Example 7.2.2 ***GI/D/1*** **Queues.** Let $\{L_t\}$ be the waiting time process of customers in a steady-state $GI/D/1$ queue. In this case (see also Example 6.2.3), the expected steady-state performance, $\ell(\boldsymbol{v}) = \mathbb{E}_{\boldsymbol{v}_1}\{L_t\}$, can be written as

$$\ell(\boldsymbol{v}) = \frac{\mathbb{E}_{\boldsymbol{v}_1}\{\sum_{t=1}^{\tau} L_t(v_2)\}}{\mathbb{E}_{\boldsymbol{v}_1}\{\tau\}} = \frac{\mathbb{E}_{\boldsymbol{v}_1}\left\{\sum_{t=1}^{\tau} \sum_{j=1}^{t-1} Z_j\right\}}{\mathbb{E}_{\boldsymbol{v}_1}\{\tau\}}, \tag{7.2.8}$$

where $Z_j = v_2 - Y_{1(j+1)}$, v_2 is the constant service time, $Y_{1j} = A_j - A_{j-1}$ is the interarrival time between customers $j-1$ and j ($A_0 = 0$ by convention), $\tau = \min\{t\colon L_{t+1} \le 0\}$, and $Y_{1j} \sim f_1(y_1, \boldsymbol{v}_1)$. We have

$$\nabla_{v_2}\ell(\boldsymbol{v}) = \frac{\mathbb{E}_{\boldsymbol{v}_1}\{\sum_{t=1}^{\tau}(t-1)\}}{\mathbb{E}_{\boldsymbol{v}_1}\{\tau\}}. \tag{7.2.9}$$

In order to estimate the gradient $\nabla\ell(\boldsymbol{v})$, $\boldsymbol{v} = (\boldsymbol{v}_1, v_2)$, we first differentiate $\ell(\boldsymbol{v})$ with respect to $\boldsymbol{v}_1$ and v_2 by taking derivatives inside the expected value (why?), and then derive a consistent estimator of $\nabla\ell(\boldsymbol{v})$, based on N regenerative cycles.

Example 7.2.3 (Example 6.2.4 continued). Consider Lindley's equation for the shortfall process,

$$L_{t+1} = \max\{0, L_t + D_t - s\},$$

of the production inventory model. Since the shortfall process $\{L_t\}$ coincides with the waiting time process in a $D/G/1$ queue having a fixed interarrival time s and service time D_t, the formula (7.2.8) holds again, provided $Z_j = v_2 - Y_{1(j+1)}$ is replaced by $Z_j = D_j - s$. Denoting $v_2 = s$, we obtain in analogy to (7.2.9)

$$\nabla_{v_2}\ell(\boldsymbol{v}) = -\frac{\mathbb{E}_{\boldsymbol{v}_1}\{\sum_{t=1}^{\tau}(t-1)\}}{\mathbb{E}_{\boldsymbol{v}_1}\{\tau\}}. \tag{7.2.10}$$

7.2.2 Push-out Technique

This section describes two techniques, called (a) push-out and (b) push-in. It will be shown that the first technique typically smoothes out the sample performance function, $L(y, \boldsymbol{v}_2)$, with respect to $\boldsymbol{v}_2$, by rendering it *independent* of $\boldsymbol{v}_2$, while the second can lead to variance reduction. Both techniques are based on the standard change-of-variable method.

Push-out Technique

To demonstrate the idea of the push-out technique, consider a simple DESS and suppose that there exist a vector-valued function $\boldsymbol{x} = \boldsymbol{x}(\boldsymbol{y}, \boldsymbol{v}_2)$, and a real-valued function $\tilde{L}(\boldsymbol{x})$, independent of $\boldsymbol{v}_2$, such that $L(\boldsymbol{y}, \boldsymbol{v}_2)$ can be represented as

$$L(\boldsymbol{y}, \boldsymbol{v}_2) = \tilde{L}(\boldsymbol{x}(\boldsymbol{y}, \boldsymbol{v}_2)). \tag{7.2.11}$$

Furthermore, suppose that $\boldsymbol{Y} \sim f(\boldsymbol{y}, \boldsymbol{v}_1)$ and that the corresponding random vector $\boldsymbol{X} = \boldsymbol{x}(\boldsymbol{Y}, \boldsymbol{v}_2)$, has a known pdf $\tilde{f}(\boldsymbol{x}, \boldsymbol{v}_1, \boldsymbol{v}_2)$. Then,

$$\ell(\boldsymbol{v}) = \int_{\boldsymbol{x} \in X} \tilde{L}(\boldsymbol{x}) \tilde{f}(\boldsymbol{x}, \boldsymbol{v})\, d\boldsymbol{x} = \mathbb{E}_{\tilde{f}}\{\tilde{L}(\boldsymbol{X})\}, \tag{7.2.12}$$

where the expectation is now taken with respect to the pdf $\tilde{f}(\boldsymbol{x}, \boldsymbol{v}_1, \boldsymbol{v}_2)$. As mentioned in Section 7.1, the term push-out derives from the fact that the parameter vector $\boldsymbol{v}_2$ is pushed out from $L(\boldsymbol{y}, \boldsymbol{v}_2)$ to an auxiliary pdf, $\tilde{f}(\boldsymbol{x}, \boldsymbol{v}_1, \boldsymbol{v}_2)$.

The derivation of $\nabla \ell(\boldsymbol{v})$ is similar, but one needs to make use of the following identity:

$$\frac{d}{dv} \int_{a(v)}^{b(v)} g(v, x)\, dx = \int_{a(v)}^{b(v)} \frac{\partial}{\partial v} g(v, x)\, dx + \frac{db(v)}{dv} g(v, b(v)) - \frac{da(v)}{dv} g(v, a(v)). \tag{7.2.13}$$

It is important to understand that, in principle, a representation of $L(\boldsymbol{y}, \boldsymbol{v}_2)$ of the form (7.2.11) and the subsequent transformation (7.2.12) are not always available; and, if available, it may be hard to calculate $\tilde{f}(\boldsymbol{x}, \boldsymbol{v}_1, \boldsymbol{v}_2)$. We shall, however, show that for many interesting applications the pdf $\tilde{f}(\boldsymbol{x}, \boldsymbol{v}_1, \boldsymbol{v}_2)$ is easy to calculate, and, thus the usefulness of the push-out technique will be established. As an example, suppose now that for every $\boldsymbol{v}_2$, the function $\boldsymbol{x} = \boldsymbol{x}(\boldsymbol{y}, \boldsymbol{v}_2)$ is one-to-one and, thus, has an inverse $\boldsymbol{y} = \boldsymbol{y}(\boldsymbol{x}, \boldsymbol{v}_2)$, which is assumed to be continuously differentiable in each component of $\boldsymbol{x}$. In this case,

$$\tilde{f}(\boldsymbol{x}, \boldsymbol{v}_1, \boldsymbol{v}_2) = f(\boldsymbol{y}(\boldsymbol{x}, \boldsymbol{v}_2), \boldsymbol{v}_1) \left| \frac{\partial \boldsymbol{y}(\boldsymbol{x}, \boldsymbol{v}_2)}{\partial \boldsymbol{x}} \right|, \tag{7.2.14}$$

where $|\partial \boldsymbol{y}/\partial \boldsymbol{x}|$ denotes the absolute value of the determinant of the Jacobian matrix of $\boldsymbol{y}(\boldsymbol{x}, \boldsymbol{v}_2)$ with respect to $\boldsymbol{x}$.

For example, suppose that $\boldsymbol{v}_2$ has the same dimensionality as $\boldsymbol{y}$ and that the function $L(\boldsymbol{y}, \boldsymbol{v}_2)$ can be represented as $L(\boldsymbol{y}, \boldsymbol{v}_2) = L(\boldsymbol{y} + \boldsymbol{v}_2)$, that is, $\boldsymbol{v}_2$ is a *location* parameter vector. Defining $\boldsymbol{x} = \boldsymbol{y} + \boldsymbol{v}_2$, one obtains

$$\tilde{f}(\boldsymbol{x}, \boldsymbol{v}) = f(\boldsymbol{x} - \boldsymbol{v}_2, \boldsymbol{v}_1). \tag{7.2.15}$$

Similar arguments can be employed when $\boldsymbol{v}_2$ is a *scale* parameter. The following simple example demonstrates the advantages of using the transformation $\boldsymbol{x} = \boldsymbol{x}(\boldsymbol{y}, \boldsymbol{v}_2)$ for the case of a location parameter $\boldsymbol{v}_2$.

Example 7.2.4 (Example 7.2.1 continued). Consider the sample performance function $L(\boldsymbol{Y}, v_1, v_4) = \max\{(v_1 + Y_2), Y_3 + v_4\}$. The corresponding transformation can then be written as $x_1 = v_1 + y_2$ and $x_2 = y_3 + v_4$. It follows that $\tilde{L}(x) = \max\{x_1, x_2\}$ and $\tilde{f}(x, v_1, v_2, v_3, v_4) = f_2(x_1 - v_1, v_2) f_3(x_2 - v_4, v_3)$. Note that here $\partial L(y, v_1)/\partial v_1$ and $\partial L(y, v_1)/\partial v_4$ are piecewise-constant functions [(see (7.2.5)] and that $\partial^2 L(\boldsymbol{Y}, v_1, v_4)/\partial v_1^2$ and $\partial^2 L(\boldsymbol{Y}, v_1, v_4)/\partial v_4^2$ vanish almost everywhere. Consequently, the associated second-order derivatives cannot be interchanged with the expectation operator in (7.2.4). On the other hand, the transformed functions are *differentiable* in $\boldsymbol{v}$ everywhere, provided $f_2(x_1 - v_1, v_2)$ and $f_3(x_2 - v_4, v_3)$ are smooth. Thus, subject to smoothness of $f(\boldsymbol{y}, \boldsymbol{v}_1)$, the push-out technique *smoothes out* the original *nonsmooth* performance measures.

At this juncture we point out that deriving smooth estimators of discontinuous expected performance functions is one of the major problems in DES. Section 7.3 discusses the push-out technique in greater detail, and in particular shows how to construct the auxiliary functions, $\tilde{L}$ and $\tilde{f}$, such that the response functions $\nabla^k \ell(\boldsymbol{v})$, $\boldsymbol{v} = (\boldsymbol{v}_1, \boldsymbol{v}_2)$, $k = 0, 1, \ldots$ are smooth; it then shows how to estimate the expected performance and associated sensitivities from *a single simulation run, for all* in $\boldsymbol{v} \in V$.

Push-in Technique

The push-in technique can be considered as a dual to the push-out technique in the sense that one searches for a transformation $\boldsymbol{y} = \boldsymbol{y}(\boldsymbol{x}, \boldsymbol{v}_1)$, such that the distribution of the corresponding random vector, $\boldsymbol{x} = \boldsymbol{x}(\boldsymbol{Y}, \boldsymbol{v}_1)$, is independent of $\boldsymbol{v}_1$. In this case, $\ell(\boldsymbol{v})$ can be represented as

$$\ell(\boldsymbol{v}) = \int \hat{L}(\boldsymbol{x}, \boldsymbol{v}) \hat{f}(\boldsymbol{x})\, d\boldsymbol{x} = \mathbb{E}_{\hat{f}}\{\hat{L}(\boldsymbol{X}, \boldsymbol{v})\}, \tag{7.2.16}$$

where $\hat{L}(\boldsymbol{x}, \boldsymbol{v}) = L(\boldsymbol{y}(\boldsymbol{x}, \boldsymbol{v}_1), \boldsymbol{v}_2)$, $\hat{f}(\boldsymbol{x})$ is the pdf of $\boldsymbol{x}(\boldsymbol{Y})$ and $\hat{f}$ is assumed to be independent of $\boldsymbol{v}_1$ and $\boldsymbol{v}_2$ in $\boldsymbol{v} = (\boldsymbol{v}_1, \boldsymbol{v}_2)$.

7.2.3 IPA Technique

This section deals with the infinitesimal perturbation analysis (IPA) method, pioneered by Ho and his co-workers (see Cao, 1994, and Glasserman, 1991). The main observation here is that IPA can be viewed as a push-in technique. Furthermore, the case where $\hat{f} \equiv 1$ on the interval [0, 1), that is, X is uniform on [0, 1), is of particular interest. In this case, the transformation $\boldsymbol{x} = \boldsymbol{x}(\boldsymbol{y}, \boldsymbol{v}_1)$ reduces to

$$x = x(y, \boldsymbol{v}_1) = F(y, \boldsymbol{v}_1), \tag{7.2.17}$$

or

$$y = F^{-1}(x, \boldsymbol{v}_1), \tag{7.2.18}$$

where $F(y, \boldsymbol{v}_1)$ is the cdf of the random variable Y (see L'Ecuyer, 1990, Ho and Cao, 1991, and Suri and Zazanis, 1988).

We point out that the original IPA approach is constrained by the fact that the transformation (7.2.17) often leads to a sample performance function, $\hat{L}(\boldsymbol{x}, \boldsymbol{v})$, which is *nondifferentiable* in $\boldsymbol{v}$ (see, e.g., Heidelberger et al., 1988, and L'Ecuyer, 1990). Consequently, the requisite interchange of expectation and differentiation is often inadmissible. Furthermore, when $\hat{L}(\boldsymbol{v})$ is not available analytically, it typically permits estimation of $\nabla^k \ell(\boldsymbol{v})$ at a fixed $\boldsymbol{v}$ *only*, whereas the SF method allows estimation of $\nabla^k \ell(\boldsymbol{v})$ for an entire range of values of $\boldsymbol{v}$. Consequently, when dealing with off-line optimization problems, the IPA approach must rely on iterative algorithms of the stochastic approximation type, requiring multiple simulation runs. In contrast, its SF counterpart can solve an entire (constrained) optimization problem from a *single simulation run*.

Example 7.2.5 Indicator Functions. Suppose that we wish to estimate $\nabla \ell(\boldsymbol{v})$, where $\ell(\boldsymbol{v}) = P_{\boldsymbol{v}_1}\{L(Y, \boldsymbol{v}_2) \leq \alpha\}$, and L is a sample performance, say, the waiting time in a $GI/G/1$ queue. The IPA approach will not work here, since the inverse transformation (7.2.18) leads to a piecewise-constant sample function (taking only the values 0 and 1). However, if we represent $\ell(\boldsymbol{v})$ as $\ell(\boldsymbol{v}) = \mathbb{E}_{\boldsymbol{v}_1}\{I_{(-\infty,0)}(L(\boldsymbol{v}_2) - \alpha)\}$, then by combining the standard SF approach with the push-out technique (see Section 8.3.4), we can estimate $\nabla \ell(\boldsymbol{v})$, $\boldsymbol{v} = (\boldsymbol{v}_1, \boldsymbol{v}_2)$, for an entire range of values of $\boldsymbol{v}$, from a single simulation run.

We now present several cases where the IPA approach is superior to the SF approach, in the sense that its derivative estimators can have *a smaller variance* than the standard SF estimators.

Example 7.2.6 (Example 7.2.1 continued). Consider again the sample performance

$$L(\boldsymbol{Y}, v_1, v_4) = \max\{v_1 + Y_2, Y_3 + v_4\}$$

where $\boldsymbol{Y} = (Y_2, Y_3)$ and $Y_i \sim F_i(y, v_i)$, $i = 2, 3$. Using the inverse transformations $Y_i = F_i^{-1}(U_i, v_i)$, $i = 2, 3$, we obtain

$$\hat{L}(\boldsymbol{U}, v_1) = \max\{v_1 + F_2^{-1}(U_2, v_2), F_3^{-1}(U_3, v_3) + v_4\},$$

where $\boldsymbol{U} = (U_2, U_3)$, U_2 and U_3 are independent $\mathscr{U}(0, 1)$ variates, and $\boldsymbol{v} = (v_1, v_2, v_3, v_4)$. In this case, the interchangeability conditions hold for the gradient,

$$\nabla \ell(\boldsymbol{v}) = \mathbb{E}\{\nabla \hat{L}(\boldsymbol{U}, \boldsymbol{v})\},$$

but not for the Hessian (why?). The IPA estimator can then be written as

$$\nabla \hat{\ell}_N(\boldsymbol{v}) = \frac{1}{N} \sum_{i=1}^{N} \nabla \hat{L}(\boldsymbol{U}_i, \boldsymbol{v}). \tag{7.2.19}$$

Note that the first-order derivatives of $\hat{L}(\boldsymbol{U}, \cdot)$ are piecewise-continuous functions with discontinuities at points where $v_1 + F_2^{-1}(U_2, v_2) = F_3^{-1}(U_3, v_3) + v_4$.

Example 7.2.7. Consider Lindley's equation for the waiting time process, $\{L_t\}$, in a $GI/G/1$ queue,

$$L_{t+1} = \max\{0,\ L_t + Z_t\}, \qquad L_0 = 0,\ t = 0, 1, \ldots, \tag{7.2.20}$$

where (see, Example 6.2.3), $Z_t = Y_{1t} - Y_{2(t+1)}$, $\{Y_{1t}\}$ is the sequence of service times, and $\{Y_{2t}\}$ is the sequence of interarrival times [Y_{2t} is the time between the arrivals of the tth and $(t-1)$st customer].

Consider separately the SF and the IPA approaches. The straightforward SF approach, based on the representation

$$\nabla^k \ell(\boldsymbol{v}) = \mathbb{E}_g\{L_t \nabla^k \tilde{W}_t\}, \tag{7.2.21}$$

where

$$\tilde{W}_t = \prod_{j=1}^{t} W_j, \qquad W_j = \frac{f(\boldsymbol{Z}, \boldsymbol{v})}{g(\boldsymbol{Z})}, \quad \text{and} \quad \boldsymbol{Z} \sim g(\boldsymbol{z}),$$

may well *fail*, since for large t, the variance of the associated SF estimators can be very large. Its IPA counterpart approach produces, however, an unbiased estimator of $\nabla \ell(\boldsymbol{v})$ with a manageable variance. The corresponding IPA algorithm for estimating $\nabla \ell(\boldsymbol{v})$ follows.

Algorithm 7.2.1

1. *Generate the output process* $\{\hat{L}_t(\boldsymbol{v})\}$, *using Lindley's equation,*

$$\hat{L}_{t+1}(\boldsymbol{v}) = \max\{0, \hat{L}_t + \hat{Z}_t(\boldsymbol{v})\}, \qquad t \geq 1, \tag{7.2.22}$$

where $\hat{L}_1 = 0$ *and* $\hat{Z}_t(\boldsymbol{v}) = F_1^{-1}(U_{1t}, v_1) - F_2^{-1}(U_{2(t+1)}, v_2)$, *for* $\mathcal{U}(0, 1)$ *sequences* $\{U_{1t}\}$ *and* $\{U_{2t}\}$.

2. *Differentiate* $\ell(\boldsymbol{v}) = \mathbb{E}\{\hat{L}_t(\boldsymbol{v})\}$ *with respect to* $\boldsymbol{v}$ *(by taking the derivative inside the expectation), and calculate* $\nabla \ell(\boldsymbol{v}) = \mathbb{E}\{\nabla \hat{L}_t(\boldsymbol{v})\}$, *where*

$$\nabla \hat{L}_{t+1}(\boldsymbol{v}) = \max\{0, \nabla \hat{L}_t + \hat{Z}_t(\boldsymbol{v})\}, \qquad t \geq 1, \tag{7.2.23}$$

3. *Simulate the* $GI/G/1$ *queue for* N *customers, and estimate* $\nabla \ell(\boldsymbol{v})$ *as*

$$\nabla \hat{\ell}_N(\boldsymbol{v}) = \frac{1}{N} \sum_{t=1}^{N} \nabla \hat{L}_t(\boldsymbol{v}). \tag{7.2.24}$$

Regenerative simulation with long regenerative cycles is another situation where the IPA estimators can outperform their standard SF counterparts. The RSF (regenerative score function) estimators are useless here, since the variance of the associated likelihood ratio process, $\{\tilde{W}_t\}$ (see Section 6.4), becomes excessive. The IPA approach will likely produce estimators with lower variance (see Glasserman, 1991, and L'Ecuyer, 1990, 1992), provided the usual interchangeability conditions hold. For additional useful IPA examples, see Glasserman (1991), Gong and Ho (1987), Ho and Cao (1991), L'Ecuyer (1990), and Wardi et al. (1992).

There are several interesting variations of the IPA method. These include the *smoothed perturbation analysis* technique of Gong and Ho (1987), which is based on the conditional Monte Carlo method; the *infinitesimal augmented perturbation analysis* technique of Gaivoronski et al. (1992), which produces unbiased estimators for the derivatives of performance measures, associated with semi-Markov processes; and the *analytic perturbation analysis* (APA) of Uryas'ev (1997), which is based on partitioning (splitting) the set of the input random vector, $\boldsymbol{Y}$, into specific subsets. Note finally, that when IPA or its variations do work, one still may use as alternatives the CSF (conditional score function) estimators of McLeish and Rollans (1992) and the DSF (decomposable score function) or TSF (truncated score function) estimators (see Section 6.4). It is, however, important to keep in mind that the IPA and CSF estimators are *consistent*, while their DSF and TSF counterparts are *asymptotically biased.*

Sections 7.3–7.5 deal with sensitivity analysis via the push out method for DESS models, queueing and production inventory models, and inventory (s, S) policy models, respectively. In particular, Section 7.3 illustrates how to construct the auxiliary functions $\tilde{L}$ and $\tilde{f}$ [see (7.2.11) and (7.2.14)], in order to estimate the quantities $\nabla^k \ell(\boldsymbol{v})$, $\boldsymbol{v} = (\boldsymbol{v}_1, \boldsymbol{v}_2)$, via the push-out technique. Emphasis is placed on the case where the sample performance function, $L(\boldsymbol{y}, \boldsymbol{v}_2)$, is not everywhere differentiable in $\boldsymbol{v}_2$. Additionally, it will be shown that using the push-out technique enables one to easily estimate the distribution and density functions of $L(\boldsymbol{Y})$. Sections 7.4 and 7.5 illustrate the application of the push-out technique to DEDS models and extend the results of Section 7.3 to queueing models and inventory models, respectively. These sections also discuss how to estimate the system performance measures $\ell(\boldsymbol{v})$ and $\nabla^k \ell(\boldsymbol{v})$, $k = 1, \dots,$ assuming the sample performance process, $\{L_t\}$, is not available analytically and can be written as $L_t = L_t(\underline{\boldsymbol{Y}}_t, \boldsymbol{v}_2)$, where $\underline{\boldsymbol{Y}}_t = (\boldsymbol{Y}_1, \boldsymbol{Y}_2, \dots, \boldsymbol{Y}_t)$ is the input sequence of iid random vectors from the pdf $f(\boldsymbol{y}, \boldsymbol{v}_1)$; sojourn times in a queueing system are a typical example. Special emphasis will be placed on deriving *smooth* estimators of the performance measures and their associated derivatives with respect to structural parameters of the model, *simultaneously* from a *single simulation run*. Note again that deriving *smooth estimators* of pdf's of noncontinuous random variables is one of the *major problems in density estimation* (see, e.g., Topia and Thompson, 1978). It will be shown that the main advantage of push-out estimators of $\nabla^k H_L(x, \boldsymbol{v})$ is that they are typically smooth in x, provided the underlying pdf is smooth (e.g., the exponential family). Finally, numerical examples will be presented to illustrate the efficacy of the push-out approach.

7.3 PUSH-OUT METHOD FOR DESS

Let $g(\mathbf{z})$ be a pdf that dominates the pdf $\tilde{f}(\mathbf{z}, \boldsymbol{v})$. Then, in analogy to (7.2.1)–(7.2.3), we have

$$\nabla^k \ell(\boldsymbol{v}) = \mathbb{E}_g\{\tilde{L}(\boldsymbol{Z})\nabla^k \hat{W}(\boldsymbol{Z}, \boldsymbol{v})\}, \tag{7.3.1}$$

where

$$\hat{W}(\mathbf{z}, \boldsymbol{v}) = \tilde{f}(\mathbf{z}, \boldsymbol{v})/g(\mathbf{z}), \tag{7.3.2}$$

and $\boldsymbol{Z} \sim g(\mathbf{z})$. Consequently, for a given sample $\{\boldsymbol{Z}_1, \ldots, \boldsymbol{Z}_N\}$ from $g(\mathbf{z})$, $\nabla^k \ell(\boldsymbol{v})$ can be estimated by

$$\nabla^k \tilde{\ell}_N(\boldsymbol{v}) = \frac{1}{N}\sum_{i=1}^{N} \tilde{L}(\boldsymbol{Z}_i)\nabla^k \hat{W}(\boldsymbol{Z}_i, \boldsymbol{v}), \qquad k = 0, 1, \ldots, \tag{7.3.3}$$

where, by convention,

$$\nabla^0 \ell_N(\boldsymbol{v}) \equiv \ell_N(\boldsymbol{v}).$$

The estimator $\nabla^k \tilde{\ell}_N(\boldsymbol{v})$ will be referred to as the POSF (*push-out score function*) estimator.

We now present two examples, illustrating the application of the push-out technique. The first example applies formula (7.3.3) to the sample performance

$$L(\boldsymbol{Y}, v_1) = \min(v_1 + Y_2, Y_3 + v_4)$$

(see Example 7.2.4), while the second example demonstrates how to estimate from a single simulation run a cdf (and pdf) of the sample performance $L(\boldsymbol{Y})$, when $L(\boldsymbol{y})$ can be represented as $L(\boldsymbol{y}) = L(\boldsymbol{y}_1) + L(\boldsymbol{y}_2)$. Such an $L(\boldsymbol{y})$ will be called an *additive sample function*.

Example 7.3.1 (Example 7.2.4 continued). Let again

$$L(\boldsymbol{Y}, v_1, v_4) = \min(v_1 + Y_2, Y_3 + v_4), \tag{7.3.4}$$

where $Y_j \sim \text{Exp}(v_j)$, $j = 2, 3$. Then, the cdf $\tilde{F}_1(y)$ of $\tilde{Y}_1 = v_1 + Y_2$ and the cdf $\tilde{F}_2(y)$ of $\tilde{Y}_2 = Y_3 + v_4$ can be written, respectively, as

$$\tilde{F}_1(y) = P(\tilde{Y}_1 \leq y) = P(Y_2 \leq y - v_1) = F_2(y - v_1)$$

and

$$\tilde{F}_2(y) = P(\tilde{Y}_2 \leq y - v_4) = F_3(y - v_4).$$

Clearly,

$$\tilde{f}_1(y, v_1, v_2) = \begin{cases} v_2 e^{-v_2(y-v_1)} & y \geq v_1 \\ 0 & \text{otherwise,} \end{cases} \tag{7.3.5}$$

and

$$\tilde{f}_2(y, v_3, v_4) = \begin{cases} v_3 e^{-v_3(y-v_4)} & y \geq v_4 \\ 0 & \text{otherwise.} \end{cases} \tag{7.3.6}$$

As a dominating pdf, take $g(\mathbf{z}) = g_2(z_2)g_3(z_3)$, where g_i is an exponential pdf with parameter v_{0i}, $i = 2, 3$. Thus,

$$\tilde{W}(\boldsymbol{v}, \boldsymbol{Z}) = \begin{cases} \dfrac{v_2 e^{-v_2(Z_1-v_1)}}{v_{02} e^{-v_{02} Z_1}} \times \dfrac{v_3 e^{-v_3(Z_2-v_4)}}{v_{03} e^{-v_{03} Z_2}} & \text{if } Z_1 \geq v_1 \quad \text{and} \quad Z_2 \geq v_4, \\ 0 & \text{otherwise.} \end{cases} \tag{7.3.7}$$

We next present a modification of the likelihood ratio (LR) (7.3.7), based on the acceptance-rejection method (see Chapter 2), which results in a more accurate estimator of $\ell(\boldsymbol{v})$ than that based on (7.3.7). Our acceptance-rejection procedure *accepts* the random vector $\boldsymbol{Z} = (Z_1, Z_2)$, if $Z_1 \geq v_1$ and $Z_2 \geq v_4$, and *rejects* it, otherwise. Let $\tilde{\boldsymbol{Z}} = (\tilde{Z}_1, \tilde{Z}_2)$ denote such a generated vector of variates. Then an alternative to the likelihood ratio (7.3.7) is

$$\hat{W}(\boldsymbol{v}, \tilde{\boldsymbol{Z}}) = p_1 p_2 \frac{v_2 e^{-v_2(\tilde{Z}_1-v_1)}}{v_{02} e^{-v_{02} Z_1}} \times \frac{v_3 e^{-v_3(\tilde{Z}_2-v_4)}}{v_{03} e^{-v_{03} Z_2}}, \tag{7.3.8}$$

where

$$p_1 = p(v_1) = \int_{v_1}^{\infty} v_{02} e^{-v_{02} y}\, dy \quad \text{and} \quad p_2 = p_2(v_4) = \int_{v_4}^{\infty} v_{03} e^{-v_{03} y}\, dy.$$

Let $\tilde{\ell}(\boldsymbol{v}, \boldsymbol{Z}) = L(\boldsymbol{Z})\tilde{W}(\boldsymbol{v}, \boldsymbol{Z})$ and $\hat{\ell}(\boldsymbol{v}, \tilde{\boldsymbol{Z}}) = L(\tilde{\boldsymbol{Z}})\hat{W}(\boldsymbol{v}, \tilde{\boldsymbol{Z}})$ be the two alternative unbiased estimators of $\ell(\boldsymbol{v}) = \mathbb{E}\{L(\boldsymbol{Y}, v_1, v_4)\}$. It is not difficult to see that although the estimator $\hat{\ell}(\boldsymbol{v}, \tilde{\boldsymbol{Z}})$, based on (7.3.8), requires the analytical calculation of the vector $(p_1, p_2) = (p_1(v_1), p_2(v_4))$ for distinct scenarios $\boldsymbol{v}$, it is more accurate than the estimator $\tilde{\ell}(\boldsymbol{v}, \boldsymbol{Z})$, based on (7.3.7) (see Exercise 8). We point out that similar acceptance-rejection procedures can be applied to other POSF estimators.

Table 7.3.1 displays the theoretical values of $\ell(\boldsymbol{v})$, the point estimators $\tilde{\ell}_N(\boldsymbol{v})$, the associated sample variance of $\tilde{\ell}_N(\boldsymbol{v})$ [denoted by $\hat{\sigma}^2\{\tilde{\ell}_N(\boldsymbol{v})\}$], and the 95% confidence intervals for $\ell(\boldsymbol{v})$ as a function of the relative perturbation $\alpha_0 = (|v - v_0|)/v_0$, and $v_1 = v_4 = 1$. Note that

$$\ell(\boldsymbol{v}) = v_1 + \frac{1}{v_2 + v_3}.$$

The reference vector, $\boldsymbol{v}_0$ was equal $\boldsymbol{v}_0 = (v_{01}, v_{02}, v_{03}, v_{04}) = (1, 2, 3, 1)$, and $\ell(\boldsymbol{v})$ was estimated using the POSF estimator $\tilde{\ell}_N(\boldsymbol{v})$, for $\alpha_0 = 0.3$ with sample size $N = 10^3$. Observe that for $\alpha_0 = 0.3$, we have $\boldsymbol{v} = (v_1, v_2, v_3, v_4) = (0.7, 1.4, 2.1, 0.7)$.

Table 7.3.2 displays similar data for $\nabla\tilde{\ell}_N(\boldsymbol{v})$. More specifically, it displays the theoretical values of $\partial\ell(\boldsymbol{v})/\partial v_j$, $j = 1, 2, 3$, the point estimators $\partial\tilde{\ell}(\boldsymbol{v})/\partial v_j$, $j = 1, 2, 3$, and the 95% confidence intervals for $\boldsymbol{v} = (v_1, v_2, v_3) = (0.7, 1.4, 2.1, 0.7)$. Note that both tables were derived with the LR (7.3.7); however, using the LR (7.3.8) yields more accurate estimators.

In deriving $\nabla\tilde{\ell}_N(\boldsymbol{v})$, we used the identity (7.2.13). Since in our model,

$$\ell(\boldsymbol{v}) = \mathbb{E}_{\boldsymbol{v}}\{L(\tilde{Y}_1, \tilde{Y}_2)\} = \int_{v_1}^{\infty}\int_{v_4}^{\infty} L(y_1, y_2)\tilde{f}_1(y_1, v_1, v_2)\tilde{f}_2(y_2, v_3, v_4)\, dy_1\, dy_2,$$

we get after some simple algebra

$$\nabla\ell(\boldsymbol{v}) = \begin{bmatrix} \mathbb{E}_{\boldsymbol{v}}[(\tilde{L} - v_1)(v_2 + v_3)] \\ \mathbb{E}_{\boldsymbol{v}}[\tilde{L}(v_2^{-1} - (\tilde{Y}_1 - v_1))] \\ \mathbb{E}_{\boldsymbol{v}}[\tilde{L}(v_3^{-1} - (\tilde{Y}_2 - v_1))] \end{bmatrix}. \tag{7.3.9}$$

Table 7.3.1 Performance of the POSF Estimator, $\tilde{\ell}_N(\boldsymbol{v})$, for Stochastic PERT Model (7.3.4)

$\ell(\boldsymbol{v})$	$\tilde{\ell}_N(\boldsymbol{v})$	$\hat{\sigma}^2\{\tilde{\ell}_N(\boldsymbol{v})\}$	95% CI
1.200	1.2910	1.7422	1.2092, 1.3728

Table 7.3.2 Performance of the POSF Estimator, $\nabla\tilde{\ell}_N(v)$, for Stochastic PERT Model (7.3.4)

$\frac{\partial\ell}{\partial v_1}$	$\frac{\partial\ell}{\partial v_2}$	$\frac{\partial\ell}{\partial v_3}$	$\frac{\partial\tilde{\ell}}{\partial v_1}$	$\frac{\partial\tilde{\ell}}{\partial v_2}$	$\frac{\partial\tilde{\ell}}{\partial v_3}$	95% CI
1.0	−0.04	−0.04	0.98	−0.038	−0.037	(0.96, 1.00) (−0.032, −0.044) (−0.033, −0.042)

Example 7.3.2 Estimating $\nabla \ell(\boldsymbol{v})$ in a Reliability Model. Consider a reliability model with sample performance

$$L(y_1, y_2, v_3, v_4) = \min\{\max(y_1, v_3), \max(y_2, v_4)\}.$$

Note that if we let $x_1 = \max(y_1, v_3)$, $x_2 = \max(y_2, v_4)$ and $\tilde{L}(\boldsymbol{x}) = \min(x_1, x_2)$, then the corresponding random variables X_1 and X_2 would take values v_3 and v_4 with nonzero probability. Hence the random vector $\boldsymbol{X} = (X_1, X_2)$ would not have a density function at point (v_3, v_4), since its distribution is a mixture of continuous and discrete ones. Consequently, the POSF method fails in its current form. To overcome this difficulty we carry out a transformation. We first write

$$x_1 = \max(y_1, v_3), \qquad x_2 = \max(y_2, v_4)$$

as

$$x_1 = v_3 \max\left(\frac{y_1}{v_3}, 1\right), \qquad x_2 = v_4 \max\left(\frac{y_2}{v_4}, 1\right),$$

and then replace $\boldsymbol{X} = (X_1, X_2)$ by the random vector

$$\tilde{\boldsymbol{X}} = (v_3 \tilde{X}_1, v_4 \tilde{X}_2), \tag{7.3.10}$$

where

$$\tilde{X}_1 = \max\left(\frac{Y_1}{v_3}, 1\right) \quad \text{and} \quad \tilde{X}_2 = \max\left(\frac{Y_2}{v_4}, 1\right).$$

The reader is asked to prove that the density of the random vector $(\tilde{X}_1, \tilde{X}_2)$ is differentiable with respect to the variables (v_3, v_4), provided $v_3 \neq 1$ and $v_4 \neq 1$.

Example 7.3.3 Estimating the cdf of a DESS for an Additive Performance Function $L(\boldsymbol{y})$. Suppose we wish to estimate the cdf

$$H_L(x, \boldsymbol{v}) = P_{\boldsymbol{v}}\{L(\boldsymbol{Y}) \leq x\} \tag{7.3.11}$$

for multiple values of $x \leq x_0$ and $\boldsymbol{v}$, where $\boldsymbol{Y} = (Y_1, \ldots, Y_n)$, $\boldsymbol{Y} \sim f(\boldsymbol{y}, \boldsymbol{v})$, and x_0 is a fixed number. Assume that the function $L(\boldsymbol{y})$ is additive in the sense that it has the representation $L(\boldsymbol{y}) = L_1(y_1) + L_2(y_n)$, where the cdf of the random variable $L_2(Y_n)$ is available. Without loss of generality, assume the particular form, $L_2(y_n) = y_n$. A procedure for estimating $H_L(x, \boldsymbol{v})$ is given below.

1. Represent H_L as

$$H_L(x, \boldsymbol{v}) = \mathbb{E}_{\boldsymbol{v}}\{I_{(-\infty,0]}(L - x)\}. \tag{7.3.12}$$

2. Define the random vector $\boldsymbol{Z} = (Z_1, \ldots, Z_n)$, where $Z_1 = Y_1, \ldots, Z_{n-1} = Y_{n-1}, Z_n = Y_n - x$, and write the cdf H_L in the equivalent representation

$$H_L(x, \boldsymbol{v}) = \mathbb{E}_{\boldsymbol{u}}\{\tilde{I}(\boldsymbol{Z})\}, \tag{7.3.13}$$

where $\boldsymbol{u} = (x, \boldsymbol{v})$, $\tilde{I}(\boldsymbol{Z}) = I_{(-\infty,0]}(L(\boldsymbol{Z}_1) + \tilde{Z}_n)$, $\boldsymbol{Z}_1 = (Z_1, \ldots, Z_{n-1})$, and the expectation is taken with respect to the pdf $\tilde{f}(\mathbf{z}, \boldsymbol{u}) = f(\mathbf{z}_1, z_n + x, \boldsymbol{v})$. Note that x is treated here as a parameter.

3. Represent the cdf H_L as

$$H_L(x, \boldsymbol{v}) = \mathbb{E}_g\{\tilde{I}(\boldsymbol{Z})\hat{W}(\boldsymbol{Z}, \boldsymbol{u})\}, \tag{7.3.14}$$

where $g(\mathbf{z})$ is a pdf that dominates the pdf $\tilde{f}(\mathbf{z}, \boldsymbol{u})$, $\hat{W}(\mathbf{z}, \boldsymbol{u}) = \tilde{f}(\mathbf{z}, \boldsymbol{u})/g(\mathbf{z})$, and $\boldsymbol{Z} \sim g(\mathbf{z})$.

4. Generate a sample $\{\boldsymbol{Z}_1, \ldots, \boldsymbol{Z}_N\}$ from $g(\mathbf{z})$, and estimate $H_L(x, \boldsymbol{v})$ by

$$\tilde{H}_{LN}(x, \boldsymbol{v}) = \frac{1}{N}\sum_{i=1}^{N} \tilde{I}(\boldsymbol{Z}_i)\hat{W}(\boldsymbol{Z}_i, \boldsymbol{u}). \tag{7.3.15}$$

Letting $g(\mathbf{z}) = f(\mathbf{z}, \boldsymbol{u}_0)$, where $\boldsymbol{u}_0 = (x_0, \boldsymbol{v}_0)$ and $x_0 \geq x$, and assuming that the components of the vector $\boldsymbol{Y} = (Y_1, \ldots, Y_n)$ are independent, the likelihood ratio $\hat{W}$ can then be written in the form

$$\hat{W}(\mathbf{z}, \boldsymbol{u}) = \frac{f_n(z_n + x, v_n)}{f_n(z_n + x_0, v_{0n})} \prod_{k=1}^{n-1} \frac{f_k(z_k, v_k)}{f_k(z_k, v_{0k})}.$$

7.3.1 Extension to Sensitivity Analysis

Suppose that the pdf $\tilde{f}(\mathbf{z}, \boldsymbol{u})$, $\boldsymbol{u} = (x, \boldsymbol{v})$, is p times continuously-differentiable with respect to both x and $\boldsymbol{v}$. Then under standard regularity conditions, the sensitivities $\nabla^k H_L(x, \boldsymbol{v})$ and their associated estimators, $\nabla^k \tilde{H}_{LN}(x, \boldsymbol{v})$, can be written as

$$\nabla^k H_L(x, \boldsymbol{v}) = \mathbb{E}_g\{\tilde{I}(\boldsymbol{Z})\nabla^k \hat{W}(\boldsymbol{Z}, \boldsymbol{u})\} \tag{7.3.16}$$

and

$$\nabla^k \tilde{H}_{LN}(x, \boldsymbol{v}) = \frac{1}{N}\sum_{i=1}^{N} \tilde{I}(\boldsymbol{Z}_i)\nabla^k \hat{W}(\boldsymbol{Z}_i, \boldsymbol{u}), \tag{7.3.17}$$

respectively, where $\nabla\hat{W}$ is the *score function*. Note that in particular,

$$h_L(x, \boldsymbol{v}) = \frac{\partial H_L(x, \boldsymbol{v})}{\partial x}$$

is the pdf of the random variable $L(\boldsymbol{Y})$, and can be estimated by

$$\tilde{h}_{LN}(x, \boldsymbol{v}) = \frac{\partial \tilde{H}_{LN}(x, \boldsymbol{v})}{\partial x} = \frac{1}{N}\sum_{i=1}^{N} \tilde{I}(\boldsymbol{Z}_i)\frac{\partial \hat{W}(\boldsymbol{Z}_i, x, \boldsymbol{v})}{\partial x}.$$

Denoting $\nabla^0 H \equiv H$ and $\nabla^0 W \equiv W$, the algorithm for estimating the cdf H_L and its sensitivities $\nabla^k H_L(x, \boldsymbol{v}),\ k = 0, 1, \ldots, p$, can be written as follows:

Algorithm 7.3.1

1. *Given* $\boldsymbol{Y} \sim f(\boldsymbol{y}, \boldsymbol{v})$ *and* $L(\boldsymbol{Y}) = L_1(Y_1, \ldots, Y_{n-1}) + Y_n$, *construct the pdf* $\tilde{f}(\boldsymbol{y}, \boldsymbol{u}),\ \boldsymbol{u} = (x, \boldsymbol{v})$.
2. *Choose a pdf* $g(\mathbf{z})$ *that dominates the pdf* $\tilde{f}(\mathbf{z}, \boldsymbol{u})$, *and generate a sample* $\{\boldsymbol{Z}_1, \ldots, \boldsymbol{Z}_N\}$ *from* $g(\mathbf{z})$.
3. *Estimate the cdf* $H_L(x, \boldsymbol{v})$ *and the associated sensitivities* $\nabla^k H_L(x, \boldsymbol{v})$, $k = 1, 2, \ldots,$ *via the SF estimator* (7.3.17).

The next two sections present more advanced material on the the push-out technique. Although the main emphasis will be placed on simple queueing and inventory models, most of the results can be easily adapted to more complex DEDS, and in particular to rather general queueing networks. It will be shown that in order to estimate the system performance measures $\nabla^k \ell(\boldsymbol{v}),\ k = 0, 1, \ldots.$ one does not have to know the analytical form of the underlying output processes.

*7.4 PUSH-OUT METHOD FOR QUEUEING AND PRODUCTION MODELS

This section applies the push-out method to queueing models and production models.

7.4.1 Queueing Models

Before proceeding further, we introduce some notation:

1. As usual, Y_{1n} denotes the service time of the nth customer, and Y_{2n} denotes the time between the arrivals of the nth and $(n-1)$st customer (by convention, $Y_{10} = Y_{20} = 0$), L_n and $\mathscr{L}_n$ denote the sojourn time and waiting time, respectively, of the nth customer.
2. X denotes a steady-state random variable with steady-state cdf

$$H_X(x, \boldsymbol{v}) = P_{\boldsymbol{v}}\{X \le x\} = \lim_{n\to\infty} P_{\boldsymbol{v}}\{X_n \le x\}.$$

It will next be shown that the push-out technique makes it possible to estimate the cdf $H_X(x, \boldsymbol{v})$, and the pdf $h_X(x, \boldsymbol{v}) = \partial H_X(x, \boldsymbol{v})/\partial x$, of random variables $X \in \{L, \mathscr{L}\}$, for multiple values of x and v from a *single simulation run*. Estimation of $H_X(x, \boldsymbol{v})$, and $h_X(x, \boldsymbol{v})$ for related random variables, such as the virtual waiting time and queue length, is treated in Rubinstein (1992).

We start with the cdf of the random variable

$$X = \sum_{t=1}^{\tau} L_t,$$

where τ is the number of customers served during a busy period in a steady-state $GI/G/1$ queue with first-in/first-out (FIFO) discipline. Let $\boldsymbol{v}$ be a vector parameter of the service time distribution, $f(y, \boldsymbol{v})$. Using standard likelihood ratios (see Chapter 6), one can represent H_X as

$$H_X(x, \boldsymbol{v}) = \mathbb{E}_g\{I_{(-\infty,0]}(X - x)\tilde{W}_\tau(\underline{Z}_\tau, \boldsymbol{v})\}, \tag{7.4.1}$$

where

$$\tilde{W}_t = \prod_{j=1}^{t} W_j, \qquad W_j = \frac{f(Z_j, \boldsymbol{v})}{g(Z_j)}, \qquad Z_j \sim g(z)$$

and $\underline{Z}_t = (Z_1, \ldots, Z_t)$. Higher-order derivatives of H_X may be similarly represented. Thus, the standard likelihood estimators of $H_X(x, \boldsymbol{v})$, based on (7.4.1), require calculation of the indicator function $I_{(-\infty,0]}(X - x)$ for each value of x, separately.

The following procedure, based on the push-out technique, permits *simultaneous* estimation of $H_X(x, \boldsymbol{v})$, for multiple values of x and $\boldsymbol{v}$, from a *single simulation run*:

1. Represent $H_X(x, \boldsymbol{v})$ as

$$H_X(x, \boldsymbol{v}) = \mathbb{E}_{\boldsymbol{v}}\{I_{(-\infty,0]}(X - x)\}.$$

Recalling that L_n and $\mathscr{L}_n$ are the sojourn and waiting time of the nth customer, it follows that

$$L_n = \mathscr{L}_n + Y_{1n}, \qquad n = 1, 2, \ldots, \tag{7.4.2}$$

so that H_X can be rewritten as

$$H_X(x, \boldsymbol{v}) = \mathbb{E}_{\boldsymbol{u}}\{I_{(-\infty,0]}(\tilde{X})\}, \qquad \boldsymbol{u} = (x, \boldsymbol{v}), \tag{7.4.3}$$

where

$$\tilde{X} = \sum_{n=1}^{\tau-1} L_n + \mathscr{L}_\tau + \tilde{Y}_\tau, \tag{7.4.4}$$

$\tilde{Y}_\tau = Y_\tau - x$ and $\tilde{Y}_\tau \sim \tilde{f}(y, \boldsymbol{u})$. Note that $\tilde{f}(y, 0, \boldsymbol{v}) = f(y, \boldsymbol{v})$.

2. Next, represent the cdf $H_X(x, \boldsymbol{v})$ as

$$H_X(x, \boldsymbol{v}) = \mathbb{E}_g\{\tilde{I}_\tau \tilde{W}_\tau(\boldsymbol{u})\}, \tag{7.4.5}$$

where

$$\begin{aligned} \tilde{W}_\tau(\boldsymbol{u}) &= \hat{W}_\tau(\boldsymbol{u}) \prod_{j=1}^{\tau-1} W_j(\boldsymbol{v}), \qquad W_j(\boldsymbol{v}) = \frac{f(Z_j, \boldsymbol{v})}{g(Z_j)}, \qquad j = 1, \ldots, \tau - 1, \\ \hat{W}_\tau(\boldsymbol{u}) &= \tilde{f}(Z_\tau, \boldsymbol{u})/g(Z_\tau), \qquad Z_k \sim g(z), \qquad k = 1, \ldots, \tau, \end{aligned} \tag{7.4.6}$$

and

$$\tilde{I}_\tau = I_{(-\infty,0]}\left(\sum_{n=1}^{\tau-1} L_n + \mathscr{L}_\tau + Z_\tau\right). \tag{7.4.7}$$

3. Finally, estimate $H_X(x, \boldsymbol{v})$ by

$$\tilde{H}_{XN}(x, \boldsymbol{v}) = \frac{1}{N} \sum_{i=1}^{N} \tilde{I}(\underline{Z}_{\tau i}) \tilde{W}(\underline{Z}_{\tau i}, \boldsymbol{u}), \tag{7.4.8}$$

where $\underline{Z}_{\tau i} = (Z_{1i}, \ldots, Z_{\tau i})$, $i = 1, \ldots, N$, is a sample from $g(\mathbf{z})$. The higher-order derivatives of $H_X(x, \boldsymbol{v})$ may be estimated similarly. In the special case of the $GI/G/\infty$ queue, $\tau = 1$ and (7.4.8) reduces to (7.3.15).

Two examples illustrate the procedure:

1. Let $f(y, v) = v \exp(-vy)$. Then $\tilde{f}(y, x, v) = v \exp(-v(y + x))$, $y \geq -x$. The dominating pdf can be chosen as

$$g(y) = v_0 \exp(-v_0(y + x_0)), \tag{7.4.9}$$

where $x < x_0$ and $y > -x_0$. Note that we could also choose $\tilde{f}(y, x, v) = f(y, v)$ for $n = 1, \ldots, \tau - 1$ and $\tilde{f}(y, x, v) = v \exp(-v(y + x))$ for $n = \tau$, and similarly for $g(y)$.

2. Let $f(y, v) = v \exp(-v(y - x_0))$, $y > x_0 > 0$. Then

$$\tilde{f}(y, v, x) = v \exp(-v(y - x_0 + x)), \qquad y > x_0 - x,$$

and the dominating pdf can be chosen as $g(y) = v_0 \exp(-v_0 y)$, $y > 0$.

It is important to note that the procedure above can be adapted to general queueing networks and, in particular, to those in which the distribution of neither L_n nor $\mathscr{L}_\tau$ is analytically available. This is so because both variates, $\sum_{n=1}^{\tau-1} L_n$ and $\mathscr{L}_\tau$, in the sample performance,

$$\tilde{I}_\tau = I_{(-\infty,0]}\left(\sum_{n=1}^{\tau-1} L_n + \mathscr{L}_\tau + Z_\tau\right),$$

can be obtained by simulation, and the variate Z_τ is generated from the dominating pdf, $g(z)$.

Arguing similarly to (7.4.2)–(7.4.8), we next present a procedure for estimating the steady-state cdf

$$H_L(x, \boldsymbol{v}) = P_{\boldsymbol{v}}(L \le x), \tag{7.4.10}$$

of the sojourn time, L, in a $GI/G/1$ queue.

1. Write $H_L(x, \boldsymbol{v})$ as

$$H_L(x, \boldsymbol{v}) = \mathbb{E}_{\boldsymbol{v}}\{I_{(-\infty,0]}(L - x)\}. \tag{7.4.11}$$

2. Represent H_L as

$$H_L(x, \boldsymbol{v}) = \mathbb{E}_{\boldsymbol{u}}\{I_{(-\infty,0]}(\tilde{L}_n)\}, \tag{7.4.12}$$

where $\tilde{L}_n = \mathscr{L}_n + \tilde{Y}_{1n}$, $Y_{1n} \sim f(y, \boldsymbol{v})$, $\tilde{Y}_{1n} = Y_{1n} - x$ and $\tilde{Y}_{1n} \sim \tilde{f}(y, \boldsymbol{u})$. Since $\{I_n = I_{(-\infty,0]}(\tilde{L}_n)\}$ is a regenerative process across busy cycles, we can write H_L as

$$H_L(x, \boldsymbol{v}) = \frac{\mathbb{E}_{\boldsymbol{u}}\left\{\sum_{n=1}^{\tau} I_n\right\}}{\mathbb{E}_{\boldsymbol{u}}\{\tau\}}. \tag{7.4.13}$$

3. Rewrite (7.3.13) as

$$H_L(x, \boldsymbol{v}) = \frac{\mathbb{E}_g\left\{\sum_{n=1}^{\tau} \tilde{I}_n \tilde{W}_n(\boldsymbol{u})\right\}}{\mathbb{E}_g\left\{\sum_{n=1}^{\tau} \tilde{W}_n(\boldsymbol{u})\right\}}, \tag{7.4.14}$$

where

$$\tilde{I}_n = I_{(-\infty,0]}(\mathscr{L}_n + Z_n), \tag{7.4.15}$$

$$\tilde{W}_n(\boldsymbol{u}) = \hat{W}_n(\boldsymbol{u}) \prod_{j=1}^{n-1} W_j(\boldsymbol{v}), \tag{7.4.16}$$

$$W_j(\boldsymbol{v}) = \frac{f(Z_j, \boldsymbol{v})}{g(Z_j)}, \qquad \hat{W}_n(\boldsymbol{u}) = \frac{\tilde{f}(Z_n, \boldsymbol{u})}{g(Z_n)}, \qquad Z_n \sim g(z),\ n = 1, \ldots, \tau.$$

It is not difficult to see that since $\tilde{Y}_1 = Y_1 + x$, we can choose the dominating pdf, $g(y)$, as $\tilde{f}(y, x_0, \boldsymbol{v}_0)$, provided $x_0 \geq x$ and $y \geq 0$.

4. Estimate $H_L(x, \boldsymbol{v})$ by

$$\tilde{H}_{LN}(x, \boldsymbol{v}) = \frac{\sum_{i=1}^{N} \sum_{n=1}^{\tau_i} \tilde{I}_{ni}(\underline{Z}_{ni}) \tilde{W}_{ni}(\underline{Z}_{ni}, \boldsymbol{u})}{\sum_{i=1}^{N} \sum_{n=1}^{\tau_i} \tilde{W}_{ni}(\underline{Z}_{ni}, \boldsymbol{u})}, \tag{7.4.17}$$

where $\{\underline{Z}_{ni}\} = \{(Z_{1i}, \ldots, Z_{ni}) \colon n = 1, \ldots, \tau_i,\ i = 1, \ldots, N\}$ is a random sample from $g(\mathbf{z})$.

In order to estimate the steady-state pdf, h, of the sojourn time process, $\{L_n\}$, we have to differentiate the ratio formula (7.4.17) with respect to x, and similarly for the estimators of the higher-order derivatives of $H_L(x, \boldsymbol{v})$.

We point out again that the procedure above can be adapted to *rather general queueing networks* and, in particular, to those where the underlying process, $\{\mathscr{L}_n\}$, is not known analytically. Again, both of the variates, $\mathscr{L}_n$ and Z_n, associated with the sample performance

$$\tilde{I}_n = I_{(-\infty,0]}(\mathscr{L}_n + Z_n),$$

can be obtained by simulation, and the variate $\tilde{Y}$ can be similarly generated from the dominating pdf, $g(y) = \tilde{f}(y, x_0, \boldsymbol{v}_0)$.

7.4.2 Production Inventory Models

Consider the *production inventory model* (PIM) of Example 6.2.5. Let $\{L_n\}$ be the associated shortfall process (equivalently, the waiting time process in a $GI/D/1$ queue). Here we show how to estimate the following from a single simulation:

1. The steady-state distribution $H_L(x, \boldsymbol{v})$.
2. The expected steady-state shortfall $\ell(\boldsymbol{v}) = \mathbb{E}_{\boldsymbol{v}}\{L\}$.
3. The expected steady-state *backlog*, defined as

$$\ell(\boldsymbol{v}) = \mathbb{E}_{\boldsymbol{v}}\{(L - S)^+\},$$

where S is the target inventory level.

4. The *fill rate*, or the expected steady-state proportion of demands met from stock, defined (see Glasserman and Liu, 1995) as

$$\ell(\boldsymbol{v}) = 1 - \frac{\mathbb{E}\{[\min(L - (S - s))^+, s]\}}{\mathbb{E}\{D\}}.$$

We now proceed to outline the estimation procedure.

1. Noting the identity (see Example 6.2.5)

$$L_n = \sum_{j=1}^{n-1}(D_j - v_2) = \sum_{j=1}^{n-2}(D_j - v_2) + (D_{n-1} - v_2), \qquad n = 1, \ldots, \tau,$$

we can write (7.4.12) as

$$H_L(x, \boldsymbol{v}) = \mathbb{E}_{\boldsymbol{u}}\{I_{(-\infty,0]}(\tilde{L}_n)\}, \qquad \boldsymbol{u} = (x, \boldsymbol{v}), \tag{7.4.18}$$

where

$$\tilde{L}_n = \sum_{j=1}^{n-2} U_j + \tilde{U}_{n-1}, \qquad U_j = D_j - v_2, \qquad \tilde{U}_{n-1} = D_{n-1} - v_2 - x, \ \ D_j \sim f(y, \boldsymbol{v}_1), \tag{7.4.19}$$

for $j = 1, \ldots, n, \ n = 1, \ldots, \tau, \ \boldsymbol{v} = (\boldsymbol{v}_1, v_2)$, where $v_2 \equiv s$ and s is the capacity of the production line.

A push-out estimator of the steady-state distribution $H_L(x, \boldsymbol{v})$ can be defined similarly to (7.4.17). Note that we need to define here two auxiliary pdf's, $\tilde{f}(y, \boldsymbol{v})$ and $\check{f}(y, x, \boldsymbol{v})$, for the variates $U_j, \ j = 1, \ldots, n-2$, and $\tilde{U}_{n-1}$, respectively.

2. Noting that for $x = 0$, we have $\tilde{f} = \check{f}$, the POSF estimator for the steady-state expected shortfall, $\ell(\boldsymbol{v}) = \mathbb{E}_{\boldsymbol{v}}\{L\}$, can be written as

$$\tilde{\ell}_N(\boldsymbol{v}) = \frac{\sum_{i=1}^{N}\sum_{n=1}^{\tau_i} \tilde{L}_{ni}(\underline{Z}_{ni})\tilde{W}_{ni}(\underline{Z}_{ni}, \boldsymbol{v}_1, v_2)}{\sum_{i=1}^{N}\sum_{n=1}^{\tau_i} \tilde{W}_{ni}(\underline{Z}_{ni}, \boldsymbol{v}_1, v_2)}. \tag{7.4.20}$$

Here, $\{\underline{Z}_{ni}\} = \{(Z_{1i}, \ldots, Z_{ni}) \colon n = 1, \ldots, \tau_i, \ i = 1, \ldots, N\}$ is a random sample from $g(\mathbf{z})$,

$$\tilde{W}_{ni}(\boldsymbol{v}) = \prod_{j=1}^{n} W_{ji}(\boldsymbol{v}), \qquad W_j = \frac{\tilde{f}(Z_{ji}, \boldsymbol{v})}{g(Z_{ji})},$$

$Z_{ji} \sim g(z)$ and $\boldsymbol{v} = (\boldsymbol{v}_1, v_2)$. The estimation of the derivatives $\nabla^k \ell_N(\boldsymbol{v})$ is similar. Observe that (7.4.20) is closely related to (7.4.17), in the sense that in order to obtain (7.4.20) from (7.4.17), we have to replace $\tilde{H}$ by $\tilde{\ell}$, v_2 by x, and $\tilde{I}$ by $\tilde{\mathscr{L}}$, with all other data remaining unchanged.

3. The mean backlog $\ell(\boldsymbol{v}) \equiv \mathbb{E}_{\boldsymbol{v}}\{(L - S)^+\}$ can be written as

$$\ell(\boldsymbol{v}) = \int_S^{\infty} (L(y) - S) f_L(y)\, dy = \mathbb{E}\{(L - S) I_{(-\infty,0]}(L - S)\}.$$

In analogy to (7.4.18), it can also be represented as

$$\ell(\boldsymbol{v}) = \mathbb{E}_{\boldsymbol{u}}\{\tilde{L}_n I_{(-\infty,0]}(\tilde{L}_n)\}, \qquad \boldsymbol{u} = (S, \boldsymbol{v}), \tag{7.4.21}$$

where $\tilde{L}_n$ in (7.4.21) coincides with $\tilde{L}_n$, in (7.4.18) but with x replaced by S.

Again a POSF estimator of the mean backlog $\ell(\boldsymbol{v}) \equiv \mathbb{E}_{\boldsymbol{v}}\{(L - S)^+\}$ coincides with that of the steady-state distribution $H_L(x, \boldsymbol{v})$ of L_n, provided $I_{(-\infty,0]}(\tilde{L}_n)$ in (7.4.18) is replaced by $\tilde{L}_n I_{(-\infty,0]}(\tilde{L}_n)$ in (7.4.21).

4. In view of the fact that the fill rate, $\ell(\boldsymbol{v})$, can be written as (see Glasserman and Liu, 1996)

$$\ell(\boldsymbol{v}) = 1 - \frac{1}{\mathbb{E}\{D\}} \mathbb{E}_{\boldsymbol{v}}\{(L + D - S)^+ - (L - S)^+\},$$

where $\mathbb{E}_{\boldsymbol{v}}\{(L - S)^+\}$ is the mean backlog, it is readily seen that an estimator of the fill rate is similar to that of the mean backlog.

We now proceed to present numerical results for simple queueing and inventory models, using the POSF method.

Table 7.4.1 displays the crude Monte Carlo (CMC) estimator, $\bar{H}_{LN}(x, v)$, the POSF estimator, $\tilde{H}_{LN}(x, v)$ [see (7.4.20)], and the associated sample variances, $\hat{\sigma}^2\{\bar{H}_{LN}\}$ and $\hat{\sigma}^2\{\tilde{H}_{LN}\}$, respectively, for the sojourn time in the $M/G/1$ queue with arrival rate $\lambda = 1$ and service time pdf

$$f(z, k, v) = v \exp(-v(z - k)), \qquad z \geq k = 0.3, \tag{7.4.22}$$

the latter being a shifted exponential pdf with shift constant k. The CMC estimator, $\bar{H}_{LN}(x, v)$, was calculated as the standard regenerative ratio estimator, requiring a separate simulation of the $M/G/1$ queue, for each parameter value $v = 4.0, 5.0, 6.0$. The dominating pdf, $g(z)$, was chosen as

$$g(z) = f(z, k_0, v_0) = v_0 \exp(-v_0(z - k_0)), \qquad z \geq k_0,$$

with $v_0 = 2.5$ and $k_0 = 0.3$. The cdf $H_L(x, v)$ was estimated from a single simulation run of $N = 10^5$ customers for multiple values of x and v. More specifically, a simulation was run for each value of v, and reliable estimators were obtained for x in the range $0.24 \leq x \leq 0.44$. Since $\mathbb{E}_g\{Z\} = 0.667$, the reference traffic intensity was set at $\rho_0 = \lambda \mathbb{E}_g\{Z\} = 0.667$.

Note that in order to extend the prediction interval of H_L, it is necessary to run an additional simulation with different values of the reference parameters, v_0 and k_0, in the dominating pdf, $g(z)$. For typical applications we found that in order to estimate the cdf H_L in a broad interval (say, in the interval $0.05 \leq H_L \leq 0.95$), it is sufficient to complete only two or three simulation runs with distinct parameter values in $g(z)$. A simple rule of thumb is to divide the interval of H_L into equal parts (e.g., $0.05 \leq H_L \leq 0.35$, $0.35 \leq H_L \leq 0.65$, $0.65 \leq H_L \leq 0.95$), and then run multiple independent simulations (in our example, three simulations with reference traffic intensities $\rho_0 = 0.35$, 0.65, 095, respectively).

Table 7.4.1 Performance of the CMC Estimator, $\bar{H}_{LN}(x, v)$, and POSF Estimator, $\tilde{H}_{LN}(x, v)$, for the $M/G/1$ Queue with Service Time pdf of (7.4.22)

v	ρ	x	$\bar{H}_{LN}(x, v)$	$\hat{\sigma}^2\{\bar{H}_{LN}\}$	$\tilde{H}_{LN}$	$\hat{\sigma}^2\{\tilde{H}_{LN}(x, v)\}$
		0.240	0.083	0.813E-06	0.083	0.328E-06
		0.280	0.157	0.152E-05	0.156	0.143E-05
4.00	0.450	0.320	0.223	0.213E-05	0.221	0.145E-04
		0.360	0.282	0.267E-05	0.284	0.217E-05
		0.400	0.337	0.308E-05	0.336	0.293E-05
		0.440	0.396	0.349E-0.5	0.404	0.325E-05
		0.240	0.113	0.106E-05	0.110	0.578E-06
		0.280	0.209	0.185E-05	0.206	0.142E-05
5.00	0.400	0.320	0.293	0.245E-05	0.288	0.245E-04
		0.360	0.364	0.290E-05	0.363	0.346E-05
		0.400	0.427	0.323E-05	0.426	0.453E-05
		0.440	0.476	0.389E-0.5	0.474	0.555E-05
		0.240	0.138	0.126E-05	0.137	0.878E-06
		0.280	0.250	0.215E-05	0.252	0.207E-05
6.00	0.367	0.320	0.343	0.266E-05	0.348	0.334E-04
		0.360	0.424	0.303E-05	0.429	0.469E-05
		0.400	0.427	0.323E-05	0.426	0.593E-05
		0.440	0.476	0.389E-0.5	0.474	0.816E-05

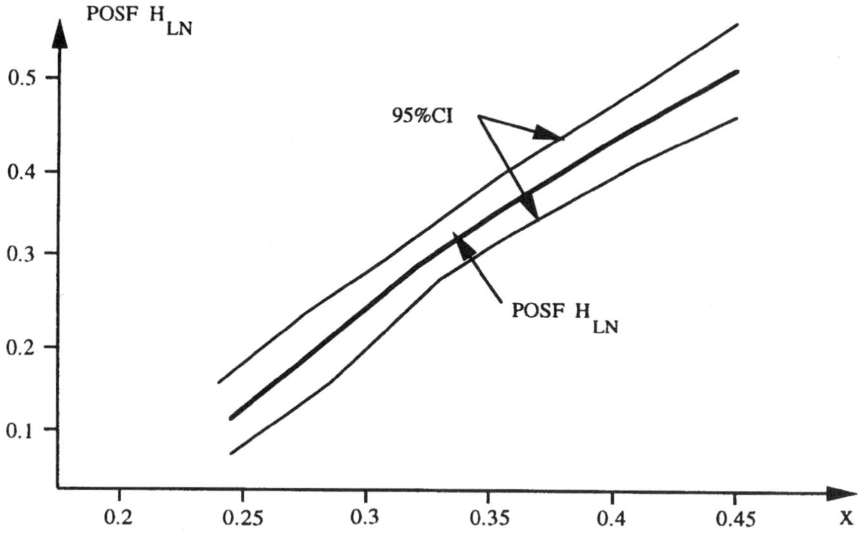

Figure 7.1. Performance of the POSF estimator, $\tilde{H}_{LN}(x, v)$, for an $M/G/1$ queue.

Figure 7.1 depicts the empirical cdf $\tilde{H}_{LN}(x, v)$ (denoted by POSF H_{LN}) as a function of x, and the associated 95% confidence intervals (denoted by 95% CI), for the POSF estimator $\tilde{H}_{LN}(x, v)$ from Table 7.4.1, with $v = 5$ ($\rho = 0.4$).

Table 7.4.2 displays the CMC estimator $\bar{\ell}(\boldsymbol{v})$, the corresponding POSF estimator $\tilde{\ell}_N(\boldsymbol{v})$ from (7.4.20) and its sample variance $\hat{\sigma}^2\{\tilde{\ell}_N(\boldsymbol{v})\}$, for the $D/M/1$ queue with $\rho = 1/v_1$. The results were obtained from a single simulation run of $N = 500{,}000$ customers. The dominating density used was $g(\mathbf{z}) = f(\mathbf{z}, v_{01}, v_{02}) = v_{02} \exp(-v_{02}(v_{01} + \mathbf{z}))$ from (7.4.9) with $v_{01} = \lambda = 1.111$, and $v_{02} = \mu = 1.0$. The expected waiting time was estimated for multiple values of v_1 in the interval $1.111 \leq v_1 \leq 10.0$.

Table 7.4.3 displays the theoretical values $\ell(\boldsymbol{v})$, the corresponding POSF estimators of $\tilde{\ell}_N(\boldsymbol{v})$ from (7.4.20), and the associated sample variance $\hat{\sigma}^2\{\tilde{\ell}_N(\boldsymbol{v})\}$, for a PIM ($M/D/1$ queue) with exponential interdemand time and interdemand rate $\lambda = 1$. The results were obtained from a single simulation run of $N = 500{,}000$ customers. The dominating density used was $g(\mathrm{z}) = f(\mathrm{z}, v_{01}, v_{02}) = v_{01} \exp(-v_{01}(v_{02} + \mathrm{z}))$ from (7.4.9) with $v_{01} = \lambda = 1$, and $v_{02} = 0.6$. The expected shortfall was estimated for multiple values of v_2 ($\rho_0 = 0.6$) in the interval $0.1 \leq v_2 \leq 0.6$.

The results of Tables 7.4.1–7.4.3 are self-explanatory. Extensive simulation studies were carried out using the POSF method for simple queueing networks as well. The results were similar to those in Tables 7.4.1–7.4.3, indicating quite good performance of the POSF estimators. For more details see Raz (1991). Note again that the above tables were derived using the likelihood ratio (7.3.7). More accurate (lower-variance) estimators could be derived with the likelihood ratio (7.3.8) based on acceptance-rejection method.

Table 7.4.2 Performance of POSF Estimator $\tilde{\ell}_N(\boldsymbol{v})$, for the $D/M/1$ Queue with $v_{01} = 1.111$ ($\rho_0 = 0.9$), and $v_{02} = \mu = 1.0$

ρ	$\bar{\ell}(\boldsymbol{v})$	$\tilde{\ell}_N(\boldsymbol{v})$	$\hat{\sigma}^2\{\tilde{\ell}_N(\boldsymbol{v})\}$
0.100	9.99E- 01	1.00E+00	1.05E-05
0.200	1.01E+00	1.01E+00	1.06E-05
0.300	1.04E+00	1.04E+00	1.14E-05
0.400	1.12E+00	1.11E+00	1.32E-05
0.500	1.24E+00	1.20E+00	1.64E-05
0.600	1.49E+00	1.41E+00	2.29E-05
0.700	1.84E+00	1.70E+00	3.74E-05
0.800	2.71E+00	2.61E+00	8.69E-05
0.900	5.14E+00	5.02E+00	7.74E-03

Table 7.4.3 Performance of the POSF Estimator $\tilde{\ell}_N(v)$, for a PIM with Exponential Interdemand Time (equivalently, an $M/D/1$ Queue with $\rho = v_2$), $\lambda = 1$ and $v_{02} = 0.6$

v_2	$\ell(v)$	$\tilde{\ell}_N(v)$	$\hat{\sigma}^2\{\tilde{\ell}_N(v)\}$
0.100	5.560E-03	5.480E-03	5.090E-09
0.150	1.320E-02	1.330E-02	2.430E-08
0.200	2.500E-02	2.553E-02	9.030E-09
0.250	4.170E-02	4.210E-02	2.080E-07
0.300	6.430E-02	6.450E-02	3.570E-07
0.350	9.420E-02	9.480E-02	4.690E-07
0.400	1.330E-01	1.330E-01	2.180E-06
0.450	1.840E-01	1.830E-01	3.100E-06
0.500	2.500E-01	2.280E-01	1.410E-05
0.550	3.360E-01	3.390E-01	3.280E-05
0.600	4.500E-01	4.530E-01	6.280E-05

7.5 PUSH-OUT METHOD FOR INVENTORY MODELS

This section deals with estimating the optimal (s, S) policy in a classic single-commodity, multiperiod, continuous review inventory model with *backlog*. A backlog is created when the system is out of stock and the requested demand is back logged and filled as soon as adequate replenishment arrives. Following established terminology in inventory theory (e.g., Sahin, 1989, Tijms, 1994), we employ the following definitions:

1. *On-hand inventory* is the inventory physically present, and immediately available.
2. *Net inventory* is the on-hand inventory minus the back logged inventory.
3. *Inventory position* is the on-hand inventory minus back-logged inventory plus on-order inventory. The latter is the inventory already requisitioned but not yet received.
4. *Lead time* is the time elapsed from placing the order until receiving it.

As in the classic (s, S) model, we make the following assumptions:

1. The lead time is a positive random variable (possibly deterministic).
2. Demands in disjoint intervals are iid random variables with a continuous pdf.
3. Inventory control is based on the *inventory position and not on the net stock.*
4. Inventory control is parameterized by a real pair (s, S). Every time the inventory position falls to or below the reorder point s, a replenishment order is placed to raise the inventory to level S; otherwise no order is placed. In other words, the quantity to be ordered is the difference between S and the inventory position at the time the order is placed.

Finally, a *cycle* is defined as the time elapsed between two consecutive epochs in which a replenishment order is received.

Let $\{\mathcal{N}_t: t \geq 0\}$ and $\{L_t: t \geq 0\}$ be the *inventory position* and the *inventory on-hand* processes, respectively (see Figure 3.3). Let A_j be the arrival time of the jth demand and define $U_j = A_j - A_{j-1}$ to be the interdemand time for $j \geq 1$, with the convention $A_0 = 0$.

Under certain mild conditions (e.g., Tijms (1994)) on the interdemand times, U_t, and the demand sizes, Y_t, the process $\{\mathcal{N}_t\}$ is regenerative with cycle length

$$\tau = \min\left\{t: S - \sum_{j=1}^{t} Y_j \leq s\right\}.$$

Note that in the case where the interdemand time is deterministic and equals unity, $\{\mathcal{N}_t\}$ can be written recursively as

$$\mathcal{N}_{t+1} = \begin{cases} \mathcal{N}_t - Y_{t+1} & \text{if } \mathcal{N}_t - Y_{t+1} \geq s \\ S & \text{if } \mathcal{N}_t - Y_{t+1} < s, \end{cases} \tag{7.5.1}$$

where $\mathcal{N}_0 = S$, and the demand random variable, Y_t, has pdf $f(y, \boldsymbol{v}_1)$.

Our goal is to estimate, from a single sample path (simulation run), the expected steady-state performances, such as the expected steady-state inventory position $\mathbb{E}\{\mathcal{N}\}$ and the expected steady-state inventory on hand, $\mathbb{E}\{L\}$, along with the associated derivatives with respect to s and S, and to find the optimal (s, S) policy that minimizes a certain cost function that takes account of ordering, shortage, and holding costs. To this end, we apply the POSF method. We also propose a new technique, called the *conditional push-out* (CPO).

The remainder of this section is organized as follows. Section 7.5.1 considers performance evaluation and sensitivity analysis of several sample performance functions associated with the steady-state processes $\{\mathcal{N}_t\}$ and $\{L_t\}$, such as the steady-state expected inventory on hand. Both the push-out and the conditional push-out methods will be employed. Finally, Section 7.5.2 identifies the optimal (s, S) policy for an optimization program of the form (5.4.3) involving functions of fixed ordering costs, shortage costs, and holding costs.

7.5.1 Performance Evaluation and Sensitivity Analysis

This subsection shows how to estimate the expected steady-state performance

$$\ell(\boldsymbol{v}) = \mathbb{E}_{\boldsymbol{v}_1}\{L(\underline{\boldsymbol{Y}}_t, \boldsymbol{v}_2)\},$$

given in (7.1.1) and the associated gradient $\nabla\ell(\boldsymbol{v}) = \nabla_{\boldsymbol{v}_2}\mathbb{E}_{\boldsymbol{v}_1}\{L(\underline{\boldsymbol{Y}}_t, \boldsymbol{v}_2)\}$, $\boldsymbol{v} = (\boldsymbol{v}_1, \boldsymbol{v}_2)$, $\boldsymbol{v}_2 = (s, S)$, *simultaneously for multiple values of* $\boldsymbol{v} = (\boldsymbol{v}_1, \boldsymbol{v}_2)$ *while using a single simulation run* of the on-hand inventory process $\{L_t\}$. Two separate cases will be considered: (a) no lead time, and (b) positive lead time.

No Lead Time

For this model $L_t = \mathcal{N}_t$, and we have

$$L_0 = S, \qquad L_t = S - \sum_{j=1}^{k-1} Y_j, \quad \text{if } A_{k-1} \le t < A_k;\ k = 1, \ldots, \tau,$$

where

$$\tau = \min\left\{t: S - \sum_{j=1}^{t} Y_j \le s\right\}$$

is the number of demands occurring in a regenerative cycle.

For convenience, we reparametrize the problem: instead of (s, S) we use (S, Q) [or (s, Q) if preferable], where $Q = S - s$. Thus, letting $E_t = \sum_{j=1}^{t} Y_j$ and $E_0 = 0$, one has

$$\tau = \min\{t: E_t \ge Q\}.$$

Two cases will be considered: (i) $\ell(\boldsymbol{v}) = \mathbb{E}_{\boldsymbol{v}_1}\{L(\underline{\boldsymbol{Y}}_t, \boldsymbol{v}_2)\}$ and (ii) $\ell(\boldsymbol{v}) = \mathbb{E}_{\boldsymbol{v}_1}\{\varphi[L(\underline{\boldsymbol{Y}}_t, \boldsymbol{v}_2)]\}$, [$\varphi(\cdot)$ is a cost function] along with the associated derivatives $\nabla\ell(\boldsymbol{v})$.

Estimation of $\ell(\boldsymbol{v}) = \mathbb{E}_{\boldsymbol{v}_1}\{L(\underline{\boldsymbol{Y}}_t, \boldsymbol{v}_2)\}$ *and* $\nabla\ell(\boldsymbol{v}) = \nabla\mathbb{E}_{\boldsymbol{v}_1}\{L(\underline{\boldsymbol{Y}}_t, \boldsymbol{v}_2)\}$

We first show that in this case one may assume that $U_t \equiv 1$ without loss of generality. Indeed, the long-run average inventory $\ell(\boldsymbol{v})$ can be written (see Figure 3.3) as

$$\ell(\boldsymbol{v}) = \frac{\mathbb{E}_{\boldsymbol{v}_1}\left\{\sum_{t=0}^{\infty} U_t(S - E_t) I_{\{E_t < Q\}}\right\}}{\mathbb{E}_{\boldsymbol{v}_1}\{\mathcal{T}\}}$$
$$= \frac{\mathbb{E}_{\boldsymbol{v}_1}\left\{\sum_{t=0}^{\tau-1} U_t(S - E_t)\right\}}{\mathbb{E}_{\boldsymbol{v}_1}\{\mathcal{T}\}}, \tag{7.5.2}$$

where $\mathcal{T} = \sum_{t=0}^{\tau-1} U_t$ is the cycle length, $\boldsymbol{v} = (\boldsymbol{v}_1, \boldsymbol{v}_2)$ and $\boldsymbol{v}_2 = (S, Q)$.

Since the random variables U_t and Y_t are mutually independent, one has

$$\mathbb{E}_{\boldsymbol{v}_1}\{\mathcal{T}\} = \mathbb{E}_{\boldsymbol{v}_1}\{U\}\mathbb{E}_{\boldsymbol{v}_1}\{\tau\},$$

and

$$\mathbb{E}_{\boldsymbol{v}_1}\left\{\sum_{t=0}^{\infty} U_t(S - E_t) I_{\{E_t < Q\}}\right\} = \mathbb{E}_{\boldsymbol{v}_1}(U)\mathbb{E}_{\boldsymbol{v}_1}\left\{\sum_{t=0}^{\infty}(S - E_t) I_{\{E_t < Q\}}\right\}.$$

Consequently

$$\ell(\boldsymbol{v}) = \frac{\mathbb{E}_{\boldsymbol{v}_1}\{\sum_{t=0}^{\tau-1}(S - E_t)\}}{\mathbb{E}_{\boldsymbol{v}_1}\{\tau\}} = \frac{\mathbb{E}_{\boldsymbol{v}_1}\{X\}}{\mathbb{E}_{\boldsymbol{v}_1}\{\tau\}}, \tag{7.5.3}$$

where

$$X = \sum_{t=0}^{\tau-1}(S - E_t). \tag{7.5.4}$$

It follows that in this simple model, $\ell(\boldsymbol{v})$ *does not depend* on the interdemand time U_t, so we may assume that $U_t \equiv 1$. For reasons of notational convenience, a regenerative cycle of length τ will henceforth be taken from 0 to $\tau - 1$ rather than from 1 to τ.

We now consider the gradient $\nabla_{\boldsymbol{v}_2}\ell(\boldsymbol{v})$ with respect to $\boldsymbol{v}_2 = (S, Q)$. Note first that

$$\ell(\boldsymbol{v}) = \frac{\mathbb{E}_{\boldsymbol{v}_1}\{\sum_{t=0}^{\tau-1}(S - E_t)\}}{\mathbb{E}_{\boldsymbol{v}_1}\{\tau\}} = S - \frac{\mathbb{E}_{\boldsymbol{v}_1}\{X_1\}}{\mathbb{E}_{\boldsymbol{v}_1}\{\tau\}}, \tag{7.5.5}$$

where

$$X_1 = \mathbb{E}_{\boldsymbol{v}_1}\left\{\sum_{t=0}^{\tau-1} E_t\right\}. \tag{7.5.6}$$

The differentiation of $\ell(\boldsymbol{v})$ with respect to S is straightforward: since X_1 and τ do not depend on S, one has

$$\nabla_S \ell(\boldsymbol{v}) = 1.$$

The differentiation of $\ell(\boldsymbol{v})$ with respect to Q is more problematic, since it involves differentiation of the indicator function $I_{\{E_t<Q\}}$. The rest of the section will be concerned mainly with sensitivity analysis and optimization with respect to Q. Two different approaches for estimating $\nabla_Q \ell(\boldsymbol{v})$ will be presented: the POSF method, based on a combination of the push-out and score function methods, and the *conditional push-out* (CPO) method, based on a combination of the *conditional Monte Carlo* and the push-out method (see Pflug and Rubinstein, 1996).

POSF Method

To apply the POSF method to the parameter Q, we first transfer (push-out) Q from the sample performance $L_t(Q)$ to an auxiliary pdf $\tilde{f}(y, \boldsymbol{v}_1, Q)$ [see (7.2.14)] by using a suitable transformation and then employing the SF method to estimate $\nabla_Q \ell(\boldsymbol{v})$. We shall now show that Q may, in fact, appear in the auxiliary pdf $\tilde{f}(y, \boldsymbol{v}_1, Q)$ either as (i) *location* or (ii) *scale* parameter.

Q as a Location Parameter. Rewrite $\ell(\boldsymbol{v})$ in (7.5.3) as

$$(\boldsymbol{v}) = S - Q - \frac{\mathbb{E}_{\boldsymbol{v}}\{X_2\}}{\mathbb{E}_{\boldsymbol{v}}\{\tau\}}, \tag{7.5.7}$$

where

$$X_2 = \sum_{t=0}^{\infty}(E_t - Q)I_{\{E_t - Q<0\}} = \sum_{t=0}^{\infty}\left[\tilde{D}_{1Q} + \sum_{j=2}^{t} Y_j\right] I_{\{\tilde{D}_{1Q}+\sum_{j=2}^{t} Y_j<0\}}, \tag{7.5.8}$$

$\tilde{D}_{1Q} = Y_1 - Q$ has a corresponding auxiliary pdf $\tilde{f}(y, \boldsymbol{v}_1, Q) = f(y + Q, \boldsymbol{v}_1)$, and $\boldsymbol{v} = (\boldsymbol{v}_1, Q)$. Note that the subscript $\boldsymbol{v}$ in (7.5.8) means that the expectation is taken with respect to the pdf $\tilde{f}(y, \boldsymbol{v})$, $\boldsymbol{v} = (\boldsymbol{v}_1, Q)$. Denoting $E_t^{(2)} = \sum_{j=2}^{t} Y_j$, we can rewrite (7.5.7) as

$$\ell(\boldsymbol{v}) = S - Q - \frac{\mathbb{E}_{\boldsymbol{v}}\{\sum_{t=0}^{\tau-1} \tilde{L}_t\}}{\mathbb{E}_{\boldsymbol{v}}\{\tau\}}, \tag{7.5.9}$$

where

$$\tilde{L}_0 = 0, \qquad \tilde{L}_t = \tilde{D}_{1Q} + E_t^{(2)}, \qquad t = 1, \ldots, \tau - 1, \tag{7.5.10}$$

and

$$\tau = \min\{t \colon \tilde{L}_t \geq 0\}.$$

The effect of this representation is to push out the parameter Q in (7.5.9) and (7.5.10) to the auxiliary pdf $\tilde{f}(y, \boldsymbol{v}_1, Q)$ of the random variable $\tilde{D}_{1Q}$. As a result, the new sample performance, $\tilde{L}_t$, *does not depend on Q*.

Next, the derivative of $\ell(\boldsymbol{v})$ w.r.t. Q is

$$\nabla_Q \ell(\boldsymbol{v}) = -1 - \left[\frac{\nabla_Q \mathbb{E}_{\boldsymbol{v}}\{X_2\}}{\mathbb{E}_{\boldsymbol{v}}\{\tau\}} - \frac{\mathbb{E}_{\boldsymbol{v}}\{X_2\}}{\mathbb{E}_{\boldsymbol{v}}\{\tau\}} \times \frac{\nabla_Q \mathbb{E}_{\boldsymbol{v}}\{\tau\}}{\mathbb{E}_{\boldsymbol{v}}\{\tau\}}\right], \tag{7.5.11}$$

with similar representations holding for the derivatives $\nabla_{\boldsymbol{v}_1}\ell(\boldsymbol{v})$ w.r.t. $\boldsymbol{v}_1$. Applying the SF method to $\nabla_{\boldsymbol{v}}\mathbb{E}_{\boldsymbol{v}}\{X_2\}$ and $\nabla_{\boldsymbol{v}}\mathbb{E}_{\boldsymbol{v}}\{\tau\}$, we obtain under standard regularity conditions,

$$\nabla_Q \mathbb{E}_{\boldsymbol{v}}\{X_2\} = \mathbb{E}_{\boldsymbol{v}}\left\{\sum_{t=0}^{\tau-1} \tilde{L}_t(\underline{Y}_t) S_1(\tilde{Y}_1, Q, \boldsymbol{v}_1)\right\} \tag{7.5.12}$$

and

$$\nabla_Q \mathbb{E}_v\{\tau\} = \mathbb{E}_v\left\{\sum_{t=0}^{\tau-1} S_1(\tilde{Y}_1, Q, \boldsymbol{v}_1)\right\}, \tag{7.5.13}$$

where $\underline{Y}_t = (\tilde{Y}_1, Y_2, \ldots, Y_t)$, $t = 1, \ldots, \tau - 1$. Here, $\tilde{Y}_1 \sim f(y + Q, \boldsymbol{v}_1)$, and

$$S_1(\tilde{Y}_1, Q, \boldsymbol{v}_1) = \frac{\nabla f(\tilde{Y}_1 + Q, \boldsymbol{v}_1)}{f(\tilde{Y}_1 + Q, \boldsymbol{v}_1)} \tag{7.5.14}$$

is the score function. Thus, $\nabla_Q \ell(\boldsymbol{v})$ in (7.5.11) can be rewritten as

$$\nabla_Q \ell(\boldsymbol{v}) = -1 - \left[\frac{\mathbb{E}_v\{\sum_{t=0}^{\tau-1} \tilde{L}_t S_1\}}{\mathbb{E}_v\{\tau\}} - \frac{\mathbb{E}_v\{\sum_{t=0}^{\tau-1} \tilde{L}_t\}}{\mathbb{E}_v\{\tau\}} \times \frac{\mathbb{E}_v\{\sum_{t=0}^{\tau-1} S_1\}}{\mathbb{E}_v\{\tau\}}\right], \tag{7.5.15}$$

and similarly for $\nabla_{\boldsymbol{v}_1} \ell(\boldsymbol{v})$.

We shall now discuss how to estimate the parameters $\ell(\boldsymbol{v})$ and $\nabla \ell(\boldsymbol{v})$, *simultaneously* for multiple values $\boldsymbol{v}$, from a single simulation run. Let G be a probability measure (distribution) on $\mathbb{R}^m$ with pdf $g(\boldsymbol{y})$, such that $dG(\boldsymbol{y}) = g(\boldsymbol{y})\, d\boldsymbol{y}$. Suppose that for every permissible value of the parameter vector $\boldsymbol{v}$, the support of $f(\boldsymbol{y}, \boldsymbol{v})$ is contained in the support of $g(\boldsymbol{y})$, that is,

$$\text{supp}\{f(\boldsymbol{y}, \boldsymbol{v})\} \subset \text{supp}\{g(\boldsymbol{y})\}, \qquad \boldsymbol{v} \in V. \tag{7.5.16}$$

We assume below that $g(\boldsymbol{y}) = f(\boldsymbol{y}, \boldsymbol{v}_0)$. The expected steady-state performance $\ell(\boldsymbol{v})$ in (7.5.9) can be written as

$$\ell(\boldsymbol{v}) = S - Q - \frac{\mathbb{E}_g\{\sum_{t=0}^{\tau-1} \tilde{L}_t \tilde{W}_t(\boldsymbol{Z}_t, \boldsymbol{v})\}}{\mathbb{E}_g\{\sum_{t=0}^{\tau-1} \tilde{W}_t(\boldsymbol{Z}_t, \boldsymbol{v})\}}, \tag{7.5.17}$$

where

$$\begin{gathered}
\tilde{L}_0 = 0, \qquad \tilde{L}_t = \tilde{Z}_1 + \sum_{j=2}^{t} Z_j, \qquad t = 1, \ldots, \tau - 1, \\
\tau = \min\{t \colon \tilde{L}_t \geq 0\}, \\
\tilde{W}_t(\boldsymbol{Z}_t, \boldsymbol{v}) = W_1(\tilde{Z}_1, Q, \boldsymbol{v}_1) \prod_{j=2}^{t} W_j(Z_j, \boldsymbol{v}_1),
\end{gathered} \tag{7.5.18}$$

$\tilde{W}_0 = W_0 \equiv 1$ is the likelihood ratio process,

$$\begin{aligned}
\boldsymbol{Z}_t &= (\tilde{Z}_1, Z_2, \ldots, Z_t), \qquad t = 1, \ldots, \tau - 1, \\
W_j &= \frac{f(Z_j, \boldsymbol{v}_1)}{f(Z_j, \boldsymbol{v}_{01})}, \qquad Z_j \sim f(y, \boldsymbol{v}_{01}), \qquad j = 2, \ldots, \tau - 1, \\
W_1 &= \begin{cases} \dfrac{f(\tilde{Z}_1 + Q, \boldsymbol{v}_1)}{f(\tilde{Z}_1 + Q_0, \boldsymbol{v}_{01})} & \text{if } \tilde{Z}_1 \geq -Q \\ 0 & \text{otherwise,} \end{cases}
\end{aligned} \tag{7.5.19}$$

and $\tilde{Z}_1 \sim f(y + Q_0, \boldsymbol{v}_{01})$.

The representation of $\nabla \ell(\boldsymbol{v})$ is similar. For example, if we let $f(y, v) = ve^{-vy} I_{\{y \geq 0\}}$, then $\tilde{f}(y, Q, v) = ve^{-v(y+Q)} I_{\{y \geq -Q\}}$, $\tilde{f}(y, Q_0, v_0) = v_0 e^{-v_0(y+Q_0)} I_{\{y \geq Q_0\}}$, and

$$W_1 = \begin{cases} \dfrac{ve^{-v(\tilde{Z}_1+Q)}}{v_0 e^{-v_0(\tilde{Z}_1+Q_0)}} & \text{if } \tilde{Z}_1 \geq -Q \\ 0 & \text{otherwise.} \end{cases} \tag{7.5.20}$$

We next present an alternative likelihood ratio W_1, based on the acceptance-rejection method [see also (7.3.8)]. The alternate acceptance-rejection procedure *accepts* the random variable $\tilde{Z}_1$, if $\tilde{Z}_1 \geq -Q$, and *rejects* it, otherwise. For example, the alternative to the likelihood ratio W_1 in (7.5.20) is

$$\hat{W}_1 = p(Q) \frac{ve^{-v(\tilde{Z}_1+Q)}}{v_0 e^{-v_0(\tilde{Z}_1+Q_0)}}, \tag{7.5.21}$$

where

$$p(Q) = \int_{-Q}^{\infty} v_0 e^{-v_0(y+Q_0)} dy = e^{-v_0(Q-Q_0)}.$$

As in (7.3.8), it is not difficult to prove that although the estimator based on $\hat{W}_1$ requires the analytical calculation of the parameter $p = p(Q)$ for distinct values of Q, that estimator is more accurate than the one based on W_1.

Consistent estimators of $\ell(\boldsymbol{v})$ and $\nabla \ell(\boldsymbol{v})$ based on (7.5.17) are

$$\bar{\ell}_N(\boldsymbol{v}) = S - Q - \frac{\sum_{i=1}^{N} \sum_{t=0}^{\tau_i - 1} \tilde{L}_{ti} \tilde{W}_{ti}(\underline{Z}_t, \boldsymbol{v})}{\sum_{i=1}^{N} \sum_{t=0}^{\tau_i - 1} \tilde{W}_{ti}(\underline{Z}_{ti}, \boldsymbol{v})}, \tag{7.5.22}$$

and

$$\nabla\bar{\ell}_N(\boldsymbol{v}) = -1 + \frac{\sum_{i=1}^{N}\sum_{t=0}^{\tau_i-1}\tilde{L}_{ti}\nabla\tilde{W}_{ti}(\underline{Z}_{ti},\boldsymbol{v})}{\sum_{i=1}^{N}\sum_{t=0}^{\tau_i-1}\tilde{W}_{ti}(\underline{Z}_{ti},\boldsymbol{v})} - \frac{\sum_{i=1}^{N}\sum_{t=0}^{\tau_i-1}\tilde{L}_{ti}\tilde{W}_{ti}(\underline{Z}_{t},\boldsymbol{v})}{\sum_{i=1}^{N}\sum_{t=0}^{\tau_i-1}\tilde{W}_{ti}(\underline{Z}_{ti},\boldsymbol{v})} \times \frac{\sum_{i=1}^{N}\sum_{t=0}^{\tau_i-1}\nabla\tilde{W}_{ti}(\underline{Z}_{t},\boldsymbol{v})}{\sum_{i=1}^{N}\sum_{t=0}^{\tau_i-1}\tilde{W}_{ti}(\underline{Z}_{ti},\boldsymbol{v})}, \tag{7.5.23}$$

where the process $\nabla\tilde{W}_{ti}(\underline{Z}_{ti},\boldsymbol{v})$ with $\tilde{W}_{0i} \equiv 1$ is the generalized score function process,

$$\tilde{L}_{0i} = 0, \qquad \tilde{L}_{ti} = \tilde{Z}_{1i} + \sum_{j=2}^{t} Z_{ji}, \tag{7.5.24}$$

$\tilde{Z}_{1i} \sim f(z + Q_0, \boldsymbol{v}_{01})$ and $Z_{ji} \sim f(z, \boldsymbol{v}_{01})$, $j = 2, \dots, \tau$.

It is readily seen that the POSF estimators (7.5.22) and (7.5.24) allow *simultaneous estimation of* $\ell(\boldsymbol{v})$ *for multiple values of* $\boldsymbol{v} = (\boldsymbol{v}_1, Q)$ *from a single simulation run*. In the special case where $\boldsymbol{v}_1$ is fixed, the POSF estimator of $\ell(\boldsymbol{v})$ reduces to

$$\bar{\ell}_N(\boldsymbol{v}) = S - Q - \frac{\sum_{i=1}^{N}\sum_{t=0}^{\tau_i-1}\tilde{L}_{ti}W_{1i}}{\sum_{i=1}^{N}[1 + (\tau_i - 1)W_{1i}]}, \tag{7.5.25}$$

where

$$W_{1i} = \begin{cases} \dfrac{f(\tilde{Z}_{1i} + Q, \boldsymbol{v}_1)}{f(\tilde{Z}_{1i} + Q_0, \boldsymbol{v}_1)} & \text{if } \tilde{Z}_{1i} \geq -Q \\ 0 & \text{otherwise.} \end{cases} \tag{7.5.26}$$

Note that using the acceptance-rejection method allows us to replace W_{1i} in (7.5.25) by

$$\hat{W}_{1i} = p(Q)\frac{f(\tilde{Z}_{1i} + Q, \boldsymbol{v}_1)}{f(\tilde{Z}_{1i} + Q_0, \boldsymbol{v}_1)}, \tag{7.5.27}$$

where

$$p(Q) = \int_{-Q}^{\infty} f(y + Q_0, \boldsymbol{v}_1)\, dy.$$

Confidence intervals for the unknown quantities, $\ell(\boldsymbol{v})$ and $\nabla\ell(\boldsymbol{v})$, can be obtained by standard statistical techniques (e.g., Rubinstein and Shapiro, 1993).

Q as a Scale Parameter. In this case we replace the *true* demand size random variable, Y, by an *auxiliary* one, $\tilde{Y} = Y/Q$. More specifically, rewriting the event $\{E_t < Q\}$ in (7.5.3) as $\{E_t/Q < 1\}$, we represent $\ell(\boldsymbol{v})$ in (7.5.3) as

$$\ell(\boldsymbol{v}) = S - Q \cdot \frac{\mathbb{E}_{\boldsymbol{v}}\{X_3\}}{\mathbb{E}_{\boldsymbol{v}}\{\tau\}}, \tag{7.5.28}$$

where

$$X_3 = \sum_{t=0}^{\infty} \frac{E_t}{Q} I_{\{E_t/Q<1\}} = \sum_{t=0}^{\infty} \tilde{E}_t I_{\{\tilde{E}_t<1\}}, \tag{7.5.29}$$

and $\tilde{E}_t = E_t/Q = \sum_{i=1}^{t} \tilde{Y}_i$, $\tilde{E}_0 = 0$. Differentiating (7.5.28) with respect to Q then yields

$$\nabla_Q \ell(\boldsymbol{v}) = -\frac{\mathbb{E}_{\boldsymbol{v}}\{X_3\}}{\mathbb{E}_{\boldsymbol{v}}\{\tau\}} - Q\left[\frac{\nabla_Q \mathbb{E}_{\boldsymbol{v}}\{X_3\}}{\mathbb{E}_{\boldsymbol{v}}\{\tau\}} - \frac{\mathbb{E}_{\boldsymbol{v}}\{X_3\}}{\mathbb{E}_{\boldsymbol{v}}\{\tau\}} \times \frac{\nabla_Q \mathbb{E}_{\boldsymbol{v}}\{\tau\}}{\mathbb{E}_{\boldsymbol{v}}\{\tau\}}\right].$$

Similarly to (7.5.12), we have

$$\nabla_Q \mathbb{E}_{\boldsymbol{v}}\{X_3\} = \mathbb{E}_{\boldsymbol{v}}\left\{\sum_{t=0}^{\tau-1} \tilde{L}_t \tilde{\boldsymbol{S}}_t\right\}, \tag{7.5.30}$$

where

$$\begin{gathered}
\tau = \min\{t\colon \tilde{L}_t \geq 1\}, \\
\tilde{L}_0 = 0, \qquad \tilde{L}_t = \frac{E_t}{Q} = \tilde{E}_t, \quad t = 1, \ldots, \tau-1, \\
\tilde{\boldsymbol{S}}_t(\underline{Y}_t, \boldsymbol{v}) = \sum_{j=1}^{t} S_j(\tilde{Y}_j, \boldsymbol{v}_1), \qquad \underline{Y}_t = (\tilde{Y}_1, \ldots, \tilde{Y}_t), \\
S_j = \frac{\nabla \tilde{f}(\tilde{Y}_j, \boldsymbol{v}_1, Q)}{\tilde{f}(\tilde{Y}_j, \boldsymbol{v}_1)}, \qquad \tilde{Y}_i = \frac{Y_i}{Q}, \qquad \tilde{Y}_i \sim \tilde{f}(y, \boldsymbol{v}_1, Q)
\end{gathered} \tag{7.5.31}$$

and

$$\tilde{f}(y, \boldsymbol{v}_1, Q) = Qf(yQ, \boldsymbol{v}_1).$$

Consider now the likelihood ratio estimators of $\ell(\boldsymbol{v})$. Noting that the dominating pdf for $\tilde{f}(y, \boldsymbol{v}_1, Q) = Qf(yQ, \boldsymbol{v}_1)$ is $g(y) = \tilde{f}(y, \boldsymbol{v}_{01}, Q_0) = Q_0 f(yQ_0, \boldsymbol{v}_{01})$, the expected steady-state performance $\ell(\boldsymbol{v})$ in (7.5.28) can be written as

$$\ell(\boldsymbol{v}) = S - Q\frac{\mathbb{E}_g\{\sum_{t=0}^{\tau-1} \tilde{L}_t \tilde{W}_t(\boldsymbol{Z}_t, \boldsymbol{v})\}}{\mathbb{E}_g\{\sum_{t=0}^{\tau-1} \tilde{W}_t(\boldsymbol{Z}_t, \boldsymbol{v})\}}, \tag{7.5.32}$$

where

$$\tilde{L}_0 = 0, \qquad \tilde{L}_t = \sum_{j=1}^{t} Z_j, \qquad t = 1, \ldots, \tau - 1, \tag{7.5.33}$$

$$\tau = \min\{t : \tilde{L}_t \geq 1\},$$

$$\tilde{W}_0 \equiv 1, \qquad \tilde{W}_t(\underline{Z}_t, \boldsymbol{v}) = \prod_{j=1}^{t} W_j(Z_j, \boldsymbol{v}_1, Q), \qquad W_j = \frac{Qf(QZ_j, \boldsymbol{v}_1)}{Q_0 f(Q_0 Z_j, \boldsymbol{v}_{01})}, \tag{7.5.34}$$

$\underline{Z}_t = (Z_1, Z_2, \ldots, Z_t)$ and $Z_j \sim \tilde{f}(y, \boldsymbol{v}_{01}, Q_0)$.

The representation of $\nabla \ell(\boldsymbol{v})$ is similar. For example, if we let again $f(y, v) = ve^{-vy} I_{\{y \geq 0\}}$, then $\tilde{f}(y, v, Q) = vQe^{-vQy} I_{\{y \geq 0\}}$, $\tilde{f}(y, v_{01}, Q_0) = v_0 Q_0 e^{-v_0 Q_0 y} I_{\{y \geq 0\}}$, and the likelihood ratio is

$$\tilde{W}_t(\underline{Z}_t, \boldsymbol{v}) = \prod_{j=1}^{t} \frac{vQe^{-vQZ_j}}{v_0 Q_0 e^{-v_0 Q_0 Z_j}} = \left(\frac{vQ}{v_0 Q_0}\right)^t e^{-(vQ - v_0 Q_0) \sum_{j=1}^{t} Z_j}, \tag{7.5.35}$$

where $Z_j \sim h(y) = v_0 Q_0 e^{-v_0 Q_0 y}$.

Note that in order to obtain accurate (low variance) estimators of $\ell(\boldsymbol{v})$ in (7.5.32), it is desirable to take $v_0 Q_0 \leq vQ$.

Remark 7.5.1. It is not difficult to see that by using Q as a scale parameter, the problem is rescaled such that $s \equiv S - 1$.

CPO Method

We now derive an alternative to the POSF method for estimating $\nabla_Q \ell(\boldsymbol{v})$ by using conditional Monte Carlo estimation. We first calculate (via conditioning) the derivatives $\nabla_Q \mathbb{E}_{\boldsymbol{v}_1}\{\tau\}$ and $\nabla_Q \mathbb{E}_{\boldsymbol{v}_1}\{X\}$ for the denominator and the numerator in (7.5.3), respectively.

Let the pdf and cdf of E_t be denoted by f^{*t} and F^{*t}, respectively. Noting that

$$f^{*t}(u) = \int f(u - v) f^{*(t-1)}(v)\, dv = \mathbb{E}\left\{ f\left(u - \sum_{j=1}^{t-1} Y_j\right)\right\},$$

we have

$$\begin{aligned}
\nabla_Q \mathbb{E}_{v_1}\{\tau\} &= \nabla_Q \mathbb{E}_{v_1}\left\{\sum_{t=0}^{\infty} I_{\{E_t<Q\}}\right\} \\
&= \sum_{t=1}^{\infty} \nabla_Q \mathbb{E}_{v_1}\{I_{\{E_t<Q\}}\} \\
&= \sum_{t=1}^{\infty} \nabla_Q F^{*t}(Q) \\
&= \sum_{t=1}^{\infty} f^{*t}(Q) \\
&= \mathbb{E}_{v_1}\left\{\sum_{t=1}^{\tau-1} f(Q - E_{t-1})\right\}.
\end{aligned} \tag{7.5.36}$$

Here, we used the convention $\sum_{j=1}^{t-1} Y_j = 0$, when $t=1$. To proceed further we need the following:

Lemma 7.5.1. *Let X and Y be two random variables, and let Y have the density* f_Y. *Then*

$$\nabla_y \mathbb{E}_v\{X I_{\{Y<y\}}\} = \mathbb{E}_v\{X|Y=y\} f_Y(y). \tag{7.5.37}$$

Proof. Letting $f_{X,Y}(x, y)$ be the joint density of X and Y, one readily obtains

$$\begin{aligned}
\nabla_y \mathbb{E}_v\{X I_{\{Y<y\}}\} &= \nabla_y \int_{-\infty}^{\infty} \int_{-\infty}^{y} x f_{X,Y}(x, u)\, du\, dx \\
&= f_Y(y) \int_{-\infty}^{\infty} x \frac{f_{X,Y}(x, y)}{f_Y(y)} f_Y(y)\, dx = \mathbb{E}_v\{X|Y=y\} f_Y(y). \quad \blacksquare
\end{aligned}$$

With the aid of Lemma (7.5.1), write

$$\begin{aligned}
\nabla_Q \mathbb{E}_{v_1}\{X\} &= \nabla_Q \mathbb{E}_{v_1}\left\{\sum_{t=0}^{\tau-1} (S - E_t) I_{\{E_t<Q\}}\right\} \\
&= \sum_{t=0}^{\infty} \nabla_Q \mathbb{E}_{v_1}\{(S - E_t) I_{\{E_t<Q\}}\} \\
&= \sum_{t=1}^{\infty} \mathbb{E}_{v_1}\{(S - E_t|E_t = Q) f^{*t}(Q)\} \\
&= \sum_{t=1}^{\infty} \mathbb{E}_{v_1}\{(S - Q) f^{*t}(Q)\} \\
&= \mathbb{E}_{v_1}\left\{s \sum_{t=0}^{\infty} f^{*t}(Q)\right\} \\
&= s \cdot \nabla_Q \mathbb{E}_{v_1}\{\tau\}.
\end{aligned} \tag{7.5.38}$$

The resulting expression for $\nabla_Q \ell(\boldsymbol{v})$ is therefore

$$\nabla_Q \ell(\boldsymbol{v}) = \frac{s \nabla_Q \mathbb{E}_{\boldsymbol{v}_1}\{\tau\}}{\mathbb{E}_{\boldsymbol{v}_1}\{\tau\}} - \frac{\mathbb{E}_{\boldsymbol{v}_1}\{X\}}{\mathbb{E}_{\boldsymbol{v}_1}\{\tau\}} \times \frac{\nabla_Q \mathbb{E}_{\boldsymbol{v}_1}\{\tau\}}{\mathbb{E}_{\boldsymbol{v}_1}\{\tau\}}, \tag{7.5.39}$$

which together with (7.5.36) can be written as

$$\nabla_Q \ell(\boldsymbol{v}) = \frac{\nabla_Q \mathbb{E}_{\boldsymbol{v}_1}\{\tau\}}{\mathbb{E}_{\boldsymbol{v}_1}\{\tau\}} [s - \ell(\boldsymbol{v})], \tag{7.5.40}$$

where

$$\nabla_Q \mathbb{E}_{\boldsymbol{v}_1}\{\tau\} = \mathbb{E}_{\boldsymbol{v}_1}\left\{ \sum_{t=1}^{\tau-1} f(Q - E_{t-1}) \right\}. \tag{7.5.41}$$

Note that the crucial step in deriving (7.5.38) is *conditioning* on the event $\{E_t = Q\}$. It is also important to note that unlike (7.5.15), the formula (7.5.40) does not rely on the push-out method when Q is *fixed*. It is not difficult to see that (7.5.40) still holds when one replaces $\nabla_Q \mathbb{E}_{\boldsymbol{v}_1}\{\tau\}$ in (7.5.41) by

$$\nabla_Q \mathbb{E}_{\boldsymbol{v}_1}\{\tau\} = \mathbb{E}_{\boldsymbol{v}}\left\{ \sum_{t=0}^{\tau-1} \tilde{\boldsymbol{S}}_t \right\},$$

given in (7.5.13). Thus, (7.5.40) combined with (7.5.13) is, in fact, a "mixture" of the POSF and CPO methods.

In order to estimate $\nabla_Q \ell(\boldsymbol{v})$ for multiple values of Q, we need to apply the push-out method. Assume for simplicity that $\boldsymbol{v}_1$ is fixed. The resulting CPO and POSF estimators, based on (7.5.40) and (7.5.25), are

$$\nabla_Q \tilde{\ell}_N(\boldsymbol{v}) = \frac{\sum_{i=1}^{N} \sum_{t=1}^{\tau_i - 1} f(s - \tilde{Z}_{1i} - \sum_{j=2}^{t-1} Y_{ji}) W_{1i}}{\sum_{i=1}^{N} [1 + (\tau_i - 1) W_{1i}]} [s - \bar{\ell}_N(\boldsymbol{v})], \tag{7.5.42}$$

and

$$\nabla_Q \bar{\ell}_N(\boldsymbol{v}) = \frac{\sum_{i=1}^{N} \sum_{t=1}^{\tau_i - 1} \tilde{L}_{ti} \nabla_Q W_{1i}}{\sum_{i=1}^{N} [1 + (\tau_i - 1) W_{1i}]} [s - \bar{\ell}_N(\boldsymbol{v})], \tag{7.5.43}$$

respectively, where $\tilde{L}_{ti}$, W_{1i} and $\bar{\ell}_N(\boldsymbol{v})$ are given in (7.5.24), (7.5.26), and (7.5.25), respectively. Note again that in both (7.5.42) and (7.5.43), W_{1i} can be replaced by $\hat{W}_{1i}$, given in (7.5.27).

Recall that the CPO estimator (7.5.42) uses Q as a location parameter. We now present an alternative to (7.5.42), which uses Q as a scale parameter. We first derive an alternative to $\nabla_Q \mathbb{E}_{\boldsymbol{v}_1}\{\tau\}$ in (7.5.36). As before, $f(y, \boldsymbol{v}_1)$ and $\tilde{f}(y, \boldsymbol{v}_1, Q)$ denote the pdf's of the true demand size, Y, and the auxiliary (demand size) $\tilde{Y} = Y/Q$, with the corresponding cdf's denoted by $F(y, \boldsymbol{v}_1)$ and $\tilde{F}(y, \boldsymbol{v}_1, Q)$. Finally, denote $\tilde{E}_t = E_t/Q$. Analogously to (7.5.36), one has

$$\begin{aligned}\nabla_Q \mathbb{E}_{\boldsymbol{v}_1}\{\tau\} &= \nabla_Q \mathbb{E}_{\boldsymbol{v}_1}\left\{\sum_{t=0}^{\infty} I_{\{E_t<Q\}}\right\} \\ &= \sum_{t=1}^{\infty} \nabla_Q \mathbb{E}_{\boldsymbol{v}_1}\{I_{\{\tilde{E}_t<1\}}\} \\ &= \sum_{t=1}^{\infty} \nabla_Q \tilde{F}^{*t}(1, \boldsymbol{v}_1, Q) \\ &= \mathbb{E}_{\boldsymbol{v}_1}\left\{\sum_{t=1}^{\tau-1} \nabla_Q \tilde{F}[(1-\tilde{E}_{t-1}), \boldsymbol{v}_1, Q]\right\},\end{aligned} \tag{7.5.44}$$

where $\tilde{F}^{*t}$ is the distribution of $\tilde{E}_t$.

As an example, let $f(y, v) = ve^{-vy}$ and $\tilde{f}(y, v, Q) = vQe^{-vQy}$, so that $\tilde{F}(y, v, Q) = 1 - e^{-vQy}$ and $\nabla_Q \tilde{F}(y, v, Q) = vQe^{-vQy} = \tilde{f}(y, v, Q)$. In view of (7.5.44),

$$\nabla_Q \mathbb{E}_v\{\tau\} = \mathbb{E}_v\left\{\sum_{t=1}^{\tau-1} \tilde{f}(1-\tilde{E}_{t-1}, v, Q)\right\} = \mathbb{E}_v\left\{\sum_{t=1}^{\tau-1} vQe^{-vQ(1-\tilde{E}_{t-1})}\right\}. \tag{7.5.45}$$

Using (7.5.40) and replacing (7.5.41) by (7.5.44) lead to an alternative estimator of (7.5.42), with the representation

$$\nabla_Q \tilde{\ell}_N(\boldsymbol{v}) = \frac{\sum_{i=1}^{N} \sum_{t=1}^{\tau_i-1} \nabla_Q \tilde{F}\left(1 - \sum_{j=1}^{t-1} Z_{ji}, \boldsymbol{v}_1, Q\right) \tilde{W}_{ti}(\underline{Z}_{ti}, \boldsymbol{v})}{\sum_{i=1}^{N} \sum_{t=1}^{\tau_i-1} \tilde{W}_{ti}(\underline{Z}_{ti}, \boldsymbol{v})} [s - \bar{\ell}_N(\boldsymbol{v})], \tag{7.5.46}$$

where

$$\tau = \min\{t : \tilde{L}_t \geq 1\},$$

$$\tilde{W}_0 \equiv 1, \qquad \tilde{W}_{ti}(\underline{Z}_{ti}, \boldsymbol{v}) = \prod_{j=1}^{t} W_{ji}(Z_{ji}, \boldsymbol{v}_1, Q), \qquad W_{ji} = \frac{\tilde{f}(Z_{ji}, \boldsymbol{v}_1, Q)}{\tilde{f}(Z_{ji}, \boldsymbol{v}_{01}, Q_0)},$$

$\underline{Z}_{ti} = (Z_{1i}, Z_{2i}, \ldots, Z_{ti})$, $\tilde{f}(y, \boldsymbol{v}_1, Q) = Qf(yQ, \boldsymbol{v}_1)$ and $Z_{ji} \sim \tilde{f}(y, \boldsymbol{v}_{01}, Q_0)$. For example, if we again let $f(y, v) = ve^{-vy}$ and $\tilde{f}(y, v, Q) = vQe^{-vQy}$, then

$$\nabla_Q \tilde{\ell}_N(\boldsymbol{v}) = \frac{\sum_{i=1}^{N} \sum_{t=1}^{\tau_i-1} \nabla_Q \tilde{F}\left(1 - \sum_{j=1}^{t-1} Z_{ji}, \boldsymbol{v}_1, Q\right) \left(\frac{vQ}{v_0 Q_0}\right)^t e^{-(vQ - v_0 Q_0)\sum_{j=1}^{t} Z_{ji}}}{\sum_{i=1}^{N} \sum_{t=1}^{\tau_i-1} \left(\frac{vQ}{v_0 Q_0}\right)^t e^{-(vQ - v_0 Q_0)\sum_{j=1}^{t} Z_{ji}}} \times [s - \bar{\ell}_N(\boldsymbol{v})] \tag{7.5.47}$$

by virtue of (7.5.35). Since conditioning always reduces variance, both CPO estimators in (7.5.46) and (7.5.47) are preferable to their POSF counterparts.

It is important to note that the POSF method first applies the push-out method and then the score function, whereas the CPO method first applies conditioning and then the push-out method.

Estimation of $\ell(\boldsymbol{v}) = \mathbb{E}_{\boldsymbol{v}_1}\{\varphi[L(\underline{Y}_t, \boldsymbol{v}_2)]\}$ *and* $\nabla\ell(\boldsymbol{v})$

Let φ be a fixed-cost function reflecting, say, inventory holding cost and shortage cost. Typical examples of $\varphi(\alpha)$ are

$$\varphi(x) = c_1 x^+ + c_2 x^- = \begin{cases} c_1 x & \text{if } x \geq 0 \\ c_2|x| & \text{if } x < 0, \end{cases} \tag{7.5.48}$$

where

$$x^+ = \max\{x, 0\}, \qquad x^- = -\min\{x, 0\}$$

and

$$\varphi(x) = \begin{cases} c_1 x & \text{if } x \geq 0 \\ 1 & \text{if } x < 0, \end{cases} \tag{7.5.49}$$

Assume further that a fixed cost K is paid for each order. In this case, the long-run costs per unit time can be written analogously to (7.5.3), as

$$\ell(\boldsymbol{v}) = \mathbb{E}_{\boldsymbol{v}_1}\{\varphi(L)\} + \frac{K}{\mathbb{E}_{\boldsymbol{v}_1}\{\mathcal{T}\}} = \frac{\mathbb{E}_{\boldsymbol{v}_1}\{X\} + \mathcal{K}}{\mathbb{E}_{\boldsymbol{v}_1}\{\tau\}}, \tag{7.5.50}$$

where

$$X = \sum_{t=0}^{\tau-1} \varphi(L_t), \qquad \mathcal{K} = \frac{K}{\mathbb{E}\{U\}}, \qquad L_t = S - E_t, \qquad \mathbb{E}\{\mathcal{T}\} = \mathbb{E}\{U\}\mathbb{E}\{\tau\}. \tag{7.5.51}$$

We next derive an expression for $\ell(\boldsymbol{v})$, when Q is a location parameter. Analogous results can be obtained when Q is a scale parameter, as well as for the associated derivatives $\nabla\ell(\boldsymbol{v})$.

Using the POSF method we readily obtain, in analogy to (7.5.17),

$$\ell(\boldsymbol{v}) = \frac{\mathbb{E}_g\left\{\sum_{t=0}^{\tau-1} \varphi(\tilde{L}_t)\tilde{W}_t(\underline{Z}_t, \boldsymbol{v})\right\} + \mathcal{K}}{\mathbb{E}_g\left\{\sum_{t=0}^{\tau-1} \tilde{W}_t(\underline{Z}_t, \boldsymbol{v})\right\}}, \tag{7.5.52}$$

where $\tilde{L}_t$ of (7.5.18) can be written as

$$\tilde{L}_0 = S, \qquad \tilde{L}_t = s - \left(\tilde{Z}_1 + \sum_{j=2}^{t} Z_j\right), \qquad t = 1, \ldots, \tau - 1. \tag{7.5.53}$$

The representations of $\nabla\ell(\boldsymbol{v})$ and $\nabla\bar{\ell}_N(\boldsymbol{v})$ are similar. It is not difficult to see that formulas (7.5.38) and (7.5.40) for $\nabla_Q \mathbb{E}_{\boldsymbol{v}_1}\{X\}$ and $\nabla_Q \ell(\boldsymbol{v})$, respectively, hold again, with s replaced by $\varphi(s)$.

Positive Lead Time

Suppose that every order arrives B units of time after being placed, where the random variable B is independent of the random variables U and Y. Let $T_0 = 0, T_1, \ldots$ be the regeneration times of the inventory position process $\{\mathcal{N}_t\}$, and let $\tau_1, \tau_2, \ldots$ be the corresponding regenerative cycles, where

$$\tau_i = T_i - T_{i-1}.$$

At first glance one might think that $\{L_t\}$ is also regenerative with regeneration times $\tilde{T}_0 = T_0 + B_0$, $\tilde{T}_1 = T_1 + B_1, \ldots$ and corresponding regenerative cycles

$$\tilde{\tau}_i = \tilde{T}_i - \tilde{T}_{i-1}.$$

Typically, however, this is not the case; $\{L_t\}$ over cycle $i+1$ depends on the demands $Y_{T_i+1}, Y_{T_i+2}, \ldots$ governing cycle i of the process $\{\mathcal{N}_t\}$. One exception occurs when the demand distribution is exponential (memoryless). In this case, $\{L_t\}$ is regenerative (with cycle length $\tilde{\tau}$ equal to the time elapsed between two consecutive upward jumps of the process $\{L_t\}$). Note that for small overshoots δ (see Figure 3.3), individual demands are small relative to Q, and the process $\{L_t\}$ can be still viewed as approximately regenerative. Clearly, when $\delta = 0$ (D is a continuous, i.e., fluid demand), then $\{L_t\}$ is regenerative, provided the process $\{\mathcal{N}_t\}$ is. Hax and Candea (1984, p. 223) remark that "large overshoots are quite improbable, and, therefore, neglecting the overshoot phenomenon is a reasonable assumption."

To fit the process $\{L_t\}$ into the regenerative framework for general demand size distributions, one can use Harris recurrence (see, e.g., Asmussen, 1987), which is beyond the scope of this book. Here we consider only the case where the interdemand times satisfy $U_t = 1$, which corresponds to $\{N_t\}$ of (7.5.1) with B a positive (deterministic) integer. In this case the process $\{L_t\}$ can be readily fitted into the regenerative framework. To this end, we argue as follows. First consider the following fundamental relationship (see Pflug and Rubinstein, 1996):

$$L_t = \mathcal{N}_{t-r} - \sum_{i=t-r+1}^{t} Y_i, \qquad t \geq r, \tag{7.5.54}$$

relating $\{\mathcal{N}_t\}$ to $\{L_t\}$, where we denote below $B \equiv r$ to distinguish between deterministic and stochastic lead times. Using (7.5.54), Pflug and Rubinstein (1996) proved that the steady-state expected performance function $\ell(\boldsymbol{v}) = \mathbb{E}_{\boldsymbol{v}_1}\{\varphi(L)\}$ can be fitted into the regenerative framework and expressed as

$$\ell(\boldsymbol{v}) = \frac{\mathbb{E}_{\boldsymbol{v}_1}\left\{\sum_{t=0}^{\infty} \varphi(S - \hat{E}_r - E_t) I_{\{E_t < Q\}}\right\}}{\mathbb{E}_{\boldsymbol{v}_1}\left\{\sum_{t=0}^{\infty} I_{\{E_t < Q\}}\right\}}, \tag{7.5.55}$$

where $\hat{E}_r = \sum_{i=1}^{r} \hat{Y}_i$ is an auxiliary sequence, independent of $\{E_j : j \geq 0\}$ and such that $\hat{Y}_i$ has the same distribution as Y_i.

From (7.5.55) we next derive (i) POSF estimators and (ii) CPO estimators for the derivative $\nabla_Q \ell(\boldsymbol{v})$, assuming for simplicity that the vector $\boldsymbol{v}_1$ is fixed.

1. Using Q as a location parameter, rewrite $\ell(\boldsymbol{v})$ in (7.5.55) as

$$\ell(\boldsymbol{v}) = \frac{\mathbb{E}_{\boldsymbol{v}_1}\left\{\sum_{t=0}^{\tau-1} \varphi(S - \hat{E}_r - E_t)\right\}}{\mathbb{E}_{\boldsymbol{v}_1}\{\tau\}} = \frac{\mathbb{E}_{\boldsymbol{v}_1}\{X\}}{\mathbb{E}_{\boldsymbol{v}_1}\{\tau\}}, \tag{7.5.56}$$

where

$$X = \sum_{t=0}^{\tau-1} \varphi(S - \hat{E}_r - E_t)$$

and

$$\tau = \sum_{t=0}^{\infty} I_{\{E_t \geq Q\}}.$$

It is readily seen that in this case, the POSF estimators (7.5.22) and (7.5.24) are valid again, provided $\tilde{L}_{ti}$ is replaced by

$$\tilde{\mathcal{L}}_{0i} = S - \hat{E}_{ri}, \qquad \tilde{\mathcal{L}}_{ti} = s - \left(\tilde{Z}_{1i} + \sum_{j=2}^{t} Z_{ji} + \hat{E}_{ri}\right), \qquad t = 1, \ldots, \tau - 1. \tag{7.5.57}$$

Depending on whether Q is a location parameter or a scale parameter, and in view of (7.5.25) and (7.5.52), one has

$$\bar{\ell}_N(\boldsymbol{v}) = \frac{\sum_{i=1}^{N} \sum_{t=0}^{\tau_i - 1} \varphi(\tilde{\mathcal{L}}_{ti}) W_{1i}}{\sum_{i=1}^{N} [1 + (\tau_i - 1) W_{1i}]} \tag{7.5.58}$$

and

$$\bar{\ell}_N(\boldsymbol{v}) = \frac{\sum_{i=1}^{N}\sum_{t=0}^{\tau_i-1}\varphi(\hat{\mathscr{L}}_{ti})\tilde{W}_{ti}}{\sum_{i=1}^{N}\sum_{t=0}^{\tau_i-1}\tilde{W}_{ti}}. \tag{7.5.59}$$

Here [see (7.5.33)],

$$\hat{\mathscr{L}}_{0i} = S - \hat{E}_{ri}, \qquad \hat{\mathscr{L}}_{ti} = S - \left(Q\sum_{j=1}^{t} Z_{ji} + \hat{E}_{ri}\right), \qquad t = 1, \ldots, \tau - 1, \tag{7.5.60}$$

while $\tilde{W}_{ti}$ and $\hat{\mathscr{L}}_{ti}$ are given in (7.5.34) and (7.5.57), respectively.

2. We first derive an expression for $\nabla_Q \mathbb{E}_{\boldsymbol{v}_1}\{X\}$. Arguing as in (7.5.38), use Lemma 7.5.1 and the independence of $\hat{E}_r$ and E_t to obtain

$$\begin{aligned}
\nabla_Q \mathbb{E}_{\boldsymbol{v}_1}\{X\} &= \nabla_Q \mathbb{E}_{\boldsymbol{v}_1}\left\{\sum_{t=0}^{\tau-1}\varphi(S - \hat{E}_r - E_t) I_{\{E_t < Q\}}\right\} \\
&= \sum_{t=0}^{\infty} \nabla_Q \mathbb{E}_{\boldsymbol{v}_1}\{\varphi(S - \hat{E}_r - E_t) I_{\{E_t < Q\}}\} \\
&= \sum_{t=1}^{\infty} \mathbb{E}_{\boldsymbol{v}_1}\{\varphi(S - \hat{E}_r - E_t | E_t = Q) f^{*t}(Q)\} \\
&= \sum_{t=1}^{\infty} \mathbb{E}_{\boldsymbol{v}_1}\{\varphi(S - Q - \hat{E}_r) f^{*t}(Q)\} \\
&= \mathbb{E}_{\boldsymbol{v}_1}\left\{\varphi(s - \hat{E}_r)\sum_{t=0}^{\infty} f^{*t}(Q)\right\} \\
&= \mathbb{E}_{\boldsymbol{v}_1}\{R\} \nabla_Q \mathbb{E}_{\boldsymbol{v}_1}\{\tau\},
\end{aligned} \tag{7.5.61}$$

where

$$R = \varphi(s - \hat{E}_r).$$

Observe that in this case, the expressions (7.5.40) and (7.5.42) for $\nabla_Q \ell(\boldsymbol{v})$ and $\nabla_Q \tilde{\ell}_N(\boldsymbol{v})$ hold again, provided s is replaced by $\mathbb{E}_{\boldsymbol{v}_1}\{R\}$, and $\tilde{L}_{ti}$ is replaced by $\tilde{\mathscr{L}}_{ti}$. Note also that both the POSF and CPO estimators require an auxiliary random variable $\hat{E}_r = \sum_{i=1}^{r} \hat{Y}_i$, and consequently an auxiliary sample $\{\hat{Y}_1, \ldots, \hat{Y}_r\}$ from the demand pdf $f(y, \boldsymbol{v}_1)$. In particular, the CPO estimators (7.5.42) and (7.5.46) generalize to

$$\nabla_Q \tilde{\ell}_N(\boldsymbol{v}) = \frac{\sum_{i=1}^{N}\sum_{t=1}^{\tau_i-1} f(s - \tilde{Z}_{1i} - \sum_{j=2}^{t-1} Y_{ji} - \hat{E}_{ri}) W_{1i}}{\sum_{i=1}^{N}[1 + (\tau_i - 1) W_{1i}]}[\bar{R} - \bar{\ell}_N(\boldsymbol{v})], \tag{7.5.62}$$

and

$$\nabla_Q \tilde{\ell}_N(\boldsymbol{v}) = \frac{\sum_{i=1}^{N} \sum_{t=1}^{\tau_i - 1} \nabla_Q \tilde{F}\left(1 - \sum_{j=1}^{t-1} Z_{ji} - \sum_{j=1}^{r} \hat{Z}_{ji}, \boldsymbol{v}_1, Q\right) \tilde{W}_{ti}(\underline{Z}_{ti}, \boldsymbol{v})}{\sum_{i=1}^{N} \sum_{t=1}^{\tau_i - 1} \tilde{W}_{ti}(\underline{Z}_{ti}, \boldsymbol{v})} \times [\bar{R} - \bar{\ell}_N(\boldsymbol{v})], \tag{7.5.63}$$

respectively, where the auxiliary random variable $\sum_{j=1}^{r} \hat{Z}_{ji}$ is distributed as $\sum_{j=1}^{r} Z_{ji}$, $\bar{R}$ is the sample mean of R, and depending on whether Q is a location or a scale parameter, $\bar{\ell}_N(\boldsymbol{v})$ is given either by (7.5.58) or by (7.5.59), respectively.

7.5.2 Optimization

Consider the unconstrained program (5.4.3),

$$\min_{\boldsymbol{v}_2} \ell(\boldsymbol{v}), \quad \boldsymbol{v} \in V, \quad \boldsymbol{v} = (\boldsymbol{v}_1, \boldsymbol{v}_2), \qquad \boldsymbol{v}_2 = (s, Q), \quad s \geq 0, \; Q \geq 0, \tag{7.5.64}$$

where the objective function is [see (7.5.56)]

$$\ell(\boldsymbol{v}) = \frac{\mathbb{E}_{\boldsymbol{v}}\{X\}}{\mathbb{E}_{\boldsymbol{v}}\{\tau\}}, \tag{7.5.65}$$

and

$$\mathbb{E}_{\boldsymbol{v}}\{X\} = \mathbb{E}_{\boldsymbol{v}}\left\{\sum_{t=0}^{\tau-1} \varphi(S - \hat{E}_r - E_t\right\} + K,$$

K being the fixed ordering cost per cycle.

The optimality condition $\nabla_{\boldsymbol{v}_2} \ell(\boldsymbol{v}) = \mathbf{0}$ [see (5.4.4) and (7.5.11)] can be written as

$$\nabla_{\boldsymbol{v}_2} \ell(\boldsymbol{v}) = \frac{\mathbb{E}_{\boldsymbol{v}}\{\tau\} \nabla_{\boldsymbol{v}_2} \mathbb{E}_{\boldsymbol{v}}\{X\} - \mathbb{E}_{\boldsymbol{v}}\{X\} \nabla_{\boldsymbol{v}_2} \mathbb{E}_{\boldsymbol{v}}\{\tau\}}{[\mathbb{E}_{\boldsymbol{v}}\{\tau\}]^2} = \mathbf{0}, \tag{7.5.66}$$

where $\boldsymbol{v}_2 = (s, Q)$. Componentwise, one has

$$\nabla_Q \ell(\boldsymbol{v}) = \mathbb{E}_{\boldsymbol{v}}\{\tau\} \nabla_Q \mathbb{E}_{\boldsymbol{v}}\{X\} - \mathbb{E}_{\boldsymbol{v}}\{X\} \nabla_Q \mathbb{E}_{\boldsymbol{v}}\{\tau\} = 0 \tag{7.5.67}$$

and

$$\nabla_s \ell(\boldsymbol{v}) = \nabla_s \mathbb{E}_{\boldsymbol{v}}\{X\} = 0. \tag{7.5.68}$$

The last equality is due to the fact that $\nabla_s \mathbb{E}_{\boldsymbol{v}}\{\tau\} = 0$.

To proceed with the solution of (7.5.66) we first present an explicit expression for (7.5.67) and for its stochastic counterpart, $\nabla_Q \bar{\ell}_N(\boldsymbol{v}) = 0$. The results below are based on the CPO method, but the treatment for the POSF method is similar.

In view of the fact that $\nabla_Q \mathbb{E}_{\boldsymbol{v}}\{X\}$ in (7.5.67) coincides with (7.5.38), with s replaced by $\mathbb{E}_{\boldsymbol{v}}\{R\} = \mathbb{E}_{\boldsymbol{v}}\{\varphi(s - \hat{E}_r\}$, the optimality condition (7.5.67) simplifies to

$$\nabla_Q \ell(\boldsymbol{v}) = \mathbb{E}_{\boldsymbol{v}}\{\tau\}\mathbb{E}_{\boldsymbol{v}}\{R\} - \mathbb{E}_{\boldsymbol{v}}\{X\} = 0. \tag{7.5.69}$$

It is important to note that unlike (7.5.67), the resulting expression (7.5.69) does not depend on $\nabla_Q \mathbb{E}_{\boldsymbol{v}}\{X\}$ and $\nabla_Q \mathbb{E}_{\boldsymbol{v}}\{\tau\}$. The stochastic counterpart for (7.5.69), is

$$\nabla_Q \tilde{\ell}_N(\boldsymbol{v}) = \bar{\tau}\bar{R} - \bar{X} + K = 0, \tag{7.5.70}$$

where

$$\bar{R} = \frac{1}{N}\sum_{i=1}^{N} \varphi(s - \hat{E}_{ri}),$$

and depending on whether Q is a location or a scale parameter, the pair $(\bar{X}, \bar{\tau})$ can be written, respectively, as [see (7.5.58)]

$$(\bar{X}, \bar{\tau}) = \left(\frac{1}{N}\sum_{i=1}^{N}\sum_{t=0}^{\tau_i - 1} \varphi(s - \tilde{Z}_{1i} - E_{ti} - \hat{E}_{ri})W_{1i} + K, \frac{1}{N}\sum_{i=1}^{N}[1 + (\tau_i - 1)W_{1i}]\right) \tag{7.5.71}$$

or as [see (7.5.59)]

$$(\bar{X}, \bar{\tau}) = \left(\frac{1}{N}\sum_{i=1}^{N}\sum_{t=0}^{\tau_i - 1} \varphi\left(S - \left[Q\sum_{j=1}^{t} Z_{ji} + \hat{E}_{ri}\right]\tilde{W}_{ti} + K\right), \frac{1}{N}\sum_{i=1}^{N}\sum_{t=0}^{\tau_i - 1} \tilde{W}_{ti}\right). \tag{7.5.72}$$

The algorithm for estimating the optimal solution $Q^*(s)$ from the stochastic counterpart (7.5.70), using Q as a location parameter, can be written as follows:

Algorithm 7.5.1 *Q* Is a Location Parameter

1. *Generate a random sample* $\{Z_{11}, \ldots, Z_{\tau_1 1}, \ldots, Z_{1N}, \ldots, Z_{\tau_N N}\}$, *based on* N *regenerative cycles, where* $Z_{1i} \sim f(z + Q_0, \boldsymbol{v}_{01})$ *and* $Z_{ji} \sim f(z, \boldsymbol{v}_{01})$, $j = 2, \ldots, \tau_i$, $i = 1, \ldots, N$.
2. *Calculate the sample output process* $\{\varphi(s - \tilde{Z}_{1i} - E_{ti} - \hat{E}_{ri}): t = 1, \ldots, \tau_i, i = 1, \ldots, N\}$ and the likelihood ratio process $\{W_{1i}\}$ [or $\{\tilde{W}_{1i}\}$] according to (7.5.26) [*or* (7.5.27)].
3. *Solve Eq.* (7.5.70).

4. *Return the solution,* $\bar{Q}_N^*$ of (7.5.70), *as an estimator of the true optimal solution* $Q^*(s)$.

The algorithm for the case where Q is a scale parameter is similar. Note that for fixed s, the optimal solution $Q^*(s)$ of (7.5.69) can also be approximated by stochastic approximation algorithms instead of the stochastic counterpart. A rigorous comparison of stochastic approximations and the stochastic counterpart is given in Shapiro (1996).

Returning to (7.5.68) and assuming that φ is differentiable, we can write

$$\begin{aligned}\nabla_s \ell(\boldsymbol{v}) &= \mathbb{E}_{\boldsymbol{v}}\left\{\sum_{t=0}^{\infty} \varphi'(S - \hat{E}_r - E_t) I_{\{E_t < Q\}}\right\} \\ &= \mathbb{E}_{\boldsymbol{v}}\left\{\sum_{t=0}^{\tau-1} \varphi'(S - \hat{E}_r - E_t)\right\} = 0.\end{aligned}$$

The associated optimal solution, $s^*(Q)$, can be approximated from the corresponding stochastic counterpart. In particular, for the examples (7.5.48) and (7.5.49), $\varphi(x)$ is differentiable almost everywhere, yielding

$$\nabla_x \varphi(x) = \begin{cases} c_1 & \text{if } x \geq 0 \\ -c_2 & \text{if } x < 0, \end{cases} \tag{7.5.73}$$

and

$$\nabla_x \varphi(x) = \begin{cases} c_1 & \text{if } x \geq 0 \\ 0 & \text{if } x < 0, \end{cases} \tag{7.5.74}$$

respectively.

If $\varphi(s)$ is not differentiable with respect to s, the POSF method can again be applied, as it was for the parameter Q. One possibility is to write $S - \hat{E}_r$ as

$$S - \hat{E}_r = s + Q - \sum_{j=1}^{r} \hat{D}_j = Q - \hat{D}_{1s} - \hat{E}_r^{(2)},$$

where $\hat{D}_{1s} = \hat{D}_1 - s$, $\hat{D}_{1s} \sim \hat{f}(y, \boldsymbol{v}_1, s)$ and $\hat{E}_r^{(2)} = \sum_{j=2}^{r} \hat{D}_j$. In so doing, we effectively push-out the (location) parameter s from the sample performance to the auxiliary pdf $\hat{f}(y, \boldsymbol{v}_1, s)$. Assuming that $\hat{f}(y, \boldsymbol{v}_1, s)$ is differentiable, for fixed $\boldsymbol{v}_1$, we can write $\nabla_s \ell(\boldsymbol{v}) = 0$ as

$$\nabla_s \ell(\boldsymbol{v}) = \nabla_s \mathbb{E}_{\boldsymbol{v}}\{X\} = \mathbb{E}\left\{\sum_{t=0}^{\tau-1} \varphi(\bar{\mathscr{L}}_t) \nabla_s W_1(\hat{Z}_1, \boldsymbol{v}_1, s)\right\} = 0.$$

The corresponding stochastic counterpart of $\nabla_s \ell(\boldsymbol{v}) = 0$ is

$$\nabla \bar{\ell}_N(\boldsymbol{v}) = \frac{1}{N} \sum_{i=1}^{N} \sum_{t=0}^{\tau_i - 1} \varphi(\bar{\mathscr{L}}_{ti}) \nabla_s W_{1i}(\hat{Z}_{1i}, \boldsymbol{v}_1, s) = 0, \tag{7.5.75}$$

where

$$W_{1i}(\hat{Z}_{1i}, \boldsymbol{v}_1, s) = \frac{f(\hat{Z}_{1i} + s, \boldsymbol{v}_1)}{f(\hat{Z}_{1i} + s_0, \boldsymbol{v}_1)},$$

$$\hat{Z}_1 \sim f(y + s_0, \boldsymbol{v}_1), \qquad \hat{D}_j \sim f(y, \boldsymbol{v}_1), \qquad j = 2, \ldots, r$$

and

$$\bar{\mathscr{L}}_{0i} = Q - \hat{Z}_{1i} - \hat{E}_r^{(2)}, \qquad \bar{\mathscr{L}}_{ti} = Q - \hat{Z}_{1i} - \hat{E}_r^{(2)} - E_{ti}, \qquad t = 1, \ldots, \tau - 1. \tag{7.5.76}$$

Confidence regions for the optimal solution, (s^*, Q^*) of the original program (7.5.64), can be obtained by standard methods simultaneously with the optimal solution ($\bar{s}_N^*$, $\bar{Q}_N^*$) of the stochastic counterpart $\nabla_{\boldsymbol{v}_2} \bar{\ell}_N(\boldsymbol{v}) = \mathbf{0}$ [see (7.5.70), (7.5.75)]. Sensitivity analysis of the optimal solution (s^*, Q^*) with respect to the parameter vector $\boldsymbol{v}_1$ (say, with respect to the expected demand) can be treated in similar fashion.

7.5.3 Numerical Results

Table 7.5.1 displays the CMC and the POSF estimators of $\mathbb{E}\{\tau\}$, $\ell(\boldsymbol{v}) = \mathbb{E}_{\boldsymbol{v}_1}\{L(\underline{Y}_t, \boldsymbol{v}_2)\}$ (without lead time), and $\ell(\boldsymbol{v}) = \mathbb{E}_{\boldsymbol{v}_1}\{\varphi[L(\underline{Y}_t, \boldsymbol{v}_2)]\}$ (with lead time), as functions of Q, based on $N = 2000$ regenerative cycles. The POSF method used Q as a scale parameter and the selected reference parameter was $Q_0 = 20$. The parameters for $\ell(\boldsymbol{v}) = \mathbb{E}_{\boldsymbol{v}_1}\{\varphi[L(\underline{Y}_t, \boldsymbol{v}_2)]\}$ were $S = 13$, $K = 10$, $c_1 = 1.0$, $c_2 = 5.0$ and $r = 5$ (deterministic lead time).

An examination of Table 7.5.1 lends support to the conclusion that the POSF method performs quite well. Similar results were obtained for the CPO method, using Q as a location parameter.

*7.6 SENSITIVITY ANALYSIS WITH AUTOCORRELATED INPUTS

This section extends some of the earlier results on sensitivity analysis of DES from independent input sequences, $\{\boldsymbol{Y}_t : t > 0\}$, to autocorrelated sequences, $\{\boldsymbol{X}_t : t > 0\}$

Table 7.5.1 Performance of CMC and POSF Estimators as Functions of Q based on $N = 2000$ Regenerative Cycles with Reference (Scale) Parameter $Q_0 = 20$

	CMC Estimators of			POSF Estimators of		
Q	$\mathbb{E}\{L(v_2)\}$	$\mathbb{E}\{\varphi[L(v_2)]\}$	$\mathbb{E}\{\tau\}$	$\mathbb{E}\{L(v_2)\}$	$\mathbb{E}\{\varphi[L(v_2)]\}$	$\mathbb{E}\{\tau\}$
1	7.61	12.80	1.94	11.75	16.42	2.14
2	7.22	10.68	2.94	10.44	13.58	3.18
3	6.78	9.46	3.87	9.21	11.55	4.28
4	6.35	8.51	4.94	8.11	9.97	5.38
5	5.86	7.78	5.96	7.12	8.69	6.40
6	5.43	7.24	6.94	6.25	7.62	7.31
7	4.93	6.82	7.97	5.48	6.71	8.16
8	4.43	6.59	8.95	4.81	6.06	9.00
9	3.93	6.58	9.93	4.21	5.97	9.86
10	3.43	6.77	10.91	3.65	6.20	10.76
11	2.92	7.35	11.92	3.12	6.69	11.71
12	2.43	8.11	12.93	2.60	7.47	12.70
13	1.91	9.15	13.92	2.09	8.58	13.72
14	1.42	10.38	14.88	1.58	9.91	14.76
15	0.97	11.76	15.93	1.06	11.38	15.81
16	0.47	13.22	16.89	0.55	12.95	16.86
17	−0.02	14.83	17.96	0.03	14.64	17.92
18	−0.52	16.57	19.00	−0.48	16.43	18.97
19	−1.03	18.34	19.99	−1.00	18.29	20.01
20	−1.51	20.23	21.03	−1.51	20.23	21.03
21	−2.05	22.11	22.07	−2.03	22.22	22.04
22	−2.54	24.17	23.08	−2.54	24.26	23.03
23	−3.07	26.23	24.07	−3.05	26.34	23.99
24	−3.53	28.23	25.13	−3.55	28.45	24.92
25	−4.02	30.35	26.07	−4.05	30.56	25.80

[see (7.1.6)]. In particular, it will be shown how to estimate the steady-state expected performance,

$$\ell(v) = \mathbb{E}_v\{L(\underline{X}_t)\}, \tag{7.6.1}$$

and the associated sensitivities, $\nabla^k \ell(v)$, $k \geq 1$, for queueing models, with special emphasis on autocorrelated TES sequences [see (7.1.9) and (7.1.10)]. Unless otherwise stated, it will be assumed throughout that $\{L_t : t > 0\}$ is the waiting time process in a stable $GI/G/1$ queue.

Before proceeding further, it is important to understand that for autocorrelated inputs, the output process, $\{L_t : t > 0\}$, is *not regenerative*. Specifically, regenerative structure is precluded owing to the dependence among cycles; in particular, the waiting time of the *first* customer in the current busy cycle and the waiting time of the *last* customer in the previous busy cycle are generally dependent. Consequently, one cannot use the direct regenerative SF estimators based on formulas (7.2.6) and (7.2.7). Instead, one can use the truncated SF (TSF) estimators, which are suitable for stationary and ergodic processes [see Section 3.3 and Melamed and Rubinstein (1992)]. Asmussen and Melamed (1994) discusses sensitivity analysis of queueing models with autocorrelated inputs, using *Harris recurrence* (see Asmussen, 1987), to obtain an embedded regenerative framework.

The rest of this section deals with applications of the push-out method to autocorrelated sequences, and in particular, it treats the estimation of the sensitivities $\nabla^k \ell(\boldsymbol{v}),\ k = 0, 1,\ \boldsymbol{v} = (\boldsymbol{v}_1, \boldsymbol{v}_2)$, where

$$\ell(\boldsymbol{v}) = \mathbb{E}_{\boldsymbol{v}_1}\{L_t(x_t(\underline{Y}_t, \boldsymbol{v}_2))\}. \tag{7.6.2}$$

For an autocorrelated sequence with members of the form $X_t = h(X_{t-1}, Y_j, \boldsymbol{v}_2),\ t = 1, 2, \ldots,$ we have $\tilde{L}(\underline{x}_t) \equiv L(\underline{x}_t)$, where $\tilde{L}(\underline{x}_t) = L(\underline{y}_t, \boldsymbol{v}_2)$ [see (7.2.11) and (7.2.12)]. Therefore, application of the push-out technique reduces to finding the joint t-dimensional pdf $\tilde{f}_t(x_1, \ldots, x_t, \boldsymbol{v})$ of the random vector $\underline{X}_t = (X_1, \ldots, X_t)$, in terms of the pdf $f(y, \boldsymbol{v}_2)$ of the variate Y. By the Markov property of $\{X_t : t > 0\}$, the joint pdf $\tilde{f}_t(x_1, \ldots, x_t, \boldsymbol{v})$ can be written as

$$\tilde{f}_t(x_1, \ldots, x_t, \boldsymbol{v}) = \tilde{f}(x_1, \boldsymbol{v})\tilde{f}(x_2|x_1, \boldsymbol{v}) \cdots \tilde{f}(x_t|x_{t-1}, \boldsymbol{v}). \tag{7.6.3}$$

We next derive the conditional pdf's, $\tilde{f}(x_t|x_{t-1}, \boldsymbol{v}),\ t = 1, 2, \ldots,$ in terms of the unconditional pdf, $f(y, \boldsymbol{v}_1)$, for AR(1) and TES processes.

Example 7.6.1 AR(1) Sequences. Consider the following simple modification of (7.1.7):

$$X_t(v_2) = \begin{cases} X_1 & \text{if } t = 1 \\ v_2 X_{t-1} + (1 - v_2) Y_t & \text{if } t > 1 \end{cases} \tag{7.6.4}$$

where $0 < v_2 < 1$. In this case it is readily seen that

$$\tilde{f}(x_t|x_{t-1}, \boldsymbol{v}) = f\left(\frac{x_t - v_2 x_{t-1}}{1 - v_2}, \boldsymbol{v}_1\right), \qquad t = 1, 2, \ldots. \tag{7.6.5}$$

In particular, let $Y \sim \text{Exp}(v_1)$. Then,

$$f\left(\frac{x_t - v_2 x_{t-1}}{1 - v_2}, \boldsymbol{v}_1\right) = v_1 \exp\left(\frac{-v_1(x_t - v_2 x_{t-1})}{1 - v_2}\right) I_{(-\infty, 0]}(x_t - v_2 x_{t-1}).$$

Example 7.6.2 TES Sequences. It is shown in Asmussen and Melamed (1994) that

$$\tilde{f}_t(x_t|x_{t-1}, \boldsymbol{v}) = 1/(\theta_2 - \theta_1) f(x_t, \boldsymbol{v}_1) I\{F(x_t, \boldsymbol{v}_1) \in C[F(x_{t-1}, \boldsymbol{v}_1), \boldsymbol{v}_2]\}, \tag{7.6.6}$$

where $t = 1, 2, \ldots, \boldsymbol{v}_2 = (\theta_1, \theta_2)$, and C is a circular interval, defined in Jagerman and Melamed (1992a).

With (7.6.5) and (7.6.6) in hand, we now show how to estimate $\nabla^k \ell(\boldsymbol{v})$, $k = 0, 1$, *simultaneously for multiple values of* $\boldsymbol{v}_1$ *and* $\boldsymbol{v}_2$, via the TSF estimator. We have (see also Rubinstein and Shapiro, 1993)

$$\nabla^k \ell^{tr}(\boldsymbol{v}, m) = \mathbb{E}_g\{\tilde{L}_t(\underline{Z}_t)\nabla^k \hat{W}_t^{tr}(\underline{Z}_t, \boldsymbol{v}, m)\} \qquad (7.6.7)$$

and

$$\nabla^k \tilde{\ell}_N^{tr}(\boldsymbol{v}, m) = \frac{1}{NT}\sum_{i=1}^{N}\sum_{t=1}^{T} \tilde{L}_{ti}(\underline{Z}_{ti})\nabla^k \hat{W}_{ti}^{tr}(\underline{Z}_{ti}, \boldsymbol{v}, m). \qquad (7.6.8)$$

Here and elsewhere, the superscript *tr* stands for truncation-related quantities. In particular,

$$\hat{W}_t^{tr}(\underline{Z}_{ti}, \boldsymbol{v}, m) = \prod_{j=t-m+1}^{t} W_j(Z_{ji}, \boldsymbol{v}),$$

m is the truncation parameter, T is the batch size, N is the number of batches, and $\{Z_{11}, \ldots, Z_{TN}\}$ is a random sample from the pdf, $g(z)$, which dominates the conditional pdf, $\tilde{f}_t(x_t|x_{t-1}, \boldsymbol{v})$, for each $t > 0$.

We next calculate $\hat{W}_t^{tr}(\underline{z}_t, \boldsymbol{v}, m)$ for the AR(1) and TES sequences, assuming that $g(\mathbf{z}) = \tilde{f}_t(x_t|x_{t-1}, \boldsymbol{v})$ and $\boldsymbol{v}_0 = (\boldsymbol{v}_{01}, \boldsymbol{v}_{02})$.

Example 7.6.3 AR(1) Sequences. In this case, we have

$$\hat{W}_t^{tr}(\underline{z}_t, \boldsymbol{v}, m) = \prod_{j=t-m+1}^{t} \frac{f(z_j - v_2 z_{j-1}, v_1)(1 - v_{02})}{f(z_j - v_{02} z_{j-1}, v_{01})(1 - v_2)}. \qquad (7.6.9)$$

In particular, letting $Y \sim \text{Exp}(v_1)$, we get

$$\begin{aligned}\hat{W}_t^{tr}(\underline{z}_t, \boldsymbol{v}, m) &= (v_1/v_{01})^m \\ &\times \exp\left\{-[(v_1 - v_1 v_2)/(1 - v_2) - (v_{01} - v_{01}v_{02})/(1 - v_{02})]\sum_{j=1}^{m-1} z_j\right\} \\ &\times \exp\{-(v_1/(1 - v_2) - v_{01}/(1 - v_{02}))z_m\} \\ &\times \prod_{j=t-m+1}^{t} I_{[-\infty,0]}(z_j - v_2 z_j - 1), \end{aligned} \qquad (7.6.10)$$

where $v_{02} \leq v_2$.

Example 7.6.4 TES Sequences. In this case, we have

$$\hat{W}_t^{tr}(\underline{z}_t, \boldsymbol{v}, m) = \left(\frac{\theta_{02} - \theta_{01}}{\theta_2 - \theta_1}\right)^m \prod_{j=t-m+1}^{t} \frac{f(z_j, \boldsymbol{v}_1)}{f(z_j, \boldsymbol{v}_{01})} I\{F(z_j, \boldsymbol{v}_1) \in C[F(z_{j-1}, \boldsymbol{v}_1), \boldsymbol{v}_2]\}, \tag{7.6.11}$$

where $\boldsymbol{v}_2 = (\theta_1, \theta_2)$. See Asmussen and Melamed (1994) for more details and Rubinstein and Shapiro (1993) for numerical results.

7.7 COMBINING THE SCORE FUNCTION METHOD WITH THE CRUDE MONTE CARLO METHOD

Although the POSF method is versatile and can be used in a broad range of real-world applications, there might exist situations in which the method cannot be applied. In such cases, alternative methods are called for.

In this section it will be assumed that for a fixed $\boldsymbol{v}_2$, the sensitivities, $\nabla^k \ell(\boldsymbol{v})$, $k = 0, 1$, can be estimated simultaneously, for multiple values of $\boldsymbol{v}_1$, via the SF method, but not vice versa; that is, both the push-in and the push-out methods fail. It will be shown below how to efficiently combine the SF and CMC methods so as to estimate $\ell(\boldsymbol{v}) = \ell(\boldsymbol{v}_1, \boldsymbol{v}_2)$ *simultaneously*, for multiple pairs $(\boldsymbol{v}_1, \boldsymbol{v}_2)$ in some set

$$\{(\boldsymbol{v}_{1i}, \boldsymbol{v}_{2j}):\ i = 1, \ldots, r_1,\ j = 1, \ldots, r_2\}. \tag{7.7.1}$$

Such general designs (configurations) will be referred to as *what-if designs*. To this end, the so-called *score function–crude Monte Carlo* (SFCMC) algorithm will be presented. Note that by (7.7.1), the CMC method requires a total of $r_1 \times r_2$ simulation runs. It is natural to expect that when the CMC method is properly combined with the SF method, the number of requisite simulation runs can be reduced to r_2. The case of estimating $\nabla\ell(\boldsymbol{v})$ is similar.

Unless otherwise stated, we assume that $g(y) = f(y, \boldsymbol{v}_{01})$, where $\boldsymbol{v}_{01}$ is the *reference parameter*, and take $\ell(\boldsymbol{v})$ to be the mean sojourn time in a $GI/G/c/m$ queue.

For $\boldsymbol{v}_2$ fixed, and $\boldsymbol{v}_1 \in \{\boldsymbol{v}_{11}, \ldots, \boldsymbol{v}_{1r_1}\}$, consider the program

$$\min_{\boldsymbol{v}_{01}} \max_{\boldsymbol{v}_1 = \{\boldsymbol{v}_{11}, \ldots, \boldsymbol{v}_{1r_1}\}} \operatorname{var}_{\boldsymbol{v}_{01}}\{\ell_N(\boldsymbol{v})\}. \tag{7.7.2}$$

It seems *natural* to choose the reference parameter, $\boldsymbol{v}_{01}$, in such a way that

$$\rho(\boldsymbol{v}_{01}) = \max_{j=1,\ldots,r_1} \rho(\boldsymbol{v}_{1j}). \tag{7.7.3}$$

Equation (7.7.3) means that $\boldsymbol{v}_{01}$ should correspond to the *highest traffic intensity* among all traffic intensities associated with the permissible values, $\boldsymbol{v}_{11}, \ldots, \boldsymbol{v}_{1r_1}$.

Example 7.7.1. Consider the $GI/G/1/m$ queue. Assume that the buffer size is *fixed* at m, and suppose we wish to estimate the expected sojourn time, $\ell(\boldsymbol{v}) = \ell(\boldsymbol{v}_1, \boldsymbol{v}_2)$, simultaneously for all $\boldsymbol{v}_1 = \boldsymbol{v}_{11}, \ldots, \boldsymbol{v}_{1r_1}$, via the what-if estimator, $\bar{\ell}_N(\boldsymbol{v})$. Let v_1 be the service rate. It readily follows from (7.7.3) that in this case, v_{01} must satisfy $v_{01} = \min(v_{11}, \ldots, v_{1r_1})$, which is the same as

$$\rho_0 = \max(\rho_1, \ldots, \rho_{r_1}), \tag{7.7.4}$$

where ρ_j corresponds to service rate v_{ij}, $1 \leq j \leq r_1$.

Consider now the general case (7.7.1). In typical applications, the traffic intensity is *monotonic* in each component of $\boldsymbol{v}_2$, in which case formula (7.7.3) is applicable again in the sense that once a "good" reference parameter, $\rho_0(\boldsymbol{v}_2)$, is chosen, it remains a good one for *all* $\boldsymbol{v}_2$ in (7.7.1). In other words, in order to find a good reference parameter $\boldsymbol{v}_{01}(\boldsymbol{v}_2)$ [and the corresponding $\rho_0(\boldsymbol{v}_2)$], suitable for *all* combinations $\{(\boldsymbol{v}_{1i}, \boldsymbol{v}_{2j})\colon\ i = 1, \ldots, r_1,\ j = 1, \ldots, r_2\}$, we have to first fix an arbitrary value, $\boldsymbol{v}_2$, from the set $\{\boldsymbol{v}_{2j}\colon j = 1, \ldots, r_2\}$, and then apply formula (7.7.3).

Example 7.7.2. Suppose we wish to estimate the expected waiting time in a $GI/G/1/m$ queue, for $r_1 \times r_2$ distinct values of the vector (v_1, v_2), where v_1 is the service rate and $v_2 = m$ is the buffer size. To this end, we may select any buffer size, m, from the set $\{m_1, \ldots, m_{r_2}\}$; then find v_{01} according to (7.7.3); and finally run r_2 simulations corresponding to the selected values $m_1, \ldots, m_{r_2}$, respectively. Clearly, in so doing, the number of runs is reduced from $r_1 \times r_2$ to r_2.

The what-if estimator of $\ell(\boldsymbol{v})$ for the jth scenario ($j = 1, \ldots, r_2$) can be written as

$$\bar{\ell}_N(\boldsymbol{v}_1, \boldsymbol{v}_{2j}) = \frac{\sum_{i=1}^N \sum_{t=1}^{\tau_i(\boldsymbol{v}_{2j})} L_{ti}(\boldsymbol{v}_{2j}) \tilde{W}_{ti}(\boldsymbol{v}_1)}{\sum_{i=1}^N \sum_{t=1}^{\tau_i(\boldsymbol{v}_{2j})} \tilde{W}_{ti}(\boldsymbol{v}_1)}. \tag{7.7.5}$$

Although τ does not depend explicitly on $\boldsymbol{v}_2$, we write $\tau_i(\boldsymbol{v}_{2j})$ rather than τ_i in (7.7.5), in order to emphasize that its distribution depends on $\boldsymbol{v}_{2j}$.

The SFCMC algorithm for estimating the response surface, $\ell(\boldsymbol{v})$, for an arbitrary what-if design $\{(\boldsymbol{v}_{1i}, \boldsymbol{v}_{2j})\colon\ i = 1, \ldots, r_1,\ j = 1, \ldots, r_2\}$ can be written as follows:

Algorithm 7.7.1 SFCMC Algorithm

1. *Specify the what-if design,* $\{(\boldsymbol{v}_{1i}, \boldsymbol{v}_{2j})\colon\ i = 1, \ldots, r_1,\ j = 1, \ldots, r_2\}$*, and then perform the following steps for* $j = 1, \ldots, r_2$.
2. *Find the reference parameter,* $\boldsymbol{v}_{01}$*, via* (7.7.3).
3. *Generate a sample* $\{\boldsymbol{Z}_1, \ldots, \boldsymbol{Z}_T\}$ *from* $f(\mathbf{z}, \boldsymbol{v}_{01})$*, where* $T = \sum_{i=1}^N \tau_i$.

4. *Generate the output processes* $\{L_t(\boldsymbol{v}_{2j})\}$ *and* $\{\tilde{W}_t(\boldsymbol{v}_1)\}$, *and calculate* $\bar{\ell}_N(\boldsymbol{v}_1, \boldsymbol{v}_{2j})$ *via* (7.7.5).

Example 7.7.3. Consider the estimation of the steady-state mean waiting time, $\ell(\boldsymbol{v})$, in an $M/M/c/m$ queue, for the what-if design $\{(\boldsymbol{v}_{1i}, \boldsymbol{v}_{2j}): i = 1, \ldots, r_1, j = 1, \ldots, r_2\}$. Here, $\boldsymbol{v}_1 = (v_{11}, v_{12})$ is the vector of the interarrival and service rates, respectively, and $\boldsymbol{v}_2 = (m, c)$ is the vector of the buffer size and the server-group size, respectively. Assume that c and v_{11} are fixed, while v_{12} and m may vary. In particular, set $c=2$, $v_{11} = 1$, while $v_{12} = 2$, 1.5, 1.4, 1.25, and $m=5$, 10, 15. According to (7.7.3), we first fix any buffer size from the set $m = \{5, 10, 15\}$, say $m=5$, and then choose the reference parameter value for v_{12} as the one that corresponds to the highest traffic intensity, ρ_0, among the values $v_{12} = 2$, 1.5, 1.4, 1.25. In our case, $v_{11} = 1$, $v_{12} = 1.25$. Finally, we make three separate runs, with respective values of $m=5$, 10, 15, to estimate $\ell(\boldsymbol{v})$ for the above $r_1 \times r_2 = 12$ scenarios. Here, the SFCMC estimator is more efficient (it requires only 3 runs instead of 12). Numerical experiments also indicate that it is more accurate than its CMC counterpart.

Remark 7.7.1. The SFCMC method above can be efficiently combined with different standard *experimental design* (ED) methods, such as the full factorial design, the central composite design, and so forth. Suppose now that we need to estimate the response surface, $\ell(\boldsymbol{v}) = \ell(\boldsymbol{v}_1, \boldsymbol{v}_2)$. Realizing that ED techniques are special cases of our general what-if design, $\{(\boldsymbol{v}_{1i}, \boldsymbol{v}_{2j}): i = 1, \ldots, r_1, j = 1, \ldots, r_2\}$, one may adopt Algorithm 7.7.1 almost verbatim, in the sense that the term *what-if design* in Step 1 need only be replaced by the term *experimental design*, with all other steps remaining intact. Such an algorithm is referred to as a *modified experimental design* (MED) algorithm.

Example 7.7.4. Consider the estimation of the steady-state expected waiting time, $\ell(\boldsymbol{v})$, for r $M/M/1/m$ queues in tandem, where $\boldsymbol{v} = (v_{11}, \ldots, v_{1r})$ and $\boldsymbol{v}_2 = (v_{21}, \ldots, v_{2r}) = (m_1, \ldots, m_r)$ are the vectors of service rates and buffer sizes, respectively. In this case, the full factorial ED method requires a total of 2^{2r} Monte Carlo experiments, while the MED method requires only 2^r such experiments. Thus, the latter is approximately 2^r times faster than the former, since the overhead of computing $\tilde{W}_t(\boldsymbol{v}_1)$ in the corresponding likelihood ratio estimators is relatively small. This reduction has been confirmed by various simulation studies.

A similar approach may be utilized in estimating $\nabla\ell(\boldsymbol{v})$. For more details see Dussault et al. (1997).

7.8 SUMMARY

This chapter treated extended versions of the model $\ell(\boldsymbol{v}) = \mathbb{E}_{\boldsymbol{v}}\{L(\underline{\boldsymbol{Y}}_t)\}$, and specifically the model $\ell(\boldsymbol{v}) = \mathbb{E}_{\boldsymbol{v}_1}\{L(\underline{\boldsymbol{Y}}_t, \boldsymbol{v}_2)\}$, $\boldsymbol{v} = (\boldsymbol{v}_1, \boldsymbol{v}_2)$, where both the underlying pdf, f, and the sample performance, L, depend on the parameter vectors, $\boldsymbol{v}_1$ and $\boldsymbol{v}_2$, respectively. The following have been shown:

1. How to estimate the performance and the sensitivities $\nabla^k \ell(\boldsymbol{v}) = \{\nabla^k_{\boldsymbol{v}_1} \ell(\boldsymbol{v}), \nabla^k_{\boldsymbol{v}_2} \ell(\boldsymbol{v})\}$, $k = 0, 1$, using the direct, push-out, and push-in (IPA) estimators.
2. That the direct estimators of $\nabla^k_{\boldsymbol{v}_1} \ell(\boldsymbol{v})$ involve differentiation of the pdf $f(\boldsymbol{y}, \boldsymbol{v}_1)$ only, calling for the standard SF estimators, whereas the direct estimators of $\nabla^k_{\boldsymbol{v}_2} \ell(\boldsymbol{v})$ involve differentiation of the sample function, $L(\boldsymbol{v}_2)$, requiring it to be both analytically available and smooth (differentiable) with respect to $\boldsymbol{v}_2$. This is often not the case.

To overcome this difficulty, a new technique, called the push-out method, was described and the following have been shown:

1. That the push-out method (based on change of variables) merely replaces the original nonsmooth sample function, $L(\boldsymbol{v}_2)$, by an auxiliary one, $\tilde{L}$, while pushing out the parameter vector $\boldsymbol{v}_2$, from $L(\boldsymbol{v}_2)$ to an auxiliary pdf, $\tilde{f}(\boldsymbol{y}, \boldsymbol{v})$. In so doing, the functions $\nabla^k \ell(\boldsymbol{v})$, $k = 0, 1$, become smooth, since the new pdf, $\tilde{f}(\boldsymbol{y}, \boldsymbol{v})$, is typically smooth and $\tilde{L}$ does not depend on $\boldsymbol{v}_2$.
2. How both the auxiliary sample performance and the auxiliary pdf can be obtained from their original counterparts, and how combining them allows one to obtain consistent estimators of $\nabla^k \ell(\boldsymbol{v})$, $k = 0, 1$.
3. That the IPA method, introduced by Ho and co-workers (1979), corresponds to the push-in technique. The latter may be viewed as a dual of the push-out method. The push-in transformation, $\boldsymbol{x} = \boldsymbol{x}(\boldsymbol{y}, \boldsymbol{v}_1)$, and its special case, $x = F(y, \boldsymbol{v}_1)$, typically lead to a nonsmooth sample performance function, $\hat{L}(\boldsymbol{u}, \boldsymbol{v})$, depending on both $\boldsymbol{v}_1$ and $\boldsymbol{v}_2$. Consequently, the interchangeability conditions of expectation and differentiation often fail to hold, a fact that tends to confine the IPA technique to more limited applications. Several examples were presented, though, where the IPA estimators outperform in accuracy their straightforward SF counterparts (e.g., when using the standard regenerative SF estimators in queueing models with long regenerative cycles).

Based on extensive studies of sensitivity analysis [estimation of $\nabla^k \ell(\boldsymbol{v})$, $k = 0, 1$, simultaneously for multiple values $\boldsymbol{v}$] and their optimization, obtained from a single-run simulation of rather general queueing networks, the following recommendations are made:

1. Use the direct SF estimators, provided (i) the variances (confidence intervals) are reasonable (e.g., when the expected number of waiting customers in a busy cycle is, say, less than 10), and (ii) the sample performance function, $L(\boldsymbol{v}_2)$, is smooth (differentiable) with respect to $\boldsymbol{v}_2$. [Clearly (ii) holds trivially when L does not depend on $\boldsymbol{v}_2$.]

2. If recommendation 1 fails (the variance is too large), then try using the low-variance DSF (decomposable score function) and TSF (truncated score function) estimators, or the CSF (conditional score function) method of McLeish and Rollans (1992). Numerical studies clearly indicate that the DSF and TSF estimators are highly efficient when optimizing a DES.
3. If recommendation 2 fails and the required transformation $\boldsymbol{x} = \boldsymbol{x}(\boldsymbol{y}, \boldsymbol{v}_2)$ is available, then try using the push-out technique.
4. If recommendation 3 fails, then try using the SFCMC technique.

7.9 EXERCISES

1. Consider Example 7.2.2, and derive

$$\nabla \ell(\boldsymbol{v}) = \left\{ \frac{\partial \ell(\boldsymbol{v})}{\partial v_1}, \frac{\partial \ell(\boldsymbol{v})}{\partial v_2} \right\}. \tag{7.9.1}$$

Take the derivatives inside the expectation and derive the associated direct estimator, $\nabla \bar{\ell}_N(\boldsymbol{v})$. Prove the consistency of $\nabla \bar{\ell}_N(\boldsymbol{v})$.

2. Consider Algorithm (7.3.1) and derive (7.5.6). Take the derivatives inside the expectation and derive the associated IPA estimator, $\nabla \bar{\ell}_N(\boldsymbol{v})$, in terms of $\hat{L}(U_i, \boldsymbol{v})$. Prove the consistency of $\nabla \bar{\ell}_N(\boldsymbol{v})$.

3. Let $\{L_t\}$ be the waiting time process in an $M/G/c$ queue (c denotes the number of servers) with arrival rate λ, and let τ be the number of customers served during a busy period. Consider the corresponding expected-value functions,

$$\ell_1(\lambda) = \mathbb{E}_\lambda \left\{ \sum_{t=1}^{\tau} L_t \right\}$$

and

$$\mathcal{L}^0(\lambda, \lambda_0) = \mathbb{E}_{\lambda_0} \left\{ \left(\sum_{t=1}^{\tau} L_t \tilde{W}_t(\lambda) \right)^2 \right\}.$$

Show that the optimal value, λ_0^*, of the reference parameter [given by the minimizer of the function $\mathcal{L}^0(\lambda, \cdot)$] satisfies $\lambda_0^* > \lambda$.

4. Repeat the results in Tables 7.3.1–7.4.2 by simulation. Use both representations, (7.3.7) and (7.3.8), of the likelihood ratio, W, and discuss the results.

5. Prove that the push-out estimator $\bar{\ell}(\boldsymbol{v}, \tilde{\boldsymbol{Z}})$, based on (7.3.8), is more accurate than the estimator $\bar{\ell}(\boldsymbol{v}, \boldsymbol{Z})$, based on (7.3.7); that is, show that $\text{var}\{\bar{\ell}(\boldsymbol{v}, \tilde{\boldsymbol{Z}})\} \leq \text{var}\{\bar{\ell}(\boldsymbol{v}, \boldsymbol{Z})\}$.

6. Derive an expression similar to (7.4.17) for the empirical cdf $\tilde{H}_{LN}(x, \boldsymbol{v})$ of the steady-state waiting time of a customer in the $GI/G/1$ queue.

7. Consider Lindley's equation (7.2.20) for the waiting time process, $\{L_t\}$, in a $GI/G/1$ queue, that is

$$L_{t+1} = \max\{0, L_t + Z_t\}, \qquad L_0 = 0, \ t = 0, 1, \ldots,$$

where as before $Z_t = Y_{1t} - Y_{2(t+1)}$, $\{Y_{1t}\}$ is the sequence of service times, and $\{Y_{2t}\}$ is the sequence of interarrival times. Assume in addition that $\{Y_{1t} = \max(a, X_{1t})\}$, where a is a constant and $\{X_{1t}\}$ is a sequence of random variables. Think about the meaning of a in the sequence $\{\max(a, X_{1t})\}$, and exhibit a POSF estimator for estimating $\nabla_a \ell(\boldsymbol{v})$ with respect to a, where $\ell(\boldsymbol{v}) = P\{L \leq x\}$ is the distribution function of L in the steady-state regime. Run a computer simulation for an $M/G/1$ queue with $X_{1t} \sim \text{Exp}(v_1 = 1)$, $Y_{2t} \sim \text{Exp}(v_1 = 0.7)$ and $a = 0.3$.
Hint: Use the relationship [see also (7.3.10)]

$$\max(a, X) = a \max(1, \tilde{X}),$$

where $\tilde{X} = X/a$.

8. Prove that the acceptance-rejection estimator $\hat{\ell}(\boldsymbol{v}, \tilde{\mathbf{Z}})$, based on (7.3.8) is more accurate than the estimator $\tilde{\ell}(\boldsymbol{v}, \boldsymbol{Z})$, based on (7.3.7).

REFERENCES

Asmussen, S. (1987). *Applied Probability and Queues*, Wiley, New York.

Asmussen, S. and Melamed, B. (1994). "Regenerative simulation of TES processes," *Acta Appl. Math.*, **34**, 237–260.

Cao, X. R. (1994). *Realization Probabilities: The Dynamics of Queueing Systems*, Springer-Verlag, New York.

Dussault, J. P., Labrecque, D., L'Ecuyer, P., and Rubinstein, R. Y. (1997). "Combining the stochastic counterpart and stochastic approximation methods," *Discrete Event Dynamic Systems, Theory and Applications*, **7**(1), 5–28.

Gaivoronski, A. A., Shi, L. Y., and Sreenivas, R. S. (1992). "Augmented infinitesimal perturbation analysis: An alternative explanation," *Discrete Event Dynamic Syst.: Theory Appl.*, **2**, 121–138.

Glasserman, P. (1991). *Gradient Estimation via Perturbation Analysis*, Kluwer, Norwell, MA.

Glasserman, P. and Liu, T. W. (1996). "Rare-event simulation for multistage production-inventory systems," *Manage. Sci.*, **42**, 1292–1307.

Gong, W. B. and Ho, Y. C. (1987). "Smoothed (conditional) perturbation analysis of discrete event dynamic systems," *IEEE Trans. Autom. Control*, **AC-32**(10), 858–866.

Hax, A. C. and Candea, D. (1984), *Production and Inventory Management*, Prentice-Hall, Englewood Cliffs, NJ.

Heidelberger, P., Cao, X. R., Suri, R., and Zazanis, M. A. (1988). "Convergence properties of infinitesimal perturbation analysis estimates," *Manage. Sci.*, **34**(11), 1281–1302.

Ho, Y. C. and Cao, X. R. (1991). *Discrete Event Dynamic Systems and Perturbation Analysis*, Kluwer, Norwell, MA.

Ho, Y. C., Eyler, M. A., and Chien, T. T. (1979). "A gradient technique for general buffer storage design in a serial production line," *Int. J. Product. Res.*, **17**(6), 557–580.

Jagerman, D. L. and Melamed, B. (1992a). "The transition and autocorrelation structure of TES processes. Part I: General theory," *Stoch. Models*, **8**(2), 193–219.

Jagerman, D. L. and Melamed, B. (1992b). "The transition and autocorrelation structure of TES processes. Part II: Special cases," *Stoch. Models*, **8**(3), 499–529.

Jagerman, D. L. and Melamed, B. (1994). "The spectral structure of TES processes," *Stoch. Models*, **10**(3), 599–618.

Kriman, V. (1995). "Sensitivity analysis of $GI/GI/m/B$ queues with respect to buffer size by the score function method," *Stoch. Models*, **39**(1), 171–194.

L'Ecuyer, P. L. (1990). "A unified version of the IPA, SF, and LR gradient estimation techniques," *Manage. Sci.*, **36**(11), 1364–1383.

L'Ecuyer, P. L. (1992). "Convergence rates for steady-state derivative estimators," *Ann. Oper. Res.*, **39**, 121–137.

Marti, K. (1990). "Stochastic optimization methods of structural design," *ZAMM*, **4**, T742–T745.

McLeish, D. L. and Rollans, S. (1992). "Conditioning for variance reduction in estimating the sensitivity of simulations," *Ann. Oper. Res.*, **39**, 157–173.

Melamed, B. (1991). "A class of methods for generating autocorrelated uniform variates," *ORSA J. Comput.*, **3**(4), 317–329.

Melamed, B. and Rubinstein, R. Y. (1992). "Sensitivity analysis of discrete event systems with autocorrelated inputs," Proc. of WSC '92, Arlington, Virginia, pp. 521–528.

Pflug, G. Ch. and Rubinstein, R. (1996). "Simulation based optimization of the (s, S)-policy inventory model by the push-out method," Manuscript, Technion, Haifa, Israel.

Raz, I. (1991). "Efficiency of the score function method for sensitivity analysis and optimization of queueing networks," Master's Thesis, Technion, Haifa, Israel.

Rubinstein, R. Y. (1992). "Sensitivity analysis of discrete event systems by the "Push out" method," *Ann. Oper. Res.*, **39**, 229–251.

Rubinstein, R. Y. and Shapiro, A. (1993). *Discrete Event Systems: Sensitivity Analysis and Stochastic Optimization via the Score Function Method*, Wiley, New York.

Sahin, I. (1989). *Regenerative Inventory Systems: Operating Characteristics and Optimization*, Springer-Verlag, New York.

Shapiro, A. (1996). "Simulation based optimization-convergence analysis and statistical inference," *Stoch. Models*, **12**(3), 425–454.

Suri, R. and Zazanis, M. A. (1988). "Perturbation analysis gives strongly consistent sensitivity estimates for the $M/G/L$ queue," *Manage. Sci.*, **34**(1), 39–64.

Tijms, H. C. (1994). *Stochastic Models: An Algorithmic Approach*, Wiley, New York.

Topia, R. A. and Thompson, J. R. (1978). *Nonparametric Density Estimation*, Johns Hopkins University Press, Baltimore.

Uryas'ev, S. (1994). "Derivatives of probability functions and integrals over sets given by inequalities," *J. Comput. Appl. Math.*, **56**, 197–223.

Uryas'ev, S. (1997). "Analytic perturbation analysis for DEDS with discontinuous sample-path functions," *Stoch. Models*, **13**(3), 457–490.

Wardi, Y., Kellmans, M. H., Cassandras, C. G., and Gong, W.-B. (1992). "Smoothed perturbation analysis algorithms for estimating the derivatives of occupancy-related functions in serial queueing networks," *Ann. OR*, **39**, 269–295.

CHAPTER 8

Response Surface Methodology via the Score Function Method

8.1 INTRODUCTION

The goal of this chapter is to rigorously treat the choice of the reference parameter, $\boldsymbol{v}_0$, in the dominating probability density function (pdf) $g(\boldsymbol{y}) = f(\boldsymbol{y}, \boldsymbol{v}_0)$, with special emphasis on the response surface methodology. If not otherwise stated, we shall assume that $\{L_t\}$ is a steady-state process representing a discrete-event dynamic system (DEDS), say, the steady-state waiting time process in a $GI/G/1$ queue. The notations $\nabla^k \bar{\ell}_N(\boldsymbol{v}, \boldsymbol{v}_0)$ and $\nabla^k \bar{\ell}_N(\boldsymbol{v})$ will be used interchangeably, as called for.

As in Chapter 5, we consider here the following two cases: (i) $\boldsymbol{v}$ is *fixed* and $\boldsymbol{v}_0$ *varies*, and (ii) $\boldsymbol{v}_0$ is *fixed* and $\boldsymbol{v}$ *varies*. As mentioned earlier, the second case is associated with the response surface methodology, that is, with estimating $\nabla^k \ell(\boldsymbol{v})$, $k = 0, 1$, from a single simulation run.

For case (i) (fixed $\boldsymbol{v}$), we show how to:

1. Choose a "good" reference vector, $\boldsymbol{v}_0$, so as to effect variance reduction in the sense that

$$\mathrm{var}_{\boldsymbol{v}_0}\{\nabla^k \bar{\ell}_N(\boldsymbol{v}, \boldsymbol{v}_0)\} < \mathrm{var}_{\boldsymbol{v}}\{\nabla^k \bar{\ell}_N(\boldsymbol{v}, \boldsymbol{v})\}, \qquad k = 0, 1. \tag{8.1.1}$$

For $k = 1$, $\mathrm{var}_{\boldsymbol{v}_0}\{\nabla \bar{\ell}_N(\boldsymbol{v}, \boldsymbol{v}_0)\}$ is defined as the trace of the corresponding covariance matrix, namely,

$$\mathrm{var}_{\boldsymbol{v}_0}\{\nabla \bar{\ell}_N(\boldsymbol{v}, \boldsymbol{v}_0)\} = \mathrm{tr}[\mathrm{cov}_{\boldsymbol{v}_0}\{\nabla \bar{\ell}_N(\boldsymbol{v}, \boldsymbol{v}_0\}]. \tag{8.1.2}$$

2. Estimate the optimal reference vector, $\boldsymbol{v}_0^*$, for the corresponding mathematical programming problem

$$\min_{\boldsymbol{v}_0} \mathrm{var}_{\boldsymbol{v}_0}\{\nabla^k \bar{\ell}_N(\boldsymbol{v}, \boldsymbol{v}_0)\}, \quad \boldsymbol{v}_0 \in V, \qquad k = 0, 1. \tag{8.1.3}$$

To fix the ideas, let $\rho = \rho(\boldsymbol{v})$ and $\rho_0 = \rho_0(\boldsymbol{v})$ be the nominal and the reference traffic intensities, respectively, in an underlying $GI/G/1$ queue, where $\boldsymbol{v}$ is either the arrival rate or service rate (or vector of both). It will be shown that for typical performance measures (e.g., those pertaining to the waiting time process), there exist traffic intensities $\hat{\rho} = \hat{\rho}(\hat{\boldsymbol{v}}, k)$, $k = 0, 1$, such that if the reference traffic intensity, ρ_0, satisfies

$$\rho \leq \rho_0 \leq \hat{\rho}, \tag{8.1.4}$$

then variance reduction is effected in the sense of (8.1.1); equivalently,

$$\text{var}_{\rho_0}\{\nabla^k \bar{\ell}_N(\rho, \rho_0)\} \leq \text{var}_{\rho}\{\nabla^k \bar{\ell}_N(\rho, \rho)\}, \qquad k = 0, 1, \tag{8.1.5}$$

with equality holding for $\rho = \hat{\rho}$. In other words, for the SF estimator, $\nabla^k \bar{\ell}_N(\rho_0)$, to perform *well* [in the sense of (8.1.5)], the reference traffic intensity, ρ_0, must belong to the interval $(\rho, \hat{\rho})$.

For case (ii) ($\boldsymbol{v}$ varies), we show how to:

1. Choose a good reference vector, $\boldsymbol{v}_0$, so as to *simultaneously* estimate $\nabla^k \ell(\boldsymbol{v})$, $k = 0, 1$, for multiple values of $\boldsymbol{v}$; or in other words, how to estimate an *entire response surface* $\{\nabla^k \ell(\boldsymbol{v}) : v \in V\}$, $k = 0, 1$, *from a single simulation run*.
2. Estimate the optimal reference vector, $\boldsymbol{v}_0^*$, for the corresponding mathematical programming problem

$$\min_{\boldsymbol{v}_0} \max_{j=1,\ldots,s} \text{var}_{\boldsymbol{v}_0}\{\nabla^k \bar{\ell}_N(\boldsymbol{v}_j, \boldsymbol{v}_0)\}, \quad \boldsymbol{v}_0 \in V, \qquad k = 0, 1. \tag{8.1.6}$$

Note that the program (8.1.3) is a special case of the program (8.1.6), with $\boldsymbol{v}_j = \boldsymbol{v}$ fixed, $j = 1, \ldots, s$.

The rest of this chapter is organized as follows. Section 8.2 deals with the choice of a good reference parameter vector, $\boldsymbol{v}_0$, for fixed $\boldsymbol{v}$, in various queueing models. Here it will be shown that the optimal solution, $\boldsymbol{v}_0^*$, for the program (8.1.3) can be approximated rather well by solving (8.1.3) with respect to the parameters of the *bottleneck queue* alone, provided such a queue exists. In other words, the optimal solution, $\boldsymbol{v}_0^*$, of the program (8.1.3) will be shown to be essentially *insensitive* to the parameters of the other (*nonbottleneck*) queues as well as the parameters of the underlying distributions, $f(\boldsymbol{y}, \boldsymbol{v})$, from the exponential family in canonical form. Section 8.3 deals with the response surface methodology for estimating an entire response surface, $\nabla^k \ell(\boldsymbol{v})$, from a single simulation run. Finally, Section 8.4 discusses how to estimate the optimal reference parameter vector, $\boldsymbol{v}_0^*$, of the programs (8.1.3) and (8.1.6) via simulation.

8.2 CHOOSING A GOOD REFERENCE PARAMETER, $\boldsymbol{v}_0$, FOR FIXED $\boldsymbol{v}$

This section discusses how to choose a good reference parameter vector, $\boldsymbol{v}_0$, for fixed $\boldsymbol{v}$. We start with static queues, such as the $GI/G/\infty$ queue, then proceed to simple queues, and conclude with non-Markovian queueing networks.

8.2.1 Static Models

Bearing in mind that static models can be considered as a special case of DEDS with $\tau = 1$, we have from Chapter 5

$$\nabla^k \ell(\boldsymbol{v}) = \mathbb{E}_{\boldsymbol{v}_0}\{L\nabla^k W(\boldsymbol{v})\}, \tag{8.2.1}$$

where

$$W(\boldsymbol{v}, \boldsymbol{v}_0) = W(\boldsymbol{Z}, \boldsymbol{v}, \boldsymbol{v}_0) = \frac{f(\boldsymbol{Z}, \boldsymbol{v})}{f(\boldsymbol{Z}, \boldsymbol{v}_0)} \quad \text{and} \quad \boldsymbol{Z} \sim f(\mathbf{z}, \boldsymbol{v}_0).$$

The score function (SF) estimator of $\nabla^k \ell(\boldsymbol{v})$ is

$$\nabla^k \bar{\ell}_N(\boldsymbol{v}, \boldsymbol{v}_0) = \frac{1}{N}\sum_{i=1}^{N} L(\boldsymbol{Z}_i)\nabla^k W(\boldsymbol{Z}_i, \boldsymbol{v}, \boldsymbol{v}_0). \tag{8.2.2}$$

Consider the program (8.1.3), which is equivalent to,

$$\min_{\boldsymbol{v}_0} \mathscr{L}^k(\boldsymbol{v}, \boldsymbol{v}_0), \quad \boldsymbol{v}_0 \in V, \qquad k = 0, 1, \tag{8.2.3}$$

where

$$\mathscr{L}^0(\boldsymbol{v}, \boldsymbol{v}_0) = \mathbb{E}_{\boldsymbol{v}_0}\{L^2 W^2(\boldsymbol{v}, \boldsymbol{v}_0)\} \tag{8.2.4}$$

and

$$\mathscr{L}^1(\boldsymbol{v}, \boldsymbol{v}_0) = \mathrm{tr}[\mathbb{E}_{\boldsymbol{v}_0}\{L^2 \nabla W(\boldsymbol{v}, \boldsymbol{v}_0)\nabla W(\boldsymbol{v}, \boldsymbol{v}_0)'\}]. \tag{8.2.5}$$

Proposition 8.2.1. *Let $\boldsymbol{Y}$ be a random vector, distributed according to an exponential family in canonical form* (5.2.24). *Then* $\mathscr{L}^k(\boldsymbol{v}, \boldsymbol{v}_0)$, $k = 0, 1$, *is a convex function with respect to* $\boldsymbol{v}_0 = (v_{01}, \ldots, v_{0n})$.

Proof. Given in Section 8.5. □

Note that if V is a convex set, then for a given $\boldsymbol{v}$, $\mathscr{L}^k(\boldsymbol{v}, \boldsymbol{v}_0)$, $k = 0, 1$, *is finite-valued on a convex subset* of V, and *strictly convex* if $L(\mathbf{y}) \neq 0$ (and thus has a *unique* minimum, $\boldsymbol{v}_0^*$, over V).

The properties of the functions $\mathcal{L}^k(\boldsymbol{v}, \boldsymbol{v}_0)$ above will be investigated in detail for the gamma pdf,

$$f(y, \lambda, \beta) = \frac{\lambda^\beta e^{-y\lambda} y^{\beta-1}}{\Gamma(\beta)}, \quad y > 0, \quad \lambda > 0, \quad \beta > 0,$$

which is a special case of the two-parameter exponential family in canonical form with

$$v_1 = \lambda, \qquad v_2 = \beta, \qquad t_1(y) = -y, \quad \text{and} \quad t_2(y) = \log(y).$$

It follows from Section 5.2 that each function $\mathcal{L}^k(\lambda, \lambda_0)$, $k = 0, 1$, is finite-valued for all $\lambda_0 \in (0, 2\lambda)$, provided β is fixed and known. It also follows from the foregoing discussion that each $\mathcal{L}^k(\lambda, \lambda_0)$ is convex and has a unique minimizer, λ_0^*, over the interval $(0, 2\lambda)$.

Proposition 8.2.2. *Let $Y \sim G(\lambda, \beta)$. Suppose that $L^2(y)$ is a monotonically increasing function on the interval $[0, \infty)$. Then*

$$\lambda_0^*(k) < \lambda, \qquad \textit{for } k = 0, 1.$$

Proof. Given in Section 8.5. □

Proposition 8.2.2 can be extended to the multivariate case, $\boldsymbol{Y} = (Y_1, \ldots, Y_m)$, where the components Y_k, $k = 1, \ldots, m$, are mutually independent, each distributed gamma with the respective parameters λ_k and β_k. Denoting $\boldsymbol{\lambda} = (\lambda_1, \ldots, \lambda_m)'$, it can be shown (see Rubinstein and Shapiro, 1993) that $\boldsymbol{\lambda}_0^* < \boldsymbol{\lambda}$, that is, *every* component of the optimal reference vector, $\boldsymbol{\lambda}_0^*$, is *less* than that of the corresponding component of $\boldsymbol{\lambda}$.

Example 8.2.1. Suppose we wish to estimate $\ell(\boldsymbol{v}) = \mathbb{E}_v\{L(Y)\}$, where $L(Y) = Y$ and $Y \sim G(\lambda, \beta)$; for example, $L(Y)$ might be the sojourn time in the $GI/G/\infty$ queue, Y being the service time random variable. From (8.2.4),

$$\mathcal{L}^0(\lambda, \lambda_0) = \frac{\lambda^{2\beta}}{\lambda_0^\beta (2\lambda - \lambda_0)^\beta} \mathbb{E}_{2\lambda - \lambda_0}\{Y^2\} = \frac{\lambda^{2\beta}\Gamma(\beta + 2)}{\lambda_0^\beta (2\lambda - \lambda_0)^{\beta+2}\Gamma(\beta)}. \tag{8.2.6}$$

The minimizer, λ_0^*, of $\mathcal{L}^0(\lambda, \lambda_0)$ over the interval $(0, 2\lambda)$ equals to

$$\lambda_0^* = \frac{\lambda\beta}{\beta + 1}.$$

Let

$$\varepsilon^0(\lambda, \lambda_0) = \frac{\mathcal{L}^0(\lambda, \lambda_0)}{\mathcal{L}^0(\lambda, \lambda)}$$

be the relative efficiency. Then

$$\varepsilon^0(\lambda, \lambda_0) = \frac{\lambda^{2\beta+2}(1+\beta)}{\lambda_0^{\beta}(2\lambda - \lambda_0)^{\beta+2}} - \beta,$$

which for $\lambda_0 = \lambda_0^*$ reduces to

$$\varepsilon^0(\lambda, \lambda_0^*) = \frac{(1+\beta)^{2\beta+3}}{(2+\beta)^{\beta+2}} - \beta.$$

Note that $\varepsilon^0(\lambda, \lambda_0^*)$ does not depend on λ. By convexity of $\varepsilon^0(\lambda, \cdot)$, there is $\hat{\lambda}$ satisfying the equality

$$\varepsilon^0(\lambda, \hat{\lambda}) = 1, \qquad \lambda \neq \hat{\lambda},$$

so that $\hat{\lambda}$ must satisfy the equation

$$\lambda^{2\beta+2} - \hat{\lambda}^{\beta}(2\lambda - \hat{\lambda})^{\beta+2} = 0.$$

To illustrate, let $\lambda = 2$, and consider the following three values of β: 3.0, 1.0, and 0.5. The corresponding values of λ_0^* are 1.5, 1.0, and 0.6667, and the corresponding variance reduction intervals, $(\hat{\lambda}, \lambda)$, are (1.04, 2), (0.3214, 2), and (0.0681, 2).

8.2.2 Simple Queues

Let $\ell(\boldsymbol{v}) = \mathbb{E}_{\boldsymbol{v}}\{L\}$ where L might be the steady-state waiting time in a simple queue, such as $GI/G/1$.

Consider first the program

$$\min_{\boldsymbol{v}_0} \operatorname{var}\{\nabla^k \bar{\ell}_{1N}(\boldsymbol{v}, \boldsymbol{v}_0)\}, \qquad \boldsymbol{v}_0 \in V, \tag{8.2.7}$$

associated with the estimator

$$\nabla^k \bar{\ell}_{1N}(\boldsymbol{v}, \boldsymbol{v}_0) = \frac{1}{N} \sum_{i=1}^{N} \sum_{t=1}^{\tau_i} L_{ti} \nabla^k \tilde{W}_{ti}(\boldsymbol{v}). \tag{8.2.8}$$

Since $\ell(\boldsymbol{v}) = \mathbb{E}_{\boldsymbol{v}}\{L\}$ does not depend on $\boldsymbol{v}_0$, it follows that analogously to (8.2.3), the program (8.2.7) is equivalent to

$$\min_{\boldsymbol{v}_0} \mathscr{L}_1^k(\boldsymbol{v}, \boldsymbol{v}_0), \qquad \boldsymbol{v}_0 \in V. \tag{8.2.9}$$

For $k=0$ and $k=1$, the functions $\mathscr{L}_1^k(\boldsymbol{v}, \boldsymbol{v}_0)$ reduce to

$$\mathscr{L}_1^0(\boldsymbol{v}, \boldsymbol{v}_0) = \mathbb{E}_{\boldsymbol{v}_0}\left\{\left[\sum_{t=1}^{\tau} L_t \tilde{W}_t(\boldsymbol{v})\right]^2\right\} \tag{8.2.10}$$

and

$$\mathscr{L}_1^1(\boldsymbol{v}, \boldsymbol{v}_0) = \mathrm{tr}\left(\mathbb{E}_{\boldsymbol{v}_0}\left\{\left[\sum_{t=1}^{\tau} L_t \nabla \tilde{W}_t(\boldsymbol{v})\right]\left[\sum_{t=1}^{\tau} L_t \nabla \tilde{W}_t(\boldsymbol{v})\right]'\right\}\right), \tag{8.2.11}$$

respectively.

Proposition 8.2.3. *Assume that the output process, $\{L_t : t > 0\}$, is nonnegative-valued and let the components of the input random vector, $\boldsymbol{Y}$, be from an exponential family in the canonical form (5.2.24). Then $\mathscr{L}_1^k(\boldsymbol{v}, \boldsymbol{v}_0)$, $k = 0, 1$, is a convex function in $\boldsymbol{v}_0$.*

Proof. An extended version of Proposition 8.2.3 is given in Shapiro and Rubinstein (1994).

We remark that the assumption that the output process, $\{L_t : t > 0\}$, must be nonnegative-valued is satisfied in most applications of interest, including queue length and waiting time processes.

Consider next the regenerative estimator, $\bar{\ell}_N(\boldsymbol{v}) = \bar{\ell}_{1N}(\boldsymbol{v})/\bar{\ell}_{2N}(\boldsymbol{v})$ [see (6.2.11)]. In this case, it is readily seen that its asymptotic variance is $1/N$ times the variance,

$$\sigma^2(\boldsymbol{v}, \boldsymbol{v}_0) = a^2 \,\mathrm{var}_{\boldsymbol{v}_0}\left\{\sum_{t=1}^{\tau} M_t(\underline{\boldsymbol{Z}}_t)\tilde{W}_t(\underline{\boldsymbol{Z}}_t, \boldsymbol{v})\right\}, \tag{8.2.12}$$

where $a = 1/\ell_2(\boldsymbol{v})$ and $M_t = L_t - \bar{\ell}_N(\boldsymbol{v})$.

Unfortunately, Proposition 8.2.3 cannot be directly extended to $\sigma^2(\boldsymbol{v}, \boldsymbol{v}_0)$, because M_t is not guaranteed to be *nonnegative*. One can, however, approximate the optimal solution, $\boldsymbol{v}_0^*$, of the program $\min_{\boldsymbol{v}_0} \sigma^2(\boldsymbol{v}_0)$, by considering, for example, the following simplified programs: $\min_{\boldsymbol{v}_0} \mathscr{L}_1^0(\boldsymbol{v}, \boldsymbol{v}_0)$, $\boldsymbol{v}_0 \in V$ or

$$\min_{\boldsymbol{v}_0}\left[a^2 \,\mathrm{var}_{\boldsymbol{v}_0}\left\{\sum_{t=1}^{\tau} L_t(\underline{\boldsymbol{Z}}_t)\tilde{W}_t(\underline{\boldsymbol{Z}}_t, \boldsymbol{v})\right\} + \ell^2(\boldsymbol{v}) \,\mathrm{var}_{\boldsymbol{v}_0}\left\{\sum_{t=1}^{\tau} \tilde{W}_t(\underline{\boldsymbol{Z}}_t, \boldsymbol{v})\right\}\right]. \qquad \square$$

Proposition 8.2.4. *Consider an output process,* $\{L_t = L_t(\underline{Y}_t)\}$, *driven by the input sequence* $\{Y_1, Y_2, \ldots\}$ *of iid random variables from* $G(\lambda, \beta)$. *Suppose that* $L_t(\underline{y}_t)$, $t = 1, 2, \ldots$, *are nonnegative-valued and monotonically increasing in every component* y_i, $i = 1, \ldots, t$, *of* $\underline{y}_t$, *and that the indicator function* $I_{[\tau \geq t]}(\underline{y}_t)$ *is nondecreasing in every component of* $\underline{y}_t$. *Then there exists a unique minimizer,* λ_0^*, *of the function*

$$\mathscr{L}(\lambda_0) = \mathbb{E}_{\lambda_0}\left\{\left[\sum_{t=1}^{\tau} L_t(\underline{Z}_t)\tilde{W}_t(\underline{Z}_t, \lambda)\right]^2\right\} \tag{8.2.13}$$

that satisfies $\lambda_0^* < \lambda$.

Proof. An extended version of Proposition 8.2.4 is given in Shapiro and Rubinstein (1994). □

If we assume that the functions $L_t(\underline{y}_t)$ are nonnegative but monotonically *decreasing* in every component of $\underline{y}_t$, and that the functions $I_{\{\tau \geq t\}}(\underline{y}_t)$ are *nonincreasing* in every component of $\underline{y}_t$, then similar arguments imply that the corresponding minimizer, λ_0^* of $\mathscr{L}(\lambda_0)$, is *greater* than λ.

To illustrate, let $L_t = L_t(\underline{\boldsymbol{Y}}_t)$ be the steady-state waiting (sojourn) time process in a stable queueing model. Suppose that the vectors $\boldsymbol{Y}_i = (Y_{i1}, Y_{i2})$ of service and interarrival times, respectively, have mutually independent components, each from a gamma distribution, that is, $Y_{ji} \sim G(\lambda_i, \beta_i)$, $j = 1, 2$, $i = 1, 2, \ldots$. Let β_1, β_2 be fixed, and consider $\mathscr{L}_1^0(\boldsymbol{\lambda}, \boldsymbol{\lambda}_0)$ as a function of the reference vector, $\boldsymbol{\lambda}_0$ [here $\boldsymbol{\lambda} = (\lambda_1, \lambda_2)$ and $\boldsymbol{\lambda}_0 = (\lambda_{01}, \lambda_{02})$]. Then, from Proposition 8.2.4 it follows that the minimizer, $\boldsymbol{\lambda}_0^* = (\lambda_{01}^*, \lambda_{02}^*)$ of $\mathscr{L}_1^0(\boldsymbol{\lambda}, \cdot)$, satisfies $\lambda_{01}^* < \lambda_1$ and $\lambda_{02}^* > \lambda_2$, implying

$$\rho_0^*(k) > \rho, \qquad k = 0, 1. \tag{8.2.14}$$

Thus, (8.2.14) shows that the optimal traffic intensity, $\rho_0^*(\boldsymbol{v}_0^*)$, is *greater* than the nominal one.

Similar propositions can be derived for other distributions belonging to the exponential family in canonical form. In particular, it is possible to show (see Rubinstein and Shapiro, 1993) that if the input sequence, $\{Y_1, Y_2, \ldots\}$, is from $N(\mu, \sigma^2)$ and the monotonicity assumptions of Proposition 8.2.4 hold, then the optimal reference vector, (μ_0^*, σ_0^*), satisfies $\mu_0^* < \mu$ and $\sigma_0^* > \sigma$.

If $\varphi(\cdot)$ is monotonically increasing on the real line and $\{L_t\}$ is an output process, satisfying the assumptions of Proposition 8.2.4, then the composite process $\tilde{L}_t = \varphi(L_t)$ will also satisfy these assumptions, and the corresponding ordering relations for the optimal value of the reference parameters will hold. In particular, one may take indicator functions of the form $\varphi(\cdot) = I_{(x,\infty)}(\cdot)$, in which case the expected performance reduces to the probability of L_t being greater than x. The case where these probabilities are very small corresponds to *rare-event* probability estimation and is the subject of Chapter 9.

From Propositions 8.2.3 and 8.2.4 it follows that for fixed traffic intensity ρ (fixed $\boldsymbol{v}$), there exist $\hat{\rho} = \hat{\rho}(k)$, satisfying

$$\rho < \rho_0 < \hat{\rho}, \tag{8.2.15}$$

such that

$$\mathscr{L}_1^k(\rho, \hat{\rho}) = \mathscr{L}_1^k(\rho, \rho), \qquad k = 0, 1. \tag{8.2.16}$$

Furthermore, if $\rho_0 \in (\rho, \hat{\rho})$, then

$$\mathscr{L}_1^k(\rho, \rho_0) < \mathscr{L}_1^k(\rho, \rho), \qquad k = 0, 1. \tag{8.2.17}$$

In other words, if the reference traffic intensity, ρ_0, satisfies $\rho < \rho_0 < \hat{\rho}$, then variance reduction will be achieved.

To provide better insight into the subsequent material, consider the steady-state expected waiting time in an $M/G/1$ queue with the two-point service time distribution (6.2.24), that is

$$f(y, v) = \begin{cases} v & \text{if } y = 1.2 \\ 1 - v & \text{if } y = 0.2 \end{cases} \tag{8.2.18}$$

where $0 < v < 1$ and $\rho = 0.6$. Table 8.2.1 depicts how various reference values of ρ_0 perform for the selected nominal value, $\rho = 0.6$. More specifically, displayed are the point estimators, $\nabla^k \bar{\ell}_N(\rho, \rho_0)$; the sample variance of $\nabla^k \bar{\ell}_N(\rho, \rho_0)$ [denoted by $\hat{\sigma}^2(\rho, \rho_0, k)$]; and the sample efficiency [denoted by $\bar{\varepsilon}^k(\rho, \rho_0)$] as functions of ρ_0 and $\Delta\delta = (\rho_0 - \rho)/\rho$ (the relative perturbation in ρ), for $k = 0$, 1. Here, $\bar{\varepsilon}^k(\rho, \rho_0)$ is defined as

$$\bar{\varepsilon}^k(\rho, \rho_0) = \frac{\hat{\sigma}^2(\rho, \rho_0, k)}{\hat{\sigma}^2(\rho, \rho, k)}. \tag{8.2.19}$$

The results of Table 8.2.1 were obtained for an arrival rate of 1 and $v = 0.4$ [so that $\rho = \mathbb{E}(Y) = 0.6$] for various reference parameters, $\boldsymbol{v}_0$, based on simulation runs of $N = 10^6$ customers. The values of $\bar{\varepsilon}^k(\rho, \rho_0)$, $k = 0, 1$, marked with an asterisk, correspond to the maximal efficiency (maximal variance reduction) in the corresponding column. Note that both $\hat{\sigma}_{\boldsymbol{v}}^2$ and $\hat{\sigma}_{\boldsymbol{v}_0}^2$ are expressed in terms of *sample variance/customer* rather than *sample variance/cycle*. Asmussen and Rubinstein (1993) gives an exact expression for $\varepsilon^2(\rho, \rho_0)$ for the steady-state waiting time in the $M/M/1$ queue.

The interval $\Delta\hat{\rho} = \Delta\hat{\rho}(\rho_0, k) = \{(\rho, \hat{\rho})\colon \rho < \rho_0 \le \hat{\rho}\}$ will be referred to as the *variance reduction interval*. From the results of Table 8.2.1 one has the following widths of the variance reduction intervals, $|\Delta\hat{\rho}(\rho_0, k)| = |\hat{\rho}(k) - \rho|$:

$$|\Delta\hat{\rho}(\rho_0, 0)| \approx 0.19,$$
$$|\Delta\hat{\rho}(\rho_0, 1)| \approx 0.30.$$

Table 8.2.1 $\nabla^k \bar{\ell}_N(\rho, \rho_0)$, $\hat{\sigma}^2(\rho, \rho_{0,k})$, **and** $\bar{\varepsilon}^k(\rho, \rho_0)$ **as Functions of** ρ_0 **and** $\Delta\delta$, **for the** $M/G/1$ **Queue with** $\rho = 0.6$

ρ_0	$\Delta\delta$	$\bar{\ell}_n(\rho, \rho_0)$ $\hat{\sigma}^2(\rho, \rho_0, 0)$ $\bar{\varepsilon}^0(\rho, \rho_0)$	$\nabla\bar{\ell}_N(\rho, \rho_0)$ $\hat{\sigma}^2(\rho, \rho_0, 1)$ $\bar{\varepsilon}^1(\rho, \rho_0)$
0.55	−0.08	7.43E-01 4.72E-05 3.08	3.68E+00 3.14E-02 8.25
0.6	0.00	7.48E-01 1.53E-05 1.00	3.65E+00 3.80E-03 1.00
0.65	0.08	7.48E-01 1.01E-05 0.66	3.61E+00 1.42E-03 0.37
0.7	0.17	7.46E-01 8.82E-06 0.58*	3.58E+00 8.28E-04 0.22
0.75	0.25	7.47E-01 1.05E-05 0.69	3.62E+00 8.10E-04 0.21*
0.8	0.33	7.47E-01 1.63E-05 1.07	3.60E+00 1.13E-03 0.30
0.85	0.42	7.38E-01 2.26E-05 1.48	3.53E+00 1.23E-03 0.32
0.9	0.50	7.55E-01 1.23E-04 8.07	3.59E+00 3.97E-03 1.04
0.95	0.58	7.19E-01 1.61E-04 10.53	3.50E+00 1.36E-02 3.57

Based on the results of Table 8.2.1, Figure 8.1 displays the sample efficiency, $\bar{\varepsilon}^k(\rho, \rho_0)$, as a function of ρ_0, for $k = 0, 1$.

From Table 8.2.1 and Figure 8.1, it follows that

1. For fixed ρ_0, the sample variance, $\hat{\sigma}^2(\rho, \rho_0, k)$, increases in k.
2. For fixed k, there exists $\hat{\rho} = \hat{\rho}(\boldsymbol{v}, \hat{v}, k)$, satisfying (8.1.4) and (8.1.5).
3. If ρ_0 satisfies (8.1.4), then (8.1.1) holds and variance reduction is attained.

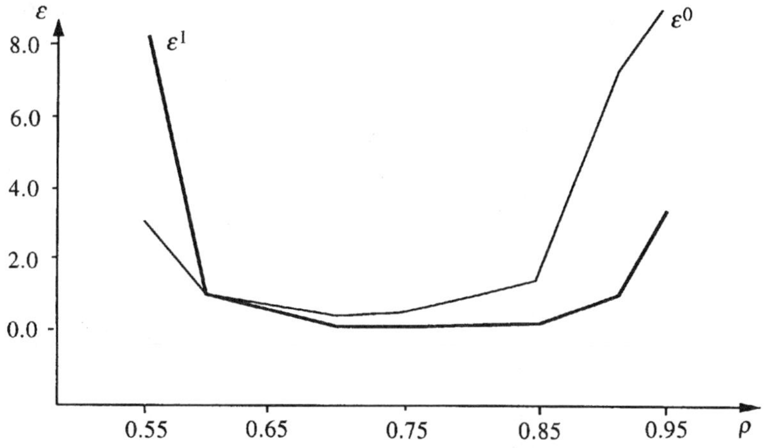

Figure 8.1. Relative efficiencies $\bar{\varepsilon}^k(\rho, \rho_0)$, $k = 0, 1$, as functions of ρ_0, for the $M/G/1$ queue with $\rho = 0.6$.

A practical guideline for variance reduction is to choose the reference traffic intensity, $\rho_0 = \rho(\boldsymbol{v}_0)$, *moderately larger* than the nominal traffic intensity, ρ. Experience shows that for $\rho_0 < \rho$, the relative efficiency, $\bar{\varepsilon}^k(\rho, \rho_0)$, increases rapidly in ρ_0. In particular, for the $M/G/1$ queue above with $\rho = 0.6$, the SF estimator, $\nabla^k \bar{\ell}_N(\rho, \rho_0)$, is practically useless (has excessive variance) when $\rho_0 < 0.5$ (see also Proposition 8.3.3).

It is reasonable to seek a simple intuitive argument, explaining why, for a given ρ, one can attain variance reduction by choosing ρ_0 to be moderately larger than ρ. One argument proceeds as follows: When estimating $\ell(\rho)$, the main contributions are due to cycles that are larger than average [e.g., in heavy traffic, only cycle lengths of order $O(1-\rho)^{-2}$ matter, while overall cycle lengths are of order $O(1)$; see the discussion by Asmussen and Rubinstein, 1993]. Thus, in choosing a reference parameter $\rho_0 > \rho$, greater weight is assigned to larger cycles.

8.2.3 Queueing Networks

We now provide some guidelines on how to choose a good vector of reference parameters, $\boldsymbol{v}_0$ (and $\boldsymbol{\rho}_0$), when estimating the system parameters $\nabla^k \ell(\boldsymbol{v}) = \nabla^k \mathbb{E}_{\boldsymbol{v}}\{L_t\}$, $k = 0, 1$, in a non-Markovian queueing network consisting of r nodes, where $\{L_t\}$ is the steady-state output process, and $\boldsymbol{v}$ is fixed. We first present simulation results for the following two simple queueing networks:

1. Two queues in tandem
2. Two queues with a feedback in the second queue

Both models are constructed with one queue being more congested than the other. Following common terminology, the more congested queue is termed the *bottleneck queue*. As will be seen later, the bottleneck queue plays a crucial role in choosing a good (optimal) reference vector.

Table 8.2.2 $\hat{\sigma}^2(\rho, \rho_0, k)$ **and** $\bar{\varepsilon}^k(\rho, \rho_0)$ **as Functions of** ρ_{01} **and** ρ_{02}**, for Two** $GI/G/1$ **Queues in Tandem with** $\rho = (\rho_1, \rho_2) = (0.3, 0.6)$

ρ_{01}	ρ_{02}	$\hat{\sigma}^2(\rho, \rho_0, 0)$	$\bar{\varepsilon}^0(\rho, \rho_0)$	$\hat{\sigma}^2(\rho, \rho_0, 1)$	$\bar{\varepsilon}^1(\rho, \rho_0)$
0.30	0.60	2.09E-06	1.000	2.72E-04	1.000
0.30	0.65	1.57E-06	0.751	1.29E-04	0.475
0.30	0.70	1.42E-06	0.680$^\diamond$	7.94E-05	0.292
0.30	0.75	1.53E-06	0.728	6.35E-05	0.234$^\diamond$
0.30	0.80	1.89E-06	0.904	6.50E-05	0.239
0.30	0.85	2.91E-06	1.387	7.59E-05	0.279
0.30	0.90	4.80E-06	2.293	1.37E-04	0.503
0.30	0.95	1.37E-05	6.529	3.31E-04	1.219
0.30	0.98	4.29E-05	20.497	1.14E-03	4.210
0.32	0.60	1.94E-06	0.928	2.46E-04	0.905
0.32	0.65	1.48E-06	0.706	1.26E-04	0.462
0.32	0.70	1.34E-06	0.639*	7.80E-05	0.287
0.32	0.75	1.42E-06	0.676	6.16E-05	0.227*
0.32	0.80	1.80E-06	0.859	6.56E-05	0.241
0.32	0.85	2.73E-06	1.302	7.58E-05	0.279
0.32	0.90	5.10E-06	2.433	1.30E-04	0.477
0.32	0.95	1.44E-05	6.852	3.27E-04	1.202
0.32	0.98	4.03E-05	19.220	1.07E-03	3.951
0.34	0.60	1.99E-06	0.949	2.68E-04	0.984
0.34	0.65	1.50E-06	0.717	1.42E-04	0.524
0.34	0.70	1.35E-06	0.646	8.46E-05	0.311
0.34	0.75	1.43E-06	0.684	6.63E-05	0.244
0.34	0.80	1.81E-06	0.863	7.08E-05	0.260
0.34	0.85	2.76E-06	1.318	8.05E-05	0.296
0.34	0.90	5.66E-06	2.703	1.34E-04	0.492
0.34	0.95	9.74E-06	4.650	2.96E-04	1.090
0.34	0.98	4.59E-05	21.930	1.44E-03	5.303

1. *Two Queues in Tandem.* Table 8.2.2 displays the sample efficiency $\bar{\varepsilon}^k(\boldsymbol{\rho}, \boldsymbol{\rho}_0)$ [the sample counterpart of $\varepsilon^k(\boldsymbol{\rho}, \boldsymbol{\rho}_0)$, where $\boldsymbol{\rho}_0 = (\rho_{01}, \rho_{02})$, $k = 0, 1$]. It also displays the sample variance of $\bar{\ell}_N(\boldsymbol{\rho})$ and the trace of the sample variance of $\nabla\bar{\ell}_N(\boldsymbol{\rho})$ [denoted by $\hat{\sigma}^2(\boldsymbol{\rho}, \boldsymbol{\rho}_0, k)$, $k = 0, 1$]. All statistics were computed as functions of $\boldsymbol{\rho}_0 = (\rho_{01}, \rho_{02})$ for two $GI/G/1$ queues in tandem with $\rho_1 = 0.3$ (light traffic) and $\rho_2 = 0.6$ (medium traffic), respectively. The interarrival times are from an exponential distribution with the two service times (in the first and the second queue) being distributed according to (6.2.24) with $v = 0.8$, 0.1, 0.4, respectively. Taking $\boldsymbol{v}_0 = (v_{01}, v_{02}) = (v_1, v_2) = (0.1, 0.4)$ as the reference vector, the network was simulated for 10^6 customers. Notice that $\boldsymbol{\rho}_0^\diamond$ corresponds to the optimal solution of (8.1.3) with respect to $\boldsymbol{v}_{02}$ *alone*, namely, when $\boldsymbol{v}_1$ (and thus ρ_1) are fixed, while $\boldsymbol{\rho}_0^*$ corresponds to the global optimum. In Table 8.2.2 and elsewhere, rows marked by diamonds and asterisks should be similarly interpreted.

Table 8.2.3 $\hat{\sigma}^2(\boldsymbol{\rho}, \boldsymbol{\rho}_0, k)$ and $\bar{\varepsilon}^k(\boldsymbol{\rho}, \boldsymbol{\rho}_0)$ as Functions of v_{01} and v_{02}, for Two $GI/G/1$ Queues with Feedback in Second Queue

v_{01}	v_{02}	$\hat{\sigma}^2(\boldsymbol{\rho}, \boldsymbol{\rho}_0, 0)$	$\bar{\epsilon}^0(\boldsymbol{\rho}, \boldsymbol{\rho}_0)$	$\hat{\sigma}^2(\boldsymbol{\rho}, \boldsymbol{\rho}_0, 1)$	$\bar{\epsilon}^1(\boldsymbol{\rho}, \boldsymbol{\rho}_0)$
0.30	0.30	7.18E-06	1.00	3.68E-03	1.00
0.30	0.36	4.87E-06	$0.69^\diamond$	1.09E-03	0.30
0.30	0.40	5.38E-06	0.75	8.78E-04	0.24
0.30	0.44	6.20E-06	0.86	7.47E-04	$0.20^\diamond$
0.30	0.50	1.07E-05	1.49	1.14E-03	0.31
0.30	0.56	1.96E-05	2.73	1.75E-03	0.48
0.30	0.63	3.98E-04	55.43	3.23E-02	8.78
0.36	0.30	7.18E-06	1.00	5.02E-03	1.36
0.36	0.36	4.85E-06	0.68*	1.25E-03	0.34
0.36	0.40	5.32E-06	0.74	1.15E-03	0.31
0.36	0.44	8.40E-06	1.17	1.20E-03	0.33
0.36	0.50	9.46E-06	1.32	1.06E-03	0.19*
0.36	0.56	2.41E-05	3.36	2.01E-03	0.55
0.36	0.63	8.36E-05	11.64	1.89E-02	5.14
0.40	0.30	1.37E-05	1.91	2.88E-02	7.83
0.40	0.36	5.83E-06	0.81	1.52E-03	0.41
0.40	0.40	5.94E-06	0.83	1.19E-03	0.32
0.40	0.44	9.58E-06	1.33	1.70E-03	0.46
0.40	0.50	2.65E-05	3.69	7.05E-03	1.92
0.40	0.56	4.06E-05	5.65	4.45E-03	1.21
0.40	0.63	1.06E-03	146.63	4.67E-02	12.69

2. *Queueing Networks with Feedback.* Table 8.2.3 displays data similar to those of Table 8.2.2 for the two-node queueing model with feedback in the second queue (the feedback probability is $p = 0.3$). The interarrival times and the two service times (in the first and the second queue) are distributed according to (6.2.24) with $v = 0.8, 0.1, 0.1$, respectively. Simulations were again run for 5×10^4 customers, with $\boldsymbol{v}_0 = (v_{01}, v_{02}) = (v_1, v_2) = (0.1, 0.1)$ being the reference vector.

Tables 8.2.2 and 8.2.3 imply that the optimal vector of traffic intensities, $\boldsymbol{\rho}_0^* = (\rho_{01}^*, \rho_{02}^*)$, and the minimal value, $\sigma^2(\boldsymbol{\rho}, \boldsymbol{\rho}_0^*, k)$ [corresponding to the minimal relative efficiency, $\varepsilon^k(\boldsymbol{\rho}, \boldsymbol{\rho}_0^*)$], are essentially *insensitive* to the parameters of the nonbottleneck queues, in the sense that both $\boldsymbol{\rho}_0^* = (\rho_{01}^*, \rho_{02}^*)$ and $\sigma^2(\boldsymbol{\rho}, \boldsymbol{\rho}_0^*, k)$ can be *approximated rather well* by minimizing the objective function, $\sigma^2(\boldsymbol{\rho}, \boldsymbol{\rho}_0, k)$, with respect to the reference traffic intensity of the bottleneck queue alone. In other words, $\boldsymbol{\rho}_0^* = (\rho_{01}^*, \rho_{02}^*)$ can be approximated rather well by $\boldsymbol{\rho}_0^\diamond = (\rho_1, \rho_{02}^\diamond)$.

Consider, for example, the statistics in Table 8.2.2, in which

$$\bar{\varepsilon}^0(\boldsymbol{\rho}_0^*) = 0.639, \qquad \boldsymbol{\rho}_0^* = (\rho_{01}^*, \rho_{02}^*) = (0.32, 0.70);$$

$$\bar{\varepsilon}^1(\boldsymbol{\rho}_0^*) = 0.227, \qquad \boldsymbol{\rho}_0^* = (\rho_{01}^*, \rho_{02}^*) = (0.32, 0.75);$$

and

$$\bar{\varepsilon}^0(\boldsymbol{\rho}_0^\diamond) = 0.680, \qquad \boldsymbol{\rho}_0^\diamond = (\rho_1, \rho_{02}^\diamond) = (0.30, 0.70);$$
$$\bar{\varepsilon}^1(\boldsymbol{\rho}_0^\diamond) = 0.234, \qquad \boldsymbol{\rho}_0^\diamond = (\rho_1, \rho_{02}^\diamond) = (0.30, 0.75).$$

Note that the first two lines above correspond to the relative efficiencies, $\bar{\varepsilon}^k(\boldsymbol{\rho}_0^*)$, $k = 0, 1$, for the optimal reference vector, $\boldsymbol{\rho}_0^* = (\rho_{01}^*, \rho_{02}^*)$, while the remaining two lines correspond to the relative efficiencies, $\bar{\varepsilon}^k(\boldsymbol{\rho}_0^\diamond)$, $k = 0, 1$, for $\boldsymbol{\rho}_0^\diamond = (\rho_1, \rho_{02}^\diamond)$, where $\rho_1 = 0.3$ is fixed. The above demonstrate that $\bar{\varepsilon}^k(\boldsymbol{\rho}_0^\diamond)$ approximates $\bar{\varepsilon}^k(\boldsymbol{\rho}_0^*)$ rather well, and consequently, $\boldsymbol{\rho}_0^\diamond = (\rho_1, \rho_{02}^\diamond)$ may be selected as a good vector of reference parameters.

Consider now a general r-node queueing network, and assume for simplicity that the first node is the bottleneck node. Our goal is again to choose a good reference vector for estimating the performance measures, $\nabla^k \ell(\boldsymbol{v})$. Arguing as before, one can solve the program (8.1.3) with respect to the parameters of the bottleneck node (i.e., with respect to $\boldsymbol{v}_{01}$), and then use the vector $\boldsymbol{v}_0^\diamond = (\boldsymbol{v}_{01}^\diamond, \boldsymbol{v}_2 \ldots, \boldsymbol{v}_r)$ as a near-optimal reference vector. Here, $\boldsymbol{v}_{01}^\diamond$ corresponds to the optimal solution of (8.1.3) with respect to $\boldsymbol{v}_{01}$ *alone*, and $(\boldsymbol{v}_2, \ldots, \boldsymbol{v}_r)$ is the vector of underlying (nominal) parameters of the remaining $r - 1$ nonbottleneck queues. Extensive simulation studies lend support to this so-called bottleneck phenomenon. Notice that when more than one bottleneck queue is present, one must increase the dimensionality of the program (8.1.3) accordingly, and solve it with respect to all bottleneck queues.

A heuristic explanation for the apparent insensitivity of the optimal vector, $\boldsymbol{\rho}_0^*$, and the optimal-value function, $\varepsilon^k(\boldsymbol{\rho}_0^*)$, with respect to the parameters $(\rho_{02}, \ldots, \rho_{0r})$ of the nonbottleneck queues may be advanced as follows. Assume, for simplicity, that $k = 0$ and solve the program (8.1.3), using the following system of equations:

$$\nabla \sigma^2(\boldsymbol{\rho}, \boldsymbol{\rho}_0) = \mathbf{0}. \tag{8.2.20}$$

Since the first (bottleneck) component of $\nabla \sigma^2(\boldsymbol{\rho}, \boldsymbol{\rho}_0)$ is *substantially larger* than any of the remaining $r - 1$ (nonbottleneck) components, the *insensitivity* result follows [see also the results of Asmussen (1992), Asmussen and Rubinstein (1993) on the variance of performance measures and their heavy traffic approximations]. Note that since $\sigma^2(\boldsymbol{\rho}, \boldsymbol{\rho}_0)$ is typically not available analytically, one may consider in lieu of $\nabla \sigma^2(\boldsymbol{\rho}, \boldsymbol{\rho}_0) = \mathbf{0}$, its stochastic counterpart,

$$\nabla \hat{\sigma}^2(\boldsymbol{\rho}, \boldsymbol{\rho}_0) = \mathbf{0}, \tag{8.2.21}$$

where $\hat{\sigma}^2(\boldsymbol{\rho}, \boldsymbol{\rho}_0)$ is the estimator of $\sigma^2(\boldsymbol{\rho}, \boldsymbol{\rho}_0)$, and similarly for $k = 1$.

8.3 RESPONSE SURFACE METHODOLOGY VIA THE SCORE FUNCTION METHOD

This section discusses the response surface methodology—the estimation of the functions $\nabla^k \bar{\ell}_N(\boldsymbol{v}, \boldsymbol{v}_0)$, $k = 0, 1$, for different values of $\boldsymbol{v}$ and fixed reference parameter vector, $\boldsymbol{v}_0$.

The following issues will be addressed:

- For which values of $\boldsymbol{v}$ is the variance of $\nabla^k \bar{\ell}_N(\boldsymbol{v}, \boldsymbol{v}_0)$ finite?
- For which values of $\boldsymbol{v}$ is the variance of $\nabla^k \bar{\ell}_N(\boldsymbol{v}, \boldsymbol{v}_0)$ relatively small, and for which is it too large to be useful?
- For which values of $\boldsymbol{v}$ (if any) does the SF estimator, $\nabla^k \bar{\ell}_N(\boldsymbol{v}, \boldsymbol{v}_0)$, lead to variance reduction, as compared with the naive estimator, $\nabla^k \bar{\ell}_N(\boldsymbol{v}, \boldsymbol{v})$?

As in Section 8.2, the treatment will start with static queueing models, proceed with simple queues, and conclude with non-Markovian queueing networks.

8.3.1 Static Models

Consider the program (8.1.6), or equivalently,

$$\min_{\boldsymbol{v}_0} \max_{j=1,\ldots,s} \mathscr{L}^k(\boldsymbol{v}_j, \boldsymbol{v}_0), \qquad \boldsymbol{v}_0 \in V, \tag{8.3.1}$$

where $\mathscr{L}^0(\boldsymbol{v}_j, \boldsymbol{v}_0)$ and $\mathscr{L}^1(\boldsymbol{v}_j, \boldsymbol{v}_0)$ are given in (8.2.4) and (8.2.5), respectively.

Proposition 8.3.1. *Let Y be a random vector from an exponential family in the canonical form* (5.2.24). *Then,* $\max_{j=1,\ldots,s} \mathscr{L}^k(\boldsymbol{v}_j, \boldsymbol{v}_0)$, $k = 0, 1$, *is a convex function in* $\boldsymbol{v}_0$.

Proof. Follows immediately from the convexity of $\mathscr{L}^k(\boldsymbol{v}_j, \boldsymbol{v}_0)$ in $\boldsymbol{v}_0$, and the fact that the maximum over a convex function is also a convex function. □

Example 8.3.1 [Example continued (λ_0 is Fixed)]. Assume that $\lambda \in (\lambda_1, \lambda_2)$, and consider (8.2.6). For $k=0$,

$$\begin{aligned}\max_{\lambda \in (\lambda_1, \lambda_2)} \mathscr{L}^0(\lambda, \lambda_0) &= \max_{\lambda \in (\lambda_1, \lambda_2)} \frac{\lambda^{2\beta}}{\lambda_0^\beta (2\lambda - \lambda_0)^\beta} \mathbb{E}_{2\lambda - \lambda_0}\{Y^2\} \\ &= \max_{\lambda \in (\lambda_1, \lambda_2)} \frac{\lambda^{2\beta} \beta(1+\beta)}{\lambda_0^\beta (2\lambda - \lambda_0)^{\beta+2}}.\end{aligned}$$

The minimizer λ_0^* of $\max_{\lambda\in(\lambda_1,\lambda_2)} \mathcal{L}^0(\lambda, \lambda_0)$ over the interval $(0, 2\lambda)$ is

$$\lambda_0^* = \frac{2(\lambda_1\lambda_2^{2\beta/(\beta+2)} - \lambda_2\lambda_1^{2\beta/(\beta+2)})}{\lambda_2^{2\beta/(\beta+2)} - \lambda_1^{2\beta/(\beta+2)}}. \tag{8.3.2}$$

It follows from (8.3.2) that λ_0^* increases in λ_2 for fixed λ_1, and it also increases in λ_1 for fixed λ_2. To illustrate, take $\beta=1$, $\lambda_1=1$, and consider the values of $\lambda_2=1$, 2, 10, 100, 1000, 10000. One has $\lambda_0^*=0.5$, 0.906, 1.412, 1.802, 1.937, 1.980, respectively. Note that for $\lambda_2 \leq 2$ one has $\lambda_0^* < \lambda_1$, while for the remaining values of λ_2 ($\lambda_2 \geq 10$), one has $\lambda_0^* > \lambda_1$.

8.3.2 Simple Queues

Consider first the estimation of the response surface $\nabla^k \ell_1(\boldsymbol{v}, \boldsymbol{v}_0)$ for different values $\boldsymbol{v}$. We are interested in the program

$$\min_{\boldsymbol{v}_0} \max_{j=1,\ldots,s} \operatorname{var}\{\nabla^k \bar{\ell}_{1N}(\boldsymbol{v}_j, \boldsymbol{v}_0)\}, \quad \boldsymbol{v}_0 \in V, \qquad k = 0, 1, \tag{8.3.3}$$

which is equivalent to the program

$$\min_{\boldsymbol{v}_0} \max_{j=1,\ldots,s} \mathcal{L}_1^k(\boldsymbol{v}_j, \boldsymbol{v}_0), \quad \boldsymbol{v}_0 \in V, \tag{8.3.4}$$

where $\mathcal{L}_1^0(\boldsymbol{v}_j, \boldsymbol{v}_0)$ and $\mathcal{L}_1^0(\boldsymbol{v}_j, \boldsymbol{v}_0)$ are given in (8.2.10) and (8.2.11), respectively.

Proposition 8.3.2. *Consider a stable queueing model, say, the GI/G/1 queue. Assume that the output process,* $\{L_t : t > 0\}$, *is nonnegative-valued, and let the components of the interarrival and service time random vector,* $\mathbf{Y}$, *be from an exponential family in the canonical form* (5.2.24). *Then,* $\max_{j=1,\ldots,s} \mathcal{L}_1^k(\boldsymbol{v}_j, \boldsymbol{v}_0)$, $k = 0, 1$, *is a convex function in* $\boldsymbol{v}_0$.

Proof. Follows immediately from the convexity of $\mathcal{L}_1^k(\boldsymbol{v}_j, \boldsymbol{v}_0)$ in $\boldsymbol{v}_0$, and the fact that the maximum over a convex function is a convex function. □

To provide better insight into the subsequent material, consider again Table 6.2.1 and Figures 6.3 and 6.4 for the $M/G/1$ queue with the two-point service time distribution (6.2.24) and $\rho_0=0.8$. It is readily seen from Table 6.2.1 that

- There exists a rather broad interval, $\rho \in (\rho_n, \rho_0)$, in which some moderate variance reduction is attained, that is, $\bar{\varepsilon}^k(\rho|\rho_0) < 1$. Here, ρ_n, called the *neutral* traffic intensity, is the solution of the equation

$$\hat{\sigma}^2(\rho_n, k|\rho_0) = \hat{\sigma}^2(\rho_0, k|\rho_0),$$

with respect to ρ, for fixed ρ_0, k, and p.

- The efficiency, $\bar{\varepsilon}^k(\rho|\rho_0)$, decreases *moderately* as we move down from ρ_0 to zero, and decreases *rapidly* as we move above ρ_0. In particular, the SF estimators, $\bar{\ell}_N(\rho|\rho_0)$ and $\nabla\bar{\ell}_N(\rho|\rho_0)$, perform reasonably well in the range $\rho \in (0.4, 0.8)$, where the relative perturbation in ρ does not exceed 200%. For large relative perturbations ($\geq$200%), the SF processes $\nabla^k \tilde{W}$ gives rise to excessive variance in the estimators $\bar{\ell}_N(\rho|\rho_0)$ and $\nabla\bar{\ell}_N(\rho|\rho_0)$.

We shall call $\Delta\rho_n = \Delta\rho_n(\rho, k) = \{(\rho, \rho_0) \colon \rho_n(\rho) < \rho \leq \rho_0\}$ *the variance reduction interval.* From Table 6.2.1 we have the following widths, $|\Delta\rho_n| = |\rho_0 - \rho_n|$, of variance reduction intervals:

$$|\Delta\rho_n(\rho, k = 0)| \approx 0.20,$$
$$|\Delta\rho_n(\rho, k = 1)| \approx 0.37.$$

Similar results were obtained for the $GI/G/1$ queue with other interarrival time and service time distributions; see Asmussen and Rubinstein, 1993.

The following proposition indicates how far one may move above ρ_0 in the $M/M/1$ queue.

Proposition 8.3.3. *The variance $\sigma^2(\rho|\rho_0)$ of the estimator $\bar{\ell}_N(\rho|\rho_0)$ for the steady-state mean waiting time in the $M/M/1$ queue is finite for*

$$\rho \leq \rho_c = \rho_c(\rho_0) = \frac{2\rho_0}{1 + \sqrt{\rho_0}}, \tag{8.3.5}$$

and is infinite for $\rho > \rho_c$.

Proof. Given in Asmussen and Rubinstein (1993). □

Table 8.3.1 displays the critical value, ρ_c, as a function of ρ_0, calculated from (8.3.5). It follows from (8.3.5) that $\rho_0 < \rho_c < 2\rho_0$. Note also that in light and heavy traffic, one has $\rho_c \cong 2\rho_0$ and $\rho_c \cong \rho_0$, respectively. To illustrate, take $\rho_0 = 0.1$ and 0.9, yielding $\rho_c = 0.152$ and 0.924, respectively.

Figure 8.2 displays the curves ρ_c (the critical value), ρ_n (the neutral value), and ρ_0^* (the optimal value) as functions of ρ_0, for the expected waiting time $\ell(\boldsymbol{v})$ in the $M/G/1$ queue, with arrival rate $\lambda = 1$, service time distribution $G(\lambda_0, \beta)$, $\beta = 2.3$, and $\lambda_0 = v_0$ being the reference parameter. Figure 8.3 displays similar curves for the derivative $\partial\ell(\boldsymbol{v})/\partial\lambda$. In both cases, separate simulations of 10^6 customers were run at $\rho_0 = 0.1, 0.2, \ldots, 0.9$.

Table 8.3.1 ρ_c as Function of ρ_0

ρ_0	0.10	0.20	0.30	0.40	0.50	0.60	0.70	0.80	0.90	0.95	0.99
ρ_c	0.15	0.28	0.39	0.49	0.59	0.66	0.76	0.84	0.92	0.96	0.99

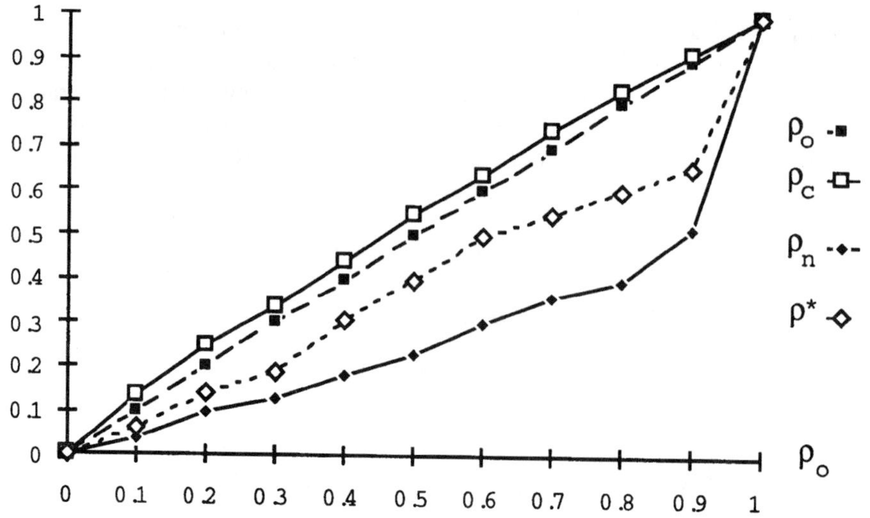

Figure 8.2. Critical ρ_c, neutral ρ_n and optimal ρ^* values as functions of ρ_0, for the expected waiting time in an $M/G/1$ queue.

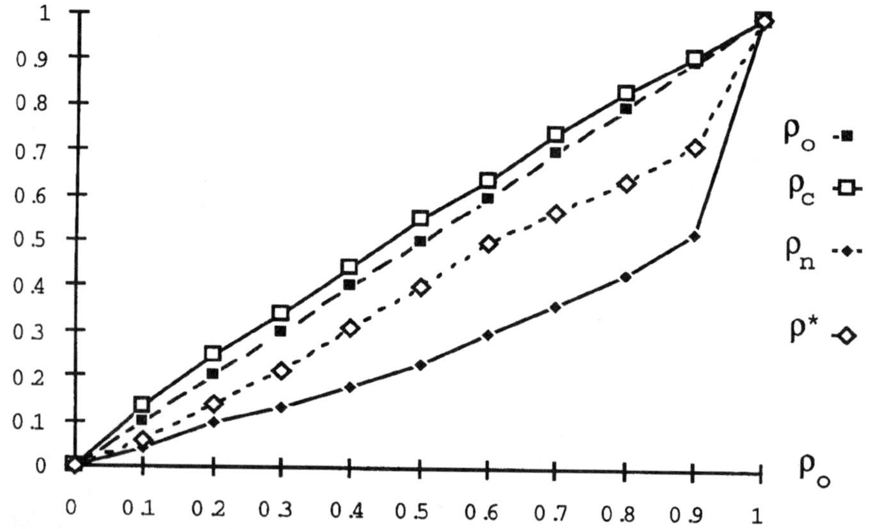

Figure 8.3. Critical ρ_c, neutral ρ_n and optimal ρ^* values as functions of ρ_0, for the derivative of the expected waiting time in the $M/G/1$ queue.

It is readily seen that the optimal traffic intensity, $\rho_0^*(\rho)$, is located inside the variance reduction interval, $\Delta\rho_n = \{(\rho, \rho_0)\colon \rho_n(\rho) < \rho \le \rho_0\}$, which in turn is quite large. This means that good performance can be obtained over a reasonably large region. Similar results were obtained for the $GI/G/1$ queue with interarrival and service times from the exponential family in the canonical form (5.2.24).

A simple practical suggestion, which follows from the above, is to choose the reference parameter, $\boldsymbol{v}_0$, such that

$$\rho(\boldsymbol{v}_0) = \{\boldsymbol{v}: \max_{j=1,\ldots,s} \rho(\boldsymbol{v}_j)\}. \tag{8.3.6}$$

Condition (8.3.6) means that the reference parameter, $\boldsymbol{v}_0$, must correspond to the *highest traffic intensity* among all those associated with the values $\boldsymbol{v}_1, \ldots, \boldsymbol{v}_s$ under consideration. Since such $\boldsymbol{v}_0$ corresponds to the "noisiest" (most variable) system configuration (regime), this in turn means that in choosing a good $\boldsymbol{v}_0$, we *ignore* all configurations except the latter. We already mentioned in Section 6.4 that through such noisiest configurations we typically gain good control over the entire system.

To illustrate, refer to the $M/G/1$ queue with data as per Table 6.2.1, and suppose we seek to estimate $\nabla^k \ell(\rho)$ for $\rho = 0.8, 0.7, 0.6$ and $k = 0, 1$, *simultaneously from a single simulation run.* It follows from Table 6.2.1 that this can be done efficiently, in the sense that $\bar{\varepsilon}^k(\rho|\rho_0) < 1$, simultaneously for all v, k, and p, by simulating a single scenario at the reference traffic intensity $\rho_0 = 0.8$. This approach is feasible since, in this case, the width of the variance reduction region, $\Delta\rho_n = \{(\rho, \rho_0)$: $\rho_n(\rho) < \rho \leq \rho_0\}$, exceeds 0.2. Note that if we need to estimate $\ell(\boldsymbol{v})$, say, for $\rho \in (0.4, 0.8)$, we could still choose $\rho_0 = 0.8$ as a reference parameter. In the latter case (see Table 6.2.1), we would attain variance reduction in the approximate interval (0.8, 0.6), but suffer a moderate increase in variance [compared to the crude Monte Carlo (CMC) method] in the interval (0.4, 0.6).

Note finally that if, in addition, we need to estimate $\ell(\boldsymbol{v})$, say, for $\rho \in (0.2, 0.8)$, we could partition the set $\rho \in (0.2, 0.8)$ into smaller subsets, say into $\rho \in (0.2, 0.6)$ and $\rho \in (0.6, 0.8)$, and use two different simulations with corresponding reference traffic intensities $\rho_{01} = 0.6$ and $\rho_{02} = 0.8$.

8.3.3 Queueing Networks

We next provide some guidelines on how to choose a good vector of reference parameters, $\boldsymbol{v}_0$ (and $\boldsymbol{\rho}_0$), for estimating the response surface $\nabla^k \ell(\boldsymbol{v}) = \nabla^k \mathbb{E}_{\boldsymbol{v}}\{L_t\}$, $k = 0, 1$, for non-Markovian queueing networks consisting of r nodes. In this section and elsewhere simulation results are presented in tables; table entries marked by an asterisk or diamond correspond to optimal values.

We begin with simulation results for two queues in tandem. Table 8.3.2 displays the sample variance, $\hat{\sigma}^2(\boldsymbol{\rho}|\boldsymbol{\rho}_0)$ of $\bar{\ell}_N(\boldsymbol{\rho}|\boldsymbol{\rho}_0)$, and the sample efficiency, $\bar{\varepsilon}^0(\boldsymbol{\rho}|\boldsymbol{\rho}_0)$ [the sample counterpart of $\varepsilon^0(\boldsymbol{\rho}|\boldsymbol{\rho}_0)$], where $\boldsymbol{\rho}_0 = (\rho_{01}, \rho_{02})$, as a function of ρ_1 and ρ_2 for two $GI/G/1$ queues in tandem with $\rho_{01} = 0.6$ and $\rho_{02} = 0.8$. The interarrival times at the first node and the service times at the two nodes were from the two-point distribution (6.2.24) with respective parameters $v = 0.8, 0.3, 0.4$. The reference parameters were set at $v_{01} = 0.3$ and $v_{02} = 0.4$, and a simulation was run for 10^6 customers.

Table 8.3.3 presents data similar to that of Table 8.3.2, but with $\rho_{01} = 0.6$ replaced by $\rho_{01} = 0.4$, while Table 8.3.4 presents data similar to that of Table 8.3.2,

Table 8.3.2. $\hat{\sigma}^2(\rho|\rho_0)$ **and** $\bar{\varepsilon}^0(\rho|\rho_0)$ **as Functions of** ρ_1 **and** ρ_2**, for Two** ***GI/G/1*** **Queues in Tandem with** $\boldsymbol{\rho} = (\rho_{01}, \rho_{02}) = (0.6, 0.8)$

ρ_1	ρ_2	$\hat{\sigma}^2(\rho\|\rho_0)$	$\bar{\varepsilon}^0(\rho\|\rho_0)$
0.25	0.70	3.94E-01	1.72E+00
		2.48E-04	3.25E-02
		4.98E+01	3.16E+01
0.40	0.70	5.45E-01	2.26E+00
		3.57E-05	1.28E-02
		5.83E+00	1.11E+01
0.25	0.60	2.55E-01	1.14E+00
		7.95E-05	3.36E-03
		4.49E+01	1.41E+01
0.40	0.60	3.69E-01	1.39E+00
		7.99E-06	4.52E-04
		3.26E+00*	1.28E+00*
0.25	0.50	1.59E-01	8.01E-01
		2.02E-05	1.99E-03
		2.77E+01	2.35E+01
0.40	0.50	2.54E-01	9.55E-01
		4.36E-06	1.73E-04
		3.76E+00	1.33E+00
0.25	0.40	9.22E-02	5.51E-01
		3.56E-06	3.83E-04
		1.17E+01	9.03E+00
0.40	0.40	1.73E-01	6.94E-01
		4.30E-06	1.24E-04
		6.99E+00	2.01E+00
0.25	0.30	4.59E-02	3.89E-01
		1.23E-06	8.64E-05
		1.10E+01	2.87E+00
0.40	0.30	1.12E-01	5.33E-01
		6.65E-06	3.21E-04
		2.93E+01	6.29E+00

but with $\rho_{02}=0.8$ replaced by $\rho_{02}=0.9$. The results of the tables are self-explanatory. Extensive simulation studies have led to a simple practical suggestion for estimating the response surface, $\ell(\boldsymbol{v})$, $\boldsymbol{v} = \boldsymbol{v}(\boldsymbol{\rho})$, $\boldsymbol{\rho} = (\rho_1, \rho_2, \ldots, \rho_r)$, for various values $\boldsymbol{v} = \boldsymbol{v}(\boldsymbol{\rho})$. It is recommended to choose the reference traffic intensity at each queue, j, as $\rho_{0j} = \max_{1,\ldots,m}(\rho_{j1}, \rho_{j2}, \ldots, \rho_{jm})$, where m is the number of what-if scenarios in the response function.

Table 8.3.3 $\hat{\sigma}^2(\rho|\rho_0)$ and $\bar{\varepsilon}^0(\rho|\rho_0)$ as Functions of ρ_1 and ρ_2, for Two $GI/G/1$ Queues in Tandem with $\rho = (\rho_{01}, \rho_{02}) = (0.4, 0.8)$

ρ_1	ρ_2	$\hat{\sigma}^2(\rho\|\rho_0)$	$\bar{\varepsilon}^0(\rho\|\rho_0)$
0.25	0.70	4.00E-01	1.96E+00
		1.73E-05	3.40E-03
		3.48E+00	3.30E+00
0.40	0.70	5.47E-01	2.25E+00
		4.85E-06	3.28E-04
		7.92E-01*	2.85E-01*
0.25	0.60	2.49E-01	1.15E+00
		3.73E-06	2.73E-04
		2.11E+00	1.15E+00
0.40	0.60	3.71E-01	1.38E+00
		3.36E-06	1.20E-04
		1.37E+00	3.40E-01
0.25	0.50	1.56E-01	7.51E-01
		1.83E-06	6.29E-05
		2.50E+00	7.44E-01
0.40	0.50	2.57E-01	9.49E-01
		3.93E-06	1.06E-04
		3.39E+00	8.13E-01
0.25	0.40	9.31E-02	5.32E-01
		1.22E-06	2.87E-05
		4.01E+00	6.77E-01
0.40	0.40	1.75E-01	7.04E-01
		6.04E-06	1.91E-04
		9.81E+00	3.11E+00
0.25	0.30	4.72E-02	3.95E-01
		4.81E-07	4.70E-05
		4.30E+00	1.56E+00
0.40	0.30	1.13E-01	5.54E-01
		1.16E-05	3.86E-04
		3.37E+01	7.56E+00

Table 8.3.4 $\hat{\sigma}^2(\rho|\rho_0)$ **and** $\bar{\varepsilon}^0(\rho|\rho_0)$ **as Functions of** ρ_1 **and** ρ_2**, for Two** ***GI/G/1*** **Queues in Tandem with** $\rho = (\rho_{01}, \rho_{02}) = (0.6, 0.9)$

ρ_1	ρ_2	$\hat{\sigma}^2(\rho\|\rho_0)$	$\bar{\varepsilon}^0(\rho\|\rho_0)$
0.25	0.70	4.07E-01	2.01E+00
		3.62E-05	3.88E-03
		7.25E+00	3.77E+00
0.40	0.70	5.45E-01	2.22E+00
		1.31E-05	5.03E-04
		2.14E+00*	4.37E-01*
0.25	0.60	2.53E-01	1.17E+00
		1.32E-05	5.59E-04
		7.46E+00	2.35E+00
0.40	0.60	3.72E-01	1.36E+00
		1.28E-05	3.58E-04
		5.22E+00	1.02E+00
0.25	0.50	1.58E-01	7.80E-01
		6.65E-06	2.30E-04
		9.10E+00	2.72E+00
0.40	0.50	2.58E-01	9.62E-01
		1.57E-05	3.67E-04
		1.35E+01	2.82E+00
0.25	0.40	9.23E-02	5.48E-01
		3.74E-06	1.15E-04
		1.23E+01	2.71E+00
0.40	0.40	1.74E-01	7.25E-01
		2.28E-05	5.05E-04
		3.71E+01	8.22E+00
0.25	0.30	4.58E-02	3.91E-01
		2.18E-06	8.72E-05
		1.95E+01	2.90E+00
0.40	0.30	1.11E-01	5.62E-01
		3.72E-05	1.00E-03
		1.08E+02	1.96E+01

*8.4 ESTIMATING THE OPTIMAL REFERENCE PARAMETER

In this section we show how to estimate the optimal reference parameter vector, $\mathbf{v}_0^*$, separately for (i) static and (ii) dynamic models.

8.4.1 Static Models

We start with the case where the parameter vector $\boldsymbol{v}$ is fixed and then extend our results to the case where $\boldsymbol{v}$ varies. Consider the function $\mathscr{L}^k(\boldsymbol{v}, \boldsymbol{v}_0)$ in (8.2.3) for $k=0$ only. The case $k=1$ can be treated similarly and is left as an exercise.

The program (8.2.3) (for $k=0$) can be written as

$$\min_{\boldsymbol{v}_0 \in V}\{\mathscr{L}^0(\boldsymbol{v}, \boldsymbol{v}_0) = \mathbb{E}_{\boldsymbol{v}_0}[L^2(\boldsymbol{Z})W^2(\boldsymbol{Z}, \boldsymbol{v}, \boldsymbol{v}_0)]\}. \tag{8.4.1}$$

or alternatively as

$$\min_{\boldsymbol{v}_0 \in V}\{\mathscr{L}^0(\boldsymbol{v}, \boldsymbol{v}_0) = \mathbb{E}_{\boldsymbol{v}}[L^2(\boldsymbol{Y})W(\boldsymbol{Y}, \boldsymbol{v}, \boldsymbol{v}_0)]\}, \tag{8.4.2}$$

where $\boldsymbol{Z} \sim f(\mathbf{y}, \boldsymbol{v}_0)$ and $\boldsymbol{Y} \sim f(\mathbf{y}, \boldsymbol{v})$, respectively.

Given a sample $\boldsymbol{Z}_1, \ldots, \boldsymbol{Z}_N$ from $f(\mathbf{z}, \boldsymbol{v}_0)$, we can estimate the optimal solution $\boldsymbol{v}_0^*$ of (8.4.1) by an optimal solution of the program

$$\min_{\boldsymbol{v}_0 \in V}\left\{\bar{\mathscr{L}}_N(\boldsymbol{v}, \boldsymbol{v}_0) = \sum_{i=1}^{N} L^2(\boldsymbol{Z}_i)W^2(\boldsymbol{Z}_i, \boldsymbol{v}, \boldsymbol{v}_0)\right\}. \tag{8.4.3}$$

To obtain a reasonable approximation of $\boldsymbol{v}_0^*$ of the program (8.4.3) we can use stochastic approximation or Newtons-type algorithms starting from $\boldsymbol{v}_0 = \boldsymbol{v}$.

We next present a slightly different approach to estimating $\boldsymbol{v}_0^*$, which is based on (8.4.2) and as will be seen below has some attractive properties. Since $\mathscr{L}^0(\boldsymbol{v}, \boldsymbol{v}_0)$ in (8.4.2) is represented under the original probability measure $f(\mathbf{y}, \boldsymbol{v})$, using the standard likelihood ratio (LR) approach we can rewrite the program (8.4.2) as

$$\min_{\boldsymbol{v}_0 \in V}\{\mathscr{L}^0(\boldsymbol{v}, \boldsymbol{v}_0) = \mathbb{E}_{\boldsymbol{v}_1}[L^2(\boldsymbol{Z})W(\boldsymbol{Z}, \boldsymbol{v}, \boldsymbol{v}_0)W(\boldsymbol{Z}, \boldsymbol{v}, \boldsymbol{v}_1)]\}. \tag{8.4.4}$$

Here, as usual, $f(\mathbf{y}, \boldsymbol{v}_1)$ dominates the pdf $f(\mathbf{y}, \boldsymbol{v})$ in the absolutely continuous sense,

$$W(\boldsymbol{Z}, \boldsymbol{v}, \boldsymbol{v}_1) = \frac{f(\boldsymbol{Z}, \boldsymbol{v})}{f(\boldsymbol{Z}, \boldsymbol{v}_1)},$$

and $\boldsymbol{Z} \sim f(\mathbf{y}, \boldsymbol{v}_1)$. Note also that the program (8.4.1) presents a particular case of the program (8.4.4) since it can be written as

$$\min_{\boldsymbol{v}_0 \in V}\{\mathscr{L}^0(\boldsymbol{v}, \boldsymbol{v}_0) = \mathbb{E}_{\boldsymbol{v}_0}[L^2(\boldsymbol{Z})W(\boldsymbol{Z}, \boldsymbol{v}, \boldsymbol{v}_0)W(\boldsymbol{Z}, \boldsymbol{v}, \boldsymbol{v}_0)]\}. \tag{8.4.5}$$

Given a sample $\{\boldsymbol{Z}_1, \ldots, \boldsymbol{Z}_N\}$ from $f(\mathbf{z}, \boldsymbol{v}_1)$, we can estimate the optimal solution (8.4.4) [and therefore that of $\boldsymbol{v}_0^*$ of (8.4.1)] by an optimal solution of the program

$$\min_{\boldsymbol{v}_0 \in V}\left\{\bar{\mathscr{L}}_N^0(\boldsymbol{v}, \boldsymbol{v}_0, \boldsymbol{v}_1) = \sum_{i=1}^{N} L^2(\boldsymbol{Z}_i)W(\boldsymbol{Z}_i, \boldsymbol{v}, \boldsymbol{v}_0)W(\boldsymbol{Z}_i, \boldsymbol{v}, \boldsymbol{v}_1)\right\}. \tag{8.4.6}$$

Note that as soon as the sample $\{\boldsymbol{Z}_1, \ldots, \boldsymbol{Z}_N\}$ is generated, $\bar{\mathscr{L}}_N^0(\boldsymbol{v}, \boldsymbol{v}_0, \boldsymbol{v}_1)$ becomes a deterministic function of $\boldsymbol{v}_0$, and consequently the program (8.4.6) can be solved by standard methods of mathematical programming. The optimal solution, $\bar{\boldsymbol{v}}_{0N}^*$, of the program (8.4.6) yields an estimator of the optimal value, $\boldsymbol{v}_0^*$, of the reference parameter vector.

From the results of Rubinstein and Shapiro (1993) it follows that when the considered pdf, $f(\boldsymbol{y}, \boldsymbol{v})$, is from an exponential family in the canonical form, then both functions, $\mathscr{L}^0(\boldsymbol{v}, \boldsymbol{v}_0)$ and $\hat{\mathscr{L}}_N^0(\boldsymbol{v}, \boldsymbol{v}_0, \boldsymbol{v}_1)$, are convex in $\boldsymbol{v}_0$; consequently, the stochastic counterpart (8.4.6) is a convex program, provided the parameter set, V, is also convex. Due to the convexity of (8.4.6), one can find its optimal solution, $\bar{\boldsymbol{v}}_{0N}^*$, from the solution of the following set of equations

$$\nabla_{\boldsymbol{v}_0} \hat{\mathscr{L}}_N(\boldsymbol{v}, \boldsymbol{v}_0, \boldsymbol{v}_1) = -\sum_{i=1}^{N} L^2(\boldsymbol{Z}_i) W^2(\boldsymbol{Z}_i, \boldsymbol{v}, \boldsymbol{v}_0) \nabla_{\boldsymbol{v}_0} W(\boldsymbol{Z}_i, \boldsymbol{v}_0, \boldsymbol{v}_1) = \boldsymbol{0}.$$

Clearly, the accuracy of the optimal solution, $\bar{\boldsymbol{v}}_{0N}^*$, depends on the particular choice of $\boldsymbol{v}_1$. Under mild regularity conditions (see Rubinstein and Shapiro, 1993) the vector $N^{1/2}(\bar{\boldsymbol{v}}_{0N}^* - \boldsymbol{v}_0^*)$ converges in distribution to multivariate normal with zero mean and covariance matrix $\boldsymbol{B}^{-1}\boldsymbol{\Sigma}\boldsymbol{B}^{-1}$, where $\boldsymbol{B}$ is the Hessian of the function $\bar{\mathscr{L}}^0(\boldsymbol{v}, \cdot)$, calculated at $\boldsymbol{v}_0 = \boldsymbol{v}_0^*$, and $\boldsymbol{\Sigma}$ is the covariance matrix given by

$$\boldsymbol{\Sigma} = \mathbb{E}_{\boldsymbol{v}_1}\{L^4(\boldsymbol{Z}) W^4(\boldsymbol{Z}, \boldsymbol{v}, \boldsymbol{v}_0^*) \nabla_{\boldsymbol{v}_0} W(\boldsymbol{Z}, \boldsymbol{v}_0^*, \boldsymbol{v}_1) \nabla_{\boldsymbol{v}_0} W(\boldsymbol{Z}, \boldsymbol{v}_0^*, \boldsymbol{v}_1)'\}, \tag{8.4.7}$$

where $\boldsymbol{Z} \sim f(\mathbf{z}, \boldsymbol{v}_1)$. Both $\boldsymbol{B}$ and $\boldsymbol{\Sigma}$ can be estimated from the generated sample, and an approximate confidence region for $\boldsymbol{v}_0^*$ can be constructed.

We shall now show that (8.4.6) has certain advantages over (8.4.3). First, note that the choice of $\boldsymbol{v}_1$ in (8.4.6) is not as crucial as that of the reference parameter vector $\boldsymbol{v}_0$ in (8.4.3), since (8.4.6) serves only as an auxiliary tool for identifying a good value of $\boldsymbol{v}_0$, and not for sensitivity analysis or optimization of $\ell(\boldsymbol{v})$. Note next that good choices of $\boldsymbol{v}_1$ and $\boldsymbol{v}_0$ in (8.4.6) and (8.4.3) would be those that render, say, the trace of their corresponding asymptotic covariance matrix, $\boldsymbol{B}^{-1}\boldsymbol{\Sigma}\boldsymbol{B}^{-1}$ of the estimator $\bar{\boldsymbol{v}}_{0N}^*$, as small as possible. Program (8.4.6) is, however, more flexible and better suited for minimizing the trace than (8.4.3). To see that, note first that the covariance matrix $\boldsymbol{\Sigma}$, associated with the optimal solution $\hat{\boldsymbol{v}}_{0N}^*$ of (8.4.3), is given in analogy to (8.4.7) by

$$\boldsymbol{\Sigma} = \mathbb{E}_{\boldsymbol{v}_0}\{L^4(\boldsymbol{Z}) W^4(\boldsymbol{Z}, \boldsymbol{v}, \boldsymbol{v}_0^*) \nabla_{\boldsymbol{v}_0} W(\boldsymbol{Z}, \boldsymbol{v}, \boldsymbol{v}_0^*) \nabla_{\boldsymbol{v}_0} W(\boldsymbol{Z}, \boldsymbol{v}, \boldsymbol{v}_0^*)'\}, \tag{8.4.8}$$

where $\boldsymbol{Z} \sim f(\mathbf{z}, \boldsymbol{v}_0)$.

Assume next that $\boldsymbol{v}_1 = \boldsymbol{v}_0^*$. By virtue of the equality (see Chapter 5)

$$\nabla_{\boldsymbol{v}_0} W(\boldsymbol{Z}, \boldsymbol{v}_0^*, \boldsymbol{v}_1) = W(\boldsymbol{Z}, \boldsymbol{v}_0^*, \boldsymbol{v}_1) \boldsymbol{S}(\boldsymbol{Z}, \boldsymbol{v}_0^*),$$

where $\boldsymbol{S}(\boldsymbol{Z}, \boldsymbol{v}) = \nabla_{\boldsymbol{v}} \log f(\boldsymbol{Z}, \boldsymbol{v})$ is the score function, we obtain for $\boldsymbol{v}_1 = \boldsymbol{v}_0^*$

$$\nabla_{\boldsymbol{v}_0} W(\boldsymbol{Z}, \boldsymbol{v}_0^*, \boldsymbol{v}_1) = S(\boldsymbol{Z}, \boldsymbol{v}_1).$$

It follows from the above that if the variability of $L(\boldsymbol{Z})$ is not large compared with the variability of the corresponding likelihood ratio functions, then $\boldsymbol{v}_1 = \boldsymbol{v}_0^*$ could be a good choice of $\boldsymbol{v}_1$ in the sense that the trace of of $\boldsymbol{\Sigma}$ is small. Clearly, the stochastic counterpart (8.4.3) does not possess such flexibility. However, $\boldsymbol{v}_0^*$ is not known. This motivates the presentation of the following iterative procedure for choosing a good parameter vector $\boldsymbol{v}_1$ (in the sense of convergence to $\boldsymbol{v}_0^*$).

Algorithm 8.4.1 Choice of a Good $\boldsymbol{v}_1$ and Optimal $\boldsymbol{v}_0^*$ for Static Models

1. *Choose an initial point $\boldsymbol{v}_1(0)$, say $\boldsymbol{v}_1(0) = \boldsymbol{v}$.*
2. *Generate a sample $\{\boldsymbol{Z}_1, \ldots, \boldsymbol{Z}_N\}$ from the pdf $f(\boldsymbol{y}, \boldsymbol{v}_1)$ and solve the stochastic counterpart (8.4.6). Denote the solution by $\bar{\boldsymbol{v}}_0^*(0)$.*
3. *Set $\boldsymbol{v}_1(1) = \bar{\boldsymbol{v}}_0^*(0)$ and repeat step 2. Denote the solution by $\bar{\boldsymbol{v}}_0^*(1)$.*
4. *Proceed with steps 2 and 3 until the desired precision is achieved.*

Empirical experience with simulation runs of rather general static models using Algorithm 8.4.1 shows that they yield reasonably good results after 2–3 iterations, in the sense that $\boldsymbol{v}_1$ converges fast to $\boldsymbol{v}_0^*$. These simulation studies used samples of size $N = 1000$ and employed the same stream of random numbers at each step of Algorithm 8.4.1.

As far as the initial point $\boldsymbol{v}_1(0)$ is concerned, one can either take $\boldsymbol{v}_1(0) = \boldsymbol{v}$ (as suggested in step 1), or again use Newton's algorithm with $\boldsymbol{v}_0(1)$ replaced by $\boldsymbol{v}_1(0)$. Alternatively, one may conduct a line search (with respect to α) in the following one-step gradient procedure

$$\boldsymbol{v}_1(0) = \boldsymbol{v} - \alpha \nabla_{\boldsymbol{v}_1} \hat{\mathscr{L}}_N(\boldsymbol{v}, \boldsymbol{v}_0, \boldsymbol{v}_1),$$

with $\boldsymbol{v} = \boldsymbol{v}_0 = \boldsymbol{v}_1$.

To provide more insight into the choice of $\boldsymbol{v}_1$, consider the following examples.

Example 8.4.1. Suppose we wish to estimate $\ell(\lambda) = P_\lambda(Y > x) = \mathbb{E}_\lambda\{I_{\{Y>x\}}\}$, where $Y \sim \text{Exp}(\lambda)$. In this case, we would like to choose λ_1 so as to make the asymptotic variance of $\hat{\lambda}_0^*$ as small as possible. Since the corresponding matrix, $\boldsymbol{B}$, reduces to a scalar, the asymptotic variance is proportional to the function

$$\begin{aligned}\hat{\mathscr{R}}^0(\lambda, \lambda_0, \lambda_1) &= \mathbb{E}_{\lambda_1}\{I_{\{L<x\}} W^4(\lambda, \lambda_0)[\partial W(\lambda_0, \lambda_1)/\partial \lambda_0]^2\} \\ &= \frac{\lambda^4}{\lambda_0^2 \lambda_1} \int_x^\infty \left(\frac{1}{\lambda_0} - z\right)^2 e^{-(4\lambda - 2\lambda_0 - \lambda_1)z} dz,\end{aligned}$$

where $\lambda_0 = \lambda_0^*$. Although even in this simple example it is not easy to write a closed-form expression for the optimal value of λ_1 which minimizes the function $\hat{\mathscr{R}}^0(\lambda, \lambda_0, \cdot)$, we can simplify the problem by minimizing the variance of the estimator $\mathscr{L}_N^0(\lambda, \lambda_0, \cdot)$. This is equivalent to minimizing the function

$$\begin{aligned}\mathscr{R}^0(\lambda, \lambda_0, \lambda_1) &= \mathbb{E}_{\lambda_1}\{I_{\{L<x\}} W^4(\lambda, \lambda_0) W^2(\lambda_0, \lambda_1)\} \\ &= \frac{e^{-(4\lambda - 2\lambda_0 - \lambda_1)x}}{\lambda_1(4\lambda - 2\lambda_0 - \lambda_1)}\end{aligned}$$

with respect to λ_1. Observe first that for fixed λ and λ_0, the function $\mathscr{R}^0(\lambda, \lambda_0, \lambda_1)$ tends to infinity as λ_1 approaches $4\lambda - 2\lambda_0$, and that the minimizer $\lambda_1^* = \lambda_1^*(x)$ of this function is given by

$$\lambda_1^*(x) = 2\lambda - \lambda_0 + \frac{1}{x} - \left((2\lambda - \lambda_0)^2 + \frac{1}{x^2}\right)^{1/2}.$$

The minimizer $\lambda_1^*(x)$ behaves similarly to the minimizer $\lambda_0^*(x)$, defined in Example 5.3.2. It satisfies $\lambda_1^* < 2\lambda - \lambda_0$, and for large x, $\lambda_1^*(x) \approx 1/x$. That is, $\lambda_1^*(x)$ coincides asymptotically with $\lambda_0^*(x)$ of Example 5.3.2. Note also that $P_{\lambda_1^*(x)}\{L > x\} = P_{\lambda_0^*(x)}\{L > x\} = e^{-1}$. Let

$$\varepsilon^0(\lambda, \lambda_0, \lambda_1) = \frac{\mathscr{R}^0(\lambda, \lambda_0, \lambda_1)}{\mathscr{R}^0(\lambda, \lambda, \lambda)}$$

be the relative efficiency. Then,

$$\varepsilon^0(\lambda, \lambda_0, \lambda_1) = \frac{\lambda^4 e^{-(3\lambda - 2\lambda_0 - \lambda_1)x}}{\lambda_0^2 \lambda_1 (4\lambda - 2\lambda_0 - \lambda_1)}.$$

In view of the fact that for large x one has $\lambda_0^* = \lambda_1^* \approx 1/x$, the efficiencies $\varepsilon^0(\lambda, \lambda_0^*, \lambda_1)$ and $\varepsilon^0(\lambda, \lambda_0^*, \lambda_1^*)$ reduce to

$$\varepsilon^0(\lambda, \lambda_0^*, \lambda_1) \approx \frac{x^2 \lambda^4}{(\lambda - \lambda_1)\lambda_1} e^{2-(3\lambda - \lambda_1)x}$$

and

$$\varepsilon^0(\lambda, \lambda_0^*, \lambda_1^*) \approx 0.25 x^3 \lambda^3 e^{3 - 3\lambda x},$$

respectively. For $\lambda = 1$ and $\lambda_0^* = \lambda_1^* = \frac{1}{12}$, one has $\varepsilon^0(1, \frac{1}{12}, \frac{1}{12}) \approx 2 \times 10^{-6}$.

Table 8.4.1 displays $\varepsilon^0(\lambda, \lambda_0, \lambda_1)$ as a function of λ_0 and λ_1, for $\lambda = 1$ and $x = 12$. Notice that values marked by an asterisk correspond to row minima, obtained as the solution of the program

$$\min_{\lambda_0} \mathscr{R}^0(\lambda, \lambda_0, \lambda_1).$$

It readily follows from Table 8.4.1 that the optimal parameter, λ_0^*, is rather insensitive to λ_1 in the sense that there exists a broad interval of λ_1 (in our case, $0.001 \le \lambda_1 \le 0.5$), such that λ_0^* may be reliably estimated. A numerical solution of the equation

$$\varepsilon^0(\lambda, \lambda_0^*, \hat{\lambda}_1) = 1$$

(see Table 8.4.1) for $\lambda = 1$ yields the following two roots: $\hat{\lambda}_{11} \approx 6.44 \times 10^{-14}$ and $\hat{\lambda}_{12} \approx 2.519$; consequently, the variance reduction interval is $(\hat{\lambda}_{11}, \hat{\lambda}_{12}) \approx (6.44 \times 10^{-14}, 2.519)$. For $\lambda = 2$, one obtains $(\hat{\lambda}_{11}, \hat{\lambda}_{12}) \approx (3.86 \times 10^{-27}, 5.40)$.

Table 8.4.1 $\epsilon^0(\lambda, \lambda_0^*, \lambda_1)$ as Function of λ_1, for $\lambda = 1$ and $x = 12$

	λ_1					
λ_0	1*E*-14	1*E*-10	0.001	0.05	$\lambda_1^* = \frac{1}{12}$	0.1
1E-18	5.8E+33	5.8E+29	5.8E+22	2.1E+21	1.9E+21*	1.9E+21
1E-14	5.8E+25	5.8E+21	5.8E+14	2.1E+13	1.9E+13*	1.9E+13
1E-10	5.8E+17	5.8E+13	5.8E+06	2.1E+05	1.9E+05*	1.9E+05
0.001	5.9E+03	5.9E-01	6.0E-08	2.1E-09	1.9E-09*	2.0E-09
0.05	7.9E+00	7.9E-04	8.0E-11	2.9E-12	2.6E-12*	2.6E-12
0.07	6.5E+00	6.5E-04	6.6E-11	2.4E-12	2.1E-12*	2.2E-12
$\lambda_0^* = \frac{1}{12}$	6.4E+00	6.4E-04	6.5E-11	2.3E-12	2.1E-12*	2.1E-12
0.09	6.5E+00	6.5E-04	6.5E-11	2.4E-12	2.1E-12	2.2E-12
0.10	6.7E+00	6.7E-04	6.8E-11	2.4E-12	2.2E-12*	2.2E-12
0.30	1.0E+02	1.0E-02	1.0E-09	3.7E-11	3.4E-11*	3.4E-11
0.50	5.0E+03	5.0E-01	5.1E-08	1.8E-09	1.6E-09*	1.7E-09
0.70	3.6E+05	3.6E+01	3.6E-06	1.3E-07	1.2E-07*	1.2E-07
0.90	3.1E+07	3.1E+03	3.1E-04	1.1E-05	1.0E-05*	1.0E-05
1.00	3.0E+08	3.0E+04	3.1E-03	1.1E-04	1.0E-04*	1.0E-04

Example 8.4.2 Suppose we wish to estimate $\ell(\lambda) = \mathbb{E}_\lambda\{L(Y)\}$, where $L(Y) = Y^{1/2}$ and $Y \sim \text{Exp}(\lambda)$. Then,

$$\begin{aligned}\mathscr{R}^0(\lambda, \lambda_0, \lambda_1) &= \mathbb{E}_{\lambda_1}\{L^4(Y)W^4(\lambda, \lambda_0)W^2(\lambda_0, \lambda_1)\} \\ &= \frac{\lambda^4}{\lambda_0^2\lambda_1(4\lambda - 2\lambda_0 - \lambda_1)}\mathbb{E}_{4\lambda-2\lambda_0-\lambda_1}\{L^4(Y)\}\,.\end{aligned}$$

Since $L^4(Y) = Y^2$ and

$$\mathbb{E}_{4\lambda-2\lambda_0-\lambda_1}\{Y^2\} = \frac{2}{(4\lambda - 2\lambda_0 - \lambda_1)^2},$$

one has

$$\mathscr{R}^0(\lambda, \lambda_0, \lambda_1) = \frac{2\lambda^4}{\lambda_0^2\lambda_1(4\lambda - 2\lambda_0 - \lambda_1)^3}.$$

It is not difficult to see that for fixed λ and λ_0, the minimizer, λ_1^* of $\mathscr{R}^0(\lambda, \lambda_0, \lambda_1)$, is given by

$$\lambda_1^*(\lambda_0) = \lambda - 0.5\lambda_0.$$

Noting that for fixed λ, the minimizer, λ_0^* of $\mathscr{L}^0(\lambda, \lambda_0)$, equals $\lambda_0^* = \frac{2}{3}$, it follows that $\lambda_1^*(\lambda_0^*) = \lambda_0^* = \frac{2}{3}$. That is, λ_1^* and λ_0^* coincide again.

The relative efficiencies are

$$\varepsilon^0(\lambda, \lambda_0, \lambda_1) = \frac{\lambda^6}{\lambda_0^2 \lambda_1 (4\lambda - 2\lambda_0 - \lambda_1)^3},$$
$$\varepsilon^0(\lambda, \lambda_0^*, \lambda_1) = \frac{(\frac{3}{2})^2 \lambda^4}{\lambda_1 (2\frac{2}{3}\lambda - \lambda_1)^3},$$
$$\varepsilon^0(\lambda, \lambda_0^*, \lambda_1^*) = (\tfrac{3}{4})^3 = 0.42,$$

the latter being independent of λ.

Table 8.4.2 displays $\varepsilon^0(\lambda, \lambda_0, \lambda_1)$ as a function of λ_1, and λ_0, for $\lambda = 1$. Here, $\mathscr{M}$ corresponds to the case $4\lambda - 2\lambda_0 - \lambda_1 \le 0$, that is, where $\mathscr{R}^0$ and ε^0 in Example 8.4.2 are undefined. It follows again from Table 8.4.2 that the optimal parameter, λ_0^*, is rather insensitive with respect to λ_1, in the sense that there exists a broad interval about λ_1 (in our case $0.3 \le \lambda_1 \le 1.0$), such that λ_0^* may be reliably estimated from the solution of the program $\min_{\lambda_0} \mathscr{R}^0(\lambda, \lambda_0, \lambda_1)$. A numerical solution of the equation

$$\varepsilon^0(\lambda, \lambda_0^*, \hat{\lambda}_1) = 1$$

(see Table 8.4.2) for the pair ($\lambda = 1$, $\lambda_0^* = \frac{2}{3}$) yields the following two roots: $\hat{\lambda}_{11} \approx 0.139$ and $\hat{\lambda}_{12} \approx 1.529$. Thus, the variance reduction interval is $(\hat{\lambda}_{11}, \hat{\lambda}_{12}) \approx (0.139, 1.529)$. For $\lambda = 2$, one has $(\hat{\lambda}_{11}, \hat{\lambda}_{12}) \approx (0.279, 3.059)$.

Consider now estimation of $\boldsymbol{v}_0^*$ for the program (8.3.1), that is, where $\boldsymbol{v}$ is allowed to vary. It is readily seen that in this case we can again apply algorithm

Table 8.4.2 $\boldsymbol{\varepsilon}^0(\boldsymbol{\lambda}, \boldsymbol{\lambda}_0, \boldsymbol{\lambda}_1)$ Function of λ_0 and λ_1, for $\lambda = 1$

	λ_0							
λ_1	0.1	0.3	0.5	$\lambda_0^* = \frac{2}{3}$	0.8	1.0	1.2	1.6
0.1	1.97	3.09	1.64	1.33	1.28*	1.46	2.06	11.4
0.2	1.07	1.70	0.91	0.75	0.73*	0.85	1.27	9.04
0.3	7.77	1.24	0.67	0.56*	0.56	0.67	1.05	10.4
0.4	6.36	1.03	0.56	0.48*	0.48	0.61	1.00	15.3
0.5	5.57	0.91	0.51	0.44*	0.45	0.59	1.04	28.9
0.6	5.09	0.84	0.48	0.42*	0.44	0.60	1.16	81.4
$\lambda_1^* = \frac{2}{3}$	4.88	0.81	0.47	0.42*	0.45	0.63	1.28	24.7
0.8	4.63	0.79	0.47	0.43*	0.47	0.72	1.70	$\mathscr{M}$
0.9	4.56	0.79	0.48	0.45*	0.51	0.83	2.25	$\mathscr{M}$
1.0	4.56	0.80	0.50	0.48*	0.56	1.00	3.22	$\mathscr{M}$
1.1	4.62	0.83	0.53*	0.53	0.64	1.25	5.05	$\mathscr{M}$
1.2	4.74	0.87	0.57*	0.59	0.75	1.63	9.04	$\mathscr{M}$
1.3	4.92	0.92	0.62*	0.67	0.90	2.24	19.8	$\mathscr{M}$

(8.4.1) with $\hat{\mathscr{L}}_N^0(\boldsymbol{v}, \boldsymbol{v}_0, \boldsymbol{v}_1)$ in (8.4.6) replaced by $\max_{j=1,\ldots,s} \hat{\mathscr{L}}_N^0(\boldsymbol{v}_j, \boldsymbol{v}_0, \boldsymbol{v}_1)$, and all other data remaining intact.

8.4.2 Queueing Models

Consider the programs (8.2.9), (8.2.10). For static models we shall use here an approach involving an auxiliary importance sampling (IS) pdf $f(\boldsymbol{y}, \boldsymbol{v}_1)$. Arguing as in (8.4.4), the stochastic counterpart, based on $f(\boldsymbol{y}, \boldsymbol{v}_1)$, can be written (similarly to (8.4.6)) as

$$\min_{\boldsymbol{v}_0 \in V} \hat{\mathscr{L}}_{1N}^0(\boldsymbol{v}, \boldsymbol{v}_0, \boldsymbol{v}_1)$$
$$= \min_{\boldsymbol{v}_0 \in V} \left\{ \frac{1}{N} \sum_{i=1}^{N} \left[\sum_{t=1}^{\tau_i} L_{ti}(\underline{\boldsymbol{Z}}_{ti}) \tilde{W}_{ti}(\underline{\boldsymbol{Z}}_{ti}, \boldsymbol{v}, \boldsymbol{v}_0) \right]^2 \tilde{W}_{\tau_i i}(\underline{\boldsymbol{Z}}_{\tau_i i}, \boldsymbol{v}_0, \boldsymbol{v}_1) \right\}, \tag{8.4.9}$$

where

$$\tilde{W}_{\tau_i i}(\underline{\boldsymbol{Z}}_{\tau_i}, \boldsymbol{v}_0, \boldsymbol{v}_1) = \prod_{t=1}^{\tau_i} \frac{f(\boldsymbol{Z}_{ti}, \boldsymbol{v}_0)}{f(\boldsymbol{Z}_{ti}, \boldsymbol{v}_1)},$$

and $\{\boldsymbol{Z}_{11}, \ldots, \boldsymbol{Z}_{\tau_1 1}, \ldots, \boldsymbol{Z}_{1N}, \ldots, \boldsymbol{Z}_{\tau_N N}\}$ is a sample of N regenerative cycles, generated from the auxiliary pdf $f(\mathbf{z}, \boldsymbol{v}_1)$.

Using the equality (see Rubinstein and Shapiro, 1993)

$$\mathbb{E}_{\boldsymbol{v}} \left\{ \left(\sum_{t=1}^{\tau} L_t \right) \left(\sum_{t=1}^{\tau} \tilde{W}_t \right) \right\} = \mathbb{E}_{\boldsymbol{v}} \left\{ \sum_{t=1}^{\tau} L_t \tilde{W}_t \right\},$$

we obtain, after some manipulation of (8.4.9), the alternative estimator

$$\min_{\boldsymbol{v}_0 \in V} \bar{\mathscr{L}}_{1N}^0(\boldsymbol{v}, \boldsymbol{v}_0, \boldsymbol{v}_1)$$
$$= \min_{\boldsymbol{v}_0 \in V} \left\{ \frac{1}{N} \sum_{i=1}^{N} \left[\sum_{t=1}^{\tau_i} L_{ti}^2(\underline{\boldsymbol{Z}}_{ti}) \tilde{W}_{ti}^2(\underline{\boldsymbol{Z}}_{ti}, \boldsymbol{v}, \boldsymbol{v}_0) \tilde{W}_{ti}(\underline{\boldsymbol{Z}}_{ti}, \boldsymbol{v}_0, \boldsymbol{v}_1) \right.\right.$$
$$\left.\left. + 2 \sum_{\substack{s=1 \\ s<t}}^{\tau_i} \sum_{t=1}^{\tau_i} L_{si}(\underline{\boldsymbol{Z}}_{si}) L_t(\underline{\boldsymbol{Z}}_{ti}) \tilde{W}_{si}(\underline{\boldsymbol{Z}}_{si}, \boldsymbol{v}, \boldsymbol{v}_0) \tilde{W}_{ti}(\underline{\boldsymbol{Z}}_{ti}, \boldsymbol{v}, \boldsymbol{v}_0) \tilde{W}_{ti}(\underline{\boldsymbol{Z}}_{ti}, \boldsymbol{v}_0, \boldsymbol{v}_1) \right] \right\}. \tag{8.4.10}$$

The relationship between the two LR estimators, $\hat{\mathscr{L}}_{1N}^0$ and $\bar{\mathscr{L}}_{1N}^0$, is discussed in Asmussen and Rubinstein (1992), where it is shown that the estimator $\bar{\mathscr{L}}_{1N}^0$ is typically more accurate than the estimator $\hat{\mathscr{L}}_{1N}^0$, in the sense that the former has a smaller variance.

With the above results we can estimate the optimal reference parameter $\boldsymbol{v}_0^*$ of the programs (8.2.9), (8.2.10) by modifying the (static) Algorithm 8.4.1 as follows: Replace (8.4.1) and (8.4.6) in step 2 of Algorithm 8.4.1 by (8.2.9) and (8.4.10), respectively, with all other data remaining intact.

Finally, consider the regenerative estimator

$$\bar{\ell}_N(\boldsymbol{v}) = \bar{\ell}_{1N}(\boldsymbol{v})/\bar{\ell}_{2N}(\boldsymbol{v}).$$

Its asymptotic variance is $1/N$ times the variance

$$\sigma^2(\boldsymbol{v}, \boldsymbol{v}_0) = a^2 \operatorname{var}_{\boldsymbol{v}_0}\left\{\sum_{t=1}^{\tau} M_t(\underline{\boldsymbol{Z}}_t)\tilde{W}_t(\underline{\boldsymbol{Z}}_t, \boldsymbol{v})\right\},$$

where $a = 1/\ell_2(\boldsymbol{v})$, and $M_t = L_t - \bar{\ell}_N(\boldsymbol{v})$. In this case, (8.4.9)–(8.4.11) hold again, but with L_t and L_{ti} replaced by $M_t = L_t - \bar{\ell}_N(\boldsymbol{v})$ and $M_{ti} = L_{ti} - \bar{\ell}_N(\boldsymbol{v})$, respectively, where

$$\bar{\ell}_N(\boldsymbol{v}) = \frac{\sum_{i=1}^{N}\sum_{t=1}^{\tau_i} L_{ti}(\underline{\boldsymbol{Z}}_{ti})\tilde{W}_t(\underline{\boldsymbol{Z}}_{ti}, \boldsymbol{v}, \boldsymbol{v}_1)}{\sum_{i=1}^{N}\sum_{t=1}^{\tau} \tilde{W}_t(\underline{\boldsymbol{Z}}_{ti}, \boldsymbol{v}, \boldsymbol{v}_1)}.$$

Observe that $\tilde{W}_t$ depends here on $\boldsymbol{v}$ and $\boldsymbol{v}_1$. The associated program can be written as

$$\min_{\boldsymbol{v}_0} \mathscr{L}^0(\boldsymbol{v}, \boldsymbol{v}_0, \boldsymbol{v}_1), \tag{8.4.11}$$

where

$$\mathscr{L}^0(\boldsymbol{v}, \boldsymbol{v}_0, \boldsymbol{v}_1) = \mathbb{E}_{\boldsymbol{v}_1}\left\{\left[\sum_{t=1}^{\tau} M_t(\underline{\boldsymbol{Z}}_t)\tilde{W}_t(\underline{\boldsymbol{Z}}_t, \boldsymbol{v}, \boldsymbol{v}_0)\right]^2 \tilde{W}_\tau(\underline{\boldsymbol{Z}}_\tau, \boldsymbol{v}_0, \boldsymbol{v}_1)\right\}, \tag{8.4.12}$$

and

$$\tilde{W}_\tau(\underline{\boldsymbol{Z}}_\tau, \boldsymbol{v}_0, \boldsymbol{v}_1) = \prod_{t=1}^{\tau} \frac{f(\boldsymbol{Z}_t, \boldsymbol{v}_0)}{f(\boldsymbol{Z}_t, \boldsymbol{v}_1)}.$$

The algorithm for estimating the optimal solution, $\boldsymbol{v}_0^*$ is similar to Algorithm 8.4.1 and can be written as follows:

Algorithm 8.4.2 Choice of a Good $\boldsymbol{v}_1$ and Optimal $\boldsymbol{v}_0^*$ for Dynamic Models

1. *Choose an initial point* $\boldsymbol{v}_1(0)$, *say* $\boldsymbol{v}_1(0) = \boldsymbol{v}$. *Generate a sample* $\{\boldsymbol{Z}_{ti}: t = 1, \ldots, \tau_i,\ i = 1, \ldots, N\}$ *from the pdf* $f(\boldsymbol{z}, \boldsymbol{v}_1)$ *and solve the stochastic counterpart*

$$\min_{\boldsymbol{v}_0} \hat{\mathscr{L}}_N^0(\boldsymbol{v}, \boldsymbol{v}_0, \boldsymbol{v}_1) = \min_{\boldsymbol{v}_0} \sum_{i=1}^{N}\left[\sum_{t=1}^{\tau_i} M_{ti}(\underline{\boldsymbol{Z}}_{ti})\tilde{W}_{ti}(\underline{\boldsymbol{Z}}_{ti}, \boldsymbol{v}, \boldsymbol{v}_0)\right]^2 \tilde{W}_{\tau_i i}(\underline{\boldsymbol{Z}}_{\tau_i i}, \boldsymbol{v}_0, \boldsymbol{v}_1) \tag{8.4.13}$$

of the program (8.4.11)–(8.4.12). *Denote the solution by* $\bar{\boldsymbol{v}}_0^*(0)$.

2. *Set* $\boldsymbol{v}_1(1) = \bar{\boldsymbol{v}}_0^*(0)$ *and repeat step 2. Denote the solution by* $\bar{\boldsymbol{v}}_0^*(1)$.
3. *Proceed with steps 2 and 3 until the desired precision is achieved.*

Extensive simulation studies show that the optimal solution, $\bar{\boldsymbol{v}}_{0N}^*$ of the stochastic counterpart (8.4.15) is rather insensitive (robust) to the choice of $\boldsymbol{v}_1$. These studies suggest that typically there exists a rather broad range of parameter vectors $\boldsymbol{v}_1$ that identify (estimate reliably) the optimal reference parameter $\boldsymbol{v}_0^*$. Consider, for example, an $M/M/1$ queue with fixed arrival rate $\lambda = 1$ and $\rho = 0.6$, and let $\{L_t\}$ be the waiting time process. In this case it follows from the analytical results of Asmussen and Rubinstein (1993) that the optimal reference parameter (service rate) is $v_0^* \approx 1.25$ ($\rho_0^* \approx 0.8$). It was found numerically that the stochastic counterpart (8.4.15) [with v_1 being the service rate in the auxiliary pdf $f(y, v_1)$] estimates the parameter $\rho_0^* \approx 0.8$ rather accurately for $v_1 \in (1.06, 2)$ [$\rho_1 = 1/v_1 \in (0.5, 0.95)$].

Finally, consider the estimation of $\boldsymbol{v}_0^*$ for the program (8.3.1), that is, the case where $\boldsymbol{v}$ varies. For static models we can again apply Algorithm 8.4.2 with $\hat{\mathscr{L}}_N^0(\boldsymbol{v}, \boldsymbol{v}_0, \boldsymbol{v}_1)$ in (8.4.15) replaced by $\max_{j=1,\dots,s} \hat{\mathscr{L}}_N^0(\boldsymbol{v}_j, \boldsymbol{v}_0, \boldsymbol{v}_1)$, with all other data remaining intact.

8.5 APPENDIX: CONVEXITY RESULTS

Proposition 8.5.1. *Let* $\boldsymbol{Y}$ *be a random vector from an exponential family in the canonical form* (5.2.24). *Then* $\mathscr{L}^k(\boldsymbol{v}, \boldsymbol{v}_0)$, $k = 0, 1$, *is a convex function in* $\boldsymbol{v}_0 = (v_{01}, \dots, v_{0n})$.

Proof. Consider first the case $k = 0$. One has

$$\mathscr{L}^0(\boldsymbol{v}, \boldsymbol{v}_0) = c^2(\boldsymbol{v}) \int \frac{L^2(\boldsymbol{y})}{c(\boldsymbol{v}_0)} \exp\left(\sum_{k=1}^{n} (2v_k - v_{0k}) t_k(\boldsymbol{y}) \right) h(\boldsymbol{y})\, d\boldsymbol{y}, \tag{8.5.1}$$

where

$$\frac{1}{c(\boldsymbol{v}_0)} = \int \exp\left(\sum_{k=1}^{n} v_{0k} t_k(\mathbf{z}) \right) h(\mathbf{z})\, d\mathbf{z}.$$

Substituting the above into (8.5.1) yields

$$\begin{aligned} \mathscr{L}^0(\boldsymbol{v}, \boldsymbol{v}_0) = c^2(\boldsymbol{v}) \int\int L^2(\boldsymbol{y}) \\ \times \exp\left\{ \sum_{k=1}^{n} [2v_k t_k(\boldsymbol{y}) + v_{0k}(t_k(\mathbf{z}) - t_k(\boldsymbol{y}))] \right\} h(\boldsymbol{y}) h(\mathbf{z})\, d\boldsymbol{y}\, d\mathbf{z}. \end{aligned} \tag{8.5.2}$$

Now, for any linear function, $a(\boldsymbol{x})$ of $\boldsymbol{x}$, the function $e^{a(\boldsymbol{x})}$ is convex. Since $L^2(\boldsymbol{y})$ is nonnegative, it follows that for any fixed $\boldsymbol{v}$, $\boldsymbol{y}$, and $\boldsymbol{z}$, the integrand of the second integral in (8.5.2) is convex in $\boldsymbol{v}_0$. This implies the convexity of $\mathscr{L}^0(\boldsymbol{v}, \cdot)$.

Consider next the case $k=1$. Then $\mathscr{L}^1(\boldsymbol{v}, \boldsymbol{v}_0)$ can be written as the trace of the covariance matrix, and one has

$$\mathrm{tr}[\mathbb{E}_{\boldsymbol{v}_0}\{L^2(\nabla W)(\nabla W)'\}] = \mathbb{E}_{\boldsymbol{v}_0}\{L^2 W^2 \ \mathrm{tr}[\mathbf{S}^{(1)}\mathbf{S}^{(1)'}]\}, \tag{8.5.3}$$

where prime denotes the transpose operator. Since $\mathrm{tr}[\mathbf{S}^{(1)}(\boldsymbol{y}, \boldsymbol{v})\mathbf{S}^{(1)'}(\boldsymbol{y}, \boldsymbol{v})]$ is nonnegative for any $\boldsymbol{y}$ and $\boldsymbol{v}$, we conclude as before that $\mathrm{tr}[\mathbb{E}_{\boldsymbol{v}_0}\{L^2(\nabla W)(\nabla W)'\}]$ is a convex function in $\boldsymbol{v}_0$.

Remark 8.5.1. Proposition 8.2.1 can be extended as follows. Let $\varphi(x_1, x_2)$ be a real-valued differentiable function of two variables, x_1 and x_2. Consider the corresponding function of expected values,

$$\ell(\boldsymbol{v}) = \varphi(\ell_1(\boldsymbol{v}), \ell_2(\boldsymbol{v})),$$

with

$$\ell_i(\boldsymbol{v}) = \mathbb{E}_{\boldsymbol{v}}\{L_i(\boldsymbol{Y})\} = \mathbb{E}_{\boldsymbol{v}_0}\{L_i(\boldsymbol{Y})W(\boldsymbol{Y}, \boldsymbol{v})\}, \qquad i = 1, 2,$$

where $L_1(\boldsymbol{Y})$ and $L_2(\boldsymbol{Y})$ are two sample functions associated with the same random vector $\boldsymbol{Y}$. In this case, the what-if estimators of $\ell(\boldsymbol{v})$ can be written as

$$\bar{\ell}_N(\boldsymbol{v}, \boldsymbol{v}_0) = \varphi(\bar{\ell}_{1N}(\boldsymbol{v}, \boldsymbol{v}_0), \bar{\ell}_{2N}(\boldsymbol{v}, \boldsymbol{v}_0)),$$

where $\bar{\ell}_{1N}(\boldsymbol{v}, \boldsymbol{v}_0)$ and $\bar{\ell}_{2N}(\boldsymbol{v}, \boldsymbol{v}_0)$ are the what-if estimators of $\ell_1(\boldsymbol{v})$ and $\ell_2(\boldsymbol{v})$, respectively.

By virtue of the delta theorem (see, e.g., Rubinstein and Shapiro, 1993), $N^{1/2}(\bar{\ell}_N(\boldsymbol{v}) - \ell(\boldsymbol{v}))$ is asymptotically normal with mean zero and variance

$$\begin{aligned}\sigma^2(\boldsymbol{v}, \boldsymbol{v}_0) &= a^2 \ \mathrm{var}_{\boldsymbol{v}_0}\{L_1 W\} + b^2 \ \mathrm{var}_{\boldsymbol{v}_0}\{L_2 W\} + 2ab \ \mathrm{cov}_{\boldsymbol{v}_0}\{L_1 W, L_2 W\} \\ &= \mathbb{E}_{\boldsymbol{v}_0}\{[aL_1 + bL_2]^2 W^2\} + R(v).\end{aligned}$$

Here, $R(v)$ consists of the remaining terms (which are independent of $\boldsymbol{v}_0$) $a = \partial\varphi(x_1, x_2)/\partial x_1$ and $b = \partial\varphi(x_1, x_2)/\partial x_2$ at $(x_1, x_2) = (\ell_1(\boldsymbol{v}), \ell_2(\boldsymbol{v}))$. For example, for $\varphi(x_1, x_2) = x_1/x_2$, one gets $a = 1/\ell_2(\boldsymbol{v})$ and $b = -\ell_1(\boldsymbol{v})/\ell_2^2(\boldsymbol{v})$.

The convexity of $\sigma^2(\boldsymbol{v}, \boldsymbol{v}_0)$ in $\boldsymbol{v}_0$ now follows similarly to the proof of Proposition 8.2.1.

Remark. This result can be extended to a function $\varphi(x_1, \ldots, x_r)$ of r variables.

Proposition 8.5.2. *Let $Y \sim G(\lambda, \beta)$. Suppose that $L^2(y)$ is a monotonically increasing function on the interval $[0, \infty)$. Then*

$$\lambda_0^*(k) < \lambda, \qquad k = 0, 1. \tag{8.5.5}$$

Proof. The proof will be given for $k=0$ only. The proof for $k=1$, using the trace operator, is similar.

Since $\mathscr{L}^k(\lambda_0)$ is convex, it suffices to prove that the derivative of $\mathscr{L}^k(\lambda_0)$ with respect to λ_0 is positive at $\lambda_0 = \lambda$. To this end, represent $\mathscr{L}^0(\lambda_0)$ as

$$\mathscr{L}^0(\lambda_0) = c\int_0^\infty \lambda_0^{-\beta} L^2(y) y^{\beta-1} \exp(-(2\lambda - \lambda_0)y\, dy,$$

where the constant $c = \lambda^{2\beta}/\Gamma(\beta)$ is independent of λ_0. Differentiating $\mathscr{L}^0(\lambda_0)$ above with respect to λ_0 at $\lambda_0 = \lambda$, one has

$$(\mathscr{L}^0)'(\lambda_0)|_{\lambda_0=\lambda} = c\int_0^\infty \left(y - \frac{\beta}{\lambda}\right)\lambda^{-\beta} L^2(y) y^{\beta-1} \exp(-\lambda y)\, dy.$$

Integrating $\int_0^\infty L^2(y) y^\beta \exp(-\lambda y)\, dy$ by parts yields

$$\begin{aligned}\int_0^\infty L^2(y) y^\beta \exp(-\lambda y)\, dy = &\frac{\beta}{\lambda}\int_0^\infty L^2(y) y^{\beta-1} \exp(-\lambda y)\, dy \\ &+ \frac{1}{\lambda}\int_0^\infty y^\beta \exp(-\lambda y)\, dL^2(y),\end{aligned}$$

provided $L^2(y) y^\beta \exp(-\lambda y)$ tends to zero as $y \to +\infty$. Putting all these facts together results in

$$(\mathscr{L}^0)'(\lambda) = c\lambda^{-\beta-1}\int_0^\infty y^\beta \exp(-\lambda y)\, dL^2(y). \tag{8.5.6}$$

Finally, since $L^2(y)$ is monotonically increasing in y, we conclude that the integral on the right-hand side of (8.5.6) is positive, and consequently, $(\mathscr{L}^0)'(\lambda) > 0$. This fact, and the convexity of $\mathscr{L}^0(\lambda_0)$ imply that $\lambda_0^*(0) < \lambda$. □

8.6 EXERCISES

1. Repeat Examples 8.2.1, 5.3.2, 8.3.1, and 8.4.2 for the Bernoulli and Weibull distributions.

2. Reproduce the data of Tables 8.2.1 and 8.3.2.

***3.** Let Y be a normally distributed random variable with mean λ and variance σ^2.

(a) Suppose that σ^2 is known and fixed. For a given λ, consider the function

$$\mathcal{L}^0(\lambda_0) = \mathbb{E}_{\lambda_0}\{L^2 W^2\}.$$

(i) Show that if the expectation $\mathbb{E}_\lambda\{L^2\}$ is finite for all $\lambda \in \mathbb{R}$, then $\mathcal{L}^0(\lambda_0)$ is convex and continuous on $\mathbb{R}$. Further, if additionally $\mathbb{E}_{\lambda_n}\{L^2\}$ does not tend to zero for any sequence $\lambda_n \to \infty$, then $\mathcal{L}^0(\lambda_0)$ has a unique minimizer, λ_0^*, over $\mathbb{R}$.

(ii) Show that if $L^2(y)$ is monotonically increasing on $\mathbb{R}$, then $\lambda_0^* < \lambda$.

(b) Suppose that λ is known, and consider the parameter σ. Note that the resulting exponential family is not of the canonical form. However, parameterizing it by $\xi = \sigma^{-2}$ transforms it into the canonical form with $t(y) = -(y-\lambda)^2/2$ and $c(\xi) = (2\pi)^{-1/2}\xi^{1/2}$.

(i) Show that

$$\mathbb{E}_{\xi_0}\{W^2\} = \frac{\xi^{1/2}}{\xi_0^{1/2}(2\xi - \xi_0)^{1/2}},$$

provided $0 < \xi_0 < 2\xi$.

(ii) Show that for given ξ, the function

$$\mathcal{L}^0(\xi_0) = \mathbb{E}_{\xi_0}\{L^2 W^2\}$$

has a unique minimizer, ξ_0^*, over the interval $(0, 2\xi)$, provided the expectation $\mathbb{E}_{\xi'}\{L^2\}$ is finite for all $\xi' \in (0, 2\xi)$ and does not tend to zero as ξ' approaches 0 or 2ξ. (Notice that this implies that the corresponding optimal value, $\sigma_0^* = (\xi_0^*)^{-1/2}$, of the reference parameter, σ_0, is also unique.)

(iii) Show that if $L^2(y)$ is monotonically increasing on $\mathbb{R}$, then $\xi_0^* < \xi$. (Notice that this implies that $\sigma_0^* > \sigma$.)

REFERENCES

Asmussen, S. (1992). "Queueing simulation in heavy traffic," *Math. Oper. Res.*, **17**, 84–111.

Asmussen, S. and Rubinstein, R. Y. (1993). "Response surface estimation and sensitivity analysis via efficient change of measure," *Stoch. Models*, **9**(3), 313–339.

Rubinstein, R. Y. and Shapiro, A. (1993). *Discrete Event Systems: Sensitivity Analysis and Stochastic Optimization via the Score Function Method*, Wiley, New York.

Shapiro, A. and Rubinstein, R. Y. (1994). "Optimal choice of the reference parameters in the score function method," Manuscript, Technion, Haifa, Israel.

CHAPTER 9

Estimating Rare-Event Probabilities and Related Optimization Issues

9.1 INTRODUCTION

Efficient estimation of rare-event probabilities, $\ell = P(A)$ (A is a rare event) is crucial to the analysis of many modern systems, such as coherent reliability systems, inventory systems, insurance risk, storage systems, computer networks, and telecommunications networks. For example, in a communications system with a buffer capability of b packets, one might be interested in selecting a buffer size so that the steady-state packet loss probability is sufficiently small, say 10^{-6}. For examples involving optimization with rare-events consider:

1. Maximizing an expected lifetime of a highly reliable coherent reliability system, subject to some cost constraints.
2. Finding an optimal (s, S) policy in a multi-item, multicommodity inventory system (under general distributions of the interdemand time and demand size), so that the backlog probability is less than 10^{-10}.

Exact analytical or "good" approximate asymptotic expressions for such rare-event probabilities are only available for a very restricted class of systems; consequently, one often has to resort to simulation. Estimation of rare-event probabilities under the original measures, via the crude Monte Carlo method, can be very time consuming or practically infeasible. In such cases, importance sampling (IS) is called for.

In the past decade, IS has been applied to a variety of problems arising in the analysis of rare events in queueing systems, inventory, and relaibility models. See, for example, Asmussen (1985, 1987, 1989, 1994), Asmussen and Nielsen (1995), Asmussen et al. (1994), Ben Letaief (1995), Bucklew (1990), Bucklew et al. (1990), Chang et al. (1994, 1995), Cottrel et al. (1983), Dembo and Zeitouni (1993), Devetsikiotis and Townsend (1992, 1993), Frater and Anderson (1989), Frater et al. (1990, 1991), Glasserman and Kou (1995), Glasserman and Liu (1996),

Gnedenko and Ushakov (1995), Goyal et al. (1992), Heidelberger (1993), Heidelberger et al. (1994), Huang et al. (1995), Kovalenko (1995), Kriman and Rubinstein (1997), Lehtonen and Nyrhinen (1992a, 1992b), Lieber et al. (1994), Lindley (1952, 1953), Mandjes (1996, 1997), Nemirovskii and Rubinstein (1993), Neuts (1977, 1981, 1989), Nicola et al. (1993), Parekh and Walrand (1989), Rubinstein and Shapiro (1993), Sadowsky (1991, 1993, 1994), Sadowsky and Bucklew (1990), Sadowsky and Szpankowski (1992a, 1992b), Shahabuddin (1994a,b,c), Shwartz and Weiss (1995), and Tsoucas (1992). For recent surveys, see Asmussen and Rubinstein (1995), Heidelberger (1993), Kovalenko (1995), and Shahabuddin (1995). The main idea of IS, when applied to rare events, is to make their occurrence more frequent, or in other words, to "speed up" the simulation. Technically, IS aims to select a probability distribution (change of measure) that reduces or minimizes the computational cost (simulation time), subject to a desired accuracy (the width of the requisite confidence interval).

It is known that choosing the IS distribution as the original distribution, given that the rare event has occurred, is not only optimal but, in fact, leads to an estimator with *zero variance*. Unfortunately, identifying such optimal IS distribution is often infeasible for the following reasons: First, it explicitly depends on $P(A)$, the unknown quantity we are trying to estimate. Obviously, if $P(A)$ were known, there would be no need to run the simulation experiment in the first place. Second, even with $P(A)$ known, it is often very difficult to sample from such a conditional distribution using a simple algorithm. To overcome this difficulty we shall use a specific IS technique, called the *exponential change of measure* (ECM), also known as *exponential twisting* and *exponential tilting*.

Before proceeding further, we review some background material. Let

$$\hat{F}[\theta] = \mathbb{E}\{e^{\theta Y_k}\} = \int_{-\infty}^{\infty} e^{\theta x} F(dx) \tag{9.1.1}$$

and

$$\Lambda(\theta) = \log \hat{F}[\theta] = \log \mathbb{E}\{e^{\theta Y_k}\} \tag{9.1.2}$$

be the *moment generating function* (mgf) and the *cummulative generating function* (cgf), also called the *log moment generating function* of the random variable Y_k; both will be assumed to be finite for all $0 \leq \theta < \theta_0$. An ECM replaces the cdf F by a cdf F_θ, given by

$$F_\theta(x) = \frac{1}{\hat{F}[\theta]} \int_{-\infty}^{x} e^{\theta y} F(dy), \tag{9.1.3}$$

where θ is a parameter. The mgf of $F_\theta(x)$ is

$$\hat{F}_\theta[s] = \frac{\hat{F}[s+\theta]}{\hat{F}[\theta]},$$

and its cgf is

$$\Lambda_\theta(s) = \log \hat{F}_\theta[s] = \Lambda(\theta + s) - \Lambda(\theta).$$

Following are some examples of ECMs, $F_\theta(x)$, for various cdf's F:

1. For $F = \mathrm{Exp}(v)$, $F_\theta = \mathrm{Exp}(v - \theta)$.
2. For $F = G(\lambda, \beta)$, $F_\theta = G(\lambda - \theta, \beta)$.
3. For $F = N(\mu, \sigma^2)$, $F_\theta = N(\mu + \theta, \sigma^2)$.
4. $F = Ge(p)$, $F_\theta = Ge(1 - (1 - p)e^\theta)$.

It is readily seen that for these examples, the original cdf, F, and the ECM, F_θ, differ only by a *single* parameter θ; in fact, the original cdf, $F \equiv F_{\theta=0}$, is a special case of F_θ with $\theta = 0$. Although this is not always the case, (e.g., the Weibull distribution), for the sake of simplicity we shall assume that the ECM F_θ ($\theta \geq 0$) dominates the original cdf, $F \equiv F_{\theta=0}$, so that θ will always be taken as the reference parameter. Moreover, in order to be consistent with the notation in previous chapters, we assume, as usual, that the original cdf F is from the exponential family in the canonical form and use both (equivalent) notations:

$$F_\theta(\boldsymbol{y}, \boldsymbol{v}) \equiv F(\boldsymbol{y}, \boldsymbol{v}_0) \quad \text{and} \quad F_{\theta^*}(\boldsymbol{y}, \boldsymbol{v}) \equiv F(\boldsymbol{y}, \boldsymbol{v}_0^*) \tag{9.1.4}$$

for ECM and OECM (*optimal exponential change of measure*), respectively, where $\boldsymbol{v}_0^*$ is the optimal solution of the program (9.1.6) (see below). For example

1′ For $Y \sim \mathrm{Exp}(v)$, one has $v_0 = v + \theta$ and $v_0^* = v + \theta^*$.
2′ For $Y \sim G(\lambda, \beta)$, one has $\boldsymbol{v}_0 = (\lambda + \theta, \beta)$ and $\boldsymbol{v}_0^* = (\lambda + \theta^*, \beta)$.

Note also that if $\{Z_1, \ldots, Z_N\}$ is a sample from $F_\theta(y, v) \equiv F(y, v_0)$, then the likelihood ratio (LR) has the following simple form:

$$W(v, v_0) = \prod_{i=1}^{N} \frac{f(Z_i, v)}{f(Z_i, v_0)} \equiv W(\theta) = \frac{\hat{F}^N(\theta)}{e^{\theta(Z_1 + \cdots + Z_N)}} = \hat{F}^N(\theta) e^{-\theta S_N}, \tag{9.1.5}$$

where $S_N = Z_1 + \cdots + Z_N$.

As in the earlier chapters, we shall minimize the variance of the associated LR estimator, say $\bar{\ell}_N(\boldsymbol{v}_0)$, with respect to the parameter $\boldsymbol{v}_0 \equiv \boldsymbol{\theta}$, and find the optimal $\boldsymbol{v}_0^* \equiv \boldsymbol{\theta}^*$, which we call the OECM parameter vector.

It appears that the ECM parameterization is extremely useful for estimation of rare-event probabilities. For static systems and reasonably simple queueing models, such as the $GI/G/1$ queue, the optimal ECM parameter vector, $\boldsymbol{v}_0^*$, which minimizes the variance of $\bar{\ell}_N(\boldsymbol{v}_0)$, leads to accurate (polynomial-time!) rare-event probability estimators. For example, estimation of packet loss probabilities on the order of 10^{-10} would require simulation of at least 10^{+10} packets by the *crude Monte Carlo* (CMC) method. Simulation results show that the OECM distribution,

$F(y, v_0^*)$, permits a very accurate estimation of such small probabilities from samples of several hundred packets only. We shall also demonstrate experimentally that under $F(y, v_0^*)$, results of similar accuracy are available for complex static systems and queueing models. Moreover, it turns out that for complex discrete-event systems (DES), both static and dynamic, there is essentially no need to parameterize the entire set of input distributions, $F(y, v)$; it is typically sufficient to parameterize only $F(y, v_{0b})$, where v_{0b} is selected from a *small subset* of v_0. The parameter v_{0b} is associated with the notion of the so-called *bottleneck cut* and *bottleneck queueing module*, for static models and queueing networks, respectively. The size of the vector v_{0b} is typically in the range 3–10, while the size of v_0 might be on the order of hundreds. To summarize, the simulation results to follow indicate that highly accurate estimators of rare-event probabilities can be derived by using a *parametric* ECM $F(y, v_{0b})$, where $v_{0b} \in V$ is selected from a small subset of v.

The theoretical framework in which we examine rare-event probability estimation is based on complexity theory, as introduced in Kriman and Rubinstein (1997). Here, IS estimators are classified either as *polynomial-time* or as *exponential-time*. We shall present necessary and sufficient conditions in various settings under which the IS estimators are polynomial-time. In fact, it will be shown that for an IS estimator, $\bar{\ell}_N(v_0, x)$ of $\ell(x)$, to be polynomial-time, it suffices that its *squared coefficient of variation* (SCV), $\kappa^2(x)$, also called the *relative error*, be bounded in x by some polynomial function, $p(x)$. For such polynomial-time estimators, the required sample size to achieve a prescribed relative error does not blow up as the event becomes rarer. In order to obtain such (polynomially bounded) $\kappa^2(x)$, we minimize $\kappa^2(x) \equiv \kappa^2(v_0, x)$ with respect to v_0, where

$$\kappa^2(v_0, x) = \frac{N \operatorname{var}\{\bar{\ell}_N(v_0, x)\}}{\ell^2(x)}.$$

This chapter treats efficient estimation of rare-event probabilities separately for static and dynamic models. ECM will be shown to play a key role in generating polynomial-time, rare-event probability estimators in both settings. In particular, a fast algorithm will be proposed for rare-event estimation, based on the OECM distribution, $F(y, v_0^*)$. Here, v_0^* is obtained as the optimal solution of the program

$$\min_{v_0} \kappa^2(v_0, x),$$

which is equivalent to

$$\min_{v_0} \operatorname{var}\{\bar{\ell}_N(v_0, x)\}. \tag{9.1.6}$$

Conditions under which the IS estimator $\bar{\ell}_N(v_0^*, x)$ is polynomial-time, where v_0^* is the optimal solution of the program (9.1.6), will also be discussed.

For queueing models, both transient simulation and steady-state rare-event simulation will be treated. Three distinct approaches to steady–state rare-event simulation may be discerned.

The first approach, proposed by Siegmund (1976a) and further developed by Asmussen (1985), relates the original rare-event queueing problem to a *ruin problem* associated with a *random walk*, which is then treated by ECM. A variety of proposed approaches to OECM will be discussed in Section 9.4. Asmussen and Rubinstein (1995) proposed a generalization based on a duality relation of Markov processes, due to Siegmund (1976b). In particular, Asmussen and Rubinstein (1995) consider queues with finite buffers, and discuss conditions under which the associated rare-event probability estimators are polynomial-time; furthermore, their approach permits *simultaneous* estimation of rare-event probabilities for different buffer sizes from a *single simulation run*.

The second approach is based on standard regenerative simulation (see, e.g., Asmussen, 1982, 1987). Under the assumption that the IS distribution is from the same parametric family as the original one, Asmussen, Rubinstein and Wang (1994) calculated explicitly the variance of $I_{\{L>x\}}$, where L is the steady-state waiting time in the $M/M/1$ queue. For more complex queueing models the paper showed empirically that in order to obtain substantial variance reduction with the regenerative LR estimator, relative to the CMC estimator, one has to choose the IS densities such that the associated traffic intensity, ρ_0 (under the IS change of measure), is moderately larger than the original traffic intensity, ρ. Asmussen and Rubinstein (1995) proved that even under the OECM distribution, the regenerative LR estimators considered in Asmussen (1987) are exponential-time.

The third approach, introduced independently by several authors (e.g., Nicola et al., 1993; Heidelberger, 1993; Rubinstein and Shapiro, 1993; Devetsikiotis and Townsend, 1992), is based on the so-called *switching regenerative* (SR) simulation method. The idea is to initially use, at each regenerative cycle, a simulation distribution that leads to accelerated occurrence of rare events (say, buffer overflow in the $GI/G/1/m$ queue) until a certain event occurs, and then complete the regenerative cycle with the original distribution. Some empirical studies of this technique are given, for example, in Goyal et al. (1992), Nicola et al. (1993), and Rubinstein and Shapiro (1993). For simple queues, such as the $GI/G/1/m$ queue, it is proved in Kriman and Rubinstein (1997) that the SR estimator is polynomial-time. Finding conditions under which SR estimators are polynomial-time for networks of queues is still an open problem. We provide, however, supporting experimental results.

This chapter also addresses the *robustness* of OECM for simple queues by exploring how far one may perturb the optimal parameter vector, $\boldsymbol{v}_0^*$, in the OECM distribution while still retaining a substantial variance reduction. Experience suggests that if the optimal parameter in the $GI/G/1$ queue is perturbed by about 20%, one only loses 2–3 orders of magnitude of variance reduction, as compared to variance reduction on the order of thousands under the optimal parameter value.

Although the focus here is on complexity and asymptotic properties (as $x \to \infty$) of OECM estimators, it is important to realize that for fixed (even large) x, one may obtain more accurate IS estimators, based on alternative parameterizations of $F(\boldsymbol{y}, \boldsymbol{v})$. Consider, for example, the estimation of a moderately small rare-event probability, say $P\{Y > x\} \approx 10^{-4}$, under a two-parameter distribution, say $G(\lambda, \beta)$.

The OECM distribution is $G(\lambda_0^*, \beta)$, where $\lambda_0^* = \lambda + \theta^*$ and θ^* can be readily found analytically. It is also not difficult to check that one might be doing better (in the sense of obtaining a more accurate estimator), by minimizing the variance of the associated LR estimator with respect to both parameters λ_0 and β_0, rather than λ_0 alone, as prescribed by the ECM method. The reason is that the OECM method is known to be asymptotically optimal, whereas $P\{Y > x\} \approx 10^{-4}$ is only moderately small. Also, optimizing in the two-dimensional space of parameters (λ_0, β_0) is at least as optimal as optimizing in the one-dimensional space of λ_0.

Asmussen and Binswanger (1995) derive polynomial-time estimators for ruin probabilities *without resorting to IS*. Their insightful procedure is based on combining order statistics with the conditional Monte Carlo method. It assumes that the original distribution, $F(y, \boldsymbol{v})$, is from the subexponential "heavy-tailed" family of distributions, such as the Raleigh distribution. In contrast, we shall be dealing with "light-tailed" distributions from the exponential family. Our basic assumption is that $\hat{F}(\theta) < \infty$, for some $\theta > 0$. This implies that the tail of the distribution F tends to zero exponentially fast, which is not the case for a heavy-tailed cumulative distribution function (cdf) F, where $\hat{F}(\theta) = \infty$ for all $\theta > 0$. Needless to say, for a fixed x, a rare event is less likely to occur under a light-tailed distribution.

The rest of this chapter is organized as follows. Section 9.2 presents a framework for time complexity of IS estimators; in particular, the concepts of *polynomial-time* and *exponential-time* Monte Carlo estimators (algorithms) will be defined. Section 9.3 deals with estimation of rare-event probabilities for static systems; a fast (polynomial-time) algorithm will be described and related numerical results will be presented. Section 9.4 deals with rare-event probabilities in simple queueing models. In particular, Section 9.4.1 combines the ideas of random walk and OECM to derive polynomial (rare-event) estimators for the $GI/G/1$ queue and classical ruin problem. Section 9.4.2 studies the time complexity of standard and switching regenerative estimators for simple queues. Section 9.4.4 investigates the robustness (stability) of IS estimators. Section 9.5 deals with estimation of rare-event probabilities in queueing networks; an adaptive algorithm will be presented for fast estimation of rare events in both transient and steady-state settings. Section 9.6 treats stochastic optimization of DES involving rare events. Here we present a two-stage procedure: in the first stage we identify (estimate) the optimal parameter vector $\boldsymbol{v}_0^*$ of the IS distribution, and in the second stage we estimate the optimal solution $\boldsymbol{v}^*$ of a constrained optimization program involving rare events. Section 9.7 presents supporting experimental results. Finally, Section 9.8 contains concluding remarks.

9.2 POLYNOMIAL-TIME ESTIMATORS

Let $\bar{\ell}_N(x)$ be an estimator of $\ell(x) = P\{L > x\}$, say a likelihood ratio estimator, where L is the steady-state sample performance and x is chosen such that $\{L > x\}$ is a rare event.

Definition 9.2.1. We say that $\bar{\ell}_N$ is an (ε, δ) accurate estimator of $\ell(x)$ for some $0 < \varepsilon, \delta < 1$, if

$$P\{|\bar{\ell}_N(x) - \ell(x)| < \varepsilon\ell(x)\} > 1 - \delta, \qquad \ell(x) \neq 0. \tag{9.2.1}$$

For example, a (0.05, 0.10) accurate estimator ensures that the relative error does not exceed 5% with probability not less than 90%.

Consider the SCV of $\bar{\ell}_N(x)$,

$$\kappa^2(x) = \frac{N \operatorname{var}\{\bar{\ell}_N(x)\}}{\ell^2(x)}. \tag{9.2.2}$$

By the central limit theorem (CLT),

$$N \approx \gamma\kappa(x)^2, \tag{9.2.3}$$

where $\gamma = [\Phi^{-1}(1 - \delta/2)]^2\varepsilon^{-2}$, and Φ denotes the standard normal cdf.

Definition 9.2.2. A Monte Carlo estimator is said to be (ε, δ) *polynomial*, if (9.2.1) is guaranteed by a sample size (computational cost) $N(x) = O(|\log \ell(x)|^p)$ for some $p < \infty$. A Monte Carlo estimator for which (9.2.1) holds for $N(x)$ of order at least $\ell^{-q}(x)$, for some $q > 0$, is called an *exponential-time* estimator. For more details see Proposition 9.2.1.

The foregoing framework is related to the basic concepts of complexity theory (see, e.g., Stockmeyer, 1992). Note, however, that in our case, the classification depends fundamentally on the way the deterministic parameter, x, is chosen. In particular, an exponential-time estimator (with respect to the parameter x) could reduce to a polynomial one, say, by choosing $\log x$. It follows that in order for the estimator $\bar{\ell}_N$ to be polynomial-time, it suffices that $\kappa^2(x)$ be bounded in x by some polynomial function, $p(x)$.

For better insight into polynomial-time and exponential-time IS estimators, consider the following simple example.

Example 9.2.1 (Example 5.3.2 continued). Suppose we wish to estimate $\ell(x) = P\{Y > x\}$, where the random variable Y has an exponential distribution with rate λ, that is, $Y \sim f(y) = \lambda \exp(-\lambda y)$. Since, in this case, $\ell(x) = e^{-\lambda x}$, it is readily seen that the SCV of the CMC estimator is

$$\kappa^2(x) \approx e^{\lambda x}, \qquad x > 0.$$

Thus the CMC estimator is exponential in x.

Let $g(y) = \lambda_0 \exp(-\lambda_0 y)$ be the IS density, and select λ_0 so as to minimize the variance of the LR estimator, $\bar{\ell}_N(\lambda_0)$. In view of (5.3.8) and (9.2.3) it is not difficult to see that the SCV of the LR estimator, $\bar{\ell}_N(\lambda_0)$, is

$$\kappa^2(\lambda_0, x) = \frac{\lambda^2 e^{\lambda_0 x}}{\lambda_0(2\lambda - \lambda_0)} - 1.$$

Suppose $x^{-1} \ll \lambda$, when $P(Y > x)$ is small, say, less than 10^{-6}. In this case it is not difficult to show that the optimal $\lambda_0^*(x)$, which minimizes $\kappa^2(\lambda_0)$, is

$$\lambda_0^*(x) \approx x^{-1}, \tag{9.2.4}$$

and κ^2 can be approximated by

$$\kappa^2(\lambda_0^*, x) \approx 0.5x\lambda e. \tag{9.2.5}$$

Thus, the SCV of the CMC increases in x exponentially $[\kappa^2(x) \approx e^{\lambda x}]$, and the optimal LR estimator increases in x linearly $[\kappa^2(\lambda_0^*, x) \approx 0.5\lambda xe]$. In other words, the CMC and IS estimators are exponential-time and polynomial-time time, respectively, with requisite sample sizes (computational costs) $N(x) = O(e^x)$ and $N(x) = O(x)$, respectively.

It is readily seen that a choice of $\lambda_0 = k\lambda_0^*$ in lieu of the optimal value, λ_0^*, results in

$$\kappa^2(\lambda_0, x) = \frac{0.5x\lambda e^k}{k}.$$

For example, if $k = 2$ (i.e., $\lambda_0 = 2\lambda_0^* \approx 2/x$), then one has

$$\kappa^2(\lambda_0, x) \approx 0.25\lambda xe^2.$$

Thus, perturbing λ_0^* $(k = 2)$ by 100% increases the variance by approximately $0.5e$ times only.

As an illustration, Table 9.2.1 displays the relative efficiency, $\varepsilon(\lambda_0, x)$, as a function of λ_0, for $\lambda = 1$ and $x = 20$.

Observe that Definition 9.2.2 does not take into account the expected CPU time, $C(x)$, of generating likelihood ratios. In the setting of Example 9.2.1, $C(x)$ would be

Table 9.2.1 Relative Efficiency, $\varepsilon(\lambda_0, x)$, of the IS Estimator as Function of λ_0, Given $\lambda = 1$ and $x = 20$

λ_0	0.6	0.4	0.2	0.1	0.075	0.05	0.025
$\varepsilon(v_0, x)$	4.0×10^{-4}	9.6×10^{-6}	3.1×10^{-7}	7.8×10^{-8}	6.2×10^{-8}	5.5×10^{-8}	6.7×10^{-8}

some constant independent of x. But if we consider, for example, a regenerative estimator, then it would be natural to take $C(x)$ as the expected regenerative cycle length, which might depend on x (see, e.g., the switching regenerative estimators in Section 9.4). Thus, for discrete-event dynamic systems (DEDS) settings, Definition 9.2.2 should be enhanced by replacing the sample size, $N(x)$, with

$$T(x) = C(x)N(x),$$

to be referred to as the *computational cost* of the estimator $\bar{\ell}_N(x)$. Notice that for regenerative estimators, $T(x)$ represents the computational cost in terms of *customers* rather than *cycles*.

The following proposition identifies conditions for an IS estimator to be either polynomial-time or exponential-time; its assumptions will serve as the typical setting for the rest of the chapter (the proof is trivial and will be omitted).

Proposition 9.2.1. *Assume that*

(i) $\ell(x) \approx Ke^{-\gamma x}$, *for some constants* $K, \ \gamma \in (0, \infty)$,
(ii) $C(x) \approx xC_0$, *for some* C_0.

Then the IS estimator is polynomial-time, if

(iii) $\mathbb{E}\{I(x)W^2(x)\} = O(x^q e^{-2\gamma x})$ *with* $q < \infty$;

and exponential-time, if

(iv) $\liminf_{x\to\infty} e^{-\delta x}\mathbb{E}\{I(x)W^2(x)\}/\ell^2(x) > 0$, *for some* $\delta > 0$.

Here W is the likelihood ratio, and I is an indicator function.

Note that (i) does not constitute a restriction, since it can be obtained by replacing x with $\varphi(x)$ for some function φ. In queueing setting, the stationary distributions under consideration typically have exponential tails, and consequently (i) holds.

Consider again Example 9.2.1. It readily follows from Proposition 9.2.1 that one can divide the interval $0 < v_0 < \infty$ into a set of subintervals, corresponding to the respective (ε, δ) exponential-time and (ε, δ) polynomial-time Monte Carlo estimators (algorithms). In particular, the expression for κ^2 readily shows that the necessary and sufficient conditions for the IS estimator to be polynomial-time are

$$\frac{C_1}{x^p} \leq v_0(x) \leq C_2 \frac{\log x}{x},$$

for all x and some constants p, C_1, C_2. Similarly, the IS estimator is exponential-time if and only if for all x, $v_0(x)$ belongs to a region of the form

$$(0, e^{-\eta x}) \cup (\beta, \infty).$$

The following motivating example is directly related to rare events in simple queues (see Section 9.4).

Example 9.2.2. Consider the *first passage time probability*

$$\ell(x) = P(\beta(x) < \infty), \tag{9.2.6}$$

where $\beta(x) = \inf\{n > 0 \colon S_n > x\}$, that is, $\beta(x)$ is the first time the process $\{S_n\}$ exceeds level x, and

$$S_n = Y_1 + \cdots + Y_n \tag{9.2.7}$$

is a random walk [sum of independent and identically distributed (iid) random variables, $Y_k \sim F(y, v)$, $k = 1, \ldots, n$]. Observe that formula (9.2.6) merely represents the probability that the random walk $\{S_n\}$ crosses level x in $\beta(x)$ steps. It is assumed that the mean drift, $\mu = \mathbb{E}\{Y_k\}$, is *negative*.

We remark that the CMC method is often useless in practice, since $\beta(x) = \infty$ with high probability. However, using OECM one transforms the random walk into one with *positive* drift, so that $\beta(x) < \infty$ with probability one.

Now, in view of (9.1.5) it is easy to check (see also Asmussen and Rubinstein, 1995) that the ECM estimator of $\ell(x)$ in (9.2.6), based on F_θ, is

$$\begin{aligned} \bar{\ell}(x) = \tilde{W}_{\beta(x)} &= \exp\{-\theta S_{\beta(x)} + \beta(x)\Lambda(\theta)\} \\ &= \exp\{-\theta x\} \exp\{-\theta \xi(x) + \beta(x)\Lambda(\theta)\}, \end{aligned} \tag{9.2.8}$$

where

$$\tilde{W}_{\beta(x)} = \prod_{j=1}^{\beta(x)} W_j, \qquad W_j = \frac{f(Z_j)}{f_\theta(Z_j)},$$

f and f_θ are the probability density functions (pdf's) corresponding to the cdf's F and F_θ, respectively, $\Lambda(\cdot)$ is the associated cgf, $Z_j \sim f_\theta(\cdot)$, and $\xi(x) = S_{\beta(x)} - x$ is the *overshoot* when crossing level x.

Suppose that θ^* is the unique positive solution of the equation

$$\hat{F}[\theta] = 1, \tag{9.2.9}$$

or equivalently,

$$\Lambda(\theta) = 0. \tag{9.2.10}$$

The following fundamental result states that applying OECM to $\bar{\ell}(x)$ leads to a polynomial-time estimator.

Theorem 9.2.1. *For the random walk* (9.2.7), *the estimator* $\bar{\ell}(x)$ *of the first passage time probability,* $\ell(x) = P\{\beta(x) < \infty\}$, *is polynomial-time if and only if the IS distribution is* $G = F_{\theta^*}$, *where* θ^* *is the unique solution of* (9.2.10).

Proof. See Asmussen and Rubinstein (1995). □

We shall see below that (9.2.7) is also of major importance in simple queueing models, due to the relation

$$P\{L \geq x\} = P\{M \geq x\} = P\{\beta(x) < \infty\}. \tag{9.2.11}$$

Here, L is the steady-state waiting time, $M = \max_{n \geq 0} S_n$, and $\{S_n\}$ is a random walk with components Y_k, where Y_k is the difference between the kth service time and the kth interarrival time [cf. Section III.7 in Asmussen (1987)].

We remark that no simple inequality is known to state that for a fixed x, the minimal variance is obtained by taking $G = F_{\theta^*}$; rather the results are asymptotic. Siegmund (1976a) showed in a slightly different setting that $G = F_{\theta^*}$ is asymptotically optimal as $x \to \infty$, provided G is chosen from the exponential family. Lehtonen and Nyrhinen (1992a) extended this result to a class of distributions, G, for which exponential moments exist. (Consequently Theorem 9.2.1 is basically a version of Lehtonen and Nyrhinen (1992a) with *slightly weaker assumptions*). Asmussen (1985) showed that the IS distribution $G = F_{\theta^*}$ is asymptotically optimal for the exponential family $\{F_\theta\}$ when the limit is not $x \to \infty$, but a diffusion (heavy-traffic) limit. Asmussen's setting is slightly different in that it specializes classical insurance risk models in continuous time, but the proof can be easily adapted. Simulation experiments confirm, however, that the choice $G = F_{\theta^*}$ works well not only for large x but also for moderate x.

Remark 9.2.1. It is known that the optimal change of measure for IS is, in fact, the conditional distribution $P\{\cdot|A(x)\}$. Therefore an obvious way to look for a good IS distribution is to try to find a simple asymptotic description of this conditional distribution, and to utilize this asymptotic description in the simulation. For the random walk setting (9.2.7), where $A(x) = \{\beta(x) < \infty\}$, it turns out that an asymptotic description of $P\{\cdot|A(x)\}$ is available; see Asmussen (1982). Roughly speaking, the results state that up to time $\beta(x)$, the random walk behaves as if its distribution changes from F to F_{θ^*}, which is precisely the type of behavior needed to infer (at least heuristically) the optimality of θ^*.

9.3 RARE-EVENT PROBABILITY ESTIMATION IN SIMPLE QUEUES

This section is based on Asmussen and Rubinstein (1995) and Kriman and Rubinstein (1994). In particular, Sction 9.3.1 extends the results of Example 9.2.2 on random walks to simple queues, with the aid of the basic formulas (9.2.9) and (9.2.11). Section 9.3.2 revisits the conventional regenerative likelihood ratio estimators (see Chapter 6) and introduces the so-called switching regenerative likelihood ratio estimator. Finally, Section 9.3.3 deals with robustness properties of switching regenerative estimators.

9.3.1 ECM and OECM for Simple Random Walks

Consider the probability of the event $A(x) = \{L > x\}$, where L is the waiting time within a cycle of length τ in a $GI/G/1$ queue. In the random walk setting of Example 9.2.2, the cycle length is

$$\tau = \inf\{n \geq 1 : S_n \leq 0\},$$

and the desired rare event is

$$A(x) = \{\beta(x) < \tau\}, \qquad \beta(x) = \inf\{n > 0 : S_n > x\}$$

[see (9.2.6)]. The IS estimator of

$$\ell(x) = P\{\beta(x) < \tau\}$$

is the average of the random variables

$$\hat{\ell}(x) = I_{\{\beta(x) < \tau\}} \tilde{W}_{\beta(x)}, \tag{9.3.1}$$

where $\tilde{W}_{\beta(x)}$ is given in (9.2.8), namely,

$$\bar{\ell}_N(x) = \frac{1}{N} \sum_{j=1}^{N} I_{\{\beta_j(x) < \tau_j\}} \tilde{W}_{\beta_j(x)}. \tag{9.3.2}$$

The following is a corollary of Theorem 9.2.1.

Corollary 9.3.1. *The IS estimator $\bar{\ell}_N(x)$, for the probability*

$$\ell(x) = P\{\beta(x) < \tau\},$$

is polynomial-time if and only if $G = F_{\theta^*}$, where θ^* is the solution of $\Lambda(\theta) = 0$.

The algorithm for estimating the probability $\ell(x) = P\{\beta(x) < \tau\}$, based on the estimator (9.3.2), can be written as follows.

Algorithm 9.3.1

1. *Select the OECM* F_{θ^*}, *where* θ^* *is obtained as the solution of* $\Lambda(\theta) = 0$.
2. *Run a simulation of the* $GI/G/1$ *queue, starting with an empty queue, until either the event* $\{\beta(x) < \tau\}$ *occurs or the queue empties again. In the former case, set* $I_{\{\beta(x)<\tau\}} = 1$; *in the latter case, set* $I_{\{\beta(x)<\tau\}} = 0$.
3. *Replicate step 2 N times, and estimate* $\ell(x)$ *via* (9.3.2).

We next compute θ^* from of $\Lambda(\theta) = 0$ for several simple queues. Let $Y_k = Y_{1k} - Y_{2k}$ be the basic random variable, where Y_{1k} and Y_{2k} are the service time and interarrival time, respectively, for the kth customer in the $GI/G/1$ queue. Assume that the Y_{1k} are iid with distribution F_1; the Y_{2k} are iid with distribution F_2; and the sequences $\{Y_{1k}\}$ and $\{Y_{2k}\}$ are independent. Then the common moment generating function of $Y_k = Y_{1k} - Y_{2k}$ is

$$\hat{F}[s] = \hat{F}_1[s]\hat{F}_2[-s]. \tag{9.3.3}$$

Let further Z_1 and Z_2 be independent random variables with corresponding ECM distributions $F_{1,\theta}$ and $F_{2,\theta}$, and let F_θ be the distribution of $Z = Z_1 - Z_2$. The moment-generating function, $\hat{F}_\theta$, of Z is

$$\hat{F}_\theta[s] = \frac{\hat{F}_1[s+\theta]}{\hat{F}_1[\theta]} \times \frac{\hat{F}_2[-s-\theta]}{\hat{F}_2[-\theta]} = \hat{F}_{1,\theta}[s]\hat{F}_{2,-\theta}[-s],$$

where $\hat{F}_{1,\theta}$ and $\hat{F}_{2,\theta}$ are the moment-generating functions of Z_1 and Z_2, respectively. Some special cases follow:

1. *The* $M/M/1$ *Queue.* Here Y_{1k} and Y_{2k} are exponential with rates μ and λ, respectively ($\mu > \lambda$). Thus,

$$\hat{F}_{1,\theta}[s] = \frac{\mu/(\mu - s - \theta)}{\mu/(\mu-\theta)} = \frac{\mu-\theta}{\mu - s - \theta},$$

and the new service time distribution, $F_{1,\theta}$, is exponential with rate $\mu_\theta = \mu - \theta$. Similarly, the new interarrival time distribution, $F_{2,\theta}$, is exponential with rate $\lambda_\theta = \lambda + \theta$. Further, in view of the identity

$$\hat{F}(\theta) = 1,$$

(9.3.3) becomes

$$1 = \mathbb{E}\{e^{\theta Y_1}\}\mathbb{E}\{e^{-\theta Y_2}\} = \frac{\mu}{\mu-\theta} \times \frac{\lambda}{\lambda+\theta}, \tag{9.3.4}$$

and its solution is $\theta^* = \mu - \lambda$. In other words, the OECM F_{θ^*} results in $\mu_{\theta^*} = \lambda$ and $\lambda_{\theta^*} = \mu$, so that the interarrival and service rates in the OECM (simulated cdf's F_{1,θ^*}, F_{2,θ^*}) are now reversed relative to the original cdf's F_1 and F_2. The OECM for the $M/M/1$ queue is no longer stable.

2. *The $M/D/1$ Queue.* Here F_1 is degenerate, say the Dirac function with impulse at 1, while F_2 is exponential with rate $\lambda < 1$. Thus, (9.3.3) becomes

$$1 = \mathbb{E}\{e^{\theta Y_1}\}\mathbb{E}\{^{-\theta Y_2}\} = e^{\theta}\frac{\lambda}{\lambda + \theta}, \tag{9.3.5}$$

which is a transcendental equation in θ, lacking an explicit solution and requiring a numerical solution instead. It is not difficult to see that for the $D/M/1$ queue (9.3.3) becomes

$$1 = \mathbb{E}\{e^{\theta Y_1}\}\mathbb{E}\{e^{-\theta Y_2}\} = e^{-\theta}\frac{\mu}{\mu - \theta}.$$

Table 9.3.1 displays the optimal parameter, θ^* (for the variate $Y_k = Y_{1k} - Y_{2k}$), and the corresponding optimal reference parameter vector, $\boldsymbol{v}_0^* = (\boldsymbol{v}_{01}^*, \boldsymbol{v}_{02}^*)$, in the OECM $\boldsymbol{F}_{\theta^*}(\boldsymbol{v}_0^*) = (F_{1,\theta^*}(\boldsymbol{v}_{01}^*), F_{2,\theta^*}(\boldsymbol{v}_{02}^*))$, for several commonly used exponential families. These are obtained as solutions of the equation $\hat{F}_\theta(0) = 1$.

The most widely used approach for estimating rare-event probabilities is based on *large deviation* (LD) theory, which involves the function

$$I(y) = \sup_{\theta \in \Theta}(\theta y - \Lambda(\theta)).$$

$I(y)$ is variously called the *LD rate function*, the *Legendre transform*, the *Legendre–Fenchel transform*, or the *Cramér transform*.

Glasserman and Kou (1995) analyze rare events for tandem Jackson queues in transient setting, namely, where a rare event corresponds to the network population starting at zero and reaching a prescribed number x before returning to zero. Using Parekh and Walrand's (1989) heuristic, based on *interchanging the arrival rate and the smallest service rate, with all other service rates remaining unchanged*, it was shown in Glasserman and Kou (1995) that in certain parameter regions, that heuristic IS estimator has linearly bounded and even constant relative error, while

Table 9.3.1 Optimal Parameter, θ*, and Optimal Reference Parameter Vector, $\boldsymbol{v}_0^* = (\boldsymbol{v}_{01}^*, \boldsymbol{v}_{02}^*)$, for Several Commonly Used Exponential Families

F_1, F_2	$\boldsymbol{v}_1, \boldsymbol{v}_2$	$\boldsymbol{v}_{01}^*, \boldsymbol{v}_{02}^*$	θ^*
Exp(·)	λ, μ	μ, λ	$\mu - \lambda$
Gamma(·,·)	$(\lambda, \beta), (\mu, \gamma)$	$(\lambda + \theta^*, \beta), (\mu - \theta^*, \gamma)$	$\lambda^\beta \mu^\gamma$
Poisson(·)	λ, μ	μ, λ	$\log(\lambda\mu)$

in other regions it is not even asymptotically efficient. [An IS estimator is said to be *asymptotically efficient* (Heidelberger, 1993), if for some c,

$$\delta \mathbb{E}_{v_0}\{IW^2\} \to 2c,$$

as $\delta \to 0$.] Also, for queues in tandem, the slowest service rate corresponds to the most congested queue. Clearly, the complexity of such a heuristic IS estimator depends on the relationship between the traffic intensity of the most congested (bottleneck) queue and the remaining ones, as well as on the rarity of the event under consideration (Glasserman and Kou, 1995). For additional references on LD theory for estimation of rare-event probabilities see Chang et al. (1994, 1995), Cottrell et al. (1983), Kriman and Rubinstein (1997), Frater et al. (1991), Bucklew et al. (1990), Sadowsky and Bucklew (1990), and Sadowsky (1991, 1993, 1994); for LD theory in general, see Bucklew (1990), Dembo and Zeitouni (1993), and Shwartz and Weiss (1995).

Although the LD approach to rare-event probability estimation is supported by a solid mathematical foundation and broad generality, it appears that it has so far enjoyed only limited success. In fact, for more complex models, LD theory typically leads to variational problems that do not have explicit solutions. For example, Frater and Anderson (1989) shows that finding the OECM for open Jackson networks using LD theory reduces to the solution of a complex minimax problem.

In Section 9.5 we shall present an alternative to the LD approach for estimating the OECM reference parameter vector $\boldsymbol{v}_0^*$ (and $\boldsymbol{\theta}^*$) in networks of queues, namely, a modified version of Algorithm 9.4.1. We first introduce, however, *conventional and switching* regenerative estimators.

9.3.2 Conventional and Switching Regenerative Estimators

Conventional Regenerative (CR) Estimators

Suppose we wish to estimate the steady-state probability $\ell(x) = P\{L > x\}$, say, the steady-state probability of excessive backlog in a $GI/GI/1$ queue, or the steady-state probability of buffer overflow in a $GI/GI/1/m$ queue ($m = x$ is the buffer size). According to the conventional regenerative LR approach (Asmussen et al., 1994), $\ell(x)$ can be written as

$$\ell(x) = P\{L_t > x\} = \frac{\mathbb{E}_g\left\{\sum_{t=1}^{\tau} I_{\{L_t > x\}} \tilde{W}_t\right\}}{\mathbb{E}_g\left\{\sum_{t=1}^{\tau} \tilde{W}_t\right\}} = \frac{\ell_1(x)}{\ell_2}, \tag{9.3.6}$$

where L_t is the number of customers in the queue just prior to the arrival of the tth customer, τ is the length of the regenerative cycle, and $\{\tilde{W}_t \colon t = 1, \ldots, \tau\}$ is the *likelihood ratio process*. For example, in a $GI/G/1$ queue with L_t being the number of customers, $\tilde{W}_t$ can be written (see Section 6.2) as

$$\tilde{W}_t = \prod_{k=1}^{t} \frac{f_1(Z_{1k}, \boldsymbol{v}_1)}{f_1(Z_{1k}, \boldsymbol{v}_{01})} \prod_{k=1}^{D(t)} \frac{f_2(Z_{2k}, \boldsymbol{v}_2)}{f_2(Z_{2k}, \boldsymbol{v}_{02})}, \tag{9.3.7}$$

where $\{Z_{1k}\}$ and $\{Z_{2k}\}$ are the sequences of interarrival times and service times with IS pdf's $f_1(\mathbf{z}, \boldsymbol{v}_{01})$ and $f_2(\mathbf{z}, \boldsymbol{v}_{02})$, respectively, and $D(t)$ denotes the serial number of the last customer being served just prior to the arrival of customer t at the queue $[D(t) < t]$.

A consistent estimator of $\ell(x)$ is

$$\bar{\ell}_N(x) = \frac{\sum_{i=1}^{N} \sum_{t=1}^{\tau_i} I_{\{L_{ti} > x\}} \tilde{W}_{ti}(\boldsymbol{v}, \boldsymbol{v}_0)}{\sum_{i=1}^{N} \sum_{t=1}^{\tau_i} \tilde{W}_{ti}(\boldsymbol{v}, \boldsymbol{v}_0)} = \frac{\bar{\ell}_{1N}(x)}{\bar{\ell}_{2N}}, \tag{9.3.8}$$

where τ_i is the length of the ith regenerative cycle, and N is the number of regenerative cycles in the simulation run.

Under the assumption that the dominating distributions, G_1 and G_2, belong to the same parametric family of distributions as $F_1(y, \boldsymbol{v}_1)$ and $F_2(y, \boldsymbol{v}_2)$, respectively, Asmussen, Rubinstein and Wang (1994) calculated explicitly the variance of the estimator $\bar{\ell}_N(x)$, for the steady-state waiting time process in the $M/M/1$ queue. They showed that in order to obtain variance reduction with $\bar{\ell}_N(x)$ relative to the CMC estimator, one has to choose G_1 and G_2 such that the associated traffic intensity, ρ_0 [under (G_1, G_2)], is moderately larger than the original traffic intensity, ρ (under (F_1, F_2)]. Asmussen and Rubinstein (1995) proved that the LR estimator, $\bar{\ell}_N(x)$, is in fact exponential-time.

Switching Regenerative (SR) Estimators

Switching estimators were introduced independently by several authors (e.g., Devetsikiotis and Townsend, 1992; Nicola et al., 1993; Heidelberger, 1993; Rubinstein and Shapiro, 1993). The underlying idea is to change the IS distribution, G, *dynamically* within the cycles. For example, one may use OECM ($G = F_{\theta^*}$) until a nonzero value of $I_{\{L_{ti} > x\}}$ occurs and then switch to the original distribution ($G = F$) for the rest of the cycle. In so doing, the process $\{L_t\}$ naturally returns to the regenerative state. We shall only treat the case where the IS distribution, G, is an ECM cdf, F_{θ_1}, during $t \le \zeta(x)$, and is replaced by F_{θ_2} during $\zeta(x) < t \le \tau$, that is,

$$G_{\theta_1, \theta_2} = \begin{cases} F_{\theta_1}, & t \le \zeta(x), \\ F_{\theta_2}, & \zeta(x) < t \le \tau, \end{cases} \tag{9.3.9}$$

where $\zeta(x)$ is a *random variable* (stopping time), defined by

$$\zeta(x) = \inf\{t\colon L_t \ge x\}.$$

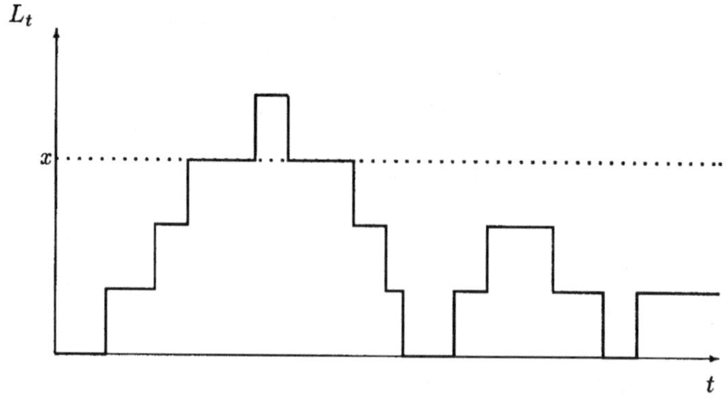

Figure 9.1. Sample path showing buffer overflow in a $GI/G/1$ queue.

In this case, the likelihood ratio can be written as

$$\hat{W}_t = \hat{W}_t(\zeta(x)) = \begin{cases} \prod_{j=1}^{t} W(\boldsymbol{Z}_j, \theta_1) & \text{if } t \le \zeta(x) \\ \prod_{j=1}^{\zeta(x)} W(\boldsymbol{Z}_j, \theta_1) \prod_{j=\zeta(x)+1}^{t} W(\boldsymbol{Z}_j, \theta_2) & \text{if } \zeta(x) < t \le \tau \end{cases} \tag{9.3.10}$$

where, as usual, $W(\boldsymbol{Z}, \theta) = f(\boldsymbol{Z})/f_\theta(\boldsymbol{Z})$ and $\boldsymbol{Z} \sim f_\theta(\mathbf{z})$. A typical sample path for the buffer overflow in a $GI/G/1$ queue is depicted in Figure 9.1.

In the special case where F_{θ_2} coincides with the original distribution ($F \equiv F_0$), equations (9.3.9) and (9.3.10) reduce to

$$G_{\theta_1,0} \equiv G_\theta = \begin{cases} F_\theta & t \le \zeta(x) \\ F_0 & \zeta(x) < t \le \tau \end{cases} \tag{9.3.11}$$

and

$$\hat{W}_t = \hat{W}_t(\zeta) = \begin{cases} \prod_{j=1}^{t} W(\boldsymbol{Z}_j, \theta) & \text{if } t \le \zeta(x) \\ \prod_{j=1}^{\zeta(x)} W(\boldsymbol{Z}_j, \theta) & \text{if } \zeta(x) < t \le \tau \end{cases} \tag{9.3.12}$$

respectively, where for convenience we denote $\theta_1 = \theta$. Formula (9.3.11) means that from the beginning of the cycle up to the first overflow one uses the ECM F_θ, and then one proceeds with the cycle, having switched to the original distribution, F.

The LR estimator (9.3.8), with $\hat{W}_t$ given as per (9.3.10) or (9.3.12), will be called the *SR estimator*. Note that the SR estimator differs from the conventional regenerative estimator *only* in the LR term, $\hat{W}_t$. Note further that some cycles in the sample path, generated under the ECM F_θ, might terminate without crossing level x (see Figure 9.1), and consequently, without switching.

Consider next the transient setting. Suppose we wish to estimate the rare-event probability

$$\ell(x) = P\{T_x < T_0\}, \tag{9.3.13}$$

where T_x is the first time the queue length reaches x, and T_0 is the first time the queue empties. This case can be viewed as a special case of the SR estimator, with F_{θ_2} in (9.3.10) corresponding to $W(\boldsymbol{Z}_j, \theta_2) \equiv 0$. Operationally, this means that if level x is reached, one switches to a server with infinite service rate. The corresponding likelihood ratio, $\hat{W}_t$, can be written as

$$\hat{W}_t = \hat{W}_t(\zeta(x)) = \begin{cases} \prod_{j=1}^{t} W(\boldsymbol{Z}_j, \theta_1) & \text{if } t \le \zeta(x) \\ 0 & \text{if } \zeta(x) < t \le \tau. \end{cases} \tag{9.3.14}$$

Assume now that the pair (θ_1, θ_2) in G_{θ_1,θ_2} has been chosen such that the variance of the SR estimator is finite.

Proposition 9.3.1. *Let* $\{L_t\}$ *be the waiting time process in the* $GI/G/1$ *queue. Consider the random walk setting* (9.2.7) *with* (9.2.6) *being used in estimating* $\ell_1(x) = \mathbb{E}\{\sum_{t=1}^{\tau} I_{\{L_t > x\}}\}$. *Then the SR estimator* [*see* (9.3.8)]

$$\bar{\ell}_{1N}(x) = \frac{1}{NT} \sum_{i=1}^{N} \sum_{t=1}^{\tau_i} I_{\{L_{ti} > x\}} \hat{W}_{ti},$$

with $\hat{W}_t$ *given in* (9.3.10), *is polynomial-time if and only if* $F_{\theta_1} = F_{\theta^*}$, *where* θ^* *is the unique solution of* $\hat{F}(\theta) = 0$.

Proof. See Asmussen and Rubinstein (1995). □

It is interesting to note that the choice of θ_2 in F_{θ_2} is irrelevant in Proposition 9.3.1, provided the variance of the SR estimator, $\bar{\ell}_{1N}(x)$, is finite. Clearly, choosing $F_{\theta_2} = F$ is not precluded.

Consider now the ratio estimator $\bar{\ell}_N(x) = \bar{\ell}_{1N}(x)/\bar{\ell}_{2N}$ in (9.3.8). Here, $\bar{\ell}_{2N}(x)$ may be defined in several ways. The two most obvious ones seem to be as follows:

(a) Estimate both $\bar{\ell}_{1N}(x)$ and $\bar{\ell}_{2N}$ from *the same sample path (simulation run)*, governed by the cdf G_{θ^*,θ_2}.
(b) Estimate $\bar{\ell}_{1N}(x)$ as above and $\bar{\ell}_{2N}$ from an *independent simulation run* using the CMC method.

Asmussen and Rubinstein (1995) proved that the SR estimator, $\bar{\ell}_N(x)$, for waiting time tail probabilities in the $GI/G/1$ queue with G_{θ^*,θ_2}, is polynomial-time in both cases (a) and (b).

Finally consider the SR estimator for excessive backlog in the $GI/G/1$ queue. In this case, $\hat{W}_t$ in (9.3.12), with $\theta = \theta^*$, can be written as

$$\hat{W}_{ti}(\theta^*, \zeta(x)) = \begin{cases} \prod_{k=1}^{t} \dfrac{f_1(Z_{1ki})}{f_{1\theta^*}(Z_{1ki})} \prod_{k=1}^{D(t)} \dfrac{f_2(Z_{2ki})}{f_{2\theta^*}(Z_{2ki})} & 1 \le t \le \zeta(x) \\ \prod_{k=1}^{\zeta(x)} \dfrac{f_1(Z_{1ki})}{f_{1\theta^*}(Z_{1ki})} \prod_{k=1}^{D(\zeta(x))} \dfrac{f_2(Z_{2ki})}{f_{2\theta^*}(Z_{2ki})} & \zeta(x) \le t \le \tau \end{cases} \tag{9.3.15}$$

where $\boldsymbol{f}_{\theta^*} = (f_{1\theta^*}, f_{2\theta^*})$, and $f_{1\theta^*}, f_{2\theta^*}$ are the IS pdf's of the service times and the interarrival times, respectively.

Under some mild regularity conditions, Kriman and Rubinstein (1997) prove that the SR estimators (9.3.8) and (9.3.15) are polynomial-time.

*9.3.3 Robustness of Switching Estimators

This section addresses the following question: How much can one perturb the parameter θ of the ECM from the optimal value, θ^*, and still ensure that the IS estimator remains polynomial-time or at least has a manageable (low) variance? The practical motivation for this question stems from the fact that the OECM may be difficult to compute exactly for more complicated queueing models.

The following two propositions address robustness for the random walk setting and the regenerative setting.

Proposition 9.3.2. *Consider the random walk setting* (9.2.7) *with* (9.2.6). *Assume that* $\theta = \theta(x)$ *varies with* x, *such that* $\theta \to \theta^*$. *Then a necessary and sufficient condition for the estimators* $\hat{\ell}(x)$, *in* (9.2.8) *and* (9.3.2), *to be polynomial-time is*

$$\theta(x) - \theta^* = O\left(\sqrt{\frac{\log x}{x}}\right) \tag{9.3.16}$$

Proof. See Asmussen and Rubinstein (1995). □

Proposition 9.3.3. *Consider the SR estimator* $\bar{\ell}_N(x)$ *in* (9.3.8) *with* (9.3.10). *Assume that* $\theta = \theta(x)$ *varies with* x, *such that* $\theta \to \theta^*$. *Then a necessary and sufficient condition for* $\bar{\ell}_N(x)$ *to be polynomial-time is* (9.3.16).

Proof. See Asmussen and Ruinstein (1995). □

Note that Propositions 9.3.2 and 9.3.3 pertain to the terminating and steady-state settings of the queue, respectively.

The robustness of the SR estimator $\bar{\ell}_N(x)$ in (9.3.8), (9.3.12) will be considered next. Kriman and Rubinstein (1997) proved that if $\theta \neq \theta^*$ in G_θ, then the estimator (9.3.8), (9.3.12) is exponential-time. They also derived explicit expressions for the computational cost, T_θ, under $G_\theta \neq G_{\theta*}$. Recall that in regenerative setting, the computational cost T is defined as

$$T = N\mathbb{E}\{\tau\}$$

rather than as $N = N(x)$ as in Section 9.2, N being the number of requisite cycles. Recall further that the former complexity measure is in terms of *time per customer* rather than *time per cycle*. When $\theta \neq \theta^*$, then the computational cost of the estimator $\bar{\ell}_N(x)$ given in (9.3.8), (9.3.12) may be characterized as follows.

Proposition 9.3.4. *Let G_θ be an IS distribution satisfying*

1. $\mathbb{E}_G\{Y_n\} > 0$.
2. $\int_{-\infty}^{\infty} \int g^2(y)/f(y)\, dy = C < \infty$, *where g is the pdf of G.*
3. *There exists a unique solution, $\theta^{**} > 0$, for the equation*

$$\int_{-\infty}^{\infty} \frac{f^2(y)}{g(y)} \exp(-\theta^{**}y)\, dy = 1,$$

such that

$$-\infty < \int_{-\infty}^{\infty} y \frac{f^2(y)}{g(y)} \exp(-\theta^{**}y)\, dy < 0.$$

Then the simulation cost, $T_\theta(x)$, of the estimator $\bar{\ell}_N(x)$ is of the form

$$T_\theta(x) = \exp(zx + O(x)),$$

where $z = \theta^{**} - 2\theta^* > 0$.

Proof. Follows directly from Theorem 5.1 in Kriman and Rubinstein (1997). □

Table 9.3.2 displays several efficiency characteristics of the SR estimator (9.3.8), (9.3.12), under both the OECM distribution, G_{θ^*}, and the ECM distribution, G_θ ($\theta^* \neq \theta$), as functions of the traffic intensity, ρ, for the probabilities of excessive backlog $\ell(x) = 10^{-15}$, in the $M/M/1$ queue with service rate $\mu = 1$. In particular, under the OECM [see (9.3.9)], $G_{\theta^*} = G_{1\theta^*, 2\theta^*}$, $\theta^* = \mu - \lambda$, this table displays $S_{\theta^*} = T_F/T_{\theta^*}$, called the *optimal speed-up factor* (T_F is the computational cost under the original cdf, F); and under the ECM $G_\theta = G_{1\theta,2\theta}$, it displays $S_\theta = T_F/T_\theta$, called the *speed-up factor*, as well as z_θ, called the *exponential rate.*

Table 9.3.2 Efficiencies S_{θ^*}, S_θ, and z_θ, of the SR Estimator (9.3.8), (9.3.12), as Functions of Traffic Intensity, ρ, for $\ell(x) = 10^{-15}$ in the $M/M/1$ Queue with $\mu = 1$

ρ	0.2	0.4	0.6	0.8
θ^*	0.8	0.6	0.4	0.2
S_{θ^*}	10^{13}	10^{12}	10^{12}	10^{11}
θ	0.64	0.48	0.32	0.16
S_θ	10^{11}	10^{10}	10^{9}	10^{8}
z_θ	1.0×10^{-1}	5.6×10^{-2}	3.1×10^{-2}	1.3×10^{-2}
x	21	36	62	107

Notice that both S_θ and z_θ were calculated for θ *less* than the optimal value, θ^*, by a factor of 20%; that is, θ was calculated from the equation $(\theta - \theta^*)/\theta^* = 0.2$. Notice also that the values x, in the last row of Table 9.3.2, correspond to the probability of excessive backlog $\ell(x) = 10^{-15}$ and $z_{\theta^*} = 0$.

Table 9.3.2 indicates that the OECM G_{θ^*} leads to a dramatic simulation speed-up (variance reduction), S_{θ^*}, and that perturbing θ^* by 20% ($\theta < \theta^*$), results in *speed-up loss* of only 2–3%.

Kriman and Rubinstein (1997) extended their complexity results on the computational cost, T, of the SR estimator (9.3.8), (9.3.12) to the following cases:

1. Rare events in the $GI/D/1$ queue (using the push-out method)
2. Rare events in the $GI/G/1$ queue with dependent arrivals
3. Rare events in the $GI/G/1$ queue with batch arrivals
4. Sensitivities (derivatives) of the probability of the excessive backlog, $\ell(x, \boldsymbol{v})$, with respect to the parameter vector, $\boldsymbol{v}$, of the interarrival time distribution.

In particular, they proved that the computational cost of the SR estimator of $\nabla\ell(\boldsymbol{v})$, obtained by differentiating (9.3.8) with respect to $\boldsymbol{v}$, is of the same order of magnitude as the computational cost of the SR estimator (9.3.8), (9.3.12). Furthermore, using the optimal parameter (OP) value, θ^*, in the estimator of $\nabla\ell(\boldsymbol{v})$ results again in a polynomial-time SR estimator. It follows that both the rare-event probability, $\ell(x)$, and its derivatives, $\nabla\ell(x)$, can be estimated *simultaneously and quickly* from a single simulation run.

We next present numerical results pertaining to the efficiency of the SR estimator (9.3.8), (9.3.12), for the $M/M/1$ queue, as well as other queues. All cases simulated 10^6 customers and estimated the probability of the excessive backlog. The SR estimator was computed for the following two cases: the optimal parameter, θ^*, corresponding to the distributions, G_{θ^*}, and the perturbed values of θ^* (denoted by θ), corresponding to G_θ. In all tables, values in the first column marked with asterisk correspond to the OECM parameter θ^*.

Table 9.3.3 displays the computational cost (simulation time), $\tilde{T}_\theta$, corresponding to the relative width of the 95% confidence interval for $\ell(x)$; the ratio $R_\theta = \tilde{T}_\theta/\tilde{T}_{\theta^*}$ [approximation of $\exp(zx)$], called the *loss factor*; and the exponential rate, z. These statistics were computed as functions of θ [and the relative perturbation $\delta = (\theta - \theta^*)/\theta^*$], while estimating the probability $\ell(x) = 5.34 \times 10^{-10}$ $(x = 40)$ in the $M/M/1$ queue with $\rho = 0.6$.

Tables 9.3.4–9.3.6 display data similar to those of Table 9.3.3. Table 9.3.4 corresponds to the $M/D/1$ queue with $\rho = 0.3$, $d = 1$, and $\ell(x) = 1.04 \times 10^{-9}$ $(x = 20)$; Table 9.3.5 corresponds to the $M/M/1$ queue with $\rho = 0.5$, $\mu = 1$, $\ell(x) = 5.34 \times 10^{-10}$ $(x = 40)$ and a (dependent) AR(1) interarrival sequence

$$A_{k+1} = \theta A_k + (1 - \theta)X_k, \quad X_k \sim \text{Exp}(\lambda), \quad \theta = 0.3;$$

and Table 9.3.6 displays results for the derivative $\nabla_\lambda \ell(x)$ with respect to the arrival rate, λ, in the $M/M/1$ queue with $\rho = 0.3$, $\mu = 1$, $\ell(x) = 1.44 \times 10^{-16}$ $(x = 30)$, and $\nabla_\lambda \ell(x) = 1.42 \times 10^{-14}$.

Tables 9.3.3–9.3.6 indicate that the RS estimators (9.3.8), (9.3.12) are robust with respect to small and moderate perturbations of θ^*, in the sense that for relative perturbations $|\delta| < 0.2$, one still attains dramatic variance reductions. Similar results were obtained for the $GI/G/1$ queue with various interarrival time and service time distributions. Extensive experimentation with SR estimators for simple queueing models also suggests that once found, the optimal parameter, θ^*, may be used for *simultaneous estimation of all probabilities*, $\ell(x)$, of order 10^{-3} or less.

Table 9.3.3 Efficiency of RS Estimator (9.3.8), (9.3.12) as a Function of θ, for the $M/M/1$ Queue with $\rho = 0.6$, $\mu = 1$, and $\ell(x) = 5.34 \times 10^{-10}$ $(x = 40)$

θ	δ	$\tilde{T}_\theta$	R_θ	exp(zx)	z
0.30	−0.25	1.20×10^6	4.5	6.1	4.53×10^{-2}
0.35	−0.125	5.87×10^5	2.2	1.7	1.33×10^{-2}
0.40*	0.00	2.76×10^5	1.0	1.0	0.0
0.43	0.075	4.27×10^5	1.6	1.3	1.36×10^{-2}
0.46	0.15	1.52×10^6	5.5	5.2	4.13×10^{-2}

Table 9.3.4 Efficiency of RS Estimator (9.3.8), (9.3.12) as a Function of θ, for the $M/D/1$ Queue with $\rho = 0.3$, $d = 1$, and $\ell(x) = 1.04 \times 10^{-9}$ $(x = 20)$

θ	δ	$\tilde{T}_\theta$	R_θ	exp(zx)	z
1.3	−0.368	1.11×10^7	39.42	27.1	1.65×10^{-1}
1.6	−0.22	1.55×10^5	5.48	3.46	6.20×10^{-2}
2.0645*	0.0	2.84×10^4	1.0	1.0	0.0
2.4	0.19	7.51×10^4	2.65	2.27	4.10×10^{-2}
2.7	0.31	1.06×10^7	37.43	48.4	1.94×10^{-2}

Table 9.3.5 Efficiency of the RS Estimator (9.3.8), (9.3.12) as a function of θ, for the $M/M/1$ Queue with AR(1) Arrival Process, $\rho = 0.5$, $\mu = 1$, $\theta = 0.3$ and $\ell(x) = 5.34 \times 10^{-10}$ ($x = 40$)

θ	δ	$\tilde{T}_\theta$	R_θ	exp(zx)	z
0.40	−0.2	1.08×10^6	4.2	3.51	1.22×10^{-2}
0.45	−0.1	5.13×10^5	2.0	1.44	4.19×10^{-2}
0.50*	0.0	2.55×10^5	1.0	1.0	0.0
0.54	0.08	5.43×10^5	2.1	1.45	1.23×10^{-2}
0.58	0.16	1.91×10^6	7.5	10.3	7.78×10^{-2}

Table 9.3.6 Efficiency of the RS Estimator (9.3.8), (9.3.12) as a Function of θ, for the $M/M/1$ Queue with $\rho = 0.3$, $\mu = 1$, $\ell(x) = 1.44 \times 10^{-16}$ ($x = 30$) and $\nabla_\lambda \ell(x) = 1.42 \times 10^{-14}$

θ	δ	$\tilde{T}_\theta$	R_θ	exp(zx)	z
0.50	−0.285	1.03×10^7	40.7	53.9	1.33×10^{-2}
0.60	−0.143	8.75×10^5	3.75	3.53	4.21×10^{-2}
0.70*	0.0	2.53×10^5	1.0	1.0	0.0
0.75	0.07	3.68×10^5	1.45	1.9	2.14×10^{-2}
0.80	0.143	3.17×10^7	124.9	171.2	1.71×10^{-1}

9.4 RARE-EVENT PROBABILITY ESTIMATION FOR DESS

This section treats the estimation of rare-event probabilities for discrete-event static systems (DESS). It will be shown here how to modify Algorithm 8.4.1 in order to find the optimal reference parameter vector $\boldsymbol{v}_0^*$ for the probability of rare events of the form $A = P\{L > x\}$.

Let

$$\bar{\ell}_N(\boldsymbol{v}, \boldsymbol{v}_0) = \frac{1}{N} \sum_{i=1}^{N} I_{\{L(\boldsymbol{Z}_i) > x\}} W(\boldsymbol{Z}_i, \boldsymbol{v}, \boldsymbol{v}_0), \tag{9.4.1}$$

be an LR estimator of $\ell(x) = P\{L > x\}$, where

$$W(\boldsymbol{Z}, \boldsymbol{v}, \boldsymbol{v}_0) = \frac{f(\boldsymbol{Z}, \boldsymbol{v})}{f(\boldsymbol{Z}, \boldsymbol{v}_0)},$$

and $\{\boldsymbol{Z}_1, \ldots, \boldsymbol{Z}_N\}$ is a random sample from the ECM pdf $f(\boldsymbol{Z}, \boldsymbol{v}_0)$. Consider the program (8.4.1) with $L^2(\boldsymbol{Y})$ replaced by the indicator function $I_A = I_{\{L(\boldsymbol{y}) > x\}}$. How should one choose a good IS density in this case? It follows directly from the expression for $\mathscr{L}(\boldsymbol{v}, \boldsymbol{v}_0)$ in (8.4.1) that in order to reduce the variance, the likelihood ratio $W(\boldsymbol{Z}, \boldsymbol{v}, \boldsymbol{v}_0)$ must be small on the event A. Since A is a rare event set, we expect that $f(\boldsymbol{y}, \boldsymbol{v})$ be of low magnitude on A. Thus, to make the likelihood ratio $W(\boldsymbol{Z}, \boldsymbol{v}, \boldsymbol{v}_0)$ small on A, one should choose the IS density $f(\boldsymbol{y}, \boldsymbol{v}_0)$ so that $f(\boldsymbol{y}, \boldsymbol{v}_0)$ is large on A. In so doing, the event A is rendered likelier to occur, that is, A is no longer a rare event under $f(y, v_0)$.

Let us revisit Algorithm 8.4.1 in a form suitable for estimating v_0^* in the standard sample performance $L(Y)$. It is readily seen that Algorithm 8.4.1 is useless for estimating v_0^* in a rare-event context, since owing to the rarity of the events $\{L(Y_i) > x\}$, the random variables $I_{\{L_i>x\}}$, $i = 1, \ldots, N$, and the associated derivatives of $\hat{\mathscr{L}}_N(v, v_0, v_1)$ [see (8.4.6)] vanish with high probability, as long as the sample size is not too large, say $N < 10^4$. With (9.4.5) in mind we can, however, modify Algorithm 8.4.1 by iterating *adaptively* on both parameters x and v_1. We start the algorithm by choosing an initial x_0 ($x_0 < x$), such that under the original pdf, $f(y, v)$, the probability $\ell(x_0) = \mathbb{E}_v\{I_{\{L>x_0\}}\}$ is not too small, say $\ell(x_0) = \alpha_0 \approx 10^{-2}$. We then proceed as follows: We start the algorithm by choosing, say $v_1(0) = v$ (see step 1) and some initial x_0 ($x_0 < x$) [see program (9.4.3)] such that under the original pdf, $f(y, v)$, the probability $\ell(x_0) = \mathbb{E}_v\{I_{\{L>x_0\}}\}$ is not too small, say $\ell(x_0) \equiv \alpha_0 \approx 10^{-2}$, and then iterate in both x and v_1.

Algorithm 9.4.1 Choice of a Good v_1 and Optimal v_0^* for $\ell(v) = \mathbb{E}_v\{I_{\{L(Y)>x\}}\}$

1. *Choose an initial point $v_1(0)$, say $v_1(0) = v$.*
2. *Generate a sample $\{Z_1, \ldots, Z_N\}$ from the pdf $f(z, v_1(0))$, and solve the stochastic counterpart*

$$\max\left\{y: \frac{1}{N}\sum_{i=1}^{N} I_{\{L(Z_i)>y\}} \geq \alpha_0\right\} \tag{9.4.2}$$

of the program

$$\max\left\{y: \mathbb{E}\{I_{\{L(Z)>y\}}\} \geq \alpha_0\right\}. \tag{9.4.3}$$

Denote the solution of (9.4.3) *and its stochastic counterpart* (9.4.2) *by x and $\bar{x}_0$, respectively. It is readily seen that, in fact,*

$$\bar{x}_0 = L_{(\lfloor(1-\alpha_0)N\rfloor)},$$

where $L_{(i)}$ is the ith-order statistics of the sequence $\{L_i = L(Z_i)$: $i = 1, \ldots, N\}$. For $\alpha_0 = 10^{-2}$, this reduces to

$$\bar{x}_0 = L_{(\lfloor 99/100N\rfloor)}.$$

3. *Estimate the optimal solution, v_0^*, of program* (8.4.1) *(with $L^2 \equiv I_{\{L(Z_i)>\bar{x}_0\}}$) by solving the stochastic counterpart* (8.4.6) *for $\bar{x}_0$. Denote the solution by $\bar{v}_0^*(\bar{x}_0)$.*
4. *Set $v_1(1) = \bar{v}_0^*(\bar{x}_0)$, and repeat step 2 with $v_1(0)$ replaced by $\bar{v}_0^*(\bar{x}_0)$. Denote the solution by $\bar{x}_1$. Repeat step 3, with $v_1(0)$ and $\bar{x}_0$ replaced by $\bar{v}_0^*(\bar{x}_0)$ and $\bar{x}_1$,*

respectively, and denote the solution by $\mathbf{v}_0^*(\bar{x}_1)$. *Denote the corresponding solutions at stage k by* $\bar{\mathbf{v}}_0^*(\bar{x}_k)$ *and* $\bar{x}_k$, *respectively.*

5. *If*

$$\bar{x}_k \geq x, \qquad k = 0, 1, 2, \ldots,$$

then set $\bar{x}_k = x$, *perform step 3 once more and take the resultant solution* $\bar{\mathbf{v}}_0^*(x)$ *as an estimate of the optimal solution of the program* (8.4.1) (*with* $L^2 \equiv I_{\{L(\mathbf{Z}_i)>x\}}$). *Otherwise go to step 2.*

It is readily seen that Algorithm 8.4.1 is a special case of Algorithm 9.4.1 in which x does not change.

To clarify the main steps of Algorithm 9.4.1, consider the following simple example.

Example 9.4.1 (Example 9.2.1 continued). Consider the estimation of $\ell(x) = P\{Y > x\}$, where $Y \sim \text{Exp}(\lambda)$. Assume without loss of generality that $\lambda = 1$, and choose the initial value x_0 corresponding to $\alpha_0 = \ell(y) = e^{-s}$. Noting that the root of $\ell(y) = e^{-y} = e^{-s}$ is $\bar{x}_0 = s$ and in view of (5.3.9), that is,

$$\lambda_0^*(x) = \lambda + x^{-1} - (\lambda^2 + x^{-2})^{1/2},$$

we obtain $\lambda_0^*(\bar{x}) = 1 + s^{-1} - [1 + s^{-2}]^{1/2} \approx s^{-1}$. This computation corresponds to step 3 of Algorithm 9.4.1.

To find $\bar{x}_1$ as per step 4 of Algorithm 9.4.1, we need to solve the program

$$\max\{y \colon \mathbb{E}\{I_{\{L(\mathbf{Z})>y\}}\} \geq \alpha_0\},$$

which in our case simply reduces to the solution of $\ell(y) = e^{-s}$ in y, with $\lambda = 1$ replaced by $\bar{\lambda}_0^*(\bar{x}_0) \approx s^{-1}$. The root of $\ell(y) \approx e^{-ys^{-1}} = e^{-s}$ is $\bar{x}_1 \approx s^2$, that is, $\bar{x}_1 = \bar{x}_0^2$. It is readily seen that $\bar{x}_r \approx s^{r+1}$. We find that if the initial and the target values are $s = 2$ and $x = 16$, 64 and 128 [corresponding to $\ell(x) \approx 10^{-7}, 10^{-30}$ and 10^{-60}], respectively, then Algorithm 9.4.1 terminates on the average after 4, 6 and 7 iterations, respectively.

A possible three-stage $(x_0, \bar{x}_1, \bar{x}_3 = x)$ realization of Algorithm 9.4.1, while estimating $\alpha = P\{Y > x\}$, $Y \sim \text{Exp}(\lambda)$, is depicted in Figure 9.2. Observe that the quantile x_0 of $\ell(x_0) = \alpha_0 \approx 10^{-2}$ can be readily estimated from a pilot run. The auxiliary (random) sequence $\bar{x}_0, \bar{x}_1, \bar{x}_2, \ldots$ was designed here to be monotonic increasing with high probability, that is,

$$\bar{x}_0 < \bar{x}_1 < \bar{x}_2 < \cdots < \bar{x}_i < \cdots = x \tag{9.4.4}$$

with high probability; furthermore, the target value x is reached after just a few iterations.

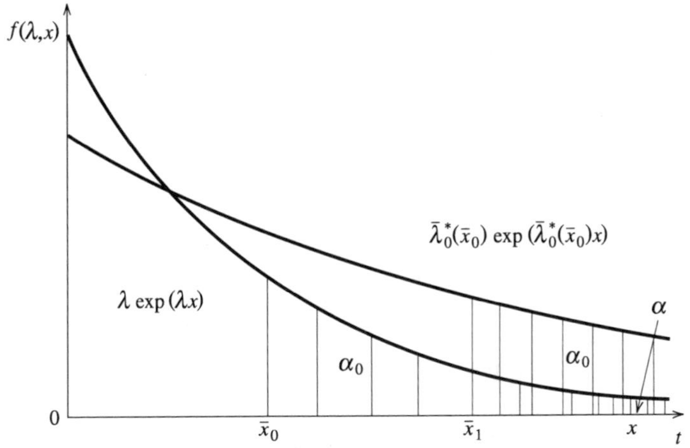

Figure 9.2. Three-stage realization of Algorithm 9.4.1.

The monotonicity of the sequence $\bar{x}_0, \bar{x}_1, \bar{x}_2, \ldots$ is an inherent part of Algorithm 9.4.1. Simulation experiments with static models, such as shortest path models, show that this is indeed the case (see also Section 9.7). The following proposition validates formula (9.4.4) for the gamma pdf.

Proposition 9.4.1. *Let $Y \sim G(\lambda, \beta)$. Suppose that $L^2(y)$ is a monotonically increasing function on the interval* $[0, \infty)$. *Then*

1. *The random sequence $\bar{x}_0, \bar{x}_1, \bar{x}_2, \ldots$ generated by Algorithm 9.4.1 is monotonic increasing, that is,* (9.4.4) *holds with high probability.*
2. *The target value x is reached with high probability in a finite number of iterations.*

Proof. We prove Part 1 of Proposition 9.4.1 first. Consider Algorithm 9.4.1, but where the stochastic counterparts (8.4.6) and (9.4.2) are replaced by the original programs (8.4.1) and (9.4.3), respectively. In this case, the random sequence $\bar{x}_0, \bar{x}_1, \bar{x}_2, \ldots$ becomes deterministic, say, $x_0, x_1, x_2, \ldots$, and by virtue of Proposition 8.2.2, the sequence $x_0, x_1, x_2, \ldots$ is monotonic increasing, since for any fixed λ and x_k, $k = 0, 1, \ldots$ [see (8.5.5)]

$$\lambda_0^*(x_0) < \lambda, \; \lambda_0^*(x_1) < \lambda_0^*(x_0), \ldots, \lambda_0^*(x_k) < \lambda_0^*(x_{k-1}), \ldots$$

The proof of Part 1 of Proposition 9.4.1 now follows, since at each stage $k = 0, 1, \ldots$ we have that asymptotically in N the optimal solutions, $\bar{\boldsymbol{v}}_k^*$ and $\bar{x}_k$, of the stochastic counterparts (8.4.6) and (9.4.2) converge almost surely to $\boldsymbol{v}_k^*$ and x_k of the original programs (8.4.1) and (9.4.3), respectively.

To prove Part 2 of Proposition 9.4.1 consider first the original programs (8.4.1) and (9.4.3). The result follows the view of formula (9.4.4) and the facts

$\ell(\bar{x}_0) = \alpha_0 < 1$, $\ell(x) = \alpha > 0$ and the finiteness of the interval (x_0, x). The proof of Part 2 using the stochastic counterparts (8.4.6) and (9.4.2) is similar to Part 1.

Proposition 9.4.1 can be readily extended to the multidimensional gamma distribution with independent components and to some other pdf's such as multidemensional normal pdf, since Proposition 4.2.2 readily extends to such cases (see Rubinstein and Shapiro, 1993).

As mentioned earlier, there is no need to solve the programs (8.4.6) and (9.4.2) very accurately, at least in the first few stages. In particular one can perform several steps of Newton's algorithm. This is the reason why Proposition 9.4.1 uses the term "with high probability" rather then "with probability one," which assumes an infinite sample.

It is interesting to note that under the IS pdf $f(y, \boldsymbol{v}_0^*)$ one has

$$P\{L(\boldsymbol{Z}^*) > x\} = \mathbb{E}_{\boldsymbol{v}_0^*}\{I_{\{L(\boldsymbol{Z}^*)>x\}}\} = O(1), \tag{9.4.5}$$

where $\boldsymbol{Z}^* \sim f(y, \boldsymbol{v}_0^*)$. This is so since at each stage of Algorithm 9.4.1 we estimate a probability $\ell(x_0) = \mathbb{E}_{\boldsymbol{v}}\{I_{\{L>x_0\}}\}$, which is of order $O(1)$.

Lieber et al. (1994) considers coherent reliability models and discusses conditions under which the LR estimator (9.4.1) with $\boldsymbol{v}_0 = \boldsymbol{v}_0^*$ is polynomial-time in x. It also gives conditions under which one may change the probability measure from $F(y, \boldsymbol{v})$ to $F(y, \boldsymbol{v}_{0b})$, where $\boldsymbol{v}_{0b}$ is a *small subset* of the $\boldsymbol{v}_0$; $\boldsymbol{v}_{0b}$ is associated with the notion of bottleneck cut. The size of the vector $\boldsymbol{v}_{0b}$ is typically in the range 3–10, while the size of $\boldsymbol{v}_0$ might be on the order of hundreds. Finally, it presents an algorithm for screening out the bottleneck cut and the associated bottleneck parameter vector, $\boldsymbol{v}_{0b}$, from the network before simulation. In Section 9.7 we shall present experimental support for the contention that Algorithm 9.4.1 screens out adaptively (during the course of simulation) the optimal subset of values, $\boldsymbol{v}_{0b}^*$, when starting the simulation procedure with the entire set of ECM distributions, $F(y, \boldsymbol{v}_0)$.

9.5 RARE-EVENT PROBABILITY ESTIMATION IN QUEUEING NETWORKS

Published work on efficient simulation for rare-event probability estimation in queueing networks (e.g., Chang et al., 1994, 1995; Devetsikiotis and Townsend, 1992, 1993; Frater et al., 1991; Glasserman and Kou, 1995; Parekh and Walrand, 1989) is far less extensive than its counterpart on single queues and simple networks. As mentioned earlier, the LD approach based on ECM typically runs into complex variational optimization problems (e.g., Frater et al., 1991). A different approach is, therefore, called for. Although it is generally believed that the ECM is still the most promising approach, finding the OECM parameter vector $\boldsymbol{v}_0^* = \boldsymbol{v}_{01}^*$, and in particular, establishing the complexity properties of relevant IS estimators is fairly difficult.

This section presents an adaptive algorithm for consistently estimating the OECM parameter vector $\boldsymbol{v}_0^*$ for queueing networks, with some supporting experimental results appearing in Section 9.7. The identification of complexity properties of the associated estimators, and in particular conditions under which they are polynomial-time, is still an open problem.

The proposed algorithm is a modified version of Algorithm 9.4.1 for static models and uses the SR estimators (9.3.8), (9.3.12) as important ingredients. If not otherwise stated, we shall assume that the rare event of interest is "the total number of customers in the network exceeding a fixed number," say x. We point out that the proposed algorithm is suitable for estimating rare-event probabilities in both transient and steady-state settings. In the latter case, we assume that the underlying process $\{L_t\}$ is regenerative, while in the former case we do not. An alternative heuristic approach to estimating rare-event probabilities in queueing networks, which is not discussed here, was developed by Devetsikiotis and Townsend (1992, 1993).

Consider the stochastic counterpart (8.4.11), where the CR estimator is replaced by the SR estimators (9.3.8), (9.3.12). Algorithm 9.4.1 for static models can be readily modified to dynamic (queueing) models. The dynamic version of Algorithm 9.4.1 calls for modifying its step 2 as follows:

2. *Generate a sample* $\{\boldsymbol{Z}_{11}, \ldots, \boldsymbol{Z}_{\tau_1 1}, \ldots, \boldsymbol{Z}_{1N}, \ldots, \boldsymbol{Z}_{\tau_N N}\}$ *of N regenerative cycles from the pdf* $f(\boldsymbol{y}, \bar{\boldsymbol{v}}_1(0))$, *and solve the stochastic counterpart*

$$\max\left\{y : \frac{1}{N\tau}\sum_{i=1}^{N}\sum_{t=1}^{\tau_i} I_{\{L(Z_{ti})>y\}} \geq \alpha_0\right\} \tag{9.5.1}$$

of the program

$$\max\{y : \mathbb{E}\{I_{\{L(\boldsymbol{Z})>y\}}\} \geq \alpha_0\}.$$

Denote the solution of (9.5.1) *by* $\bar{x}_0$. *It is readily seen that, in fact,*

$$\bar{x}_0 = L_{(\lfloor(1-\alpha_0)N\rfloor)},$$

where $L_{(i)}$ *is the ith-order statistic of the sequence* $\{L_i = L(\boldsymbol{Z}_i) : i = 1, \ldots, N\}$.

The rest of Algorithm 9.4.1 remains intact. The resultant algorithm will be referred to as the *modified Algorithm* 9.4.1. Note that the sequence $\bar{x}_0, \bar{x}_1, \bar{x}_2, \ldots, x$ corresponds here to different level crossings by the SR estimator. Two possible sample path realizations, $L_t(\boldsymbol{v})$ and $L_t(\bar{\boldsymbol{v}}^*(\bar{x}_0))$, of Algorithm 9.4.1 with three level crossings $\bar{x}_0, \bar{x}_1$ and $\bar{x}_2 = x$ are depicted in Figure 9.3.

Notice also that the modified Algorithm 9.4.1 estimates the optimal reference parameter vector $\boldsymbol{v}_0^*$ for the numerator $\ell_1(\boldsymbol{v}, \boldsymbol{v}_0, x)$ in the switching estimators (9.3.8), (9.3.12) alone and then uses it for estimating rare-event probabilities $\ell(x)$ in the ratio formula (9.3.6). Thus, once found, the estimate of the optimal reference

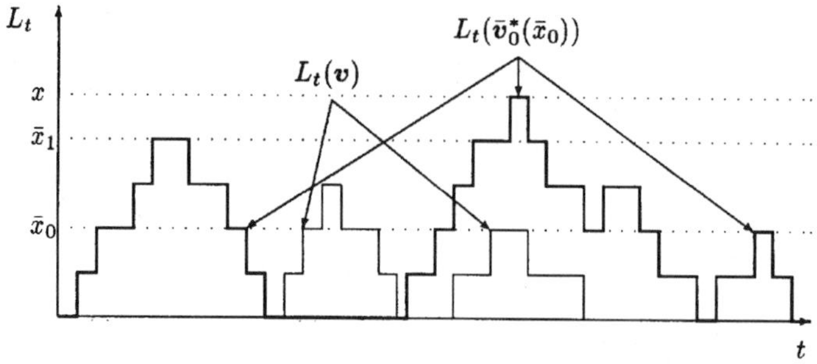

Figure 9.3. Three-stage realization of the modified Algorithm 9.4.1.

parameter $\boldsymbol{v}_0^*$ for the numerator $\ell_1(\boldsymbol{v}, \boldsymbol{v}_0, x)$ is used to estimate the entire ratio $\ell(x)$ in (9.3.6). A possible alternative is to estimate the optimal reference parameters from two simulations: one for the numerator $\ell_1(\boldsymbol{v}, \boldsymbol{v}_0, x)$ and another for the denumerator $\ell_2(\boldsymbol{v}, \boldsymbol{v}_0)$, and then combine them together into the switching estimators (9.3.8), (9.3.12). (Recall from Proposition 9.3.1 that for the $GI/G/1$ queue both alternatives result in polynomial-time estimators, provided one uses the optimal reference parameter vector, $\boldsymbol{v}_0^*$). Our numerical results suggest that the second alternative is preferable to the first one when the probability of a rare event is of order 10^{-6} and less. The reason is that for very small probabilities the switching estimator of the denumerator $\ell_2(\boldsymbol{v}, \boldsymbol{v}_0) = \mathbb{E}_{\boldsymbol{v}}\{\tau\}$ is less accurate then its standard LR counterpart.

As for static models, the optimal reference parameter vectors, $\boldsymbol{v}_0^*$, in the standard regenerative and switching regenerative estimators might be quite different in their order of magnitude. To see that, consider the following example.

Example 9.5.1. Let $\mathbb{E}_{\boldsymbol{v}}\{I_{\{L_t>x\}}\}$ and $\mathbb{E}_{\boldsymbol{v}}\{L_t\}$ be two performance measures. Consider first $\mathbb{E}_{\boldsymbol{v}}\{I_{\{L_t>x\}}\}$, where $\{L_t\}$ is an output process, say, the waiting time process in the $M/M/1$ queue. It is well known (e.g., Heidelberger, 1993) that the optimal $\hat{\boldsymbol{v}}_{01}^*$ in the SR estimators (9.3.8) and (9.3.12) corresponds to the interchange of λ and v, that is, it equals $\hat{\boldsymbol{v}}_{01}^* = (v, \lambda)$; furthermore, such an RS estimator is polynomial-time. For example, if $\lambda = 1$ and $v = 2$, then $\hat{\boldsymbol{v}}_{01}^* = (2, 1)$.

Consider next $\mathbb{E}_{\boldsymbol{v}}\{L_t\}$, with the same $\lambda = 1$ and $v = 2$ ($\rho = 0.5$). Assume that v_0 is the reference parameter. From Asmussen, Rubinstein and Wang (1994) one has $v_0^* \approx 1.4$ ($\rho_0^* \approx 0.72$). Thus, the optimal reference parameter vectors which minimize the variances of the corresponding CR and SR estimators are quite different.

We remark again that the purpose of the modified Algorithms 8.4.1 and 9.4.1 (when applied to Example 9.5.1) is to consistently estimate the optimal vectors $\boldsymbol{v}_0^*$, which for $\hat{\boldsymbol{v}}_{01}^* = (v, \lambda)$ in Example 9.5.1 become $\boldsymbol{v}_0^* = \hat{\boldsymbol{v}}_{01}^* = (2, 1)$ and $v_0^* \approx 1.4$, respectively.

Finally, note that since the reference parameter vectors $\boldsymbol{v}_0^*$ in the CR and SR estimators (for fixed $\boldsymbol{v}$) are quite different, one might use *two* separate simulations:

one for the standard performance function $\mathbb{E}_v\{L_t\}$ and another for the performance function $\mathbb{E}_v\{I_{\{L_t>x\}}\}$, using rare-event probability estimation. For example, separate simulations might be appropriate for $\mathbb{E}_v\{L_t\}$ and $\mathbb{E}_v\{I_{\{L_t>x\}}\}$ in the performance

$$\ell(\boldsymbol{v}) = \mathbb{E}_v\{L_t\} + \mathbb{E}_v\{I_{\{L_t>x\}}\}.$$

Simulation experiments with queueing models suggest, however, that a *single* simulation run with the reference parameter $\hat{\boldsymbol{v}}_{01}^* = (v, \lambda)$ (optimal value for $\mathbb{E}_v\{I_{\{L_t>x\}}\}$) is sufficient, provided $\mathbb{E}_v\{I_{\{L_t>x\}}\} < 10^{-6}$. That is, if $\mathbb{E}_v\{I_{\{L_t>x\}}\} < 10^{-6}$, then one may use the same SR estimator with the same optimal reference parameter, $\hat{\boldsymbol{v}}_{01}^*$ corresponding to $\mathbb{E}_v\{I_{\{L_t>x\}}\}$, for both the standard sample performances, $\mathbb{E}_v\{L_t\}$ and $\mathbb{E}_v\{I_{\{L_t>x\}}\}$, and to ensure that the accuracies of both runs, the single run and the double run, result in LR estimates of comparable magnitudes. Returning to Example 9.5.1, this would imply replacing the optimal value, $v_0^* \approx 1.4$ for $\mathbb{E}_v\{L_t\}$, by $\hat{\boldsymbol{v}}_{01}^* = (v, \lambda)$. Note again that as far as the $GI/G/1$ queue is concerned, Proposition 9.3.1 ensures that both alternatives (single and double simulation scenarios) result in polynomial-time estimators.

9.6 OPTIMIZATION WITH RARE EVENTS

Consider the optimization program (6.4.1), (6.4.2) and its stochastics counterpart (6.4.3), (6.4.4) for dynamic (queueing) models. Assume that the output processes $\{L_{jt}\}$, $j = 0, 1, \ldots, M$ (say, the steady-state waiting times at successive queues) are regenerative. Recall that in this case the functions

$$\ell_j(\boldsymbol{v}) = \mathbb{E}_v\{L_j(\boldsymbol{Y})\}, \qquad j = 0, 1, \ldots, M,$$

are approximated by their sample equivalents $\bar{\ell}_{jN}(\boldsymbol{v})$, given in (6.4.4), as

$$\bar{\ell}_{jN}(\boldsymbol{v}) = \frac{\sum_{i=1}^{N}\sum_{t=1}^{\tau_i} L_{jti}\tilde{W}_{ti}(\boldsymbol{v}, \boldsymbol{v}_0)}{\sum_{i=1}^{N}\sum_{t=1}^{\tau_i} \tilde{W}_{ti}(\boldsymbol{v}, \boldsymbol{v}_0)}.$$

For the case of rare-event probability estimation, one need only replace L_{jti} by the indicator function $I_{\{L_{jti}>x\}}$ with all other data remaining intact. It is crucial to realize that the likelihood ratio processes vary from case to case: for the standard processes $\{L_{jti}\}$ one has

$$\tilde{W}_{ti}(\boldsymbol{v}, \boldsymbol{v}_0) = \prod_{j=1}^{t} W_{ji}(\boldsymbol{v}, \boldsymbol{v}_0) = \prod_{j=1}^{t} \frac{f(\boldsymbol{Z}_{ji}, \boldsymbol{v})}{f(\boldsymbol{Z}_{ji}, \boldsymbol{v}_0)}, \quad \boldsymbol{Z}_{ji} \sim f(\boldsymbol{z}, \boldsymbol{v}_0), \tag{9.6.1}$$

while for the indicator function $I_{\{L_{jti}>x\}}$ (x large), one has $\hat{W}_{ti}$ as per (9.3.10). To summarize, we use the CR estimator (6.4.4), (9.6.1) and the SR estimator (6.4.4), (9.3.10), for L_{jti} and $I_{\{L_{jti}>x\}}$, respectively.

Consider the choice of optimal (good) parameter vectors $\boldsymbol{v}_0$ for the program (6.4.1), (6.4.2), (6.4.5) using the stochastic counterpart (6.4.3), (6.4.3), (6.4.5). There are three separate cases: (i) *None* of the functions $\ell_j(\boldsymbol{v})$, $j = 0, 1, \ldots, M$, are computed over rare events; (ii) *all* functions $\ell_j(\boldsymbol{v})$, $j = 0, 1, \ldots, M$, are computed over rare events; and (iii) *some* of the functions $\ell_j(\boldsymbol{v})$, $j = 0, 1, \ldots, M$, are computed over rare events.

(i) This case is discussed rigorously in Section 6.4.1.

(ii) In this case, we estimate the optimal solution, $\boldsymbol{v}^*$ of the program (6.4.1), (6.4.2), (6.4.5), using the SR estimator (6.4.4), (9.3.10) with the reference parameter vector $\boldsymbol{v}_0 = (\boldsymbol{v}_{01}, \boldsymbol{v}_{02})$. As in case (i), for standard performance functions, it is suggested to estimate the optimal reference parameter vector $\boldsymbol{v}_0^* = (\boldsymbol{v}_{01}^*, \boldsymbol{v}_{02}^*)$ using the modified Algorithm (9.4.1) but with $\boldsymbol{\rho}$ replaced by $\boldsymbol{\rho}^+$, where $\boldsymbol{\rho}^+$ corresponds to the highest traffic intensity, defined in (6.4.5). Thus, the estimator of the optimal reference parameter vector $\boldsymbol{v}_0^* = \boldsymbol{v}_0^*(\boldsymbol{\rho}^+) = (\boldsymbol{v}_{01}^*(\boldsymbol{\rho}^+), \boldsymbol{v}_{02}^*(\boldsymbol{\rho}^+))$ will correspond to a configuration with the highest traffic intensity $\boldsymbol{\rho}^+$. Notice that as in the modified Algorithm 9.4.1, we estimate $\boldsymbol{v}_{01}^*$ from the program (8.4.11) (but with $\boldsymbol{\rho}$ replaced by $\boldsymbol{\rho}^+$), and take $\boldsymbol{v}_{02}^*$ as in the SR estimator (9.3.8), (9.3.10), but with $\boldsymbol{\rho}$ replaced by $\boldsymbol{\rho}^+$.

The above considerations lead to the following two-stage procedure for estimating the optimal solution, $\boldsymbol{v}^*$, using the stochastic counterpart (6.4.3), (6.4.4) (6.4.5).

Algorithm 9.6.1 Two-Stage Procedure for Estimating the Optimal Solution $\boldsymbol{v}^*$

1. *Apply the modified Algorithm 9.4.1 to estimate the optimal reference parameter vector* $\boldsymbol{v}_0^* = \boldsymbol{v}_0^*(\boldsymbol{\rho}^+) = (\boldsymbol{v}_{01}^*(\boldsymbol{\rho}^+), \boldsymbol{v}_{02}^*(\boldsymbol{\rho}^+))$, *corresponding to the highest traffic intensity* $\boldsymbol{\rho}^+$, *defined in (6.4.5).*
2. *Solve the program (6.4.3), (6.4.4), (6.4.5) using the derived estimator* $\bar{\boldsymbol{v}}_{0N}^* = \bar{\boldsymbol{v}}_{0N}^*(\boldsymbol{\rho}^+)$ *and deliver the optimal solution* $\bar{\boldsymbol{v}}_N^*$ *as an estimate of* $\boldsymbol{v}^*$.

How reliable is such an estimator of $\boldsymbol{v}^*$, when using the reference parameter vector $\bar{\boldsymbol{v}}_{0N}^*(\boldsymbol{\rho}^+)$ of Algorithm 9.6.1? Some suggestive information may be found in Kriman and Rubinstein (1994) (see also Section 9.3), which reports that for fixed $\boldsymbol{v}$, the RS estimator (9.3.8), (9.3.10) is *robust* in the following sense: perturbing the optimal reference vector $\boldsymbol{v}_0^*$ by 20–30% results in a loss of only 2–3 orders of magnitude of variance reduction as compared to a corresponding loss of some 10 orders of magnitude under the optimal value $\boldsymbol{v}_0^*$. Thus, if one perturbs $\boldsymbol{v}_0^*$ by 20–30%, one still gets dramatic variance reductions as compared to the crude Monte Carlo method. It is important to note that since $\boldsymbol{\rho}^+$ corresponds to the highest traffic intensity, the perturbations in $\boldsymbol{\rho}^+$ must be toward lower traffic intensities, which corresponds to (6.4.5). This implies that Algorithm 9.6.1 should work reasonably well, if $(\boldsymbol{\rho}^+ - \boldsymbol{\rho}^-)/\boldsymbol{\rho}^+ \leq 0.3$ component-wise (see also supporting numerical results below). The above considerations must be carefully factored in when defining the so-called trust optimization region, particularly the constraint (6.4.5).

(iii) In this case, step 1 of Algorithm 9.6.1 is modified as follows: Take the highest traffic intensity, $\boldsymbol{\rho}^+$, as a reference parameter vector, $\boldsymbol{\rho}_0$, for standard performance functions of the type $\mathbb{E}_{\boldsymbol{v}}\{L_t\}$, and take the parameter vector $\boldsymbol{v}_0^* = (\boldsymbol{v}_{01}^*, \boldsymbol{v}_{02}^*)$ from Algorithm 9.6.1 for performance functions with rare events of the type $\mathbb{E}_{\boldsymbol{v}}\{I_{\{L_t>x\}}\}$. Such an algorithm will be referred to as the *modified Algorithm* 9.6.1. Clearly such an algorithm requires two simulations: one for standard performance functions and another for performance functions with rare events.

To emphasize again that the reference vectors, $\boldsymbol{v}_0^*$, might be quite different in CS and SR estimators when solving the program (6.4.3), (6.4.4), (6.4.5), consider the following example.

Example 9.6.1. Let $\ell_0(\boldsymbol{v}) = \mathbb{E}_{\boldsymbol{v}}\{L_t\}$ and $\ell_1(\boldsymbol{v}, x) = \mathbb{E}_{\boldsymbol{v}}\{I_{\{L_t>x\}}\}$ be two performance functions in the program (6.4.1), (6.4.5), say, the objective function and the constraint functions, respectively. Let $\{L_t\}$ is the waiting time process in a $M/M/1$ queue with $\lambda = 1$, and assume that the parameters ρ^+ and ρ^- in (6.4.5) are $\rho^+ = 0.7$ and $\rho^- = 0.5$, respectively, so that $0.5 \le \rho \le 0.7$. Finally, let the decision variable be the service rate v. For the $\ell_0(\boldsymbol{v})$ part, take the highest traffic intensity $\rho_0 = \rho^+ = 0.7$ ($v_0 = \frac{1}{0.7} \approx 1.4$) as the reference parameter ρ_0 (see Section 8.3). For the $\ell_1(\boldsymbol{v})$ part, use the switching regenerative estimator (9.3.8), (9.3.12), and following the modified Algorithm 9.6.1, set $\boldsymbol{v}_0^* = (\boldsymbol{v}_{01}^*, \boldsymbol{v}_{02}^*)$, where $\hat{\boldsymbol{v}}_{01}^* = (v, \lambda) = (1.4, 1)$ and $\boldsymbol{v}_{02}^* = (\lambda, v) = (1, 1.4)$. Observe that in $\boldsymbol{v}_{01}^*$ we interchanged λ and v $[v = v(\rho^+)]$, while in $\boldsymbol{v}_{02}^*$ we switched them back. Finally note that $\boldsymbol{v}_{02}^* = (1, 1.4)$ in $\ell_1(\boldsymbol{v})$ coincides with the reference parameter vector $\boldsymbol{v}_0 = (1, 1.4)$ in the standard performance $\ell_0(\boldsymbol{v})$.

Since the reference parameter vectors, $\boldsymbol{v}_0$ in the CR and SR estimators, might be quite different, we could deploy again *two separate simulations*: one for standard performance functions and another for performance functions computed over rare events. Empirical evidence suggests that the modified Algorithm 9.6.1, based on a *single simulation* run with SR estimators, is in fact sufficient, provided $\mathbb{E}_{\boldsymbol{v}}\{I_{\{L_t>x\}}\} < 10^{-6}$. More precisely, when the rare-event probabilities are not too small, simulating the entire system just once, with reference parameters $\boldsymbol{v}_0^* = (\boldsymbol{v}_{01}^*, \boldsymbol{v}_{02}^*)$ corresponding to performance functions computed over rare events, gives rise to estimators of the optimal solution, $\boldsymbol{v}^*$ of the program (6.4.1), (6.4.5), which are as good (accurate) as estimators based on the two aforementioned independent simulations. In other words, ignoring standard performance functions when estimating the optimal reference parameter vector $\boldsymbol{v}_0^*$, does hardly affect the accuracy of the estimator of the optimal decision vector $\boldsymbol{v}^*$. This observation also means that taking $\boldsymbol{v}_0^* = \boldsymbol{v}_0^*(\boldsymbol{\rho}^+) = (\boldsymbol{v}_{01}^*, \boldsymbol{v}_{02}^*)$ as reference parameters for both the standard performance functions and those computed under rare events, affords good control over the entire system with the decision vector set $\boldsymbol{v} \in V$ given in (6.4.5). This suggestive (but still empirical) evidence needs to be established on solid mathematical and statistical grounds, and thus merits further investigation.

9.7 SIMULATION RESULTS

This section presents simulation results concerning the estimation of the optimal reference parameter $\boldsymbol{v}_0^*$ and the optimal solution $\boldsymbol{v}^*$ of a simple unconstrained program.

9.7.1 Estimating the Optimal Parameter Vector

This section presents two simulation studies: one for static systems and another one for dynamic models. The first example is a stochastic shortest path model and the second is a queueing model.

Static Models

Consider the double-bridge stochastic shortest path model containing 10 elements, depicted in Figure 9.4.

Table 9.7.1 displays the point estimators (PE) $\bar{\ell}_N(\boldsymbol{v}, \bar{\boldsymbol{v}}_0^*)$, and the 95% relative width of the confidence interval (CW) for the probability $\ell = P\{L > x\} = 10^{-3}$, with components $Y_i \sim \text{Exp}(v_i)$, $i = 1, \ldots, 10$. A sample size of $N = 10^5$ from $\boldsymbol{Z} \sim \text{Exp}(v_{0i})$, $i = 1, \ldots, 10$, was used to estimate the optimal vector of reference parameters, $\boldsymbol{v}_0^*$, according to Algorithm 9.4.1, and an additional sample size of

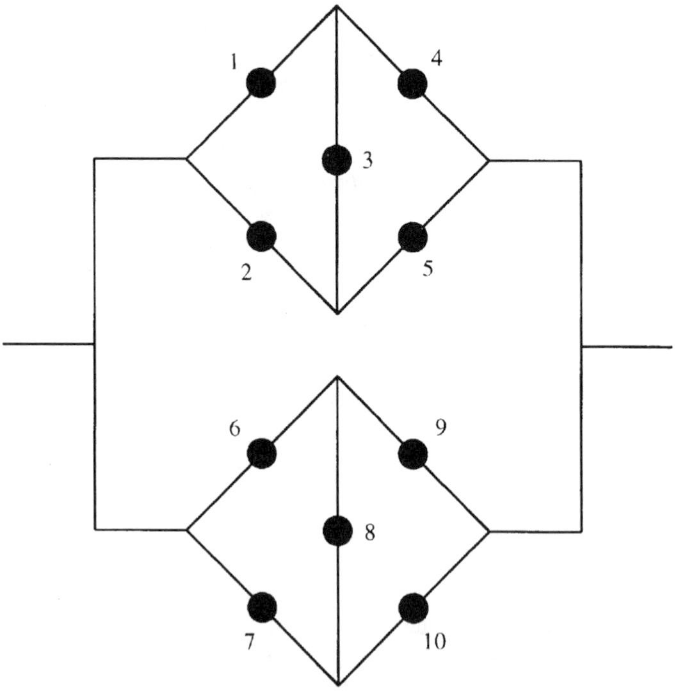

Figure 9.4. Double-bridge structure.

Table 9.7.1 Performance of the LR and CMC Estimators for the Probability $P\{L > x\} = 10^{-3}$ in the Double-Bridge Structure of Figure 9.7.1

			$\bar{\ell}_N(\boldsymbol{v}, \bar{\boldsymbol{v}}_0^*)$		$\bar{\ell}_{N_{\text{CMC}}}(\boldsymbol{v})$	
i	v_i	$\bar{v}_{0i}^*$	PE	CW	PE	CW
1	1.0	0.43				
2	1.0	0.45				
3	2.0	1.71				
4	2.0	1.34				
5	2.0	1.65	1.1E-3	0.17	9.8E-4	0.51
6	1.0	0.35				
7	1.0	0.43				
8	2.0	2.00				
9	2.0	1.75				
10	2.0	1.47				

$N = 5 \times 10^5$ was used to estimate the rare-event probability $\ell(\boldsymbol{v}, x) = \mathbb{E}_{\boldsymbol{v}}\{I_{(L>x)}\}$. Algorithm 9.4.1 was started with $\boldsymbol{v}_0 = \boldsymbol{v}$ and x_0 corresponding to $P_{\boldsymbol{v}}\{L > x_0\} \approx 10^{-2}$, and stopped after two iterations at the requisite point x.

It is readily seen that the LR estimator, $\bar{\ell}_N(\boldsymbol{v}, \boldsymbol{v}_0^*)$, outperforms its CMC counterpart, $\bar{\ell}_{N_{\text{CMC}}}(\boldsymbol{v})$ by a factor of approximately 3. The efficiency of the estimator $\bar{\ell}_N(\boldsymbol{v}, \boldsymbol{v}_0^*)$, relative to $\bar{\ell}_{N_{\text{CMC}}}(\boldsymbol{v})$, was found to increase with the rarity of the event $\{L > x\}$. Furthermore, the optimal parameters $\bar{v}_{0i}^*$, $i = 1, 2, 6, 7$, are less than half the corresponding values of the original parameters, v_i, while the remaining six elements of $\bar{v}_{0i}^*$ differ little from their original values, v_i. In the terminology of Mandjes (1996), the first set of elements (numbered $i = 1, 2, 6, 7$) defines the bottleneck cut of the network.

Table 9.7.2 displays simulation results with Algorithm 9.4.1 for the same input parameters as Table 9.7.1, but for the probability $P\{L > x\} = 10^{-8}$ instead of $P\{L > x\} = 10^{-3}$. In addition we performed the following experiment: we took the four-dimensional optimal bottleneck parameter vector $\bar{\boldsymbol{v}}_{0b}^*$ (the optimal parameters $\bar{v}_{0i}^*$ corresponding to $i = 1, 2, 6, 7$) from Table 9.7.2, while the remaining six optimal nonbottleneck elements $\bar{v}_{0i}^*$ in Table 9.7.2 were set to their corresponding original values v_i. The results in Table 9.7.2 clearly indicate that the estimator $\bar{\ell}_N(\boldsymbol{v}, \bar{\boldsymbol{v}}_{0b}^*)$, based solely on the optimal bottleneck parameter vector $\bar{\boldsymbol{v}}_{0b}^*$ is nearly as accurate as the estimator $\bar{\ell}_N(\boldsymbol{v}, \boldsymbol{v}_0^*)$. A heuristic explanation of this observation is that the nonbottleneck components contribute very little to Algorithm 9.4.1 optimization, and as the rarity of the underlying event increases, they only contribute "noise" to the estimator $\bar{\ell}_N(\boldsymbol{v}, \boldsymbol{v}_0^*)$. Thus, if available, the ECM and Algorithm 9.4.1 may be applied to the set of bottleneck parameters only.

Table 9.7.3 presents the behavior of the parameters $\bar{x}_k$ and $\bar{v}_{0i}^*$ $(\bar{x}_k)$ in Algorithm 9.4.1 for the input parameters of Table 9.7.2. It is readily seen that Algorithm 9.4.1 stopped after four iterations, corresponding to $\bar{x}_0$, $\bar{x}_1$, $\bar{x}_2$, and $\bar{x}_2 = x$; furthermore,

Table 9.7.2 Comparison of Efficiency of Estimators $\ell(v, v^*_{0b})$ and $\ell_N(v, v^*_0)$ for $x = 5.0$ $[\ell(x) \approx P\{L > x\} = 10^{-8}]$ in the Double-Bridge Structure of Figure 9.4

				$\bar{\ell}_N(v, \bar{v}^*_{0b})$		$\bar{\ell}_N(v, v^*_0)$	
i	v_i	$\bar{v}^*_{0bi}$	$\bar{v}^*_{0i}$	PE	CW	PE	CW
1	1.0	0.210	0.210				
2	1.0	0.165	0.165				
3	2.0	2.000	1.625				
4	2.0	2.000	1.336				
5	2.0	2.000	1.240	1.5E-8	0.36	1.3E-8	0.31
6	1.0	0.155	0.155				
7	1.0	0.157	0.157				
8	2.0	2.000	1.879				
9	2.0	2.000	1.322				
10	2.0	2.000	1.820				

Table 9.7.3 Behavior of Parameters $\bar{x}_k$ and $\bar{v}^*_{0i}$ $(\bar{x}_k)$ in Algorithm 9.4.1 for the Probability $P\{L > x\} \approx 10^{-8}$ in the Double-Bridge Structure of Figure 9.4

		$\bar{x}_0 = 1.927$	$\bar{x}_1 = 2.909$	$\bar{x}_2 = 4.012$	$\bar{x}_3 = x = 5.000$
i	v_i	$\bar{v}^*_{0i}(\bar{x}_0)$	$\bar{v}^*_{0i}(\bar{x}_1)$	$\bar{v}^*_{0i}(\bar{x}_2)$	$\bar{v}^*_{0i}(x)$
1	1.000	0.492	0.365	0.296	0.210
2	1.000	0.424	0.376	0.251	0.165
3	2.000	1.791	2.000	2.000	1.625
4	2.000	1.817	1.608	1.245	1.336
5	2.000	1.765	1.404	1.217	1.240
6	1.000	0.467	0.285	0.245	0.155
7	1.000	0.482	0.354	0.159	0.157
8	2.000	1.687	1.747	2.000	1.879
9	2.000	1.836	1.575	2.000	1.322
10	2.000	1.770	1.621	1.021	1.820

the optimal values of the bottleneck parameters, $\bar{v}^*_{0i}$, increase as the $\bar{x}_i$, $i = 0, 1, 2, 3$, increase. (The magnitude of the $\bar{x}_i$ corresponds to the rarity of the events $\{L > \bar{x}_i\}$).

Table 9.7.4 presents data similar to Table 9.7.3, but with the exponential components $Y_i \sim \text{Exp}(v_i)$, $i = 1, \ldots, 10$, replaced by Poisson components with corresponding parameters v_i, $i = 1, \ldots, 10$. The results of Table 9.7.4 are self-explanatory.

Finally, Table 9.7.5 displays data similar to Table 9.7.1, but with two equal bottleneck cuts: one cut with elements (1, 2, 9, 10) and the other with elements (4, 5, 6, 7). As expected, the optimal parameter values in both bottleneck cuts are substantially different from the original values, while for the remaining two

Table 9.7.4 Behavior of Parameters $\bar{x}_k$ and $\bar{v}^*_{0i}(\bar{x}_k)$ in Algorithm 9.4.1 for the Probability $P\{L > x\} \approx 3.510^{-8}$ in the Double-Bridge Structure of Figure 9.4 with Poisson Components

		$\bar{x}_0 = 4.000$	$\bar{x}_1 = 6.000$	$\bar{x}_2 = x = 7.000$
i	v_i	$\bar{v}^*_{0i}(\bar{x}_0)$	$\bar{v}^*_{0i}(\bar{x}_1)$	$\bar{v}^*_{0i}(x)$
1	2.000	3.836	4.231	5.249
2	2.000	3.561	4.586	5.253
3	1.000	1.052	1.245	1.620
4	1.000	1.631	2.525	2.374
5	1.000	2.117	2.166	2.547
6	2.000	3.798	4.630	4.866
7	2.000	3.301	4.249	5.011
8	1.000	1.000	1.364	1.232
9	1.000	1.547	2.254	2.597
10	1.000	1.774	2.252	2.254

Table 9.7.5 Performance of LR and CMC Estimators for the Probability $P\{\gg x\} = 10^{-3}$ in the Double-Bridge Structure of Figure 9.4 with Two Equal Bottleneck Cuts

			$\bar{\ell}_N(\boldsymbol{v}, \bar{\boldsymbol{v}}^*_0)$		$\bar{\ell}_{N_{\text{CMC}}}(\boldsymbol{v})$	
i	v_i	$\bar{v}^*_{0i}$	PE	CW	PE	CW
1	1.0	0.47				
2	1.0	0.54				
3	2.0	1.58				
4	1.0	0.54				
5	1.0	0.57	1.1E-3	0.17	1.0E-5	0.50
6	1.0	0.60				
7	1.0	0.57				
8	2.0	1.63				
9	1.0	0.48				
10	1.0	0.40				

elements (3, 8), the optimal parameter values are much closer to their original counterparts.

Queueing Models

Consider two $M/M/1$ queues in tandem with routing as depicted in Figure 9.5. Here v_1 denotes the customer interarrival rate; v_2, v_3 denote the service rate of the first and second queue, respectively; and v_4 is the routing probability to exit the network from node 1.

Consider the estimation of the optimal reference parameter vector $\boldsymbol{v}^*_0 = (v^*_{01}, \ldots, v^*_{04})$, using the modified Algorithm 9.4.1 and the associated rare-

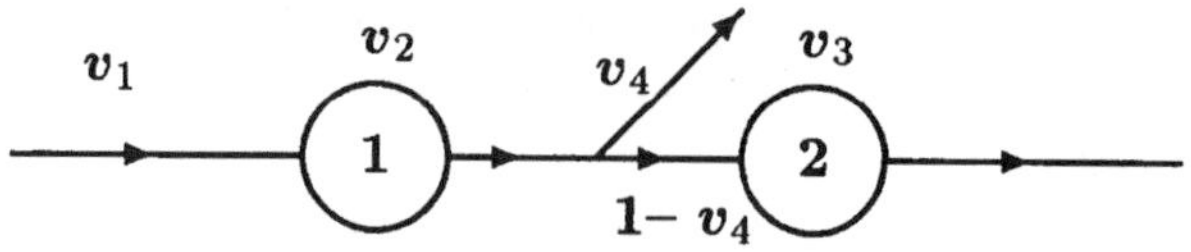

Figure 9.5. Two $M/M/1$ queues in tandem with routing.

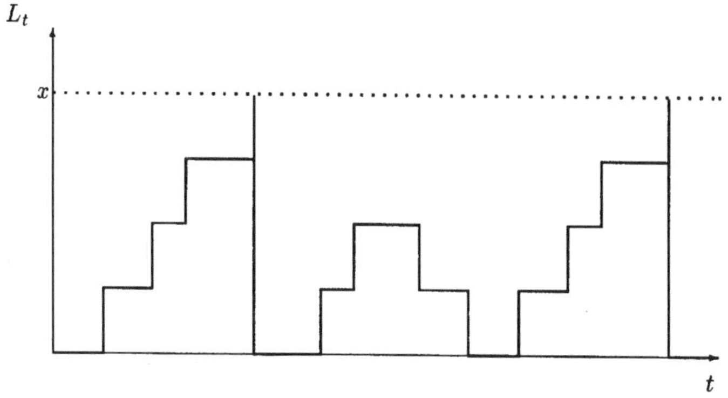

Figure 9.6. Sample path showing buffer overflow in a $GI/G/1$ queue.

event probability $\ell(x) = P(L_t > x)$ in a *transient* setting [see (9.3.13)]. Recall that a cycle is defined as a time duration starting with an empty system and ending the instant the system either becomes empty again or the number of customers in the system, L_t, reaches x (see Figure 9.6).

Letting

$$V_k = I_{\{L_t \text{ reaches } x \text{ in cycle } k\}},$$

one has $\ell(x) = P\{V_k = 1\}$—the desired excessive backlog probability (see Parekh and Walrand, 1989). For the CMC estimator $\tilde{\ell}_N(x)$ and the IS estimator $\bar{\ell}_N(x)$, one has

$$\tilde{\ell}_N(x) = \frac{1}{N} \sum_{i=1}^{N} V_i$$

and

$$\bar{\ell}_N(x) = \frac{1}{N} \sum_{i=1}^{N} V_i \hat{W}_i, \tag{9.7.1}$$

where $\hat{W}_i$ is the likelihood ratio during cycle i [see (9.3.14)].

Experimental results using the modified Algorithm 9.4.1 will be next compared with those of Parekh and Walrand in (1989), which are based on large deviation theory.

Tables 9.7.6 and 9.7.7 display the results of two simulation experiments with the model. Both experiments were conducted for $x = 20$. In the first case $\ell(x) = 3.269 \times 10^{-4}$, while in the second $\ell(x) = 2.390 \times 10^{-8}$. The relevant statistics were obtained from 20 experiments, each consisting of 1500 cycles in the first case and 2000 in the second. Both experiments used 2 iterations of the modified Algorithm 9.4.1, each starting at $x_0 = 7.0$ and $\boldsymbol{v}_0 = \boldsymbol{v}$; a total of 2×10^4 cycles were simulated to estimate $\hat{v}_0^*$, and the same number of cycles to calculate $\bar{\ell}_N(\hat{v}_0^*, x)$.

The true parameters, v_i, of the model are given in the first column of Tables 9.7.5 and 9.7.6. The optimal values of the reference parameters v_{0i}^*, $i = 1, 2, 3, 4$ [calculated analytically in Parekh and Walrand (1989)] and their counterparts $\hat{v}_{0i}^*$, $i = 1, 2, 3, 4$ (estimated via Algorithm 9.4.1) are given in the third and fourth columns, respectively.

Tables 9.7.6 and 9.7.7 show that the accuracies (variances) of estimators obtained via the modified Algorithm 9.4.1 are close to those of Parekh and Walrand in (1989). Similar results were obtained for more complex models and in steady-state settings. It is important to bear in mind that although the large deviation approach is based on solid mathematical foundations, this approach is limited to simple settings, while the approach in this section is rather general.

Table 9.7.6 Performance of Modified Algorithm 9.4.1 for Two $M/M/1$ Queues in Tandem with Routing, for the Probability $\ell(x) = 3.269 \times 10^{-4}$

				Parekh and Walrand (1989)		Algorithm 9.4.1	
i	v_i	v_{0i}^*	$\hat{v}_{0i}^*$	$\bar{\ell}_N(\boldsymbol{v}_0^*, x)$	$\hat{\sigma}$	$\bar{\ell}_N(\hat{v}_0^*, x)$	$\hat{\sigma}$
1	0.200	0.300	0.310				
2	0.300	0.200	0.205	3.194×10^{-4}	1.370×10^{-5}	3.286×10^{-4}	1.218×10^{-5}
3	0.500	0.500	0.479				
4	0.100	0.100	0.099				

Table 9.7.7 Performance of Algorithm 9.4.1 for Two $M/M/1$ Queues in Tandem with Routing, for the Probability $\ell(x) = 2.390 \times 10^{-8}$

				Parekh and Walrand (1989)		Algorithm 9.4.1	
i	v_i	v_{0i}^*	$\hat{v}_{0i}^*$	$\bar{\ell}_N(\boldsymbol{v}_0^*, x)$	$\hat{\sigma}$	$\bar{\ell}_N(\hat{v}_0^*, x)$	$\hat{\sigma}$
1	0.100	0.220	0.228				
2	0.700	0.700	0.687	2.405×10^{-8}	5.110×10^{-10}	2.378×10^{-8}	6.391×10^{-10}
3	0.200	0.080	0.082				
4	0.200	0.090	0.086				

9.7.2 Estimating the Optimal Solution

Consider the following unconstrained program:

$$\min_{v \in V} \ell(\boldsymbol{v}) = \min_{v \in V}[c\mathbb{E}_{\boldsymbol{v}}\{L_1\} + b\mathbb{E}_{\boldsymbol{v}}\{I_{\{L_2>x\}}\} + d\|\boldsymbol{v}\|], \qquad \boldsymbol{v} \in V, \tag{9.7.2}$$

where L_1 and L_2 are some random variables in steady state (say, sojourn times and queue lengths in a $GI/G/1$ queue), c is the cost of a waiting customer, b is the unit cost associated with the probability $P\{L_2 > x\}$ (x large), and d is the cost associated with the service rates.

From Chapter 8 it is readily seen that for the exponential family in the canonical form, $\ell(\boldsymbol{v})$ is a strictly convex differentiable function with respect to $\boldsymbol{v}$. Thus, $\boldsymbol{v}^*$ is a unique minimizer of (9.7.2) over the convex region V. This observation allows us to estimate the optimal solution $\boldsymbol{v}^*$ of the program (9.7.2) from the following stochastic counterpart:

$$\nabla\bar{\ell}_N(\boldsymbol{v}) = \mathbf{0}, \qquad \boldsymbol{v} \in V.$$

Furthermore, in analogy to (6.4.11), the estimator of $\nabla\ell_1(\boldsymbol{v}) = \nabla\mathbb{E}_{\boldsymbol{v}}\{L_1\}$ can be written as

$$\nabla\bar{\ell}_{1N}(\boldsymbol{v}) = \frac{\sum_{i=1}^N \sum_{t=1}^{\tau_i} L_{ti1}\nabla\tilde{W}_{ti}}{\sum_{i=1}^N \sum_{t=1}^{\tau_i} \tilde{W}_{ti}(\boldsymbol{v}_0)} - \frac{\sum_{i=1}^N \sum_{t=1}^{\tau_i} L_{ti1}\tilde{W}_{ti}(\boldsymbol{v}_0)}{\sum_{i=1}^N \sum_{t=1}^{\tau_i} \tilde{W}_{ti}(\boldsymbol{v}_0)} \times \frac{\sum_{i=1}^N \sum_{t=1}^{\tau_i} \nabla\tilde{W}_{ti}(\boldsymbol{v}_0)}{\sum_{i=1}^N \sum_{t=1}^{\tau_i} \tilde{W}_{ti}(\boldsymbol{v}_0)}, \qquad \boldsymbol{v} \in V, \tag{9.7.3}$$

and similarly for the estimator of $\nabla\ell_2(\boldsymbol{v}) = \nabla\mathbb{E}_{\boldsymbol{v}}\{I_{\{L_2>x\}}\}$.

Example 9.7.1 The *M/M/1* Queue. Let λ and v be the interarrival and service rates, respectively, and let v be the decision parameter. Assume that L_1 and L_2 are the steady-state random variables of the sojourn time and the queue length, respectively, and consider the program (9.7.2) with $v \in V$, corresponding to

$$0.2 = \rho^- \leq \rho(v) \leq \rho^+ = 0.8.$$

Since $\mathbb{E}_{\boldsymbol{v}}\{L_1\} = 1/(v - \lambda)$ and $\mathbb{E}_{\boldsymbol{v}}\{I_{\{L_2>x\}}\} = (\lambda/v)^x$ (e.g., Assmusen, 1987), it is readily seen that the optimal v^* that minimizes the performance measure given in (9.7.2) is the solution of the following system of equations:

$$\left(\frac{\lambda}{v}\right)^x + d = 0$$

$$-\frac{c}{4(v-\lambda)^{5/4}} - \frac{bx}{v}\left(\frac{\lambda}{v}\right)^x + d = 0.$$

Table 9.7.8 Theoretical Values $\ell(\rho)$ and $\nabla\ell(\rho)$ and Corresponding Estimated Values, $\bar{\ell}_N(\rho)$ and $\nabla\bar{\ell}_N(\rho)$, as Functions of ρ for $b=5\times10^5$, $c=1$, $d=5$, $\lambda=1$, and $x=20$

	N=5,000		N=50,000		N=500,000		Theoretical Value	
ρ	$\bar{\ell}_N$	$\nabla\bar{\ell}_N$	$\bar{\ell}_N$	$\nabla\bar{\ell}_N$	$\bar{\ell}_N$	$\nabla\bar{\ell}_N$	ℓ	$\nabla\ell$
0.1	50.1	4.98	50.6	4.98	50.6	4.98	50.6	4.98
0.2	25.4	4.96	25.6	4.96	25.7	4.96	25.7	4.96
0.3	17.2	4.91	17.3	4.91	17.5	4.91	17.5	4.91
0.4	13.0	4.81	13.2	4.79	13.3	4.81	13.4	4.81
0.5	6.41	0.11	6.44	0.06	6.47	0.00	6.48	0.00
0.6	26.9	−196	27.1	−206	27.7	−214	27.7	−216
0.7	382	-5×10^4	396	-5×10^4	409	-6×10^4	407	-6×10^4
0.8	6×10^3	-8×10^4	6×10^3	-8×10^4	6×10^3	-9×10^4	6×10^3	-9×10^4

Table 9.7.8 displays the theoretical values $\ell(\rho)$ and $\nabla\ell(\rho)$, and their stochastic analogs $\bar{\ell}_N(\rho)$ and $\nabla\bar{\ell}_N(\rho)$, as functions of $\rho = \rho(v)$ ($\rho = \lambda/v$) and the sample sizes N. The parameters chosen were $\lambda=1$, $b=5\times10^5$, $c=1$, $d=5$, $x=20$ [these correspond to $v^* \approx 2$, $\rho^* = \rho(v^*) \approx 0.5$, and $\mathbb{E}_v\{I_{\{L_2>x\}}\} = (\lambda/v)^x \approx 10^{-6}$]. The experiment used two separate simulations: one for the $\mathbb{E}_v\{L_1\}$ part and the other for the $\mathbb{E}_v\{I_{\{L_2>x\}}\}$ part (the latter involving rare events). More precisely, the conventional regenerative and switching regenerative estimators, (6.4.4), (9.6.1) and (9.3.8), (9.3.12), were used respectively. Since $\rho^+ = 0.8$, the corresponding reference parameters (for $\mathbb{E}_v\{L_1\}$ and $\mathbb{E}_v\{I_{\{L_2>x\}}\}$) were set at $\boldsymbol{v}_0 = (v_0, \lambda) = (1.25, 1)$ and $\hat{\boldsymbol{v}}_0^* = (\hat{\boldsymbol{v}}_{01}^*, \hat{\boldsymbol{v}}_{02}^*)$, where $\hat{\boldsymbol{v}}_{01}^* = (v_0, \lambda) = (1.25, 1)$, $\hat{\boldsymbol{v}}_{02}^* = (\lambda, v_0) = (1, 1.25)$, respectively. This corresponds to $\rho_0 = \rho^+ = 0.8$ for the $\mathbb{E}_v\{L_1\}$ part, and to the interchange of λ and v_0 (in $\hat{\boldsymbol{v}}_{01}^*$) followed by a switch back (in $\hat{\boldsymbol{v}}_{02}^*$) for the $\mathbb{E}_v\{I_{\{L_2>x\}}\}$ part [see (9.3.12)].

Table 9.7.8 indicates that as the sample size N increases the estimated values, $\bar{\ell}_N(\rho)$ and $\nabla\bar{\ell}_N(\rho)$, converge to their respective theoretical counterparts, $\ell(\rho)$ and $\nabla\ell(\rho)$.

Table 9.7.9 displays the theoretical values of v^*, the point estimators, $\bar{v}_N^* = \bar{v}_N^*(b)$, and the 95% confidence intervals of v^* (denoted by 95% CI), as

Table 9.7.9 Performance of CR and SR Estimators, Based on Two Separate Simulations

ρ^*	b	α	$\bar{v}_N^*(b)$	v^*	95% CI
0.8	21.34	0.000	1.261	1.255	(1.15, 1.37)
0.7	385.6	0.143	1.433	1.429	(1.33, 1.54)
0.6	1.0×10^4	0.333	1.654	1.667	(1.58, 1.73)
0.5	5.0×10^5	0.600	1.971	2.000	(1.91, 2.03)
0.4	5.5×10^7	1.000	2.467	2.500	(2.42, 2.51)
0.3	2.3×10^{10}	1.667	3.324	3.333	(3.25, 3.39)
0.2	1.2×10^{14}	3.000	4.947	5.000	(4.74, 5.15)
0.1	2.5×10^{20}	7.000	9.741	10.00	(9.42, 10.06)

functions of b, $\rho^* = \lambda/v^*$ and $\alpha = (\rho_0 - \rho^*)/\rho^*$. Note that *all* estimators $\bar{v}_N^*(b)$ were obtained *simultaneously*, for different values of b, using two separate simulations (one for $\mathbb{E}_v\{L_1\}$ and the other for $\mathbb{E}_v\{I_{\{L_2>x\}}\}$), each based on $N = 10{,}000$ cycles (approximately 50,000 customers). It is readily seen that the estimator $\bar{v}_N^*$ performs reasonably well.

9.8 CONCLUDING REMARKS

This chapter presented a methodology for efficient estimation of rare-event probabilities for both static and dynamic models. For static models it has been shown how to estimate consistently the OECM parameter vector $\boldsymbol{v}_0^*$, using Algorithm 9.4.1, and how to get fast (polynomial-time) estimators. For dynamic (queueing) models we distinguished between single-node queues and networks of queues. In the former case, LD principles were used to calculate analytically the OECM $\boldsymbol{v}^*$, and then the complexity of corresponding IS estimators was discussed. In the latter case, the modified Algorithm 9.4.1 was proposed to estimate consistently the OECM $\boldsymbol{v}_0^*$. Proving or disproving that for queueing networks the corresponding estimator is polynomial-time under $\boldsymbol{v}_0^*$ is still an open problem. An important feature of Algorithm 9.4.1 and its modification is that regardless of the parameterization type used for $F(\boldsymbol{y}, \boldsymbol{v})$ (ECM or alternatives), it estimates consistently the optimal parameters $\boldsymbol{v}_0^*$, some of whose components might eventually turn out to be redundant.

It was also shown how to optimize complex DES involving rare events from a single simulation run. More specifically, the two-stage Algorithm 9.6.1 was shown to be suited for optimization with the stochastic counterpart (6.4.3), (6.4.4), (6.4.5): In the first stage the OECM reference parameter vector $\boldsymbol{v}_0^*$ is estimated, and in the second stage the optimal solution $\boldsymbol{v}^*$ of the program (6.4.1), (6.4.2), (6.4.5) is estimated. The efficiency of the two-stage Algorithm 9.6.1 has been supported by extensive simulation experiments that indicate that in many interesting cases it suffices to change the probability measure to $F(\boldsymbol{y}, \boldsymbol{v}_{0b})$ rather than to $F(\boldsymbol{y}, \boldsymbol{v}_0)$, where $\boldsymbol{v}_{0b}$ is chosen from a small subset of $\boldsymbol{v}_0$. This subset is associated with the notion of the so-called bottleneck cut for static models, and bottleneck queueing module for queueing networks. Finding (screening out) the bottleneck cuts and the bottleneck modules before running a simulation, and finding conditions under which sampling under $F(\boldsymbol{y}, \boldsymbol{v}_{0b})$ is as good as under $F(\boldsymbol{y}, \boldsymbol{v}_0)$ are still open problems. Similar ideas may be applied to other networks, such as coherent reliability networks, stochastic shortest path, computer communications networks, and inventory systems.

The area of estimating rare-event probabilities and stochastic optimization over rare events is still in its infancy and considerable progress can be expected, especially in the domain of dynamic (queueing) networks. Some further topics for investigation are listed below.

1. Optimizing computer simulation models involving rare events with respect to both distributional and structural decision variables. In the latter case it would

be interesting to explore the capabilities of the push-out method (see Chapter 7).

2. Combining efficiently recursive and nonrecursive procedures, such as stochastic approximation and the stochastic counterpart.
3. Investigating complexity properties of IS estimators based on the modified Algorithm 9.4.1 for dynamic models.
4. Extending the above simulation-based optimization methodology to such important fields as inventories, reliability, and insurance risk modeling.
5. Investigating the bottleneck phenomenon for fast estimation of rare-event probabilities, and in particular identifying the bottleneck cut for static systems and bottleneck modules in dynamic systems, thereby identifying redundant elements.

9.9 EXERCISES

1. Produce a table similar to Table 9.3.1 for normal, geometric, and Bernoulli distributions.

2. Consider the double-bridge structure in Figure 9.4. Assume that all 10 elements $i = 1, \dots, 10$, are mutually independent, each distributed Poisson with the corresponding parameters, v_i, as in Table 9.7.1. Run a simulation for Algorithm 9.4.1, and display data similar to that in Table 9.7.1.

3. **Switching estimators for the performance $\ell(\boldsymbol{v}) = \mathbb{E}_{\boldsymbol{v}}\{L\}$.** Let $\{L_t\}$ be a standard sample performance, such as the number of customers or the waiting (sojourn) time in a queueing system. Simulation studies show that the switching estimator (9.3.8), (9.3.15) can be useful not only in estimating rare-event probabilities, but in standard performance estimation as well.

   In the former case (rare events), one chooses $\zeta(x) = \inf\{t\colon L_t \geq x\}$, while in the latter case, one can choose $\zeta(x^*) = \inf\{t\colon L_t \geq x^*\}$, where x^* is an *auxiliary level*, obtained from the solution of the problem

$$\min_x \mathscr{L}_1^0(x),$$

where

$$\mathscr{L}_1^0(x) = \mathbb{E}_{\boldsymbol{v}_0}\left\{\left[\sum_{t=1}^{\tau} L_t \tilde{W}_t\right]^2\right\} = \mathbb{E}_{\boldsymbol{v}_0}\left\{\left[\sum_{t=1}^{\tau}\left(\sum_{y=0}^{\infty} I_{\{L>y\}}\right)\tilde{W}_t(\zeta(x))\right]^2\right\}. \quad (9.9.1)$$

The last equality in (9.9.1) can be justified, since for $L > 0$, one has

$$\mathbb{E}\{L\} = \sum_x P\{L > y\} = \sum_{y=0}^{\infty} \mathbb{E}\{I_{\{L>y\}}\}.$$

Estimate x^* from the stochastic counterpart of (9.9.1) and provide numerical evidence that

$$\mathscr{L}_1^0(\boldsymbol{v}, \boldsymbol{v}_0, x^*) < \mathscr{L}_1^0(\boldsymbol{v}, \boldsymbol{v}),$$

where $\mathscr{L}_1^0(\boldsymbol{v}, \boldsymbol{v})$ corresponds to the crude Monte Carlo method (which does not employ switching).

4. Designing a buffer size using stochastic approximation. Suppose we wish to find the root x of the equation

$$\ell(x) = P\{L > x\} = \mathbb{E}\{I_{\{L>x\}}\} = \varepsilon,$$

where ε is a fixed number. As an example, think of designing a buffer size such that the associated buffer overflow probability is bounded by a fixed quantity, ε, corresponding to a rare event.

The stochastic counterpart of $\ell(x) = \varepsilon$, based on the SR estimators (9.3.8), (9.3.15), is

$$\bar{\ell}_N(x) = \frac{\sum_{i=1}^N \sum_{t=1}^{\tau_i} I_{\{L_{ti}>x\}} \hat{W}_{ti}(\zeta(x))}{\sum_{i=1}^N \sum_{t=1}^{\tau_i} \hat{W}_{ti}(\zeta(x))} = \varepsilon.$$

The corresponding stochastic approximation algorithm for estimating the root, x, of $\ell(x) = \varepsilon$ is

$$x_{i+1} = x_i - \gamma_i[\bar{\ell}_{N_i}(x_i, \zeta_i(x_i)) - \varepsilon], \qquad \gamma_i > 0,$$

where $\zeta(x_i) = \inf\{t\colon L_t \geq x_i\}$. Discuss the convergence of the stochastic approximation (SA) algorithm above.

5. Consider the estimation of an expected performance of the form

$$\ell(v) = \mathbb{E}_v\{Y\} + \mathbb{E}_v\{I_{\{Y>x\}}\}, \tag{9.8.2}$$

where $Y \sim \text{Exp}(v)$. Recall that the optimal values of the reference parameter for $\mathbb{E}_v\{Y\}$ and $\mathbb{E}_v\{I_{\{Y>x\}}\}$ (x large) are $v_0^* = 0.5$ and $v_0^* \approx 1/x$, respectively. (For $v = 1$ and $x = 100$, which corresponds to $\ell(x) = e^{-100}$, one has $v_0^* = 0.5$ and $v_0^* \approx 0.01$, respectively).

Due to the dissimilarity of the reference parameter vector $\boldsymbol{v}_0^*$ (for fixed $\boldsymbol{v}$) in the standard performance function, $\mathbb{E}_v\{Y\}$, and the rare-event performance function, one might use two separate simulations (double run). Show that, in fact, a *single* simulation run with the reference parameter $v_0^* \approx 1/x$ (optimal value for $\mathbb{E}_v\{I_{\{Y>x\}}\}$) is sufficient. More precisely, show that if one replaces the optimal value $v_0^* = 0.5$ (for $\mathbb{E}_v\{Y\}$) by $v_0^* \approx 1/x$, then the accuracies (relative errors) of the *single* run and the *double* run LR estimators, are of the same order of magnitude.

Hint: Take note of (9.2.5).

6. Run a simulation of the (s, S) policy inventory model given in Table 7.5.1. Use the modified Algorithm 9.4.1 to estimate the following rare-event probabilities:

(a) *Stock out probability*:

$$\ell_1(\boldsymbol{v}) = P_{\boldsymbol{v}}\{L < 0\} = \mathbb{E}\{I_{\{L<0\}}\}.$$

(b) *Inventory unavailability* during a cycle,

$$\ell_2(\boldsymbol{v}) = \frac{\mathbb{E}\left\{\sum_{t=1}^{\tau} I_{\{L_t<0\}}\right\}}{\mathbb{E}\{\tau\}},$$

which represents the proportion of time stock is out during a cycle. Assume that both demand sizes and interdemand times are independent exponentially distributed with parameters $\lambda_1 = 10$ and $\lambda_2 = 0.2$, respectively, while L is the steady-state on-hand inventory process.

Hint: Take λ_{10} and λ_{20} as reference parameters.

REFERENCES

Asmussen, S. (1982). "Conditioned limit theorems relating a random walk to its associate, with applications to risk reserve processes and the $GI/G/1$ queue," *Adv. Appl. Probab.*, **14**, 143–170.

Asmussen, S. (1985). "Conjugate processes and the simulation of ruin problems," *Stoch. Proc. Appl.*, **20**, 213–229.

Asmussen, S. (1987). *Applied Probability and Queues*, Wiley, New York.

Asmussen, S. (1989). "Risk theory in a Markovian environment," *Scand. Actuarial J.*, **1989**(2), 69–100.

Asmussen, S. (1994). "Siegmund duality, Markov dependence and applications," Manuscript, Aalborg University, Denmark.

Asmussen, S. and Binswanger, K. (1995). "Simulation of ruin probabilities for subsexponential claims," Manuscript, ETH Zurich and University of Lund, Lund, Sweden.

Asmussen, S. and Nielsen, H. M. (1995). "Ruin probabilities via local adjustment coefficients," *J. Appl. Probab.*, **32**(3), 736–755.

Asmussen, S. and Rubinstein, R. Y. (1995). "Complexity properties of steady state rare-events simulation in queueing models," in *Advances in Queueing: Theory, Methods and Open Problems*, J. Dshalalow, Ed., CRC Press, Boca Raton, pp. 429–462.

Asmussen, S., Rubinstein, R. Y., and Wang, C. L. (1994). "Regenerative rare-events simulation via likelihood ratios," *J. Appl. Probab.*, **31**(3), 797–815.

Ben Letaief, K. (1995). "Performance analysis of digital lightwave systems using efficient computer simulation techniques," *IEEE Trans. Commun.*, **43**(2/3/4), 240–251.

Bucklew, J. A. (1990). *Large Deviation Techniques in Decision, Simulation, and Estimation*, Wiley, New York.

Bucklew, J. A., Ney, P., and Sadowsky, J. S. (1990). "Monte Carlo simulation and deviations theory for uniformly recurrent Markov chains," *J. Appl. Probab.*, **27**, 44–59.

Chang, C. S., Heidelberger, P., Juneja, S., and Shahabuddin, P. (1994). "Effective bandwidth and fast simulation of ATM in tree networks," *Perf. Eval.*, **20**, 45–65.

Chang, C. S., Heidelberger, P., and Shahabuddin, P. (1995) "Fast simulation of packet loss rates in a shared buffer communications switch," *ACM Trans. Mod. Computer Simul.*, **5**(4), 306–325.

Cottrel, M., Fort, J. C., and Malgoures, G. (1983). "Deviations and rare events in the study of stochastic algorithms," *IEEE Trans. Automat. Contr.*, **AC-28**, 907–918.

Dembo, A. and Zeitouni, O. (1993). *Deviation Techniques*, Jones & Bartlett, Boston.

Devetsikiotis, M. and Townsend, K. R. (1992). "A dynamic importance sampling methodology for the efficient estimation of rare-events probabilities in regenerative simulations of queueing systems," *Proceedings of IEEE Globecom '92*, IEEE Computer Society Press, Los Alamitos, pp. 1290–1297.

Devetsikiotis, M. and Townsend, K. R. (1993). "Statistical optimization of dynamic importance sampling parameters for efficient simulation of communication networks," *IEEE/ACM Trans. Network.*, **1**(3), 293–305.

Frater, M. R. and Anderson, B. D. O. (1989). "Fast estimation of the statistics of excessive backlogs in tandem networks of queues," *Australian Telecommun. Res.*, **23**, 49–55.

Frater, M. R., Walrand, J., and Anderson, B. D. O. (1990). "Optimality and efficient estimation of the buffer overflow in queues with deterministic service times," *Australian Telecommun. Res.*, **24**, 1–8.

Frater, M. R., Lennon, T. M., and Anderson, B. D. O. (1991). "Optimally efficient estimation of the statistics of rare events in queueing networks," *IEEE Trans. Automat. Contr.*, **AC-36**, 1395–1405.

Glasserman, P. and Kou, Sh.-G. (1995). "Analysis of importance sampling estimator for tandem queues," *ACM Trans. Mod. Computer Simul.*, January, 1995.

Glasserman, P. and Liu, T. W. (1995). "Rare-event simulation for multistage production-inventory systems," *Manage. Sci.*, **42**, 1292–1307.

Gnedenko, B. V. and Ushakov, I. A. (1995). *Probabilistic Reliability Engineering*, Wiley, New York.

Goyal, A., Shahabuddin, P., Heidelberger, P., Nicola, V. F., and Glynn, P. W. (1992). "A unified framework for simulating Markovian models of highly dependable systems," *IEEE Trans. Comput.*, **C-41**, 36–51.

Heidelberger, P. (1993). "Fast simulation of rare events in queueing and reliability models," *ACM Trans. Mod. Computer Simul.*, **5**(1), 43–85.

Heidelberger, P., Shahabuddin, P., and Nicola, V. F. (1994). "Bounded relative error in estimating transient measures of highly dependable mon-Markovian systems," *ACM Trans. Mod. Computer Simul.*, **4**(2), 137–164.

Huang, Ch., Devetsikiotis, M., Lambadaris, I., and Kaye, A. R. (1995). "Fast simulation for self-similar traffic in ATM networks," *Proc. IEEE ICC '95*, Seattle, Washington, 1995, pp. 438–444.

Kovalenko, I. (1995). "Approximations of queues via small parameter method," in *Advances in Queueing: Theory, Methods and Open Problems*, J. Dshalalow, Ed., CRC Press, Boca Raton, pp. 481–509.

Kriman, V. and Rubinstein, R. Y. (1997). "Polynomial and exponential time algorithms for estimation of rare events in queueing models," in *Frontiers in Queueing: Models and Applications in Science and Engineering*, J. Dshalalow, Ed., CRC Press, Boca Raton,

Lehtonen, T. and Nyrhinen, H. (1992a). "Simulating level crossing probabilities by importance sampling," *Adv. Appl. Probab.*, **24**, 858–874.

Lehtonen, T. and Nyrhinen, H. (1992b). "On asymptotically efficient simulation of ruin probabilities in a Markovian environment," *Scand. Actuarial J.*, **1992**(1), 60–75.

Lieber, D., Rubinstein, R. Y., and Elmakis, D. (1994). "Quick estimation of rare events in stochastic networks," Manuscript, Technion, Haifa.

Lindley, D. (1952). "The theory of queues with a single server," *Proc. Cambr. Philos. Soc.*, **48**, 277–289.

Lindley, D. (1959). "Discussion of a paper of C.B. Winsten," *J. Roy. Statist. Soc. Ser. B*, **21**, 22–23.

Mandjes, M. (1996). "Rare event analysis of communication networks," PhD Thesis, Vrije Universiteit, Amsterdam.

Mandjes, M. (1997). "Fast simulation of bloking probabilities in loss networks," *Europ. J. OR*, to appear.

Nemirovskii, A. and Rubinstein, R. (1993). "Robust stochastic approximation algorithms and their applications to discrete event systems," Manuscript, Technion, Haifa, Israel.

Neuts, M. F. (1977). "A versatile Markovian point process," *J. Appl. Probab.*, **16**, 764–779.

Neuts, M. F. (1981). *Matrix-Geometric Solutions in Stochastic Models*, Johns Hopkins University Press, Baltimore.

Neuts, M. F. (1989). *Structured Stochastic Matrices of the M/G/1 Type and Their Applications*, Marcel Dekker, New York.

Nicola, V. F., Shahabuddin, P., Heidelberger, P., and Glynn, P. W. (1993). "Fast simulation of steady-state availability in non-Markovian highly dependable systems," in *Proceedings of the Twentieth International Symposium on Fault-Tolerant Computing*, IEEE Computer Society Press, Los Alamitos, pp. 491–498.

Parekh, S. and Walrand, J. (1989). "A quick simulation method for excessive backlogs in networks of queues," *IEEE Trans. Automat. Contr.*, **AC-34**, 54–66.

Rubinstein, R. Y. and Shapiro, A. (1993). *Discrete Event Systems: Sensitivity Analysis and Stochastic Optimization via the Score Function Method*, Wiley, New York.

Sadowsky, J. S. (1991). "Deviations theory and efficient simulation of excessive backlogs in a $GI/GI/m$ queue," *IEEE Trans. Automat. Contr.*, **AC-36**, 1383–1394.

Sadowsky, J. S. (1993). "On the optimality and stability of exponential twisting in Monte Carlo simulation," *IEEE Trans. Inform. Theory*, **IT-39**, 119–128.

Sadowsky, J. S. (1994). "Monte Carlo estimation of deviations probabilities," Manuscript, Arizona State University, Tempe, AZ.

Sadowsky, J. S. and Bucklew, J. S. (1990). "On deviations theory and asymptotically efficient Monte Carlo estimation," *IEEE Trans. Inform. Theory*, **IT-36**, 579–588.

Sadowsky, J. S. and Szpankowski W. (1992a). "The probability of queue length and waiting times in a heterogeneous multiserver queue. Part I: Tight limits," Manuscript, School of Electrical Engineering and Department of Computer Science, Purdue University, West Lafayette, IN.

Sadowsky, J. S. and Szpankowski, W. (1992b). "The probability of queue length and waiting times in a heterogeneous multiserver queue. Part II: Positive recurrence and logarithmic limits," Manuscript, School of Electrical Engineering and Department of Computer Science, Purdue University, West Lafayette, IN.

Shahabuddin, P. (1994a). "Fast simulation of packet loss rates in communications networks with priorities," in *Proceedings of the 1994 Winter Simulation Conference, Orlando, Florida*, IEEE Press, New York, pp. 274–281.

Shahabuddin, P. (1994b). "Importance sampling for the simulation of highly reliable Markovian systems," *Manage. Sci.*, **40**(3), 333–352.

Shahabuddin, P. (1995). "Rare event simulation of stochastic systems," in *Proceedings of the 1995 Winter Simulation Conference, Washington, DC*, IEEE Press, New York, pp. 178–185.

Shwartz, A. and Weiss, A. (1995). *Deviation for Performance Analysis: Queues, Communication and Computers*, Chapman-Hall, London.

Siegmund, D. (1976a). "Importance sampling in the Monte Carlo study of sequential tests," *Ann. Statist.*, **4**, 673–684.

Siegmund, D. (1976b). "The equivalence of absorbing and reflecting barrier problems for stochastically monotone Markov processes," *Ann. Probab.*, **4**, 914–924.

Stockmeyer, L. J. (1992). "Computational complexity," in *Handbooks in OR & MS*, Vol. 3. (E. G. Coffman et al., editors), pp. 455–517, North Holland, New York.

Tsoucas, P. (1992). "Rare events in series of queues," *J. Appl. Probab.*, **29**, 168–175.

Index

WILEY SERIES IN PROBABILITY AND STATISTICS
ESTABLISHED BY WALTER A. SHEWHART AND SAMUEL S. WILKS

Editors
Vic Barnett, Ralph A. Bradley, Noel A. C. Cressie, Nicholas I. Fisher, Iain M. Johnstone, J. B. Kadane, David G. Kendall, David W. Scott, Bernard W. Silverman, Adrian F. M. Smith, Jozef L. Teugels, Geoffrey S. Watson; J. Stuart Hunter, Emeritus

Probability and Statistics Section

*ANDERSON · The Statistical Analysis of Time Series
ARNOLD, BALAKRISHNAN, and NAGARAJA · A First Course in Order Statistics
BACCELLI, COHEN, OLSDER, and QUADRAT · Synchronization and Linearity: An Algebra for Discrete Event Systems
BASILEVSKY · Statistical Factor Analysis and Related Methods: Theory and Applications
BERNARDO and SMITH · Bayesian Statistical Concepts and Theory
BILLINGSLEY · Convergence of Probability Measures
BOROVKOV · Asymptotic Methods in Queuing Theory
BRANDT, FRANKEN, and LISEK · Stationary Stochastic Models
CAINES · Linear Stochastic Systems
CAIROLI and DALANG · Sequential Stochastic Optimization
CONSTANTINE · Combinatorial Theory and Statistical Design
COVER and THOMAS · Elements of Information Theory
CSÖRGŐ and HORVÁTH · Weighted Approximations in Probability Statistics
CSÖRGŐ and HORVÁTH · Limit Theorems in Change Point Analysis
DETTE and STUDDEN · The Theory of Canonical Moments with Applications in Statistics, Probability, and Analysis
*DOOB · Stochastic Processes
DRYDEN and MARDIA · Statistical Analysis of Shape
DUPUIS and ELLIS · A Weak Convergence Approach to the Theory of Large Deviations
ETHIER and KURTZ · Markov Processes: Characterization and Convergence
FELLER · An Introduction to Probability Theory and Its Applications, Volume 1, *Third Edition,* Revised; Volume II, *Second Edition*
FULLER · Introduction to Statistical Time Series, *Second Edition*
FULLER · Measurement Error Models
GELFAND and SMITH · Bayesian Computation
GHOSH, MUKHOPADHYAY, and SEN · Sequential Estimation
GIFI · Nonlinear Multivariate Analysis
GUTTORP · Statistical Inference for Branching Processes
HALL · Introduction to the Theory of Coverage Processes
HAMPEL · Robust Statistics: The Approach Based on Influence Functions
HANNAN and DEISTLER · The Statistical Theory of Linear Systems
HUBER · Robust Statistics
IMAN and CONOVER · A Modern Approach to Statistics
JUREK and MASON · Operator-Limit Distributions in Probability Theory
KASS and VOS · Geometrical Foundations of Asymptotic Inference
KAUFMAN and ROUSSEEUW · Finding Groups in Data: An Introduction to Cluster Analysis

*Now available in a lower priced paperback edition in the Wiley Classics Library.

Probability and Statistics (Continued)

KELLY · Probability, Statistics, and Optimization
LINDVALL · Lectures on the Coupling Method
McFADDEN · Management of Data in Clinical Trials
MANTON, WOODBURY, and TOLLEY · Statistical Applications Using Fuzzy Sets
MARDIA and JUPP · Statistics of Directional Data, *Second Edition*
MORGENTHALER and TUKEY · Configural Polysampling: A Route to Practical Robustness
MUIRHEAD · Aspects of Multivariate Statistical Theory
OLIVER and SMITH · Influence Diagrams, Belief Nets and Decision Analysis
*PARZEN · Modern Probability Theory and Its Applications
PRESS · Bayesian Statistics: Principles, Models, and Applications
PUKELSHEIM · Optimal Experimental Design
RAO · Asymptotic Theory of Statistical Inference
RAO · Linear Statistical Inference and Its Applications, *Second Edition*
*RAO and SHANBHAG · Choquet-Deny Type Functional Equations with Applications to Stochastic Models
ROBERTSON, WRIGHT, and DYKSTRA · Order Restricted Statistical Inference
ROGERS and WILLIAMS · Diffusions, Markov Processes, and Martingales, Volume I: Foundations, *Second Edition;* Volume II: Îto Calculus
RUBINSTEIN and SHAPIRO · Discrete Event Systems: Sensitivity Analysis and Stochastic Optimization by the Score Function Method
RUZSA and SZEKELY · Algebraic Probability Theory
SCHEFFE · The Analysis of Variance
SEBER · Linear Regression Analysis
SEBER · Multivariate Observations
SEBER and WILD · Nonlinear Regression
SERFLING · Approximation Theorems of Mathematical Statistics
SHORACK and WELLNER · Empirical Processes with Applications to Statistics
SMALL and McLEISH · Hilbert Space Methods in Probability and Statistical Inference
STAPLETON · Linear Statistical Models
STAUDTE and SHEATHER · Robust Estimation and Testing
STOYANOV · Counterexamples in Probability
TANAKA · Time Series Analysis: Nonstationary and Noninvertible Distribution Theory
THOMPSON and SEBER · Adaptive Sampling
WELSH · Aspects of Statistical Inference
WHITTAKER · Graphical Models in Applied Multivariate Statistics
YANG · The Construction Theory of Denumerable Markov Processes

Applied Probability and Statistics Section

ABRAHAM and LEDOLTER · Statistical Methods for Forecasting
AGRESTI · Analysis of Ordinal Categorical Data
AGRESTI · Categorical Data Analysis
ANDERSON, AUQUIER, HAUCK, OAKES, VANDAELE, and WEISBERG · Statistical Methods for Comparative Studies
ARMITAGE and DAVID (editors) · Advances in Biometry
*ARTHANARI and DODGE · Mathematical Programming in Statistics
ASMUSSEN · Applied Probability and Queues
*BAILEY · The Elements of Stochastic Processes with Applications to the Natural Sciences
BARNETT and LEWIS · Outliers in Statistical Data, *Third Edition*

*Now available in a lower priced paperback edition in the Wiley Classics Library.

Applied Probability and Statistics (Continued)

BARTHOLOMEW, FORBES, and McLEAN · Statistical Techniques for Manpower Planning, *Second Edition*
BATES and WATTS · Nonlinear Regression Analysis and Its Applications
BECHHOFER, SANTNER, and GOLDSMAN · Design and Analysis of Experiments for Statistical Selection, Screening, and Multiple Comparisons
BELSLEY · Conditioning Diagnostics: Collinearity and Weak Data in Regression
BELSLEY, KUH, and WELSCH · Regression Diagnostics: Identifying Influential Data and Sources of Collinearity
BHAT · Elements of Applied Stochastic Processes, *Second Edition*
BHATTACHARYA and WAYMIRE · Stochastic Processes with Applications
BIRKES and DODGE · Alternative Methods of Regression
BLOOMFIELD · Fourier Analysis of Time Series: An Introduction
BOLLEN · Structural Equations with Latent Variables
BOULEAU · Numerical Methods for Stochastic Processes
BOX · Bayesian Inference in Statistical Analysis
BOX and DRAPER · Empirical Model-Building and Response Surfaces
BOX and DRAPER · Evolutionary Operation: A Statistical Method for Process Improvement
BUCKLEW · Large Deviation Techniques in Decision, Simulation, and Estimation
BUNKE and BUNKE · Nonlinear Regression, Functional Relations and Robust Methods: Statistical Methods of Model Building
CHATTERJEE and HADI · Sensitivity Analysis in Linear Regression
CHOW and LIU · Design and Analysis of Clinical Trials
CLARKE and DISNEY · Probability and Random Processes: A First Course with Applications, *Second Edition*
*COCHRAN and COX · Experimental Designs, *Second Edition*
CONOVER · Practical Nonparametric Statistics, *Second Edition*
CORNELL · Experiments with Mixtures, Designs, Models, and the Analysis of Mixture Data, *Second Edition*
*COX · Planning of Experiments
CRESSIE · Statistics for Spatial Data, *Revised Edition*
DANIEL · Applications of Statistics to Industrial Experimentation
DANIEL · Biostatistics: A Foundation for Analysis in the Health Sciences, *Sixth Edition*
DAVID · Order Statistics, *Second Edition*
*DEGROOT, FIENBERG, and KADANE · Statistics and the Law
DODGE · Alternative Methods of Regression
DOWDY and WEARDEN · Statistics for Research, *Second Edition*
DUNN and CLARK · Applied Statistics: Analysis of Variance and Regression, *Second Edition*
ELANDT-JOHNSON and JOHNSON · Survival Models and Data Analysis
EVANS, PEACOCK, and HASTINGS · Statistical Distributions, *Second Edition*
FLEISS · The Design and Analysis of Clinical Experiments
FLEISS · Statistical Methods for Rates and Proportions, *Second Edition*
FLEMING and HARRINGTON · Counting Processes and Survival Analysis
GALLANT · Nonlinear Statistical Models
GLASSERMAN and YAO · Monotone Structure in Discrete-Event Systems
GNANADESIKAN · Methods for Statistical Data Analysis of Multivariate Observations, *Second Edition*
GOLDSTEIN and LEWIS · Assessment: Problems, Development, and Statistical Issues
GREENWOOD and NIKULIN · A Guide to Chi-Squared Testing
*HAHN · Statistical Models in Engineering
HAHN and MEEKER · Statistical Intervals: A Guide for Practitioners

*Now available in a lower priced paperback edition in the Wiley Classics Library.

Applied Probability and Statistics (Continued)

HAND · Construction and Assessment of Classification Rules
HAND · Discrimination and Classification
HEIBERGER · Computation for the Analysis of Designed Experiments
HINKELMAN and KEMPTHORNE: · Design and Analysis of Experiments, Volume 1: Introduction to Experimental Design
HOAGLIN, MOSTELLER, and TUKEY · Exploratory Approach to Analysis of Variance
HOAGLIN, MOSTELLER, and TUKEY · Exploring Data Tables, Trends and Shapes
HOAGLIN, MOSTELLER, and TUKEY · Understanding Robust and Exploratory Data Analysis
HOCHBERG and TAMHANE · Multiple Comparison Procedures
HOCKING · Methods and Applications of Linear Models: Regression and the Analysis of Variables
HOGG and KLUGMAN · Loss Distributions
HOLLANDER and WOLFE · Nonparametric Statistical Methods
HOSMER and LEMESHOW · Applied Logistic Regression
HØYLAND and RAUSAND · System Reliability Theory: Models and Statistical Methods
HUBERTY · Applied Discriminant Analysis
JACKSON · A User's Guide to Principle Components
JOHN · Statistical Methods in Engineering and Quality Assurance
JOHNSON · Multivariate Statistical Simulation
JOHNSON and KOTZ · Distributions in Statistics
Continuous Multivariate Distributions
JOHNSON, KOTZ, and BALAKRISHNAN · Continuous Univariate Distributions, Volume 1, *Second Edition*
JOHNSON, KOTZ, and BALAKRISHNAN · Continuous Univariate Distributions, Volume 2, *Second Edition*
JOHNSON, KOTZ, and BALAKRISHNAN · Discrete Multivariate Distributions
JOHNSON, KOTZ, and KEMP · Univariate Discrete Distributions, *Second Edition*
JUREČKOVÁ and SEN · Robust Statistical Procedures: Aymptotics and Interrelations
KADANE · Bayesian Methods and Ethics in a Clinical Trial Design
KADANE AND SCHUM · A Probabilistic Analysis of the Sacco and Vanzetti Evidence
KALBFLEISCH and PRENTICE · The Statistical Analysis of Failure Time Data
KELLY · Reversability and Stochastic Networks
KHURI, MATHEW, and SINHA · Statistical Tests for Mixed Linear Models
KLUGMAN, PANJER, and WILLMOT · Loss Models: From Data to Decisions
KLUGMAN, PANJER, and WILLMOT · Solutions Manual to Accompany Loss Models: From Data to Decisions
KOVALENKO, KUZNETZOV, and PEGG · Mathematical Theory of Reliability of Time-Dependent Systems with Practical Applications
LAD · Operational Subjective Statistical Methods: A Mathematical, Philosophical, and Historical Introduction
LANGE, RYAN, BILLARD, BRILLINGER, CONQUEST, and GREENHOUSE · Case Studies in Biometry
LAWLESS · Statistical Models and Methods for Lifetime Data
LEE · Statistical Methods for Survival Data Analysis, *Second Edition*
LEPAGE and BILLARD · Exploring the Limits of Bootstrap
LINHART and ZUCCHINI · Model Selection
LITTLE and RUBIN · Statistical Analysis with Missing Data
MAGNUS and NEUDECKER · Matrix Differential Calculus with Applications in Statistics and Econometrics
MALLER and ZHOU · Survival Analysis with Long Term Survivors
MANN, SCHAFER, and SINGPURWALLA · Methods for Statistical Analysis of Reliability and Life Data

*Now available in a lower priced paperback edition in the Wiley Classics Library.

Applied Probability and Statistics (Continued)

McLACHLAN and KRISHNAN · The EM Algorithm and Extensions
McLACHLAN · Discriminant Analysis and Statistical Pattern Recognition
McNEIL · Epidemiological Research Methods
MILLER · Survival Analysis
MONTGOMERY and PECK · Introduction to Linear Regression Analysis, *Second Edition*
MYERS and MONTGOMERY · Response Surface Methodology: Process and Product in Optimization Using Designed Experiments
NELSON · Accelerated Testing, Statistical Models, Test Plans, and Data Analyses
NELSON · Applied Life Data Analysis
OCHI · Applied Probability and Stochastic Processes in Engineering and Physical Sciences
OKABE, BOOTS, and SUGIHARA · Spatial Tesselations: Concepts and Applications of Voronoi Diagrams
PANKRATZ · Forecasting with Dynamic Regression Models
PANKRATZ · Forecasting with Univariate Box-Jenkins Models: Concepts and Cases
PIANTADOSI · Clinical Trials: A Methodologic Perspective
PORT · Theoretical Probability for Applications
PUTERMAN · Markov Decision Processes: Discrete Stochastic Dynamic Programming
RACHEV · Probability Metrics and the Stability of Stochastic Models
RÉNYI · A Diary on Information Theory
RIPLEY · Spatial Statistics
RIPLEY · Stochastic Simulation
ROUSSEEUW and LEROY · Robust Regression and Outlier Detection
RUBIN · Multiple Imputation for Nonresponse in Surveys
RUBINSTEIN · Simulation and the Monte Carlo Method
RUBINSTEIN and MELAMED · Modern Simulation and Modeling
RYAN · Statistical Methods for Quality Improvement
SCHUSS · Theory and Applications of Stochastic Differential Equations
SCOTT · Multivariate Density Estimation: Theory, Practice, and Visualization
*SEARLE · Linear Models
SEARLE · Linear Models for Unbalanced Data
SEARLE, CASELLA, and McCULLOCH · Variance Components
STOYAN, KENDALL, and MECKE · Stochastic Geometry and Its Applications, *Second Edition*
STOYAN and STOYAN · Fractals, Random Shapes and Point Fields: Methods of Geometrical Statistics
THOMPSON · Empirical Model Building
THOMPSON · Sampling
TIJMS · Stochastic Modeling and Analysis: A Computational Approach
TIJMS · Stochastic Models: An Algorithmic Approach
TITTERINGTON, SMITH, and MAKOV · Statistical Analysis of Finite Mixture Distributions
UPTON and FINGLETON · Spatial Data Analysis by Example, Volume 1: Point Pattern and Quantitative Data
UPTON and FINGLETON · Spatial Data Analysis by Example, Volume II: Categorical and Directional Data
VAN RIJCKEVORSEL and DE LEEUW · Component and Correspondence Analysis
WEISBERG · Applied Linear Regression, *Second Edition*
WESTFALL and YOUNG · Resampling-Based Multiple Testing: Examples and Methods for *p*-Value Adjustment
WHITTLE · Systems in Stochastic Equilibrium
WOODING · Planning Pharmaceutical Clinical Trials: Basic Statistical Principles
WOOLSON · Statistical Methods for the Analysis of Biomedical Data
*ZELLNER · An Introduction to Bayesian Inference in Econometrics

*Now available in a lower priced paperback edition in the Wiley Classics Library.

Texts and References Section

AGRESTI · An Introduction to Categorical Data Analysis
ANDERSON · An Introduction to Multivariate Statistical Analysis, *Second Edition*
ANDERSON and LOYNES · The Teaching of Practical Statistics
ARMITAGE and COLTON · Encyclopedia of Biostatistics: Volumes 1 to 6 with Index
BARTOSZYNSKI and NIEWIADOMSKA-BUGAJ · Probability and Statistical Inference
BERRY, CHALONER, and GEWEKE · Bayesian Analysis in Statistics and Econometrics: Essays in Honor of Arnold Zellner
BHATTACHARYA and JOHNSON · Statistical Concepts and Methods
BILLINGSLEY · Probability and Measure, *Second Edition*
BOX · R. A. Fisher, the Life of a Scientist
BOX, HUNTER, and HUNTER · Statistics for Experimenters: An Introduction to Design, Data Analysis, and Model Building
BOX and LUCEÑO · Statistical Control by Monitoring and Feedback Adjustment
BROWN and HOLLANDER · Statistics: A Biomedical Introduction
CHATTERJEE and PRICE · Regression Analysis by Example, *Second Edition*
COOK and WEISBERG · An Introduction to Regression Graphics
COX · A Handbook of Introductory Statistical Methods
DILLON and GOLDSTEIN · Multivariate Analysis: Methods and Applications
DODGE and ROMIG · Sampling Inspection Tables, *Second Edition*
DRAPER and SMITH · Applied Regression Analysis, *Third Edition*
DUDEWICZ and MISHRA · Modern Mathematical Statistics
DUNN · Basic Statistics: A Primer for the Biomedical Sciences, *Second Edition*
FISHER and VAN BELLE · Biostatistics: A Methodology for the Health Sciences
FREEMAN and SMITH · Aspects of Uncertainty: A Tribute to D. V. Lindley
GROSS and HARRIS · Fundamentals of Queueing Theory, *Third Edition*
HALD · A History of Probability and Statistics and their Applications Before 1750
HALD · A History of Mathematical Statistics from 1750 to 1930
HELLER · MACSYMA for Statisticians
HOEL · Introduction to Mathematical Statistics, *Fifth Edition*
JOHNSON and BALAKRISHNAN · Advances in the Theory and Practice of Statistics: A Volume in Honor of Samuel Kotz
JOHNSON and KOTZ (editors) · Leading Personalities in Statistical Sciences: From the Seventeenth Century to the Present
JUDGE, GRIFFITHS, HILL, LÜTKEPOHL, and LEE · The Theory and Practice of Econometrics, *Second Edition*
KHURI · Advanced Calculus with Applications in Statistics
KOTZ and JOHNSON (editors) · Encyclopedia of Statistical Sciences: Volumes 1 to 9 wtih Index
KOTZ and JOHNSON (editors) · Encyclopedia of Statistical Sciences: Supplement Volume
KOTZ, REED, and BANKS (editors) · Encyclopedia of Statistical Sciences: Update Volume 1
KOTZ, REED, and BANKS (editors) · Encyclopedia of Statistical Sciences: Update Volume 2
LAMPERTI · Probability: A Survey of the Mathematical Theory, *Second Edition*
LARSON · Introduction to Probability Theory and Statistical Inference, *Third Edition*
LE · Applied Survival Analysis
MALLOWS · Design, Data, and Analysis by Some Friends of Cuthbert Daniel
MARDIA · The Art of Statistical Science: A Tribute to G. S. Watson
MASON, GUNST, and HESS · Statistical Design and Analysis of Experiments with Applications to Engineering and Science
MURRAY · X-STAT 2.0 Statistical Experimentation, Design Data Analysis, and Nonlinear Optimization

*Now available in a lower priced paperback edition in the Wiley Classics Library.

Texts and References (Continued)

PURI, VILAPLANA, and WERTZ · New Perspectives in Theoretical and Applied Statistics
RENCHER · Methods of Multivariate Analysis
RENCHER · Multivariate Statistical Inference with Applications
ROSS · Introduction to Probability and Statistics for Engineers and Scientists
ROHATGI · An Introduction to Probability Theory and Mathematical Statistics
RYAN · Modern Regression Methods
SCHOTT · Matrix Analysis for Statistics
SEARLE · Matrix Algebra Useful for Statistics
STYAN · The Collected Papers of T. W. Anderson: 1943–1985
TIERNEY · LISP-STAT: An Object-Oriented Environment for Statistical Computing and Dynamic Graphics
WONNACOTT and WONNACOTT · Econometrics, *Second Edition*

WILEY SERIES IN PROBABILITY AND STATISTICS

ESTABLISHED BY WALTER A. SHEWHART AND SAMUEL S. WILKS

Editors
Robert M. Groves, Graham Kalton, J. N. K. Rao, Norbert Schwarz, Christopher Skinner

Survey Methodology Section

BIEMER, GROVES, LYBERG, MATHIOWETZ, and SUDMAN · Measurement Errors in Surveys
COCHRAN · Sampling Techniques, *Third Edition*
COX, BINDER, CHINNAPPA, CHRISTIANSON, COLLEDGE, and KOTT (editors) · Business Survey Methods
*DEMING · Sample Design in Business Research
DILLMAN · Mail and Telephone Surveys: The Total Design Method
GROVES · Survey Errors and Survey Costs
GROVES, BIEMER, LYBERG, MASSEY, NICHOLLS, and WAKSBERG · Telephone Survey Methodology
*HANSEN, HURWITZ, and MADOW · Sample Survey Methods and Theory, Volume 1: Methods and Applications
*HANSEN, HURWITZ, and MADOW · Sample Survey Methods and Theory, Volume II: Theory
KASPRZYK, DUNCAN, KALTON, and SINGH · Panel Surveys
KISH · Statistical Design for Research
*KISH · Survey Sampling
LESSLER and KALSBEEK · Nonsampling Error in Surveys
LEVY and LEMESHOW · Sampling of Populations: Methods and Applications
LYBERG, BIEMER, COLLINS, de LEEUW, DIPPO, SCHWARZ, TREWIN (editors) · Survey Measurement and Process Quality
SKINNER, HOLT, and SMITH · Analysis of Complex Surveys

*Now available in a lower priced paperback edition in the Wiley Classics Library.

Rethinking University Teaching 2nd Edition

There have been extensive changes in the technologies available for learning over the last decade. These technologies have the potential to improve radically the way students engage with knowledge and negotiate ideas. However, this book argues that the promises made for e-learning will only be realised if we begin with an understanding of how students learn, and design the use of learning technologies from this standpoint.

This new edition has been updated in view of recent technological advances and provides a sound theoretical basis for designing and using learning technologies in university teaching. The author argues that although the new learning technologies are not individually capable of matching the effectiveness of the one-to-one teacher, together they can support the full range of student learning, both efficiently and effectively.

This book is essential reading for all academics and academic support staff concerned with improving the quality of teaching in Higher Education.

Diana Laurillard is Professor of Educational Technology and Pro-Vice Chancellor for Learning Technologies and Teaching at The Open University.

Rethinking University Teaching
2nd Edition

A conversational framework for the effective use of learning technologies

Diana Laurillard

London and New York

First published 2002 by RoutledgeFalmer
11 New Fetter Lane, London EC4P 4EE

Simultaneously published in the USA and Canada
by RoutledgeFalmer
29 West 35th Street, New York, NY 10001

RoutledgeFalmer is an imprint of the Taylor & Francis Group

Typeset in Baskerville by GreenGate Publishing Services,
Tonbridge, Kent
Printed and bound in Great Britain by Biddles Ltd, Guildford and King's Lynn

British Library Cataloguing in Publication Data
A catalogue record for this book is available from the British Library

Library of Congress Cataloging in Publication Data
A catalog record for this book has been requested

ISBN 0-415-25679-8 (pb)
ISBN 0-415-25678-X (hb)

To Brian, Amy and Anna, from whom I continue to learn.

Contents

Plates

Figures

Tables

Acknowledgements

I owe my thanks, as ever, to the staff and students of the Open University. In preparation of this second edition I have drawn, as before, on the work of my colleagues: in the Institute of Educational Technology, the Knowledge Media Institute, the faculties, and the academic support units, who create the learning experiences from which I try to learn. Open University students are an extraordinary community of scholars and practitioners, from all walks of life, who find some place in their full lives not only to study, but also to reflect on their study. Without their willingness to do this, we would not have the means to rethink our teaching.

I am especially grateful to those of my colleagues, at the OU and elsewhere, whose work is represented in the examples used to illustrated the concepts in the book: Professor Ference Marton, Dr Shirley Booth, Professor Paul Ramsden, Professor Tony Bates, Dr Josie Taylor, Dr Rose Luckin, Dr Lydia Plowman, Dr Peter Wright, Dr Joel Greenberg, Dave Meara, Dr Rod Moyse, Nicola Durbridge, Professor Tom Vincent, Dr Jon Rosewell, John Naughton, Dr David Johnson, Karen Shipp, Professor Shirley Alexander, Professor Ray Ison, Professor Robin Mason, Dr Gilly Salmon, Professor Marc Eisenstadt, Dr Simon Buckingham Shum, Dr Peter Scott. We have all been blessed with a Vice Chancellor, Sir John Daniel, who created the environment in which new technologies could be explored and exploited. His benign and visionary leadership in this field has enabled us to keep raising our ambitions for the value that learning technologies can bring to the learning experience.

I must also pay tribute to the late Professor Gordon Pask as the original inspiration for the ideas developed here. His special contribution was to bring a new rigour to educational technology, and for all his technical wizardry, a deep humanity as well.

Finally, I give heartfelt thanks to my Personal Assistant, Alison Nash, whose professional commitment to making my life manageable has made this book possible.

Preface to the 2nd edition

Since the first edition of this book, there have been extensive changes in the technologies available for learning. The Web has become established, interface design has matured, and PC access has become widespread. The demands of technological change have hindered the theory and practice of its application, however. Learning technologies are unfamiliar and complex. Few of the current generation of academics have ever learned through technology, so practice develops slowly, and theory hardly at all. Fortunately, the Conversational Framework introduced in the first edition has proved to be remarkably robust in the face of the new technologies. Its development has benefited from application, and from discussions with many academics, and their critiques have contributed to elaboration of the original theory.

The revisions to the first edition are extensive because the general principles of learning design are communicated most convincingly through the detail of example. And illustrative examples change with the technology. Part I updates the research studies on students' learning needs, creating the challenge that new technologies must meet. Part II extends the Conversational Framework to test how well new media contribute to academic learning. Five different types of learning media are illustrated by examples drawn from recent innovative learning materials. Part III revises the design methodology for the course material and its programme context.

As before, this edition finishes with a blueprint for a university infrastructure that is not sidetracked by the uncertain notion of an 'e-university' or an 'online university'. The integrity of the academic institution is paramount. Throughout the book there remains the fundamental assumption that a university is defined by the quality of its academic conversations, not by the technologies that service them.

Diana Laurillard
London, February 2001

Introduction

My first lecture as a student was a wretched experience. With 199 other students I counted myself lucky that I was in the main lecture theatre and not in the overspill room receiving closed circuit television. The lecturer was talking formulae as he came in, and for fifty minutes he scribbled them on the board as he talked, and we all scribbled more, in a desperate attempt to keep up with his dictation.

My first lecture as a teacher was no better. For this group, I had a syllabus listing thirty or so topics, and a timetable of three lectures a week. Fresh from finals and desperate not to bore the seventy-odd engineering students with the trivia of introductory complex analysis I prepared reams of notes from several textbooks and my own scribbled lecture notes, entered the room talking formulae, and scribbled them on the board as I went.

One lucky thing happened. At the end of the lecture, I asked if there were any questions, and one brave student asked a question of such breathtaking 'stupidity' that it was clear he could not have understood anything beyond my first sentence. Did anyone else have that problem? Yes, they all had that problem. I learned a lot more than they did from that lecture. Their stupidity or mine? Who has the greater responsibility for that situation?

This book starts from the premise that university teachers must take the main responsibility for what and how their students learn. Students have only limited choices in how they learn: they can attend lectures or not; they can work hard or not; they can seek truth or better marks – but teachers create the choices open to them. The students in my lecture could only choose to concentrate hard, they could not choose to understand. It is the teacher's responsibility to create the conditions in which understanding is possible, and the student's responsibility to take advantage of that. Students have little control over their access to knowledge. The university operates a complex system of departments, curricula, teaching methods, support facilities, timetables, assessment – all of which determine the possible ways in which students may learn. Yes they have libraries and the Web, giving them access to alternative resources. But university teachers make heavy demands of student time: reading around and browsing the Web are luxuries they can ill afford. Our responsibility as teachers is commensurate with the degree of control we exert over the learners.

It would be quite possible to argue that students should take responsibility for their own learning, that they should use the university as a set of resources largely under their control. This is the most attractive vision of academic learning as a community of scholars pursuing their own course towards knowledge and enlightenment, inspired but not directed by their teachers. Universities still aspire to this at postgraduate level and, at its best, this model is indeed attractive and highly productive. It is essentially a minority provision, however. To support properly students in their own exploration of what is known in a field, where its frontiers are, and how they might be extended, is extremely costly in staff time. Guidance is a labour-intensive process, which means that any one academic can service only a small number of students. Assessment is also labour-intensive, as each case must be judged on its own merit, not in terms of a pre-defined 'model answer'. Moreover, working at the frontiers of knowledge is essentially a lonely task performed by individuals and very small groups; such a task is not suited to any form of mass education or support. It is the proper model of postgraduate education, but that is where it must be confined. At undergraduate level, students are exploring an already known field of knowledge, they are explicitly not breaking new ground, except at a personal level. Although we often argue that in university education students should develop their own point of view within a subject, not accept spoon-feeding, and be critical, we nonetheless expect right answers. It is perfectly permissible to criticise an authority's argument, but students must give an accurate account of it, and their critique must be well argued. No matter how democratic we are about respecting the student's point of view, there is always a pre-defined standard of answer. That is why our model of education at undergraduate level is more often didactic than negotiated, teaching methods are many-to-one rather than one-to-one, and we control rather than offer resources. And that is why as teachers we have the major responsibility for what and how our students learn.

So are students just puppets, dancing to the tunes of their various teachers, helplessly buffeted by the forces around them? This is a model that university teachers strongly resist, remembering perhaps their own heightened sense of personal responsibility for what they learned, and anxious to preserve the joy of exploration and discovery for their own students. We particularly value those students who, in Bruner's phrase go 'beyond the information given'. Yet, the individual learner's sense of breaking new ground makes learning something personal, peculiar to that individual, and therefore not so amenable to the mass treatment that a didactic education system tends to adopt. This is the paradox that challenges the teaching profession: we want all our students to learn the same thing, yet we want each to make it their own.

As teachers relating to individual students, it is possible to adopt and live by the values of a community of scholars, and that can be a common experience of postgraduate teaching. Nevertheless, at undergraduate level, while teaching and assessing en masse, teachers are as embedded in a system outside their control as their students. My first lecture may have been worse than most – it does not have to be that bad – but it brought home to me the farcical nature of the system I was caught up in. Consider what the lecturer, meeting a class for the first time, has to do: they must guide this collection of individuals through territory they are unfamiliar with towards a common meeting point, but without knowing where they are starting from, how much baggage they are carrying, and what kind of vehicle they are using. This is insanity. It is truly a miracle, and a tribute to human ingenuity, that any student ever learns anything worthwhile in such a system.

The academic system must change. It works to some extent, for some students, but not well enough. As higher education expands, we cannot always rely on human ingenuity to overcome its inadequacies. It is always possible to defend the inspirational lecturer, the importance of academic individuality, the value of pressurising students to work independently, but we cannot defend a mode of operation that actively undermines a professional approach to teaching. Teachers need to know more than just their subject. They need to know the ways it can come to be understood, the ways it can be misunderstood, what counts as understanding: they need to know how individuals experience the subject. However, they are neither required nor enabled to know these things. Moreover, our system of mass lectures, examinations, and low staff:student ratios ensures that they will never find them out.

Higher education cannot change easily. Traditions, values, infrastructure all create the conditions for a natural inertia. It is being forced to change, and the pressures wrought upon it have nothing to do with traditions and values. Instead, the pressure is for reduced costs, for greater scale and scope, and for innovation through technology. Academics are facing an unprecedented challenge to the traditions and values of the profession. There is an appetite for reform from within higher education in many countries now, but it moves slowly as we all scurry about in response to the increasing external pressures which exercise their own peculiar forms of change. Academics are going on courses on management training and marketing methods. Reform of an education system might progress faster if they went on courses on how to teach better.

Higher education should be reformed through pressure from within. Academics share a number of important traditions, some of which should be preserved: the pursuit of research and scholarship (OECD, 1987), the advancement of learning (Robbins, 1963), the freedom to conduct a radical critique of knowledge claims (Barnett, 1990). The Dearing Report in the UK, the first comprehensive review of higher education since Robbins, reconsidered the aims of higher education, and defined them in very similar ways, preserving the valued traditions:

> To inspire and enable individuals to develop their capabilities …
> To increase knowledge and understanding for their own sake …
> To serve the needs of a knowledge-based economy …
> To play a major role in shaping a democratic, civilised, inclusive society.
>
> (Dearing, 1997: 72)

Only the last differed significantly from Robbins' view that universities should 'transmit a common culture and common standards of citizenship', which does not reflect the diverse, multicultural values of our society now. The consensus was, however, that we should preserve the traditional academic values, while seeking change in the means of addressing them. We need to rebuild the infrastructure that will enable a fit between the academic values we wish to preserve and the new conditions of educating larger numbers.

I see the solution as being found in a new organisational infrastructure, not in guidelines on how to teach. There is no body of knowledge out there on how to teach thermodynamics, as there is on how to cure headaches. Given that the human mind is probably at least as complicated as the human body, it will be a long time before we understand the many ways in which it can fail. The time, energy and money spent on medical science will never be spent on instructional science, so the outlook for knowing very much about how to teach is bleak. But we can at least take the right approach to the task. The organisational infrastructure will be a series of mechanisms, tasks, and responsibilities that together ensure a benign process, one that will be progressive, in much the same way as research methodology aims to ensure that knowledge progresses. Methodology is generative and ultimately more productive than prescriptive guidelines. We may have only limited knowledge of how to teach well, but at least we can use a productive methodology that helps us build our knowledge.

This book discusses how to think about teaching. It works towards an analysis in the final chapter of what the infrastructure of universities should be if the effectiveness of teaching and therefore the quality of student learning is to improve and go on improving. The final chapter attempts to describe the structures and mechanisms constituting a system that would not have the farcical effects we experience as students and later as teachers. The system should support both sides in an approach to learning and teaching that fits the academic values we proclaim. The idea is to find an infrastructure that enables university teachers to be as professional in their teaching as they aspire to be in their research.

The chapters in Part I build up to this. Each one takes an aspect of learning and teaching that the professional teacher needs to know about, and gives a critical account of the key research studies to establish both what we know and how it comes to be known. Methodology plays a part all the way through, therefore, as this will inform our approach to the task of teaching. The argument begins in Chapter 1 with an exploration of the nature of academic learning. University teachers have a rich but largely unarticulated experience of what it means to learn their subject. In this first chapter, I attempt to tap that experience and develop a description of it that will provide the basis for motivating the rest of the book. The central idea is that academic learning is different from other kinds of learning in everyday life because it is not directly experienced, and is necessarily mediated by the teacher. Undergraduates are not learning about the world directly, but about others' descriptions of the world, hence the term 'mediated'. This view is developed from and contrasted with other prominent views on the nature of learning in the current literature, and raises the question of how teachers are to perform this mediating role.

Chapters 2 and 3 address the issues I referred to in my partial analogy of teacher as travel guide, as needing to know what students bring to their learning and how they do it. Here, the sum total of what we know is negligible in comparison with what there is to know. This is not because every student is individual and infinitely variable, but because there is so much for them to know, and so many ways it can be known. The research literature tells us enough about a few key concepts in certain subject areas to make it clear that the ways in which a concept can be understood is an empirical question, not a logical one. As we shall see in Chapter 2, we cannot deduce from the definition of a concept the range of misconceptions students will exhibit; we have to discover them. The research methodologies that produce these results are generalisable, however, and that will enable us in later chapters to look at how they might extend our knowledge base. Chapter 3 moves away from students' epistemology into the even more uncertain area of trying to understand how they come by what they know. Here we are trying to see the learning process from their point of view, to see their ways of understanding not as wilful perversity but as something explicable and rational. These studies provide the basis we need for thinking about how to teach.

By Chapter 4, we return to territory that is more familiar for the teacher. Having looked in some detail at the inner life of the learner, we should be in a better position to see the implications of this for teaching strategies. There is a long tradition of instructional design, particularly in the US, and there are current contrasting theories of instruction, and the chapter critically analyses and draws on some of the principal ones. At this stage of the book, the problem is still being treated analytically. I do not wish to suggest that teaching is a science that can determine precisely how a topic should be taught, and in later chapters, I come to the creative side of designing teaching. But first I think it is possible, given the grounding from Chapters 2 and 3, to take an analytical approach to the relationship between the curriculum-defined goals of teaching, the specification of the learning activities students must therefore carry out, and the formative assessment appropriate to these goals and activities.

So far, there is nothing in the argument that implicates the use of any particular teaching method or medium. Chapter 4 leads us towards thinking about how to specify appropriate learning activities for students, which raises the question of how we might facilitate them. Up to this point, the idea of teaching as 'mediating between the world and the learner' has been used to define a particular way of viewing the teaching–learning process. It gives access to some interesting research findings, and a more elaborated description of the role of the teacher. Now we come to the practicalities of what it means to mediate, and the ways we can do this via educational media. Teachers are familiar with teaching methods that constitute their contact hours – lectures, seminars, tutorials – and they will be aware of a range of educational media – print, audiovisual, computer-based learning, teleconferencing, and Web access. Part II begins by describing an analytical framework for classifying these media, the Conversational Framework. It is based on the specification developed at the end of Part I for the activities the media must foster if all aspects of the learning process are to be supported. Chapters 5 to 9 define the different categories of learning media, though not from the point of view of their apparent characteristics, which is the standard approach of books on educational media. They are described instead in terms of the nature of the learning activities they support: narrative, interactive, adaptive, communicative, and productive. The last is an addition to those discussed in the first edition, in recognition of the new forms of user-controlled software that support creativity. These should surely begin to take their place among the learning media. Describing the different media in terms of the Conversational Framework, allows us to develop a comparison of what they contribute to learning.

Part III is more practical. Chapter 10 outlines learning design for the media described in Part II, and uses the analysis of learning activities arrived at in Part I to deduce how to combine the media to facilitate learning. The details of this analysis are different for every topic, so the aim is to describe the design process in general terms, for application to specific content.

The most brilliantly designed educational materials can fail completely if they are not used with the same care. Research and development projects on educational media pay quantities of hard cash for development, lip-service to evaluation, and no attention to implementation. There is rarely enough cash to equip a decent programme of piloting, dissemination, and staff training. Development projects trust to luck and the dedication of enthusiasts to carry them through. Learning technologies have progressed well on the backs of enthusiasts, but cannot achieve their potential this way. Chapter 11 covers the use of learning technologies as a fully integrated part of everyday academic life. Students respond primarily to the institutional context and its demands, so these must be congruent with the demands of the technology. The same is true for teachers, who are no less subject to the institution's demands. Full integration is vital for optimising any investment in learning technology. We can draw on evaluation studies of the implementation of new media, and on studies of institutional contexts, and use these to define the aspects of institutional life that will influence the success or failure of new technologies.

A new medium or method rarely works well in its first implementation, but the academic community is failing to learn the lessons of experience. Too few academics build on each others' previous work in the field; journal articles do not critique others' work, they only mention it; research and development projects do not build on what has gone before, so the same conclusions are continually repeated. Innovation in the teaching of a subject does not match the standards of innovative research in the subject itself. We should be building a body of knowledge of how best to use learning media, and creating a teaching profession that knows what it is doing and why. Chapter 12 discusses how we might do this, focusing on the organisational system in a university. The book does not offer 'how to do it' advice in the other chapters because teaching is not a normative science. My strategy, therefore, is to offer a way of thinking about teaching and the use of learning technology that is informed by a more elaborated understanding of what students do when they learn. The assumption is that when teachers think differently, they can act differently. Thinking differently is not a sufficient condition for acting differently, however. We must also be enabled to act differently. The institutional context must afford and encourage the actions we need. For that reason, the final chapter turns to 'how to do it', not just at the level of teaching, but at the level of defining a 'blueprint' for an organisational infrastructure that enables good teaching to be delivered.

As the impact of learning technology begins to bite, many universities discover that the economics of this innovation fit poorly with their institutional structures. Technology, as ever, requires standardisation, project management and teamwork. Universities are used to these disciplines in a research context, but not in teaching. Bringing e-learning to most universities will mean giving them access to the production and delivery infrastructure that learning technology requires, but that only distance teaching universities have developed.

In Chapter 12, we have to address the full context within which the professional teacher is operating to design an effective infrastructure. All universities' external relationships are becoming more complex, but the nature of the academic mission must remain paramount. It must not be lost within e-business models and the mesmerising effects of changing technologies. Every professional academic has a responsibility to their students and their discipline. The technologies, the new organisational structures, and the re-cast business models are subservient to that end.

The book concludes, then, with a suggestion for how higher education should operate – a blueprint for enabling academics to use learning technologies effectively, in the widest context. Universities operate from within national boundaries, but education as an ideal does not recognise national boundaries. A sense of nation has no place in the sense of vocation that an academic feels in wanting others to experience the delight of a true understanding of their subject. National and cultural differences play a part in curriculum design, applied learning, and the logistics of implementation, but at the level of affording understanding of a subject, all academics come together in a common purpose. That is why the international context has to be part of the analysis.

There is a boundary to all this, however. Rethinking university teaching is not the responsibility of government. It is the task of professional educators to change education, not politicians. The book ends with a systemic blueprint for an academic institution as the logical conclusion from the premise that university teachers must take responsibility for what and how their students learn.

Part 1

What students need from learning technologies

Chapter 1

Teaching as mediating learning

INTRODUCTION

What we believe to be of practical help to lecturers depends upon how we define the aim of teaching, so the greater part of this chapter is concerned with clarifying this basic issue. If you were to believe that teaching is about imparting knowledge, then the main requirement of the lecturer would be the possession of that knowledge. For some time, this has been the prevailing view of university teaching, and therefore academics are appointed on the basis of their qualifications in subject matter knowledge. There is probably also an implicit requirement that they should be capable of imparting the knowledge as well as knowing it. However, since this is done through lectures, and they can all talk, the requirement has not been dignified with any sort of qualification.

Of course, 'imparting knowledge' has not usually been a very successful teaching aim, as every essay and examination paper testifies. Academics have always been well aware of this, but while higher education was an élitist enterprise, it was possible to make this failure the responsibility of the student, reified in the 'fail' grade. This is not now the prevailing view. As higher education has become less élitist, and has taken on the task of educating anyone who wishes to pursue their studies, many institutions of higher education have developed an approach to teaching that has a higher ambition: 'The aim of teaching is simple: it is to make student learning possible' (Ramsden, 1992: 5). Changes in approach are important. However, it is changes in practice will make the real difference to students, and we are still a long way from defining and requiring professional practice for university teachers.

What might that be? If it is not simply imparting knowledge, what is it? 'Making student learning possible' places much more responsibility with the teacher. It implies that the teacher must know something about student learning, and what makes it possible. This is what I have characterised in the chapter title as 'mediating learning'. Since this is the idea that motivates the approach taken in the remainder of the book, I should begin by explaining it. An analysis of the nature of academic learning of the kind done by students at university level should reveal what it might mean to 'make student learning possible'.

THE CHARACTER OF ACADEMIC LEARNING

There is no professional training requirement for university academics in terms of their teaching competence, as there is for school teaching. Possibly for this reason, there is comparatively little research on student learning at university level. Of the many books and journals concerned with teaching, the great majority relate to school level. The Dearing Report on higher education in the UK acknowledged this:

> While higher education has increased its class sizes, reduced its teaching time, modularised, accepted students without traditional academic preparation, refocused programmes to prepare students for employment, and so on, it has done so on the basis of little evidence of the consequences, and with little strategic research in place to monitor them.
>
> (Dearing, Main Report, 1997: 126)

Advice to university teachers has to draw on other fields to supplement the meagre information we have from direct research. This book is directed specifically at university teaching, so it is worth deciding what kind of transformation has to be wrought on the available data to make it applicable to this context. Is learning at university different from learning at school, or learning outside formal education?

Academics have ambitious definitions for student learning. When asked to define the nature of learning in their subject area they produce descriptions of high-level thinking, such as 'critically assessing the arguments', 'compiling patterns to integrate their knowledge', 'becoming aware of the limitations of theoretical knowledge in the transfer of theory to practice', 'coming to accept relativism as a positive position'. Course descriptions and syllabuses inevitably tend to focus on the subject content that students will be learning, but clearly, in reflecting on what it is really about, academics are fascinated by the process itself. They see learning not simply as a product, but as a series of activities, and developing skills and capabilities as much as formal knowledge. How students approach their subject is as important as what they end up knowing. If we were to eavesdrop on academics' discussions in an examiners' meeting, the point would be confirmed. Missing out some key points will be forgiven if the argument is good; high praise is offered not just for accuracy, but more often for evidence of integrating lectures with background reading; accuracy is the *sine qua non*, perhaps, but more is needed. Evidently, student learning is not just about acquiring high-level knowledge. The way students handle that knowledge is what really concerns academics.

If academic learning is not just about imparting knowledge, is it really different from the acquisition of everyday knowledge? We learn a great deal about the world very successfully without academic institutions, and with no help from any didactic process. There is a tradition of pedagogy that stretches back to John Dewey's rejection of the classical mode of passing on knowledge in the form of unchangeable ideas. This strand of educational theory has always argued for the learner to be actively engaged in the formation of their ideas. More recent exponents of the latter tradition are Vygotsky, Piaget, Bruner, Papert, all of whom argue for the active engagement of the learner rather than the passive reception of given knowledge. These psychologists have had an effect in schools, especially at primary level. However, in universities, with their continued reliance on lectures and textbooks, the classical tradition of 'imparting knowledge' still flourishes in the forms through which we teach, if not in the rhetoric of individual academics.

The idea of academic knowledge as an abstract Platonic form had a new impetus from the development of an information-processing model of cognition. It used the metaphor of knowledge structures, or conceptual structures, to describe mentalistic entities that can be changed through instruction, or even represented in a computer program. Computational models of cognition now form the mainstream of cognitive psychology, and where psychology leads, educational theorists like to follow. There is an undeniable attraction in the rigour that computational modelling can bring to the description of learning. Lecturers are also likely to be attracted by the idea of a conceptual structure as a stable and well-defined entity abstracted from the contexts in which the concept was experienced. The notion sits well with the ideal of 'discipline' knowledge. However, it does not address the reality that all teachers surely recognise – that students do not transfer their knowledge across different settings, that they often find it difficult to relate theory to practice, that knowledge does seem to be context-dependent. University teachers are not aided by the representation of knowledge as a formal structure if they prefer to see learning as an activity that develops capabilities, and knowledge as an aspect of that activity. They need a description of academic knowledge that is more realistic than a stable mental model.

The next section presents a recent critique of educational tradition and its emphasis on decontextualising knowledge. This is followed by a critique of the critique, and the chapter ends with a synthesis of what I take to be the essential character of academic learning that provides the basis for discussion in the rest of the book.

A CRITIQUE OF ACADEMIC LEARNING AS IMPARTED KNOWLEDGE

The recent interest in the idea of 'situated learning' expresses dissatisfaction with the idea of formal knowledge, and with the computational models of mainstream cognitive psychology. The origins of this approach lie in ethnographic studies and in Vygotsky's theory of the social character of learning (Vygotsky, 1962). The idea is to recognise that learning must be 'situated', in the sense that the learner is located in a situation. Therefore, what they know from that experience they know in relation to that particular context:

> Situations might be said to co-produce knowledge through activity. Learning and cognition, it is now argued, are fundamentally situated.
>
> (Brown *et al.*, 1989a: 32)

The article outlining the approach was published in *Educational Researcher*, and provided a well-articulated statement of the position, based on several research studies of learning. The article attracted a great deal of comment, and had the benefit of further discussion through critiques from others and a reply by the authors, so it makes a good focus for our analysis of the nature of academic learning. The detail of the argument, rather than a general summary, is the best way to see how the perspective defines learning, and what it means for the practising teacher. Going through the detail makes it easier for the lecturer to relate the broad generalities to their own subject.

The argument begins with a demonstration that knowledge has a contextualised character, which means that we cannot separate knowledge to be learned from the situations in which it is used. The idea of 'situated knowledge' invites the analogy of knowledge as tool:

> We should abandon once and for all any notion that a concept is some sort of abstract, self-contained substance. Instead, it may be more useful to consider conceptual knowledge as in some ways similar to a set of tools.
>
> (Ibid. 5)

A corollary of this argument is that the acquisition of inert concepts (e.g. algorithms, routines, decontextualised definitions – that is the stuff of many university courses) is no use if the student cannot apply them. The analogy they use for students having inert concepts is those people who have a Swiss Army knife with a device for getting stones out of horses hooves: they can talk knowledgeably about it, but would not know what to do if they saw a limping horse. We have to be careful with analogies. Many engineering students have no idea how to do a Laplace transform within a week or so of passing finals, but knowing of its existence and its function they can reassemble the heuristic knowledge they need when necessary. If they know about the device, and can recognise a 'limping horse', it is easy to look up the heuristics of 'removing stones' they once knew. However, academic knowledge is not just the heuristics of 'removing stones', or 'doing Laplace transforms'; it has a broader and deeper functionality than that. The far greater problem is that students can exhibit competence in doing Laplace transforms without having any idea of when to use them or why. They are good at removing stones, but too often, they cannot recognise a limping horse. The distinction is important. As Brown *et al.* (1989a) argue, we have to use our knowledge in authentic activity, i.e. genuine application of the knowledge, which allows us to build an increasingly rich understanding of the tool itself and how it operates. The reason for unpacking the analogy is that many lecturers would argue that they do indeed give students the opportunity to do 'authentic activity': to understand Laplace transforms you have to do lots of examples of them and use them in different problems. This is common practice in every engineering course and has its parallel in every other kind of course. The problem arises from the scope of 'authentic', the degree of embeddedness in the social and physical world. We have to help students not just to perform the procedure, but also to stand back from it and see why it is necessary, where it fits and does not fit, distinguish situations where it is needed from those where it is not, i.e. carry out the authentic activities of the subject expert. But these remain implicit objectives in most course descriptions, and that implicitness persists all the way through to the activities we prescribe for students. One conclusion we can draw is that learning must be situated in the domain of its objective. If you want students to be able to recognise a limping horse, you must situate their learning activity within the domain of that objective, not simply in the domain of removing stones. We shall return to this point.

As a further example of the value of situating learning, rather than decontextualising it, Brown *et al.* (1989a) demonstrate the unity between problem, context and solution when the problem is experienced, rather than given. A weight-watcher was trying to serve the correct amount of cottage cheese, and worked it out as three-quarters of the two-thirds of a cup he was allowed. After muttering about his college calculus course, he suddenly brightened, and certain that he had found the solution, proceeded to dump two-thirds of a cup of cheese onto a board, flatten it into a circle, cut it in four and serve three of the quarters.

> This sort of problem solving is carried out in conjunction with the environment and is quite distinct from processing solely inside heads that many teaching practices implicitly endorse.
>
> (Ibid. 35)

This example gives rise to consternation among academics because it looks as though the weight-watcher's achievement is valued as more important than the more abstract knowledge of the arithmetic of fractions. But the example only demonstrates the 'sense-making' nature of naturally embedded activities; the weight-watcher is not being applauded. The point is that if formal education provided more naturally embedded activities, students could do their own sense-making. The authors are arguing against the decontextualising of knowledge by teaching abstractions:

> Our argument is that to the degree that abstractions are not grounded in multiple contexts, they will not transfer well. After all, it is not learning the abstraction, but learning the appropriate circumstances in which to ground the abstraction that is difficult.
>
> (Brown *et al.*, 1989b: 12)

There is a distinction made, therefore, between teaching abstractions and enabling students to learn abstractions from multiple contexts. The latter stands between the extreme of the weight-watcher's purely situated knowledge, which is clearly not academic, and the purely abstract, which academic knowledge is often thought to be. The implication is that academic learning should occupy the middle position of an activity that develops abstractions from multiple contexts.

A CRITIQUE OF ACADEMIC LEARNING AS SITUATED COGNITION

Teaching practices that encourage abstraction from experience do not have to subscribe to an epistemology that places knowledge 'solely inside heads'. It is legitimate and necessary for teaching to go beyond the specific experience, to offer the symbolic representation that allows the learner to use their knowledge in an unfamiliar situation. Situated cognition is attractive in well-chosen situations, but one of the reasons that education has evolved the way it has over the centuries is that situated cognition is not enough. Suppose the weight-watcher were trying to work out his share of a discounted car hire with a couple of friends and had to figure out the logically equivalent problem of one-third of 5 per cent off the total cost? The unity between problem, context and solution is not quite so apparent here. The point of an academic education is that knowledge has to be abstracted, and represented formally to become generalisable and therefore more generally useful. It then empowers people like the weight-watcher to deal with quantities of things other than cottage cheese.

Can students be taught to acquire an abstraction from multiple contexts without it being taught directly as an abstraction? Some of the illustrations of situated cognition come from real teachers, and demonstrate what most practising teachers know, that concepts need to be grounded in experience and practice before they can be abstracted. It is common pedagogical practice at all levels of education to start with concrete examples, or to provide illustrative examples of general principles, but the way the teacher conducts this process is crucial for its success. Again, the only way we can see how the idea of situated cognition applies is to go through an example in detail, and analyse the extent to which it provides an adequate account of academic learning.

The teaching of multiplication may seem a rather elementary example to use as an illustration for university teaching, but it works well for two reasons: it reveals some interesting aspects of what it means to acquire an abstraction from multiple contexts, and it is an abstraction that everyone is familiar with.

Multiplication does not have to be taught as an abstraction. Brown *et al.* describe the approach of a teacher whose method is to make mathematical exploration continuous with everyday knowledge. She sets out to help the learners towards the abstract algorithm in the context of real world problems and the stories the group creates about them:

> Lampert helps her students explore their implicit knowledge. Then in the second phase, the students create stories for multiplication problems. They

> perform a series of decompositions and discover that there is no one magically 'right' decomposition decreed by authority, just more and less useful decompositions [e.g. 24 = 8 × 3 or 6 × 4], whose *use* is judged in the context of the problem to be solved.
>
> (Brown *et al.*, 1989a: 38)

It is clear from this example that situated cognition in the context of education is not concerned simply with learning about the world, but with learning about a way of looking at the world. This is important if it is to describe academic learning. Look at the dialogue they are describing:

Teacher: Can anyone give me a story that could go with this multiplication … 12 × 4?
Student 1: There were 12 jars and each had 4 butterflies in it.

Clearly the learner has already acquired a way of interpreting 12 × 4, and the usage of language such as 'story' and 'go with this multiplication'. This is not everyday language; it is already academic language, describing the notion of interpreting a symbolism. The teacher then draws a picture to represent jars, and another to represent butterflies.

Teacher: Now it will be easier for us to count how many butterflies there are altogether if we think of the jars in groups. And as usual the mathematician's favourite number for thinking about groups is?
Student 2: 10

Grouping in order to count is a fundamental aspect of the nature of multiplication, but the learners are not grounding this aspect of their knowledge in the activity: it is being handed on as a precept, as the best way to do this kind of task. The choice of decomposition is not theirs but the teacher's. The focus here is on different ways of decomposing:

Teacher: Is there any other way I could group them to make it easier to count all the butterflies?
Student 6: You could do 6 and 6
Teacher: Now how many do I have in this group [of six jars]?
Student 7: 24
Teacher: How did you figure that out?
Student 7: 8 and 8 and 8 [He puts the 6 jars into 3 pairs intuitively finding a grouping that made the figuring easier for him]

Now a student has offered a different grouping, and the teacher can use this to show that the total is still the same.

Teacher: That's 3×8. It's also 6×4. Now how many are in this group [the other group of six jars]?
Student 6: 24. It's the same. They both have 6 jars.
Teacher: And how many are there altogether?
Student 8: 24 and 24 is 48
Teacher: Do we get the same number of butterflies as before? Why?
Student 8: Yeah, because we have the same number of jars and they still have 4 butterflies in each.

This is not strictly an example of students discovering 'that there is no one magically right decomposition'. They are being led through a reasoning process planned by the teacher. The final student comment is reasoning from conservation of matter, not from equivalence of decomposition. They can understand that whichever way you group the jars, the total is the same, but the leap of mathematical reasoning required is to see that there are two symbolic forms that describe the groupings, and that also match the real-world properties of those groupings. In fact, the more accurate interpretation is that 'authority decrees that there is no one magically right decomposition' and that this is the idea the teacher is trying to get across. Brown *et al.* (1989a) are justified in using this to exemplify good teaching. The teacher is clearly aiming to make the idea of decomposition meaningful, but she is also directing the way the students are to think about this activity. She carefully constructs the situation as a benign environment for learning about an abstract description of the world. But this extract does not demonstrate that the students do think about the activity in the way she requires, nor that they have yet abstracted this knowledge from the multiple contexts being set up for them. It demonstrates that students can do their own sense-making in this naturally embedded activity, but does not demonstrate the process of abstraction that is essential for academic learning.

Using situated learning as a metaphor for academic learning is interesting and powerful as an idea, because it analyses successful naturally-occurring learning to understand how that operates, and then transfers that analysis to the academic context to see how it should be applied there. The problem with the analogy is that the learner stands in a different relation to the content of what is learned in the two cases. Learning in naturalistic contexts is synergistic with the context; the learning outcome is an aspect of the situation, an aspect of the relation between learner, activity and environment, so it is learning about that world and how it works within that relationship. Those naturalistic contexts afford learning through situated cognition.

On the other hand, learning in educational contexts requires learning about descriptions of the world, or about a particular way of looking at the world. The learner cannot relate to a description, nor to someone else's perspective, as they can to an object. The weight-watcher could deal with hundreds of piles of cottage cheese successfully and never abstract the principle of proportional reasoning needed to deal with the car-hire problem. Academic learning requires him to take a different perspective on those activities, to generalise from them to obtain an abstraction, a description of the world that does not consist in doing the activity alone.

Brown *et al.* (1989a) argue that we have to recognise the situated character of learning, and use it to devise ways of constructing a situation that is benign with respect to what we want students to learn. That is what Ms Lampert does, and what their extract demonstrates very well. However, the analysis does not go far enough for the purposes of academic knowledge because it also has to address how the process of abstraction is to be done by the student. Multiple contexts may be necessary but they are not sufficient. Those children could be taken through hundreds of examples by the teacher, but while the teacher does all the planning, hands out precepts, and asks the questions, the students can easily fail to engage actively with her way of thinking. The authentic activity of the mathematician is not grouping jars in alternative ways, but exploring the relationships between the real-world activities and the symbolic descriptions of them. A more authentic activity, given the teacher's objective, would be to ask the class if they could generate different ways of describing the groupings of butterflies and jars. They then focus their attention on the relationships between theory and practice, and engage in a more authentic activity than the grouping task offers. For this to be an adequate account of academic learning, the detail of the teaching process should have shown us how those learners were to engage, not just with their own experience, but with knowledge derived from someone else's experience. The concept of 'authentic activity' is nonetheless a valuable one because of its implication for design. Applied correctly, it defines how learners must engage with content.

ACADEMIC LEARNING AS A WAY OF EXPERIENCING THE WORLD

Some years ago, I contributed to a debate about the relationship between psychology and education, and made a specific request to the psychologists:

> Our problem is that at present cognitive psychology produces generalized, not content-specific principles and theories of learning ... A general principle that describes, for example the importance of active manipulation, or of relating new knowledge to existing knowledge, is no help because it does not clarify the logic of the relationship between the cognitive activity and the content to be learned. We need cognitive psychology to tell us, in a content-specific way, how a natural environment affords learning. Then, perhaps, we can construct the means of access that will turn an unnatural environment into one that affords learning.
>
> (Laurillard, 1987a: 206)

The psychology of situated cognition does provide a content-specific account of how a natural environment affords learning. It is valuable because it sees learning as essentially situated, and as requiring a non-dualistic epistemology. Analyses of how natural environments afford learning are used to construct descriptions of how an academic environment can do so too. The work on situated cognition does not, however, illuminate the essential difference between academic knowledge and everyday knowledge, and I want to pursue that point a little further.

In that earlier debate I drew a distinction between natural environments which afford the learning of 'percepts' in everyday life, and unnatural environments which are constructed for learning 'precepts' in education. Situated cognition makes the same distinction in arguing that the one type of environment should emulate the other, but does not elaborate on the nature of the difference. I argued that learning precepts is different from learning percepts because our means of access to them are so limited:

> We cannot experience structuralism in the same way as we experience good table manners. We cannot experience molecules in the same way as we experience dogs. Because we have to rely on the artificial structuring of our experience of precepts, via academic texts, for example, it is unlikely that the mechanisms we use in the natural environment will transfer directly to this unnatural environment. Thus, our means of access to precepts becomes critical to our success in learning them.
>
> (Laurillard, 1987a: 202)

The distinction between learning percepts and learning precepts is important for my subsequent argument about the nature of academic teaching, but it is a difficult one to make as Eysenck and Warren-Piper pointed out in the debate.

> Choosing 'molecule' as an example of a precept which cannot be experienced seems to overlook the degree to which people, when they imagine such entities, call upon what they have experienced so as to give such abstractions substance. Ping-Pong balls and gravity are dragooned into service to explain the molecule ... And while to Laurillard structuralism cannot be experienced like table manners, others might argue that that is precisely the way in which it is experienced, both being a protocol for going about a ritual task ... One is led to conclude that there is no clear division between the natural and unnatural environments.
>
> (Eysenck and Warren-Piper, 1987: 209–210)

It is true that molecules can be experienced, but the very fact that we have to dragoon other such disparate experiences into service in order to experience them demonstrates how different that is from the way a dog is experienced. The point about structuralism is interesting because it probably could be experienced as a protocol, but I think academics want students to understand structuralism in a deeper sense than in being able to perform a ritual procedure. If the teacher simply acculturates students into 'performing' structuralism, then it will not also be available to them as an articulated idea, accessible to comparison with other approaches, and open to criticism. Academics want students to learn more than that which is already available from experiencing the world.

The point about academic knowledge is that, being articulated, it is known through exposition, argument, interpretation. It is known through reflection on experience and represents therefore a second-order experience of the world. Knowledge derived from experiencing the world at one remove must be accessed differently from that known through a first-order experience. We need a specific example to clarify what this actually means for a teacher, and the classic one that emerges sooner or later in every discussion of student learning is the problem of understanding Newton's concept of force. Like any illustrative example it suffers from the fact that to appreciate what it tells us you really have to understand the concept, but it gains from the fact that it has been extensively researched, so we know a lot about how it is misunderstood. We all experience force as an aspect of daily life, and we have multitudinous contexts from which to abstract a general idea about its nature. We learn the use of the word 'force' in a number of different contexts, and learn to distinguish its use, as in 'police force', 'force it open', 'force of gravity', 'force them to do it', etc. Our knowledge of 'force' is situated and we have no great problem with it. The physics lecturer then offers some new ways of thinking about force, and using the word 'force'. We meet the idea of 'force acting at a distance', which is curious, but not unlike forcing someone to do something. We hear about a falling apple interpreted as 'the force of the earth acting on the apple' which makes a kind of sense if you accepted action at a distance. However, there is also 'the force of the apple acting on the earth', which makes no sense at all and had better be ignored. A reaction like this latter one dooms us never to understand Newton's idea of force (more of this later).

We certainly use our everyday experience to help interpret the meaning of the physics lecture, and to an extent that helps. But it is important to go beyond that to attain the true scientific meaning. The physics lecture cannot, however, offer any new experience of the world to match this new idea. It offers only a different way of thinking about apples falling, of seeing them as being essentially similar to planets orbiting the sun, or atoms orbiting an electron. Every academic subject faces this same kind of challenge, to help students go beyond their experience, to use it and reflect on it, and thereby change their perspective on it, and therefore change the way they experience the world. That is why education must act at the second-order level of 'reflecting on' experience. Everyday knowledge is located in our experience of the world. Academic knowledge is located in our experience of our experience of the world. Both are situated, but in logically distinct contexts. Teaching may use the analogy of situated learning of the world, but must adapt it to the learning of descriptions of the world. I have termed this 'mediated learning', after Vygotsky:

> A scientific concept involves from the first a 'mediated' attitude towards its object.
>
> (Vygotsky, 1962: 102)

Teaching as mediating learning involves constructing the environments which afford not only learning of the world, but also learning of descriptions of the world. The means of access to the two types of knowledge is different. The one is direct, the other mediated.

Because academic knowledge has this second-order character, it relies heavily on symbolic representation as the medium through which it is known. This is usually language, but may also be mathematical symbols, diagrams, musical notation, phonetics, or any symbol system that can represent a description of the world, and requires interpretation. Students must learn the representation system as well as the ideas it represents. The difficulty of this has attracted a fair amount of attention at the level of school mathematics, but surprisingly little has been done on how students interpret teachers' language, how they read academic texts, and how they interpret graphical and symbolic information. Roger Säljö makes the same point in his analysis of 'the written code' as a medium for learning, fittingly subtitled as 'observations on the problems of profiting from somebody else's insights'. The problems arise from the fact that the two worlds, of everyday knowledge and academic knowledge, are not always compatible:

> In scientific texts, new 'versions of the world', or fragments of such, are offered, and the act of learning through reading may thus be seen as containing an implicit commitment to transcend assumptions vis à vis reality for which we have a firm basis in terms of our own previous daily experiences. Our knowledge gained by personal experience and therefore 'true' in our everyday realm of life, may in our culture have to yield to an alternative mode of conceptualisation that links with a scientific 'version of the world'.
>
> (Säljö, 1984: 31)

A similar dichotomy has been explored by Gibbons *et al.* (1994) as a contrast between the formal, codified 'Mode 1' knowledge of the traditional disciplines, and the informal, implicit 'Mode 2' knowledge created by communities of practice. However, their argument is not that we must transcend everyday experiential knowledge to acquire the formal scientific knowledge, but that experiential knowledge is more valuable than formal knowledge. Gibbons *et al.* argue that university teaching must also address itself to experiential knowledge if it is to remain relevant to the way knowledge is actually used in our society. But what is the scope of relevance? In a further development of their earlier account of situated learning, Brown and Duguid carry the same dichotomy through to its logical conclusion that practice knowledge is highly contextualised. Informal, experiential, situated knowledge, developed through communities of practice, becomes fully contextualised, to the extent that it is no longer functional beyond that community:

> The tasks undertaken by communities of practice develop particular, local, and highly specialized knowledge within the community … communities develop their own distinct criteria for what counts as evidence … the division

> of labour produces the division of knowledge … Within communities, producing, warranting, and propagating knowledge are almost indivisible … Hence, the knowledge produced doesn't turn readily into something with exchange value or use value elsewhere.
>
> (Brown and Duguid, 1998)

Brown and Duguid use this argument as a warning to organisations that the cumulative knowledge developed within the workforce cannot be commodified and communicated easily. There is a danger that organisations are unable to know what they know, and therefore under-utilise the knowledge being accumulated. The argument against the primacy of formal knowledge, advanced by Gibbons *et al.*, and by Brown and Duguid, thus comes full circle to an acknowledgement that without the processes of decontextualisation, and formalisation, knowledge remains situated and uncommunicable. The dialectic process will probably lead us to a resolution of these two aspects of knowledge, and an acknowledgement that neither can predominate. Academic knowledge will necessarily address both aspects of knowledge.

Before summarising the arguments made in this chapter, it will be useful to reconsider the concerns of university teachers: 'critically assessing the arguments', 'compiling patterns to integrate knowledge', 'becoming aware of the limitations of theoretical knowledge in the transfer of theory to practice', 'coming to accept relativism as a positive position'. If the analysis here locates academic knowledge between our first-order and second-order experience of the world, and the teacher must mediate the latter, I feel we have not strayed too far from the focus of those concerns and the assumptions that underlie them. A computational model would not so easily embrace the sense of action they describe. Situated cognition certainly gives a sense of action, but not the sense of 'standing back' from the content that is implicit in what teachers want of their students. Academic knowledge is not like other kinds of everyday knowledge. Teaching is essentially a rhetorical activity, seeking to persuade students to change the way they experience the world through an understanding of the insights of others. It has to create the environment that enables students to embrace the twin poles of experiential and formal knowledge.

SUMMARY

The chapter began by accepting that the aim of university teaching is to make student learning possible. Because academics are concerned with how their subject is known, as well as what is known, teaching must not simply impart decontextualised knowledge, but must emulate the success of everyday learning by situating knowledge in real-world activity. However, academic learning has a second-order character, as it concerns descriptions of the world. Whereas natural environments afford learning of percepts through situated cognition, teaching must create artificial environments that afford the learning of 'precepts', i.e. descriptions of the world. The implications for the design of teaching are that:

- academic learning must be situated in the domain of the objective, and learning activities must match that domain;
- learning environments must be designed with features that afford the learning of precepts, the affordances for academic learning;
- academic teaching must help students reflect on their experience of the world in a way that produces the intended way of representing it.

Thus teaching is a rhetorical activity: it mediates learning, allowing students to acquire knowledge of someone else's way of experiencing the world. With that analysis of the nature of the task, we can now attempt to see how this might be done.

Chapter 2

What students bring to learning

INTRODUCTION

There is a many-to-one relationship between where students are at the start of a course, and where teachers want them to be by the end, not because teachers want to turn out identical replicas of themselves, but because there is a consensual aspect to the didactic process without which academic life fails in its responsibility to progress knowledge. We expect students to use their knowledge in a variety of ways, and to contribute personal and even original ways of thinking about their subject, but we expect them also to exhibit some point of contact with the consensus view of a subject: if they cannot agree on the substantive content, then they must be able to provide an acceptable argument for the opposing point of view. It follows from this combined personal and consensual character of academic knowledge that there are many ways of knowing a topic, and also that there are many ways of failing to know it.

The knowledge that students bring to a course will necessarily affect how they deal with the new knowledge being taught. Because this relationship has always been understood, the progress through an academic career has typically been governed by a student's acquisition of pre-requisite knowledge; each new course builds on an assumption about what the student has already mastered. This is a dangerous assumption, as we shall see in this chapter. Mastery of the art of taking examinations designed to test knowledge is more prevalent than mastery of the knowledge itself. The teacher will often be building on sand. As teaching and assessment techniques improve, we could look forward to a gradual lessening of this problem, perhaps, were it not for the fact that other changes exacerbate the problem. Increased intake to university courses increases the likelihood that students will not have fully mastered all the pre-requisite ideas in a subject area; greater modularity in courses decreases the likelihood that they will have acquired those concepts that used to be considered pre-requisite. It will continue to be necessary, therefore, for academics to understand not only where students should get to, but also where they are as they begin a course.

How can we know this? The educational system offers academics only one source of information, the examination result. It is hardly sufficient. Even the topmost grade obscures a multitude of sins of omission and commission in the student's knowledge and understanding of the detail of the subject. What are the different 'ways of knowing' that a whole class of students might bring to a topic? There are a number of studies of university students that tell us something about where they might be at the start of a course. In this chapter I shall introduce these studies as alternative sources of information for the academic to make use of. Like the examination system, they each use a particular methodological approach and therefore offer a particular kind of description of the student useful at different stages of the educative process. The examination system may be good for the selection process, but not for the academic who needs to know what kind of teaching the students will need if they are to cope with new ideas.

There are two fundamentally different ways of investigating what students bring to their learning of a topic. One approach considers student-specific characteristics, such as approach to study, epistemological belief, and intellectual development; the other illuminates the task-specific aspects important for understanding: conceptions, reasoning processes and representational skills.

QUESTIONNAIRE STUDIES OF STUDENTS' CHARACTERISTICS

Do students have individual learning styles or approaches to study which we should take into account? This is a question that intrigues many academics who think about how to design their teaching, and there are a number of studies that set out to answer it.

Individual student characteristics are explored in questionnaire studies as though they are independent of the context of particular learning tasks. The methodology is to survey or interview a sample of students, asking them questions about how they approach learning, how they define learning, how they organise their study, etc. Factor analysis of survey data, or content analysis of interview data into categories of similar responses, enables the researcher to sort the student sample into different individual types and to correlate some characteristics with others, e.g. learning style with motivation. The methodology invites the emergence of individual characteristics which are necessarily independent of context, because that is the way they were collected, and which are therefore presumed to be present in the context of any learning task. The methodology cannot determine how important these characteristics are in the learning process.

The main difficulty in interpreting findings of this type is to decide how far we consider the characteristics discovered to be fixed and immutable for an individual. In Chapter 1 I argue that we should consider the learning process holistically, and knowledge as being, in part, situated and contextualised. This sits unhappily with the notion that students might have personality characteristics that determine the way they think, irrespective of the context. On the other hand, there is undoubtedly an expectation and an intuition on the part of academics that there are identifiably different ways of thinking, often linked to the type of subject being studied: course aims may be defined as being to help students 'think like a social scientist', or 'think like a technologist'. This expresses an intention to acculturate the student, rather than a belief in a personality type, but what about phrases like 'first class mind', or 'scatterbrain'? These express the idea that some aspects of thinking perseverate across a variety of contexts, and constitute an individual style of thinking. Clearly if there are such characteristic styles they would affect the way a learner responds to a particular task, and they would therefore be of interest to us.

Entwistle decribes a study which asked students to indicate the extent of their agreement with a series of statements about their normal academic work, for example:

> I try to relate ideas in one subject to those in others, whenever possible.
> I like to be told precisely what to do in essays or other set work.
> It's important to me to do really well in courses here.
> When I'm reading I try to memorize important facts which may come in useful later.
>
> (Entwistle, 1981: 57)

Factor analysis of the results (e.g. for one study 767 first year students from three British universities) then links together several of these items as related to each other. For example, students who agree with the second statement above (indicating extrinsic motivation) are likely also to agree with the last one (indicating a superficial approach to study). Another factor links a deep approach to intrinsic motivation, and a third links organised study methods to achievement motivation. In this way, Entwistle is able to identify three types of motivation with these three factors, which he characterises as 'personal meaning', 'reproducing', and 'achieving'.

The idea of pigeonholing students is a convenient simplification of the vast diversity of those idiosyncratic individuals we meet when we teach. It is always salutary, however, to try to pigeonhole oneself in one of these categories – do you think you are achievement-oriented rather than meaning-oriented, or vice versa? The idea has a certain face validity when applied to other people, preferably people you don't know very well, but applying it to oneself illuminates the crudity of the classification. There is a strong temptation to respond 'both', or 'it all depends'. That is probably closer to the reality. As Entwistle (1981) and Ramsden (1992) pointed out, students are capable of variation as well:

> It is possible to accept that there can be both consistency and variability in students' approaches to learning. The tendency to adopt a certain approach, or to prefer a certain style of learning, may be a useful way of describing differences between students. But a more complete explanation would also involve a recognition of the way an individual student's strategy may vary from task to task.
>
> (Entwistle, 1981: 105)

> Although it is abundantly clear that the same student uses different approaches on different occasions, it is also true that general tendencies to adopt particular approaches, related to the different demands of courses and previous educational experiences, do exist. Variability in approaches thus coexists with consistency.
>
> (Ramsden, 1992: 51)

In an early study of students learning through problem-solving this variation within individual students was apparent (Laurillard, 1979). The study was conducted with a group of students carrying out a series of problem-solving tasks set by their lecturers in different subjects. The students submitted their assignments and were interviewed about their approach to the tasks. The outcomes and approaches were analysed in terms of Marton's 'deep' and 'surface' approaches to learning, and Pask's 'operation' and 'comprehension' styles of learning (where 'operation' refers to procedures and 'comprehension' refers to descriptions). It was clear that each student's choice of deep or surface approach, and of operation or comprehension learning, was dependent to some extent on the nature of the problem set and to some extent on their perception of the teacher's requirements (Laurillard, 1997). A similar point is made by Prosser and Trigwell, who explain the variation in individual acts of learning in terms of the learner's awareness of the learning context and prior experiences (Prosser and Trigwell, 1999).

We do not have strong enough evidence of the existence of stable individual learning characteristics, whether motivation, learning style, or study pattern, to abandon the idea that a student's approach interacts with particular learning situations, and is therefore context-dependent. There may still be some antecedent influence on what a student does during learning. The entire pre-history of their academic experience up to the time of a learning session can affect what they do. Each individual student probably has a repertoire of approaches of which one will be salient for a particular learning task. Moreover, part of Entwistle and Ramsden's research programme showed that students' approaches could also be influenced by their perceptions of teaching and assessment (Entwistle and Ramsden, 1983).

The characteristics found in questionnaire studies describe the population as a whole, therefore, with all of the characteristics potentially available to all students as aspects of their learning. We do not need to make the much stronger assumption that they are stable characteristics of individuals.

EXPLORATORY STUDIES OF THE STUDENT POPULATION

Exploratory studies attempt to describe the characteristic ways of conceptualising and learning a topic that can be found in the student population, without identifying the characteristics with individuals. Some of these studies are also known as 'phenomenographic', because they set out to find characteristics in the form of students' descriptions of the phenomena, in contrast with studies that set out to explain student behaviour by finding relations between pre-defined characteristics.

I have begun this section with a conscious parallel of the description of methodology given in the previous section, with differences italicised to aid comparison.

Population characteristics are explored as though they are *dependent* on the context of particular learning tasks. The methodology is to survey or interview a sample of students *working on a particular task, either given by the researcher or occurring within their normal study*. The students are asked questions about how they approach this learning task, how they think about it, *why they do what they do*, etc. *Content* analysis of interview or open-ended questionnaire data produces categories of similar responses which enable the researcher to sort the protocols into different types and *to find common patterns of internal relations between characteristics of each protocol*. The methodology invites the emergence of *characteristics of the learning process* which are necessarily *contextualised*, because that is the way they were collected, and which are therefore presumed to be applicable to the context of any learning task. The methodology elicits characteristics *that are important* in the learning process, but it *cannot* determine how consistent these are for individual students.

The unit of phenomenographic research is 'a way of experiencing something' and the focus is the variation in ways of experiencing something among a population of students. The output is a set of characteristic ways of experiencing something for that population. Using phenomenography for a study of academic learning, therefore, will yield a set of characteristic ways of experiencing learning for a population of university students. A feature of these studies is that they are conducted through an interview about the learner's experience of a specific learning event, not through surveys about descriptions of learning in general.

For educational purposes, we need to look at the variation in ways of experiencing a particular idea, which might be:

> a common-sense conception of a phenomenon on the one hand, and the conception used within a scientific framework for understanding the very same phenomenon on the other hand.
>
> (Säljö, 1988: 38)

The study generates a set of variations, for example in conceptions of 'force', which map the range of ways the idea can be known. But there is no attempt at a reductionist explanation of these conceptions in terms of underlying psychological processes. A conception is not a property of an individual in the way a nose is; it is an aspect of their behaviour in the world and their experience of it. With this relational epistemology, it is impossible to expect that we can discover anything worthwhile about conceptions by looking at traces of how people carry out tasks, such as their written performance on subtraction problems. This gives us access only to their behaviour, not their experience. Without careful interviewing and observation, it is impossible to interpret correctly students' actions. The power of the phenomenographic methodology is that it sets out to discover precisely what teachers need to know – the range of conceptions of an idea that students may already have.

The phenomenographic method uses critical tasks within the topic concerned, probing interviews with students about their experience of carrying out those tasks, and comparative analysis of the protocols to reveal the main forms of conception. There is no expectation of being able to identify an individual with a particular type of conception, and it can happen that in the course of an interview a student will exhibit more than one conception. The analysis is not by individual, therefore, but is carried out in terms of the meaning of the conceptions invoked in the course of a student's explanation. Interviewing many students within a population will result in several forms of conception of one idea or topic. The relationship between these different conceptions is not clear, however. Some researchers see them as inclusive – a more sophisticated conception will logically include the lower ones. Others, myself among them, see them as being related not to each other, but to the history of the students' experiences with the idea. And others see them as defining a developmental progression, where each successive conception is better, in a similar way to the progression defined for scientific theories: they explain more, they are more productive. It is an intriguing fact of research on students' conceptions that they may sometimes bear a strong resemblance to earlier scientific theories – students have Aristotelian conceptions of motion, Lamarckian conceptions of evolution, phlogiston theories of combustion, Eysenckian theories of intelligence, for example, suggesting the delightful idea that the intellectual development of the individual recapitulates the development of the history of ideas (see Säljö, 1988; Brumby, 1984; Champagne *et al.*, 1982). There are studies that support all these views. I do not propose to decide among these just now. The important point, for my purpose, is that such an approach can illuminate for the teacher what their students might already know.

The remarkable conclusion from all these studies is that what students know can be described in a relatively concise way, as long as you penetrate to the level of what the concept means to the student. Brown and Van Lehn, in their study of subtraction procedures, found 89 different ways of doing it wrongly (Brown and Van Lehn, 1980). However, by going to a different level of description, at the level of understanding, Resnick and Omanson (1987) found just two ways of misconceptualising subtraction. Looking at procedures attacks the problem at the wrong level. If a student borrows across zero incorrectly, we want to teach him not 'how to borrow across zero', but what 'borrowing' means. Subtraction is not just the formal manipulation of a procedural skill. The skill is inseparable from the knowledge it invokes. The most commonly identified problem with the teaching of arithmetic is that it divorces the process of subtraction from its meaning in action, as Brown himself argued in his later work, already discussed in Chapter 1 (Brown *et al.*, 1989a). It makes no sense to remediate a faulty procedural skill with reference to the procedure alone; we have to appeal to the conceptual apparatus that supports it as well. We have already established that knowledge is situated in action. Similarly, action manifests knowledge; and 'buggy' behaviour manifests an underlying conceptualisation that itself needs remediation. This means we must know how the student conceptualises all the aspects involved in the procedure: the action of borrowing, the representation of the task, the representation of the procedure, the concept of number, the concept of 'difference' and the concept of 'subtraction'. Students need practice in the interpretation and manipulation of the formal representations found in any academic subject.

Resnick and Omanson made use of the systematic definition of Brown and Van Lehn's 'buggy algorithms' (i.e. flawed procedures), but concluded that it is best not to remediate them, but to take them as revealing a fundamental flaw in the way children think about number in the context of these tasks. It is this fundamental flaw that should be remediated, thus pre-empting the formation of meaningless algorithms in the first place:

> ... if we look beyond the symbol manipulations of written arithmetic to what the symbols represent, the buggy algorithms look much less sensible ... It seems reasonable to suggest ... that a major reason that children invent buggy algorithms so freely is that they either do not know or fail to apply to calculation problems the basic principles relevant to the domain. If so, instruction focused on principles and on their application to calculation ought to eliminate or at least substantially decrease buggy performances.
>
> (Resnick and Omanson, 1987: 49)

If you remediate one of the eighty-nine wrong procedures, you have another eighty-eight to contend with; but if you remediate one of the misconceptions, you avoid all the inherited bugs and faulty procedures as well. A methodology that searches for the fundamental misconceptions will yield data that are far more valuable.

There is remarkably little research on how students interpret formal representations. Although many disciplines confine themselves to specialist language, there is increasing use of other forms, such as diagrams, symbols, pictures, tables, and equations, used to explicate an idea, or formalise the knowledge. Students have to learn how to interpret these standard forms, and academics need to use them without creating a further barrier to understanding. Formal representations mediate our experience of the world, and embody a particular way of describing it. Students must be able to apprehend both the form and the content, but neither is unproblematic. Tabachneck-Schijf and Simon offer a nice example from economics of a form of representation that obscures its interpretation. Supply and demand curves are typically represented with the quantity bought on the x-axis and the price offered on the y-axis. This invites us to read the graph from left to right in terms of changes in quantity demanded affecting changes in price. However:

> this way of reading the graph does not reflect our normal causal thinking about the underlying economic relations, i.e. the economic reasons why the curves slope this way. If we were asked 'why?' about these slopes, we might want to read the graphs from bottom to top: as the price increases, the quantity demanded decreases ... It would be easier to answer the 'why' question if the curves were graphed so that ... we could read them 'causally' from left to right.
> (Tabachneck-Schijf and Simon, 1996: 34)

As a way of mediating the world of economic relations, this form of representation lacks fidelity. If it more faithfully reflected the students' natural way of thinking about economic relations, then it would be easier for them to situate their interpretation of the graph in the real-world events it is representing. Without that, the formalism becomes a barrier.

The same argument applies not just to symbols and diagrams, but to specialist language as well. An appropriate topic for detailed analysis here is Newton's Third Law. Many people misunderstand Newton's Third Law (including many of the people who teach it), and it is a good example of the difference between experiential and academic knowledge. This discussion is based on an extensive study of students' conceptions in mechanics, described more fully elsewhere (Bowden *et al.*, 1992; Laurillard, 1992).

Using a phenomenographic approach, physics students in their first year of university were interviewed about their solutions to five or six carefully selected problems. For one of them, they are asked to state Newton's Third Law (if they cannot remember it they are given a prepared statement of its canonical form: 'every force has an equal and opposite reaction'). They are then shown a diagram of a box resting on a table, and a box in mid-air and asked to use the law to describe the forces acting in the two situations (see Figure 2.1).

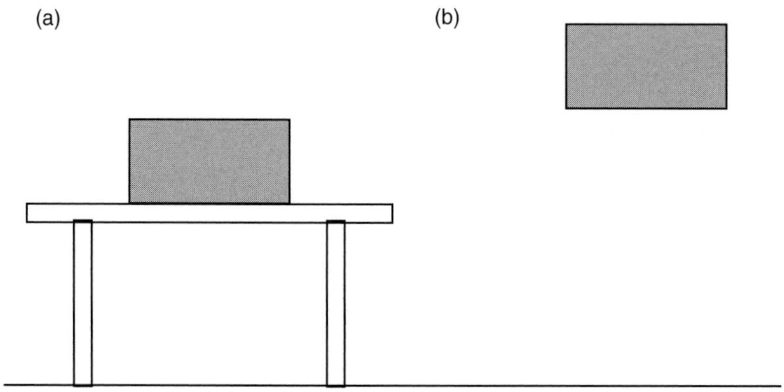

Figure 2.1 Applying Newton's Third Law: students are asked to explain the application of the law for (a) the box on the table, and (b) the box falling to the ground.

The interviewer probes to find out why they give the answer they do and how they define their terms, trying to obtain as complete a picture as possible of the way the student thinks about the problem. Of course, this kind of interview is a learning experience in itself and, not surprisingly, students often change the way they reason about the situation in response to mild but probing questions. To give a flavour of this kind of interview, I have reprinted a longish extract. The student begins with a definition of the law applied to the box on the table, which he later finds difficult to apply to the box in mid-air:

Interviewer: And in the case when it's falling, how does Newton's Law apply there?
Student 1: Well umm it's not at a constant velocity and it's not at rest ...
Interviewer: Mm ...
Student 1: so there is a a net force, um, which is which is present and that's what is causing the acceleration.
Interviewer: And what is the net force?
Student 1: Um, it's the force of the gravity which is the mg force ...
Interviewer: Mm ...
Student 1: minus the the component which is the resistance.

This is correct, but does not relate to the Third Law as it is not relating paired forces. Another student displayed the correct conception:

Student 2: The force of the earth, uh, box falling towards the ground, yeah, the earth on the box, is equal to the force of the box on the earth, but because the box's mass is so much less that the earth, it is ... the box moves towards the earth.

Without the idea of the pairing of forces – box on earth, earth on box – the other student remains in difficulty:

Student 1: This case isn't applicable because the body is moving ...
Interviewer: Mm ...
Student 1: and and it's not constant velocity.
Interviewer: So does Newton's Law sometimes work and sometimes not?
Student 1: Well the [laugh] you can't sort of say the law only exists for certain bodies. Um – it's gee [laugh] – yeah. I think I'll just, I can just say that, that the case which is just stated on this side is – um. See it's not the central situation so that specialised case doesn't hold on that side [the block in mid-air].
Interviewer: Mm.
Student 1: It doesn't mean that the forces don't exist.
Interviewer: Mm.
Student 1: It's just that, it's just that the the cancellation doesn't occur.

Interviewer: And because the cancellation doesn't occur?
Student 1: The, the, yeah, the effect that the cancellation has is to have made the body at rest or moving at a constant speed or velocity and because that hasn't occurred …
Interviewer: Mm …
Student 1: then the, then the body is not in that situation.
Interviewer: And so?
Student 1: Yeah um well that, that case doesn't apply there.

Without knowing any physics, it should be possible to see that this student is having difficulty in reconciling different bits of knowledge about physics. He cannot reconcile his definition of the law and its application to an instance, with his belief that the law should hold for all situations. Many of these interview sessions end up being a powerful learning experience for students. It should also be clear that analysis is not straightforward. Defining this student's conception of the law would be very complex, especially as it appears here as rather fluid in form, certainly not stable and bounded. We can discern within this dialogue, however, an aspect of thinking about the problem that is present in other interviews as well, namely the idea of the forces cancelling out:

> [The forces in the second case] are just the weight of the box acting down and there's air resistance acting up … The force acting down is bigger so that's why it falls down to the ground.
> It [the law] applies to when they're in equilibrium and at rest, but when the actual system is trying to reach equilibrium it doesn't apply.

The idea of forces cancelling out to give equilibrium is not relevant to the application of Newton's Third Law. The phraseology of its most common form, 'equal and opposite', sounds like the balancing of forces on a body, and leads inevitably to confusion with the concept of equilibrium. Furthermore, this common version neglects the additional idea contained in Newton's original formulation, that the forces are acting on different bodies ('the mutual action of two bodies upon each other'). The law expresses the idea that a force cannot exist that does not have its counterpart. It is not expressed as a property of the world: it is a definition of what he means by force. (It is rather like defining a mirror through the definition 'for every mirror image there is an equal and opposite mirror image', in the sense that it is a tautology if you already know what a mirror image is.) To apply the law to the falling box, you have to recognise that gravity, the force of the earth on the box, presupposes the equal and opposite force of the box on the earth. What makes the box accelerate is its tiny mass in comparison with that of the earth. The earth is also accelerating towards the box, but its great mass makes its acceleration tiny in comparison with that of the box. The two forces are the same, however, and that is where 'equal and opposite' is applicable, not in any notion of balance which implies lack of motion.

The specialist language creates an inevitable confusion between 'equilibrium' on the one hand and 'equal and opposite' on the other.

This aspect of the students' conception has no logical relation to an expert conception. Once described we can begin to see its etiology in our everyday experience of force, in the language used to describe the law, in the problems set to students. Few textbooks quote Newton's Law in its original form, and in simplifying it, they often omit the very phrases that would help students see that their everyday conception of force is quite different. This is a 'pedagogenic error', comparable to 'iatrogenic disease', and is avoidable if we become sufficiently aware of the contaminating effects of everyday language.

In addition to the conceptualisation of the law in terms of equilibrium, we also found some students who conceptualised it correctly, using all three key components of the law – that all forces are paired, that paired forces are equal and opposite, and that they act on different objects – and they were able to apply this successfully to both situations. We also found a third form of conception, which did not insist on finding a pair of forces, and therefore explained the box in midair in terms of an unbalanced force of gravity:

> The second one, no table, so, ah, the box has still got gravity acting on it … and seeing there's no [other] forces coming from anywhere … it's just going to fall towards the ground with constant acceleration.

When challenged about how the equal and opposite forces mentioned in the Third Law could be applied to this situation, the student had little option but to reject Newton altogether:

> I'd only apply Newton's Third Law where there's no resulting acceleration for a thing, whereas this box is accelerating as it comes downwards, so I don't know, I wouldn't use Newton at all here.

This was not an isolated example, and given the lack of any sense of intellectual struggle in these cases, the interviewers were constrained to prolong these interviews a lot further to avoid this unfortunate conclusion becoming the learning outcome of the session.

This completes the analysis for this question, and leaves us with an outcome space of just three main conceptions of the law, as (1) including all three components, (2) neglecting the condition that the forces described act on different objects, (3) also neglecting the requirement that forces are paired. This is a complete description of the outcome space for this population of students: it can account for all the explanations and reasoning processes students used. The three conceptions cannot be related to each other except via the definition of the law. They are well-ordered, in the sense that being wrong in type (2) is better than being wrong in type (3), but it is not obviously a developmental progression, where every individual has to go through each one. The outcome space is empirically defined, therefore, and leaves open the possibility that a different population of students could add to the number of ways of thinking about Newton's Third Law.

This basic method can be applied to any subject area to clarify the alternative possible ways of thinking about complex concepts. Particular examples are: the law of diminishing returns in economics (Dahlgren and Marton, 1978), the mole in chemistry (Lybeck *et al.*, 1988), clinical diagnosis in medicine (Whelan, 1988), essay writing (Hounsell, 1984), recursion in programming, (Marton and Booth, 1997), and a much deeper analysis of the concept of subtraction (Neuman, 1987). Richardson questions the value of a research methodology that focuses only on the 'product of learning rather than the process of learning' (Richardson, 2000: 36), but it should be clear from these studies and from the detailed analysis of one example above, that it is important for teachers to know how their students think. Without this, we build on sand. Teachers must address and challenge those fundamental misconceptions, but first they need to know what they are. The methodology of phenomenography will tell us, but it is a labour-intensive task to undertake that kind of research for one topic at a time. In Chapter 4 we will look at the extent to which studies of this type can be generalised.

LONGITUDINAL STUDIES OF DEVELOPMENTAL CHANGE IN UNIVERSITY STUDENTS

One further aspect of what the student brings to learning derives from long-term studies of students that reveal the changes they go through over time. William Perry's study of Harvard undergraduates, carried out from the vantage point of the academic counsellor, documents the long intellectual journey from a basic dualism – regarding knowledge as facts that are right or wrong, dispensed by authority – to a generalised relativism and a commitment to personal values (Perry, 1970). This study was carried out via long open-ended interviews with students, one for each year of their degree. The analysis looked at variations between students to generate the existence of different categories of epistemological belief, and at the changes within students to describe a developmental pattern. He documented both epistemological and ethical aspects, the former developing from the dualistic position of knowledge as right or wrong, to a multiplicity of possible correct explanations, to the relativism of contextualised knowledge. There is a parallel in the ethical development, from seeing authority as responsible for what is known, to the solipsism of everyone being equally correct, to a personal responsibility for one's own set of values.

Epistemological and ethical development do not necessarily stay in step with each other. When I analysed interviews with British Open University students studying the first year social science course, it appeared that some, being mature students, had already established a sense of a personal system of values and did not want to be told what to think. At the same time, their epistemology told them that there were independent facts:

Student 1: They're going to tell me eventually what capitalism is ... Well, I suppose they can leave it open for me to disagree, but I'd always assumed we were in a mixed economy, and the fact they're now saying are we a mixed economy surprised me, because I thought it was a mixed economy within capitalism ... but it seems to me they're telling us all the time.

Interviewer: So what do you want?

Student 1: I don't know. I suppose what I want is to be presented with the facts and make up my own mind, which ultimately is what will happen, isn't it?

For this student the course focuses her attention on relativistic conceptions – whether our society is mixed economy or capitalist, not which it is – and yet she has no basis for deciding the issue herself. She is beginning to recognise the idea of relativism, but is unsure how to handle it:

Student 1: It seems to me that – I always get confused over things like this – that they're producing a model to explain how something works, but that the definition depends on how they explain it. So how am I to know? I suppose I look at capitalism and see if that explains the definition. Is that how it's supposed to work?

Students have to work out where they can locate themselves in this academic debate: on what basis can they decide their view? They do not want to be told, but they want to know how they can know. It is a problem common to students in every subject area.

This kind of description of the student population differs from 'personality' studies in that all students are expected to go through all stages at some point in their academic career, though the pace of change may differ. Longitudinal studies differ from 'conceptions' studies because at any one time a particular student is likely to be at a particular stage. In this sense, Perry's scheme identifies individual differences between students. What should the teacher do with developmental stages? Do we have to wait until a student is ready for a new way of thinking, or can we push them forward? Perry is clear:

> We cannot push anyone to develop, or 'get them to see' or 'impact' them. The causal metaphors hidden in English verbs give us a distracting vocabulary for pedagogy. The tone is Lockean and provocative of resistance. We can provide, we can design opportunities. We can create settings in which students who are ready will be more likely to make new kinds of sense.
>
> (Perry, 1988: 159-160)

Sometimes we can see this happening. Another student, who began the social science course with the belief that individuals are responsible for their actions, was naturally resistant to structural explanations of vandalism:

> I thought some of the explanations that were given, I didn't accept them all – because they were bored – I don't necessarily think that's a good reason to kick in a telephone box. I think it's possibly just because they're vandals.
>
> (Laurillard, 1982: 16)

The course tried to steer students away from individualistic explanations of this sort, but at the same time recognised them and addressed them. In discussing the question raised by the course 'is social science really just providing excuses' the same student was able to acknowledge his earlier view and also recognise how it had changed:

> Yes this is an attitude that certainly I had very strongly when I just started up on the course. I think I'm gradually beginning to see the social sciences' point of view.
>
> (Ibid. 16)

In this case the course created the setting in which this student was able to start making a new kind of sense. For mature students like him there can be an uncomfortable dislocation between the epistemological and ethical stages of Perry's developmental scheme. They have already acquired a personal ethical stance with respect to, say, vandals, but have not yet acquired a relativistic point of view. The course has to help them realign the two – to engage with alternative ways of seeing the world, but to use that analysis to decide on their personal knowledge, and the evidence that supports it. As university teaching incorporates more project work, collaborative working, and independent research by students, it is possible that we will see students enabled to progress more easily to the higher levels identified by Perry than they could within the transmission model of teaching.

Finally, we should consider Säljö's identification of five conceptions of learning, which also mark a developmental progression. His study was again interview-based but, instead of being longitudinal, took interviewees from all age groups and stages of learning. Säljö's five stages (Säljö, 1979) are compatible with Perry's nine, but they bring out what is implicit in Perry's analysis with respect to how students conceptualise the process of learning itself, as:

- the increase of knowledge;
- memorising;
- the acquisition of facts, procedures for use in practice;
- the abstraction of meaning;
- an interpretive process for understanding reality.

Their conception of learning is an important manifestation of a student's epistemology, being, quite literally, the way they believe they can come to know. There is a world of difference between students who take learning to be 'an increase in knowledge' or 'memorising':

> Accumulation of knowledge.
> Filling my head with facts.
> Drumming it into the brain and reeling it off.
> Learning it up for exams and reproducing it.
>
> (Marton and Booth, 1997:36)

and those for whom it is an interpretive process:

> All the time it keeps cropping up, you might have seen it in one way before, you sort of see it in different ways.
> Opening your mind a little bit more so you see things in different ways.
> Being able to look at things, from all sides, and see that what is right for one person is not right for another person.
>
> (Ibid. 37)

It is important for teachers to know which conception of learning their students incline to, and to take responsibility for helping them develop this most fundamental aspect of what they bring to their learning.

A further longitudinal study of OU students by Marton *et al.* (1993), building on Säljö's study in 1979, generated a sixth conception of learning in addition to replicating the original five. Perhaps because OU students experience a struggle to align the ethical with the epistemological, found in the earlier study, the sixth conception concerns the changes they experience in themselves:

Student 1: I suppose it's what lights you ... It's something personal and it's something continuous ... You should be doing it not for the exam but for the person before and the person afterwards.

Student 2: Expanding yourself ... you tend to think that life just took hold of you and did what it wanted with you ... You should take hold of life and make it go your way.

> This is *learning as change as a person*, the most extensive way of understanding learning in that it embraces the learner, not only as the agent of knowledge acquisition, retention and application, and not merely as the beneficiary of learning, but also as the ultimate recipient of the effects of learning.
>
> (Marton and Booth, 1997: 38)

It is appropriate that a population of mature students should have expanded the outcome space for conceptions of learning with a more extensive conception, but its form is surprising. The fact that these learners are capable of changing as a person through academic study makes the ideal of 'lifelong learning' seem at least plausible.

We shall return to how students' epistemological beliefs might affect teaching strategy in Chapter 4, and Chapter 3 shows how students at different developmental stages deal with intellectually challenging subject matter.

SUMMARY

In this chapter, I have made a division between studies of individuals' characteristics, characteristics of a student population, and longitudinal studies. The methodologies produced different kinds of data that will operate at different levels of description of the teaching process. Some can inform the curriculum planning level (e.g. how to address students' epistemologies) whereas others can suggest the language to be used in teaching (e.g. how to talk about 'force'). All the studies describe aspects of what students bring with them to learning a new topic. In summary these are:

- conceptions of the topic – teachers need descriptions of the ways students conceptualise a topic to be able to challenge their fundamental misconceptions;
- representational skills – students need explicit practice in the representation of knowledge of their subject, in language, symbols, graphs, diagrams, and in the manipulation and interpretation of those representations;
- an epistemology – teachers must enable students to develop their epistemological and ethical beliefs, and in particular, their conceptions of learning.

These issues at least must be addressed later in devising teaching strategies, and will contribute to the generation of a teaching strategy in Chapter 4. Meanwhile, having considered what students bring to their study, Chapter 3 goes on to consider what they do when they study.

Chapter 3

The complexity of coming to know

INTRODUCTION

This is the point of the book at which we come as close as possible to what goes on while a student is learning. It is not easy to penetrate the private world of someone coming to an understanding of an idea, and much of this chapter will discuss the ways this can be done, as well as what is found out. I once caught myself wishing I could attach electrodes to students' heads to see what goes on when they learn. Never mind humanitarian principles of research investigation, or anti-reductionist beliefs about the nature of learning; it would be so wonderful to be able to see how their sense-making cognitive apparatus arrives at some of those weird outcomes. Retrospective interviews are a very unsatisfactory substitute. The fantasy deserves to be nothing more than that, but it does convey that sense of wanting to see the learning process from the students' perspective, in all its complexity, and in such a way that we can make sense of it.

An insight into the student's view of the learning process would give us some basis for deciding on a teaching strategy. Chapter 2 elaborated the relevant features a student might bring to a learning session. We now consider what goes on within it. The teacher has to encourage 'mathemagenic' activities in the students. This is a term originally coined by Rothkopf to refer to those activities that 'give birth to learning', such as 'systematic eye fixations' while reading. The term defines 'truly, a student-centred approach' to instruction (Rothkopf, 1970: 334), but it is a shame to confine it to the realm of such minute behaviours as eye fixations. The context of predominantly behavioural psychology within which Rothkopf was working constrained the application of his idea. He acknowledged the importance of cognitive processing but did not then have the means to take it further than one brief paragraph.

In the last twenty years, psychologists have done a great deal of research on processing and, with a different epistemological orientation, we can extend Rothkopf's idea to its proper domain. The concept of mathemagenic activities expresses exactly the idea that there are activities the learner can carry out that will result in their learning. Encouraging these activities is the proper focus of a teaching strategy. So our task in this chapter is to consider what kinds of activities could be mathemagenic.

FINDING OUT WHAT HAPPENS IN LEARNING

The approach must be to look at what happens during the learning process and relate this to learning outcome. We need a methodology that provides a deep level of description of what is happening for the student when they learn, linking the way they think about the content to what they achieve as an outcome. Because of the focus on content, these studies have to investigate students working on particular learning tasks, and because of the requirement to illuminate their perspective on the topic, the methods have to include observation, interview and a trace of students' performance (written protocols, input to a program, dialogue, etc). The interviews are not 'introspective', and the protocols are not 'think-aloud'. Both techniques derive from a different epistemological tradition in psychology: they require meta-level monitoring by the subject, which presupposes that this gives them access to an accurate account of the object-level activities involved in the task, and that meta-level monitoring is not itself used in the task. I had a graphic demonstration of the fallacy of the latter assumption when I first tried using think-aloud protocols for problem-solving tasks. The typical pattern produced plenty of talk while the subject was figuring out how to go about the task, but the point at which they said something like 'Aha …', was followed by total silence until either they completed their plan of action or they got stuck again. It was as if the point at which the really productive thinking was happening did not allow them spare capacity for a meta-level account.

A better method is to allow the student to complete the task undisturbed, and to give a retrospective account of how they experienced it, much as one might describe an event witnessed. The student's account is not taken as an objective description of a psychological process, but as being itself a phenomenon which is to be analysed. The student performance protocols (e.g. worked problems or written explanations of a concept) are used by the interviewer to focus students' explanations of why they did what they did, and to provide a stimulus to recall their activity. The combination of protocol and retrospective interview is then analysed by the researcher in relation to other students' data and to the content of the topic discussed to produce an account of what they learned and how. This procedure provides the kind of detailed insight we need into what constitutes the learning process.

There are several aspects of learning that have been investigated in enough detail to admit a general account that can inform teaching. Given everything I have said so far about the integrative nature of the learning process, the inseparability of knowledge and action, and of process and outcome, there is no logical ordering of parts of the process, as each part is constituted in its relation to the other parts. The particular aspects I want to focus on are organised according to some of the most important findings in the literature: apprehending structure, integrating parts, acting on the world, using feedback, reflecting on goals. The division is simply a convenience to make discussion more manageable. An integrative whole can be divided up in many ways, none of them more correct than the other; the difference is only in their utility.

APPREHENDING STRUCTURE

The most common method of learning in higher education, is via acquisition, especially through lectures and reading. In Chapter 1 I argued that a peculiarity of academic learning is to focus, not on the world itself, but on others' views of that world. The idea that people can learn through listening to lectures most clearly expresses the fact that teaching is a rhetorical activity, seeking to persuade students of an alternative way of looking at the world they already know through experience. This way of learning presupposes that students must be able to interpret correctly a complex discourse of words, symbols, and diagrams, each bearing a specific meaning that must be interpreted correctly if the student is to learn what is intended. How do students deal with this?

Meaning is given through structure. The Gestalt psychologists gave a clear demonstration of this using the famous picture that organised one way meant young girl, and organised a different way meant old woman. The same information structured differently has a different meaning. This is why I have begun by focusing on apprehending structure. In Chapter 1 academic teaching was linked to a didactic process that has as its goal a consensual viewpoint on the world, a particular meaning. For students to interpret a complex academic discourse as having a specific intended meaning, they must be able to apprehend the implicit structure of that discourse. A number of studies show that they fail to do this. Since deciding on the structure and how it is to be displayed is part of the teacher's instructional strategy, this needs elaborating.

Phenomenography is particularly successful at illuminating how students deal with structure and meaning because these studies focus on content. They have led to the identification of two contrasting approaches to studying a text: one known as the 'deep approach', where the student looks for meaning, and processes the text in a 'holistic' way, preserving the original structure of the discourse and therefore preserving its intended meaning; the other known as the 'surface approach', where the student focuses on key words or phrases and processes the text in an 'atomistic' way, distorting the original structure and therefore changing its meaning. There have been many studies demonstrating this contrast. Marton and Säljö (1976a and b) and Svensson (1997) documented the earliest studies. More recent books by Entwistle (1981), Marton *et al.* (1997), Ramsden (1988 and 1998), and Marton and Booth (1997) document the many later ones, spanning a range of educational contexts and topic areas.

Extracts from a study I carried out on students reading a social science text illustrate their different approaches to structure and meaning. Students were given an article by Bertrand Russell ('Can a scientific society be stable?'), in which he argues that a scientific society is not stable, and adduces a number of reasons for this. The connections made between science, population increase and instability are moderately complex, and Russell discusses several reasons for instability, so that there is a degree of information overload. Since the argument is complex enough to extend over the whole article, a deep approach is necessary to apprehend its structure and combine its parts to give the intended meaning. A surface approach would result in, at most, a list of unconnected points in the argument. Some students took a 'deep' approach to discerning the intended meaning:

Student 1: I was trying to remember the main points he was arguing. I tried to find out first what it's about from the introduction, and then went on to his reasons, which was what I was looking for, relating it to his title.

Student 2: I tried to understand his argument, see where it's leading, see if it makes sense.

and in their summaries of the text were able to preserve the original meaning by linking scientific progress to social factors:

Student 1: Because science is progressing and we can support large populations, the population growth will overtake scientific growth.

Student 2: He's basically advocating that in its present form the scientific society is unstable unless there is drastic population control and control of resources.

Other students described a 'surface' approach:

Student 3: I didn't read it deeply ... I tend when I'm reading to forget what went before. I take it in at the time, but if nothing really strikes me I forget it.

Student 4: I just read straight through … I found I would think about it and carry on reading and find I'd have read the last few sentences again because I hadn't been concentrating on it ... some bits go in easily, others don't.

and their summaries reflected this, being more disjointed and failing to preserve the original links between science and social factors:

Student 3: It was about whether a scientific society could be stable.

Student 4: It's basically about the ethics of science and how he doesn't reckon we will survive much longer unless man's wisdom increases.

Without attempting to discern the structure inherent in the text, these students were unable to unravel its complex argument, and were left with isolated statements lacking any clear relationship to each other. As we shall see repeatedly in these examples, the internal structure of academic ideas, arguments and conceptions tends to be complex, and usually more complex than the everyday conception of the same phenomenon. With an example from economics,

Dahlgren and Marton demonstrate that the main difficulty with understanding the law of diminishing returns lies in its second-order character, that it involves 'decrement of increment, i.e. change of change' (Dahlgren and Marton, 1978: 28). Brumby shows that students of biology assign adaptive characteristics to changes in the individual rather than to the more complex mechanism of changes to the species via natural selection among individuals (Brumby, 1984). Booth shows that computer science students are likely to view the concept of recursion as being about repetition, rather than the more complex idea of self-referential repetition (Booth, 1992). The typically greater complexity of the academic conception makes it extremely important that students attend to the full scope of the discourse structure.

The same problem of appropriately apprehending structure will occur in the interpretation of discourse in any medium. To clarify some aspects of the argument an author will often appeal to experience and use a specific example to illustrate an idea, but the description of that example will have its own complex internal structure embedded within the structure of the text as a whole. Discerning the structure is difficult not just because of its complexity but also because it is rarely explicit. It is conveyed via syntax, conjunctions, and expressions such as 'in order to', 'not only ... but also ...', 'instead', etc. For many of the ideas students have to grapple with, their only access to them is via text. Academic knowledge does not present itself through experience with the world. The link is more tenuous than that. Academic knowledge relates to the experience of the world it describes, but it requires also a great deal of contemplative reflection on that experience. Furthermore, the perspective described in an academic text is not a clear glass window onto the world. A closer analogy would be looking through the wrong end of some grubby binoculars adjusted to someone else's eyes. The student has to do a lot of work to discern the point being made. The principle–example structure, which is a common feature of teaching texts, is missed by many students whose attention is captured by the intriguing example (Marton and Wenestam, 1979). The relational argument structure, also important for expressing a complex idea, will be unpacked to its constituent components but may never be reassembled.

Marton and Booth (1997), in their analysis of Roger Säljö's study of students reading a text (1979), argue that only when learners are aware of the different structural levels within a text can they read it as it is meant to be read, and discern its intended message. In Säljö's study, the text was meant to be read as being about a certain topic, which embraced two different ways of looking at learning, classical conditioning and instrumental conditioning, and gave examples of each. Some students understood the text as a hierarchical structure and interpreted its message as being about forms of learning. Others took the linear format of the medium as the form of the content, and 'horizontalised' the structure. Not having discerned the structure, they therefore distorted the meaning, and read the text as being about Pavlov's dogs and Skinner's rats:

> If the former understanding better reflects how the text *should* be read – as we argue it does – what it takes to understand the text is a discernment and simultaneous awareness of the different levels.
>
> (Marton and Booth, 1997: 102, original italics)

Figure 3.1 shows how they represent the internal structure of the text, and the corresponding levels of awareness the reader must maintain.

The results of their study, differentiating students' success at maintaining this awareness, suggest that the features of the text which were designed to afford discernment of its structure were too subtle for some of the students. Such features might be the title, sub-headings, connecting phrases such as 'this example illustrates our main point that …', and other helpful markers of structure. If they are absent or too subtle, then students are less likely to discern the internal structure, and hence the intended meaning.

I have focused so far on teaching techniques that assume learning through acquisition, but the issue is just as important for other kinds of learning method, such as problem-solving. The point of problem-solving as a method is to enable the student to manipulate the internal relations within their conceptual knowledge, such as definitional relations, causal relations, forms of representation, mathematical relations, sign-signifier relations, etc., much as they would manipulate the world

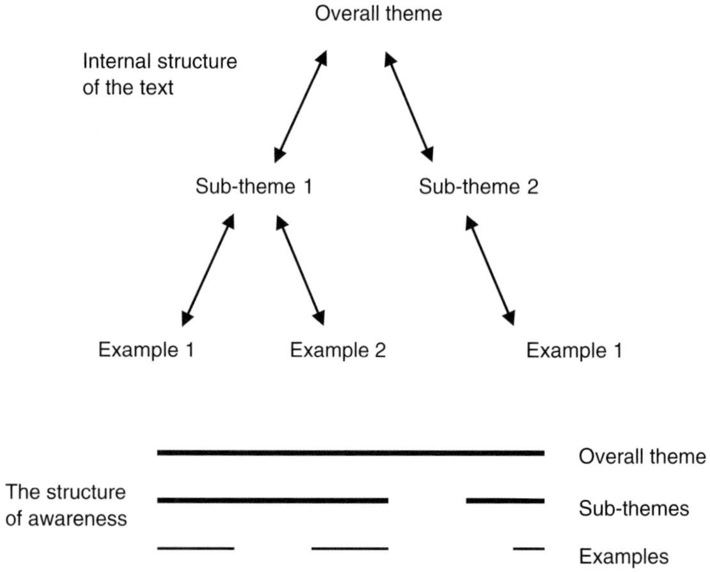

Figure 3.1 Levels of awareness in reading a text: the significance of the imposed structure for awareness of the point of the text as overarching theme and examples. *Source*: based on Marton and Booth, 1997, Fig 5.6

in order to learn about it. As a method of learning, the focus is not the solution, but the relations between the problem statement, the solution, and all the intervening steps. The problem-solving exercise has a structure and embodies a meaning, a description of the world. In this sense, the apprehension of structure is just as important in problem-solving as in interpreting discourse.

In a study of how students carry out problem-solving (Laurillard, 1979), different approaches again distinguished those who addressed the structure as a whole:

> First I had to decide on the criteria of how to approach it, then drew a flow diagram, and checked through each stage. You have to think about it and understand it first.
> You have to make a basic assumption to work through, then you work backwards to check your input, then forwards again.
>
> (Laurillard, 1979: 399–400)

and those who made no attempt to deal with the overall structure:

> You don't need to look at the system, you don't have to interpret it.
> I looked up the formulae and made the calculations from those.
>
> (Ibid. 399)

For some students, the focus of a problem-solving exercise is getting the answer out. That approach will help them develop a facility for mathematical manipulation, but will not do very much to enrich their understanding. The process of selecting the equation that fits the variables given in the problem does not involve the student in thinking about the meaning of the equation, nor about the relation it expresses. In more general terms, without looking at the structure as a whole in relation to their task, the student will be unable to appreciate the meaning of the answer they have produced, whether they are reading philosophy, watching a social science programme, or doing a maths problem.

The mathemagenic activities relevant to apprehending structure are already well defined in the literature in terms of the 'deep' or 'holistic' approach, which characterises them as follows:

> Focus on 'what is signified' (e.g. author's argument, or the concepts applicable to solving the problem).
> Relate and distinguish evidence and argument.
> Organise and structure content into a coherent whole.
>
> (Ramsden, 1992: 42)

Students may understand what it means to take a deep approach, and still find it difficult to organise the content into a coherent whole. Undoubtedly some examples of academic discourse, whether lecture, book or television programme, seriously obfuscate their meaning by pursuing a muddled or over-complex structure. Through understanding students' different approaches to structure and meaning we can design teaching to encourage the mathemagenic activities they need, summarised by the characteristics of a deep approach to apprehending structure.

INTERPRETING FORMS OF REPRESENTATION

The importance of the integrative aspect of learning is already clear from the discussion of structure. The view of academic knowledge presented so far in this book constantly stresses its relational nature. Learning academic knowledge requires activities that address and deal with relations. One of the most important is the sign-signified relation, which concerns the interpretation of symbol systems, whether linguistic, symbolic or pictorial. It plays an essential role in the study of any academic subject because it requires students to make sense of the theoretical in terms of the practical, and vice versa. The forms of representation adopted by a discipline embody both a way of looking at the world and a description of it from that perspective. Students have to perform an interpretive and integrative process if they are to master the ideas. This is the focus of the following section.

Academic study cannot do without special forms of representation – language, mathematics, diagrams, symbols – but how do students make sense of them? No subject area escapes the problem, because they all use at least language to represent ideas. However, there are few studies of the problem at university level. Two examples from the humanities and science may serve to illustrate how it occurs.

The first example describes a study of formal representation in economics. By investigating students' responses to different ways of displaying the same data, Tabachneck-Schijf and Simon show that:

> Visual saliency of particular features of a display may direct attention to these features independently of their relevance for the problem at hand.
>
> (Tabachneck-Schijf and Simon, 1996: 45)

The problem is to interpret a supply-demand diagram. The increase of $1 changes the supply curve from S to S' in Figure 3.2. Students are asked to derive the change in equilibrium price resulting from the change in tax.

Most students were able to describe the change by comparing the two intersection points on the diagram, concluding that the initial equilibrium price and quantity will be $7.50 for 55,000 knives, changing to $8.00 for 50,000. This form of representation enabled students to think about the problem appropriately. Two further forms did not. The same data represented as equations, in Figure 3.3, created confusion because students could not see which equations should be paired.

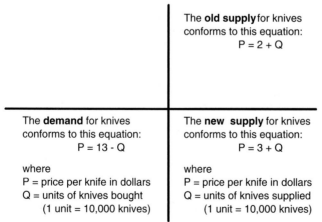

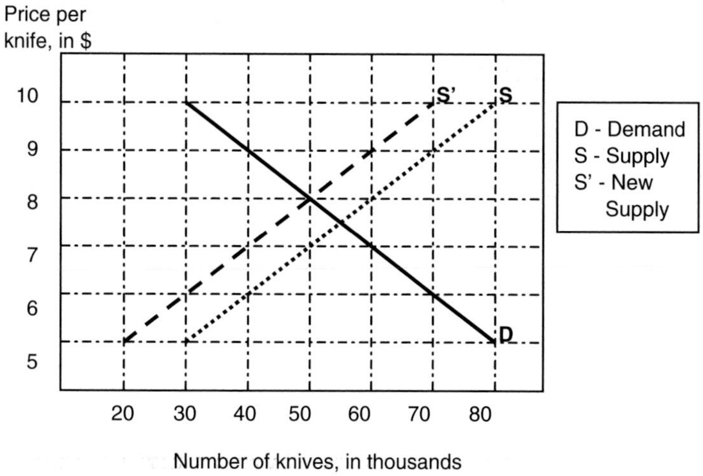

Figure 3.2 Graphical representation of the supply–demand problem: students are asked to derive the change in equilibrium price resulting from the increase in tax of $1, which changes the supply curve from S to S'. *Source*: based on Tabachneck-Schijf and Simon, 1994, Figure 9B

These equations are valid for 4 < P < 11

The **old supply** for knives conforms to this equation:

P = 2 + Q

The **demand** for knives conforms to this equation:

P = 13 - Q

where
P = price per knife in dollars
Q = units of knives bought
(1 unit = 10,000 knives)

The **new supply** for knives conforms to this equation:

P = 3 + Q

where
P = price per knife in dollars
Q = units of knives supplied
(1 unit = 10,000 knives)

Figure 3.3 Algebraic representation of the supply–demand problem: students are asked to derive the change in equilibrium price resulting from the increase in tax of $1. *Source:* based on Tabachneck-Schijf and Simon, 1994, Figure 9A

Comparing the two supply equations on the right-hand side, for example, they concluded that the change in equilibrium price was $1. The authors conclude that 'there was nothing in the visual display that signalled the appropriate pairing of equations to them' (Ibid. 44). By contrast, numerical tables, as in Figure 3.4, focused attention on inappropriate elements.

Comparing the first line of the table for S with the second line of the table for S', students noted a $1 difference in price for the same quantity, and concluded that the equilibrium price would increase by $1. They ignored the relationship between supply and demand that affects the equilibrium quantity as well.

It is clear from the latter two examples that the students were not taking a deep approach to their interpretation of the information. Those using the diagram carried out an interpretation that was compatible with a deep approach, apprehending the full complexity of the structure, and interpreting and integrating the data correctly. We would probably judge a student as having a full understanding only if they could correctly handle all three forms of representation. So we cannot conclude, as Tabachneck-Schijf and Simon do, that:

> To avoid misleading the users of a display, it must be designed so that irrelevant cues will not be salient and that salient cues to the relevant features are provided.
>
> (Ibid. 44)

		This is the **old supply** schedule:	
		at this price:	this many would be supplied:
		$5.-	30,000
		$6.-	40,000
		$7.-	50,000
		$8.-	60,000
		$9.-	70,000
		$10.-	80,000
This is the **demand** schedule:		This is the **new supply** schedule:	
at this price:	this many would be bought:	at this price:	this many would be supplied:
$5.-	80,000	$5.-	20,000
$6.-	70,000	$6.-	30,000
$7.-	60,000	$7.-	40,000
$8.-	50,000	$8.-	50,000
$9.-	40,000	$9.-	60,000
$10.-	30,000	$10.-	70,000

Figure 3.4 Tabular representation of the supply–demand problem: students are asked to derive the change in equilibrium price resulting from the increase in tax of $1.
Source: based on Tabachneck-Schijf and Simon, 1994, Figure 9C

They have shown that students can interpret a well-designed diagram, but cannot interpret information that challenges their understanding of the topic. It is important for the teacher to be aware of these differences, and to design forms of representation accordingly, but if students are to secure a deep understanding of the idea, they should be able to handle distracting features of the data presented. A more appropriate conclusion would be to use this study to generate exercises for students in exploring the relationship between the three forms of representation, progressing to generating their own versions to describe a similar situation. This would give them further practice in mapping between the situation described and its representation.

The second example describes a study I carried out with students learning about crystallographic projection. Three-dimensional mathematical diagrams represent the different shapes of crystalline matter, and their complexity gave plenty of opportunity to see how students cope with a new formalism. One strategy was to treat it strictly as a procedure that does not need to be interpreted:

Student 1: It's about, um, representation of a unit cell of a close-packed hexagonal structure. It's a way of representing, well I'm not actually sure, but all I know is, it's a way of representing the atoms in the unit cell by means of sixty-degree graph paper. And they just refer to the positions of the atoms. That's all I know, really, but that's all you need to know to do it. You don't need to know anything else.

Other students were working hard at the subject, often devoting hours of study to trying to figure out how to interpret and draw these diagrams. It is a little like trying to understand a foreign language when you have only a cursory knowledge; once you miss the odd phrase, maintaining the sense of what the discourse is about becomes ever harder. What to the lecturer seems a logical progression through successively abstract diagrams that allow ever more complex crystals to be represented, seems to the student like utter confusion:

Student 2: There are so many ways of describing one crystal, it seems illogical. We draw it naturally, the way you see it, then we're told to draw it in three-dimensional projection to see it that way. Now we're told to draw it in a circle. Totally illogical. Then we have to see not only how the crystal fits in the circle – and that looks nothing like a crystal to me – we have to see how it works in that diagram by drawing another diagram and another circle. It's very confusing, all the different terminologies for one crystal. It would be nice if we had one thing now that brought all these planes, this stereographic projection and this [diagram] and tried to relate them all and show exactly how they fitted, in a sort of sequence of events, whereas we've been given them totally separately.

This student demonstrates an awareness of what a deep approach would be, but he does not have the intellectual means to carry it out. He wants to have a sense of a coherent whole. He is trying to make sense of the diagrams in terms of the crystal it is supposed to represent, but the sign-signified relation is unintelligible so far. The student seems to have no idea that the point of the whole process is precisely to build a formalism that makes it easy to represent complex crystals and their orientations in the world. The lecturer, by the way, was unusually committed and the students reckoned him an excellent teacher. This is a non-trivial and persistent problem throughout higher education: students need help in practising the mapping between world and formalism, the ways of representing academic ideas and their interrelations.

Another student hints at what they need in order to grasp the meaning of these complex representations. He also felt it was not available to him in the worked examples offered in the handouts:

Student 3: This doesn't fall into my particular method of learning, this handout business, because I don't learn things photographically. I can't look at something and remember it. I can't read something and remember it. The only way I can learn something is to do it. It's all done for us here and it's not fully educational for me. Instead of all these being drawn in, if these were exercises and we had to do work, then I'd get it straight away.

He recognises the need to practise the mapping process between the formalism and the reality it represents. It is not sufficient to follow someone else's practice. For the representation to be intelligible, they need to practise the translation in both directions. In doing so, they will begin to see how the abstraction works, which aspects of the reality this perspective attends to, and how it can be generalised beyond specific instances. It is the process that takes situated learning beyond the situation, and is therefore critical to an academic understanding.

ACTING ON THE WORLD (OF DESCRIPTIONS)

One of the most often quoted maxims about learning is the one which concludes 'I do and I understand'. All teachers recognise the importance of learning as an activity done by the learners. Teaching methods in use in universities therefore include many examples of learning through practice or imitation of practice (laboratory practicals, demonstrations, fieldwork, seminars, essays, problems, exercises, etc). Action as an aspect of learning is not in dispute. But what are learners acting on when they are learning academic knowledge? Is it the same world they are acting on when they are learning experiential knowledge? I have already begun the argument in Chapter 1 that academic knowledge is importantly different from experiential knowledge, and this distinction becomes unavoidable when we consider this aspect of the process of learning. I contrasted second-order academic knowledge with first-order experiential knowledge as being knowledge of descriptions of the world rather than knowledge of the world. The distinction is particularly important when we consider how learners are to access the knowledge. When the 'what' that is being learned is objects, behaviours, sensations, then experience serves as the access; when the 'what' is theories, descriptions, viewpoints, then the access can only be through some form of representation: language, symbols, diagrams, pictures. The actions learners must carry out can only be usage of language and symbols, therefore. Learning about dogs can be done through actions on the object: observation, touch, smell, interaction (offering a biscuit, throwing a ball), comparing these experiences with the same actions on other animals, all done without recourse to any form of representation or use of language. Learning about molecules cannot be done without recourse to representation of some kind.

Eysenck and Warren-Piper in their paper debating this issue suggest that students can imagine molecules as ping-pong balls, and in this sense, they are experienced (Eysenck and Warren-Piper, 1987). Yes, but not in the way ping-pong balls are experienced. The access to molecules is via an analogy, and this is a difficult trick, because setting up the correct analogy for a particular exploratory action on a molecule presupposes an understanding of molecules. Physics is notorious for its alluring concrete analogies that lead you falsely. Electrical current is a telling example. Most people feel they have a rough understanding of current flow, using water flow as an analogy, but try using it to predict the answer to the following question to an undergraduate physics class (McDermott, 1991): rank in order of brightness (1) a bulb in a simple circuit, (2) two bulbs in series, (3) two bulbs in parallel.

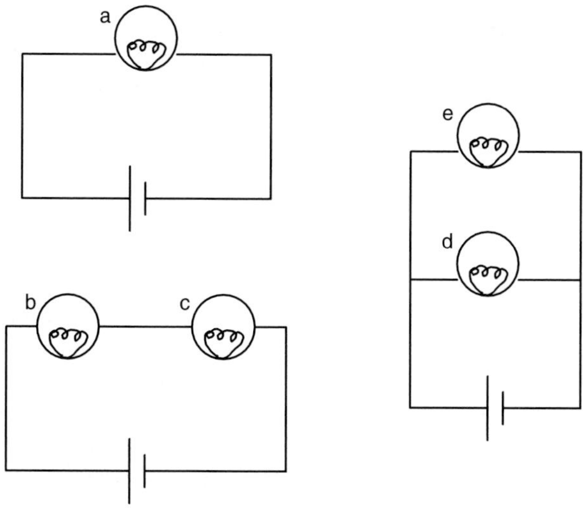

Figure 3.5 The current flow problem: the bulbs in the circuits must be ranked in order of brightness.

One form of the analogy assumes that the battery acts like a kind of waterfall, the greater the height the greater the voltage. The bulb acts as a filter slowing the flow, so the faster the water flows the brighter the bulb. This means that the water flows at the same rate through a and b, but by the time it gets to c it has already slowed down a bit, so $a = b > c$. In the third situation, the same amount of water is divided, so the rate of flow through d and e is half what it is for a, so $a > d = e$. That is wrong: the right answer is $d = e = a > b = c$. The difficulty occurs in setting up the analogy. The rate of flow depends upon the relation between the voltage and the resistance. Therefore, an ever-replenishing waterfall whose rate of flow is not governed just by what you put in its way, but by the quantity of water available, is an inappropriate analogy.

A more appropriate analogy would be to imagine the circuit as a route through a room in an art gallery with a new acquisition on display (this is the resistance, slowing the flow of people). To avoid a build-up of people, the guard is letting them in at a rate of 10 per minute (this is the current flow), even though there is a queue that is always about 50 people long (this is the voltage). If another new acquisition is put in the same room, he will have to slow the rate of people even further (because of the greater resistance), even though the queue is the same. However, if the second new acquisition is put in a parallel room, then he can keep 10 people per minute flowing through each room without a build-up in either room (this is the parallel circuit).

In case you find it confusing, only one-eighth of the physics undergraduates got the correct answer, and used a variety of unsuitable analogies (McDermott, 1991: 308). The study makes the point that imagining concrete analogies is not a reliable way of experiencing academic knowledge. Setting up the correct analogy is highly dependent on a good understanding of the concept being learned.

There is no equivalent of 'water play' for learning about electric current. An experience of water flowing faster as the bucket tips more steeply, of the conservation of amount, of watching the river flow, all enable us to develop an elaborated understanding of the concept of water flow, and to make reliable predictions about effects of actions on it. Our access to what electric current does when a circuit divides is available to us only via equations, diagrams, definitions, and language. Laboratory experiments are intended to provide 'current play' – and it is a fair bet that every person who reads this book will probably have carried out a school science experiment that put an ammeter in circuits wired up as in Figure 3.5 to demonstrate the effect above. But how did that trembling needle relate to our mental model of electric current? It is a very tenuous link, which hardly compares with the physical model of a water wheel spinning faster as you pour more water on it. The thought experiments we carry out as we think through the problem in Figure 3.5, as if we were acting on that world, are operating on the analogy we have built. Therefore, everything depends, for the success of this activity, on having an appropriate analogy.

This kind of analysis is reminiscent of discussions in the philosophy of science about how scientists decide between theories. It is generally accepted that a decision about which theory is to be preferred can only reasonably be made when there is some basis for agreement about observables. Schoolchildren and physics teachers looking at an ammeter in a laboratory are rather like the Pope and Galileo looking through a telescope at the moons of Jupiter. Is there agreement about what is observed? In what sense is this a shared experience of the behaviour of electricity? Even if pupils could play with different arrangements of circuits they would still need to acquire somehow the art gallery analogy (or similar) rather than the water-flow analogy to interpret their findings correctly.

Students carry out many learning activities during an academic course that appear to consist in learning directly about the world. Students of literature read novels and see plays; students of science look down microscopes to see substances in more detail or investigate the behaviour of electricity; students of management visit organisations. However, these apparently direct experiences of the world are mediated by the teacher, contextualised within the course, and encountered within the way the subject is being taught. When students engage with those worlds by interpreting a novel, or identifying a substance, or critiquing an organisation, they are generating further descriptions, or representations, which do not themselves engage directly with the world, only with the world of the teacher. That is why the proper subject of this section is 'acting on the world of descriptions'. The great majority of study time is not spent on activities in the field, but in working with analogies, historical accounts, critiques, statistics, case studies, diagrams, etc., which is straightforwardly 'acting on descriptions of the world', acting on a mediated world.

USING FEEDBACK

Action without feedback is completely unproductive for a learner. As we learn about the world through acting on it, there is continual feedback of some kind. If we can make the right connection between action and feedback, then we can adjust the action accordingly and this constitutes an aspect of learning. Receiving feedback is important. Being able to use it is also important. To the child who reads 'struct*ur*al' we do not say 'accentuate the antepenultimate'; we say 'you mean *struc*tural', because they can make an immediate connection between that feedback and their action. Feedback has to be meaningful to be useful.

There are two easily distinguished types of feedback, intrinsic and extrinsic, and both play an important role in learning. 'Intrinsic' feedback is that which is given as a natural consequence of the action; the feedback is intrinsic to the action. Clear examples of this are abundant in water play, as the physical world responds to the child's actions of filling, pouring, emptying, etc. We experience it every time we move a mouse, and adjust the movement to its manifestation on the screen. Correcting pronunciation is another example: although not a physical response to the action, it is a socially natural response. Pronunciation is a social norm and feedback of this type is natural and expected in a social situation.

These examples contrast with 'extrinsic' feedback, which does not occur within the situation but as an external comment on it: right or wrong, approval or disapproval. It is not a necessary consequence of the action, and therefore is not expressed in the world of the action itself. Extrinsic feedback is the feedback that operates at the level of descriptions of actions, and is therefore common in educational contexts. It may or may not be helpful or meaningful. A simple 'right' or 'wrong' gives the learner no information at all about how to correct their performance, only that correction is needed. It may also not be obvious which aspect of the performance is wrong: are they saying 'wrong' because I mis-read it or because I mis-pronounced it? A more helpful form of extrinsic feedback would give the learner information about how to adapt their performance. An elaborated comment like 'accentuate the antepenultimate' offers a generalised rule from which the action '*struc*tural' can be derived. It is a description of an action, unlike intrinsic feedback, which is a response from the system in which the student is acting, whether physical or social.

The key feature of extrinsic feedback is that it is external to the context of action. It is feedback that is not 'situated'. As an example, I shall draw on a study by Stevens, Collins and Goldin, who use a tutor–student dialogue to characterise tutor behaviour in response to students' 'bugs'. The dialogue concerns the causes of rainfall. The student demonstrates an understanding that rainfall comes from moist air cooling and condensing, but then exhibits a misconception, which the tutor corrects:

Student: … the moist air cools so the clouds can't hold the water so it rains.
Tutor: OK, what causes the moist air to cool?
Student: It cools when the wind blows it and it lowers from the sky.
Tutor: What happens to the temperature of moist air when it rises?
Student: It gets warm.
Tutor: No. Warm air rises, but as it rises, it cools off.

(Stevens *et al.*, 1979: 150)

The student has demonstrated a common misconception that the higher the air the warmer it is. It is common knowledge that warm air rises, but he interprets this as an attribute of air that is high up. The tutor offers an alternative description of what happens, more complex in the sense that it describes a relation: as height increases, temperature decreases. This is a description of what could have been an experience for the student, but here it is extrinsic feedback on their description. If geography students were able to practise 'air-play' as children can practise 'water-play', then they could develop a situated understanding of the relation. It would be an equivalent to the generalised rule that 'as the angle of pouring increases the water flows faster', along the lines of 'as air rises it cools'. The difference is that obtaining the feedback as a description, not as an experience, means that from the student's point of view this result is not connected to any goal or action on their part. It is situated in the dialogue only, as feedback on their description.

The dialogue offers further opportunity for action in the form of description, after the tutor asks how mountains could cool moist air:

Student: As it rises, or even if it doesn't rise, the cold mountains could cool it off.
Tutor: No, contact with a cold object does not provide enough cooling to an air mass to cause rain.

(Ibid. 150)

The student appears to be aware of the possibility that the mountains could make the air rise, but is sufficiently uncomfortable with that conception at the moment to prefer the security of another common misconception, that the cooling is done by contact with something cold. The next opportunity to rehearse the newly-learned action of invoking the 'as it rises it cools' rule comes later in the dialogue when the tutor is trying to achieve transfer to the new situation of a non-mountainous region:

Tutor: What happens when the warmer moist air is blown off the water and hits the stationary cold air mass?
Student: It makes it rise.
Tutor: Right, why?
Student: Because warm air rises, and when this warm air rises with the cool air on top of it, then the air will cool and it will rain.
Tutor: Almost. The warm air rises over the cool air.

(Ibid. 150–151)

Perhaps the student has now adopted the tutor's view that warm air rising is sufficient to cool it, but not to the extent of dissociating it completely from the cool air mass, which from the student's point of view still seems to have the function of helping the warm air to cool rather than helping it rise. The tutor makes the appropriate adjustment to the student's description.

The to and fro of this kind of dialogue makes a good match with the to and fro of interaction with the world, where action elicits feedback in the form of some event or behaviour, and adjustments to the action in the light of feedback elicit further feedback, enabling a refinement of the action to match what the world requires. In the dialogue above what is being refined is not actions in the world, but descriptions of the world, also a kind of action, perhaps, and undoubtedly experienced, but the experience is not of rainfall: it is of descriptions of rainfall. The only way this student could get intrinsic feedback on their actions is by carrying out experiments on rainfall and air temperatures in the range of different contexts discussed in the dialogue. With good experimental technique, they would then discover that the 'cooling by contact' hypothesis is flouted for certain conditions.

The nature of feedback on academic learning will reappear again in the following chapters. At this stage I hope to have established, from the student's point of view, the unity between, on the one hand, action with intrinsic feedback in the world as experienced directly, and on the other hand, action as description of the world with extrinsic feedback in the form of redescription. To use feedback, students must be able to make sense of it. The teacher has to devise situated actions that elicit meaningful intrinsic feedback for the student, or redescribe the student's description in a way that gives meaningful extrinsic feedback to the student.

REFLECTING ON GOALS–ACTION–FEEDBACK

It has been unavoidable in the previous sections of this chapter to include mention of the goal of the learning process. The presence of a goal is prefigured in the unity between action, feedback and integration; these aspects of the process only make sense if there is also direction, provided by a goal. The link between them is only made if the learner can reflect on the relationships between them all: on what the feedback means for the action in relation to the goal to be achieved; on what the goal means for the action now to be set up in the light of feedback on the last action, etc. Reflection is not confined to the goal, but as an aspect of the learning process it must always attend to the goal.

There is not a great deal of work in higher education, nor indeed at other levels of learning, that focuses specifically on the way learners handle the goals of a learning situation. Teachers have traditionally used assessment to act as both goal and feedback for the learner, and thereby promote the kind of reflection needed to master the material. But the use of assessment has been largely unreflective practice by the teaching profession.

The interviews conducted within the 'phenomenographic' method focus on how the student perceived the goal, and how they used this in their execution of the task. Some of the early work on reading, which established the deep–surface dichotomy, used retrospective accounts by students of what they were doing. This provided direct evidence of intentions such as 'looking for the meaning', 'trying to discover what the author wanted to put across', and conversely the absence of these intentions in a surface approach. This link between intention, process and outcome is an empirical one, and demonstrates the importance for the learning process of the way the student interprets the goal of the task.

But whose goal is it? We keep returning to the essential unity of the learning process that requires mutual interaction between its various aspects. In learning about the world through experience it is relatively straightforward to interpret the individual's actions as being goal-directed. The goal is itself a product of the individual's interaction with the world, inextricable from the individual learning in that situation. This does not transfer very well to the academic learner, however. The goal of an academic learning situation is generally set by the teacher. The students may be aware of it, and may even share it, but stand in a different relation to it in comparison with their goal-directed actions in the world.

A goal may be apparently agreed and shared, but the execution of actions directed to that goal may betray subtle differences in interpretation of the goal. Wertsch *et al.* studied two different groups of adult–child pairs given the task of reproducing a model. They found that although all pairs achieved the match, the way they did it was dependent on the group they were in: in mother–child pairs in a domestic context the adult was more likely to direct the child's actions than in teacher–child pairs in a classroom context (Wertsch *et al.*, 1984). They interpret this result to suggest that although goal and execution are logically independent of each other, in the sense that a goal can be executed many ways, and conversely, an action can serve many goals, the regularity observed in the way a common goal is executed in two different ways in two different groups must be explained in terms of the way the task goal is seen by the two groups. In the case cited, the mothers see the task as being to reproduce the model correctly, whereas the teachers see the task as being to instruct the children, hence the difference in the way the task was carried out.

What happens when students are solving problems set by a lecturer? In what sense are they able to maintain a unity between the nature of the task, the goal and their actions? I have reported elsewhere a study of how twelve students on a micro-electronics course set about a problem-solving exercise. The students reported retrospectively on their approach to a problem in microelectronics which asked them to write a device control program. The analysis of those protocols showed that the students were united in their perception of the task as being about providing the teacher with what he required of them, rather than as being about designing a program (Laurillard, 1984b). This was evident at the initial planning stage:

> I have to sort through the wording very slowly to understand what he wants us to do.
> I read through the notes to see what was familiar from the lecture, i.e. phrases or specific words that were repeated.
>
> (Ibid. 130)

It was evident also at the operational stage, where they might be expected to focus purely on the content of the task:

> I thought of a diagram drawn in a lecture and immediately referred back to it. Then I decided which components were wanted and which were not and started to draw it out, more or less copying without really thinking.
> I decided since X was setting the questions block diagrams were needed.
>
> (Ibid. 131)

and again at the final stage of checking back over the solution:

> I don't think the finished product was right but I decided it would do.
> I drew what I thought seemed logical although [I] was not satisfied as I couldn't really see how it fitted in … I didn't really do this exercise with a view to

> getting anything out of it. I felt it was something to copy down and nothing to understand really.
>
> (Ibid. 132)

These were model students, who worked hard and conscientiously on a tough course, taught by an enthusiastic teacher. They cannot be dismissed as out of the ordinary in any way. But their perception of the task in hand is intriguingly contrary to what the teacher supposed was going on. The point of these exercises was to familiarise the students with the intricacies of this kind of program, to give them a feel for the way the control of the electronics device could be analysed. The teacher saw the exercise as a challenging logical problem in linking the features of the device to the capabilities of the microprocessor via the medium of a set of coherent and unambiguous instructions. The students saw it as a problem in matching the demands of the teacher, as defined in the exercise, to the information available, as encoded in the linguistic and pictorial forms of representation he used in the lecture, via the medium of symbols and diagrams. At every point the task, the goal and the operations are seen differently by teacher and students. The focus of the students' attention is the 'problem-in-context', rather than the problem itself. This is similar to Wertsch's analysis of the theory of activity (1984): the same task can be perceived differently by teacher and students, and therefore operationalised in a way the teacher may not expect.

This does not destroy the essential unity between goal and action. It is preserved in the mutual shift of focus of both goal and operations on the part of the students from the substantive problem to the problem-in-context. The teacher's goal is not their goal, so reflection on their actions in relation to their goal produces a different analysis than it would if they were concerned about getting a functional program written.

As a mathemagenic activity, reflecting on action in a learning task in relation to its goal is known to be important from the work on deep and surface approaches. We have seen from the above discussion that the teacher has some additional work to do, not just in setting the goal, but in helping to form students' perceptions of what is required and what is important in the task set, as well as encouraging students to do the reflecting.

SUMMARY

In this chapter we have looked at students' learning activities in terms of five interdependent aspects of the learning process. Students must address all these mathemagenic activities if learning is to succeed:

- apprehend the structure of the discourse – e.g. focus on the narrative line, distinguish evidence and argument, organise and structure the content into a coherent whole;

- interpret the forms of representation – e.g. practise mapping between the concept, system, event or situation and its representation, practise using the forms of representation of an idea, represent the discourse as a whole as well as its constituent parts;
- act on descriptions of the world – e.g. combine descriptions and representations to generate further descriptions of the world, manipulate the various forms of representation of the world;
- use feedback – e.g. use both intrinsic and extrinsic feedback to adjust actions to fit the task goal, and adjust descriptions to fit the topic goal;
- reflect on the goal–action–feedback cycle – e.g. relate the feedback to the goal or message of the discourse, reflect on how the link between action and feedback relates to the structure of the whole.

The division of the learning process into five aspects does not make them in any sense independent. Whichever way the process is divided up, it will always be necessary to see one aspect in relation to the others. Throughout this chapter I have repeatedly invoked one in the discussion of another, and this is inevitable. The five aspects chosen enabled me to make use of the research literature in an orderly way, and provides a framework for further discussion, but they are not meant to be seen as logically distinct. It would be like trying to divide a society into mutually exclusive families. 'Family' is a useful category, but not an analytical one; each aspect of the learning process identified is constituted by its relation to the others.

Mathemagenic activities have been defined as those that give birth to learning, and encouraging these is an appropriately student-centred way of thinking about the teacher's task. We have looked at what count as mathemagenic activities for each of the five aspects of the learning process, and considered also the non-mathemagenic activities that students engage in. An awareness of both types will give us a grounding for devising teaching strategies in the next chapter.

Chapter 4

Generating a teaching strategy

INTRODUCTION

This chapter addresses the task of forming the bridge between what we know about student learning and what we should therefore do as teachers. That 'therefore' contains the assumption that there is some kind of logical link between the two. At the end of every study of student learning, and indeed of instructional psychology, educational psychology, and even sometimes cognitive psychology, there is an 'implications for teaching' section, which sets out the supposed link. In this chapter, we shall look at some of these links and their resultant implications. However, I feel I should issue a warning at the start that although this can be a respectable analytical process – going from what we know about student learning to what this means for teaching – it is not a logical one. It is clearly important to base a teaching strategy on an understanding of learning, but the relationship is fuzzy. The character of student learning is elusive, dependent on former experiences of the world and of education, and on the nature of the current teaching situation. What we learn from this will have an uncertain relation to what will happen in a new teaching situation. The dialectical character of the teaching–learning situation means that the connection will not transfer exactly to the different context of a new teaching strategy. We cannot tweak the teaching without altering the way that learning relates to it. The nature of student learning described in all the previous studies embodies within it the nature of the teaching situation the students were experiencing. That is why it was important not to decouple the description of learning from its content. However, it was usually decoupled from its context. In the one example I quoted in Chapter 3, p.59, where the context was taken into account (students on the microelectronics course), it became clear that there was a dissociation between the content and the context of the learning process (Laurillard, 1984b). The students' problem-in-context had little relation to the substantive problem set by the teacher. This remains an unresolved issue for educational design, and I believe it is an important one. The epistemological position laid out in Chapter 1, and everything that has followed, requires a relational view of knowledge and of learning, and emphasises the situated character of all types of learning. The bulk of the research we have to call upon, if it adopts this epistemology at all, does so in relation to content, rather than context. I do not wish to suggest that with funds and enough time we could establish complete and reliable connections between learning, content and context that would enable us to define reliable prescriptions for teaching strategies. Rather, the absence of research on the context of learning gives us an over-simplified view of student learning. Therefore, we are basing the design of a teaching strategy on a minimal analysis of student learning. It can still be principled, however, and in this chapter I hope to clarify what makes it principled.

Chapter 2 showed that a teaching strategy has to address three key aspects of the content of the students' learning experience:

- conceptions of the topic
- representational skills
- epistemological development.

Chapter 3 showed that a teaching strategy also has to assist students in the process of learning, in terms of the following mathemagenic activities:

- apprehending the structure of academic discourse
- interpreting forms of representation
- acting on descriptions of the world
- using feedback
- reflecting on the goal–action–feedback cycle.

These, together with the subject matter content, are the principal empirical basis for generating a teaching strategy. They were arrived at by considering the empirical evidence gathered from a selection of studies that investigated the outcomes and the process of learning in particular contexts. None of the studies pretends to completeness of description of the learning process, nor do they produce complementary coverage of what there is to be known about learning. The two lists therefore constitute a collection of things we ought to include, rather than an analysis of everything needed to generate a teaching strategy. Given the lack of any logical relation between learning and a teaching strategy, and this incomplete analysis of what a teaching strategy must include, it will be useful to look first at other attempts to derive strategies for effective teaching to see if they find a principled way of doing it.

I can identify in the current literature three distinct ways of handling this problem, deriving from different scientific traditions:

- instructional design, deriving originally from behavioural psychology but increasingly incorporating findings from cognitive psychology;
- constructivist psychology, deriving from developmental psychology;
- phenomenography, deriving from phenomenological psychology.

Each of these provides a link between an empirical base and a principle for design, so we can compare the nature of their empirical base, and the nature of the link made.

INSTRUCTIONAL DESIGN

The undisputed father of the field of instructional design is Robert Gagné, whose book *The Conditions of Learning*, first published in 1965 and now in its fourth edition, forms the precursor to all the current work. A more recent analysis given in *The Selection and Use of Media* (Romiszowski, 1988), acknowledges the influence of his work, so it is worth looking at as an example of a principled approach to generating teaching strategies.

Since it was first published, Gagné's analysis has shifted from a grounding in behavioural psychology to using information-processing theory as its empirical base. The system itself underwent only relatively minor revisions and elaborations, however. This is because it has only a tenuous link to any empirical base. Gagné's approach is essentially a logical analysis of what must be the case, rather than an empirically grounded theory. He begins with definitions of the general types of human capabilities that are learned: intellectual skills, cognitive strategies, verbal information, etc., a common-sense classification of what there is. He then describes the 'learning events' for each capability. These are derived from theoretical constructs generated by experimental studies in cognitive psychology, and based on information-processing theory. The constructs include, for example, 'short-term memory storage', based on studies of telephone number retrieval, and 'encoding', based on studies of memory of short passages of text. These 'learning events', together with the desired outcomes already defined as capabilities, are then used to generate the internal (mental) and hence external (situational) conditions for learning. For example, for 'defined concept learning', a sub-category of intellectual skills, the internal conditions are that the learner should:

1 have access in working memory to the component concepts;
2 have acquired the intellectual skill of being able to represent the syntax of the statement of the definition, i.e. distinguish subject from verb and object.

The external conditions 'usually consist in the presentation of the definition of the concept in oral or printed form' (Gagné, 1977: 134). That completes the analysis, and all the remaining combinations of capabilities and learning events are analysed in the same way to produce the same kinds of 'external conditions', i.e. the design of instructional events. The complete list of instructional events to be carried out by the teacher is:

- activating motivation
- informing learner of the objective

- directing attention
- stimulating recall
- providing learner guidance
- enhancing retention
- promoting transfer of learning
- eliciting performance
- providing feedback.

They seem unobjectionable and have an intuitive logical appeal, which is probably why the approach has been so influential. However, its empirical base is constituted in the theoretical constructs of another empirically based discipline. Cognitive psychology has an empirical foundation, but one that is built for its own purposes. These studies of, for example, short-term memory are carried out in experimental situations, and in isolation from all the other components Gagné includes in the learning process. They are used to infer possible constructs to describe how the human brain works. These are then transferred to the context of an academic learning task, as though the transfer were unproblematic. The empirical base is insufficient, therefore, to provide a holistic understanding of student learning. There is no data in the theoretical development of this approach that derives from students learning in an instructional context. The theory may be used to generate teaching which is then evaluated, but this does not test the approach, only its instantiation in that piece of instruction.

A further problem with instructional design of this type is that the analysis into components of the teaching–learning process is not followed by any synthesis. Any relationship between cognitive strategies and motor skills, for example, is not considered. Gagné himself has recognised this recently in a paper with another of the key figures in instructional design, David Merrill. They begin by outlining what they see as the value of their approach:

> The procedure of working backwards from goals to the requirements of instructional events is one of the most effective and widely employed techniques. This approach requires the initial identification of a category of instructional objectives, such as *verbal information, intellectual skill, cognitive strategies* ... From each of the single categories of learning outcome, the designer is able to analyze and prescribe the instructional conditions necessary for effective learning.
>
> (Gagné and Merrill, 1990: 2, original italics)

This analysis deals with one objective at a time, so that the designer must plan for instruction 'at the level of an individual topic'. However, they acknowledge that this is sometimes an inadequate level of analysis:

> When instruction is considered in the more comprehensive sense of a module, section or course, it becomes apparent that *multiple objectives* commonly

> occur … When the comprehensiveness of topics reaches a level such as often occurs in practice, instructional design is forced to deal with multiple objectives and the relationship among these objectives.
>
> (Ibid. 24, original italics)

Their solution is to add 'integrative goals' to the existing design theory, though without any perceivable shift in the underlying approach:

> We propose that integrative goals are represented in cognitive space by *enterprise schemas* whose focal integrating concept is the integrative goal. Associated with the integrative goal is an enterprise scenario and the various items of verbal knowledge, intellectual skills and cognitive strategies that must be learned in order to support the required performances … a consideration of enterprises as integrated wholes may lead to a future focus on more holistic student interactions.
>
> (Ibid. 29, original italics)

However, it is not possible to effect a synthesis of those analytical components simply by drawing a circle round them, as the diagram in the paper does, and then naming it. 'Integrative goals', and 'holistic student interactions' have to be derived from studies that look at interactions holistically. Their enterprise is word games; it is not science.

The influence of this kind of instructional design is enormous, however, which is why we must consider the approach. Perhaps it is the blandness of its conclusions that has permitted the largely uncritical acceptance of this way of tackling the task. Whatever the reasons, it is not a progressive force. It does not find out how the world is, it merely supposes. It is rather like reading a treatise on mediaeval physics, where theories, if they were built on anything other than supposition, were built on other theories, rather than on descriptions of the phenomena themselves. Gagné and Merrill begin their paper with these words:

> One of the signal accomplishments of contemporary doctrine on the design of instruction … is the idea that design begins with the identification of the goals of learning.
>
> (Ibid. 23)

This may seem rather obvious for an idea dignified as a 'signal accomplishment', but its complete absence from much educational planning shows that it was worth saying. And achieving widespread acceptance of such an idea is a worthwhile accomplishment. My argument is not so much against their conclusions as against their method. A progressive force in educational design theory would be one that cumulatively builds our knowledge of the phenomena concerned, and this does not. I think we can do better.

CONSTRUCTIVIST PSYCHOLOGY

The focus of teaching has to be on the way the individual interacts with their world, as we have already seen in Chapter 3. Constructivist psychology is valuable because it provides an account of how the individual learns through interaction with their world. This understanding then provides a principled approach to formal teaching which can be designed to manage the interaction in such a way that it optimises the learning process. That is the kind of analysis we need.

Constructivism is a broad church, encompassing all educators who reject the 'transmission' model of teaching or anything that sounds non-cognitive. A recent overview of current views of constructivism corrals the wide range of ideologies into two common tenets, that:

> (1) learning is an active process of constructing rather than acquiring knowledge, and
> (2) instruction is a process of supporting that construction rather than communicating knowledge.
>
> (Duffy and Cunningham, 1996: 171)

Duffy and Cunningham clarify the disparity of views of constructivism in the contrast they draw between 'cognitive constructivist' and 'socio-cultural constructivist' versions. The former derives originally from Piaget's work, describing children's development of increasingly abstract constructions of their world. The latter derives from Vygotsky's description of the development of knowledge through social interaction and the later idea of 'situated cognition' discussed in Chapter 1. It posits knowledge as a social construct, with cultural practices 'acting on and transforming reality within the context of those practices' (Ibid. 176). Duffy and Cunningham regard the two points of view as contradictory rather than complementary. Their valuable analysis of the range of key concepts offered by this cognitive approach to instructional design leads to a framework that synthesises cognitivism in 'problem-based learning' as the instructional model of choice. The design of the problem 'as a stimulus for authentic activity' is carried out via the processes of:

- task analysis;
- problem generation from the syllabus content;
- the learning sequence of collaborative and self-directed learning;
- the definition of the facilitator's role as challenger;
- the assessment grounded in the context of the problem.

This is a useful checklist for a teacher planning their lesson, but does not focus on the student's role, on what they must do to learn, despite the constructivist origins of cognitivism. There is no focus here on empirical findings on student learning, and no means to build further understanding of the learners. A teaching strategy, especially one that acknowledges the importance of the nature of the learner's interaction with their world, should build on our understanding of how that works.

By contrast, Burge's use of constructivism goes deeper into exactly how the concept might inform teaching design. While she lists the teacher's tasks, they all presuppose actions and attitudes of the learner:

> To teach constructively is to provide opportunities for complex information processing related to a learner's needs and knowledge of the world, design relevant and real world (authentic) tasks, help to identify conflicting ideas and attitudes, provide complex and controversial stimuli, challenge the learner's existing knowledge structures and values, acknowledge vague structures in knowledge, help learners revisit material in greater depths, confirm the learning identified by learners, and guide learners to generate correct solutions.
>
> (Burge, 1995:156)

These guidelines more clearly reveal a sense of how the teacher might encourage constructive learning activities. However, they do not prescribe a principle for the design of an independent learning environment.

Biggs is more direct about how to develop a constructivist teaching strategy. He uses the idea of 'constructive alignment' to describe the link to be made between the curriculum objectives and the corresponding activities by students. The latter are defined in terms of appropriately 'high-level verbs', which means: 'you get students to do the things that the objectives nominate' (Biggs, 1999: 26). High-level verbs such as theorise, or reflect, are contrasted with low-level verbs such as recognise, or memorise. These are derived from empirical studies of student learning such as those described in the previous two chapters. However, the teacher needs a principled teaching strategy that goes beyond the definition of high-level active verbs. Biggs defines teaching and learning activities as either teacher-directed, peer-directed, or self-directed. Teacher-directed activities are those that ensure the presentation is clear. Peer-directed are those that involve discussion, although he suggests that prior training in generic questions is required to promote productive discussion. We might expect self-directed activities to define those we try to elicit in educational design. However, these focus only on study skills, such as 'note-taking' and 'reading for main ideas', without suggesting how the teacher might elicit either these or the other higher-level verbs. A principled strategy for designing a constructive learning environment for the individual learner is not easily derived from this approach.

Marton and Booth (1997) also identify two forms of cognitivism, 'individual cognitivism' and 'social cognitivism', and demonstrate that the two forms offer quite different analyses of the learning process. Whereas the former situates the explanation of learning in the learner's cognitive mental acts, the latter situates it in their social, external behaviour. Marton and Booth argue that we have to transcend the person–world dualism assumed in both forms of constructivism, and accept that the world that we experience is constituted as an internal relation between the world and the learner. This brings us to the third type of approach to generating a teaching strategy.

PHENOMENOGRAPHY

The methodology of phenomenography, described in previous chapters, derives its empirical base from discovery rather than hypothesis testing. It uses qualitative rather than quantitative data, and its output is categories of experience, rather than relational explanations. It cannot aim to be prescriptive in defining the implications of its findings, because it does not define a relationship between aspects of teaching and consequent learning outcomes. In being descriptive of how students experience learning, however, it provides an empirical base that can inform our approach to teaching.

Marton and Ramsden (1988) list six implications for the design of a learning session, which derive from phenomenographic studies.

1 Present the learner with new ways of seeing.
2 Focus on a few critical issues and show how they relate.
3 Integrate substantive and syntactic structures.
4 Make the learners' conceptions explicit to them.
5 Highlight the inconsistencies within and the consequences of learners' conceptions.
6 Create situations where learners centre attention on relevant aspects.

The first two suggest using the variation in students' conceptions that are revealed through phenomenographic studies. The third focuses on integrating forms of representation with the event or system represented. The next two use what Marton later refers to as the 'architecture of variation' to help learners change from one conception to another, and the sixth suggests using the relevance structure for the topic to focus students' attention appropriately (Marton and Booth, 1997: 185).

Marton and Ramsden's recommendations show how the empirical base generates the strategy they define. Implicit in this discussion are two distinct ways of linking research results to implications for teaching:

- from descriptions of the internal structures of different conceptions, deduce how teachers and students should make their conceptions explicit so that they can be compared and contrasted;
- from descriptions of the differences between successful and unsuccessful teaching, deduce the characteristics of successful teacher–student interactions.

Phenomenographic studies clarify the variation in conceptions, such as the three conceptions of Newton's Third Law discussed in Chapter 2. From these we should be able to deduce how the teacher can make them explicit, using, perhaps, the same task used in the phenomenographic study. The difference between students' internal structures is the 'architecture of variation', such as that described in Chapter 2, p.32, for the Newton example. This becomes the focus of the interaction. In that example we could see that for some students the relevance structure would involve contrasting the Third Law with the concept of equilibrium. For others, it would involve focusing on the notion of a scientific law.

The descriptions of differences in teaching are based on studies of teachers teaching students, so the derived characteristics are at the level of the relation between teacher–student–subject, and not at the greater remove of conditions–person–task, as in instructional design.

In their more recent analysis of teaching strategies, Marton and Booth discuss at length the ways in which phenomenography can contribute to better learning. Their approach begins with a definition of pedagogy in which:

> teachers mold experiences for their students with the aim of bringing about learning, and the essential feature is that the *teacher takes the part of the learner* ... becomes aware of the experience through the learner's awareness.
>
> (Marton and Booth, 1997: 179, original italics)

This is compatible with their non-dualistic position that situates the learning experience as inclusive of the learner and the object of learning. The teacher's strategy must therefore focus on the learner's experience of the object of learning. They can do this in two ways: by 'building a relevance structure' for the topic in question, and by using the 'architecture of variation' in conducting the dialogue, as elaborated in the six recommendations above. This neither specifies how something is to be taught, nor what methods are to be used: 'there is never one way of teaching something' (Ibid. 179). However, it does specify the conditions that any method must address if it is to elicit meaningful learning. The research prescribes not the action the teacher must do to the student, but the form of the interaction that must take place between teacher, student and subject matter. Prosser and Trigwell have reconceptualised this approach as a 'constitutionalist' perspective:

> In any act of learning, students simultaneously engage in three successive phases – acquiring, knowing, and applying ... From the constitutionalist perspective, we consider students' prior experiences, perceptions, approaches and outcomes to be simultaneously present in their awareness.
>
> (Prosser and Trigwell, 1999: 17)

The learning process is constituted in the succession of expectations, perceptions, approaches, and outcomes. The approach contrasts with the deterministic box-and-arrows models that abound in psychology, and expresses the learning experience in a more holistic, iterative form. As the learner iterates through the learning sequence, there is an opportunity for development of perceptions and approaches, creating new experiences that become background for the next in the sequence. For this to be possible, the learning process must be designed to elicit awareness of inconsistencies in conception, variation in conception, etc., such as those identified above. This acknowledgment of the necessary iteration between teacher, student and content is more realistic than the cause–effect models of instructional design and cognitivism. This is why I believe phenomenography offers the best hope for a principled way of generating teaching strategy from research outcomes.

A PRINCIPLED APPROACH TO GENERATING TEACHING STRATEGY

Returning to the list of findings to be addressed by a teaching strategy, I want to reconsider these in the light of the principle, expressed above, of using them to deduce the form of interaction between teacher, student and content. This shift in focus from what the teacher should do, to how they must set up the interaction, reflects the fact that we cannot generalise these findings, only the methodology (Marton, 1988). We cannot claim to have sorted out once and for all what students need to be told if they are to make sense of topic X. No matter how much detailed research is done on the way the topic is conceptualised, the solution will not necessarily be found for new ways of putting it across. The new way of telling may sort out one difficulty, but it may well create others. All we can definitely claim is that there are different ways of conceptualising the topics we want to teach. So all we can definitely conclude is that teachers and students need to be aware that there are such differences and they must have the means to resolve them within the learning situation. The only prescriptive implication from our analysis here is that there must be:

- a continuing iterative dialogue between teacher and student, which reveals the participants' conceptions, and the variations between them, and these in turn will determine the focus for the further dialogue.

There is no escape from the need for dialogue, according to this analysis. There is no room for mere telling, nor for practice without description, nor for experimentation without reflection, nor for student action without feedback. This very 'prescriptive' implication from phenomenographic studies is compatible with the analysis of the nature of academic knowledge in Chapter 1. If you accept that academic knowledge is knowledge of descriptions of the world and will become known through operations on descriptions, then teaching must be a dialogic process.

The findings on students' epistemologies tell us that teaching should focus on the nature of the learning process, encouraging students to take a reflective, interpretive approach to their learning. Marton and Booth describe several studies that attempted to do this, but all of which failed (Ibid. 168–171). They all share the technique of focusing the students' awareness on the act of learning itself, e.g. by including in a text instructions on how to read it, to reflect on it, and to summarise it. In all cases, the students responded by focusing on the guidance rather than the content of the text, thereby undermining any meaningful outcome they might otherwise have derived. A learning skills programme for history students found a more successful strategy. It made use of history materials as the focus of reflection. This integration of content with process resulted in a more advanced conception of learning, compared with a similar programme that used generic materials. By contrast, the separation of content and process in the other studies had served merely to technify the learning process, making the instructions themselves the object of learning. Deriving teaching strategy from research findings is not straightforward.

The findings on productive learning activities (see Chapter 3) will be the best source for a teaching strategy about how to conduct an interactive dialogue that fully supports the learning process. Table 4.1 elaborates each aspect of the process, following the organisation of Chapter 3, to show what roles student and teacher should play in the interaction.

Table 4.1 Student and teacher roles in the learning process

Aspects of the learning process	*Student's role*	*Teacher's role*
Apprehending structure	Look for structure. Discern topic goal. Relate goal to structure of discourse.	Explain phenomena. Clarify structure. Negotiate topic goal. Ask about internal relations.
Interpreting forms of representation	Model events/systems in terms of forms of representation. Interpret forms of representation as events/systems.	Set mapping tasks between forms of representation and events/systems. Relate form of representation to student's view.
Acting on descriptions	Derive implications, solve problems, and test hypotheses, to produce descriptions.	Elicit descriptions. Compare descriptions. Highlight inconsistencies.
Using feedback	Link teacher's redescription to relation between action and goal, to produce new action on description.	Provide redescription. Elicit new description. Support linking process.
Reflecting on goal–action–feedback cycle	Engage with goal. Relate to actions and feedback.	Prompt reflection. Support reflection on goal–action–feedback cycle.

We can check the validity and utility of this way of describing the learning–teaching dialogue by applying it. I have selected two learning problems for which the literature describes the teaching strategy and records something of its success. In each case, we look at how the teacher–student interaction is conducted, in terms of the five aspects of the process listed above.

Teaching the process of rainfall

Taking the dialogue between tutor and student already outlined in Chapter 3, p.56, as an example of an interaction designed to help the student change their conception, how would the above analysis be applied? In terms of the prescription offered at the beginning of this section, it meets the criteria:

- there *is* a continuing iterative dialogue between teacher and student;
- it *does* reveal the participants' conceptions, and the variation between them;
- this *does* in turn determine the focus for the continuing dialogue.

It is not just conducting a dialogue that is important, but how it is conducted. This particular dialogue failed to address several of the essential aspects of the learning process, as listed at the end of Chapter 3, p.60, and in Table 4.1.

First, there was rather little opportunity for the student to interpret forms of representation, as there is no form other than language, and there is no specialist use of the language involved. The difficulty arises in understanding the system, not its representation.

Other aspects of the process are more evident. The tutor certainly provides an interactive environment that allows the student to generate descriptions; he elicits descriptions from the student relating in different ways to the descriptions he offers. The student is asked to explain phenomena ('it rains, why? – because the moist air cools and the clouds can't hold the water'), to make predictions about new situations ('Can you guess what the average rainfall is like on the other side of the mountains? – It's probably heavy') to compare analogous situations ('what is the relation between mountains and cold air mass? – the cold air mass stays low'). This is exactly what that peculiar phrase 'acting on descriptions' is about: making connections between propositions, offering re-articulations, deducing new propositions.

There is also feedback, in the form of extrinsic feedback on the student's hypothesis, a new description ('the cold mountains could cool it off – no, contact with a cold object does not provide enough cooling'). The student then has to link this feedback to the goal and action to produce a new description. It is clear what his action was – his hypothesis about cooling by contact – but not so clear what the topic goal is, because this has not been explicitly negotiated. The current tutor-set goal is to explain the role of the mountains in cooling the air. If the student shares this goal he now has a reason to look for an alternative hypothesis, since 'cooling by contact' has been rejected as inadequate. The tutor follows this with feedback relevant to his current goal: 'rainfall is almost always the result of cooling due to rising air', and then sets up an opportunity for the student to apply this ('how do you think the mountains might affect the rising of the moist air?'). So the tutor is making good connections between the student's description (action), feedback and goal, but the student is not in control. He may not be following those same crucial links, which may be why the 'cooling by contact' bug appears to surface again later in the dialogue. The tutor does not support the student's reflective process of using the feedback to modify their description in relation to the goal.

Finally, there was no opportunity for the student to 'apprehend the structure' of the tutor's discourse. It would be a difficult exercise from the transcript, and next to impossible in the cut and thrust of a conversation, to discern the totality of the tutor's point of view, the key planks in his argument, and the nature of the connections between them. The representation of the tutor's knowledge structure in the original paper involves nine propositional nodes and seven connecting relations of three different types. It remains only implicit in the dialogue quoted in Chapter 3. It is difficult for the student to relate the goal to the structure, or to integrate the different parts of the structure, because the tutor is directing the dialogue according to his goals. They remain un-negotiated with the student, never subject to reflection.

The myth of Socratic teaching

This lack of explicit focus on the goal and its relation to the components of the structure is a common feature of the 'Socratic dialogue'. It is worth taking time to analyse a Socratic dialogue, because it is a respected teaching strategy, and it takes the form of dialogue, so it should fit my purpose well. However, interestingly, it fails the application of the above principles.

Brown and Atkins (1991), in their discussion of effective teaching, offer Socrates as 'the great proponent' of small group teaching. The illustrative example they use is from Plato's *The Symposium* (Hamilton, 1951) in the dialogue with Agathon. They quote, with approval, an interaction in which Socrates engages in a kind of rhetorical bullying:

Socrates: You said, I think, that the troubles among the gods were composed by love of beauty, for there could not be such a thing as love of ugliness. Wasn't that it?

Agathon: Yes.

Socrates: Quite right, my dear friend, and if that is so, Love will be love of beauty, will he not, and not love of ugliness?

Agathon agrees.

Socrates: Now we have agreed that Love is in love with what he lacks and does not possess.

Agathon: Yes.

Socrates: So after all, Love lacks and does not possess beauty?
Agathon: Certainly not.
Socrates: Do you still think then that Love is beautiful if this is so?

They omit Agathon's immediate admission of humiliation:

Agathon: It looks, Socrates, as if I didn't know what I was talking about when I said that.

as well as the remainder of the dialogue which shows Socrates is apparently magnanimous in victory, but condescending, nonetheless:

Socrates: Still, it was a beautiful speech, Agathon. But there is just one more small point. Do you think that what is good is the same as what is beautiful?
Agathon: I do.
Socrates: Then if Love lacks beauty, and what is good coincides with what is beautiful, he also lacks goodness.
Agathon: I can't find any way of withstanding you Socrates. Let it be as you say.

Is this really the kind of response we want from our students?

Socrates: Not at all, my dear Agathon. It is truth that you may find it impossible to withstand; there is never the slightest difficulty in withstanding Socrates. But now I will leave you in peace.

(Hamilton, 1951: 78–79)

Is that a fitting conclusion for a tutorial? This is hardly an interactive style to be emulated by tutors. Hamilton, in his introduction, has a more realistic assessment of the Socratic method, pointing out that he employs upon Agathon:

> … the instrument of philosophical inquiry that is peculiarly his own, the method of question and answer, of which the first stage consists in reducing the interlocutor to 'helplessness', the admission that his own existing views upon the subject under discussion are completely mistaken.
>
> (Ibid. 18)

The role of the teacher is to mediate the person–world relationship and ensure that it can change over time in the direction of the desired learning outcome. The Socratic dialogue is unlikely to achieve this because it does not invite the person to relate to their world, only to highly localised descriptions within the tutor's world. This is part of the sequence of responses by Meno's slave (the full transcript of this section of the dialogue is in Appendix 1):

Boy: True.
Boy: Yes.
Boy: There are.
Boy: I do not understand.
Boy: Yes.
Boy: Four.
Boy: Two.

It is a completely one-sided dialogue in which there is no attempt to mediate the person–world relationship by eliciting the boy's conception of geometry. He can remain focused on the internal logic of each question in order to frame his answer, and need never see the overall structure of the goal and its relation to the component parts. The dialogue ends with the ultimate leading question:

Socrates: And that is the line which the learned call the diagonal. And if this is the proper name, then you, Meno's slave, are prepared to affirm that the double space is the square of the diagonal?
Boy: Certainly, Socrates.

(Jowett, 1953: 283–284)

However, the dialogue provides evidence only of the boy's knowledge of counting and logic. He produces not one statement about geometry. The goal for Socrates is Truth, to be achieved through philosophical inquiry. That is not the same as a goal of enhancing the intellectual skills and understanding of others. In essence it is a strategy designed to reduce his interlocutor to helplessness, when they are ready to capitulate to anything he says: 'let it be as you say', 'certainly, Socrates'. It is extremely authoritarian. The Socratic method is not, as it is often described, a tutorial method that allows the student to come to an understanding of what they know. It is a rhetorical method that gives all the responsibility to, and therefore achieves all the benefit for, the teacher. To appreciate the true value of a dialogic interaction for the student, we have to look at the totality of what the student says. In both Socrates' original, and in the rainfall dialogue, removal of the teacher's role reveals just how minimal the student's role is. They engage actively at a localised level only, the overall structure remaining inaccessible to them, and therefore the overall meaning in danger of being lost to them.

Again, in terms of the prescription offered at the beginning of this chapter, this kind of dialogue does not meet all the criteria:

- there *is* a continuing iterative dialogue between teacher and student;
- but it *does not* reveal all the participants' conceptions, nor the variation between them, and therefore it is only Socrates' personal narrative that can determine the focus for the continuing dialogue.

If a tutorial dialogue is to be successful from the student's point of view, it must be carefully managed to address all the mathemagenic activities listed in Table 4.1, p.72. The successful tutorial dialogue is the means by which the tutor resolves Meno's paradox (how can we learn from the world what we do not already know?). The tutor must mediate the process of successive focused iterations in which the student attempts to capture experience of the world in descriptions, or forms of representation. That is how they elicit from the student a new way of experiencing a concept, which is constituted in the person–world relationship.

SUMMARY

This chapter has sought a way to generate a principled teaching strategy, given what we know about the characteristics of student learning. We considered three very different approaches.

Instructional design theory is logically principled, not empirically based, and therefore unable to build teaching on a knowledge of students.

In the first edition of this book, I included intelligent tutoring systems as another logically principled approach. I concluded then that it did not offer a principled derivation of a teaching strategy because, like instructional design, it did not attempt to link teaching design to empirical data about students learning. Its explicit rejection of empirical data, together with its failure to instantiate theories about learning and teaching have since led to its demise as a field of inquiry.

Constructivist approaches have focused more on the teacher–student interaction but without offering a detailed link between teaching, student activity and interaction with the subject.

I found phenomenography a more fitting approach. The co-operative style is more democratic, giving full representation to students' as well as teachers' conceptions, and if it prescribes anything, it does so at the level of how the iterative dialogue should be conducted.

The best expression of an empirically based teaching strategy so far, therefore, is as an iterative dialogue between teacher and student focused on a topic goal. The responsibilities of both teacher and student, generated throughout the chapter, can be grouped as four distinct aspects of the progression of the dialogue.

Teaching strategy

Discursive:

- teacher's and student's conceptions should each be continually accessible to the other;
- teacher and student must agree learning goals for the topic;

- the teacher must provide a discussion environment for the topic goal, within which students can generate and receive feedback on descriptions appropriate to the topic goal.

Adaptive:

- the teacher has the responsibility to use the relationship between their own and the student's conception to determine the task focus of the continuing dialogue;
- the student has the responsibility to use the feedback from their work on the task and relate it to their conception.

Interactive:

- the teacher must provide a task environment within which students can act on, generate and receive feedback on actions appropriate to the task goal;
- the students must act to achieve the task goal;
- the teacher must provide meaningful intrinsic feedback on their actions that relates to the nature of the task goal.

Reflective:

- the teacher must support the process in which students link the feedback on their actions to the topic goal for every level of description within the topic structure;
- the student must reflect on the task goal, their action on it, and the feedback they received, and link this to their description of their conception of the topic goal.

The strategy is undeniably prescriptive, but aspires to prescribe a form of interaction between teacher and student, rather than action on the student. In this way, it provides a structure capable of its own improvement. The claim for this higher level of prescriptive teaching strategy is the strong one, that it should not fail. It will be difficult to apply, and might be misapplied, but it should result in improved quality of learning. The chapters in Part II use the requirements of the teaching strategy to challenge the extent to which the new learning media are capable of supporting academic learning.

Part II

Analysing the media for learning and teaching

A framework for analysis

INTRODUCTION

Part II has the task of examining what the various media have to offer learning and teaching. Having arrived at a perspective on learning and teaching that sees the process as essentially a dialogue, this may appear to rule out any contribution from teaching methods other than the one-to-one tutorial. Whatever you may think of the approach developed over the last four chapters, it has something to recommend it if it derives the one-to-one tutorial as the ideal teaching situation. Sadly, the one-to-one tutorial is rarely feasible as a method in a system of rapid expansion beyond a carefully selected élite, so we look to other methods to provide the same effect more efficiently.

The familiar methods of teaching in higher education are there to support learning as it is commonly understood to occur:

- through acquisition, so we offer lectures and reading;
- through practice, so we set exercises and problems;
- through discussion, so we conduct seminars and tutorials;
- through discovery, so we arrange field trips and practicals.

These methods, if practised in combination, are capable of satisfying most of the requirements of the teaching strategy derived at the end of Chapter 4. Feedback on students' actions is the weakest link, because there is only a small amount relative to their learning actions. Feedback is handled within the assessment procedures adopted for set work, and within supervised practicals and tutorials, but is not guaranteed, is usually not closely associated with the actions, and tends to be only extrinsic, rarely intrinsic.

I do not accept, however, that these methods are essentially unable to yield the ideal form of the teaching–learning process. Paul Ramsden, in his book on teaching in higher education, seems more pessimistic. Having developed an extensive analysis of what must be required of the best teaching methods – that they must involve students in actively finding knowledge, interpreting results and testing hypotheses – he notes the sharp contrast between these and the methods that traditionally place authoritative information before students and leave the rest to them:

> The reader will now I hope be able to see one step ahead in the argument and confront the inevitable truth that many popular methods, such as the traditional lecture–tutorial–discussion–laboratory–class method of teaching science and social science courses, do not emerge from this analytical process unscathed. In fact, not to put too fine a point on it, many teaching methods in higher education would seem, in terms of our theory, to be actually detrimental to the quality of student learning.
>
> (Ramsden, 1992: 152)

On the other hand, he certainly does not see any salvation in the technological media:

> Computers and video in higher education have so far rarely lived up to the promises made for them ... No medium, however useful, can solve fundamental educational problems.
>
> (Ibid. 159–161)

In the remainder of the book, he retrieves the position for many of the more traditional methods, or at least for a combination of them, by describing better ways of doing them. He gives examples of how the traditional methods can, with careful planning, meet the requirements of good teaching:

> In short, a teacher faced with a series of classes with a large group of students should plan to do things that encourage deep approaches to learning; these things imply dialogue, structured goals, and activity … Teaching is a sort of conversation.
>
> (Ibid. 167–168)

It is possible, then, to examine the ways traditional methods can meet the requirements derived from the research on student learning, and I believe it can be done in the way Ramsden suggests. I also agree with his point that no one medium can solve the problems, as will become clear. However, given that we agree on the essentially conversational character of the teaching–learning process, what kind of role could the various media possibly play, since most of them cannot support conversations at all? Moreover, media are sometimes defined as transmitters, the very opposite of what we need:

> We define 'media' as the *carriers of messages, from some transmitting source* (which may be a human being or an inanimate object) *to the receiver of the message* (which in our case is the learner).
>
> (Romiszowski, 1988: 8, original italics)

How can the use of media possibly fit with an epistemology, such as the one explored in this book, that argues against a transmission model for education, and against the idea that knowledge is an entity separable from knower and known? It will mean a redefinition of 'media' at the very least.

PEDAGOGICAL CATEGORIES FOR CLASSIFYING MEDIA

There are many attempts in the literature to categorise and classify the forms of media, none of which is very illuminating or useful for our purpose here. Classification of forms is a notoriously difficult task, even when we can expect there to be some guiding principle inherent in their existence and formation. The development of educational media has an odd mix of engines driving it, technological pull, commercial empire-building, financial drag, logistical imperatives, pedagogical pleas, and between them they generate a strange assortment of equipment and systems from which the educational technologist must fashion something academically respectable. None of the media to be discussed in Part II was developed as a response to a pedagogical imperative, and it shows. They do not easily lend themselves to a pedagogical classification.

The point of a good classification system is that it should be powerful enough to embrace the ideal as well as a recognisable reality, and thereby make the shortcomings of our realities apparent. A classification system that starts by classifying what there is will fail to address a pedagogical ideal, and that is why the current attempts are unsatisfying. Chapter 4 ended with principles for generating a teaching strategy, and that is where a classification of educational media should begin.

The categories defined at the end of Chapter 4, p.77, reflect the interdependent relationships between all the aspects of the learning process previously defined. On that basis, educational media should be classifiable in terms of the extent to which they support the interpersonal and internal dialogue forms, the 'discursive', 'adaptive', 'interactive' and 'reflective' processes:

Classification of educational media

Discursive:

- teacher's and students' conceptions are each accessible to the other and the topic goal is negotiable;
- students must be able to generate and receive feedback on descriptions appropriate to the topic goal;
- the teacher must be able to reflect on student's descriptions and adjust their own descriptions to be more meaningful to the student.

Adaptive:

- the teacher can use the relationship between their own and the student's conception to set up and adapt a task environment for the continuing dialogue, in the light of the topic goals;
- the student must be able to use their existing conceptual knowledge to adapt their actions in the task environment in order to achieve the task goal.

Interactive:

- the students can act within the task environment to achieve the task goal;
- they should receive meaningful intrinsic feedback on their actions that relate to the nature of the task goal;
- something in the environment must change in a meaningful way as a result of their actions;

Reflective:

- teachers must support the process by which students link the feedback on their actions to the topic goal, i.e. link experience to descriptions of experience;
- the pace of the learning process must be controllable by the students, so that they can take the time needed to reflect on the task goal–action–feedback cycle in order to develop their conception in relation to the topic goal.

To illustrate these different processes as they occur in real teaching, I have applied it to a one-to-one tutorial in what I hope is an accessible topic. This is not, I'm afraid, another excursion to the primary classroom. It is an edited extract of a remedial maths session for UK undergraduate technology students.

Topic goal: to represent correctly cancellation of algebraic quotients.
Task goal: simplify (ca + b)/c.

Student:	$\frac{ca + b}{c} = a + b$	*Action on task*
Teacher:	Can we just look at this bit again? Let's turn this into a real problem. Suppose you've got six apples and six bananas and you divide them between six people, what does each person get?	*Indirect extrinsic feedback on action.* *Adapts task goal to imagined real-world task.*
Student:	One apple and one banana.	*Implicit intrinsic feedback from imagining action in world.*
Teacher:	OK. And we can write "six a's plus six b's shared among six" down as	*Extrinsic feedback.* *Redescription of action*

	(6a + 6b)/6, and that comes to a + b.	*as mathematical representation.*
	OK?	*Checks description is shared.*
Student:	Yes.	*Agrees description.*
Teacher:	Now, can you write down something similar for dividing four apples and one banana between four people?	*Adapts task goal to check that S does share topic goal of the principle for describing real-world action. Selects example to match form of original task.*
Student:	[writes] (4a + 1b)/4. Is that it?	*Reflects on T's earlier redescription. Presents own redescription of imagined action as mathematical representation.*
Teacher:	Yes, terrific.	*Extrinsic feedback.*
	So four apples, and a banana, what would they get each?	*Adapts task goal to carry out imagined action in the world.*
Student:	One apple and a quarter of a banana … Ahh.	*Reflects on interaction, and on relationship between descriptions of events.*
Teacher:	Right. Can you write that down? Write down the whole thing.	*Adapts task goal in light of topic goal to check that redescription of event as mathematical representation is shared.*
Student:	[writes] $= a + \frac{b}{4}$	*S redescribes own representation of action.*
Teacher:	Good.	*Extrinsic feedback on redescription.*
	So you can't just cancel one of them. The four divides both the terms on the top line.	*Redescription of T's conception in relation to task goal.*

The teacher's focus throughout the dialogue is not on the real-world task. The student's ability to share out real-world objects equally is not in doubt. The teacher is utilising this by translating the problem into a real-world equivalent, enabling the student to imagine the intrinsic feedback they would receive in that context. The point is to help the student learn to represent the real world mathematically – to acquire the academic knowledge of how to represent experience. The 'thought experiment' with apples and bananas is an imagined interaction with the world. The student's reflection on that imagined interaction, together with their discursive interaction at the level of description, enables them to arrive at the teacher's way of looking at the world of apples and bananas.

In any dialogue of this kind, the student is learning how to represent the world, not how to act on the world. They use their knowledge of real-world experience to make sure there is a fit between the representation and those actions, and that therefore the representation is correct. Description with extrinsic feedback, and imagined action with its intrinsic feedback, enables the student to make the link between the world of experience and the world of academic representations of experience. Once this link is established, the student should then be able to enhance their future actions in the world by referring to the academic theory it relates to, whether it is performing mathematical calculations, or predicting rainfall.

A FRAMEWORK FOR ANALYSING EDUCATIONAL MEDIA

The previous chapter generated a teaching strategy from the findings of research studies of student learning, and based on an epistemology that situates learning as a relationship between the learner and the world, mediated by the teacher. The teaching strategy has been refined into a set of requirements for any learning situation:

- it must operate as an iterative dialogue;
- which must be discursive, adaptive, interactive and reflective;
- and which must operate at the level of descriptions of the topic;
- and at the level of actions within related tasks.

This is the framework against which we now evaluate the extent to which the various media support the full specification. Because of its essentially dialogic form, I have termed it a 'Conversational Framework'.

Figure II.1 offers an alternative form of representation of the above descriptions. Teacher and student are represented as interacting through some medium – it may be a face-to-face tutorial, it may be conducted entirely through correspondence, or it may employ a combination of several media. Teacher and student each operate at the level of descriptions of the topic goal, and actions on a task environment. The arrows represent learning and teaching activities that constitute the dialogic relationships within and between the two participants.

- The discursive process is represented as a series of activities by teacher and student at the level of descriptions of the topic goal: describing and redescribing each participant's conception of it (activities 1–4).
- The adaptive process is represented as activities (5 and 10) internal to both teacher and student, each of whom adapts their actions at the task level in the light of the discursive process at the description level.
- The interactive process is represented as a series of activities (6 to 9) by teacher and student at the level of the task environment, setting and aiming to achieve the task goal, giving and acting on feedback in the light of the task goal.
- The reflective process is represented as activities (11 and 12), internal to both teacher and student, each of whom reflects on the interaction at the task level in order to redescribe their conceptions at the level of descriptions of the topic goal.

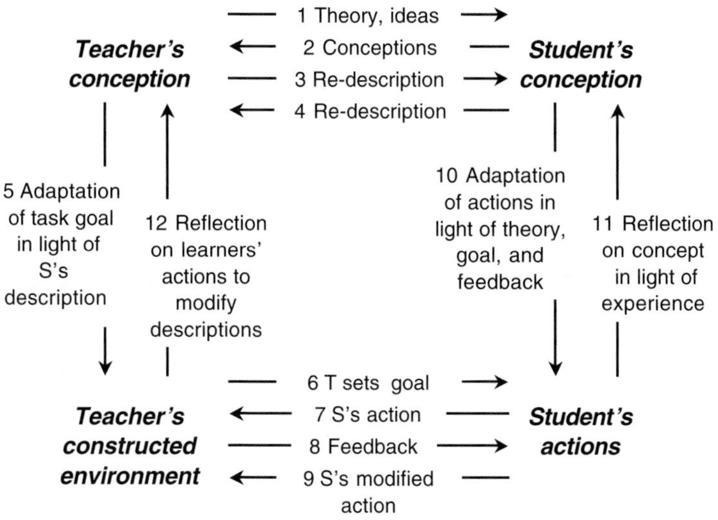

Figure II.1 The Conversational Framework identifying the activities necessary to complete the learning process.

This Conversational Framework for describing the learning process is intended to be applicable to any academic learning situation: to the full range of subject areas and types of topic. It is not normally applicable to learning through experience, nor to 'everyday' learning.

The characterisation of the teaching–learning process as a iterative 'conversation' is hardly a new idea. I have already quoted Paul Ramsden's statement that teaching is a sort of conversation. Kolb's "learning cycle" (Kolb, 1984) states that learning occurs through an iterative cycle of experience followed by feedback, which is reflected on, and then used to revise action (equivalent to activities 6-7-8-11-10-9 in Figure II.1). Gordon Pask formalised the idea of learning as a conversation in Conversation Theory (Pask, 1976), which included the separation of 'descriptions' and 'model-building behaviours', and the definition of understanding as 'determined by two levels of agreement' (Ibid. 22). Vygotsky drew the same kind of distinction between the 'spontaneous' concepts of everyday learning, and the 'scientific' concepts of the classroom:

> The inception of a spontaneous concept can usually be traced to a face-to-face meeting with a concrete situation, while a scientific concept involves from the first a 'mediated' attitude towards its object.
>
> (Vygotsky, 1962: 108)

Most interesting ideas have their counterparts in the culture of Ancient Greece, as does this one in the 'Socratic dialogue'. It is still referred to as epitomising the tutorial process, although Chapter 4 questions its value as a teaching strategy. A Conversational Framework as a representation of the learning process has at least face validity, therefore, and serves both to clarify the second-order character of academic learning, and to define its essential components.

The Conversational Framework outlined here defines the core structure of an academic dialogue and relates it to content in terms of a topic goal. Any particular dialogue, where the topic focus shifts as the conversation proceeds, would be mapped by a series of conversational frameworks, where the topic goal breaks down into nested sub-topics, or switches to a parallel topic before returning to the main topic. The dialogue may never actually include action-in-the-world; it may only refer to former experience or 'thought experiments' as in the remedial maths dialogue above, but the core structure remains two-level. Similarly, the dialogue may never take place explicitly between teacher and student. It could be a purely internal dialogue with the student playing both roles. This kind of process is manifest in the research interviews described in Chapter 2, where students talk themselves into realising that they fail to understand the point. In clarifying this fact, of course, they sometimes see their way past the cognitive block. Figure II.2 shows how the Conversational Framework could be interpreted for learning from lectures, where there is little opportunity for the teacher to do anything other than deliver the theory. The remaining activities to complete the learning process must come from the student's own internal dialogue.

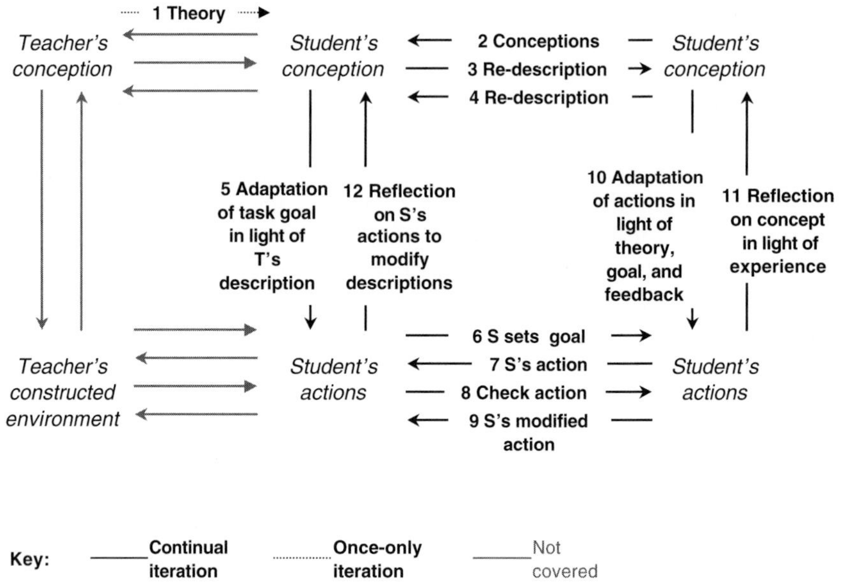

Figure II.2 The Conversational Framework interpreted for learning through lectures.

For learning to take place, the core structure of the Conversational Framework must remain intact in some form: the dialogue must take place somewhere, the actions must happen somewhere, even if it is all carried out by the student. That is what is needed for learning from lectures.

The question before us now is the extent to which educational media can support the Conversational Framework and thereby assist the learning process.

FORMS OF EDUCATIONAL MEDIA

In the following chapters each type of educational medium is analysed in terms of the Conversational Framework to see how far it serves the needs of a principled teaching strategy, using evidence from evaluation and design studies in the literature where possible. The chapters are organised according to the main types of educational media. This allows us to focus on their essential pedagogical characteristics and to identify the unique contribution made by each one. We also need a way of characterising the main types of educational media in advance of the pedagogical classification.

It is most useful to organise the discussion of media around their logistical properties, because these can affect access, cost, and durability. The logistics of media development and delivery are salient issues for the producer, and these will be discussed in Part III. In Part II, we organise the prior pedagogical analysis around a classification of the media in terms of their logistics.

The media of text, talk, visuals, or interaction can be delivered via meetings, print, cassette, disc, or link to a network. The different formats for delivery support different kinds of learning experience, and require different kinds of production and presentation resources. For example, print, television, video and DVD all require prior design and development, and relatively little labour-intensive presentation support. By contrast, seminars, discussion groups and online conferences require relatively little prior preparation, but do need labour-intensive presentation support in the form of tutors, or discussion leaders. These logistical differences will be important when we consider resources in Part III. For the analysis of pedagogical characteristics of media in Part II we need to group the media forms by the learning experiences they support. There are probably many ways of grouping the learning activities in the Conversational Framework. By combining learning experience with logistical characteristics, it is possible to focus on just five media forms, which cover all the key activities identified above. Table II.1 shows how each media form, characterised as narrative, interactive, communicative, adaptive, and productive, identifies with particular kinds of learning experience and delivery method.

These five media forms provide a reasonable way of organising the following chapters, which consider the pedagogical characteristics of each form.

Table II.1 Five principal media forms with the learning experiences they support and the methods used to deliver them

Learning experience	*Methods/technologies*	*Media forms*
Attending, apprehending	Print, TV, video, DVD	Narrative
Investigating, exploring	Library, CD, DVD, Web resources	Interactive
Discussing, debating	Seminar, online conference	Communicative
Experimenting, practising	Laboratory, field trip, simulation	Adaptive
Articulating, expressing	Essay, product, animation, model	Productive

Chapter 5

Narrative media

INTRODUCTION

Narrative media are the linear presentational media that include print (text and graphics) audio, usually audiocassette, audiovision (an audiocassette talk accompanied by some separate visual material), broadcast television or film, and videocassette or digital disc. These presentational media share the core common property that they are non-interactive, which distinguishes them from all the computer-based media. It is a feature that Socrates recognised as a failing in an educational medium, by comparison with interactive dialogue:

> I cannot help feeling, Phaedrus, that writing has one grave fault in common with painting; for the creations of the painter have the attitude of life, and yet if you ask them a question they preserve a solemn silence. And the same may be said of books. You would imagine that they had intelligence, but if you require any explanation of something that has been said, they preserve one unvarying meaning.
>
> (Jowett, 1953: 185)

And the same may be said of television, audio and video media. They cannot respond to their audience's enquiries, and the learner must make what they can of them.

In distinguishing narrative media from the computer-based media discussed in the following chapters, we should be able to discern some significant pedagogical consequence in the use of narrative. Books have been established as the supreme educational medium for several centuries, despite their non-interactive form, and they clearly support at least some of the essential activities students must engage in during the learning process.

The traditional educational methods and media, such as lectures, books, films, and television programmes, are all narrative in form, and for good reason. Narrative provides a structure that creates global coherence in a text that contains many component parts. The structure provides a linear dynamic that links the components to each other via relationships, which may be causal, temporal, or motivational, depending on the content. In an educational context, print, audio and video all use a variety of structural cues, such as headings, textual signposts, paragraphing, captions, locations, and camera movement, to allow learners to maintain a sense of the overall structure of the narrative, and hence understand its meaning. Narrative is fundamentally linked to cognition by providing the structure that enables the reader to discern the author's meaning.

However, in terms of the learning activities discussed in the last chapter, it is clear that these media offer only descriptions of the teacher's conception, with no opportunity for iteration through the remaining learning activities. The requirements of the Conversational Framework suggest that if the narrative presentational media are to move beyond the limits of the solemnly silent, uninterrogatable text to meet the demands of the learning process, then they have to structure the narrative to engage the learner in reflecting and articulating at the discursive level, and in playing some vicarious part in adapting and acting at the experiential level. Here we examine the extent to which they do this.

LECTURE

The lecture is under consideration here only to provide a baseline for comparison, as the traditionally favoured university teaching method. It is designed to be presentational and employs the narrative form of the ancient oral cultures. Only the teacher is able to articulate their conception. It therefore puts a tremendous burden on the students to engage in the full range of mathemagenic activities. They must do the work to render the implicit structure explicit to themselves, must reflect on the relationship between what the lecturer is saying and what they previously understood, and decide if it is different and how the difference is to be resolved. They must then check that this is compatible with everything else the lecturer said, initiating their own reflective activities, retrospectively, using their notes of the lecture. Their personal redescriptions are then articulated in tutorial discussions or essays which later elicit feedback from the teacher to complete the 'discursive' loop. It can be done, but opportunities for breakdown or failure are numerous.

Some lecturers acknowledge these limitations, and use techniques designed to address the essential learning activities omitted from the traditional form of one person talking to many for fifty minutes. Questions to students encourage them to reflect, and their answers allow the lecturer to refine the descriptions and explanations offered. Questions from students provide further opportunity at the discursive level for the lecturer to gain an insight into how students are thinking about the topic. Buzz groups encourage students to articulate descriptions and redescriptions of their understanding in interactive discussion with each other. The experiential level is addressed only rarely. There are examples of experiments and demonstrations, usually by heroic physics lecturers whose enthusiasm somehow sustains them through the logistical challenges of setting them up. But even these are essentially presentational, allowing the students only vicarious experience of the goal–action–feedback loop, in which they cannot test their own conceptually-generated action. The more usual way of linking experience to theory within a lecture is to appeal to the students' own previous experiences, using analogies, or illustrative examples. The remembered experience becomes an interaction on which to reflect and build their conceptions. Techniques such as these restore the lecture to something a little closer to the ideal of the one-to-one tutorial, but its inevitable one-to-many format maintains its position as very far from the ideal.

Why aren't lectures scrapped as a teaching method? If we forget the eight hundred years of university tradition that legitimises them, and imagine starting afresh with the problem of how best to enable a large percentage of the population to understand difficult and complex ideas, I doubt that lectures will immediately spring to mind as the obvious solution. Their success depends upon the lecturer knowing very well the capabilities of the students, and on the students having very similar capabilities and prior knowledge. Lectures were defensible, perhaps, in the old university systems in which students were selected through standardised entrance examinations. Open access and modular courses make it most unlikely that a class of students will be sufficiently similar in background and capabilities to make lectures workable as a principal teaching method. The economic pressures forcing open access to universities generate higher student numbers and, while universities remain designed around lectures, therefore dictate larger classes. Yet, the open access that creates a highly diverse audience makes lectures hopelessly inefficient for the individual student, in terms of pedagogical needs.

Academics will always defend the value of the 'inspirational' lecture, as though this could clinch the argument. But how many inspirational lectures could you reasonably give in a week? How many could a student reasonably absorb? Inspirational lectures are likely to be occasional events. Academics as 'students' typically think little of the method. It is commonplace to observe that the only valuable parts of an academic conference are the informal sessions. Students often defend the lecture system as a way of finding out what the curriculum is. There must be better ways. The lecture is a very unreliable way of transmitting the lecturer's knowledge to the student's notes. When I was teaching as a maths lecturer I once looked at a student's lecture notes, and saw reference to a '∂∂∂ function'. Intrigued by this I asked him what it meant. He had no idea but claimed that I ought to know as I had written it on the board. It turned out to be my badly written 'odd function'. The implications of this were horrifying. Not only did the transmission of my knowledge fail, it was also clear that he did not even expect it to succeed, and moreover, knowing it had failed did nothing to remedy the fact, and moreover accepted his fate. It was probably around this time that I began to question the whole idea of the transmission model of education, although my immediate solution was the one that many lecturers adopt routinely: distributing prepared notes. This combination of lecture and print has almost become the standard form of the 'lecture'. The point of the lecturer's presence, if not to deliver the ideas, must therefore be to use their oral presentation skills to enable the student to see the subject from their perspective, to see why they are enthusiastic about it. They must see what is elegant or pleasing, and see how it makes sense of the world. Good writing can put all that into print, however, so it remains difficult to see the point of having lectures, beyond providing a shared sense of community of scholarship, of like minds interested in the topic of discussion. At least the printed notes are accurate, and are more easily controlled by the student than the lecture.

For the individual learner, the lecture is a grossly inefficient way of engaging with academic knowledge. For the institution, it is very convenient, and so, despite the inconvenience to the students, who have to fit to its logistical demands, and despite its questionable pedagogical value, it survives.

Alternatives to the predominance of the lecture method at university level have been practised successfully for years in distance-learning universities such as the Open University. These have relied on a combination of media-based learning, occasional tutorials, and individualised support from tutors via mail, telephone, and now email. For the campus-based university the balance could be similar, but with the advantage of more opportunity for contact with the tutors and with other students.

In the remainder of this and the following chapters, we shall test the range of educational media against the Conversational Framework to see how far they can support the required activities for students to learn. Then we can deduce an appropriate balance of media and methods for a university not enfeebled by tradition.

PRINT

Print is easily the most important educational medium, in terms of proportion of teaching delivered that way, in both distance teaching and campus universities. It owes its predominance to logistical rather than pedagogical advantages. It satisfies only one of the pedagogical requirements of the Conversational Framework – that the teacher can describe their conception – but logistically, it shines. It is the easiest medium to design (single author), to produce (established publishing mechanisms), to deliver (bookshops and libraries), to handle (light and portable), to use (random access, contents, indexes). Logistics change with technological and cultural changes, however, so we have to be clear about the true extent of the pedagogical characteristics of print to be able to judge these against its changing comparative logistics.

Print is similar to the lecture in that it can support only the description of the teacher's conception, but has the key advantage that, like most educational media, it is controllable by the students. They can control the topic focus: they can re-read, skip, browse, go to another topic via the index or contents page, and in doing so control the pace of delivery of the material. For a cohort of students with diverse academic backgrounds control over the pace of study is essential.

Print still has the disadvantages of failing to be interactive, adaptive or reflective, and this has been a particular concern of academics in distance learning institutions such as the Open University. To counter these essential deficiencies of the printed format, a number of design features have been adopted:

- the statement of learning objectives as a way of clarifying the topic goal;
- wide margins to encourage students to make their own annotations on the text;
- the use of in-text questions and activities to encourage action, e.g. students are asked to write down their point of view on a topic before reading on to compare the author's point of view with their own; students are set analytical tasks, or calculation tasks, as appropriate to the material;
- the provision of supplementary texts to make the material adaptable for students who need to spend more time on some aspects of the work;
- the use of self-assessment questions (SAQs) to help students to reflect on what they know, and to check their performance against a given answer.

The combination of activities and SAQs enables print to be more discursive, by inviting the student to describe and even redescribe their conception in the light of further reading. It is not fully discursive, of course, because it is not possible for the teacher, as author, to redescribe their conception in response to the student's description. Some texts do this pre-emptively, by predicting possible misconceptions and addressing those, which is an excellent way to write a teaching text, and diminishes the constraints of the medium.

The print medium can be improved considerably over its standard form, therefore, and although it still fails to satisfy all the requirements for an ideal teaching strategy, the students are given some support for what they have to contribute themselves.

The structure of the discourse for both lectures and print remains essentially implicit. There have been attempts, following the investigations of the 'surface approach' to text, to help students take a 'deep approach' to apprehending the structure, in an attempt to negotiate a shared understanding of what the topic goal is. The 'in-text activities' referred to in the list above sometimes take this form. Evaluation studies of these design features have not been particularly encouraging (see, for example, Lockwood, 1992; Marton and Booth, 1997). They appear to suggest that the solution does not lie in a design fix alone, but depends also upon the student's appreciation of the idea of the 'deep approach' itself, their conception of learning, and their perception of the learning context. The addition of in-text activities and all the other add-on features discussed above do not themselves change the format of the medium. It is still print, and only print, and therefore open to the same distortions as the original simple text. The students have to imbue these activities with a different status from the activity of reading, have to acknowledge that it invites them to stand back from the text and reflect upon it, and then do that:

> Reading means approaching something that is just coming into being.
>
> (Calvino, 1979)

An active approach to reading transcends the passive and becomes a creative act by the reader. If reading is to be a productive learning activity, then it must be approached with that expectation. In his immensely scholarly history of reading, Manguel describes the change in teaching from the scholastic method, which is similar to the transmission model of teaching:

> … the teacher would copy the complicated rules of grammar onto the blackboard – usually without explaining them, since, according to scholastic pedagogy, understanding was not a requisite of knowledge … Following the scholastic method, students were taught to read through orthodox commentaries that were the equivalent of our potted lecture notes … The merit of such a reading lay not in discovering a private significance in the text but in being able to recite and compare the interpretations of acknowledged authorities.
>
> (Manguel, 1997: 76–77)

to what we would now call a constructivist model, inspired by the humanist philosophers:

> in the mid-fifteenth century, reading, at least in a humanist school, was gradually becoming the responsibility of each individual reader. Previous authorities … had established official hierarchies and ascribed intentions to the different works. Now the readers were asked to read for themselves, and sometimes to determine value and meaning on their own in light of those authorities … the scholastic methods were questioned and then gradually changed.
>
> (Ibid. 82)

Manguel attributes the change in part to the wider availability of books soon after the invention of the printing press. They were no longer rarities entrusted only to the teacher as guide, but objects in the hands of the students, to be interrogated for their personal perspective. The book is no longer the medium of solemn silence with 'one unvarying meaning', but a text that speaks to many readers in different ways, depending on what each reader brings into being as they approach it.

However, there is nothing in the format of the print medium that requires students to take this active approach. And many of them choose not to. Ference Marton's studies of students reading demonstrated this through the discovery of the surface approach (see Chapter 3, p.43), and he further confirmed the technification of reading when the text included instructions to take a more active approach (Marton and Booth, 1997: 169). Only a small proportion of students actually write something down when asked to do so in an activity (Lockwood, 1992). However, all of them produce an essay for assessment, which is another way of getting them to reflect upon the text. The essay is more successful in terms of the proportion who do it because it represents a structural change to the format of the medium. Linking the reading to a marked essay allows the print medium to establish more links in the chain of 'display teacher's conception (text)' – 'set task (essay question)' – 'action (write essay)' – 'feedback (marks and comment)'. The chain has to endure over a long time-span, and the dialogue between teacher and student that this represents is similarly attenuated. It cannot match the cut-and-thrust of the face-to-face tutorial, where the student can interrogate their teaching resource and expect a response. Nonetheless, the design enhancements listed above, which address the requirements of other activities in the Conversational Framework, can render a book closer to being a 'tutorial in print' (Rowntree, 1992), as Figure 5.1 shows. When these are combined with the very superior logistical characteristics of a printed text, print inevitably takes its place as the predominant medium for learning.

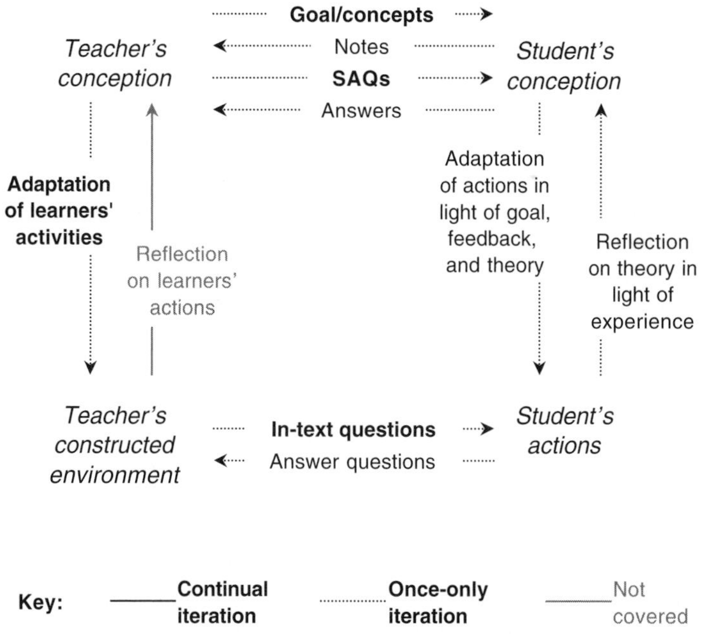

Figure 5.1 Interpretation of the Conversational Framework for print material. *Students may choose not to carry out any of their activities, but they are encouraged to do so through the provision of in-text questions and self-assessed questions (SAQs).*

AUDIOVISION

The audiocassette as a learning medium is underrated by the textbooks on educational media. Its principal contrast with the lecture is that it is more controllable, though less so than print, being difficult to browse or index. Its principal contrast with print is that it uses the auditory channel rather than the visual, which means it has the tremendous potential for students who cannot easily read, that it makes the world of print available to them. The lecture loses little, pedagogically, by being transferred to audiocassette, and gains in giving greater control to the student. Moreover, the audiocassette can offer at least a vicarious experience of discussion, such as a recorded tutorial, or an academic debate. The disadvantage of audio for sighted students – that it provides nothing for the visual channel to focus on – is what makes it logistically advantageous. It is the ideal medium for the lifelong learner, whose study can be done in parallel with other necessary activities such as travel, gardening, shopping, and ironing. When this audio-only activity releases the student's focus of attention to the auditory channel, this is a highly efficient medium in terms of material covered. In terms of material learned, it is less efficient. Unless the material is unchallenging, it requires a considerable feat of memory to sustain an understanding of the full meaning until it can be reflected upon and tied into other activities at a later stage.

For this reason, 'audiovision' is a more acceptable medium, as well as offering more scope pedagogically. The hybrid 'audiovision', uses the auditory channel in combination with something for the visual channel to focus on, usually print. Thus, it creates an additional representation in print of the descriptions being given in sound (Durbridge, 1984a). Since print is not just text, but also pictures and diagrams, the print can provide an iconic or graphic version of the verbal description. The 'vision' part does not have to be print, it may also be material. One example from a geology course is a piece of rock, where the audiocassette talks the student through an examination of its look and feel. Another example from a technology course is a computer program, where the audio talks the student through their actions on the computer and provides an interpretation of the screen at each stage.

Audiovision is not usually adaptive. If the audio is being used to set tasks that extend and enhance students' experience of the world, then the medium achieves a degree of adaptivity. I define 'adaptive' as a medium in which something changes in the state of the 'system' as a consequence of the student's action. Clearly, if they are operating a piece of equipment, or cutting a piece of rock, then they are changing the state of the world and seeing the consequences. Moreover, since the audio commentary is designed to interpret these (presumably) known consequences, the student is receiving tuition at the levels of both experience and description of experience, making the medium a surprisingly powerful one. In general, however, the medium does not link to real-world actions, but to actions on descriptions, e.g. text, diagrams, pictures.

Print and audio, in their standard forms cover only a fraction of the Conversational Framework. They cannot be discursive, in the sense of being able to comment on the student's representation of the topic, and even in combination, they cannot easily incorporate adaptivity or reflection by the teacher. These have to come from the student.

TELEVISION

Broadcast television has been a solution to special educational conditions, such as widely distributed campuses in Australia, Canada, the Philippines, or widely distributed students in distance-learning universities. With more widespread introduction of cable television and satellite broadcasting as the communications infrastructure develops, there has been an increase in this form of delivery of the lecture. It extends also to training and continuing education as companies with widely distributed organisational networks find it worthwhile to use the medium. Like the lecture, it is neither discursive, interactive, adaptive nor reflective, and is not self-paced. Its principal contrast with lectures is the form of representation it can use: dynamic images as well as language.

Television has the frequently underestimated power to assist in the difficult trick of conveying a particular viewpoint or idea. Academic knowledge consists in descriptions of the world, and these descriptions represent a particular way of experiencing the world (see Chapter 1). Much of the work a lecturer has to do involves finding ways of conveying the peculiar characteristic viewpoint of their subject. Television (and film, which I take to be equivalent for this discussion) is peculiarly able to convey a way of experiencing the world. It provides a vicarious experience through dynamic sound and vision, and uses a number of technical devices to manipulate that experience. Salomon has called these devices 'supplantation', in the sense that they supplant a cognitive process (Salomon, 1979). For example, a 'zoom' from long shot to close-up supplants the process of selective attention; a 'pan' supplants the process of shifting attention; a 'montage' supplants the process of association of ideas. These are powerful rhetorical devices. Add to these the production decisions about what to film, where to point the camera, or how to edit a sequence of images, and the potential for establishing a point of view is clear.

For the academic who wants to convey a complex theoretical idea, television can offer a way of supplanting the process the student must follow in order to understand the meaning. I would have great difficulty in trying to describe a Riemann surface to non-mathematicians, but if you were to see the sequence where trick photography is used to make a man seem to get smaller as he walks along a radius crossing concentric circles which gradually get closer together, then you would know it in a way you could not from words alone. The sociologist trying to get students to take an objective look at the world, and see vandalism not just as something perpetrated by youths, but as an aspect of the way we all live, uses a series of shots of industrial waste, ugly hoardings hiding a beautiful tree, a house covered in stone cladding, the destruction of a cottage to make way for a by-pass. These are all representations of what the academic sociologist means by describing vandalism as an aspect of the structure of society rather than the product of agents. These sequences extend and enhance the way the students experience the world, and good educational television frequently achieves that. By bringing the world to the student's study it becomes possible for them to experience vicariously a variety of actions on the world: fieldwork (climbing a volcano and inspecting samples), experimentation (add another chemical and watch the reaction), interpretation (compare one part of a painting with another). However, these define purely logistical, delivery roles for television, whereas given enough resources the students would engage in these experiences directly. The 'supplantation' devices are convergent with the way we see in that medium. They develop over many years as the cinematic medium shapes and is shaped by our cultural responses. 'Supplantation' allows our perception of the world through television to imitate our perception of the real world. As television offers a 'vicarious perception' of the world, it acts as a solution to the logistical problem of enabling large numbers of students to experience that aspect of the world directly.

The more interesting role for television, as a unique pedagogical medium, exploits its rhetorical power. Television as a public information medium necessarily has its rhetorical power constrained, in the interests of appearing to be balanced and objective. In educational broadcasting, given my position that academic knowledge is essentially rhetoric anyway, the medium can legitimately fulfil its potential.

There are not many studies of the rhetorical aspect of educational television. From all that has gone before, it follows that it should not be seen as primarily a means of transmitting information. It is a poor informational medium anyway, because it is not controllable, so the viewer is too easily swamped with information; alternatively, the information is meted out in digestible quantities, which then makes it inefficient. It hardly matters if students fail to remember some constituent item within a sequence or programme. If the medium is being used as it should be, to persuade the viewer of a line of argument, or a way of seeing the world, then the important question is whether they got the point being made.

In a study of students learning from social science programmes, for example, I found that often they did not. The internal structure of the programme was elaborate and yet obscured from the students, so they found it difficult to discern the overall meaning conveyed through that structure (Laurillard, 1991). Their summaries of the programmes focused on local meanings of particular sequences, especially those represented most evocatively through vicarious experience, instead of talking heads.

The study built on Marton and Wenestam's work (see Chapter 3, p.45) on students' understanding of texts with a principle–example structure, and applied a similar approach to the medium of educational television. Like print, the linear format of the medium nonetheless contains an internal narrative structure that is hierarchical in form. The structural levels within five Open University television programmes were described in a similar way to that used for text in Marton and Booth's (1997) analysis. The overall theme was referred to as 'the main point', which was made up from 'component points', each of which were illustrated with 'examples of components'. Students' understanding of the programmes was judged from their summaries. There was considerable variation across the programmes in terms of the level at which students pitched their summaries:

> ... some students fail to perceive the underlying message structure of a television programme, in the sense that they perceive the main point correctly when it is made, but fail to accord it the appropriate status. It is possible that this may be because they assume a linear rather than a hierarchical structure, make no effort to discern the structure that is present, and tend to ignore the cues that point to it.
>
> (Laurillard, 1991: 15)

One programme, on political theory, generated summaries at the main point level from only 46 per cent of the students; another, on social integration, generated summaries at the main point level from 80 per cent of the students. Why was the latter so much more successful?

The programmes were analysed for their respective design features to see if there was some relation between programme design and perception of its underlying structure. All the programmes exhibited the same principle–example structure as the text examples in the earlier study, but a key discriminating feature was the amount of time a programme devoted explicitly to the main point, component points, and the examples. Figure 5.2 shows the analysis for the two programmes mentioned above.

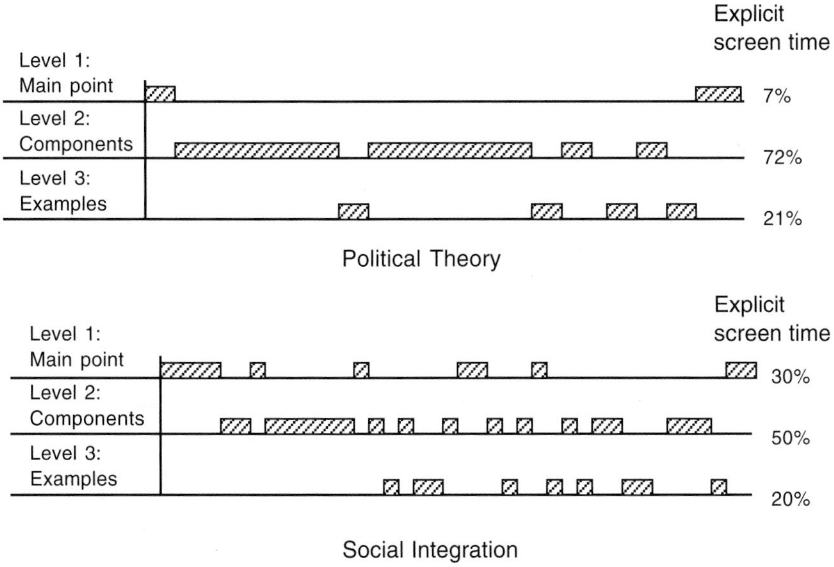

Figure 5.2 Structural analysis of the content of two television programmes. *Source:* based on Figure 2 in Laurillard, 1991

Could there be a relation between comprehension of the main point and the amount of screen time devoted to it? The analysis in Figure 5.2 shows that the programme on Social Integration had a relatively high proportion of screen time on the main point, repeatedly clarifying the relation between it and the evidence cited at the example level. A feature of this kind would assist those students for whom the internal structure might otherwise be 'horizontalised'. The politics programme made the main point that there are alternative political theories to explain how society changes. The total time spent on this was very small in comparison to the time spent on the two contrasted theories, of Marxism and Pluralism and their respective examples. The 54 per cent of student summaries not focusing on the main point instead described these two theories, rather than the nature of the contrast between them, which the programme was attempting to illuminate.

The implication of such findings is not that all educational programmes should devote a third of their screen time to explicit reference to the main point, but that they should offer ways of helping students discern the main point. The key data in these studies is not the quantitative analysis of the constituent activities. It is the variation in ways of experiencing the programmes that illuminates how students use the cues provided for discerning the different structural levels, whether in a text or in a programme.

The programmes for which the students' summaries were very similar to the producer's message were identified as programmes that had an 'image–argument synergy' for the overall message (Ibid. 19). The term is meant to express the closeness of correspondence between the academic's description of the world, and what the viewer experiences through the on-screen images. Television can provide an analogue representation of an idea normally expressed through language, e.g. 'states can be violent' can also be expressed as images of war, riot police, capital punishment. When it does that, either within a sequence or at programme level, the 'supplantation' achieved is of a different kind from the 'vicarious perception' I described earlier. Here it supports the students' cognitive efforts to discern the meaning embedded in the implicit structure of the discourse itself. 'Image–argument synergy' ties the experience (the image) to the description (the argument), synthesising both levels of the academic discourse, and giving the students a 'vicarious conception', i.e. offering an insight into the way that the teacher thinks about the topic.

When this more elaborate kind of 'supplantation' succeeds, the medium scarcely needs the other rhetorical props of interactivity and adaptivity to bring the teacher–student dialogue to a consensus. Of course, the same can be true of a lecture. An idea may not be so difficult that students need such props; alternatively, the inspired lecturer finds a way to convey the idea well enough through language alone. However, if the idea is too complex, or unfamiliar, then its alternative representation as some televisual analogue may help, where supplantation via image–argument synergy attempts to replace the entire rhetorical cycle necessary for learning to take place.

Television is engaging and powerful, and those advantages can be exploited effectively to assist student learning, but it is not a reflective medium, partly because it is not controllable by the student. Reflection has to come later, in a tutorial discussion, or prompted by printed notes with SAQs for the individual learner. In its standard form, however, television covers at most three of the required activities within the Conversational Framework: the teacher's description, the teacher's set task, and intrinsic feedback on the teacher's actions.

VIDEO

The principal contrast between broadcast television and video is the relative controllability of the latter, making it adaptive by the student. Some researchers have referred to videocassette plus exercises as 'interactive', but I believe this overstates the case. The term 'interactive' has already been emasculated in its application to media that offer open access to resources, such as the Web, or video-on-demand. At least these are open presentational media, which do allow the user to be responsive to what they find, even if the medium is not responsive to their actions. A video, on the other hand, is essentially a linear presentational medium. Nothing in the video changes when a student rewinds it, just as nothing in a book changes when you turn a page. The epithet 'interactive' is applied to video because a cassette allows students to carry out activities in between watching sections, and to carry out analytical exercises on the video material itself. These are excellent ways of using a videocassette, and of exploiting its controllability, but they are not interactive in the strong sense. They are essentially the same kind of activities as reading a book, re-reading it, analysing passages, doing activities between reading, etc. The medium is unvarying and cannot adapt itself to meet a variety of student needs. It is 'active video' perhaps, but not 'interactive video'.

Video has the same ability as television, however, to bring together experience and description of that experience and, being self-paced, can enhance this further with the opportunity for students to reflect on what they are doing. Nicola Durbridge, in an evaluation of video use at the Open University, observed this in the way a set of videos of children doing mathematics were used in a course for teachers:

> Thus, the video can be described as having two aspects to its full meaning. One is the *sense* of the problems of doing mathematics, the other involves a *critical appreciation* of these problems. Students need to respond in two ways to understand the whole; they need to be receptive to the stimulus of the 'real-life' sound plus vision and to show a sympathetic but instinctive understanding of it; they also need to pursue a rational enquiry into its fuller meaning along the lines prompted by the notes and voice-over elements.
>
> (Durbridge, 1984b: 234, original italics)

In fact, she found that the voice-over technique was less successful than the notes, because students were 'more engrossed with the action' on screen, and felt the simultaneous instructions to focus on particular content were distracting and off-putting. This accords with the point above about the technification of a text that includes instructions on how to read it. For video, the synergy between image and argument may only work when the image is given time to be 'sensed', or the event experienced, and there is separate time for the argument to be 'critically appreciated', or the concept described.

Although this form of 'active' video gives students set task goals at the end of short sections of video, there is no feedback on their actions. Durbridge highlights students' sense of frustration with this aspect of work on videocassettes:

> There is also clear evidence that if questions and directives are highlighted … they will need to be supported by some indication of the answers or observations students might make. Without such support many students felt both frustrated and anxious about the quality of their learning.
>
> (Ibid. 240)

This is the disadvantage of a medium that is neither fully discursive (giving extrinsic feedback) nor fully interactive (giving intrinsic feedback). However, Durbridge does suggest ways in which 'pre-emptive' extrinsic feedback can be offered, e.g. where the academic's version of the answer, or their comment on an expected wrong answer, is written at the end of the notes. It may be summarised at the beginning of the next video section, in much the same way as print may comment of what a student is presumed to have done in an activity. Students need to know what they are meant to be learning, and need to have a sense of when they have achieved what is expected. The non-interactive media must attend to this aspect of the learning process, even though they cannot support it fully.

The main advantage of video over television is in the self-pacing provided by greater learner control, which at least allows students to reflect on the interaction they have witnessed. Their reflection is then available to the activity of modifying their description, should they be invited to do this by additional instructions or notes. Other than that, video retains all the pedagogical advantages of broadcast television as a medium, and loses only that shadowy sense of belonging to a synchronised scholarly community.

DIGITAL VERSATILE DISC (DVD)

DVD can be seen as having exactly similar pedagogical properties to a videocassette, except that it offers easier access to the video material, as did the now obsolete interactive videodiscs. It can embrace all the pedagogical qualities of television when used as a narrative medium, as well as an even greater degree of learner control over sequence and pace. It has one interesting and critical property that distinguishes it from both television and video, however: it can be delivered through a PC. In this mode, it inherits expectations of interactivity. In the next chapter, we will see the extent to which this could affect the learning experience associated with DVD. In the meantime, as a delivery system for narrative television it embodies far greater user control than all the others, so its logistical value is likely to make it a more popular medium for learning than either broadcast or video.

SUMMARY

Table 5.1 summarises the characteristics of the media discussed in this chapter. I have included SAQs because they offer a way of enhancing any of the other media, providing no less than four of the required activities. The table enables us to see how combinations of media can cover the Conversational Framework more fully than the standard forms.

The table can be read as a way of deciding how to cover the range of activities required by the Conversational Framework, but it does not decide between the media. It does not say 'choose television rather than print because it gets more

Table 5.1 Summary of narrative media characteristics

		Print	*AV*	*TV*	*Video*	*P/SAQs*
1	T can describe conception	✔	✔	✔	✔	✔
2	S can describe conception	O	✔	O	✔	✔
3	T can redescribe in light of S's conception or action	O	O	O	O	O
4	S can redescribe in light of T's redescription or S's action	O	✔	O	✔	✔
5	T can adapt task goal in light of S's description or action	O	O	O	O	O
6	T can set task goal	O	✔	✔	✔	✔
7	S can act to achieve task goal	O	✔	O	✔	✔
8	T can set up world to give intrinsic feedback on actions	O	O	O	O	O
9	S can modify action in light of feedback on action	O	O	O	O	O
10	S can adapt actions in light of T's description or S's redescription	O	O	O	O	O
11	S can reflect on interaction to modify redescription	O	O	O	O	O
12	T can reflect on S's action to modify redescription	O	O	O	O	O

ticks'. The decision on media choice is more complex than that, involving both the obvious presentational properties of the medium (e.g. that television presents dynamic visuals better) and the logistics of development and distribution, to be discussed in Chapter 11. The table should rather be read as a way of indicating which activities are unsupported by a particular medium. It clarifies the nature of the responsibility such media place on students, requiring that they sustain a tenuous link across these and other learning sessions in order to complete the learning process. Once this is clear, the teacher can decide on how best to deal with it – by adding another medium, by offering tutorial support, or by assuming that students can provide the additional activities for themselves. Analysing the audiovisual media in terms of the Conversational Framework allows the academic to design their teaching with a more realistic expectation of success.

Chapter 6

Interactive media

INTRODUCTION

Interactive media are the presentational media that include hypertext, hypermedia, multimedia resources, Web-based resources and Internet-delivered television. They share the core common property that they are essentially linear media delivered in an open, user-controlled environment, either by disc or over a network. Being essentially linear, they offer a given text, in its widest sense, that remains unchanged by the user. The environment in which they are delivered, offering open access to any part of the material, in any sequence, lends them a degree of user-responsiveness that has earned these media the epithet 'interactive'. The term was formerly applied to media which supported reciprocal action, implying an equality between the participant and the medium which these media cannot aspire to. However, the word has now become a term of art, and its meaning has moved on. 'Interactive' now refers to a medium in which the user can navigate and select content at will. The content may be text, graphics, audio, video, or any combination.

The important features of interactive media, from a pedagogical point of view, are the scope of the access and the nature of the user control. It may seem illogical to group together media that are delivered in such different ways – via discs or networks – and most analyses of educational media separate them. However the motivation here is to begin with the pedagogical analysis, for which the mode of delivery is irrelevant by comparison with the mode of engagement with the content. In any case, as delivery systems converge – text is delivered on television screens via WebTV, while television is delivered over the Internet – they make an unreliable basis for any categorisation. We return to logistical issues in later chapters. For now we consider only the pedagogical characteristics of hypermedia, Web resources, and interactive television.

There is little in the literature to help with the definition of interactive media. In the early days of interactive video there were attempts to define 'levels of interactivity' to distinguish between actions such as the selection of media sections, and the selection of answers to multiple choice questions. It is clear that interactive media offer different types of learner control: over the sequence of content, over the type of learning activity, and over input to content questions. However, it takes no more than a moment's thought to establish that these are likely to be basic features, and they do not advance our understanding of how best to use interactive media. Barker suggests using a 'basic principle of interactivity' based on interaction in dynamic systems theory: the mechanism of interaction between two dynamic processes (e.g. a student and a computer program) works through successive messages sent between them. Each receiver undergoes a change of state on receipt of the message, and generates a new message. Each thereby learns more about the other. At this level of generality, the principle is applicable to either student–tutor interactions or student–program interactions:

> This is important because knowledge, ideas and experience obtained with one type of system can often be beneficially 'carried across' to the other.
>
> (Barker, 1994: 6)

But without some analysis of the nature of the messages, or the nature of the state-changes, the principle is so general that it could even be applied to program–program interactions. It is not clear how we could apply such a principle meaningfully to the design of interactive media for learning. Barker describes a case study that apparently applied the principle to the creation of multimedia courseware, but does not explain how it informed the design. If they are to inform courseware design, we must establish principles of interactivity that give a detailed analysis of the nature of the medium, and students' experience of the interaction. That is what this chapter tries to do.

HYPERMEDIA

Hypertext is the original form of hypermedia, and is probably best defined through an understanding of its historical origins. John Naughton, in his fascinating account of the origins of the Internet, credits Vannevar Bush with its invention in a magazine article in 1945 titled 'As we may think' (Naughton, 1999: 212). The title is an important clue to the nature of hypertext systems. Bush wanted to create the means for an information retrieval system to mirror the associative retrieval characteristic of human memory. The aim was to go beyond the static forms of paper-based index and retrieval systems to a more dynamic form, 'whereby any item may be caused at will to select immediately and automatically to another'. His idea of a 'trail' is remarkably close to what we now understand by hyperlinking:

> When the user is building a trail, he names it, inserts the name in his code-book, and taps it out on the keyboard. Before him are two items to be joined … The user taps a key and the items are permanently joined … Thereafter, at any time, when one of these items is in view, the other can be instantly recalled.
>
> (Bush, quoted in Naughton, 1999: 214)

The hypertext tool that Bush invented was designed to act as an aid to thinking that would work better because it more closely matched the associative linking we naturally use in managing large amounts of information. There is a nice irony in the fact that as computers became more fully developed and more widely understood, psychologists began to describe human memory in terms of the organisation of computers. The information processing theory of cognition was the result. It was the Bush project in reverse: nature was apparently imitating technology.

There is an important feature of Bush's idea and the way he expresses it, which has implications for the pedagogical power of hypertext: for him it is a tool for thinking. He talks in terms of what the user will do in building their system. It is the means by which the thinker organises the information available in a way that makes it easier for personal retrieval. Comparing this account with the kinds of hypermedia resources we are familiar with now, we notice the vital missing ingredient: we do not typically create the links. We follow the links created for us. There could be a system allowing us to do the creating, and there once was. In the late 1980s a Macintosh system called HyperCard allowed the user to create their own associations, and build their own information environment with no knowledge of programming necessary. It was meant to open up the world of personal computing to the non-programmers:

> The sad truth is that it didn't – and for one very simple reason. It was rooted in the notion that the computer was a standalone device, complete in itself … it assumed that all the connections worth making were on your hard disk.
>
> (Naughton, Ibid. 227)

Naughton's point is that it was overtaken by the Web. However, the Web merely extends the connections worth making, as we shall see in the next section. Bill Atkinson's HyperCard gave us creativity, the ability to create the links ourselves, not merely follow the links created for us. The two should not be in competition. Only fashion, and timing, made them so. This crucial difference between the two systems, following links and creating links, puts the Web in this chapter, and HyperCard in Chapter 9 (Productive media, p.161).

We will now discuss how the properties of hypermedia, the open access to navigable links between text, graphics, and multimedia, relate to the requirements of the Conversational Framework.

Discursive iteration

The discursive iteration between lecturer and student cannot be a continual loop because the system cannot respond to the student's questions with other than the same pre-scripted reply to a particular question. Like print, a hypermedia system cannot be interrogated. It can offer alternative perspectives on the same question, but there is no 're-articulation' in the light of the student's performance or puzzlement. Its strength at the discursive level is that it offers open access to a range of statements of the lecturer's conception, and uses a range of media. Its presentational qualities can be impressive. Its multiple hyperlinking offers freedom of navigation. However, the user exerts control in this medium, which has the effect of reducing the amount of time they are likely spend on the node at the end of each link. Unlike print or television, or even video, where there is an implicit surrender of pace to the control of the author, a user-controlled medium creates the expectation that the user will not have to submit to author control for long. If a hyperlinked video clip lasts longer than thirty seconds there is a sense of the user having ceded control, and they revert to being the viewer, rather than active participant (Laurillard, 1984a). Ten to twenty seconds is more comfortable. User control is fundamental to the 'sit-forward' interactive media, and the user expects to be doing something every few seconds, in contrast with the 'sit-back' narrative media of print and television. This fact seriously limits the presentational capability of hypermedia, therefore. It would not be appropriate to use it for a complex account or explanation where the author needs to hold the learner's attention over a period of many minutes. That is a narrative, and is rightly confined to the narrative media. The presentational qualities of hypermedia are better suited to the focused, goal-oriented gathering of information and ideas by the student who has their own narrative in mind. Nevertheless, there is no obvious opportunity within a hypermedia environment for the students to articulate their own analysis of the material. Without this, the discursive level cannot generate a student response, and therefore cannot iterate.

Interactive iteration

The interactive iteration offered by hypermedia is limited in the sense that the tasks set cannot be developed in response to the student's performance at the discursive level, as they would be in a class or tutorial. However, once the task is set, the medium does allow continual iteration of the student's action, a response to that action, and then a further response by the student. For a completely open resource environment, such as an encyclopedia CD, the adaptive–reflective iteration for the lecturer is non-existent. The adaptive–reflective iteration for the student depends for its success on the degree to which the student has a particular goal in mind, as this drives their adaptation of their actions, and their reflection on the interactive experience. Without a clear personal goal, students will tend to iterate through the resources without either reflection or adaptation. Interactive hypermedia do not necessarily offer a productive learning environment.

There is research evidence for this from the MENO project (Multimedia, Education, and Narrative Organisation).[1] A classroom-based study of learners using history resources on an encyclopedia CD showed that the lack of pedagogical support left them learning very little. Students were observed in groups of three, using the CD to investigate a topic set by the teacher. We analysed their dialogue by looking for critical sections of discourse and interaction between the learners, and between the learners and the material. We were looking for evidence of productive learning activity, but usually found evidence of clearly unproductive activity: lack of engagement by the students, and a consistent focus on the operational aspects of the task in hand, rather than its content or meaning. The following is an illustration of this kind of data. The example comes from a project set by the teacher, for a group of three 14-year-old students using a history disc on the Second World War to investigate the topic of nuclear bombing. There is a varied set of resource materials on the disc, all well indexed, including documents, speeches from war leaders, newsreel documentaries of the time, video, and audio, all fully controllable by the user.

The dialogue excerpt below begins at a point about ten minutes into the session when the students are using the index and media controls to find their way to the resource material they need (numbers indicate different students speaking).

Student 1: There's no film there, is there?
Student 2: No.
That one has no film there, either.
Student 3: It's the last one.
Student 2: Is there any text to go with it? (Reading from text "Cities in Hiroshima").
One more.

This is an example of dialogue that we interpreted as entirely operational. The focus of their attention is all on the navigational aspects of the interface. A similar kind of dialogue is found in another group, at about the same point in the session:

Student 1: What else shall we do?
Student 2: Go back to the index … See if we can..
Student 3: Go down, keep going … 2: Hiroshima.
Just type it in at the top. Fill in the box. You only have to put the first couple of letters in for it anyway.
Student 2: Alright.
Student 3: You see, 'Dropping the bomb', and 'Using the bomb'.
This is what we did before. OK. Go back because that's the one we did before.

Focusing on the task form is appropriate for some parts of the exercise, but it seemed there was little else in the data. The same approach is evident even when they find some relevant material. This extract continues with them watching a newsreel clip of about one minute, showing the aftermath of the Hiroshima bombing. The voiceover is emphasising the devastation and the human tragedy of the event. As might be expected, it is very shocking, emotionally charged footage. Despite the nature and relevance of the material, however, the students remain focused entirely on the process, on the operational aspects of the task in hand.

Student 2: So …
Student 1: Did you stop that?
Student 2: No it just finished.
Student 3: It didn't finish properly, actually.
Student 1: So what do we do now?

The material they found was highly relevant to their overall goal, to investigate nuclear bombing, it was highly engaging material, and yet it appears to have afforded no productive response of any kind. There is no sense of a storyline to their investigation, no goal in sight, no progression towards it, no sense of achievement. There were many such episodes in our observation of learners using these kinds of discs (Plowman, 1996).

In our analysis of student dialogue we were looking for evidence of each aspect of the Conversational Framework in operation, with students:

- interpreting the overall goal set by the teacher;
- deciding on the sub-goals of the material they need to collect;
- adapting their actions to finding this;
- reflecting on the material they find in relation to their sub-goal;
- articulating their conclusions in terms of the overall goal.

However, the design of the CD did not afford such activities. We concluded that within an educational experience provided by a non-linear narrative medium, such as hypermedia, we must take care to help learners maintain their own narrative line (Laurillard *et al.*, 2000). The design has to embody affordances for the activities required by the Conversational Framework, as learners do not generate these for themselves. Figure 6.1 shows the meagre coverage of the Conversational Framework that is achieved by interactive media design of this kind.

The minimal coverage of the Conversational Framework shows why this basic form of hypermedia is so unsuccessful, but it can also suggest the design enhancements necessary to give learners the support they need.

ENHANCED HYPERMEDIA

To what extent can we enhance hypermedia to support learners in their interactive exploration of a set of resources? The list above outlines the activities learners must undertake, as they iterate through the cycles in the Conversational Framework in order to maintain their narrative line. Figure 6.2 unravels the iterative cycles as a timeline, and illustrates the different levels at which students operate as they work through the activities – describing the overall goal and articulating the concept, defining and evaluating the sub-goals to determine tasks, deciding actions to carry out the tasks, and interpreting the feedback they receive in order to adapt the next action.

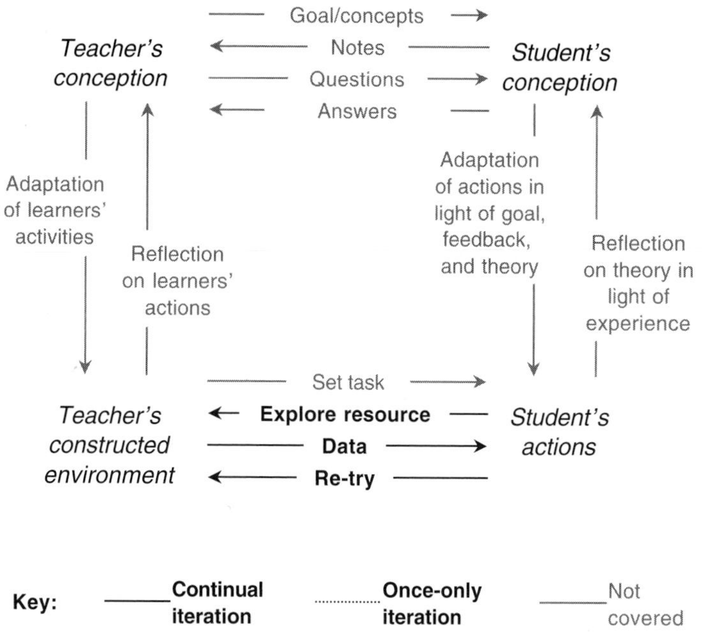

Figure 6.1 Interpretation of the Conversational Framework for hypermedia, e.g. interactive multimedia resources on CD or DVD.

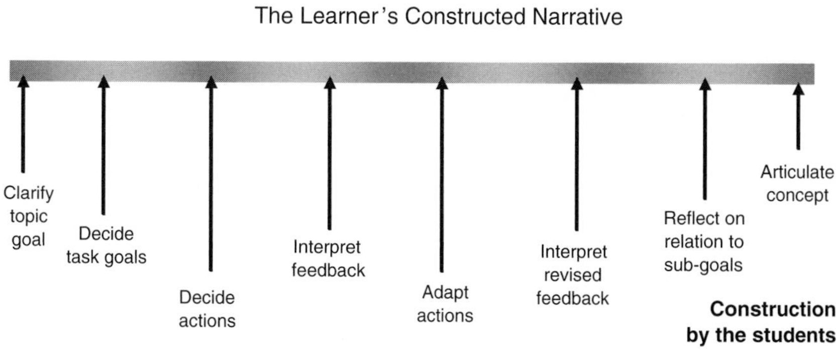

Figure 6.2 Learning activities needed to construct and maintain the learner's narrative line.

Each activity helps to drive forward and build the internal structure of their own narrative, relating to their own topic goal. This simplified version shows the kind of sequential iterations they need to go through to construct the different levels in the structure of their narrative: the overall conceptual goal, the sub-goals, the evidence for these, the refinement of the concepts and actions to obtain evidence, and so on, building to the final outcome. However, as the MENO project showed, learners find it hard to do this (see p.112). The design properties of a resource disc, offering just selections from an index, do nothing to encourage the activities needed. We can use the Conversational Framework to deduce the kinds of design features that will prompt the activities needed.

Tom Carey uses a similar approach to instructional design. He uses the Conversational Framework as a means of visualising the sequence of learner activities that must be supported (Carey *et al.*, 1999):

> Using such a diagram in the design toolkit requires designers to focus on high-level issues like the balance between expert and novice activities and between building and applying concepts.
>
> (Carey *et al.*, 1999: 21)

His design software represents a timeline for the student as a sequence of learning activities within the Conversational Framework. He defines the timing and activities precisely, as this is a design tool. Figure 6.3 is not a design tool, so does not focus on precise timing, nor on the detail of the activities. It does show how each of the productive learning activities might be elicited from the student through guidance features built into the program.

An example of such a design is an interactive CD on the poetry of Homer, based on material from an Open University course *Homer: Poetry and Society* (Open University, 1993). Students have to select a topic goal from a series of investigations, e.g. 'Compare the mortal characters in the Iliad and the Odyssey'. On selecting their investigation, the student is advised of what kinds of tasks to carry out to complete it, e.g. to search for three or four occurrences of a character in both texts (there would be several hundred references for each character altogether), and to use the Note Pad to describe them. The CD design provides an environment that is suitable for exploring these topics, and is adapted for the interactive level of explorative activities. Figure 6.4 (Plate 1) shows the screen layout of the environment. The student is reminded of the topic under discussion within the Note Pad. They begin work on the constituent tasks, searching for Nestor in this case, and making notes in the Note Pad on what they find.

Other material on the disc includes other war poetry, archaeological maps, video walk-throughs of the sites of Ancient Greece, and museum artefacts. The interactive iteration offered by the disc is therefore extensive, allowing the student to explore a wide range of material in their investigation of other topics such as 'Investigate the values of the society represented in the Iliad', 'Investigate the evidence for what kind of society existed at Mycenae'.

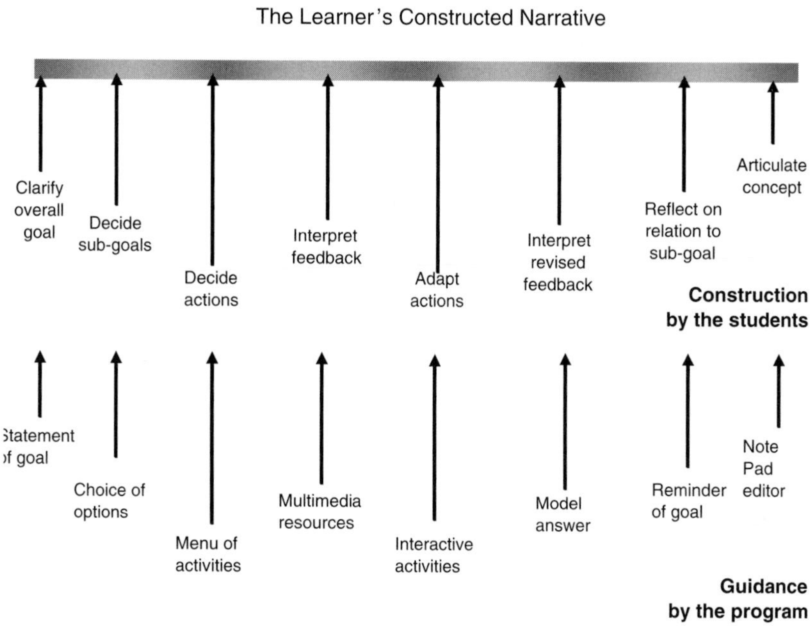

Figure 6.3 Design features needed to support the learning activities that will construct and maintain the learner's narrative line.

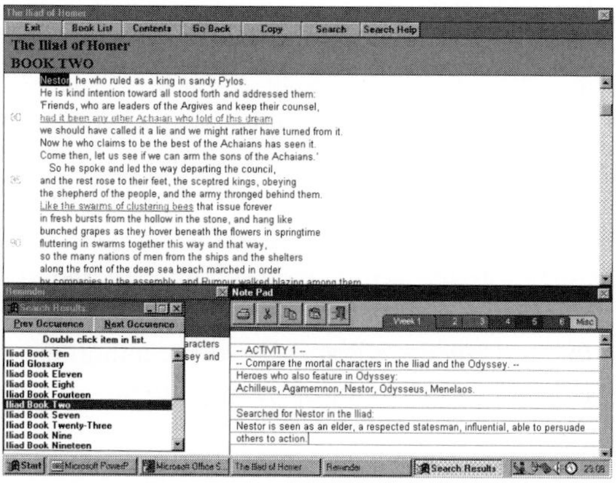

Figure 6.4 (Plate 1) An interactive program on the Homeric poems. *The search window shows occurrences of the item (Nestor) being searched. The text window displays the extract selected with the item highlighted. It also shows hyperlinks to further notes in the Companion Guide. The Note Pad shows the current activity and the student's notes.* © the Open University.

In this enhanced form of hypermedia the adaptive–reflective iteration for the lecturer is limited to the adaptive part only, i.e. the one-off design of the environment, and the tasks recommended within it, based on previous experience of teaching the course. For the 'Homer' CD students needed freedom to explore the material, but needed also guidance. The original version of the course included only print, audio and videocassette, which made investigative searches difficult and time-consuming. With all the material in electronic form exploration became easy. The design of the disc environment was based very closely on the printed text. This, of course, was written in narrative form, explaining the argument, adducing evidence, and setting the student the task of thinking through their own interpretation of the evidence before reading on to see the author's view. The design of the disc transferred this narrative line to the interactive environment, but in doing so ceded control to the learner. This was done by deconstructing the text into its essential structure – in this case, a series of hypotheses or interpretive ideas, the evidence for them, and the conclusions, building to reveal the synthesis of all these conclusions as the dénouement. The disc revealed the entire narrative structure in the list of investigations offered to the student, drawn from the succession of hypotheses. The lecturer also designed the resources and navigation devices to fit the evidence students would need to build their conclusions. The adaptive role of the lecturer was significant, albeit once off, in creating this environment for investigation and discovery. The reflective role of the lecturer would mean using the student's performance in this interactive environment to determine some more suitable redescription of the topic. This was non-existent, as it would require a highly sophisticated analysis of an open-ended activity. Figure 6.5 shows how each iteration can be interpreted for this example of hypermedia.

The adaptive–reflective iteration for the students is very well supported in this CD. Given the topic goal, and existing conceptions acquired from their previous reading of the texts (provided in print form for this very different kind of reading), students are able to adapt their actions within the interactive environment, and then reflect on how these findings relate to the overall topic goal. The structure of some activities requires students to articulate their findings in the Note Pad before being able to gain access to the author's commentary on the topic. In contrast with the print version, students are therefore motivated to generate their own ideas before comparing them with the lecturer's ideas. This is an important aspect of the design of the disc, which drives the reflective part of the cycle for the student. Comments from the evaluation studies testify to the importance of this delayed access to the expert's view, as well as the open access to the resources:

> I like it. I do have a tendency to be a bit superior and tend to skip the activities. Being forced to do it – I like that.
>
> I've done a lot more on the Companion, because it's easy to do – just click without interrupting the flow, but getting there is more difficult in print so I don't do it.

I did find it easier to click on the Commentary. It's far superior to books in that way.

(Laurillard, 1998)

It reinforces information better. In the Units if I reach a bit I'm not particularly interested in I'll skip through it. This encourages you to sit and complete the activity before you pass on.

You can switch from the literature to the archaeology and still continue on the same line of thought. It's the interactivity – it's exciting.

(Chambers and Rae, 1999)

The control given to the students appears to elicit more active processing of the ideas because the learner is driving the narrative line. The context of the interactive resource affords learner-construction of their own narrative line in contrast with the narrative media in which they accede to the author's narrative. These students are aware that the interactive medium gives them instant access to the next link in their own narrative line, whereas the print medium imposes the sequence of the author's narrative. We must expect very different types of learning outcome from these two highly contrasted activities – both are valuable and necessary, but interactive media elicit distinctly different responses from the students, and are therefore likely to yield different learning outcomes.

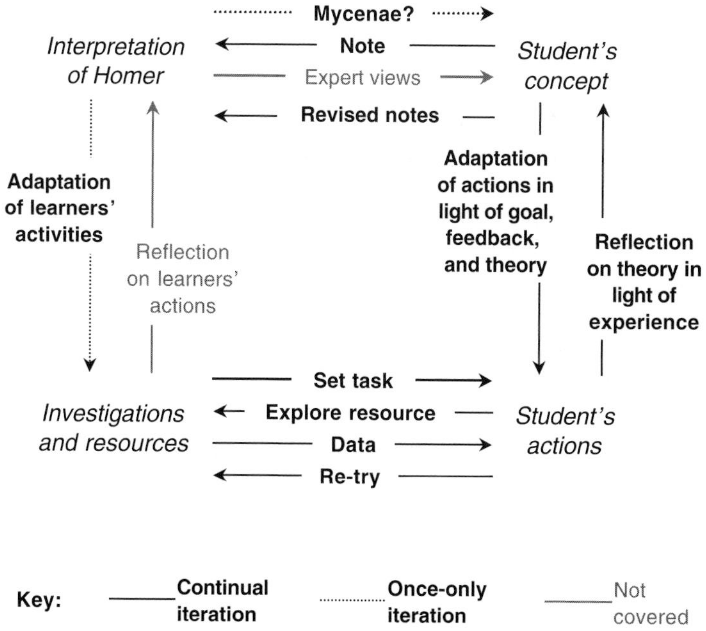

Figure 6.5 Interpretation of the Conversational Framework for the 'Homer' CD.

The responsiveness of the interactive medium is limited, however. Hypermedia environments, enhanced or otherwise, are not adaptive to the student's needs at either the discursive or the interactive level. It would not be possible for the student to tell if they had made an inappropriate interpretation of the resources, as the system remains neutral and unvarying with respect to anything they do. The feedback it offers at the interactive level is intrinsic in the sense that it offers a representation of the student's request for information about that world. In that sense they can test their idea about a character, say, against the references to him in the text. But there is no way of the student being able to test whether their interpretation is correct, except by comparing it with the various expert views then made available in the form of model answers.

Nonetheless, hypermedia can offer considerable coverage of the Conversational Framework in a stand-alone medium, if constructed with the kinds of features that are exemplified in the 'Homer' CD. Without these additional features, hypermedia environments offer nothing more than a series of navigable resources.

The claims made for the educational potential of hypermedia should be examined with care. On the one hand a basic hypertext is nothing more than a small albeit beautifully connected library. On the other hand, by its very nature, it undermines the structure of the 'texts' it uses and reduces knowledge to fragments of information. Jonassen claims that "learners in college courses can browse through interconnected knowledge bases in lieu of textbooks" (Jonassen, 1991: 84). The phrase 'interconnected knowledge bases' lends a spurious status to this series of associative links between components of a structured text. It suggests that the student, by browsing through these links, can thereby acquire knowledge in a similar, but more user-controlled way, than by reading a book. This is unlikely. Academic knowledge is complex, highly structured, cast in terms of specialised forms of representation. Knowledge of rainfall is not adequately expressed as an associative network of fragments of information: even a simple statement such as 'as air rises it cools' cannot be expressed as an association between two component fragments. Knowledge of rainfall will be developed through using that relation in a variety of contexts, as the dialogue we looked at in Chapter 3, p.56, tried to do. An academic book will take the learner through the narrative structure that tries to mirror the Conversational Framework by including in the account both overarching theory and its interpretation in practical events. It calls on the reader's own experience to assist their understanding. The learner finding their own way through the associated components in the story may find the meaning distorted, because each component has more than one simple associative link to others. The value of the narrative is that it helps the reader to keep track of the multiple links that make up a complex idea. A hypertext cannot do that on behalf of a student navigating their own route through the components. Remember the origin of the hypertext: it was a system for enabling experts to move within material through links they had constructed. They used the system to construct their own narrative line, which they wished to preserve. A student browsing a hypertext is not in a position to construct their own links, because this is not a feature these systems now offer. In following someone else's they are in danger of being unable to build the complex structure they need from a series of paired components.

The problem is that most of the ideas we are concerned with in education are more complex than can be expressed by an associative network. When knowledge networks are constructed to represent some kind of academic knowledge the links defined between the nodes are many and various. Even within the information-processing paradigm, association alone cannot do it. One example quoted by Jonassen uses the relation 'is a component process of' to link 'needs assessment' to 'instructional systems development'. But the link is much more complex than that. The original text linking the two would undoubtedly take a strong line about the importance of 'needs assessment'. It would discuss where in the 'development' process it should come, and how it would relate to the other components. It would express the full complexity of all the important links between the two 'nodes'. In the hypertext system, these points may be made within the documents associated with each node, but then the true internal structure is not explicit. Perhaps it could be, but unpacking that complex structure into an explicit form generates an extremely complex network that would be difficult to navigate, and even more difficult to keep track of as you do so. The display of a network makes it explicit but does not make it known. The student still has to do a great deal of work to internalise its structure and interpret its meaning, just as they do with the implicit structure of text.

So what sense can we make of 'interconnected knowledge bases in lieu of textbooks'? Textbooks are already interconnected knowledge bases. The interconnections they use cannot be represented as simple links. Hypertext cannot replace textbooks. Shoehorning a textbook into hypertext format could easily distort the internal structure of its argument so that the discourse loses its meaning. The process must carefully deconstruct and rebuild the internal structure, as in the 'Homer' example, revealing the narrative line as a series of investigations, and supporting the student through the exploration and synthesis of what they discover. The strength of hypermedia is the range of material it makes available for exploration: several books, pictures, graphics, audio and video together on one disc. Scholars delight in the capabilities of these systems, because prohibitively time-consuming research tasks become feasible, and therefore accessible, and therefore accessible also to the undergraduate. Hypermedia can change the curriculum. If students can now, with a few key presses, call up every instance of Homer's references to 'Helen', or every paragraph in which 'Helen' occurs as well as 'beauty', then comprehensive textual analysis becomes possible. However, what must concern the lecturer, is not so much the information retrieved by the student, but the use of that information – the transformation wrought by the student to render it as knowledge. The number of references they use in their analysis is far less important than the quality of the analysis, if what you are teaching is how to do textual analysis. Once it indicated diligence, perhaps; now it indicates access to a powerful system.

Hypermedia offer students the chance to do what a scholar does, to do the extensive library work that enables them to explore a wide range of material in a comprehensive way. As one student in the 'Homer' study commented:

> I wish I'd had it before. The ability to search the text, there's no other way to pick up certain things. You can be like a scholar.
>
> (Laurillard, 1998)

With the appropriate support within the design of the material, yes. Hypermedia are fascinating and motivating for students able at last to act like researchers in their field. However, the responsibility is with the lecturer and the designer to build in the features students need if they are to avoid producing extensively documented rubbish.

WEB RESOURCES

Web resources refer to hypermedia resources made available via the Web, rather than a disc. The distinction is important because the connections made possible by the Web are unconstrained. A discussion of the Conversational Framework that extends the 'teacher's conceptual knowledge' to include 'all conceptual knowledge' reminds us of the dilemma this creates for the student:

> Digital depositories such as the WWW ... put the learner directly in touch with the source material, but without the teacher ... to recommend one book over another or show where to find it. Our learner ends up lost among the virtual stacks, easily distracted by trivia and irrelevancies on the way. So quality and relevance of resources remain, as ever, the issues.
>
> (Murison-Bowie, 1999: 148)

The Web is the medium that most obviously supports the contextual, transdisciplinary, and socially distributed form of knowledge that is emerging alongside disciplinary knowledge (Gibbons *et al.*, 1994). It supports the needs of the lifelong learner, who has learned how to learn and has the skills needed to explore and evaluate the multiply-connected network of knowledge in their own and related fields. For the student who is a novice in their field (which, incidentally, includes the lifelong learner who is exploring a field that is new to them), the scale and scope of this online library requires a kind of 'reading list'. The Web equivalent of the reading list is the 'gateway'. A gateway is a high-quality resource discovery service designed for a target community. One example is the ROUTES services supplied by the Open University Library, and based on the ROADS service, which provides the tools needed to do this (see Web References, p.260). ROUTES provides a database of internet resources designed to support students registered on a particular OU course. All the links are accompanied by searchable descriptions and keywords, which allow users to understand the scope of a website before connecting to it. The system enables a course team or academic to set up a subset of approved websites for students to use as background or research material that takes them beyond the essential requirements of their course. A stringent set of criteria are used to ensure that the database of resources is optimised for student use:

Quality: information content is approved by the Library staff and by members of relevant course teams; sites meet the University guidelines for web pages (e.g. no illegal or offensive material.)

Reliability: resources are maintained by established organisations or a level of durability can be ensured; links are checked regularly by Library staff to maintain accuracy.

Access: to all Internet users; or any information about rights, costs, exclusivity or special software requirements is part of the description of the resource, with further contextual help made available where necessary.

Rights: any permission needed to create a link to the resource is done through the ROUTES Manager.

Through systems such as this, university libraries are able to offer students both the freedoms of the scholar, and the guidance they need as learners. The service extends the reading list to a much wider range of resources, not normally sustainable within a single institution. At the same time, through the involvement of both academics and librarians, it ensures good quality and relevant resources. New subject-specific gateways are appearing daily on the Web, and these will support the undergraduate in their continuing role as lifelong learner throughout a range of interests.

With careful adherence to the criteria of good quality provision, learning-oriented gateway services open up campus and distance universities to extensive resource provision for their students. The design criteria need to extend beyond the quality of the resources, however, if students' learning experiences are to be optimised. Web resources support no more of the Conversational Framework than a library does, less if you consider the personal service offered by the librarian on duty. A study of chemistry students' use of the Web as a resource for problem-based learning found that the role of the teacher was crucial to the students' ability to use it effectively (Dobson and McCracken, 1997). By contrast with a disc, the Web provides an admittedly large, but often bizarrely connected library. Academics are naturally excited by a medium that enables students to explore widely and follow their own research pathways, but students are working under time pressures. Full-time campus students are fitting jobs around their studies, and part-time distance learners are fitting studies around their jobs. If they are to use precious learning time efficiently, then the academic must design their use of Web resources in the light of the whole Conversational Framework. The Web provides an environment that offers hyperlinked access to a range of resources, but gives no further support than that. Figure 6.6 shows its limited coverage of the Conversational Framework.

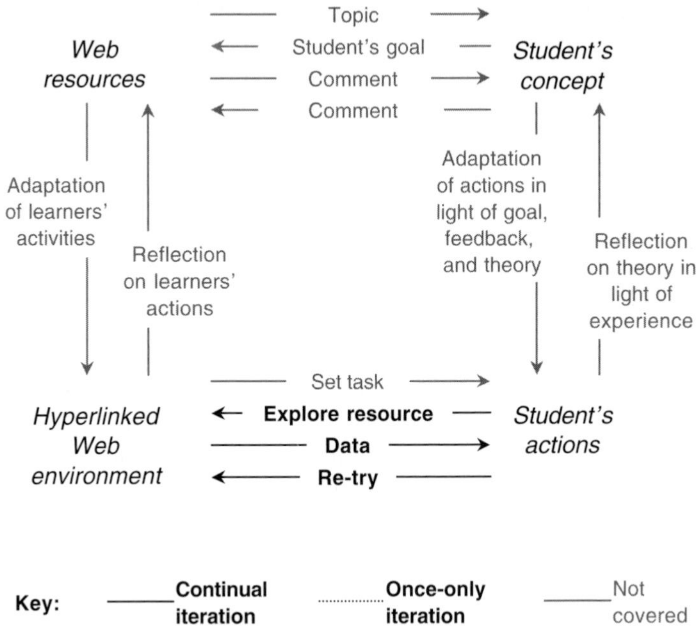

Figure 6.6 Interpretation of the Conversational Framework for web resources with no additional guidance.

However, the Web could support student learning as efficiently as hypermedia. The gateway would act like the search space on the disc. The academic's website could provide the additional design features discussed in the previous section: investigation topics, suggested task goals, a way of collecting and submitting notes on findings, and access to expert analyses of those topics, with reference to evidence from the linked sites. These additional support devices reduce the degree of uncertainty for students. They need to be protected from the tyranny of choice offered by the Web. They can easily escape the protection, should they wish to, but it is our responsibility to make the material learnable. The additional design features suggested by the Conversational Framework transform the Web into a supported learning environment.

INTERACTIVE TELEVISION

The final media type in this section is interactive television, the narrative medium of television made digital, and therefore available through a user-controlled interactive network. Interactive television is internet plus television. I have already contrasted the sit-forward interactive media with the sit-back narrative media, but in interactive television the two converge. The medium is being created through technological and business pressures, not from user pressure, and it makes an uncomfortable hybrid medium:

> Watching television is a shared, passive experience; using the Web is an active, personal one … The successful integration of television and the Web is a goal which many organisations are trying to bring about in different ways. However … it is a marriage which cannot be forced, because they are entirely different media. Television is passive; the Web is interactive.
>
> (Thompson, 1999: 45, 52)

The major television networks are losing audience share to home computers, as well as to other digital providers, and see the internet as a source of added value for their programmes that will help to stem the flow. Along with every other information provider, they also see education as an expanding consumer market. The challenge will be to find the form of interactive television that bridges the gulf between its incompatible constituent media, and also raises the quality of online education. As Thompson says, mere choice, even with 500 channels, is not interaction. He lists three forms of interactive television identified to date:

- Enhanced television – a data broadcasting system which offers news, information, advertising;
- Walled garden – restricted internet access and a store of magazine-like content;
- Portal television – internet access via a gateway leading to easily navigable predefined areas.

All of these offer little more than 'mere choice'. The latter two are most clearly applicable to an educational context, where subject-specific programmes for informal learning on educational channels can be linked to further services on the Internet. Examples can be found on the Open University's broadcasting website (see Web References, p.260). A link from a broadcast television programme via the Web address takes the viewer to further interactive material on that programme, and to taster material and short courses on the topic. Given this access to interesting resources, in terms of the Conversational Framework, interactive television is identical with Web-based multimedia resources, or disc-based DVD. There is no marriage of the two different media: narrative television leads to interactive multimedia via a website acting as the bridge. They remain separate pedagogical forms. Interactive television is interesting more for its logistics. Because it is sited on an interactive network, digital television will create new educational channels that are capable of adding value to the broadcast programmes, and respond more directly to a developing community of learners. It is worth further discussion in the context of the communicative media in Chapter 8, therefore.

SUMMARY

Having considered hypermedia, Web resources, and interactive television in relation to the Conversational Framework defined in Part II, we can broadly summarise the learning activities they can support. In Table 6.1, the three have been grouped together, given their similar pedagogic characteristics. The first column shows hypermedia and interactive television. The second column shows the Web, which although interactive, does not offer any particular topic for exploration in the way that a hypermedia CD or an interactive television programme does. The third column shows how the additional design features discussed above in the hypermedia section enable these media to support more of the learning activities in the Conversational Framework.

The essence of the interactive media is to offer resources for students to explore. Multimedia CDs and DVDs and Web resources enable students to make their own links between topics, and follow their own line of investigation, and this is valuable. However, as we have seen, it is not sufficient to ensure that the student is fully supported. The enhanced features used in programs such as

Table 6.1 Summary of interactive media characteristics

		Hypermedia, TV-Web	*Web resources*	*Enhanced hypermedia*
1	T can describe conception	O	O	✔
2	S can describe conception	O	O	✔
3	T can redescribe in light of S's conception or action	O	O	O
4	S can redescribe in light of T's redescription or S's action	O	O	✔
5	T can adapt task goal in light of S's description or action	O	O	O
6	T can set task goal	O	O	✔
7	S can act to achieve task goal	✔	✔	✔
8	T can set up world to give intrinsic feedback on actions	✔	✔	✔
9	S can modify action in light of feedback on action	✔	✔	✔
10	S can adapt actions in light of T's description or S's redescription	O	O	✔
11	S can reflect on interaction to modify redescription	O	O	✔
12	T can reflect on S's action to modify redescription	O	O	O

'Homer' contributed to much more efficient learning. Alternatively, combining the interactive media, with the other media forms will also help to complete the Conversational Framework.

NOTE

[1]The MENO project (1993–97) was conducted as part the Economic and Social Research Council's Cognitive Engineering Programme, grant no. L127251018 (see Web References, p.260). The aim of the project was to investigate the role of narrative in non-narrative educational interactive media.

Chapter 7

Adaptive media

INTRODUCTION

The adaptive media are the computer-based media capable of changing their state in response to the user's actions. This does not imply any kind of reciprocity, as 'communicative' does, where there is an equality of responsiveness between both parties in a communicative interaction, each changing and being changed by the other. An adaptive program is one that uses the modelling capability of computer programs to accept input from the user, transform the state of the model, and display the resulting output. In this sense, it 'knows' what the user has done in its world, and can therefore provide direct intrinsic feedback on their action. Feedback is critical to the learning process, as every theory of learning acknowledges, from behaviourist to social constructivist. The ability to offer intrinsic feedback is unique to the computer, and forms the core of any understanding of the contribution that ICT can make in education. This characteristic marks out the adaptive media from other computer-based media, which exploit other capabilities of ICT.

Since this chapter focuses on intrinsic feedback, it is worth considering exactly what it means. In Part IIa, I argued that feedback on students' actions is the weakest link in the traditional educational process. For the learning process to be fully supported, students should receive meaningful intrinsic feedback on their actions that relate to the nature of the task goal. The goal–action–feedback cycle constitutes the core of the interactive level of the Conversational Framework. The *Shorter Oxford Dictionary* defines 'intrinsic' as 'inherent, belonging to the thing in itself', and 'extrinsic' as 'not inherent, lying outside the object under consideration'. The two forms are distinct:

- intrinsic feedback is feedback that is internal to the action, that cannot be helped once the action occurs;
- extrinsic feedback is feedback that is external to the action, which may occur as a commentary on the action.

These definitions reveal their pedagogical significance. The former is inherent in the action and unlike the latter, requires no third party judgement on the quality of the action. The requirement for the interactive level of the Conversational Framework is that the feedback should be meaningful to the student. This is important because although it occurs as a necessary part of the action, they must find it easy to interpret in relation to the goal they are trying to achieve. With all those conditions in place, it is possible for the student to use the intrinsic feedback to improve their performance.

Extrinsic feedback, as a comment on the action, is usually confined to the quality of the action – 'very good', 'should try harder', etc. Good quality feedback of this kind does its best to emulate intrinsic feedback – 'you have offered good evidence for your arguments here', 'you would have achieved a better introduction to this essay by including some historical background to the field', etc. Comments of this kind relate the student's action to the goal, and to how they need to change their action in order to meet the goal. By its nature, comment on essays is extrinsic feedback, essentially external to the actions of the student. However, it should always attempt to emulate the more valuable information content of intrinsic feedback.

The informational content of intrinsic feedback is extremely valuable to the learner. It enables them to know how close they are to a good performance, and what more they need to do. It is individualised, private, formative feedback, which helps to build their understanding of the internal relations between theory and practice. Like the teacher–student dialogue, it is fundamental to the Conversational Framework.

SIMULATIONS

A computer-based simulation is a program that embodies some model of an aspect of the world, allows the user to make inputs to the model, runs the model, and displays the results. The model could take several forms: a system of equations, for example, for describing coexisting plant populations (Golluscio *et al.*, 1990); a set of procedures, for example, for guiding a rocket (Brna, 1989); semi-quantitative models to support reasoning about the direction of change in a system (Ogborn, 1990); an operational simulation using experimental performance data, for example, of an engineering plant (Edwards, 1996); a set of condition–action rules, for example, for operating a nuclear power plant (see Figure 7.1).

For all these types of simulations, the program interface will allow the student to make inputs to the model. They may take the form of selecting parameters to change, choosing parameter values within a range, or choosing when to change parameters. The students' inputs to the model determine its subsequent behaviour, which is then displayed, either as numerical values, a diagram, a picture, an animation, or as a verbal description of its new state. From this very general description, it should be clear that a simulation is possible for anything that can be implemented as a manipulable model.

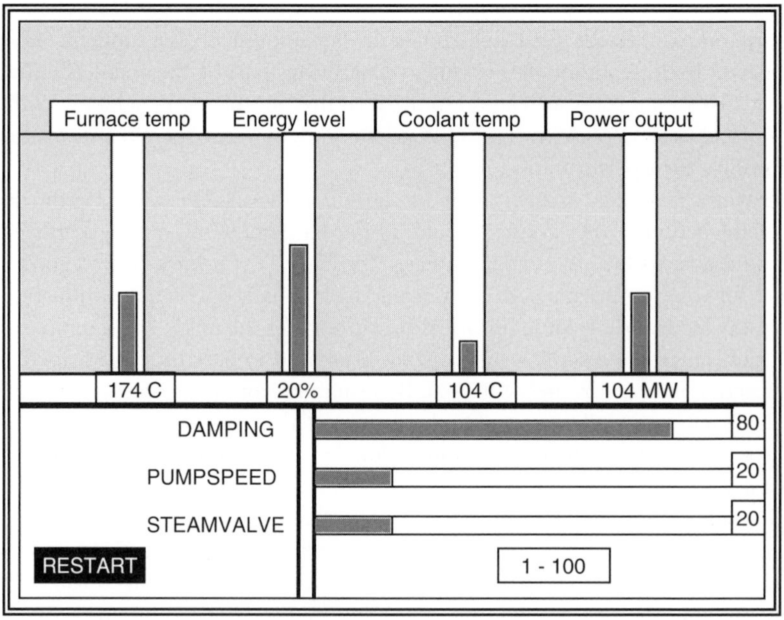

Figure 7.1 The screen for the simulated power system. Users click in the horizontal bars to set the related control. Readouts of the system state are given in the vertical bars. *Source:* Moyse (1991:26)

Simulations are useful for representing complex relations. There would be little point, for example, in simulating a model of an aspect of the economy, such as 'increasing inflation leads to increasing unemployment', as the relation is simple enough to understand from the description alone. A more complex relation, such as 'there is a phase difference of 12 months in the rate of change of inflation and that of unemployment', might be better understood by non-mathematical students if they could see what happened to unemployment figures, in either numerical or graphical form, as they change the inflation figures. A simulation is a representation of the actions and events of the real world in a simulated world.

The computer-based simulation is the first medium we have considered that is adaptive, in the sense that it gives intrinsic feedback on students' actions. The actions are inputs to a model, so the simulation is allowing the student to have a direct experience of the (simulated) world; it is not operating at the level of descriptions of experience; it offers direct experience, albeit simulated. The design of a simulation determines the extent to which the task goal is controllable by the student. With maximum freedom, they can select parameters to change, and thereby determine the task focus. For example, in a plant population simulation they can decide when to stop investigating the effects of rainfall and move on to looking at effects of pollution. Alternatively, within their investigation of rainfall, they can look at conditions for stable equilibrium, or effects of competition below the equilibrium point (see Golluscio *et al.*, 1990). The decision about task focus is the student's, not the teacher's. The power plant simulation in Figure 7.1 allows the student to explore any combination of parameter values, or to test the system to destruction, to find the lowest values at which it still operates, or to find the values which give the optimum read-outs for all measures – to explore the system at will.

Locus of control is important. If the task goal is controllable by the student, then the program cannot comment on how well they have achieved it. The power plant simulation offers exploratory freedom, but because of this cannot supplement the direct intrinsic feedback with any extrinsic comment on what their performance means in terms of achieving an overall goal. The designer has to decide on the appropriate trade-off between freedom for the learner to define their own task goal and richness of feedback from a program that defines the task goal. In the latter case, the simulation can be programmed to supplement its intrinsic feedback with extrinsic feedback giving some commentary on the extent to which the goal was achieved.

The explicitness of the goal is critical: with it, the simulation can encourage reflection on the goal and its relation to the action and the feedback, and can comment on the student's action. Students appreciate the feedback offered by simulations, and may even make favourable comparisons with laboratory work in this respect (Edwards, 1996: 48). Without an explicit goal, where the student explores their own 'what if' questions, reflection is not encouraged because there is nothing specific to aim for, and comment is not possible. Freedom to explore system behaviour was 'intensely motivating' in one study, whereas for another, students found it 'too time-consuming' (Edwards, 1996: 51).

Simulations support a kind of experiential learning, but the students' actions are confined to quantitative representations. Much of their reasoning will be quantitative because the only form their actions take is to determine the quantities of parameters. However, it can also be qualitative, as Moyse has shown. He found an important dichotomy in the way the use of a simulation was set up for students. Those given a 'structural' model describing 'the flow of energy through the system' were far more likely to reason qualitatively, referring to real-world knowledge. Those given a 'task-action mapping' model describing 'a list of control movements which would achieve operational goals' reasoned more quantitatively (Moyse, 1991: 25). Qualitative reasoning, incorporating knowledge of real-world objects or processes produced comments like this:

> I have increased the speed of the coolant going round the system and it's bringing down the temperature again.
> I'm going to try and stop the furnace from cooling too much, by clicking on the damping allowing it a bit more air by going down towards zero.
> (Moyse, 1991: 27–30)

On the other hand, students in the task-action mapping group justify their decisions using instrumental reasoning referring only to quantities and processes explicitly represented on the screen, so that the actions remain uninterpreted:

> Power is going down, we need to increase the steam valve.
> We are getting nothing like the power we need. Open steam valve up, about, put it up to fifty.
>
> (Ibid. 29)

For this latter group, the content was irrelevant; only the quantities were figural. The structural description given to the other group, with the focus on the behaviour of the system rather than quantitative control, elicited more real-world, qualitative reasoning. Purely quantitative reasoning is not inevitable in simulations, therefore, but we may need to encourage a more interpretive approach, as it is not elicited when students simply operate the simulation. I reported a similar finding from an evaluation study in engineering, where students spent much of their time 'number-hunting', trying to find the exact value of a parameter that produced the critical effect (Laurillard, 1987b). The dialogue was of the pure 'up-a-bit, down-a-bit' form that Moyse reported for his task–action mapping group.

The internal relations between the parameters in many simulations are too complex for the underlying model to be determined with any accuracy from these kinds of numerical experiments. Instead, lecturers using simulations must provide detailed notes that set out the derivation of the model. Students may then refer to the symbolic description to help them make their decisions, especially if it is available on screen (Edwards, 1996). Checking the formal representation enables the student to tell, for example, that one parameter is affected by an exponential, and needs to be increased a great deal before it has a noticeable effect on another. This is a valuable educational tool for encouraging something like Resnick's 'mapping strategy' (Resnick and Omanson, 1987), because it enables students to relate the mathematical symbolism to the behaviour of the system. If this works, then they begin to achieve better coverage of the learning process, as they are now focusing on formal descriptions of the simulated world, as well as actions within it.

This kind of access to the teacher's conception within a simulation is a matter of design decision. Simulations are based on a model, and in most the model remains hidden in the depths of the program, inaccessible to inspection by the student. It is common for a teaching program to be issued with accompanying notes that state the model, and even, if the students are supposed to have some mathematical competence, its derivation. The model is the topic structure, so it exists in an explicit form in the program code and on paper somewhere. The complexity of the explicit form is usually the main reason for creating a simulation, so that students can become familiar with it by investigating the behaviour it models rather than by inspecting its explicit form. But having both forms of access – explicitly via the equations or rules, and implicitly via the behaviour of the model – gives the student a better chance of relating their experience of the world (actions on the model) to descriptions of the world (the formal statement of the model).

The examples discussed here all come from the quantitative disciplines because those subjects have developed models as ways of representing the theoretical ideas. Such well-defined models and rule systems do not exist in the humanities and arts in the same way. Nonetheless, intrinsic feedback is possible for certain kinds of action, especially those with visual output. One example is an art history task on constructing a cubist collage (Perkins, 1995). Another is the Art Explorer environment (Durbridge and Stratfold, 1996). This is designed for art history students to explore their own ways of looking at paintings prior to a more theoretical approach to the subject. The task is to generate their own categories of description for a given set of paintings, and then to sort them into those that satisfy a particular category and those that do not. If they repeat the task for several different categories, the program can then analyse their groupings and tell them which categories are similar in terms of their sorting. They can then inspect other students' or experts' categories, sort according to those, and see if they did it the same way. The sequence of activities works well as a way of engaging students in thinking about ways of looking at paintings. The only direct intrinsic feedback is their visual grouping of the paintings. However, there is also commentary on what they have done at that interactive level, which helps them expand their categories, and hence their ways of seeing the paintings. With this experience, and an enhanced set of constructs for paintings, students are better prepared to engage with theories of genre in paintings. Figure 7.2 shows that this combination enables the environment to cover much of the Conversational Framework.

With no model of how the student-defined category should sort the paintings, however, the program cannot comment on the student's actions, so the teacher's reflection on the interaction is missing. Nor does it support the student's description at the discursive level – there is no point at which they have to articulate their findings, as they do in the 'Homer' program discussed in Chapter 6, p.114. A simulation tends to be an interactive environment that the student can explore by acting on it in some way, thereby experiencing some aspect of the practice of the discipline. In practice, therefore, simulations are often embedded in a teaching context that supplies these other aspects of the Conversational Framework in other ways:

- students use it in pairs, so use dialogue to interact at the level of descriptions;
- the equivalent of the 'lab sheet' provides the teacher's pre-emptive task focus for what they do with the program;
- students describe their conception in the form of a write-up on their work on the simulation, rather like a lab report on an experiment;
- the teacher gives extrinsic feedback in the form of comments on the write-up.

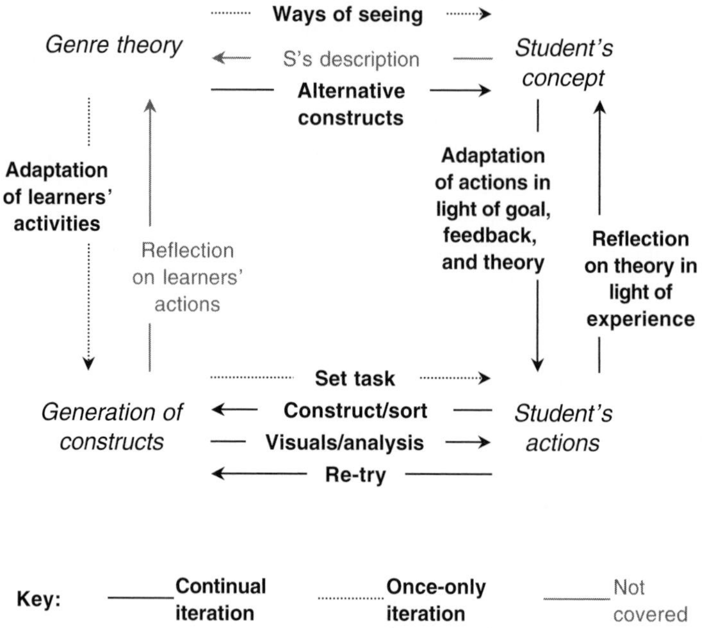

Figure 7.2 Interpretation of the Conversational Framework for Art Explorer.

In this way teachers improve the basic form of the simulation by adding to it the medium of print, and student discussion. Moyse describes a design that can prompt reflection by the student, and that could inform a student–program discussion:

> A partial interpreter, or simulator … is used to produce an execution history in terms which allow its interpretation through any of the required [available] viewpoints. This facilitates a range of tutorial interactions and allows the student to choose a viewpoint which is suited to their current [topic] goals.
>
> (Moyse, 1992: 207)

As the program has a model of the system, and can record all the student's input, it can use that information (the execution history) to help the student reflect on the interaction. This a sophisticated design, but its complexity means that, unfortunately, it is not often emulated. The difference that collaborative work makes, whether student–program or student–student, is very important, as we shall see in Chapter 8. It can lift a very limited educational medium to one that covers many aspects of the learning process.

VIRTUAL ENVIRONMENTS

Virtual environments are an interesting form of simulation, which differ in the nature of the representation of the reality they are simulating. Whereas simulations use a generative mathematical model of the system, virtual environments use a graphical model to display the visual and positional properties of the system, rather than its behaviour. Examples would be virtual art galleries, or virtual field trips, where the user can explore a representation of a three-dimensional environment. They may be aiming to understand the positional relationships, which would be important for a geology course, for example, or the visual properties of chemical explosions, or the detail and interrelationships of components of paintings, or buildings. A particularly successful virtual environment, the virtual microscope (see Figure 7.3 and Plate 2) was developed at the Open University initially for disabled students (see Web References, p.260). It simulates the views through a microscope of slides displaying different kinds of materials.

The value of such a simulation for disabled students is clear: it provides access to key data for students with motor disabilities who cannot visit laboratories, or students who are partially-sighted and find use of microscopes difficult. Materials

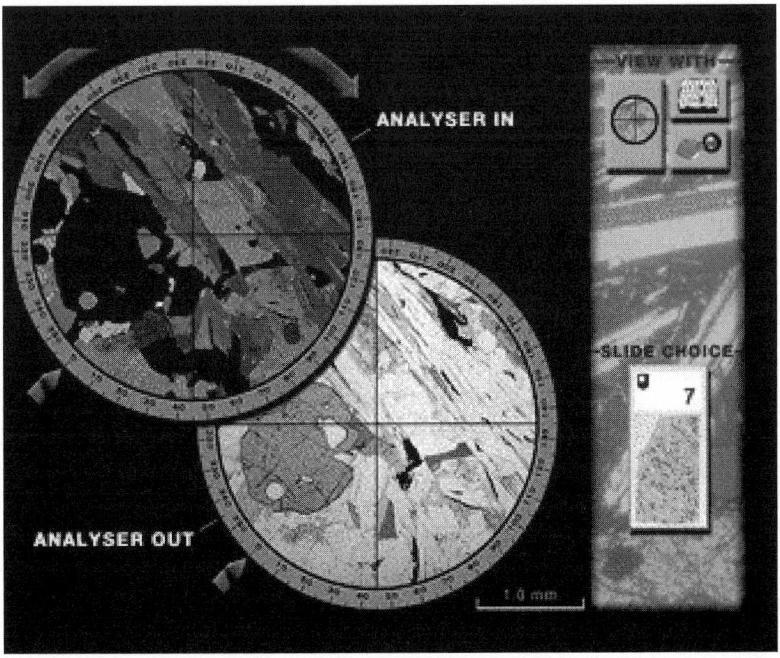

Figure 7.3 (Plate 2) The Virtual Microscope, showing the two views through the simulated microscope, and the icon for selecting different materials for investigation. © the Open University.

designed for students with disabilities invariably add value for other students as well, in this case because the interpretation of microscopic data is difficult when the view is not being shared. This virtual environment enables tutors and students to point unambiguously to the same section for discussion and interpretation.

Do virtual environments cover the Conversational Framework in the same way as simulations? Interestingly, they are closer to interactive media than to adaptive media. Like interactive media, they provide new information according to the user's selection; they are environments for exploration and discovery. They do not provide intrinsic feedback on actions, as simulations do, because they do not model the behaviour of a system. A virtual environment that is adaptive to a student's actions would be one whose positional or visual properties were manipulable in relation to some goal. An example would be a program that offers the components of a picture and asks students to assemble them as a meaningful composition. The composition resulting is intrinsic feedback on their actions, and this is then comparable with a simulation of the type described above. Access to the original composition would provide further feedback, in the form of the 'model answer'. When additional features of this kind are designed in, they can overcome the limitations of the otherwise unsupportive virtual environment (Dobson *et al.*, in press). Most virtual environments are not of this type, and would need additional media forms to complete the other aspects of the Conversational Framework.

TUTORIAL PROGRAMS

Tutorial programs differ from all the previous media forms because they embody an explicit teaching strategy. They are premised on the idea that it is possible for a computer program to emulate a teacher. In Chapter 4, I drew the conclusion that the ideal teaching system was a one-to-one teacher–student dialogue. We now begin to consider whether it is possible to achieve that ideal without very high staff:student ratios, with a computer program acting as 'teacher'. Given the aspirations of this type of medium, we must expect it to come close to covering all aspects of the learning process earlier identified as being essential.

The ideal tutoring program offers extrinsic feedback on the student's actions, and an adaptive task focus related to previous actions and the overall goal.

Tutorial programs do not necessarily set out to express either the teacher's or the student's conceptions fully. Because of the limitations of the computer as a presentational medium, it should only be used to offer initial teaching that can be expressed in a few words and simple diagrams. Tutorial programs tend to assume some previous initial teaching of the topic, and to focus instead on the practice of related tasks. The teacher's conception is implicit in the form of the feedback given, but this does not make it available to the student to inspect in its totality.

The teaching strategy embodied in the tasks put to the student is usually designed to elicit known misconceptions. In that sense there is an intention to make the student's conception available to the program. However, because of the limitations of computer interface styles, these are extremely constraining on how students are allowed to express their ideas. The most risky style is also the most common – the multiple choice question (mcq) technique. This is perfectly acceptable in those cases where all possible answers to the question can be listed for the student to select from, e.g. yes/no questions, or questions of the form:

> How do increases in government spending and in private investment compare with respect to their effect on aggregate demand:
> A. Only government spending shifts aggregate demand.
> B. Only private investment shifts aggregate demand.
> C. Both shift aggregate demand.
> D. Neither shifts aggregate demand.
>
> (Saunders, 1991)

As there is no other possible answer to the question as phrased – logically it has to be one or the other – the mcq technique works well as a way of communicating the student's answer to the question. Compare this with another way of asking about the same topic:

> In comparing an increase in government spending to an increase in private investment, we can correctly say that in the short run:
> A. they will both shift aggregate supply.
> B. they will both shift aggregate demand.
> C. government spending is inflationary; private spending is not.
> D. government spending must equal taxes; private investment must equal saving.
>
> (Ibid. 1991)

This is much more hazardous as a way of gauging the student's answer, because there are many possible ways of answering this more open question – the particular ones chosen do not cover all logical possibilities. This means the student's own way may not be represented. Therefore, they must consider each answer in turn for its plausibility. In doing so, they will be likely to postulate a reason why each answer might be correct – 'government spending must equal taxes' sounds sensible, so that must be the right answer. Even if they end up selecting the correct answer, they cannot expunge that reasoning. It could be what they remember best, even though they may never have thought of it without prompting. The mcq technique therefore runs the great pedagogical risk of inviting students to make sense of wrong answers. At least with the first example above it does not introduce ideas they could not have thought of.

On the other hand, the first example fails to identify those students who have the conception that spending increases aggregate supply. The only way to elicit this conception without also inviting it is to rephrase the question, and use keyword identification of an open-ended answer, a technique known as a 'concealed multiple choice question'. This would work as follows. Ask the question 'What else increases when government spending increases?' and compare the string of letters input by the student to the strings 'supply' and 'demand'. If any part of the input matches 'supply', then the program assumes that they believe that supply increases with government spending. The matching algorithm may be more sophisticated, e.g. allow mis-spellings of 'supply'; or allow certain synonyms. However, with this method it is always possible for the student to get a right answer that the program cannot recognise. The feedback should therefore be cautious about right/wrong judgements. A common solution is to say something non-judgemental, but making explicit the correct answer, such as 'In fact it increases demand'.

The questionable pedagogic value of mcqs is raised in an extensive evaluation of the WinEcon software for economics. A key criticism is the transmission model of teaching embodied in both its presentational, non-interactive design, and in the mcq form of assessment it uses:

> There is the predominant use of multiple-choice questioning, which is typical of the transmission approach as it serves to check that the message has been received. Such assessment is based on how much, and how accurately, information is known rather than what is understood, whereas a more student-centred approach focuses on what is understood.
>
> (Brooksbank *et al.*, 1998: 51)

Whatever form of elicitation of student conception is used, it should be designed to allow the student to express what they think as closely as possible.

The presentational qualities of multimedia allow tutorial programs to offer brief introductions to the content being studied, but these are unlikely to use any lengthy narrative form. The style of study required for apprehension of the structure of a narrative and through that, of its meaning, is not compatible with the style of study required for ICT. Because computer-based media are minutely controllable and interactive, the student inevitably expects continual prompting, whereas a passage of text or a video sequence requires sustained attention, but no action. I do not want to term this contrast 'active/passive', as students reading or watching a video are not passive. They are necessarily active in using these media if they are to experience anything at all. However, moving from an adaptive medium of continual activity, to a narrative medium of continual receptivity is a disquieting jump for students. The sit-forward/sit-back media do not make a happy combination. For that reason, the video used in conjunction with tutorials is usually shown in very small clips, of a few seconds or so. Text as well, aside from the difficulties of reading text on a screen, is kept to short passages in well-designed programs. Otherwise, there is an observable tendency for students to ignore it, or become impatient with it, a point confirmed in an evaluation of the WinEcon materials (Brooksbank *et al.*, 1998: 49). Students expect to be able to consult explanatory material as they need it within a tutorial, but in general would prefer to study its narrative in the more appropriate medium of a video or a book.

The main difference between simulations and tutorial programs is that the latter necessarily embody an explicit teaching strategy. For a simulation there is an implicit teaching strategy in the choice of model, and in the way the interface operates to support the student's exploration. But there may be no explicit goal. This constrains the feedback that can be given to a student. In a tutorial, there is an explicit teaching goal, and this fact enables the program to comment on the student's performance in relation to that goal. This type of adaptivity means that the program uses the student's performance on previous tasks to decide what feedback or subsequent task should be offered. For example, in a tutorial on chemical periodicity (Figure 7.4 and Plate 3), the goal is to identify the noble gases from their reactive properties. The student can explore the properties of a range of gases and is then tested on what they know. In the illustration, they have mis-identified chlorine, and the program branches to a re-run of the demo showing the violence of its reactive properties.

This embodies a teaching strategy of the form described at the beginning of this section. There is a default sequence built into the program structure, such that the program moves through topics that progress in complexity. Based on the frequency of the student's errors, it may suggest they do more before moving on to a different topic, or taking a test. Thus, adaptivity acts at the level of deciding what task to set, and how much practice to offer on each one. However, it is crucial to allow the student to override the default sequence. This is a user-control medium, and they will wrest that control somehow, to the extent of abandoning the program if it does not give them the freedom appropriate to the medium. For

Figure 7.4 (Plate 3) A tutorial program on chemical periodicity. *The student has previously selected reactions with each of the gases, to see which ones yield a violent reaction. They are now being tested on which gases are 'noble' and do not react. On selecting chlorine, they are given the feedback of a repeat of the demo showing how chlorine reacts, together with a brief comment.* © the Open University.

the program in Figure 7.4, the pull-down menus offer free navigation among all the sections of the material, at any stage. Tutorial programs should be fully controllable by the student because although the teacher's adaptive strategy may be generally effective, it may not be so on every occasion as far as the student is concerned. In a study of students' control preferences, I found that, given the option, some students did far more exercises than any teaching strategy would ever dare to suggest. Others would abandon an exercise as soon as they got something wrong, but return to it later (Laurillard, 1984a).

Because a tutorial is a succession of explicitly designed tasks, those tasks can be interactive at the level of actions, or discursive at the level of descriptions – it is a matter of design, rather than the nature of the medium. At the level of action, it can ask students to perform exercises whose input the program can analyse in order to supply feedback. Too many tutorial programs use the mcq format to define the task set for the student, so that the input is easy to analyse. They provide only extrinsic feedback, of the form 'Yes, because ...' where the reason is stated just in case the student made a guess and did not know the actual reason for the correct answer, or for wrong answers 'No, because ...', or sometimes 'Try again' in case it was a trivial error. This is not intrinsic feedback, but extrinsic feedback with more teaching attached. It provides information, and will assist memorisation of a procedure, and this may well be sufficient in many cases, but it will not do much to develop conceptual understanding if the student is having conceptual difficulties. For that, intrinsic feedback on action is necessary. In the case of chemical periodicity, the learning objective is a very simple relation between name and property, and the illustration of the property enhances the experience of what it means, and may assist memorisation. The format is appropriate for such an objective. Objectives that are more complex need the combination of tutorial with simulation, to provide intrinsic feedback.

TUTORIAL SIMULATIONS

It is quite feasible for a tutorial to offer intrinsic feedback, but only if it has some kind of model of the task it sets. This defines the 'tutorial-simulation', and being a combination of two complementary media, one offering extrinsic feedback and the other intrinsic, it is an importantly different medium from either on their own.

Tait makes a similar point based on his study of students using a simulation of homeostasis in the kidney. He found that they often need help in making sense of their actions on the model:

> Simulations in themselves are not necessarily effective learning tools unless used to perform appropriate tasks...This emphasises the need for further learning support, for example in the form of explanation similar to that which might be given over the shoulder by a human tutor or by different learning tasks.
>
> (Tait, 1994: 122)

The simulation program must therefore embed the model itself within a supportive learner interface. His 'Discourse' environment, for example, proposes the appropriate tasks (e.g. manipulation and diagnosis), encourages students to make predictions (e.g. by sketching a graph of the expected result), and provides explanations of the events (e.g. by organising the knowledge the program has about the state of the model and linking it to qualitative annotations to generate a verbal explanation). This way of offering explanations enables the simulation to supplement its intrinsic feedback with extrinsic feedback, elaborating what has happened to ensure that it is meaningful to the student. The provision of learning tasks and goals, as in tutorials, enables the simulation to be supportive of the student's process of learning.

The power of the combination is evident in programs such as the geology example in Figure 7.5 (Plate 4). The task goal is set by the program: to find the shift in the geological formation which gives rise to a given surface feature. It creates for the student an environment in which they can drive an operation in the geological process itself, namely the direction and amount of movement of a rock formation along a fault line.

Because the program provides both the task goal, and an interactive environment, it can offer feedback at two levels. The model offers intrinsic feedback on the student's action: in this case, we can see that the right side of the formation has moved up and the top layer has eroded to expose the dark green rock layer (see Plate 4). In order to match the surface combination on the left of the screen, it should have moved up more and eroded two layers to expose the bright green layer. Furthermore, because the program knows what the goal is, it can give extrinsic

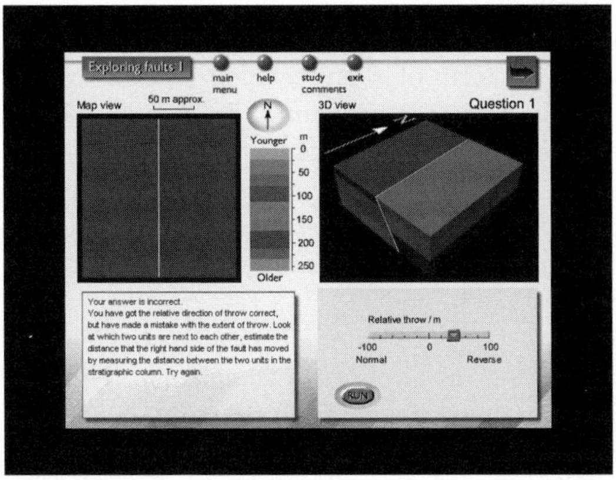

Figure 7.5 (Plate 4) A tutorial-simulation program on geological formations. *The student must specify the way the rock formation on the right has to move and erode in order to expose the surface features represented on the left. The direction and amount of the movement is controlled by the slider. The result of their first move is shown on the right, with commentary below left. The order of the sedimentary rock layers is shown in the middle.* © the Open University.

feedback on the action that emulates the intrinsic: they have correctly defined the direction of the relative throw, but not the correct amount. This helps to ensure that the student has correctly interpreted the intrinsic feedback. Students of geology experience great difficulty in visualising the processes involved in these three-dimensional changes over time, as a recent study has shown (McCracken and Laurillard, 1994). Despite carefully crafted printed materials with illustrations, students need practice in thinking about how the formations move and change, and hence need a more dynamic environment. The intrinsic feedback from the control over the environment provided here will help the visualisation. It provides meaningful feedback on actions in such a way that students can see what they need to do in order to correct their input. The extrinsic feedback is possible because the program knows what the goal is and that they have not reached it. The hint provided in the extrinsic feedback goes beyond what the intrinsic feedback offers, because it relates the movement to the underlying rock structure.

This simulation model in the program gives students an experiential sense of how the system behaves. With this highly constrained way of experiencing the world of stratigraphs, surface areas, and rock formation movements, similar to the idea of 'supplantation' in learning from video, students begin to see the system as the geologist would wish them to. The tutorial part of the program provides extrinsic feedback to complement the intrinsic, in the form of a canned text comment on the interaction. This does not make it fully discursive, as there is no provision for the student to articulate their own description. The program is capable of offering a redescription of the topic in the light of the student's action, tailored to that event, and therefore capable of helping the student interpret the intrinsic feedback. Figure 7.6 shows the extent to which the program covers the Conversational Framework.

The most general way of defining the aim of the program here is to help the student understand the form of representation of a real-world system, i.e. the concept of 'throw' in geological structures. For this kind of aim, a computer-based tutorial-simulation can approximate quite closely to supporting the discursive mode, because it 'knows' about the correct interpretation of the simulated event via the model it uses, just as Tait's learner interface does. The interpretation has to be plugged into canned text, but is clearly focused on the event just experienced by the student.

Another example is shown in Figure 7.7 (Plate 5), which is essentially a tutorial, giving remedial teaching on algebraic manipulation. The teaching strategy is adaptive, building from simple manipulations to more complex ones, as the student progresses. The interactive environment is analogous to a game, which defines a goal and allows only certain rules of manipulation from the initial conditions to reach the goal. The rules are introduced at the beginning of the program in terms of a tiddlywinks game. This simple initial engagement in a concrete instantiation of the rules helps students gain familiarity with the environment and its behaviour. In successive exercises, the objects become more abstract, to be replaced by letters, but still operating under the same rules, with the same interface.

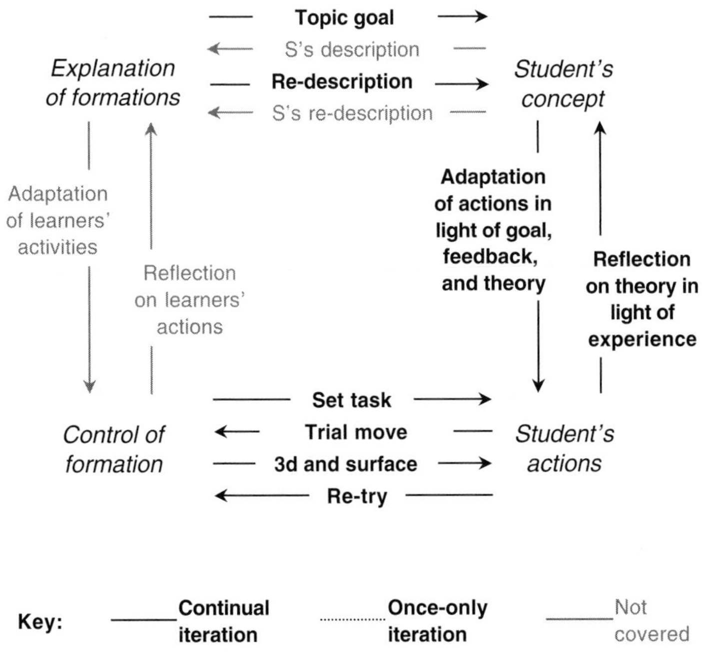

Figure 7.6 Interpretation of the Conversational Framework for the geology simulation.

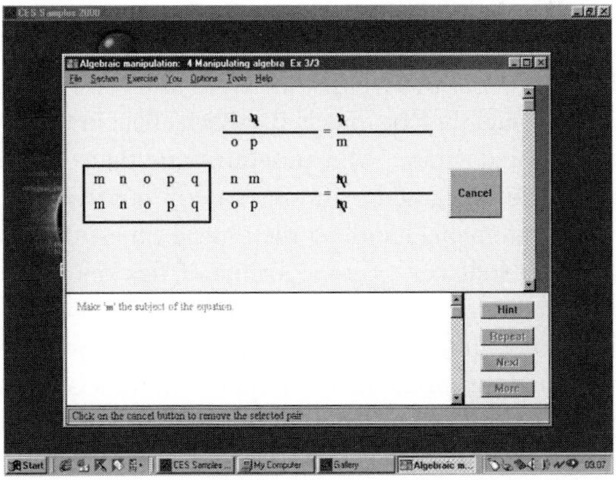

Figure 7.7 (Plate 5) A tutorial-simulation program on algebraic manipulation. *Each step is driven by the student. In the previous step illustrated they clicked on each 'n' and then on the Cancel button. In the current step (the lower one), they had to click on two m's in the box, and then drag each one to its current position, and then click on the two right-hand m's and the Cancel button to cancel them.* © the Open University.

As in the geology program, the environment is designed to allow the student to drive the process. Because they are here learning about the appropriate sequence, they must decide on each successive manipulation in order to reach the appropriate final goal. All the moves are defined in relation to their mathematical equivalence. For example, to make a letter in the denominator the subject of an equation, they must multiply it through both sides. This move can only be accomplished by selecting two of the same letter from the box (m in the example), and dragging each one to the top line on each side. They can perform a cancellation either horizontally or vertically by clicking on the two letters and then on the Cancel button. The manipulations in the environment therefore provide intrinsic feedback by allowing only mathematically correct moves, and by showing the result of each successive manipulation. They cannot go wrong, but it is easy for a novice to go round in circles if they are unsure of the procedure. Intrinsic feedback may not be sufficient, therefore, and extrinsic feedback is provided either from the Hint, or from the option to see the next step done. There is no control of the student: the repeat, next and more buttons, together with drop-down menus to move around the exercises, provide complete navigational freedom for the student.

Again, like the geology program, students receive both intrinsic feedback from the model in the form of permissible moves, and extrinsic feedback in the form of canned text hints to advise them on the nature of the next move.

The tutorial-simulation is a powerful combination. Once we recognise that anything that can be modelled can be programmed as a simulation, it follows that it can also be designed as an even more effective tutorial-simulation. The model does not have to be mathematical. The economics example we looked at on p.135, could be modelled as a series of causal relations: e.g. increase in government spending → increase in aggregate demand; increase in private investment → increase in aggregate demand; increase in aggregate demand → increase in aggregate supply; and so on. A statement by the student that 'supply increases as a consequence of increased government spending' could then be matched to each effect in the cause–effect relations to discover its stated cause. The program can then output the intrinsic feedback 'No, that would be caused by an increase in demand'. A more accurate model would include statements handling the crucial timescale refinements of 'in the short run' and 'in the long run', but the format of this kind of model should be clear. The difficulty lies in the complexity of producing a model that is complete enough and accurate enough to handle the range of student responses to the questions put. Designing an appropriate model requires an understanding of both the subject matter and the ways students typically conceptualise it. This is discussed further in Chapter 10. For now, it is sufficient to note that such models are possible elements of tutorial-simulations for many subject areas.

Without a model of the system, tutorial programs are only weakly discursive. Programs built around multiple-choice questions do not show students what happens in the world as a result of their actions. To operate intrinsic feedback, the basic tutorial must be augmented with a simulation, or dynamic environment. Then it becomes a very powerful teaching medium, because it provides both adaptation of the environment to the student's actions, and reflection on that interaction at the discursive level in the form of extrinsic feedback in relation to the overall goal. In terms of the Conversational Framework, Figure 7.6 on p.141 shows that the tutorial-simulation can address almost all of the learning activities, except the iteration around the student's own articulation of the topic.

EDUCATIONAL GAMES

In the first edition of this book, I discussed intelligent tutoring systems (ITSs) at this point. They promised to offer everything we could need to cover the full Conversational Framework, had they existed, but failed in practice because they were driven more by the cognitive science research agenda than by pedagogy. Sadly this trend continued and the current educational research agenda is not seeking to enhance the capability of adaptive programs with AI-based research.

Instead, we turn to another chimera: educational games. They ought to exist, and like ITSs, they would constitute the acme of educational media if they did. However, here again, the research agenda has diverged from education, driven by more rewarding markets than education can ever aspire to. Computer games have developed usable virtual reality environments, and a variety of forms of user control of those environments. They offer primarily intrinsic feedback, although some will also offer advice and hints on how to achieve specific outcomes. The goal is sometimes program-defined (e.g. to reach a target performance) and sometimes user-defined, as in construction environments where the user builds the hospital or city of their choice, though it behaves according to the rules of the game. A key feature is the real-time nature of the interaction, because this requires close attention and responsiveness from the user, whether it is a combative game, or an environment that changes over time. The intrinsic feedback on the user's actions from the environment is usually meaningful enough to enable them to adjust their actions in relation to the current goal. In terms of their form, computer games offer exciting and motivating learning environments. Aligned with the communicative interface of the Web, they can also create social interactive environments with multi-player games. It should be possible to harness their content to the educational agenda. Where educational objectives fit the form of these kinds of environments and activities, especially where there is a certain tedium experienced in mastering a skill or procedure, for example, a gaming environment has much to offer. While the development communities of games and education remain as separate as they are now, however, there is little prospect of this. The theoretical demands of the Conversational Framework suggest that there should be convergence of the form of games with the function of education. Perhaps this will materialise as the commercial sector enters the education market, though it will be at the expense of academic control.

SUMMARY

This chapter has considered simulations, tutorial programs and tutorial-simulations in relation to learning activities in the Conversational Framework. Those they can support are summarised in Table 7.1.

The main difference between the simulation and the tutorial lies in the fact that the teacher's conception and goal is expressed explicitly in the latter. The combination of these in the tutorial-simulation naturally has an additive effect on what is covered. None of the three is genuinely discursive. However, the modelling component, incorporated to allow the student to express their conception in description language, makes them the only media so far to offer adaptation and redescription by the teacher, based on students' actions. These media come closest to covering the range of essential learning activities we defined in Chapter 4 and, especially in combination with other presentational media, are potentially effective alternatives to the one-to-one teacher–student dialogue.

Table 7.1 Summary of adaptive media characteristics

		Simulations	*Tutorials*	*Tut-Sims*
1	T can describe conception	O	✔	✔
2	S can describe conception	O	✔	O
3	T can redescribe conception or action	O	✔	✔
4	S can redescribe in light of T's redescription or S's action	O	✔	O
5	T can adapt task goal in light of S's description or action	O	✔	✔
6	T can set task goal	O	✔	✔
7	S can act to achieve task goal	✔	✔	✔
8	T can set up world to give intrinsic feedback on actions	✔	O	✔
9	S can modify action in light of feedback on action	✔	O	✔
10	S can adapt actions in light of T's description or S's redescription	✔	✔	✔
11	S can reflect on interaction to modify redescription	O	O	✔
12	T can reflect on S's action to modify redescription	O	✔	✔

Chapter 8

Communicative media

INTRODUCTION

The communicative media are those that serve the discursive level of the Conversational Framework, having the specific task of bringing people together to discuss. The discussion may be between tutor and student, or between students. The medium of communication is either text/graphics, audio, video, or any combination of the three.

The communicative media are designed to provide a solution to a logistical problem, rather than a pedagogical one, and were only ever used in education to communicate with students who are geographically distributed. Email, telephone and videoconferencing have only been seen as desirable media by distance-learning universities, unlike the other media we have considered. Since the first edition of this book, two developments have changed the importance of communicative media for higher education as a whole: the increase in lifelong learning, and the Web. There is a clear recognition now that the undergraduate population is changing:

> Fewer than one-fourth of the students on college campuses today are between the ages of eighteen and twenty-two and attending full-time – our definition of a traditional undergraduate.
>
> (Palloff and Pratt, 1999: 3)

The Dearing Report for the UK (Dearing, 1997) put the same figure at one-third, but the trend is going in the same direction. In addition, as universities find that the majority of their students are part-time, mature learners, often returning to university study, the demands of the student population inevitably change:

> Our understanding of how people learn is growing, suggesting that increased individualization of the learning process is the way to respond to the diverse learning styles brought by our students as they enter and re-enter the world of higher education.
>
> (Twigg, 1994: 1)

This is a highly discerning student population, relatively affluent, mobile, and hard working. They will demand a lot from the university they return to, and as Palloff and Pratt (1999) point out, the campuses will respond by working hard to create learning communities among these groups. The key environment for this group is the online community. The Web would have been an attractive medium for campus-based students in any case, because unlike the earlier forms of communication over the Internet, it facilitates a much wider range of communicative forms. For distance-learning students, it becomes a lifeline.

A medium that can support discussion immediately addresses the two types of learning activity that we have so far found it most difficult to cover: interaction at the level of descriptions, and reflection on action, feedback and goals. The use of communications media in education is based on the assumption that students can learn through discussion and collaboration, even at a distance and asynchronously. This is the assumption we examine in this chapter.

Communications media take two forms: synchronous – where participants are together in time, communicating through text, audio or video via a network; and asynchronous – where participants use the system at different times. Tutors and students may be engaged in a one-to-one conversation via email, audio link or desktop video, or more usually in many-to-many conversations. All these media forms allow an eavesdropping audience of other students participating vicariously in observing the discussion.

Exciting claims have been made for the significance of the Web for education. Collis defines a new educational paradigm as 'interconnectiveness', being able to connect to experts and resources beyond one's local possibilities:

> Through interconnectivity, we can not only access perspectives on a topic, but also alternative perspectives on that topic, not only those chosen by our 'local experts'. Even more powerfully, we can access the authors themselves, casually, instantaneously, we can initiate a communication. We can sit ourselves at the feet of a master, in the way of disciples in the second paradigm [the mediæval university], limited not by the radius of her voice, but her willingness to return an email message.
>
> (Collis, 1996: 582–583)

Quite. The value of interconnectivity may be perceived differently by the expert with an overloaded email system. The interconnectivity provided by snail-mail and telephones quickly led to ways of limiting access by the many to the few, and the Web will be no different. Access to alternative perspectives on a topic is hardly new, and is usually thought to be one of the benefits of libraries. Nonetheless, when the networks are functioning, there is no question that the Web provides ease of search and access.

The design of the Web interface, together with improved networking speeds, is transforming the capability of the communicative media. Text and graphics are the basic mode, but these can now be combined with audio and video, all three available through the same network link. Digitised audio-video information is encoded in much larger files than the equivalent text, and therefore requires fast transmission speeds to make the communication acceptable in real time. Humans use minute detail of sound and visuals in communication, and this requires very high information content in the transmission of the audio and video data. At the beginning of the twenty-first century, there are still major technical constraints on the communication that is possible over networks. The technology exists to carry high-information content across the world, but the infrastructure needed to support it is expensive and physically difficult to install. Nonetheless, the pressure to communicate at a distance will only increase, and the demands for higher quality transmission will only increase, so the infrastructure will eventually follow. As the technology increasingly supports richer forms of telecommunication, education will be able to exploit this valuable discussion and collaboration medium to the full. The rollout of the technology will vary in rapidity in different parts of the world. This is an implementation issue that we will consider in Chapter 11. In this chapter, the key issue is the quality and type of learning activity the communications media can support, and the role they play in the learning process as a whole.

COMPUTER-MEDIATED CONFERENCING

A conferencing system supports an online discussion environment in which remote users send and receive text messages, usually reading and creating messages offline, and then connecting to the system to upload their messages and download new ones. In synchronous mode they remain online, and send and receive messages with just a few seconds delay, depending on the speed of the connection. The system supplies a structured environment which groups messages in separate conferences according to topic, and allows the user to identify their message as a comment on another.

The standard way to use the system is for the student to join the conference they are interested in, and the system to display all the messages sent into it that they have not already looked at. The student can display each message in turn, and may either contribute comments relating to a particular message, or add a message making a new point. Their message will then appear on everyone else's system next time they join the conference. Computer conferencing is therefore a little like taking part in a normal conference discussion, but via text alone, and over a much longer span of time, as the discussion is asynchronous.

The claims made for the educational value of CMC rest on the assumption that students learn effectively through discussion and collaboration: 'the digital learning environment will probably be the most efficacious "enabler" of independent and self-determined learning' (Peters, 2000: 16); 'In the online classroom, it is the relationships and interactions among people through which knowledge is primarily generated' (Palloff and Pratt, 1999: 15). However, this is not a well-tested assumption as far as the research literature is concerned. It remains a strong belief, given new impetus from the significance of 'communities of practice' (Wenger, 1999), and 'Mode 2 knowledge' (Gibbons *et al.*, 1994), but as we know from the studies of student learning reported in Part I, the properties of a medium do not determine the quality of learning that takes place. Collaborative learning is undeniably important, and the communicative media are powerful enablers that match what is needed for discussion and collaboration, but to what extent do they succeed in enabling learning?

Several studies suggest ways in which we can use the media to support collaboration and discussion. We can learn some key lessons from them. Palloff and Pratt, from extensive experience of online teaching in the field of management, offer evidence of students clearly recognising the benefits of their online community (Palloff and Pratt, 1999). Mason documents a long-term study of an online postgraduate course in educational technology. It showed that students can engage in authentic tasks directly relevant to their work, that these experienced students can bring valuable contributions to the community, and that the burden of student support can be shared among tutors and students alike (Mason, 1998). Another extensive evaluation at the Open University was carried out for an online maths course. It showed that students valued the discussion environment for the alternative perspectives and explanations they encountered, for the opportunity to learn from others' mistakes and insights, and for the sense of community it offered (Petrie *et al.*, 1998). Together these studies show that a collaborative discussion environment is highly valued by students, in ways that affirm its value for the discursive and reflective activities in the Conversational Framework:

- students have access to an expert whom they can question to clarify the expert's description;
- students can articulate and re-articulate their descriptions of the topic in response to others' ideas and comments';
- students can reflect on the discussion to clarify their own understanding.

An obvious pedagogical advantage over the normal face-to-face tutorial is that students can take time to ponder the various points made, and can make their contribution in their own time. Topic negotiation is possible, as in face-to-face discussion, and a tutor may pursue several lines of discussion with different groups of students in sub-conferences, as the topic develops. Student control is therefore relatively high for this medium.

However, the pedagogical benefits of the medium rest entirely on how successfully it maintains a fruitful dialogue between tutor and students, or between students. This is determined to a great extent by the role the tutor plays. In practice, the relationship is asymmetrical, as it is in any face-to-face tutorial, and the tutor is more likely to be responsible for establishing the ground rules of the interaction. Tutors generally have little trouble in articulating their own view, whatever the medium; that is their art. The more difficult trick for them is to give the student the space to express theirs, and to encourage them to elaborate it sufficiently for the tutor to make sense of any points of departure. One factor that diminishes the student's opportunity to express their view is time pressure. Analysis of audioconferencing within a satellite television programme has shown, for example, that during the broadcast transmission, the ratio of tutor:student airtime in conferencing was 3:1. Once the programme was finished and the audioconference alone continued off-air, the ratio was 2:1. This illustrates the value of an asynchronous format such as CMC, which allows the student unlimited time to compose what they want to say, and to say it. Analysis of tutor and student messages in a computer conference running during an Open University course showed that the average length of student contribution was 200 words, equivalent to over a minute of continuous speech. This would be rare indeed in the standard face-to-face tutorial.

The skill of conducting a fruitful dialogue via conferencing is as important here as it is in face-to-face situations, perhaps more so, as there is less information from body language and facial expression to help the interlocutors. If we combine the results from evaluation studies in many different educational contexts (see also Hawkridge, 1998; Jones, 1999; in addition to those above), it becomes clear that the moderator (the conferencing equivalent of a chair) must take responsibility for the success of the community as follows:

- negotiate goals and schedules for the programme of work within the conference;
- define norms and a clear code of conduct to regulate expectations;
- provide access to an expert for a defined period;
- set up new branches and topics as the discussion progresses;
- nurture group collaborative processes to carry out specific tasks;
- encourage students to draw on their own experience in making contributions;
- ensure that adequate responses and reactions are given to all relevant contributions.

Keeping abreast of the conference, which can proceed apace when the moderator turns away just for a day or so, is difficult enough for a busy lecturer. The work required to play an adequate role as moderator, to ensure that the interaction will indeed be successful, is considerable. The first three points above all help to constrain student expectations of the amount of time the tutor will spend in the conference. This is a socially unconstrained medium, compared with a one-hour tutorial conducted within a formal timetable on the tutor's territory. All the studies of online tutoring emphasise the time-consuming nature of the medium. However, this is not an essential feature of communicative media. It is a useful trick in judging the essential qualities of innovations to imagine the change process in reverse. If we were now converting from electronic meetings to place-based meetings, tutors would find it immensely difficult by comparison to adjust to the travel, the strain of responding immediately to questions, the problem of how to end a discussion in reasonable time. We have developed formal mechanisms to protect tutors in the traditional place-based mode – offices, status symbols, and timetables. We will in time develop similar mechanisms for the electronic world, and many of these studies have useful suggestions that others can build on. Nonetheless, at present, the online discussion environment needs careful planning and management to be both pedagogically successful and economically feasible (Salmon, 2000).

The fact that conferencing has a one-to-many form, rather than one-to-one, leads to an expectation of the value of eavesdropping by other students. This is largely borne out by what students say about the medium, and evaluation studies report many expressions of delight at hearing others expressing the same worries or confusions or criticisms. As Figure 8.1 shows, the Conversational Framework is covered only at the discursive level, but among students, as well as between tutor and student.

A dialogue between a tutor and a student can stand for many such dialogues if that student is indeed representative of many others. In the course of the dialogue, the tutor's viewpoint is also likely to be re-expressed or elaborated, which then benefits all students.

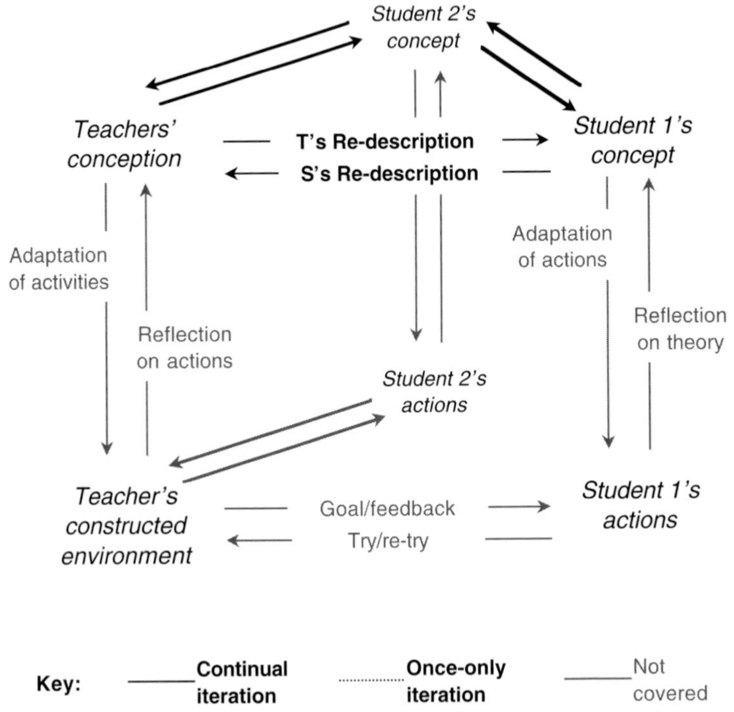

Figure 8.1 Interpretation of the Conversational Framework for a computer-mediated conferencing environment. Only two students are shown, but there are potentially many in the same kind of networked relationship.

The logistical advantages of computer conferencing are obvious for distance-learning universities, as it can be a lifeline for students otherwise cut off from any form of discussion with tutors and fellow students. The advantages for campus-based students will be apparent in courses with low populations, or with distant experts, or with tutors and students who cannot easily arrange to meet, as is frequently the case in teaching hospitals, for example. For all students, the sense of belonging to an engaged and supportive community is highly valuable in itself. This student quoted by Palloff and Pratt summarises the point:

> It seems that we as students have been more willing to talk and discuss the issues at hand than we probably would inside the classroom. I feel this is so for two reasons. One is that we have time to concentrate on the question and think, whereas in the class you are asked and an immediate response is in need. Two, we can discuss openly, and not have to worry about failure as much. If you post something that is not quite right, no one has said this is wrong but instead we give encouragement and try to guide each other to find the right answer.
>
> (Palloff and Pratt, 1999: 31–32)

The conferencing media, as we can see from all the studies reported above, contribute as much to pedagogy as to logistics. With the appropriate planning and moderating, text-based computer conferencing offers an opportunity for articulation, and for reflection on participants' contributions, and helps to build a sense of a scholarly community. The success is totally dependent on a good moderator, however, and this is likely to be as time-consuming as any other form of face-to-face tutoring. None of the existing studies suggests that this is the kind of medium where students can be left to work independently.

DIGITAL DOCUMENT DISCUSSION ENVIRONMENT (D3E)

Conferencing alone does not support any task-based activity other than the description and redescription of the student's view. The availability of conferencing on the Web, however, makes possible other, augmented forms of communication. A discussion environment can be linked to other 'documents', where the document may be a text, or could also be a Java applet running a simulation or animation. Figure 8.2 (Plate 6) shows an example of a discussion environment linked to a paper. Each section of the paper is shown in the contents list at the left-hand side, which helps navigation. If a user wishes to make a comment on a particular section, they click on the associated comment button to get access to that part of the discussion environment, where they can add a new point, or reply to someone else's.

The digital document discussion environment, known as D3E at the Open University where it was developed, creates an asynchronous network, a close equivalent of the reading group, or seminar. Every member of the group has

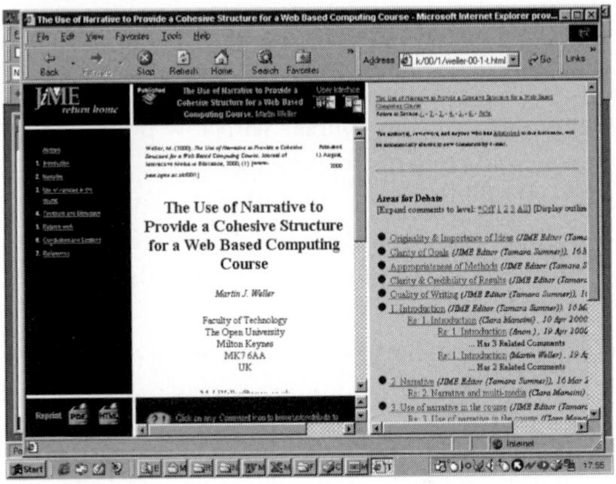

Figure 8.2 (Plate 6) A digital document discussion environment for an article. *The index to sections in the article is listed on the left for navigation of the text shown in the middle. The right-hand page shows the outline of the discussion, with comments linked to each section of the digital document.* Digital Document Discourse Environment (D3E). Knowledge Media Institute, Open University, UK. Available at: http://d3e.open.ac.uk. © the Open University.

access to the same material, and each can comment and debate the text in detail. The digital environment also offers the considerable advantages of facilitating a more orderly coverage of the whole text. Being asynchronous, it creates time for reflective responses, free of the cut-and-thrust of face-to-face discussion. The opportunity for detailed commentary on a text introduces an interactive-level task to the discursive-level topic. The tutor uses students' comments to offer feedback on their interpretation, or on the way they have linked one part of the text to another. It enables comment on their practice, not just on their descriptions of their understanding. In this sense the environment can emulate intrinsic feedback, and offers an additional dimension of learning activity to the discussion environment of conferencing alone.

The 'document' can also be a runnable program. The example in Figure 8.3 (Plate 7) is taken from the electronic Journal of Interactive Media in Education (www-jime.open.ac.uk). The journal enables authors to link their papers to dynamic versions of the software they are describing. A sequence from an economic modelling task in a paper on WinEcon, is one of several examples among the papers now available online (see Buckingham Shum and Sumner, 1998). The same format could be used for an online student group.

Students could be set a task to achieve within a simulation environment, and link their practice here to a comment or question to the group. The combination of discussion and task environments enables students and tutor to link their dialogue at the discursive level to their actions at this interactive task level. The D3E therefore offers an extremely powerful learning environment. Figure 8.4 shows the extent to which it addresses the activities within the Conversational Framework.

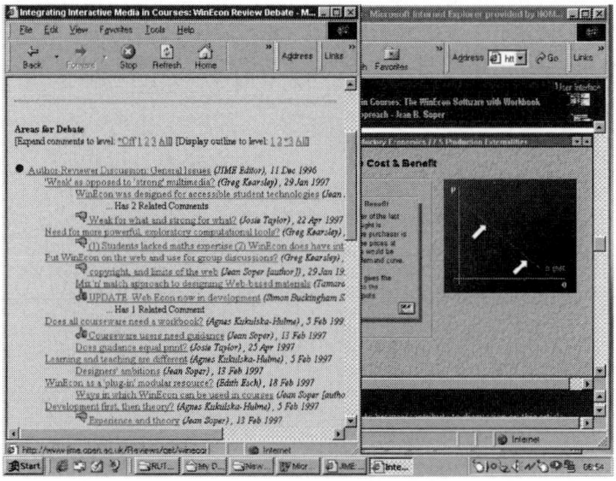

Figure 8.3 (Plate 7) A digital document discussion environment for a runnable simulation. *The top page shows the discussion, linked to the digital document behind it. This example is a scholarly debate within an electronic journal. The format could equally well be an online class discussion of an interactive simulation.* Digital Document Discourse Environment (D3E). Knowledge Media Institute, Open University, UK. Available at: http://d3e.open.ac.uk. © the Open University.

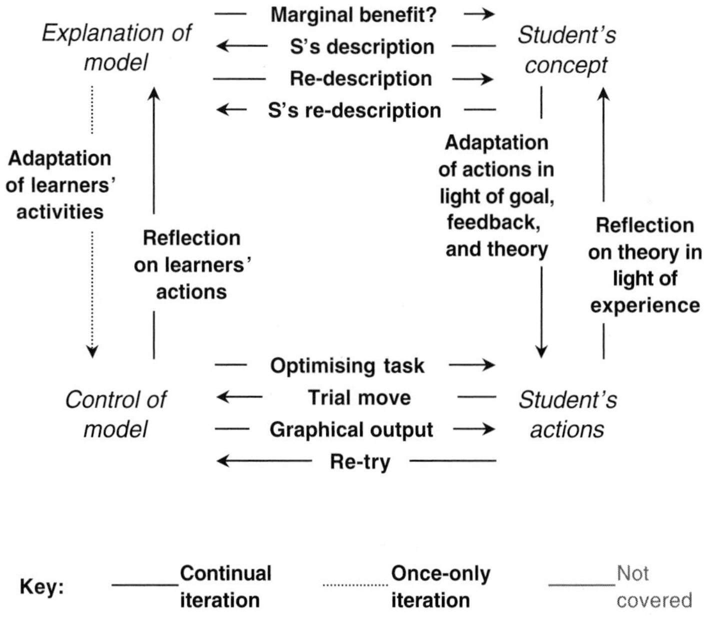

Figure 8.4 Interpretation of the Conversational Framework for a digital document discussion environment on marginal private cost benefit in economics.

There are no examples of its application in an educational context to date, but this is surely a future direction. We could imagine the form transferred to an online economic modelling tutorial, in which students are given an optimisation task to demonstrate their understanding of a particular concept. Then, in terms of the Conversational Framework, they would be operating the model interactively at the task level, and discussing their reflections on that goal–action–feedback cycle at the discursive level. Figure 8.4 illustrates that this kind of medium supports almost the full range of the iterative activities. The only non-iterative activity is the adaptation of the model in the light of student needs discerned at the discursive level. In most cases, it would be difficult for the tutor to re-program the model responsively.

Conferencing on the Web should not always be confined to the discursive level. Using a hybrid medium like D3E enables students to be supported through a much more intensive learning process, iterating through communication and interaction in both theory and practice.

AUDIOCONFERENCING

Audioconferencing is group discussion by telephone, and has been useful in distance-learning contexts to support remote discussion for small tutorial groups. The interface characteristics of the technology make it an uncomfortable medium to use, as there are only inadequate sound cues to distinguish who is talking and who wishes to talk (see Mason, 1994 for a fuller discussion). With carefully designed ground rules, a tutor can make this work, but neither students nor tutors enjoy the medium, and it is used only *in extremis*.

Again, the Web has wrought a radical transformation. The availability of audio on the Web brings audioconferencing into the limelight of significant educational media. Like text-based conferencing, the Web medium offers convergence between audio discussion and a range of other media, requiring a new epithet: 'audiographics' has some currency already as a generic term, so I will use that. There are two key differences between audiographics and D3E: it is synchronous, and it transmits voices, not just text. Digitised voice and other data are both transmitted via the same network line, whether cable or dial-up modem. All participants arrange to join a session at the same time, wherever their location. They wear headphones with microphone attached, leaving them hands-free to input data or text to the shared screen, on which each participant sees the same data, dynamically updated as they contribute to it. Figure 8.5 (Plate 8) shows how a screen looks to a student group working in the Lyceum environment developed at the Open University (see Scott and Eisenstadt, 1998).

A well-designed interface will use simple visual devices to support the progress of the audio-only discussion: e.g. names, photos, indicators of current speaker and those wishing to speak. The shared area can contain dynamically created diagrams, as shown, or can use a cut and paste tool to import an existing picture or

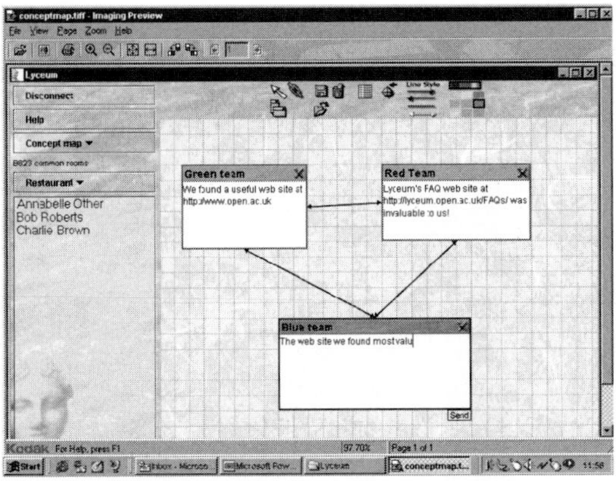

Figure 8.5 (Plate 8) An audiographics conferencing environment on the Web. *The left-hand column shows who is online. When someone speaks, a microphone icon appears beside their name. The shared screen on the right is a concept mapping tool being used by the group to organise contributions from the three participants. Each one may be simultaneously typing into a box. When one hits the Send button, their contribution appears on all screens.* © the Open University.

Web page. This feature brings greater flexibility to the interactive task level than in the D3E example on p.152. It means that the tutor can adapt the task to the needs of the students, in light of the discussion, as shown in the imagined example in Figure 8.6 (Plate 9).

Voice communication alone will not always have the power to elicit the precision of expression that can be achieved by asking someone to draw or add to a diagram. Audiographics is therefore more effective at allowing the teacher, and especially students, to express their point of view, through both language and diagrams. Similarly, the students can introduce their own material or create diagrams online. This also gives students some control over the direction the discussion takes, as it gives them an additional way of expressing an idea, or asking a question. Student control is dependent on tutor restraint, however. The disadvantage is that we give the academic a presentational device via this system. They can all too easily make use of it for delivering new material, rather than allowing a student-led discussion to develop.

Audiographics constitutes the potentially most powerful medium so far in terms of coverage of the Conversational Framework. There is relatively little evaluation data on usage because it is still a very innovative technology, not yet developed to its full potential. However, as Figure 8.7 shows, the medium can support many of the iterations of the learning activities defined by the Conversational Framework.

There are two key pedagogical differences between D3E and Lyceum: whereas the former cannot offer adaptation of the task environment the latter can; and where the former can offer intrinsic feedback at the interactive level, the latter

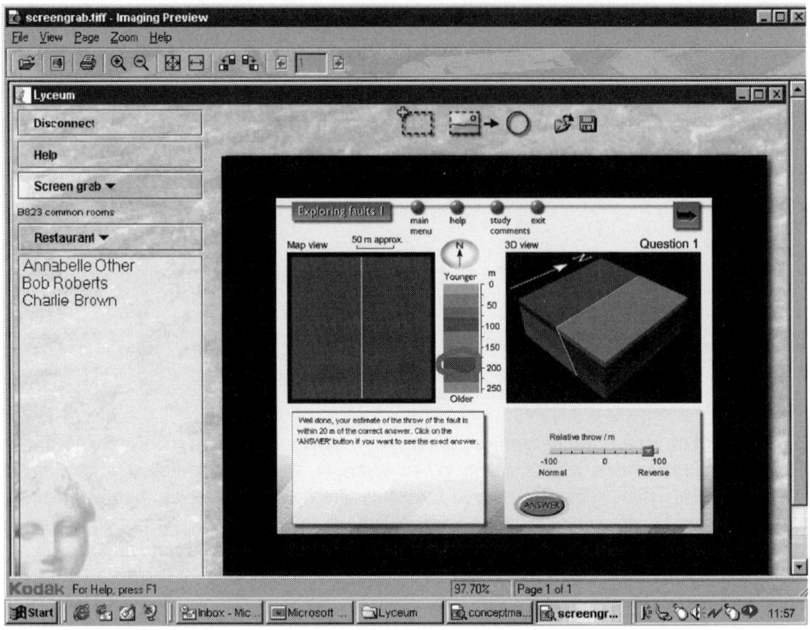

Figure 8.6 (Plate 9) An audiographics task-based environment on the Web. *The discussion concerns a difficulty one student has with a task on the geology course CD, already seen in Chapter 7, Figure 7.5 (Plate 4). The student has copied the troublesome screen to the shared window, and the tutor has asked them to indicate which rock layer they should be aiming to expose in the task. The student has responded by circling the appropriate layer in red (see Plate 9). The discussion can now proceed in light of the student's action.* © the Open University.

cannot. When Lyceum can import a runnable model to the shared area, then it will fulfil the requirements of the whole framework.

VIDEOCONFERENCING

Videoconferencing is a one-to-many medium, making it a sensible way to provide access to a remote academic expert. The availability of video on the Web enables desktop videoconferencing between individuals, though this is less likely to be used in education, except in special cases.

At either end of a videoconferencing link there is a camera focused on an individual or group. This carries their picture via a network to the screen at the other end. The feasibility of good quality communication due to the additional visuals is wholly dependent on having the appropriate bandwidth of the network for the high information content needed to transmit humans talking. The lecturer is usually given control over what is transmitted via a console governing the various cameras.

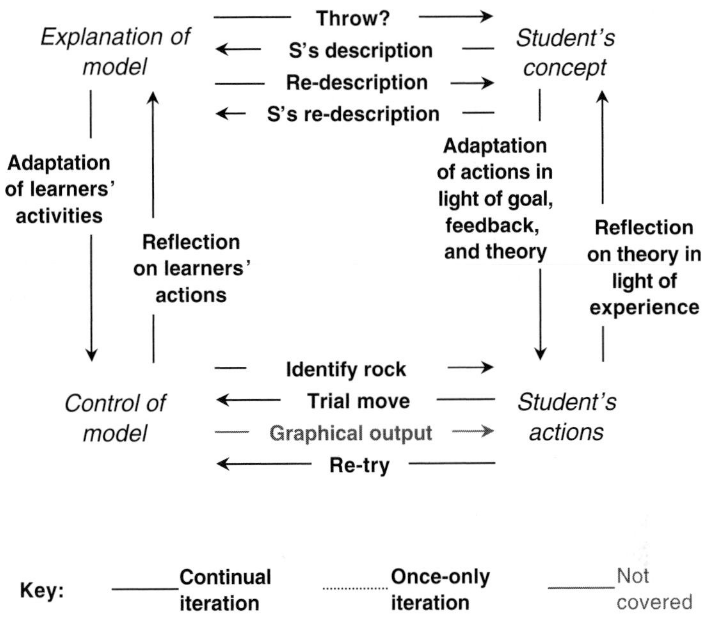

Figure 8.7 Interpretation of the Conversational Framework for a audiographic task-based environment on concepts in geological structures.

Cameras may also film live action elsewhere, such as an operating theatre, an experimental set-up, or an interesting object. Each local site is furnished with a microphone, or with several if the group is large, to enable individuals to communicate spontaneously. Participants need to be able to signal their intention to speak to the lecturer, who then activates that line if they wish. The degree of student control over the communication is therefore similar to a large lecture, i.e. not very great. It is further diminished by the barrier of a largely unseen audience. As in a lecture, there is little opportunity for social negotiation.

Videoconferencing invites the delivery of lectures. It is essentially a presentational medium as well as being a minimally discursive one. A two-way visual link can be hard to justify in an educational context. As one evaluation study found, students were reluctant to make use of the facility to ask questions themselves, and found the best use of the medium was often the traditional didactic lecture (Bollom *et al.*, 1989). Of course, the student may ask an academic question, or may be pounced on to broadcast their answer to one, but this does not make the medium truly discursive in reality. The current technology makes it an uncomfortable way to negotiate a shared conception. As a way of transmitting a didactic lecture, a video would be cheaper and easier. As a way of allowing communication with the tutor, a series of small-group audio-graphic links would be more effective. The link to a remote expert will always be a valued application. The two-way link between small groups learning from each other will be valued, though it needs additional support from text and audioconferencing to be fully useful. Videoconferencing as a medium offers less than the lecture in terms of pedagogy, and wins mainly on the logistical value of bringing people together across a distance.

Another Web-based video technology could have a future within education, however. The remote Web-Cam, giving participants access to a remote event, could be used in the context of the vicarious field trip. Embedded within a discussion environment, it is possible to imagine a tutor taking a group of students to an expensively-located art gallery, or a high-tech scientific laboratory. They could use a local camera linked to a shared area within a Lyceum environment, talking students through what they see, responding to their commands to control the camera to ask about what they are interested in. This exploits the virtual reality capability of video, but embeds it within a highly interactive discussion environment. The convergence of television and the Web should offer us some imaginative new hybrids to consider (Thompson, 1999). If they rise to the challenge of the Conversational Framework, and set out to meet the full range of students' learning needs, then we could see some very interesting developments around a new form of 'interactive video'.

STUDENT COLLABORATION

One of the great untested assumptions of current educational practice is that students learn through discussion. In the UK, the idea of 'learning through discussion' dominates many of the National Curriculum documents. It has always been acknowledged as important at university level, where seminars are a key teaching method. There is increasing research on collaboration between students using computers, but this work is only just beginning to look at the nature of the student–student discussion that results. Student–student discussion is certainly valued by students, as we saw in the quote above, but in comparison with the aspects of the learning process I suggested were essential, it addresses rather few. It supports the communication of the student's point of view. It is controllable by the student. It supports interaction at the level of description, although the fact that the feedback offered on a student's description is from another student, and not from a teacher is a significant difference. Argument between students about a topic can be an extremely effective way of enabling students to find out what they know, and indeed what they do not know, but it does not necessarily lead them to what they are supposed to know. As a mathematics lecturer I often found it difficult to get students to express what they had trouble with beyond 'I just don't understand it'. The most successful technique was to ask them to role-play teaching each other some small chunk of theory. After five minutes of this they were able to formulate (a) some profound questions I was unable to answer immediately, and (b) some points of such startling banality that I had assumed they were obvious after the first few minutes of the first lecture. The technique was excellent for alerting me to the deficiencies in my teaching, and for unblocking our communicative impasse, but their discussion alone was not always sufficient to sort out their problem. Discussion between students is an excellent partial method of learning that needs to be complemented by something offering the other characteristics, if students are not to flounder in mutually progressive ignorance. To avoid this, student–student discussions need to be able to consult a tutor, or should be required to summarise some kind of articulated output for monitoring by the tutor.

Studies of student–student interaction are universal in their enthusiasm for the richness of the interactions produced, and the potential they offer for learning to take place. They are all carried out as observations of what happens to take place as students interact in the pursuance of some task, whether generated by video, paper and pencil tasks, tutorial program, computer simulation, microworld or modelling task. They are equally universal in their recognition that the interactions are not always successful (see for example Durbridge, 1984b; McMahon, 1990; Hoyles *et al.*, 1991; Mason, 1994; Palloff and Pratt, 1999; Jones, 1999; Mason, 1998). Studies typically identify as sources of failure those aspects of the learning process that are missing from the media combination concerned, i.e. feedback from the teacher, and reflection by students on the goal–action–feedback cycle.

This is a field of research that has yet to produce a practice-oriented consensus on how we should support student–student dialogue to engender successful learning. From these early beginnings of recording the phenomena will eventually emerge some patterns of interactions, and some relations between these and the contextual characteristics of their occurrence. Some contexts seem to support productive interactions better than others do. All the studies mentioned here make recommendations for ways in which the tutor can foster the community, and help students make their collaboration productive. Studies like this will give us the means to develop computer and other environments that provide better support for students working together and unsupervised, both at the task level and at the description level.

SUMMARY

However the teacher–student discussion is managed, it is a vital part of the learning process. Without it, students have no opportunity to stand back from their experience, articulate the academic knowledge they are acquiring, and receive feedback on how they are expressing it. This is why misconceptions persist and remain resistant to the most concerted efforts of presentational teaching. Teaching has to be interactive and communicative to overcome misconceptions; the students need individualised responses to how they express what they know. The academic has to provide the learning environment in which this kind of interaction can take place: not just interaction with the world, but interaction also with the world of ideas and descriptions.

We would expect communicative media to be able to handle the discursive aspects of the learning process well. Audio and computer conferencing do support discussion, but offer far better support to the learning process when combined with interactive and adaptive media on the Web. Videoconferencing tends to approximate more to the lecture than the conversation. Student–student collaboration is highly valued, but without the teacher's roles of redescription and adaptation, the method remains at risk of failing to support learning. The contrasts are summarised in Table 8.1.

Table 8.1 Summary of communicative media characteristics

		CMC	*D3E*	*Audio-graphics*	*Video conf'g*	*Student collab'n*
1	T can describe conception	✔	✔	✔	✔	O
2	S can describe conception	✔	✔	✔	✔	O
3	T can redescribe in light of S's conception or action	✔	✔	✔	✔	✔
4	S can redescribe in light of T's redescription or S's action	✔	✔	✔	✔	✔
5	T can adapt task goal in light of S's description or action	O	O	✔	O	O
6	T can set task goal	O	✔	✔	O	O
7	S can act to achieve task goal	O	✔	✔	O	✔
8	T can set up world to give intrinsic feedback on actions	O	✔	O	O	O
9	S can modify action in light of feedback on action	O	✔	O	O	O
10	S can adapt actions in light of T's description or S's redescription	O	✔	✔	O	✔
11	S can reflect on interaction to modify redescription	O	✔	O	O	✔
12	T can reflect on S's action to modify redescription	O	✔	✔	O	O

Chapter 9

Productive media

INTRODUCTION

At the end of the introduction to Part II, I introduced the idea of 'productive media' construed entirely from the demands of the Conversational Framework. It makes repeated reference to action by the student, and articulation of their conceptions, hence the need for educational media that enable students to produce their own contributions via paper, disc, cassette or network. Paper has always been, and always will be an important productive medium for learners, still significant in schools, less so in universities, now that production of words is almost entirely electronic. The constructive areas of the curriculum, such as fine arts, media studies, design and technology, have a range of imaginative ways of enabling their students to produce work in a variety of media. The more theoretical areas have been confined to the written essay, report, or project. Electronic media have radically extended the range of expression for these areas with some rich and varied tools for building instantiations of ideas. HyperCard, mentioned in Chapter 6, p.109, is one example of a tool for building a network of ideas. The animation capabilities of PowerPoint could be a way of enabling a student to express their view of how a system works. But what are we actually using as the key enabler for student expression? Microsoft Word. Given what is possible in the electronic world, it might as well be a quill pen. In this chapter, we consider what might be, rather than what is, because there is very little in reality that exploits the productive capability of electronic media to allow the student to be the author. The impetus comes entirely from the predictive capacity of the Conversational Framework. It is, after all, one of the useful properties of a theoretical framework that it creates an expectation of what might be, rather than classifies what is.

MICROWORLDS

A terminological difficulty attends the usage of 'microworlds' to describe one form of productive media. Some authors refer to simulations as microworlds, an understandable confusion, because it is a feature of simulations that they allow the user to act within a 'little world'. Microworlds made their biggest impact in education in the form of Logo, Seymour Papert's programming language for geometry. In his book *Mindstorms*, Papert describes the reasoning behind the development of a Newtonian microworld, and in doing so, expresses exactly the difference between a microworld and a simulation. What makes Papert's microworlds interesting is that they appear to address explicit descriptions of the student's point of view:

> Direct experience with Newtonian motion is a valuable asset for the learning of Newtonian physics. But more is needed to understand it than an intuitive, seat-of-the-pants experience. The student needs the means to conceptualize and 'capture' this world ... The Dynaturtle on the computer screen allows the beginner to play with Newtonian objects. The concept of Dynaturtle allows the student to think about them. And programs governing the behaviour of Dynaturtles provide a *formalism in which we capture our otherwise too fleeting thoughts.*
>
> (Papert, 1980, 124, my italics)

The formalism provided as an essential feature of a microworld allows the student to express their description of some aspect of the world in a form interpretable by the program itself. The simulation offers no such means of representation, only actions encoded as option choices of parameter changes. In a microworld, the student is building their own runnable system, whereas in a simulation they are controlling a system that someone else has built. The mode of interaction with the subject matter is very different.

The microworld provides a mediating mechanism for acting in its world, e.g. a programming language. This provides a level of description of what is happening in that world. To use Papert's physics microworld, a student has to describe their actions in the form of a set of commands, then run them as one would a program, and the result is either the intended behaviour or something unexpected. The feedback operates at the level of the description. In one version of Logo geometry, they control the movements of the Dynaturtle. For example, the student types in a set of commands to draw a square (e.g. 'forward 100, right 90, forward 100, right 90, forward 100, right 90'), then runs it, and the computer draws three sides of a square (see Figure 9.1).

The program provides intrinsic feedback on their description of the action, i.e. that their description was incomplete and they need to add another command similar to the first. Having perfected the description of a square (by adding another 'forward 100'), they can adapt and develop that to produce more elaborate outcomes in the microworld. If they create a more general model, i.e. a program with variables (e.g. 'repeat X (forward Y, right Z)'), they can then investigate its behaviour under different initial inputs, using the model as one does in a simulation. Some create star patterns, others polygons, others spirals, and so on. Through experimentation of this kind, students will gradually understand the

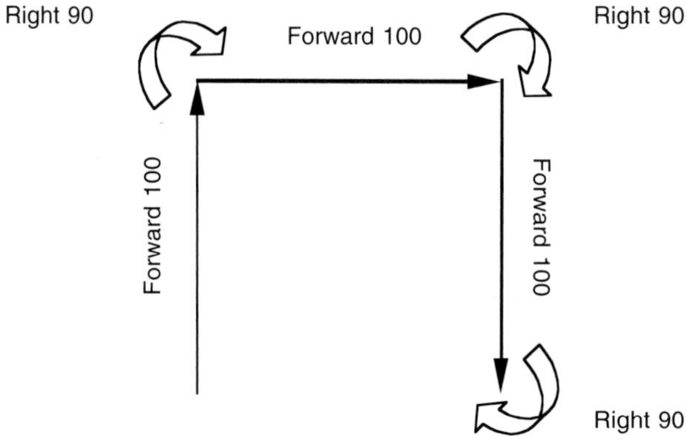

Figure 9.1 A microworld for geometry. *The program gives visual intrinsic feedback on an incomplete representation of a square as the sequence of commands: 'forward 100; right 90; forward 100; right 90; forward 100; right 90'.*

relationship between the behaviour of the system and the form of its representation, to the point where they can control it at will.

In creating the model, students clearly have a different kind of learning experience from those who only use a model created for them. The difference lies in whether the action exists as a description, and can be 'captured' for inspection, reflection and revision resulting from feedback, as in a microworld, or whether it remains a fleeting thought captured only as part of the memory of the action, as in a simulation.

Microworlds are productive media, therefore, in the sense that they enable the learner to create and produce a system of their own, designed to achieve a specific end. Simulations are adaptive, and the student can only explore and investigate, not create and produce.

A microworld is a very limited productive medium, however. The little world the user inhabits is one that has been highly constrained by the designer. In this sense, microworlds are similar to simulations. Using the power plant simulation discussed in Chapter 7 is quite unlike acting in the real situation, where the student might be disposed to investigate, say, the effects of alternative forms of coolant. In the simulation, the user has no choice but to use the one offered by the program. The students are unable to experience aspects of this world that are not part of the simulation, such as varieties of coolant. They cannot avoid looking through the designer's spectacles. Similarly, in the design of the Dynaturtle and the programming language that governs its behaviour, Papert constrained what students could do with it, so that their view of this microworld incorporated the perspective he wished them to take. The microworld is a productive medium that incorporates a learning objective into its design. It is importantly different, therefore, from other productive media, such as pen and paper, wordprocessor, spreadsheet, which are too generic to incorporate a specific learning objective.

The key structural difference between simulations and microworlds that classifies simulations as adaptive media, and microworlds as productive media is the nature of the intrinsic feedback they support.

Feedback at the level of description, instead of just action, is important if we want an educational medium to address all aspects of the learning process. We have seen the value of this in the communicative media, in which people provide feedback on descriptions through language. The microworld does it differently. The form of description is a formalism, designed to be easier to understand and use than a mathematical formalism:

> It bypasses the long route (arithmetic, algebra, trigonometry, calculus) into the formalism that has passed with only superficial modification from Newton's own writing to the modern textbook.
>
> (Papert, 1980: 124)

Papert's primary concern is not in fact to provide the means for describing actions in the world, but to provide the means for students to enter the Newtonian way of thinking without having to use mathematics as the medium:

> We shall design a microworld to serve as an incubator for Newtonian physics. The design of a microworld makes it a 'growing place' for a specific species of powerful ideas or intellectual structures.
>
> (Ibid. 125)

Papert wants to give students direct access to the physicist's way of experiencing the world, enabling them to develop an intuitive grasp of the correct Newtonian conceptions. This is more compatible with the idea of academic knowledge as 'situated cognition' than as second-order descriptions of the world. It seeks experiential knowledge of the world, just as a simulation does, rather than articulated knowledge. Nonetheless, surely the formalism of the microworld fits the requirement I developed earlier for a teaching strategy that can interact at the level of descriptions? Papert's concern was that the formalism should be intelligible to the student, and he has made it so. He created it to be a medium of expression through which the student can create and explore a simulated world, aided by intrinsic feedback, just as a mathematical physicist uses mathematics as a medium of expression for exploring the physical world. But is the formalism of a microworld a formal description at the discursive level or a situated description at the interactive level? It is an interesting challenge to the Conversational Framework to be sufficiently well defined that it can adequately account for the descriptions used in a microworld.

I previously referred to academic descriptions as though they were purely language-based. The characterisation of academic knowledge as being second-order – standing back from experience of the world, articulating what is known – all presuppose language as the vehicle of expression. And yet academic knowledge is not only represented through language. The more usual alternatives are mathematical symbolism and diagrams, carried through the medium of print. When computers become as ubiquitous a medium as the book, why should academic knowledge not also be expressed through the medium of a program?

This is close to the mission that Papert expresses in his book, that learners of all ages should be taught computational modelling as a powerful intellectual tool. He argued that being able to program a system is as good a way of knowing it as being able to describe it mathematically or in language. Is it though?

Papert quotes a teacher as saying 'I love your microworlds, but is it physics?' and he seriously considers the question it raises about how far one should attempt to reconceptualise classical domains of knowledge as microworlds (Ibid. 140ff). His main argument is that turtle physics is closer to the spirit of real physics than the physics of the stereotyped classroom where formulae are meaningless rituals. It is a good defence, but is it close enough to real physics? Or to take the discussion beyond just physics: is computational modelling within a microworld an adequate representation of the academic knowledge in a discipline? The reader can probably sense my implicit 'no' in answer to these questions. The reason lies in what is embedded in the design of the microworld. That is where the real physics is. Reading the description of the development of any microworld, be it physics (Sellman, 1991) or music (Holland, 1987), it is apparent that tremendously hard thinking about the subject has to be built into the way the program objects are designed. The user will use these building blocks of programmable objects, or commands, or rules, to create a system that models some theory of the real world; they are not generic. They are theoretical constructs, peculiar to the theory of the world that is being built. Once they are designed, students can build with them and explore how they work. Then they have access to that special world, defined by the theoretician, and can learn about it in the way we learn about the real world by experiencing and acting on it. The computational model they devise brings them closer to an intuitive understanding of what goes on in that world, but does not help them express it fully. The programmable objects are a truncation of the real physics into manipulable chunks. So my answer to the teacher's question is no, I do not think this is physics, at least not academic physics, any more than playing with real bricks is academic physics. I do accept Papert's argument that the microworld provides students with the experience of this perspective on the world that will be useful for thinking about academic physics.

The intrinsic feedback in a microworld is at the experiential, interactive level, not the theoretical discursive level. This is reminiscent of the video analogue of an academic idea – it affords a particular way of experiencing the world. The microworld covers a wider range of learning activities than the narrative medium, as Figure 9.2 shows.

The microworld is controllable by the student and therefore supports interaction, adaptation and reflection. Students construct something, see how well it works, use this intrinsic feedback to improve the construction in terms of their

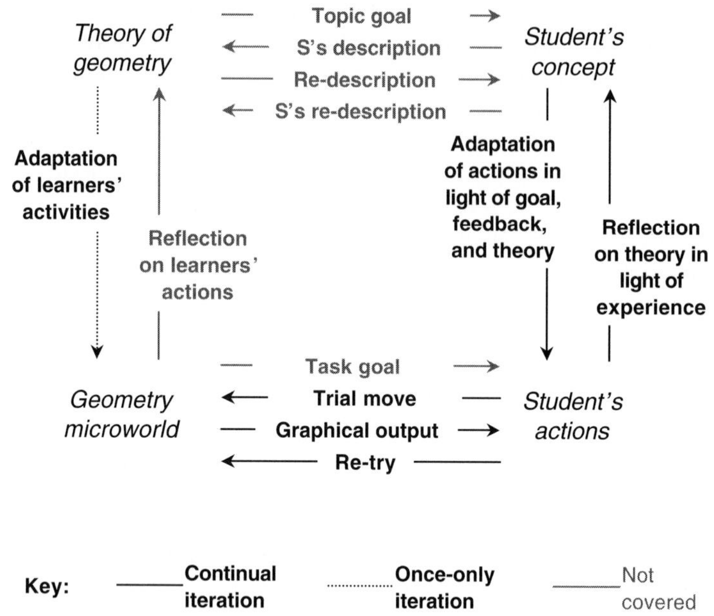

Figure 9.2 Interpretation of the Conversational Framework for a microworld on concepts in geometry.

immediate goal, and because they can create their own goal, are able to test their own conceptions. But there is no iteration at the theoretical discursive level.

It is important to be clear about the nature of the learning experience each style of medium offers, because in the consideration of individual examples it is very easy to be distracted by the particular content, or the particular implementation. Many of these systems are extremely attractive, especially to teachers and experts who already understand the subject very well, and see immediately the potential for discovery and play that they offer. But we are considering them here from the point of view of benefit to learners who have all the usual conceptual difficulties. What is the real pedagogical significance of a microworld for them? It is designed, remember, to help them become familiar with a world which is normally only accessible through mathematics. A Logo-type microworld makes exploring Newtonian objects and motion as much as possible like exploring the behaviour of building bricks in the real world. That is the whole point, as defined by Papert. The metaphor is entirely apt. But to what extent is the child playing with bricks, even if noting down a record of the moves made with them, doing physics? This is coming to an intuitive understanding of the world, it is learning about the world through experience, and the analogy follows through to the physics students using the Dynaturtle – they are learning about the Newtonian world through experiencing it. So it is not academic knowledge they are acquiring, but experiential knowledge of an academically defined world. I would not deny its importance, nor its value to the student who then uses this in building their academic knowledge, but the two are not the same. This is what accounts for Papert's teacher's anguish – constructing systems in a microworld is not what it takes to do academic physics. The reasoning students are doing while operating in a microworld is helping to build their personal theory of that world, just as the child builds theories of the physical world by playing with bricks. But they must also learn the generic formalism of mathematics and the way that physics theories are expressed in language, to be able to take their own exploration and understanding further than the confines of the microworld. Microworlds do not operate at the level of formal descriptions of the subject.

To summarise: simulations and microworlds are similar in the sense that they both operate at the experiential level. They are different because microworlds are also productive, allowing students to go beyond exploration of a given model to creating their own model. Does this structural difference have any pedagogical significance in terms of the Conversational Framework? I think it does, and it lies in the fact that a microworld enables the student to define their own topic goal in the interaction, and therefore encourages them to reflect upon the interaction. A simulation supports a more limited number of possible goals, depending on what parameters are available to the student. It needs to be augmented with a teacher-defined goal, to ensure that the student addresses a goal, and does not simply play with the parameters to 'see what happens if …', and only see, and not reflect on what they see. Without reflection, the simulation, for all its interactive adaptivity, contributes little to new understanding. The microworld presupposes that the student will define their own goal. This may be more motivating than working to a teacher-defined goal, and is also more likely to encourage reflection on how well their interaction is meeting their conceptual, topic level goal.

COLLABORATIVE MICROWORLDS

Simulations and microworlds have always been used collaboratively in school classrooms, where sharing of computing equipment is inevitable. The collaborative mode can add to students' motivation and enjoyment, but unless the design of the program takes account of group use, it can be frustrating for students who are participating only vicariously while one student operates the controls. The program must support and encourage collaborative decision-making if it is to succeed in this mode (Luckin *et al.*, 2001). Group collaboration around computer-supported media has not been a focus of design for university teaching, but this is changing with the increase in Web-based collaboration (Winer *et al.*, 2000). Whalley gives a historical overview of these productive media, and describes the aim of linking an interactive modelling environment for mechanics to a communicative discussion environment for networked classrooms:

> Our aim was simply to create an environment in which the procedural aspects of the transformation of observation to data, and data to symbolic representation could be brought out clearly, and allow them to be enriched by collaborative discussion of what appears to be happening, and the children's prediction of what will be happening.
>
> (Whalley, 1998: 57–58)

This environment does follow through to symbolic representation, though not as expressed by the learners – the program carries out the transformation for them:

> The plot mode ... dynamically illustrates how each pair of data points combine to make a single plot point, and is designed to extend conceptually the 'what if' thinking that children may already carry out with spreadsheet packages ... All users share the same model space, and if in the same mode are locked together and see the result of each other's actions.
>
> (Ibid. 58)

Collaboration over the network enables the students to comment on each other's experimental actions and results, and to discuss predictions from further actions. There is relatively little research on students' use of these collaborative microworld environments as yet, but with more distance learning, and more use of the Web, it is likely that this will be a an area of rapid growth. In terms of the Conversational Framework, their value is to extend the microworld to the discursive level, either with other students, who then collectively produce something that a teacher evaluates independently, or by incorporating the teacher's comments as part of the discussion environment, as in Figure 9.3.

The Conversational Framework is reproduced for each student, but in contrast to the microworld alone, there is additional interaction at the discursive level between students, as well as with the teacher. Each student is acting on the microworld and receiving feedback on the actions of both. As with the combined media discussed in the previous chapter, the inclusion of the discussion environment completes the Conversational Framework for the productive media. In comparison with the adaptive media, they have the additional motivational benefit of allowing students to set their own goals in building a model. The constraint of the microworld design still obtains, however, and this is no more adaptive in the light of discussion than is the simulation.

MODELLING

A modelling program invites the learner to create their own model of a system, which it then runs, allowing the output to be compared with stored data of a real-world system, or the program's own model. It contrasts with a simulation because

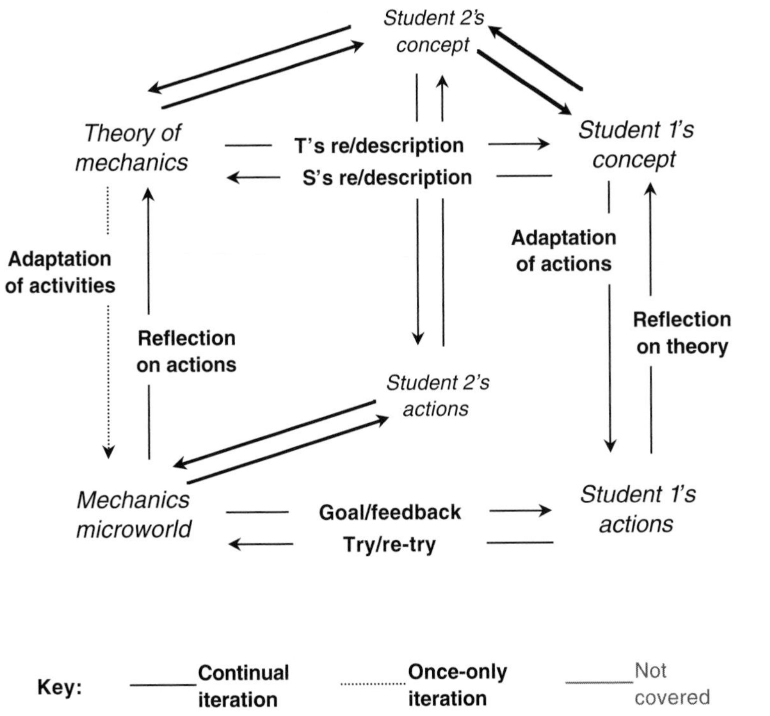

Figure 9.3 Interpretation of the Conversational Framework for a collaborative microworld on concepts in mechanics.

the student manipulates the model itself, not just parameters within a given model. For example, in a program on the concept of moments of inertia, students are asked to define the equation to be plotted in order to find the best fit to a given set of case study data (Laurillard *et al.*, 1991). The topic goal is set by the teacher within the program, at the discursive level. The student uses the modelling environment to test their model, and their inputs to the model, against a task goal set in terms of the case study data. The student's articulation of their understanding is therefore carried out in terms of the formalism of the subject matter, i.e. mathematical equations. Figure 9.4 shows the nature of the fit with the Conversational Framework.

There is no iteration with the teacher's description because they are not present, and their reflection is once only – in terms of the case study data. But the student does have the means to redevelop their own description because of the intrinsic feedback from the model.

A modelling program contrasts with a microworld in the sense that the student defines their model directly. It is not buried within the design of objects. In a modelling program, the program merely interprets formulae (or rules): it knows nothing about any subject matter, unlike the microworld. A physics microworld

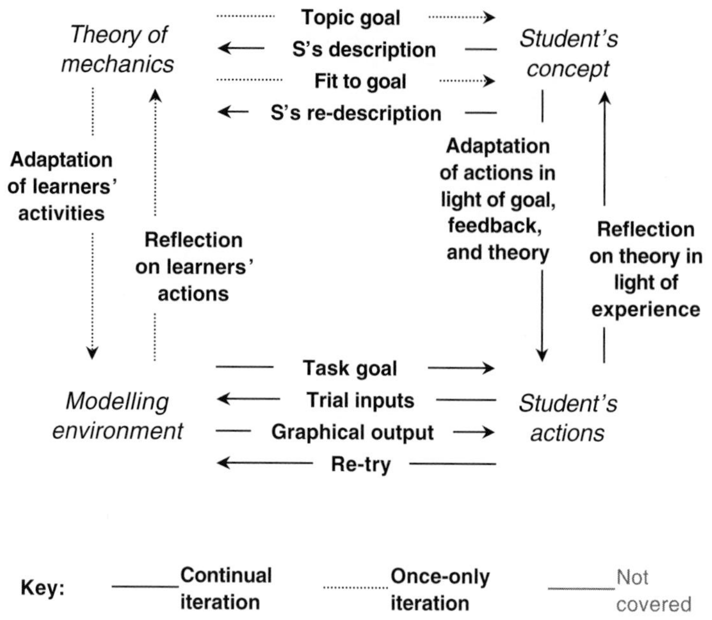

Figure 9.4 Interpretation of the Conversational Framework for a modelling environment for mechanics.

can only be used for physics; a modelling program can be used for anything that can be modelled. There is a considerable interface design problem in getting the program to interpret the learner's description of the model, and this is the clever part of designing such a tool. Perhaps the best known commercial modelling program is 'Stella', although students find it difficult to operate (Mandinach and Cline, 1996). Defining a suitable description language or representational system for students to express their conception of an idea, or model a problem, is a complex design task (see for example Reusser, 1992). Spreadsheets, which offer a template for defining an equation, provide a much simpler form of productive tool. The concept-mapping tool shown in Figure 8.5, and other tools of a similar kind, also offer useful ways of enabling students to articulate their own ideas.

The program may contain its own model, or data, for a particular topic, which it can compare with what the student has done. If so, it can prompt some reflection on the student's outcome in comparison with the goal, as in the mechanics example above. Since the program has access to the student's model, it could also be programmed to offer canned text as extrinsic comment on particular characteristics of its relationship to the known model, assuming it knows the topic goal. This is a plausible extension, but certainly not part of the standard form. It supports an explicit representation of the student's model, but in mathematical and graphical form only. It gives feedback on the student's action by running the model, and is fully controllable by the student.

The structural form of a modelling system is more iterative than a microworld. The difference is in how the student's conception is expressed: in a microworld as representations of actions in the world; in a modelling system as a representation of the world in which those actions take place. Here there would be no doubt in the teacher's mind that this is 'doing physics', and that is because the focus of the students' talk is on how to express the behaviour of a system mathematically. If the aim is not merely to experience the world but also to explain it, then the modelling program is the closest so far to supporting the learning of academic knowledge. The modelling environment is more generic, and would allow the student to create their own kind of model, rather than their own model of a particular type. It is the most unconstrained form of productive medium.

SUMMARY

The productive media include microworlds, productive tools, and modelling environments. Their key properties are to provide an electronic context in which:

- the learner can build something;
- they engage with the subject by directly experiencing its internal relationships;
- they learn to represent these relationships in some general formalism.

At the end of Chapter 7, I suggested that educational games would become an important form of interactive learning environment. They could also provide the basis for productive learning environments. Several forms of computer game offer creative or productive opportunities to users; the 'simulation worlds' are an obvious example. Collaborative games have become a key feature of Web use for whole communities of games players. But these types of productive media develop in entirely separate universes from that of education, not even converging in research labs. It is a lost opportunity, because, as we have seen, the collaborative microworld and the modelling environment, as shown in Table 9.1, are powerful media for learning. Collaborative gaming environments take this form. Harnessed to serve an educational end, they would be extremely valuable.

Table 9.1 Summary of productive media characteristics

		Microworlds	*Collaborative microworlds*	*Modelling environment*
1	T can describe conception	O	✔	✔
2	S can describe conception	O	✔	✔
3	T can redescribe in light of S's conception or action	O	✔	O
4	S can redescribe in light of T's redescription or S's action	O	✔	✔
5	T can adapt task goal in light of S's description or action	O	O	O
6	T can set task goal	O	✔	O
7	S can act to achieve task goal	✔	✔	✔
8	T can set up world to give intrinsic feedback on actions	✔	✔	✔
9	S can modify action in light of feedback on action	✔	✔	✔
10	S can adapt actions in light of T's description or S's redescription	O	✔	✔
11	S can reflect on interaction to modify redescription	O	✔	✔
12	T can reflect on S's action to modify redescription	O	✔	O

Summary of Part II

COMPARING THE MEDIA

The five chapters in Part II have covered most of the technological media likely to be used in the service of education. The analysis at the end of each chapter showed the extent to which each one can support the learning process as defined by the Conversational Framework at the beginning of Part II. The Media Comparison chart in Table II.2 compares all the different forms of media – narrative, interactive, adaptive, communicative and productive – in their non-enhanced forms. Each chapter has discussed ways in which the media forms can be combined to produce better coverage, e.g. the narrative medium of audio is enhanced with exercises, interactive Web resources are enhanced by inclusion of a communicative environment, and so on. Here we compare the basic forms.

We can see from the comparison that none of the current learning media covers the full iteration between reflective and adaptive discussion and interaction in the way that a teacher in a practical session could. However, they cover the majority of learning activities, and in combination, they cover all the essential activities in the learning process, as defined by the Conversational Framework. Figure II.3 represents the same point graphically.

An analysis of this kind is not fine-grained enough to differentiate all the contrasts possible. It would become unwieldy if it attempted to, but it does show how each type of medium needs further support, and which other media might provide it. The columns are at least additive: a combination of the two media represented by two columns will inherit the combined characteristics of both, as the enhanced media show in the tables at the end of each chapter in Part II. They may even be multiplicative, as in the case of tutorial and simulation. Neither on its own allows the student to 'reflect on interaction to modify description', the tutorial because there is no interaction, and the simulation because there is no facility for the student to describe their conception. In combination, however, each provides the facility the other lacks, and so produces a multiplicative effect of better coverage of the learning process.

This kind of analysis does not determine the selection of media; it is not a prescriptive process. However, it does show how to integrate a range of media in

Table 11.2 Media comparison by degree of fit to the Conversational Framework

		Narrative	*Interactive*	*Adaptive*	*Commun'tive*	*Prod'tive*
1	T can describe conception	✔	O	O	O	O
2	S can describe conception	O	O	O	✔	O
3	T can redescribe in light of S's conception or action	O	O	O	✔	O
4	S can redescribe in light of T's redescription or S's action	O	O	O	✔	✔
5	T can adapt task goal in light of S's description or action	O	O	✔	O	O
6	T can set task goal	O	O	✔	O	O
7	S can act to achieve task goal	O	✔	✔	O	✔
8	T can set up world to give intrinsic feedback on actions	O	✔	✔	O	✔
9	S can modify action in light of feedback on action	O	✔	✔	O	✔
10	S can adapt actions in light of T's description or S's redescription	O	O	✔	O	✔
11	S can reflect on interaction to modify redescription	O	O	✔	O	✔
12	T can reflect on S's action to modify redescription	O	O	✔	O	O

Key
Narrative: print, TV, videocassette
Interactive: CD, DVD or Web-based resources
Communicative: Web-based conferencing, asynchronous or synchronous
Adaptive: manipulable model on disc or Web
Productive: tools for student to create models or descriptions, on disc or Web

order to exploit the strengths of each. The clear conclusion is that improvements in university teaching are more likely to be achieved through 'multiple media', appropriately balanced for their pedagogic value, than through reliance on any one learning technology.

BALANCING THE MEDIA

It is hard to predict the optimal balance of time a student should spend in working on learning materials, participating in discussion, reading, writing, listening, and practising. It will vary from one subject to another, and according to the way teachers design their courses. The optimal balance evolves with practice. The more attentive teachers and students are to evaluating and reflecting on the practice, the more effective it will become. There is the danger that without this, the pressures of

Table II.3 Distribution of study time across media forms and modes of study

Media forms	*Lecturer*		*Group*		*Student*		*Total*
	Standard	*ICT*	*Standard*	*ICT*	*Standard*	*ICT*	
Narrative	–	–	–	–	–	–	30
Interactive	–	–	–	–	–	–	20
Communicative	–	–	–	–	–	–	20
Adaptive	–	–	–	–	–	–	10
Productive	–	–	–	–	–	–	20
Totals		**21**		**21**		**58**	**100**

Key

Narrative	print, TV, videocassette
Interactive	CD, DVD or Web-based resources
Communicative	Web-based conferencing, asynchronous or synchronous
Adaptive	manipulable model on disc or Web
Productive	tools for student to create models or descriptions, on disc or Web
Standard	standard methods: lectures, tutorials, essays
ICT	information and communications technologies

institutional demands will determine the balance, which would not necessarily be optimal for student learning. Careful planning of student workload, and the distribution of their time across different kinds of learning, does not often feature prominently in course design. This is worth doing even at the basic level of the balance across media forms, to clarify for both staff and students how the optimal balance is construed. The notion of balanced media becomes figural. It is part of the debate about how to support students, and feeds into resource planning discussions (see Chapter 11). One way of looking at the balance is shown in Table II.3, which considers the distribution of study time across the five media forms, and across the three modes of lecturer-led, group work, and individual study. As long as all five media forms are included then it is likely that there is complete coverage of the learning activities in the Conversational Framework. For each dimension, there should be some optimal distribution, which will then break down into delivery via normal or ICT-based media. But what would be optimal?

In a hundred hours of study, over, say, two weeks of full-time study, the student would be unlikely to have more than ten contact hours per week with a tutor or in a lecture, and a similar number of hours spent on group work with other students. The majority of their study time is likely to be spent on individual self-study: reading, practising, producing. Comparing rows, notice that narrative forms are assigned the highest proportion for undergraduate study, as the learner is engaged in trying to understand what is already known in their field. Proportions would vary for different stages of education, and types of course, but probably not by more than 10% for each line. The proportion for narrative media would reduce for postgraduate study, for example, where a greater proportion of time would be spent in practising through adaptive forms (such as experiments) and producing (reports and ideas). Once the general distribution is agreed for a particular type of course, the academic designers can then decide how they would assign these hours to normal and ICT methods within the media forms and modes of study. Table II.4 suggests one such breakdown.

Narrative is delivered mainly through lectures and books, as there is little point in using ICT for such forms. Interactive forms would be partly tutorials with tutor as interactive resource, and partly self-study interactive materials. Communicative forms would be tutor-led seminars and student discussion groups, face-to-face and online. Adaptive forms would be partly tutor-led practical classes, workshops, and fieldwork, and partly self-study practice. Productive forms would involve both group projects with other students, essay writing or the equivalent, and feedback from the tutor on the work produced. With this distribution, even attempting to maximise ICT, only 52% of the work is ICT-based. To increase this proportion would mean less staff contact, less place-based work with other students, and inappropriate use of ICT for narrative forms, all of which would be less than optimal. With an analysis of this kind, it becomes possible to see the extent to which the idea of a wholly electronic university is an extremely sub-optimal solution. Even for distance learning universities, the maximum proportion of ICT-based work is similar, as Table II.5 shows.

Table II.4 Breakdown of study time across media forms and modes of study

Media forms	*Lecturer* *Standard*	*ICT*	*Group* *Standard*	*ICT*	*Student* *Standard*	*ICT*	*Total*
Narrative	2	2	–	–	26	–	30
Interactive	4	8	–	–	–	8	20
Communicative	2	–	6	12	–	–	20
Adaptive	2	–	–	–	2	6	10
Productive	1	–	3	–	–	16	20
Sub-totals	*11*	*10*	*9*	*12*	*28*	*30*	
Totals		**21**		**21**		**58**	**100**

Table II.5 Breakdown of study time for distance learning

Media forms	*Lecturer* *Standard*	*ICT*	*Group* *Standard*	*ICT*	*Student* *Standard*	*ICT*	*Total*
Narrative	–	3	–	–	27	–	40
Interactive	1	2	–	–	–	12	15
Communicative	1	–	2	12	–	–	15
Adaptive	–	–	–	–	1	9	10
Productive	–	1	–	3	–	16	20
Sub-totals	*2*	*6*	*2*	*15*	*28*	*37*	
Totals		**8**		**17**		**65**	**100**

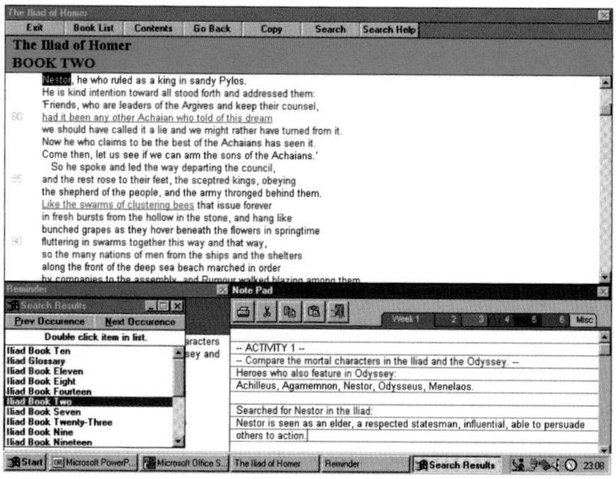

Plate 1 (Figure 6.4) An interactive program on the Homeric poems. *The search window shows occurrences of the item (Nestor) being searched. The text window displays the extract selected with the item highlighted. It also shows hyperlinks to further notes in the Companion Guide. The Note Pad shows the current activity and the student's notes.* © The Open University.

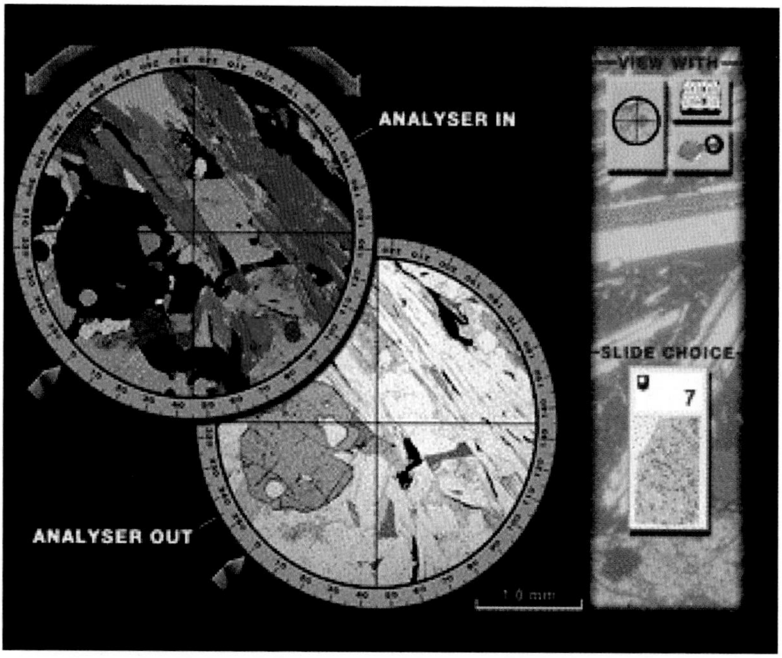

Plate 2 (Figure 7.3) The Virtual Microscope, showing the two views through the simulated microscope, and the icon for selecting different materials for investigation.
© The Open University.

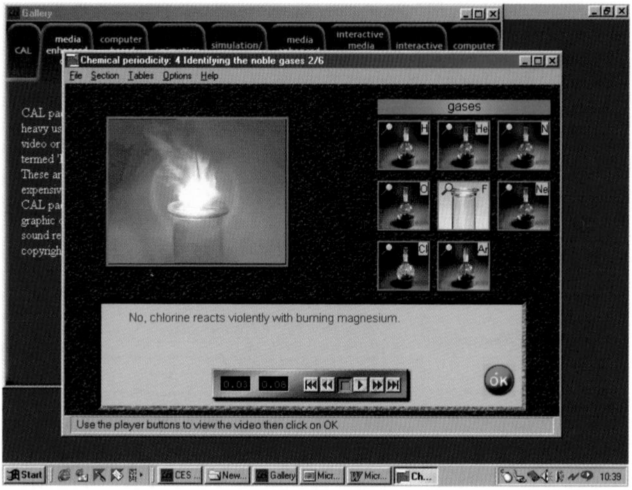

Plate 3 (Figure 7.4) A tutorial program on chemical periodicity. *The student has previously selected reactions with each of the gases, to see which ones yield a violent reaction. They are now being tested on which gases are 'noble' and do not react. On selecting chlorine, they are given the feedback of a repeat of the demo showing how chlorine reacts, together with a brief comment.* © The Open University.

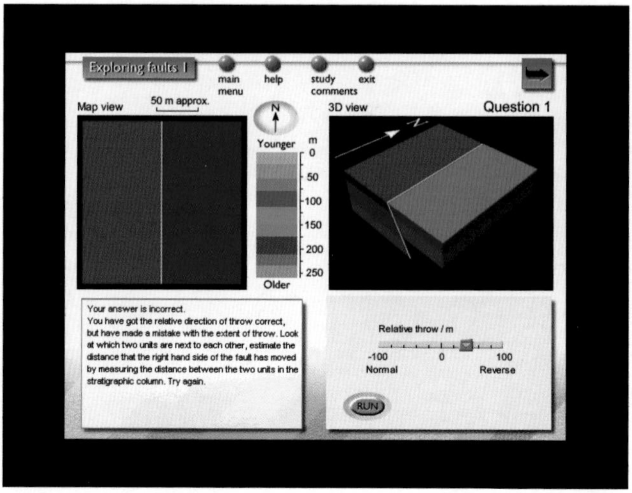

Plate 4 (Figure 7.5) A tutorial-simulation program on geological formations. *The student must specify the way the rock formation on the right has to move and erode in order to expose the surface features represented on the left. The direction and amount of the movement is controlled by the slider. The result of their first move is shown on the right, with commentary below left. The order of the sedimentary rock layers is shown in the middle.* © The Open University.

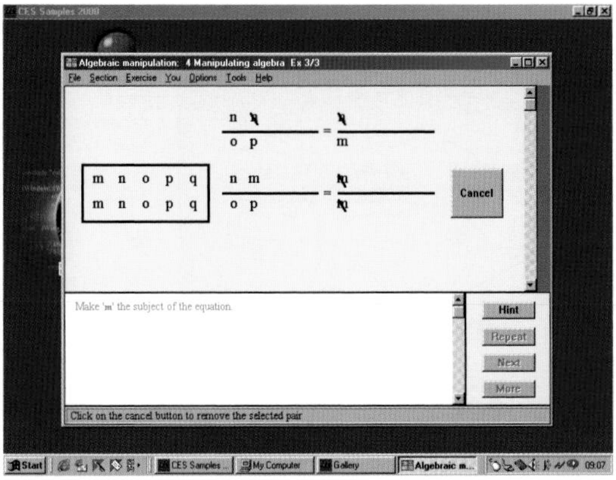

Plate 5 (Figure 7.7) A tutorial-simulation program on algebraic manipulation. *Each step is driven by the student. In the previous step illustrated they clicked on each 'n' and then on the Cancel button. In the current step (the lower one), they had to click on two m's in the box, and then drag each one to its current position, and then click on the two right-hand m's and the Cancel button to cancel them.* © The Open University.

Plate 6 (Figure 8.2) A digital document discussion environment for an article. *Digital Document Discourse Environment (D3E). Knowledge Media Institute, Open University, UK. Available at: http://d3e.open.ac.uk.* © The Open University.

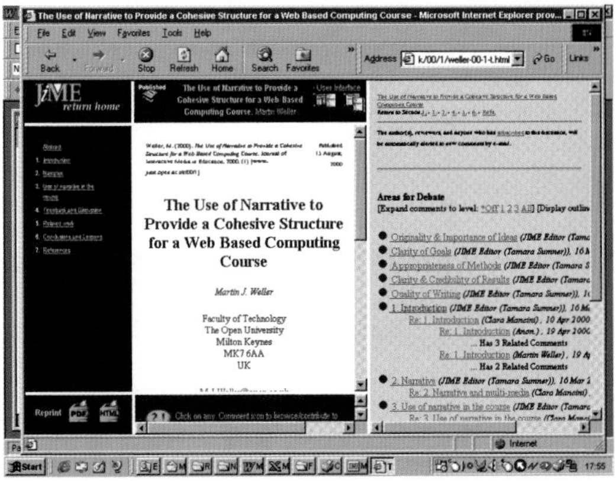

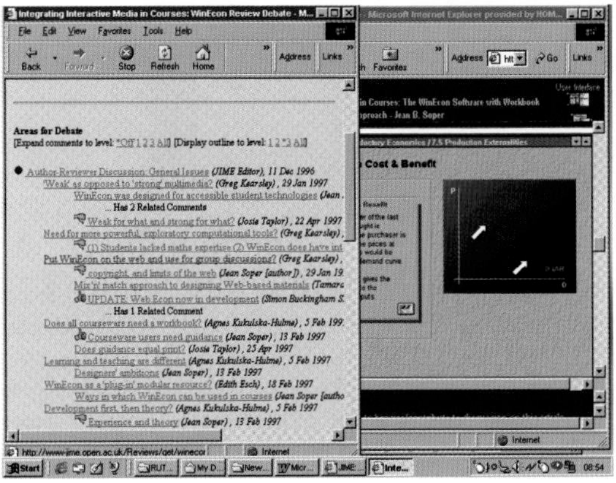

Plate 7 (Figure 8.3) A digital document discussion environment for a runnable simulation. *Digital Document Discourse Environment (D3E). Knowledge Media Institute, Open University, UK. Available at: http://d3e.open.ac.uk.* © The Open University.

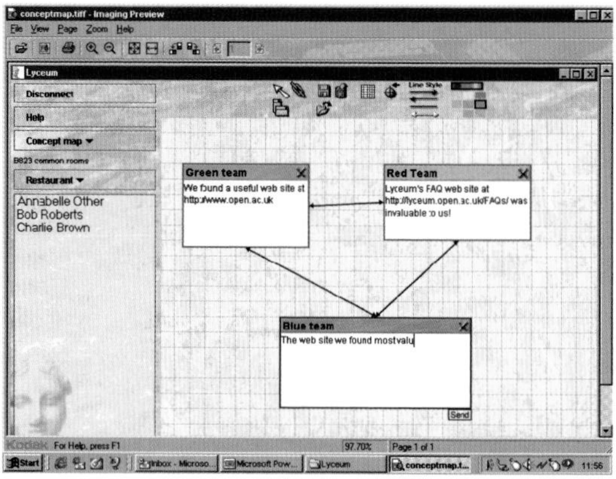

Plate 8 (Figure 8.5) An audiographics conferencing environment on the Web. *The left-hand column shows who is online. When someone speaks, a microphone icon appears beside their name. The shared screen on the right is a concept mapping tool being used by the group to organise contributions from the three participants. Each one may be simultaneously typing into a box. When one hits the Send button, their contribution appears on all screens.* © The Open University.

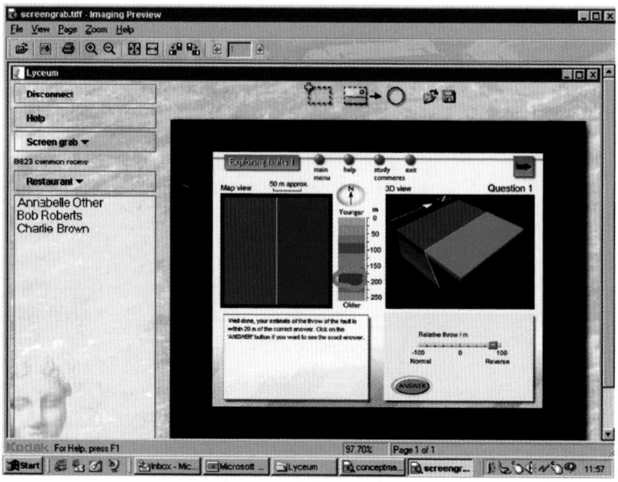

Plate 9 (Figure 8.6) An audiographics task-based environment on the Web. *The discussion concerns a difficulty one student has with a task on the geology course CD, already seen in Chapter 7, Figure 7.5 (Plate 4). The student has copied the troublesome screen to the shared window, and the tutor has asked them to indicate which rock layer they should be aiming to expose in the task. The student has responded by circling the appropriate layer in red (see Plate 9). The discussion can now proceed in light of the student's action.*
© The Open University.

The proportion of time spent on narrative forms is higher, with less opportunity for interactive and communicative forms, but the introduction of ICT communications has improved these proportions over earlier forms of distance learning. Contact with the lecturer or tutor is necessarily lower, but again is improved with the introduction of ICT. Nonetheless, the proportion of ICT-based learning is still only 58%. The wholly electronic university is not likely to be optimal, and is certainly not the only model for an online course (Mason, 1998). A significant increase in use of ICT would be possible if narrative forms were optimised for ICT. If electronic paper became a commonplace reality, then we would be coming very close to the genuine e-university.

Given that academic knowledge is a consensual description of experience, it follows that discussion between teachers and students should play a very important part. It should be the mode of learning that drives everything else a student does, even if it is allocated only a small part of the total study time. It should not be vanishingly small, however, and there is an increasing danger that it will be. The continuing increase in student numbers in universities makes it ever more unlikely that individual students will have more than the briefest conversation with an academic during a course. Careful planning of study time for a course will clarify the extent to which students are receiving a genuinely discursive education. Without this element of debate and discussion around academic ideas, universities will become training camps, unable to do more than expose their students to what there is to be known, and to rehearse them in the ability to reproduce it. The learning technologies will not overcome the problem of worsening staff:student ratios. Each medium has its strengths, so they can help, but every learning environment needs to embrace a teacher–student dialogue, and that is undeniably labour-intensive.

Part III

The design methodology

Chapter 10

Designing teaching materials

INTRODUCTION

This chapter begins the practical section of the book. Chapter 1 laid the foundations for an underlying philosophy of what academic education is trying to do. Chapters 2 to 4 used our knowledge of how students learn to establish a principled approach to the selection and design of media-based methods. This formed the basis of a Conversational Framework for the learning process, developed at the beginning of Part II as a basis for analysing the main technology-based educational media. Chapters 5 to 9 used existing examples and studies of each medium to clarify their respective capabilities, and to establish the extent to which each supports the essential activities involved in the learning process.

Now there has to be a slight dislocation in the line of argument. We cannot simply deduce the design of learning activities from the media capabilities. First we return to a consideration of what the student needs, and then use the Conversational Framework to bring the two together. The needs as defined will challenge the media, and clarify the extent to which they fail to deliver what pedagogy requires. We may as well know it.

The design of learning materials for any medium should begin with the definition of objectives and analysis of student learning needs. Objectives will usually be given via the curriculum aims that determine what students need to know or be able to do for a particular subject area. The curriculum aims are defined in terms of the topic – the perceived priorities and values from the academic's point of view, and from the point of view of market demand – in terms of future learning needs, and the knowledge and skills appropriate for those graduates. The student is considered primarily as a future expert at this stage.

Curriculum aims are general, and need to be specified in more detail as learning objectives if they are to assist the design process. Without clearly defined objectives, educational design becomes mere exposition. The academics must know what they wish the student to achieve if they are to bring rational planning to design and development, and to recognise when the end has been achieved.

We must also address students' learning needs. Without an appreciation of students' learning difficulties, the teacher risks talking over students' heads, or bypassing them completely, or at worst, creating such confusion that they are incapable of rational judgement. Students' misconceptions are typically 'pedagogenic errors', the equivalent of iatrogenic diseases (the kind that are caused by doctors' actions), born of poor teaching rather than ignorance. Teachers must address students' current conceptions if they are to guide them towards the consensus conception.

This analysis of where students are, and where teachers wish them to be, will reveal a clear logical relation between the two. Defining the learning activities that will bridge that gap is not a simple logical problem, however. We do not have a learning theory or instructional theory complete enough to perform that trick, and I even doubt that such a thing is possible. That is why this book stresses methodology more than theory. We may not be able to determine the appropriate educational design to meet a learning objective, but we can optimise the design. The methodological approach discussed here builds on a principled teaching strategy, using the Conversational Framework to link educational media to learning activities. The bulk of this chapter describes what this design methodology looks like in practice.

DEFINING LEARNING OBJECTIVES

Defining learning objectives sounds to many academics like a fearsome constraint on their creative teaching aspirations. As you delight in the intricacies and excitement of the ideas you want to promulgate, it can seem like an unwelcome intrusion to have to consider what your students will be able to do as a result. It is not about doing – the protest goes – it is about understanding, appreciating, seeing in a new way. However, the point of having learning objectives is to answer the question: how will you know if the students do understand, appreciate, or see in a new way? What would count as evidence that they understand? Without knowing this, the teacher remains ignorant of the effect of their teaching. Hence the proliferation of pedagogenic errors.

Academics most easily approach the definition of objectives via the definition of an aim. A teaching aim is couched in terms of what the teacher is trying to do, grounded in what the subject demands. Then, having clearly articulated what this piece of teaching is about, it is a little easier to approach the task of defining what this means for what students must be able to do. A teacher can easily produce aims such as the following diverse examples, where the teaching should help students to:

- understand Newton's Third Law of Motion;
- appreciate the civic origins of architectural designs;
- see gestures as having communicative value.

These are not learning objectives, however, as they do not define precisely how the teacher would know whether the aim had been achieved. What would count as the student understanding the law? Turning an aim into a series of objectives is a challenging analytical process. The objectives have to be precise, challenging and complete. Anyone who has ever designed a marking scheme for an examination paper will have gone through this analysis, albeit implicitly. The procedure can be summarised as in Table 10.1.

Table 10.1 Defining learning objectives

1 State the aim.
2 Define what actions by the student would count as demonstrating to you that they had achieved this aim.
3 Are the actions in 2 defined precisely enough to allow you to agree with a colleague about whether a student has achieved them? If not, return to 2, and refine the precision of the definitions.
4 Do the actions in 2 differentiate students who have achieved the aim, from those who have not?
5 Does the list generated so far cover everything implicit in the aim? If not, then return to 2 and generate the further objectives needed for completeness.
6 List the aim and objectives so defined.

Working through the procedure should lead the lecturer to a more thorough analysis of what their teaching has to do. That is the point of it. It will draw mainly on their knowledge of the subject, but probably also on their experience of teaching it. A lecturer would be aware that students frequently fail a particular kind of task, so including it would be a suitable challenge to their real understanding. At the end of the analysis, there should be a sense of having elaborated and operationalised the meaning of the aim, and of the scale of the task the teaching has to accomplish. It should not be seen as a task completed, however. It is most unlikely that this kind of means–end analysis will have pre-empted all the difficulties a student is likely to have in studying the topic. This analysis has stayed within the legitimate logical moves of the expert's understanding of the topic. It has not attempted to predict the illegitimate moves and prior misconceptions which a student may well bring to their study. That is why we need the further analytical step of identifying students' needs, considered in the next section.

IDENTIFYING STUDENTS' NEEDS

It must be clear by this stage of the book that it is impossible for teaching to succeed if it does not address the current forms of students' understanding of a subject. It is always hard for academics to empathise with a learner's sense of bewilderment in encountering a new idea, for the obvious reason that they either never experienced it that way or have long since forgotten it, which is why they are where they are now. The only way subject matter experts can hope to enter into the students' world is by setting out to understand it. We used to think it was unnecessary to try. The privilege and élitism that once determined cultural norms in higher education allowed academics to argue that it was their job to make the knowledge available to students, and the students' job to make of it what they could. Hence the uni-directional, transmission model of teaching epitomised by the lecture method. Students had to be both highly motivated and clever enough to puzzle out for themselves the obscurities of a discourse that rarely set out to be communicative, merely expository. As higher education becomes less élitist, and academics recognise the importance of inculcating academic knowledge in students who find the challenge too great to succeed unaided, the teaching enterprise has to aspire to something better than mere exposition. The subject matter expert must remember their own boredom and bafflement in subjects outside their chosen domain; they must ask whether their lack of motivation was essential or circumstantial. Can you recall your attention lapsing during some lesson or lecture, as you felt that this was something that held no interest for you? And was the topic essentially uninteresting? Can that really be said about anything? Or is it that the speaker failed to engage you, failed to speak your language, too quickly moved on before your basic uncertainties were addressed? That is what happens to students every day in the thousands of lectures taking place in universities all over the world. But if their lecturers better understood their point of view, and addressed that, it would happen less. There are three obvious sources for a better understanding: experience, students, and research.

Evidence from experience

Is it possible for a lecturer ever to address all the concerns of a class of students? Are there not as many ways of understanding a topic as there are students? In Chapter 2 we went through a number of studies of what students bring to a subject, and from all these it was clear that there are usually rather few ways of misunderstanding any one idea. In some areas, the ways are so few, and they bear such a close resemblance to the various historical conceptions of the topic, that some researchers have even suggested that misconceptions might be predictable, i.e. identical to former expert views of the subject. In physics, students appear to exhibit Aristotelian conceptions of force; in biology, they have Lamarckian views of evolution; in chemistry they have a phlogiston theory of change; in sociology they assign explanations to individual behaviour rather than social forces, and so on. It is an attractive thought that the history of ideas might be recapitulated in every student's personal development. It suggests that a lecturer who understands the development of ideas in their field may be able to recognise and pre-empt some of the plausible misconceptions, or naïve approaches in their students.

The empirical result that possible misconceptions are few in number reflects the experience of most teachers that the student errors they encounter in tutorials, assignments and examinations are the same every year. It is rare for a student to come up with a wholly new way of getting it wrong. So if the forms of error are relatively few in number, why are they not documented so that we can address them in future teaching? It has not been done because it has not been seen as a necessary or proper thing to do. Tutorials, assignments and examinations are there as feedback to the student on how they failed to learn, not to the teacher on how they failed to teach. Yet, what the student produces in these encounters could easily be used as data. Through these devices, every teacher has access to extensive fieldwork data, capable of analysis into categories of misconceptions, which they could later address in the design of future teaching.

This convenient solution, of teachers becoming their own educational researchers, is not straightforward, however. As in any discipline area, good research cannot be done by untrained amateurs. Academics, in spite of their competence in their own field, have a poor record as reflective practitioners. There are now enough studies of student learning to show that the level of subject matter understanding indicated by examination results is rarely achieved by the same students put to more challenging tests. Students can achieve very good examination results and still exhibit fundamental misconceptions (see Brumby, 1984, in biology; Dahlgren *et al.*, 1978, in economics; Bowden *et al.*, 1992, in physics; McCracken and Dobson, 1999 in geology). The problem is that what begins as a performance indicator soon becomes an end in itself. Inevitably, students learn what assessment assesses. However, if there is a careful analysis of the objectives, then appropriately challenging assignments are possible. This makes it more likely that students will reveal a fundamental misunderstanding, or an over-simplification of an issue. The reflective practitioner, however, will see this as formative evaluation of their own teaching. There is an excellent summary of the issue in Bowden and Marton:

> If we want to use assessment questions to find out what the students actually learn… this information must be found out from their answers, and this is by no means a trivial undertaking … In order to reveal the variation in students' understanding the teacher has to discern and focus on critical aspects of the students' understandings, and these can be discerned precisely due to the variation in the answers … A hidden world of varying ways of thinking about the phenomena dealt with in teaching is revealed. The teacher learns from and about the students.
>
> (Bowden and Marton, 1998: 185)

It is possible, therefore, for lecturers to conduct their own educational research, using the data available to them from their experience of tutorial and assessment contacts with students. We should not underestimate how difficult this is. It will take a cultural shift in our definition of professionalism in university teaching to legitimise this approach for the majority of academics, a point to follow up later in this section.

Evidence from students

Students can sometimes make clear to a lecturer what it is they find difficult, which is why lecturers find it fruitful to ask if there are any questions, and why tutorials are supposed to be useful. The only problem is that the ones who are really struggling cannot even frame a question. Students have to be coaxed towards an awareness of what it is they fail to grasp. This is, in effect, what research interviews frequently achieve as a by-product. By asking students to articulate and explain their perspective on a topic, the researcher is engaging them in a rhetorical dialogue that helps to disambiguate expressions, to expose and resolve internal contradictions, and to frame questions. This role does not have to be confined to researchers; it can be taken as well by other students. One of the most productive activities a teacher can suggest to students is that they engage in a kind of teacher–student role-play, where one spends, say, five minutes trying to teach the other a particular theory or concept. The one acting as student undertakes to ask whatever questions are necessary to clarify the explanation. Five minutes is usually enough to generate some very fundamental questions from the one acting as 'teacher', who has now discovered more precisely what it is they don't know. In my own experience with the technique, as a mathematics lecturer struggling to understand precisely what my students found confusing, it generated some absurdly basic questions that had been covered in the first lecture of the course, and some profoundly difficult ones that I had to think hard about. This technique lacks the analytical rigour and generality of the phenomenographic method. However, it does have the advantage of mimicking its immediate pedagogical benefits, and it defines the problem in the students' own terms, which is the principal concern of any teacher.

The self-help group is also a valuable source of revealing student questions. Students can use each other to clarify their confusion, and to reinforce their sense that the confusion is not entirely their fault. If the teacher consciously builds in an encouragement to form self-help groups with the explicit intention of generating precise questions to the teacher, then students are given the responsibility and the means to ensure they understand. The process can be conducted in class or online. The online provision of 'frequently asked questions' is the ideal tool, but the format is too often populated by experts' invented questions. When they are genuine products of self-help groups honing the question they really need the answer to, they are much more effective. The answers created then become formative development of the initial teaching design.

Evidence from the research literature

A relatively low proportion of academics read the research journals on teaching in their subject. Reading is now a luxury for academics, and the precious time there is must be for research, or at best scholarship, never teaching itself. In fact, many of the journals of subject teaching reflect this concern. They are devoted entirely to informing the teacher about developments in the subject, and about teaching strategies based on experience in the classroom. They rarely include any analysis of what students are likely to need. Chapter 2 documented research on the particular problems students bring to their learning. Two stand out as being widely applicable to all subject areas: alternative conceptions, and difficulties in generating and interpreting representational forms. However, the particular instantiations of these general forms are peculiar to the subject area. Because of the importance of keeping abreast of current thinking in the pedagogy of each academic discipline, the Learning and Teaching Support Network has been set up in the UK to facilitate this (see Web References, p.260). Referring to the existing work of close colleagues will alert teaching designers to what has already been done.

Combining the sources

Academics for whom the role of teacher is as important as the role of researcher will use all the above sources of information to help them identify their students' learning needs. Teaching must be communication, not just exposition, so the teacher must know something about their interlocutors. In social conversation, we adjust our language and our arguments according to what we know about the person we are talking to, because we know that communicative success depends on it. Similarly, physics lecturers must know that when they talk about 'force', students imagine not the Newtonian action at a distance, but the kind of 'oomph' it takes to move a table, and the two are not compatible. The psychology lecturer must be aware that students naturally interpret 'short-term memory' as the kind of memory span they are aware of, not the theoretical concept of a process that spans only fractions of a second. The mathematics lecturer must remember that as students become deeply enmeshed in the intricacies of a proof, they tend to forget the meaning of the manipulations they are undertaking, so that each successive stage becomes increasingly meaningless.

Many lecturers will claim that they do know their students, that they talk to them in tutorials, they take note of performance in assignments and examinations, and this informs their subsequent teaching. This is important and no doubt accounts for the many pedagogic successes that higher education can claim. However, it can only be a relatively superficial analysis, and will not always reveal the cognitive links that have to be made if the student is to progress to the expert view. The lecturer may be able to discern incorrect terminological usage, but ensuring correct usage does not necessarily dismantle the conceptual construction already built. Another kind of misconception is the simplification of the internal logical structure of a concept. It will be clear to the lecturer that the student has a misconception of some kind, but its precise logical relation to the correct one may not be obvious. The naïve conception, or the lay person's everyday model of a system, is likely to be the bedrock on which the academic is attempting to erect a wholly new perspective. Students' attempts to reconcile the two can lead to bizarre distortions that will be difficult for the lecturer to unravel. Shifting that original conception is the first step in communicating a new idea, but it can only be done if you know what it is.

Some knowledge of where students are conceptually, as well as where we wish to get them, is therefore essential to good pedagogic design. However the designer does it, whether through basic phenomenographic research, role-play by students, teachback exercises, assignments, or via the existing literature, some initial analysis is important to motivate the design of the learning activities the student must undertake. Even if it is only guesswork based on experience of teaching, the students' supposed prior conceptual state should be articulated, as then it can be challenged and refined in the light of further experience.

Combining the three sources of knowledge about students' learning needs, we can summarise the key activities to be undertaken at this stage of the design process as in Table 10.2.

Table 10.2 Key activities in identifying learning needs

1. Which naïve (historical?) conceptions or simplifications might be prevalent in this topic?
2. What are the standard forms of error or misrepresentation that occur in assignments and exams?
3. In what ways might the internal logical structure of the main concept be distorted?
4. Which technical terms have everyday meanings that could lead to their misinterpretation?
5. What do students' questions and discussion reveal about their learning needs?
6. Which educational research results in this field identify learning needs?
7. Which forms of representation (linguistic, notational, diagrammatic, graphical, symbolic, iconic, numeric) are difficult for students?

It is not easy to second-guess the inventiveness students can bring to their attempts at comprehension of a subject. There is no substitute for proper investigation of these issues, but the prior analysis of students' needs will pre-empt some of the problems.

DECIDING THE BALANCE OF LEARNING OBJECTIVES

Refining the curriculum aims to specific learning objectives helps to structure the course or programme into manageable sections. A critical stage in the design process is then to decide on how to balance students' workload across the range of objectives. The needs analysis will help to determine this because it will suggest which objectives are likely to be most problematic. Estimating time needed can only be done from experience of teaching the subject to the expected target students. Experts underestimate how long novices need, and experience does not transfer easily across institutions, or topics, or even within institutions and topics. Furthermore, there is almost no research evidence on time taken for study. There are a number of developmental testing studies at the Open University, conducted on trial materials, which invariably show that students will need more time than was estimated. On the other hand, students invariably report that they have too little time available for study, so simply lengthening course times is no solution.

The best option for all concerned is to be explicit about time estimates for study, both teacher-directed, and private study. Students need realistic estimates of how much time it is appropriate for them to spend on materials, activities, discussion, group work, projects, electronic searches, as well as on their own self-paced study. An explicit expectation helps students to plan their time, and sets a target against which the course can be evaluated – enabling the academic to check on the quality of the estimates. Gradually, experience builds to generate better estimates at the design stage. Every course or programme has an expected study time associated with it, so that it must always be possible to estimate the general breakdown of student workload. Table 10.3 suggests the key activities for this stage of design.

Table 10.3 Key activities in estimating the balance of objectives

1	What is the total formal and informal study time needed for the course?
2	What are the key learning objectives defined for the course/programme?
3	Given the needs analysis, what is the appropriate breakdown of study time, formal and informal, across the key objectives?

Greater clarity about how much student time can be spent on each of the key objectives enables better planning and well-targeted design of the specific learning activities linked to each objective. Course planning is more usually carried out in relation to curriculum topics and the academic time needed for presentation, or class contact. It is an entirely provider-centric perspective that takes no account of students' academic or logistical needs. The majority of students at university are part-time, and necessarily careful with their time. Students will increasingly opt for the university that is genuinely student-centred, that structures study time around student needs, not institutional needs.

Designing specific learning activities

Once the learning objectives and student needs are articulated, by whatever means, it becomes easier to plan what the student must therefore do to achieve the desired learning outcome, in the time available. This next stage is largely creative. From the first two stages we may well have a neat logical description of the relation between where students are and where they need to be, but this does not define the psychological pathway between the two. That is why it is a mainly creative process.

Objectives generate actions such as 'distinguish types', 'give examples', 'name parts', 'interpret', 'state relations', 'make predictions', but what does a student have to do in order to be able to 'distinguish types of X', 'give examples of Y', etc? What does it take to learn these things? Which mathemagenic activities will yield these objectives as outcomes? Although designing learning activities is mainly a creative process, with the analysis we have done so far, it should be possible to build an analytical tool to assist this process.

The features of the successful learning environment will be the affordances for academic learning first defined in Chapter 1, i.e. the design features that invoke successful learning activities by the students. The template below sets out a distillation of what we know from the previous chapters. It links the learning activities needed to achieve the more challenging objectives to the affordances for these, and to the media forms that best support such activities. This analytical tool for the design process can direct the activity, and pre-empt some of the students' problems and thereby reduce design time.

Teaching is most effective if it can avoid creating the pits we know students are likely to fall into. In thinking through the best way to teach a topic, the academic, who is extremely knowledgeable, will have great difficulty in stepping into the students' shoes to accomplish the feat of pre-emptive adaptation. The template in Table 10.4 is meant to support that process, given what we know from preceding chapters, and is constructed using the Conversational Framework developed on pp.86–89.

The design template begins the detailed design with an initial analysis of what it will take for the student to learn, and how the teaching can best support this. The most difficult part is designing the actions that have intrinsic feedback for comparison against target goals, as this is the most unfamiliar for academics. It relates mainly to the adaptive media, which are exclusively computer-based. The essence of it is to find ways of enabling students to emulate the scholar. Give them the interactive environment in which they perform the activities of the scholar but with feedback related to a goal in such a way that it exposes the internal relation to them, and makes it meaningful. We saw examples of this in the programs on the power plant, geology, and algebraic manipulation in Chapter 7. The particular form it takes will depend on the context and the learning objectives; beyond this, it is hard to generalise. Some of the most useful techniques to build into the interface the student operates are listed in Table 10.5. These enable an adaptive or interactive medium to give maximum support to the student.

These are all necessary if the program is to support students adequately, but they are difficult to implement. This is why ICT designers promulgate the idea of the importance of student control over their learning, and there is a sudden interest in 'student-centred learning'. It has a lot more to do with the difficulty of program design and the complexity of learning than it does with pedagogical high-mindedness.

It is a time-consuming process to address students' needs: far easier to make the material available and give them the navigation tools to find their own way

Table 10.4 Designing affordances for learning

Learning activities needed	*Affordances (design features that afford those activities)*	*Media forms*
Attending	Describe the narrative in terms of an overall topic goal.	Narrative
Apprehending	Clarify structure of argument, nature of evidence.	Narrative
Experiencing	Offer vicarious experiences or supplantation of experiences of ideas, concepts.	Narrative
Discriminating	Offer alternative forms of description, based on misconceptions and misrepresentations identified in learning needs.	Narrative
Articulating	Encourage student's articulation of conceptions and perspectives.	Communicative Productive
Challenging conceptions	Generate the questions and exercises that will elicit likely misconceptions, or representational difficulties.	Communicative
Clarifying internal relations	Create environment for actions with intrinsic feedback.	Adaptive
Experimenting	Define the goals against which students can compare the intrinsic feedback to modify their next action.	Adaptive
Relating experience to theory	Refer to prior experiences of interacting with the world that students should reflect on to appreciate the points being made.	Narrative
Investigating, analysing	Offer student the means to select or negotiate their own task goal.	Interactive
Reflecting on experience	Generate questions on topic goal that require students to use their experience at the interactive task level.	Communicative Productive
Relating theory to practice	Develop goals and activities at interactive level that require students to use their knowledge of the theory.	Interactive Adaptive
Synthesising	Ask students to reflect on the comparison between theirs and the teacher's conceptions, and on goal–action–feedback cycle.	Productive

Table 10.5 Interface techniques for ICT-based activities

- Optional investigations to encourage students to construct their own narrative to drive their navigation of resources.
- Keyword analysis algorithm to interpret student descriptions, requests, or answers.
- Matching algorithm to provide extrinsic feedback on students' constructed answers.
- Algorithm to generate repeatable tasks.
- Manipulable model of system that can provide intrinsic feedback in the form of graphical, pictorial, or textual output.
- Categorisation of student's actions or descriptions to support interpretation of their performance, to provide extrinsic feedback.
- Record of student's actions to enable categorisation and interpretation.
- Editable Note Pad, to encourage students to articulate their findings.
- Model answers for students to compare against their own descriptions.

through it. This is why we have seen a proliferation of resource-based CDs and interactive Web resources. However, beneath the rhetoric of 'giving students control over their learning' is a dereliction of duty. We never supposed students could do that with a real library, or a real laboratory. Why should they be able to it with an electronic one?

There is now a further excuse for avoiding this more complex type of design. The communicative capabilities of the Web offer the illusion of human support. Learning through discussion is supposed to be beneficial, and therefore we need only provide the communication links. However, human support is still highly labour-intensive, and cannot be there whenever the student needs it. This is not a plausible excuse for avoiding the design of student support within an interactive stand-alone environment. We can use high-cost human support more efficiently in the design time that is needed for doing the preparatory work: to discover what students need, to devise the diagnostic strategies, and to specify the generative tasks.

DESIGNING THE LOCUS OF CONTROL

The different ICT media have the capability to support the learning process very well, but will only do this if we fully exploit their properties. The key issue is the locus of control in the program – does it rest with student, program, or both?

Control by the student is important because we cannot possibly predict the exact sequence and pacing that each individual student needs. To adapt their actions, and to reflect on the goal–action–feedback cycle students need control over what they do and when. The control features that should be available in the interface to support each type of activity are shown in Table 10.6. All these control features should normally be available on-screen throughout any ICT material, as icons, buttons or pull-down menus. Their design and functionality should follow current good practice found in the most popular commercial sites and programs, because these will define the current universal grammar of the medium. Immediate intelligibility is crucial.

Table 10.6 Control features needed for ICT interface design

Discursive	A structured map of the content to allow access at any time to all aspects of the teacher's description of their conception.
	Concealed multiple choice questions (cmcq) with keyword analysis to allow student to express their conception and obtain extrinsic feedback on it.
Adaptive	Ability to sequence and select/construct their own task goal, enabling them to generate the experiences they feel they need.
	Access to statement of objectives for program and for sections of content, so that they know what counts as achieving the topic goal.
Interactive	Clear task goals, so that they know when they have achieved them.
	Intrinsic feedback that is meaningful, accompanied by access to extrinsic feedback (such as a 'help' option) that interprets it.
Reflective	An indication of the amount of material in each section to allow planning for self-pacing.
	Requirement to test a new conception by offering a description of it for comment (e.g. via cmcq).

The omission of any one of these control features impairs the student's ability to maintain control of their learning. The reflection–adaptation cycle is extremely important. It is the key to successful learning and must be supported by teaching materials, not sabotaged at every turn because the materials cannot adapt to the student's needs. None of the above features is especially difficult to offer; they are the minimum requirements for good design.

The student's reflection must be focused on the content of learning, on the meaning of their interaction, not on how to operate the program. This means the interface must be operationally transparent, so that they do not waste time trying to figure out how to work it. Computer environments have been the breeding ground for a new strain of learning activities, which I can only describe as 'anathemagenic' – activities that give birth to loathing. Here are some of the most common forms:

- looking for how to get started;
- wondering why nothing is happening;
- discovering you are unable to get back to the page you just left;
- being told you are wrong when you know you are right;
- wondering how long this is going on;
- trying to guess the word the program is waiting for;
- wondering what you are supposed to do next;
- coming upon the same feeble joke for the fifteenth time.

You will have your own additions to the list. The considerable improvements in human–computer interface design in recent years have brought operational transparency to many types of computer tool. Educational programs must aspire to the same standards.

DEVELOPMENTAL TESTING

The design process is not complete without evaluation, beginning as early as possible in the process. Educational design is not a precise science, and there is too little secure knowledge learned and shared from existing experience for academics and designers to be able to build on experience. The design makes decisions of different types, as we have seen: objectives, learning activities, and interface. Each of these must be tested through developmental testing, or formative evaluation, although the traditional media are well established in terms of user interface, and need little testing of this kind. Testing with students is needed to give feedback on both the interface design and the learning activities.

The quality of the interface for ICT media is most easily judged from observation of target students trying to operate the software, recording options chosen, mouse moves, time taken to complete an operation, etc. The aim is to design a user interface that never intrudes on the task in hand.

The design of the learning activities has to be evaluated in terms of the thinking they elicit. Do students indeed use theory to adapt their actions to the task goal, do they focus on the task goal, do they reflect on the action–feedback cycle to relate this to the topic goal? Students cannot talk aloud as they work, as this intrudes on their cognitive processing of the task in hand, but they can talk to each other. This level of evaluation therefore has to use pairs of students, working together to achieve the topic goal, and discussing their reactions as they work. The data combining both actions and talk will help to reveal the kind of thinking elicited by the design of the learning activities. In the context of communicative media using the Web, the discussion is captured by the medium itself, and is therefore directly available for evaluation analysis. Where confusion or uncertainty remains in the way students discuss the material, there is an opportunity for improvement in design.

The main task of the developmental testing is to judge the extent to which the design of the learning activities enables students to achieve the intended objectives. This will be revealed through observation of their actions and discussion, but is better judged in interview, using questions of the type that would be used in an essay or exam. Alternatively, asking the students to write a brief summary may be sufficient. The close attention paid to the way target students deal with the subject matter in these studies will sometimes affect the design of the objectives themselves – perhaps they are pitched at the wrong level, perhaps another objective emerges. It should be a legitimate outcome of the evaluation process that the objectives may also be challenged.

There is little help in the literature for academics and designers wishing to use evaluation tools and methodology, but there are some available on the Open University website for the Programme on Learner Use of Media (see Web References, p.260). This provides both a plan for a study, and a set of formative evaluation tools designed for learning technologies. The approach assumes that there is little resource for evaluation. The techniques require no more than a few pairs of students, observed and interviewed for a session length appropriate to the material (see for example Laurillard and Taylor, 1994). By (1) targeting the weaker students, (2) observing two or three pairs working collaboratively on the materials, and (3) maintaining the belief that the materials must adjust to the students rather than vice versa, the design team will be able to produce good teaching materials and document the generalisable lessons for others to learn from. All three conditions are necessary, but the third is the most important, and the most difficult to achieve. Even at prototype stages of design, the data that emerges from such a study is rich and informative, and capable of contributing to great improvement in the materials.

Piloting should allow the team to assess the success of the materials in a new environment. The context within which the materials are used helps to determine their success, and each needs to adjust to the other. It is important to use the piloting phase to discover, through observation and interviewing, the contextual conditions that enable the courseware to work most effectively.

Student time commitment will be necessary for developmental testing. This should be carried out in the context of real usage of the materials within a course, in order to study the conditions under which it can be made to work optimally. Staff teaching the course must be prepared for this and closely involved with the way the development team intends the materials to be used. They must be involved in both developmental testing and piloting. If use of the materials is a genuine part of the course, then students will expect their work with them to be included in the assessment of the course.

The pilot study should fully integrate the materials into the course, linking them to other teaching, following up on what students did with them, and assessing their work. The study should enable the design team to identify any necessary changes in how they organise the use of the materials.

COMPARATIVE DEVELOPMENT COSTS

The output from the initial design process will be a set of aims and key objectives, together with their associated media prototypes, and the time allotted to each one. This should be sufficient information to carry out a preliminary check on student workload and development costs, assuming that costing information is available. This should be done as early as possible in the development process, so that adjustments can be made if necessary. Table 10.7 shows an extract from the kind of spreadsheet tool in use at the Open University to model the initial design of a course in terms of aims, balance of media, and workload for students. Each of the course aims generates objectives that require combinations of media. The study time allocated to each medium is entered for each aim, and the total time allotted for each aim and each medium is calculated. This enables designers to check that the total workload is not too high, and there is the right balance between the course aims.

From the study time allotted to each medium the spreadsheet calculates the academic design time and production time, for each one, and presents the results. A tool such as this will make apparent to the design team the sensitivity of staff workload to the distribution of student workload – for example, that it costs much more staff time per study hour supported to increase use of video than use of print. The data used for this is derived from experience, and changes with time, so the data shown will be relevant only for a specific context, and are not generalisable. Design for print is well practised, and efficient production mechanisms are available to deal with it, so the total costs will be less volatile than those for the ICT media. Design time has to account for inexperience; production time has to account for frequent changes in response to developmental testing. Design time data should be monitored, so that estimates can be made, but it changes year on year.

There have been studies to test the comparative efficiency of different media, but there can be no definitive result because too much depends on local circumstances and on the quality of the particular material developed. It should be clear from the complexity of the discussions in the previous five chapters that a question like 'which is the most cost-effective medium?' can only expect an unhelpful answer – it all depends. A cost-modelling tool such as that in Table 10.7 can only be an aid to decision-making, and will build understanding over time, but it is not exact. The costing data included in development costs will include those listed in Table 10.8. Universities collect and record costs of activities in different ways, so local tools will be more useful. Nonetheless, Table 10.8 categorises the key elements in a realistic costing.

Table 10.7 Extract of worksheet to model study hours for each aim and medium

	Print/SAQs	*Video*	*D3E*	*Audiographics*	*Total*
Aim 1	25		10		35
Aim 2	5	5	10		20
Aim 3	30			15	45
Total time	*60*	*5*	*20*	*15*	*100 hrs*
Academic time	250	50	60	40	400
Production time	150	230	200	20	600
Total time	*400*	*280*	*260*	*60*	*1000 days*

Table 10.8 Categories of development costs for learning technologies

Academic staff	*Production staff*
pedagogic design	production design
travel (for collaboration, material collection, etc.)	prototype production and revisions
developmental testing	developmental testing
drafting and revision of materials	copyright costs
quality assurance procedures (reviewing, discussing)	quality assurance procedures (editing, testing)
project management	project management
administrative support	technical support
	final production

Staff responsible for these activities must be responsible also for estimating and recording the time needed, and for managing the time more productively where possible. Differences between departments within an institution can be greater than those between institutions, so any attempts at general estimates are impossible.

This is as far as it is reasonable to take a comparative cost analysis at a general level of description. The greatest expense is in allowing time for careful design, and the quality assurance mechanisms and developmental testing required to produce high quality materials, and that is the same for all media.

SUMMARY OF THE LEARNING DESIGN PROCESS

This chapter has outlined the design process needed for optimising the use of a range of educational media. The design process has progressed from the curriculum aims, through objectives, learning activities, media forms and evaluative feedback. The full process is summarised in Figure 10.1, which shows the iterations between the different stages. The specification for the media prototype is the beginning of the development process, which continues with repeated design–test–redesign cycles.

There is no simple prescriptive rule connecting the analysis of learning activities to the required medium. However, the elaboration of what the teaching is trying to achieve, and how, will inform media selection by clarifying which learning activities are most likely to need support.

However, it is not possible to conclude that some particular combination would be significantly better than the others without a consideration of the logistics of the teaching–learning context, which we consider in Chapter 11. It is only logistics, after all, that rule out the ideal form of teacher and student discoursing on Newton's Laws of Motion while punting on the river.

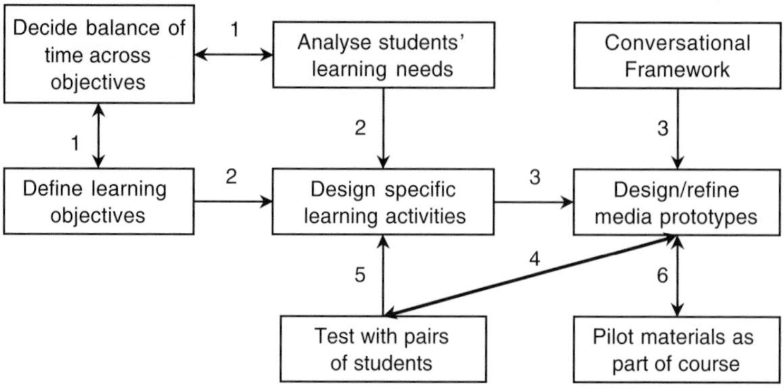

KEY
1 Objectives and needs analysis suggest the appropriate balance of time across objectives.
2 The design of a learning activity uses the needs analysis for that objective; the balance between objectives defines the time allotted to it.
3 The activities needed are analysed in terms of the Conversational Framework to design the media prototype.
4 Successive refinements of the prototype are tested, and refined to the final form in an iterative process.
5 Analysis of student talk and actions further defines the learning activities needed, and feeds back into the media design.
6 The prototypes for the learning materials developed are piloted to ensure they integrate with each other, and fit the learning context, with further feedback to the design.

Figure 10.1 The sequence of stages in the learning design process.

Throughout this chapter, the discussion has been focused at the topic level, at how to get students to think constructively about particular ideas, and how to engage with those ideas in a productive way. The understanding of an idea or concept does not occur in isolation from the other aspects of a student's university life. It takes place in the context of a course, a department, and an institution, and these contextual factors will have an effect on student learning, and must be attended to if the materials are to work. As well as the pedagogical issues we have considered in this chapter, the logistics of these different institutional contexts will also affect the teacher's judgement about the most appropriate media to use, and how to combine them in teaching the subject. The next chapter therefore looks at the institutional context that envelops the student as they learn their subject.

Chapter 11

Setting up the learning context

INTRODUCTION

There is a folk wisdom in academic circles that educational technologies come and go, leaving expensive machines to lie in cupboards, gathering dust. The main reason for this, when it occurs, is neglect of the organisational context for the learning process, not just, as is often supposed, the poor quality of teaching the machines provide. There is plenty of traditional teaching on offer in universities that is poor in quality, sustained nonetheless by its fit with the learning context. Educational technologies, especially new ones, demand effort and ingenuity in the development of materials, but rarely is this extended to the embedding of those materials in their educational niche. This is one of the key reasons why they have made relatively little impact in higher education, despite their potential, and why we need to devote the two final chapters of this book to the organisational context for learning.

In this chapter we remain with the student's perspective, and document the contextual factors that affect how they learn. In the next chapter we move to considering what this means for the institutional infrastructure.

Students are not simply learners of an academic subject; they are social beings. Like everyone else, they respond to the social, political and organisational context around them, and this directly affects what they do in their day to day work:

> Approaches to learning are intimately connected to students' perceptions of the context of learning. Perceptions of assessment requirements, of workload, of the effectiveness of teaching and the commitment of teachers, and of the amount of control students might exert over their own learning, influence deployment of different approaches.
>
> (Ramsden, 1998: 48)

All these points apply equally to the context of use of learning technologies, and this chapter addresses each one.

A few years ago, in a study of students' problem-solving, I was looking for evidence that students use heuristic methods of the kind advocated for problem-solving, e.g. understanding the problem, doing a means–end analysis, creating sub-goals, working backwards from the solution needed, checking back, etc. An interesting finding emerged (Laurillard, 1984b). Based on written protocols and interviews about their approach, it was difficult to credit many students with the use of these heuristics in relation to the substantive problem. In a problem about writing a device control program for a microprocessor, for example, there was no focusing on the nature of the device, or which instructions might be needed, or what form the final solution should take to do the job. However, it was certainly possible to ascribe the use of these heuristics to the solving of 'a problem set by a particular teacher in a particular course' – they were solving the problem-in-context. As a consequence, the information they considered as relevant to solving the problem was: what was done in the lecture, the teacher's diagram, the wording of the question, the relation to similar examples done by the teacher, what the teacher gave high marks for. In the checking back stage, they used as a criterion their own level of commitment to the course, rather than the accuracy of the solution. This is all perfectly rational behaviour. But it means that in setting work for students we must think of them not necessarily as grappling with the intriguing ideas we have put before them, but as trying to second-guess what we want of them. It follows from this rigid orientation to what the teacher requires that the teacher has a great responsibility to require the sort of thing that will help them learn.

This argument does nothing to diminish the importance of students taking responsibility for their own learning. The point is that they will inevitably respond to the demands of the context, so the teacher must be sure that the demands of the context are compatible with their pedagogic intentions.

The following sections list the contextual factors that affect the quality of student learning. In each case, we consider where the teacher's responsibility lies, and how this affects the introduction of learning technologies.

STUDENT PREPARATION

As students approach each new learning session in their course, they need to be oriented towards the ideas or skills they are about to encounter. This is true for all media-based materials. Learning, when it happens within a taught course, is not a voyage of discovery with the student in control. Academics never want to spoon-feed their students, but since they generally take control of what is to be learned, and when, and how it is to be judged, students are very much at their mercy. If students are to have any control over their learning, then they need some information. The voyage of discovery does not have to be a mystery tour. To be well equipped to get the most out of the learning session, they need to know why this topic is important and interesting, the prerequisite knowledge or skills, the learning objectives in view and how they are assessed, how much time to allot to it, and how to approach it.

Students' approach to a topic is affected by the way they are invited to engage with it. Whether the material is Web resources or adaptive simulation, the right kind of preliminary exercise will enable students to feel some sense of ownership of what they are studying. They will attend to a case study with more interest if they have been challenged by a question about their own experience. They will watch a video-clip on social interaction with more attention if they have already tried to list all the different forms of questioning they can think of. They will watch a simulation of the manufacture of netting with more interest if they have already tried to figure out how it might be done. If they have access to a database resource, they will use it better if they have learning objectives to aim for than if they are asked to see what they can find out.

It is the academic designer's responsibility to include all these considerations as part of the design of each course component.

Summary

Prepare briefing for the use of materials to:

- orient students to why this topic is important and interesting;
- describe what they already need to know to make best use of it;
- define the learning objectives, and how they are assessed;
- offer a diagnostic pre-test to orient them to what they should focus on;
- suggest the time to be allotted to formal and informal study of the topic;
- provide preliminary exercises that alert them to the challenges of the topic.

INTEGRATION WITH OTHER MEDIA

New learning materials will be likely to change aspects of the existing teaching, so the need to revise all the teaching must be kept under review. For example, computer simulations, or database resources, can give students access to sophisticated material for doing their own analysis. In this case, they may need additional teaching on analytical procedures if they are to make good use of the new material. Access to information databases gives students a wealth of material to work from. However, this is of no value to them if they are not able to make selective judgements about what to use, and critical judgements about the content of what they find. The teaching that surrounds students' use of new learning technologies will need to address this kind of issue. It is essential that academics taking on new material clarify its relationship to other course components, the learning objectives it is meeting, and the pre-requisite skills it entails.

The learning students achieve within each component should then be followed through in subsequent teaching, to ensure its integration with the rest of the course. The work should have a natural place in the course and its role should be clear to students. Several evaluation studies of videos and computer packages have looked at retention of learning over time. However, these can tell us nothing about the teaching method itself. Retention is dependent upon whether what is learned is followed up soon after by applying it in other learning sessions, or practice, or assignments. Unless students use what they learn on a package, they will soon forget it, no matter how good it is, or how well they learned initially. It has to become embedded in the way they think, before retention can be expected, and that means repeated use, not just an isolated event, no matter how impressive it appears.

For students to feel they have control over their learning, a well-integrated course will offer ease of navigation through the different components. They benefit greatly from a unifying device, such as a study guide, Web page, or course calendar, with hyperlinked access to brief descriptions of all course components, including non-ICT elements.

Summary

Review each course component to:

- clarify its relationship to other components of the course or programme;
- check that pre-requisite knowledge and skills are covered;
- decide how and when to follow up on what students have learned;
- offer a map of the course components and how they relate to each other.

EPISTEMOLOGICAL VALUES

As we saw in Chapter 2, students bring their own epistemological values to studying a course. They will have been nurtured throughout all the students' previous educational encounters. Every teacher plays a part in nurturing their students' epistemological values – their conception of how we come to know – and hence their conception of what learning is, and how it should be done. None of this features very much in course syllabuses, because they tend to be concerned with the content to be learned, rather than its epistemological status. It is often implicit, however, in the way academics talk about the aims of university education, and in the discussions that ensue at examination boards, as I suggested in Chapter 1. It also links to the synergy between teaching and research at the core of university activity. So we have to consider it.

Inculcating an appropriate conception of learning, or a desirable epistemology, is not an issue peculiar to the use of learning technologies. Clearly, it is fundamental to any teaching. It is particularly important to confront it in the use of new technology, however, because it presupposes a diminution of teacher–student contact. Part of the great value of the tradition of teacher–student contact is that, in the interstices between content-related talk, the academic can stand back from the task in hand and encourage the student to look at the nature of the academic enterprise itself. It will probably be in such a discussion that the student is treated to the sudden revelation that getting the right answer may not always be the most important goal. That kind of sentiment never gets written on the board, never appears in course syllabuses, nor in lecture handouts, but it needs to be made explicit to students. Whatever the academic feels is an appropriate way to approach the acquisition and manipulation of knowledge in a subject should itself be a topic for discussion with students. When discussion time reduces, as student:teacher ratios worsen, then some treatment of this issue has to be included consciously in the course materials.

The importance of this issue is demonstrated by Ramsden's work on students' perceptions of teaching in different departments. From questionnaire studies in a range of institutions and academic departments, he found:

> differences in students' orientations and attitudes to study which are only explicable in terms of the powerful effects of contexts of learning… [and] … associations between [students'] approaches and the perceived quality of teaching in first and second year university level study.
>
> (Ramsden, 1992: 80)

The quality of learning relates strongly to the quality of the academic context provided. And quality of teaching is not judged here by the clarity of the lectures given. Ramsden lists the characteristics of the learning context that are associated with a 'deep approach', among them:

> teaching that addresses the nature of the subject and its relevance;
> the lecturer's personal commitment to the subject;
> opportunities for students to choose their methods of studying.
>
> (Ibid. 81)

The consequences of inattention to the epistemological values of a university education are well documented by William Perry. In a revealing study of approaches to reading among these top undergraduates, he found that:

> What they seem to do with almost any kind of reading is to open the book and read from word to word, having in advance abandoned all responsibility in regard to the purpose of the reading to those who had made the assignment.
>
> (Perry, 1959: 195)

Perry's classic study of Harvard undergraduates (1970) explored the relationship between their intellectual and ethical development. Students do not necessarily take responsibility for their learning, nor for what they know.

The status of knowledge, one's personal commitment to it, and the appropriate ways of approaching the study of it, are all topics that should be figural in any course. They equip students to take personal responsibility for their knowledge and their learning. With the greater distance introduced by learning technologies, it is more difficult for academics to convey these informal, and more personal aspects of their teaching.

Summary

In order to create an environment for students to develop their conception of learning and their own appropriate epistemology:

- use the communicative media to demonstrate your own commitment to the subject, and your way of approaching it;
- give students opportunities to debate their methods of study, and to defend their choices;
- provide opportunities for discussion of the status of the knowledge in the subject, how it can be known, and how it may be learned.

ASSESSMENT

The first section of this chapter argued for the importance of clarifying for students the nature of the assessment, as this is inevitably a strong influence on the way they approach their study. It was linked to the setting of objectives, as it must logically be. It is a task of some importance, however, to decide exactly what kinds of questions or assignments will adequately test the achievement of the more interesting learning objectives. Bowden and Marton (1998) offer an extensive account of research and evaluation studies of university assessment. There is an ongoing debate about whether we should assess what students know, or what they can do. The traditional modes of assessment of knowledge are seen as inadequate because they fail to assess students' capability in the authentic activities of their discipline. The authentic assessment movement would instead reflect the complex performances that are central to a field of study – e.g. writing a position paper on an environmental issue, investigating a mathematical concept. The debate continues, questioning the validity of the claim that authentic assessment is a true measure of students' capacity to generalise their learning to new situations. Given that students orient their study towards their perception of the assessment, the solution offered is to find more challenging forms of assessment. They must link to the learning aims and reveal what students have learned at a general level, rather than simply assess the technicalities, which leads to a more instrumental form of learning:

> In many current assessment systems, the form of the question is such that the relevant aspects are given and the capability of the student to discern them is not tested ... we are arguing for using assessment to define learning aims, for

> revealing students' capability for discerning critical aspects of certain classes of situation, and to find out what students have learned in general.
>
> (Bowden and Marton, 1998: 167, 175)

Their conclusion is that assessment should be:

- open to different perspectives, which requires students to discern the critical features;
- non-technical, going beyond the focus on specific facts or procedures that constrain the student, and do not test their capacity to take responsibility for what they know;
- conceptual, focusing on:

> the phenomena, concepts and principles that are central to the field of knowledge studied and which are vital to the students' capabilities for handling situations in the future, which is what we are trying to prepare them for.
>
> (Ibid. 184)

This is a critique of traditional assessment methods, but new technology methods create further challenges for academics designing assessments to fit the ambitious aims of university teaching. Too frequently, teachers introduce learning technologies to students on an experimental, pilot basis, without properly integrating them into the teaching. Students therefore see them as peripheral to the real teaching, and invest in them less effort than they otherwise would. The only real test of any learning material is its use under normal course conditions. This means it must be integrated with other methods, the teacher must build on the work done and follow it through, and most important, the work students do with ICT media must be assessed. This may require new standards to be set.

The best way of using some ICT-based methods may be in small groups, or in pairs. Use of discussion environments often means that work produced is collaborative, and must be assessed in a different way from work produced individually. Mason suggests several strategies, appropriate for different objectives:

> - Some part of the mark for individual effort and some part for the group effort;
> - One mark applicable to all participants regardless of their individual input;
> - Some form of negotiated mark, in which either the individual or the group decides, in consultation with the tutor, what individual mark each participant deserves.
>
> (Mason, 1994: 33)

Palloff and Pratt also suggest asking students to submit a self-evaluation with their own grading, asking a group to appoint a leader to determine grades for its members, allowing groups to negotiate their own grade, and referring students to assessment guidelines (Palloff and Pratt, 1999). Chalmers and Fuller emphasise the value of involving students in the setting of questions and marking schemes against given standards. This informs students of the value placed on critical thinking, use of the subject matter, and a reflective approach (Chalmers and Fuller, 1996). Salmon suggests a halfway house, in which students carry out a peer review of each others' work, which is also subject to assessment by the teacher, and providing extra marks for participation (Salmon, 2000). Greater openness about assessment and grading has its own benefits in allowing students to understand the process better, and rarely does negotiation or collaboration in marking lead to complaints or argument.

The kind of work students do using learning technologies is necessarily different from what they do in learning via other methods. Therefore, the teacher has to decide what counts as a good performance, and what counts as useful feedback to students on what they did. If they have used a database package to obtain information for example, are they to be assessed on the basis of the results they obtained, or on the imaginativeness of their exploration of it? When comprehensive and detailed bibliographic research is feasible through new technology methods, the criteria for judging this work must change. Academics will be too easily impressed by the result of a few key presses, if they equate it with days of hard slog among the library catalogues. The assessment must require students to rework the information they find. Chapter 9 on productive media suggested that we can now offer more imaginative ways of assessing students by asking them to use tools such as spreadsheets, PowerPoint animation, website design, or an annotated collection of multimedia resources, as ways of presenting their ideas. The criteria will be similar to the traditional ones for essays: coherence, accuracy, originality, good use of evidence in support of an argument, etc., but we will have to learn how to apply and interpret these in new contexts.

Part of the point of new teaching methods is that they change the nature of learning, and of what students are able to do. It follows that the teachers then have the task of rethinking their assessment of what they do.

Part of the value of learning technologies is that they can carry out some limited forms of automated assessment. Chapter 7 discussed ways in which adaptive media can offer more constructive assessment than the ubiquitous multiple-choice questions. It is possible to give students feedback on their work in several ways, already discussed in Chapters 7 and 10: using keyword matching for user-constructed input, offering possible model answers for students to compare with their own, using manipulation of a model to achieve a particular output. A recent study found that some of these more innovative forms were valued by students, and were pedagogically beneficial:

> The findings confirmed the value of innovative assessment strategies such as the electronic delivery of model answers, marking schemes and peer review as a way of enhancing formative feedback to students, in assisting

> the development of critical and analytical skills, and in demonstrating alternative approaches to written work.
>
> (MacDonald, 1999: 241)

Many such formats can be adequately assessed by a program, which can thereby provide constructive feedback to students.

Whatever forms of assessment are decided upon, it is vital that these are communicated to students clearly. One of the greatest dissatisfactions with student performance, most commonly expressed in examiners' meetings, is that students did not appear to understand what was required of them. The greatest service teachers can do for themselves and their students is to take time to clarify assessment requirements and check that they are understood, and take steps to make them understood better. It is reasonable to maintain a continuing dialogue about this, so important is it for the success of any teaching method.

Summary

To ensure materials are properly embedded into a course:

- design assessment in terms of objectives;
- design questions to be open, non-technical and conceptual;
- ensure that learning through new media is assessed and accredited;
- design group assessment to fit objectives and modes of collaborative learning;
- involve students in the design of assessment and marking;
- reinterpret assessment criteria explicitly for learning from new media;
- use the productive media to test the new learning activities that are being encouraged;
- communicate assessment requirements clearly.

LOGISTICS

This section concerns all the conditions that significantly affect the quality of learning the students can achieve, but which are determined more by institutional context than pedagogic considerations. They concern: the amount of material covered in a course; the sequence of courses; the time and duration allocated for the course; the amount of teacher contact; the scheduling of contact hours; the means of access students have to relevant material, equipment, and activities for their study; the timing of assessment; the form of assessment; the administrative and technical support given to students, etc. Most of these aspects of a student's experience are effectively out of the hands of the individual academic planning their course or media design, and act as constraints within which they work. However, they can all have a significant effect on how students study:

- Amount of material covered – pressure of time leads students to cut corners, cover breadth rather than depth, use superficial study methods.
- Access to relevant material, equipment and activities for study – scarcity of library resources and ICT equipment creates barriers to students being able to study as they need to, especially for the many students who are part-time, or distance learners.
- Technical support given to students – as the use of learning technology increases students need support for networking and for running unfamiliar software, on a variety of personal systems, without which there is a risk that they will be unable to study.

These effects have been reported in many evaluation studies at all levels of education. Because these aspects of the teaching–learning context are so often beyond the control of the individual academic planning their course, the solutions are more likely to be found in changes to the institutional context, which we come to in Chapter 12. On the other hand, because institutional changes can sometimes occur through the action of individual academics demanding better organisational conditions for their teaching, it is worth considering them here as well. They usually require additional resource, which is the subject of the next section. The 'hygiene factors', the logistics of running new technology, are often underestimated in terms of the staff time and expertise they require, but without this even the best ICT materials will be unusable.

Summary

Ensure that the logistics of the academic context allow students to study effectively and efficiently:

- inform students about the importance of material covered in a course, whether it is, e.g. essential, important, or optional, to encourage depth of study rather than breadth of study;
- ensure all students have good access to relevant material, equipment and activities for their study, with particular attention to the needs of part-time and distance learners;
- ensure all students have good online and telephone-based technical support for their study, available at all hours.

THE VIRTUAL LEARNING ENVIRONMENT

This chapter has clarified the importance of the 'context of delivery' of learning and teaching. The development of learning materials is important, but delivery is paramount. The most stunning educational materials ever developed will fail to teach if the context of delivery fails. The 'context of delivery' encompasses the support system needed to help students achieve the maximum benefit from their study. Learning technologies can create their own context of delivery through a 'virtual learning environment' (VLE) – a Web-based environment that provides for the online student all the support facilities that a good campus would provide. The following list of key features for a VLE is derived from a distillation of the best of current practice (Britain and Liber, 1999; Ryan *et al.*, 2000).

Noticeboard

A noticeboard enables tutors and course organisers to keep students in touch with ongoing arrangements, updates to the materials, topical events, etc. Students should also be able to contribute short items. The noticeboard needs daily management to ensure it is up-to-date, and that only appropriate material is being displayed.

Course outline

The course outline or schedule provides an overview of the course structure with pre-requisites, contents, objectives, study calendar with suggested workload times, critical dates, e.g. for assignments, assessments, synchronous online conferences, etc. It will provide hyperlinks to each ICT component, or to a brief description of a non-ICT component. This could act as the course homepage, and be the principal site from which the rest are accessible. A more visual map of the course may give better support to navigation of the materials for some students.

Students' personal pages

This feature enables students on a course to get to know each other. The system would provide a standard personal page for each student to edit, linked to the list of students on the course. Students should also be able to upload a page of their own design instead of the standard format.

Narrative media

Many courses running a VLE will include print and video materials as well as those available online. If students are asked to download and print themselves, they will be receiving poorly presented material at great cost. It is better to mail properly produced printed books. Video material takes a long time to download – mailing a disc is better. Print material should also be available, via the course outline, in electronic form, for ease of indexing, browsing and searching.

Adaptive media

The multimedia materials available for a course can be offered for downloading via the VLE, but in many cases would be better sent via CD or DVD. Such programs are large and slow to load via the Web. The course outline should hyperlink to a runnable 'taster' of these materials, and a brief description, but an option to download should state the likely time for given modem speeds.

Web resources

The environment should include a facility for tutors to add to a managed list of Web resources relevant to the course. These should be managed by library staff, just as the book collection is managed for a reading list – monitoring availability and appropriateness.

Conferencing tools

Asynchronous conferencing tools for discussion groups provide the means for students to engage in collaborative exchange about topics on the course. Synchronous environments are valuable for small groups up to six, using audio and graphics on the Web. Conferencing tools such as Lyceum and D3E have been discussed in Chapter 8.

Assessment formats

Diagnostic pre-tests should be available to help prospective students test themselves on pre-requisites for a course. Interactive computer-marked assessment should go beyond multiple choice questions to offer more challenging 'concealed mcq's', open-ended questions with access to model answers, manipulation of interactive models and simulations to achieve target output, etc.

Assignment handling

This feature provides the means for students to send their completed assignments in electronic form to the tutor. The document is returned electronically, once annotated with comment, feedback, and marks. The system also automatically records and accumulates the marks for all students.

Student notebook

A notebook facility would allow students to annotate and link to the material they are currently studying. If they are studying online, the Web page address should be stored so that when they return to the page their previous notes are available, or when they look at their notes there is a link to the appropriate page.

Student contributions

Students should be able to upload their own materials to a shared area for other students to use. This may take the form of simply contributing links to useful websites, or it may include other useful files such as spreadsheets or design tools. The inter-student exchange that is part of campus life should be feasible in a VLE through such a feature.

Bookmarking

A bookmarking facility can decrease the time spent navigating to frequently used places or items within the environment. Some systems include a more sophisticated version of bookmarking that allows participants to build up their own individual resource base.

Email

An email system can be used to email the tutor and individual students on the same course, and others in the same institution.

Student's homepage

This should feel like home to a student. The homepage would take a standard form, and would include access to data on their assignments, individual information about their own progress through the material, access to their email messages, access to their personal bookmarks, notebook, and online library, etc. It would also offer a link to other useful institutional sites, such as the Student Union, the Library, the technical Helpline, and the Administration.

Navigation

The navigation features affect the usability of the environment. Good design practice will ensure that students can always return to the homepage from any point, that this has a well-structured hierarchical index to the whole site, and that there is a keyword search facility. The institution has to decide whether the course homepage or the student homepage is the default option for when a student logs in – or may offer the student the option.

Metadata

Each course component should include information about author, date, who holds copyright, the target audience, etc., just as printed materials do. Such information should be in a standardised format such as IMS (see Web References, p.260).

Tutor support

The VLE holds information about students' names, their marks, and how they have accessed the system. This can be available to the tutor to enable them to check progress, and target help as necessary. It should also offer tutors a structured FAQ format that enables them to compose answers to genuine frequently-asked questions, as they come up, for the benefit of all students.

Student support

Students on every course will need generic support, just as they do in a campus university. The VLE should therefore include access from all course or student homepages to support and guidance services and documents on: induction, study preparation and learning skills development, course choice and study planning, careers guidance, advice on issues affecting progress, support for students with disabilities.

Universities and commercial providers are developing VLEs, but few include all the features covered here as essential to support students fully in an online study environment. The above list could act as a reasonable benchmark against which to test the learning support aspects of any system under consideration.

SUMMARY

This chapter has outlined some of the key factors in the learning context which are likely to affect the way students learn. We began with the assertion that learning technologies depend for their success upon being embedded properly into the existing learning context. Innovation will necessarily require changes in what exists already, and if this is not acknowledged and accommodated then the innovation will not succeed. Students respond primarily to the institutional context as they perceive it. The demands and constraints it imposes, in terms of the issues discussed in this chapter, will have a greater effect on what students know than will any ingenious pedagogic design. These are the issues every academic must attend to for their teaching, or use of media, to succeed:

- student preparation for studying from the new materials;
- integration of the new materials with the rest of the course;
- discussion of epistemological values;
- pedagogical support to complete the materials' coverage of the learning process;
- revision of the form of assessment in line with revisions to the teaching;
- logistical conditions that will allow students to study effectively;
- a supportive learning context that will help students to study effectively.

Revolutionary improvements in the quality of teaching do not usually succeed in the context of one course, however. Many of the changes necessary need to occur at departmental or institutional level, or beyond. That is why, in the final chapter, we consider not just the student's learning context, but the academic's teaching context as well, to the widest context of the academic profession as a whole.

Chapter 12

Designing an effective organisational infrastructure

INTRODUCTION

It is not feasible to ensure effective teaching through ICT methods by promulgating prescriptive guidelines on how to design materials. Our use of new media over the last few years has been prodigious but is not matched by our understanding of it, because the emphasis has been on development and use rather than research and evaluation. This book has used what we do know from studies of student learning and from what few evaluation studies there are to develop a methodology for the design of ICT-based teaching that builds on what is known and enables that knowledge base to continue to be developed. This chapter takes that approach to its logical conclusion by applying the methodology to the whole academic system for a university. The implementation of new technology methods cannot take place without the system around it adjusting to the intrusion of this new organism. The biological metaphor is apt. The academic system has to learn, has to be able to respond to its environment, which is a hostile one in most countries now, and respond also to its internal changes, which again in most countries are radical ones. If academe is to preserve what is good in its traditions and also preserve its mission to develop knowledge and educate others, then the higher education system needs a more robustly adaptive mechanism than it has had to develop hitherto. This chapter postulates what that system must look like, if we are to make best use of the new technologies.

CONDITIONS FOR A UNIVERSITY TO BE A LEARNING ORGANISATION

What kind of university organisational system is capable of being adaptive to the changing environment universities find themselves in? In a paper for a systems conference discussing the Dearing Report (Dearing, 1997), I argued that it has to be a learning organisation (Laurillard, 1999). It is an attractive, but overused concept, with many meanings for different contexts. But there is a very straightforward way of thinking about what a learning organisation has to be.

Like any organism adapting to its environment in order to survive, an organisation has to be capable of adaptive learning. If we think we know what it takes for an individual to learn, and we do have a usable framework for this now, perhaps this is applicable to an analysis of what it takes for an organisation to learn. The difference is that in the academic context for the individual, there is a mediating teacher. In the experiential context for an individual and for any other organism, where the learning is done from the environment directly, without mediation, the model becomes reliant entirely on the internal conversation, similar to that needed for learning from the non-dialogic medium of lectures (see p.88). A learning organisation, therefore, is one that attempts to conduct an internal learning conversation that allows it to learn from experience, and adapt to its environment. And if the logical structure of a learning organisation is congruent with the logical structure of a learning individual, then its internal structure ought to mirror the Conversational Framework for an individual learning.

One test of whether this model of an educational institution is coherent or useful is to interpret each part, and use that interpretation to challenge constructively the way we run our universities. It means reinterpreting the Conversational Framework defined for an individual learning through an internal dialogue around their experience. Figure 12.1 reinterprets the framework for a learning organisation, in terms of the activities carried out at each node, and the kind of information that passes between the nodes.

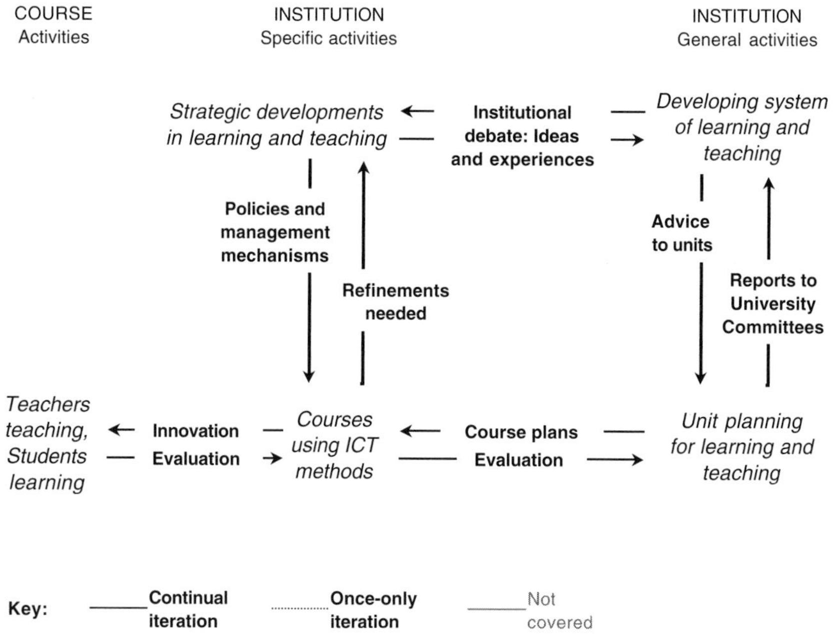

Figure 12.1 The Conversational Framework for the learning organisation.

The diagram represents the internal conversation in which the lessons from the specific activity context are combined with other specific lessons to inform the general approach, at the level of institutional policy, and at the level of teaching and learning practice. At the level of practice, there is interaction with the world of students learning, through the activities of course teams, academic designers, tutors and lecturers. The direct experience afforded by the conduct of teaching is evaluated to inform academic unit plans for further courses, constituting the internal dialogue at the level of practice. Reflection on the lessons learned from these specific developments feeds back to the strategic development policies that can then put in place the improved policy or management mechanisms needed. Similarly, the experience of each unit is reported to university committees through the planning process, to inform the further development of the university's system of teaching and learning, which further informs the unit planning process. Insitutional strategies for learning and teaching, for curriculum development and for institutional planning would be developed at this node, informed by the lessons fed back through the quality assurance, reporting and planning processes that match the form of the Conversational Framework. Many of the pilot experiments on learning technologies in universities have been conducted in isolation from the institutional management process. This analogy with the individual learning shows how their integration into an organisational learning process could enable such experiments to contribute continually to the long-term reconceptualisation of the institution's learning and teaching strategy.

The appeal to the Conversational Framework as an organising principle for a university suggests, therefore, that its organisational infrastructure must be cyclical to ensure improvement in its learning and teaching. High academic standards are assured partly through setting up mechanisms that are capable of monitoring, learning, and changing. The goal–action–feedback–revise action cycle should be evident at every point in the organisational process, and this includes management actions (Elton and Middlehurst, 1992). As in any learning process, there has to be a meta-level function that reflects on the process at the next level down in order to set up improvements to it. Therefore, in thinking about how development and implementation should be organised, we must be aware that every level of operation presupposes a higher level that is monitoring and reflecting on the way the lower level carries out its tasks. The same people may be operating on both levels; the two levels define different aspects of their activity: focusing on the operational aspects at the lower level, and reflecting on the strategic lessons from that operation at the higher level.

Following the principle of 'self-similarity' of learning systems, the same structure will necessarily be mirrored at each level of description of the organisation, such as 'unit', 'department', 'programme board', etc. This recursive form was alluded to in Beer's description of a viable management system – like fractals, whichever level of the organisation you describe, the structure should be the same (Beer, 1985). A complete picture of this organisational structure would look like a fractal picture of nested Conversational Frameworks, operating at each organisational level. A committee responsible for project funding, for example, would ensure its own progressive learning by defining the proposal template. Similar to a research proposal, developers must show what existing knowledge and development they are building on, define the objectives to which their outcomes must converge, explain how they will evaluate their work, and describe how they will articulate and disseminate the results. The research project, after all, is itself the microcosm of a learning system.

Universities as learning organisations are best described in terms of the Conversational Framework for experiential learning, rather than mediated learning. With respect to a university's strategy for learning and teaching, there is no strong external agency playing the guiding and supporting role of the teacher, and none could, if universities are to maintain their academic freedom. In some universities there is a strong internal agency that plays the role of enabling the organisation to learn about its teaching. In the case of the Open University the Institute of Educational Technology played this role, joined later by the Knowledge Media Institute, with respect to the new media specifically. Both academic agencies help to innovate and test new ideas, and to conceptualise what is being learned in order to generate future enhanced action. Both report to the senior management office for learning technologies and teaching. Internal academic agencies of this kind can be extremely valuable to university management that is consciously setting out to create a learning organisation.

From the institution's point of view, the core conversation is between teacher and student. At the next level of description, we should expect to find the Conversational Framework for the individual teacher, or teaching team. The link between the bottom left-hand nodes of Figure 12.1 define the experiential learning that the academics use to learn through the experience of teaching. The full representation of the Conversational Framework at this level is shown in Figure 12.2. The teacher's own unmediated learning from experience uses the iteration at the level of practice to identify and attempt to respond to students' learning needs. They then attempt to improve on their own practice, through reflection and adaptation in the light of their improving understanding.

The top left-hand node would complete the framework for mediated learning if the academic were engaged in a professional development course, providing advice and resources on the theory and practice of teaching and learning. In this case, there would still be no link with the course environment. In the application of theory to local practice, as in any professional development course, the providers cannot govern the local environment in which the learner is working.

Applying the Conversational Framework in this way suggests that we should diminish the distinction between teaching and research as essentially separate activities. The academic should be seen not just as researcher and teacher of their subject, but also as researcher into the teaching of their subject, providing the bridge between the two activities that effectively blurs their distinction. With the academic playing this kind of role, problematising their teaching, and learning from the student, the university will be a learning organisation right through to

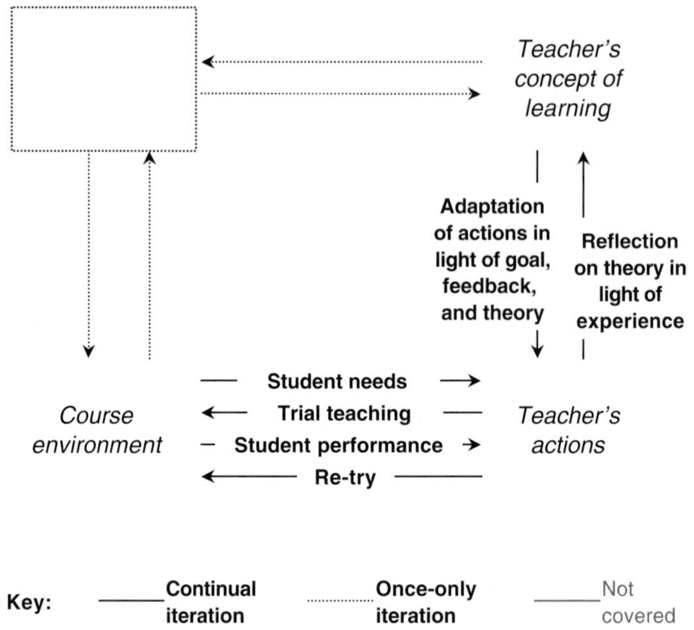

Figure 12.2 Interpretation of the Conversational Framework for learning through experience for academic teachers.

the course level. Ison goes further, and argues that a university cannot be a centre of learning while the teaching–research partition remains, and advocates instead:

> ... action-research as a means of integrating research and learning, leaving 'teaching' to wither from its place as the dominating paradigm ...
>
> (Ison, 1994: 379)

Certainly, it is clear from the arguments advanced throughout this book that teaching at universities must be linked to the origination and negotiation of knowledge. Academic knowledge is distinct from experiential knowledge (see Chapter 1). Academic knowledge has an integrative character. It is a reflection on experience, rather than being synonymous with knowledge of experience *per se*. It also includes knowledge of how that knowledge comes to be known. William Perry (1970) described the Harvard students' struggle to achieve this epistemological perspective on their knowledge. It remains an important part of what counts as being a graduate, that they should understand the limitations of knowledge, and what it takes to generate new knowledge in their field. Academics who wish to enthuse their students and help them keep abreast of developments in the field will strive to make even complex ideas intelligible. Keeping the curriculum close to the areas most salient for application in future working environments will be valuable for all students, including those who do not work in that field. By treating teaching as an extension of their research interests, academics will increase their own and their students' motivation. The fundamentals of the subject can be taught as well in relation to the latest findings and become the more obviously relevant at the same time. The link to research helps to clarify the nature of the knowledge being learned, its origins and limitations, and the relevance of even the most uninspiring aspects of theory. The close synergy between research and teaching ensures that a university remains a true centre of learning.

ESTABLISHING AN APPROPRIATE ORGANISATIONAL INFRASTRUCTURE

The now extensive literature on knowledge management draws our attention to the importance of continual innovation, if an organisation is to remain competitive. A learning organisation is:

> continually expanding its capacity to create its future … "adaptive learning" must be joined by "generative learning" – learning that enhances our capacity to create.
>
> (Senge, 1993: 14)

Senge's quote captures the twin tasks of both generating new knowledge, and monitoring existing activities, to ensure adaptive change in response to the external environment. Nonaka made the link between knowledge creation and competition in his seminal paper on organisational knowledge, and his model draws attention to the relationship between individual learning and organisational learning (Nonaka, 1994). Organisational knowledge creation is seen as a continual dynamic process of conversion between tacit (experiential) and explicit (articulated) knowledge, iterating between the different levels of the individual, the group and the organisation. The principles of iteration between practice and theory, in a dialogic process between individuals and groups at different levels of description of the organisation, are very similar to the principles embodied within the Conversational Framework. Nonaka's organisational knowledge process successively iterates through 'enlarging individual knowledge', 'sharing tacit knowledge', 'conceptualisation', 'crystallisation', 'justification', and 'networking knowledge'. The evaluation and validation of innovations combine in the 'justification' process, which evaluates the knowledge produced in relation to the management requirements. In practice it is valuable to separate the two into (1) the iterative formative evaluation of projects to the point where they appear to meet the objectives, and (2) the summative validation of an implementation to test whether the product as a whole works in the marketplace. The complete process for organisational learning can then be characterised as a succession of activities:

- expanding knowledge
- sharing
- innovating
- evaluating
- implementing
- validating

linked through the iterative flow of information between them. Figure 12.3 shows the kinds of activities undertaken within each node, and the kind of information that flows between them.

Beginning at the bottom left-hand node, the responsibility for expanding knowledge rests at every level of the organisation. The individual undertakes literature searches and keeps abreast of new developments and ideas. The department regularly appoints new staff, and visits similar departments elsewhere. The institution collects market intelligence and environmental analyses. This is where the institution exposes itself to new ideas from wherever in the world they may originate. The ideas and awareness produced from these activities are difficult to communicate within an institution, and this is elaborated in the following section.

Sharing tacit knowledge, which is itself derived from these activities, happens within the different kinds of working groups in a university. It will in turn generate further explicit knowledge in the form of ideas and plans that begin the design process. This is where innovation becomes explicitly embodied in prototypes of learning materials and services. The iterative evaluation loop allows the team to develop and refine the ideas in relation to the intended objectives, and this process produces the explicit specification for how the actual implementation should work. Finally, collecting performance data will enable the team and others to validate the implementation against the market response, and articulate the lessons learned, to contribute to the further expansion of our knowledge of learning and teaching.

The activities and information flows defined in Figure 12.3 are further elaborated in the following sections. Each task is assigned either to academic management, including pro-vice-chancellors, vice-presidents, deans, unit directors, etc., or to academic teaching, including individual academics, course teams, programme boards, curriculum planning groups, teaching committees, etc. High academic standards are assured in the use of new media in teaching if evaluation and validation mechanisms are in place for both design and implementation. Organisational learning is necessary for renewal and survival, and it requires the kind of iterative knowledge creation process outlined here. The rest of this chapter considers the implications of this for the way the academic system must operate, in terms of the constituent tasks to be carried out at each academic level.

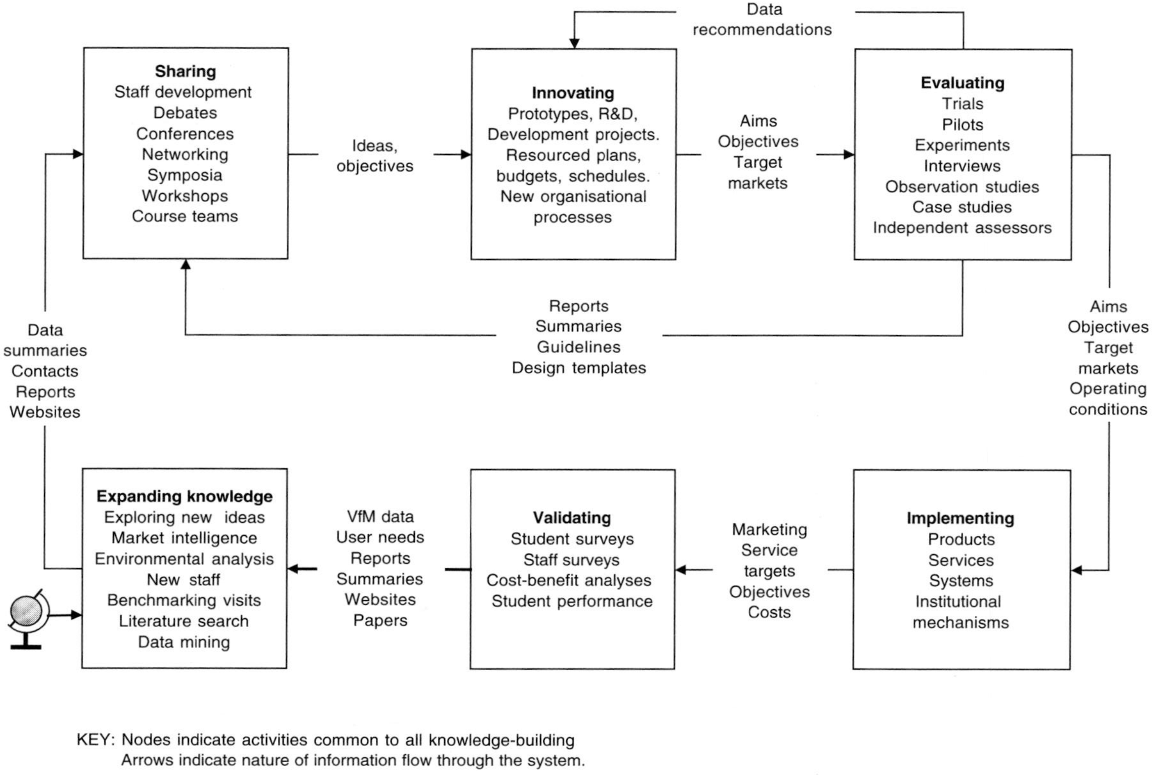

Figure 12.3 Knowledge management activities to assure competitive advantage through innovation.

EXPANDING KNOWLEDGE

The important activities here support staff and groups at all levels in the university in their exposure to new ideas, and existing knowledge. It is the natural task of any academic to do this in the context of research, but not teaching. Higher education must build its knowledge of learning technologies, and that process begins with exposure to the, albeit meagre, existing knowledge in the field. However, this sits in the context of market intelligence more generally: environmental analysis of what others are doing, and what students are looking for. The main tasks for management and teaching are as follows.

Academic management

Provide access to an institutional database of learning materials and evaluation reports

As universities move to creating more learning materials, in either traditional or new media, they will be developing an implicit archive of learning materials and documentation about their success or failure. The archive should be explicit, managed and promoted. An institution's information strategy, or knowledge management strategy, would enshrine such activities, and enable academics and course teams to build on existing knowledge and ideas.

Promote access to national databases of learning materials

Access to national databases, if they exist, should be available through the website for the national funding council. An institution's IT strategy should aim to support staff members in their use of national and international databases, and help to legitimise the migration of good design ideas across institutions.

Provide funding for journals, travel

Funding should be available for progressing the quality of teaching as well as the quality of research. Network access can provide for some degree of collaboration, but travel will be essential occasionally, so this has to be provided for within academic teaching budgets.

Convene teachers' forum to identify key areas for development

The responsibility for developing the best uses of new technology in each subject area will rest primarily with the discipline area itself. The usual fora of academic journals, the academic conference, the professional institute, the professors' conference, etc., will enable this debate to be pursued, but each of these must recognise their responsibility to promote such debate, not once, but continually.

New technology can be a barrier to progress, and needs to face a political as well as a pedagogic critique (Hawkridge, 1993). Academics must be aware of the politics of new technology if they are not to be misled by it. Each discipline area will have its particular challenges and vulnerabilities, which is why the debate must be reworked for each one.

Recruit new staff

Maintaining a healthy turnover of staff has always been valued in research areas, and is equally important for ensuring innovation in teaching. The recruitment and appointing process is an ideal opportunity to define the importance of teaching innovation, and to demonstrate how excellence in teaching is valued by the academic management.

Academic teaching

Analyse course provision at universities

Teaching through learning technologies is only economically feasible if it is used across a wide range of institutions. Any materials developed must have a large potential market. An analysis of course prospectuses for universities would determine common topic areas. Further information from course convenors is necessary to determine the likely characteristics of the student audience, and the approach to the subject. This kind of survey is necessary to guarantee the largest possible audience for the materials developed.

Access national database of courseware, courseware reviews

If there is to be progress in the field, it is vital that courseware developers build on what has gone before. They need access to it, therefore, not just via reviews, but also in demo versions, available through websites, so that they can move forward from the best of what already exists. National and international databases already exist, at least for the exchange of ideas and good practice, and preferably for the migration of the materials themselves, if design is to achieve maximum efficiency.

Analyse market needs

Universities have recognised the need to be less provider-led in their design of curricula, programmes of study, and course offerings. There is still an imperative for the curriculum to be driven in part by developments in the discipline if future graduates are to be abreast of key developments. However, it is equally important for universities to be responsive to students' academic interests, and to their logistical needs. If more people wish to study part-time, or at a distance, or with interrupted periods throughout a long working life, then universities must respond. Learning technologies support these new modes of study, and a greater shift to more flexible forms would demonstrate greater responsiveness to students' needs.

Carry out literature search in key journals

Many subject areas have their own academic journals of research and development in teaching the subject, and library staff should make these a priority in their collection management. They will be a valuable source of information on students' needs, and teaching design ideas. They also alert academic designers to where their own work might be published.

Build on previous evaluation reports

Teachers and designers should use the output from previous validation stages, at least within their own department or institution. If an institutional database exists, then this is the obvious place to begin. The university library should house copies of such reports. Wherever dissemination from the validation stage has taken place, this stage should make use of it.

Analyse exam scripts

Critical problems in students' understanding of a subject area will often be apparent from assessment scripts and assignments. Analysis of these would give rich material for deciding what kind of learning activities students need in order to attain the defined objectives. Examination scripts should be seen as an evaluation of the quality of teaching, as well as the quality of learning. Data mining of the institutional records on assessment and examination across departments, courses, types of student, types of teaching, would enable a course team to begin understanding the relationships between some of these critical variables.

SHARING KNOWLEDGE

The outcomes from the stage of expanding knowledge will sometimes be in the form of formal reports, e.g. of conferences or academic visits, or of environmental analyses, usually made available on websites, or as strategic papers to key committees. Much of it will be tacit knowledge built up by individuals who keep up to date with current developments. It is important for a university to create the opportunities for ideas and experiences to be shared and debated, so that they become part of the institutional awareness of what works, what more is needed. The activities at this stage will include informal gatherings, discussion groups, and planning groups, as well as formal symposia, staff development activities and institutional debates. Both management and teaching have key roles to play.

Academic management

Optimise the deployment of staff resources

As academic staff begin to be more involved in the development of teaching materials, rather than face-to-face teaching, there will be considerable logistical implications for the way learning and teaching is organised within the institution. It changes the way space is used, and the way teaching is timetabled. There has to be a clear understanding in all university departments of the nature of the changes being made, so that they can decide how best to organise staff and student time. Such a plan would decide how the timetable would gradually change from mainly face-to-face teaching to more resource-based teaching with small group work and mentors. It would also plan the use of staff resource: how the department would use some staff for traditional methods and some for development of new technology methods, who should receive a staff development programme, and over what period skill in the new methods would be developed for all staff. Appendix 2 in the Dearing Report suggests one approach to this kind of planning (Dearing, 1997).

Optimise the organisation of teaching

The new technology revolution cannot happen in one department alone. Students should have an integrated system to work in. If they attend courses offered by different departments, the workstations and the software they use should be compatible. If the optimal way of managing large numbers using new technology turns out to be within a block teaching format – where students study just one subject for a few weeks, then move onto another, rather than studying several at once – then this has to be a cross-disciplinary decision. Remedial or foundational courses, such as basic maths, IT, report writing, etc., should be available as generic learning materials, in a form that is customisable to each different subject. They must therefore be acceptable to all departments. It will be the responsibility of the academic administration at both central and departmental levels to devise and monitor the optimal teaching organisation for their university.

Encourage use of good materials developed elsewhere

In tandem with a national funding council commitment to encouraging transfer of courseware between institutions, there should be an institutional level of commitment to acquiring courseware. Selection and acquisition should be carried out by academics, but may be encouraged by the academic administration if funds are available for importing materials. As standards develop, the transfer of courseware should become easier, to the point where it can even threaten the providers:

> The development of marketable and interchangeable course units has the potential for taking decisions about the shape of courses out of the hands of the educators.
>
> (McInnes, 1995: 50)

The Instructional Management Systems project in the US (see Web References, p.260) offers an example of how course materials could be accessed not just by educational providers, but by the students themselves. The form of the Conversational Framework suggests that this would not constitute good educational practice, but a more commercial market-oriented approach to education would nonetheless become feasible. If the educators remain appropriately responsive to students, and to potential student needs, then they are the more likely to remain in control of the shape of courses and curricula.

Establish a programme of staff development

Staff who are to be involved in implementing new technology methods will need an induction programme which includes the objectives of:

- raising awareness of current teaching practice and use of new technology in their field;
- elaborating their understanding of how students learn through different media;
- developing their expectations of, and critical approach to, new technology;
- developing formative evaluation skills for improving learning design;
- increasing the likelihood that they will make their own contribution to the field.

The programme will be more successful if it is subject-specific, as generalisations about learning can be difficult to follow through to the particular concerns of academic teachers, without careful mediation. It will also be more successful if it takes account of the needs of practitioners, who will not wish to give up time to sustained courses. McNaught and Kennedy argue the importance of faculty-based support groups, because they can propagate new skills among colleagues far more rapidly than centrally planned provision does (McNaught and Kennedy, 2000). The resistance of academics to educational courses is remarkable for a profession that lives by them, but none the less real. Staff development for academics has to use the most extreme form of work-based professional updating possible. Induction presentations will be needed at timely moments in course design and planning, linked to the current uppermost concerns. Familiarity with the key technologies is a pre-requisite for academics to think through how to use them in courses, with access to good practice sample materials provided through the library. Academic management should formally encourage new and existing staff to gain teaching credentials of some kind (in the UK this would be the ILT at present – see Web References, p.260).

Set up multi-skilled development teams

Collaborative development is crucial for learning technology because of the range of skills needed. An appropriately balanced team is the first step in the quality assurance of the materials and services produced. At some level of academic administration, there must be a judgement about the capability of the team to produce the planned materials. Evidence of experience and previous success will be important. It helps if such evidence is available to inspection, as this in turn helps to promote the documentation of the lessons learned. Where experience is lacking, which is often the case for new technology, then there should be evidence of knowledge of existing materials to build on, and formative evaluation skills needed to refine the new development. Project management experience and success is critical for the team leader.

Set up forum for teachers to discuss ideas, experience

Academics and designers need to discuss their experience of learning technologies, and the academic issues surrounding the balance of learning methods. This forum would receive and debate evaluation reports on developments. It would probably range over topics such as the design of media-based teaching in the subject, the success and failure of ways of supporting students in their use of media, the problems of integrating learning materials with other teaching methods, and teachers' requirements for future technological development.

Academic teaching

Share awareness of current developments in design

Many subject areas have conferences on teaching, increasingly with software exhibitions, which allow academic designers to see what has gone before them. Keeping abreast of developments in the teaching of their subject will become as important as doing so in research, as university teaching becomes more professional. The analogy with research follows through into all modes of updating academics' knowledge of current developments, including seminars, the library and the Web. Teaching seminars are as uncommon, however, as research seminars are common in a thriving academic department. A rapid and radical contribution to awareness of teaching design, and recognition of its importance, would be to inaugurate just such a seminar series.

INNOVATING

Innovation is at the core of a university's competitive advantage, in both research and teaching. In one sense, it cannot be managed. Creativity is an uncertain process. However, the conditions for innovation can be managed. The previous stages of expanding and sharing knowledge will have laid the foundations. Teaching could not be an activity that once honed to perfection could remain static, any more than research could be. There is a continual interaction between knowledge and the way it comes to be known, whether the knowledge is being taught, or being discovered. Innovation in learning and teaching will include introducing new ideas to be taught, new ways of combining topics, new ways of engaging students' thinking, new media, new forms of assessment, new ways of organising teaching, new modes of learning. Continual innovation should be a natural part of an academic department, not necessarily in all areas of teaching at once, but nothing is sacred. Creating the right conditions for innovation to be both feasible and successful, without undermining what remains unchanged, is the responsibility of both management and teaching.

Academic management

Agree development resources and costing

Innovation is expensive and needs protected resource. The introduction of new technology is uncertain, and therefore needs to operate within agreed limits. Universities do not have the resources to afford the massive cost overruns that are common in software projects in the commercial sector. Resource for development should be planned and managed at institutional and at departmental level. Academic management needs to operate a well-defined approval process for development projects. From the previous analysis, a departmental teaching committee should be able to assure itself that the project has attended to:

- market needs analysis;
- use or re-use of existing materials;
- skills audit of the team;
- development/induction programme for the team;
- staff resource needed;
- bought-in resources needed;
- rights clearance process;
- risk analysis;
- development and production schedule;
- developmental testing programme.

Costs can be managed more easily if resources are committed to full-scale implementation only after a first-stage prototype has proved successful. Hence the importance of scheduling, and of a testing programme. The development team should be making resource-related decisions in the light of their cost, and it is management responsibility to make sure that this information is available to them. The greatest cost, inevitably, is staff time.

Agree staff time commitment

Academic staff time in universities is rarely fully costed in relation to specific areas of their work. This will be an important part of planning if lecturers begin to spend less time on lectures and large classes, and more time on materials development and small-group mentoring. Teaching timetables have to be planned around this well in advance. Figure 12.4 shows the dramatic shifts in staff time that result when moving from traditional modes of teaching to the kind of open and distance learning that the new technology offers. Using learning technology materials means that staff must find time for their development, which can only come from student contact time. The time spent on presentation of the curriculum through lectures disappears, to be replaced by much greater design time for materials. Time for marking assignments reduces in the expectation that some formative, and even summative, assessment can be computer-based. Time needed for research and development in this innovative and fast-moving field, for every discipline, must be planned.

For many academic staff, the introduction of new technology has been a nightmare of overwork and lack of support. Advance planning and project management will help to constrain ambitions for what can be completed within any one project, and will also enable valuable data to be collected on the real costs of innovation of this kind. It is essential to approach innovation with realistic expectations of the resource needed, if it is not to be undermined by the sheer exhaustion of the enthusiasts needed to make it happen.

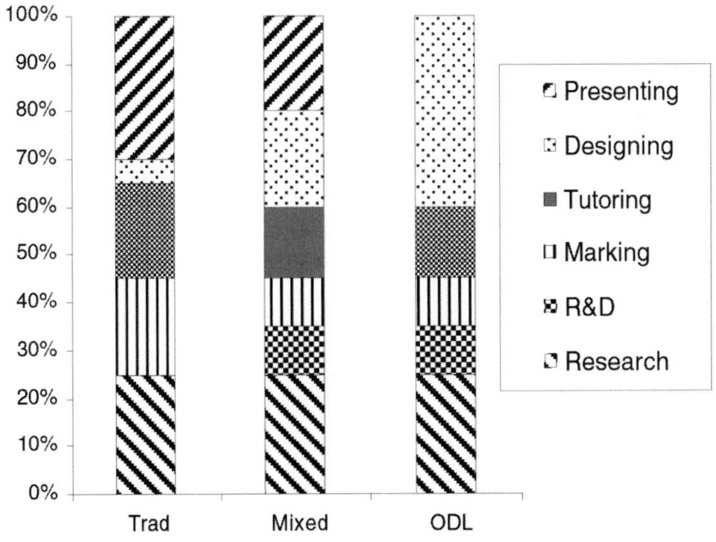

Figure 12.4 The change in the distribution of staff time across different modes of teaching; contrasting traditional with open distance learning, and with mixed mode combining campus and distance learning.

Monitor project management

The courseware development procedure is itself a form of quality assurance mechanism. As John Daniel points out in his analysis of the Open University, the mechanism of the course team, through the collective work of the specialists it brings together, is considered to give the content of the materials they produce a special authority (Daniel, 1991: 24). For innovative multimedia materials, however, continual monitoring of quality is necessary, because innovation is uncertain. The academic administration, in the form of a teaching committee, should:

- receive reports from course teams on project management and progress;
- ensure that the team is working to the resources available;
- take an active approach to risk management;
- take responsibility for changes to procedures as necessary.

The uncertainties of innovation will not be removed, but they can be contained.

Establish policy on reversioning support, productivity tools, and standards

Efficient use of resource depends on optimising the productivity of academic staff in the development process, and on maximising the value of the output. The ability to reversion existing learning technology materials should be an essential benefit of the use of electronic media. Editing, updating, and recombining is feasible, and materials for new objectives and contexts can be generated at relatively low cost. However, the technical infrastructure for creation in a form that allows reversioning for archiving, storage, and retrieval, is non-trivial, and requires specialist support, including systems development, copyright, and design experts. There has to be a clear policy on commitment to reversioning, the use of productivity tools (see p.231) and adoption of industry standards. These efficiency measures will be seen as an unwelcome constraint on creativity, although they are likely to enhance the resource available for creativity, and provide a catalyst for innovative ideas.

Match innovation in assessment to innovation in learning and teaching

New technology changes both the curriculum and the way content is learned. If assessment is to match what students have learned, it is likely that assessment processes and requirements will change (see Chapter 11). Responsibility for assessment policy lies with all levels from individual academic to institution, but policy will tend to change in response to bottom-up demand, which can be a slow process. Management must be prepared to innovate at policy level to avoid undermining learning innovation with the adverse effects of inappropriate assessment. Students optimise their effectiveness by responding first to assessment requirements, so all learning innovation must include design of assessment as well.

Academic teaching

Address innovation to the critical aspects of learning the subject

The use of learning technology is difficult, time-consuming and expensive, and should therefore focus on the critical aspects of learning for each course or programme. There may be foundational concepts or skills, or new curriculum topics reflecting changes in the subject area due to ICT, or aspects of the field that are traditionally difficult for students. The earlier stages of origination and sharing of knowledge should have clarified what these are. All these are good reasons for attracting innovative ways of teaching. There is little point in wasting valuable creative effort in areas where traditional or cheaper methods are already effective. Before investing resource in technological innovation there must be a clear justification in terms of how it will deliver better quality learning. Two examples from the Open University are (1) numeracy software that taught students far more effectively than print materials had ever succeeded in doing, and (2) the collection of students' history project data on CDs to act as an additional resource for later students. Both could demonstrate in advance that there was good justification for ICT, later vindicated by students' enthusiastic evaluation.

Use productivity tools, reversioning principles, and standards

The production members of a course team should be familiar with all the productivity tools available for media development. The team must be aware of the reversioning principles and standards that design must respect if the materials are to be maximally useful. This is partly a technical issue that the production staff should be able to advise on, but is also a pedagogic issue. Materials are designed as coherent wholes for the immediate context, and it is difficult to deconstruct them into separable parts that could be re-used in new contexts. This is a nascent skill at present, with little experience or success to build on, but this principle is critical to the sustainable development of learning technologies: *learning materials should be designed to be customisable by others*. If they are designed with re-use in mind, then we begin to migrate the good practice lessons learned through the design of customisable models, and we begin to reduce the overall cost of these necessarily expensive developments (Twining *et al.*, 1998). Similarly, the copyright for use of existing materials should be investigated at the earliest opportunity, to ensure that the materials produced can be used as widely as possible.

Receive and act on evaluation reports

This stage of the creative process receives information also from the evaluation stage, given its iterative nature. The course team responsible for developing a set of materials should be in receipt of the evaluation data from developmental testing. Discovering student reaction to early designs can be highly illuminating for designers, as fond expectations are frequently destroyed by the reality of the learner learning. But the process is also a catalyst for invention and creativity. The detailed study of learners learning provided by the observation and interview studies in a developmental testing programme enhances academics' and designers' understanding of the conceptual difficulties they have, which in turn contributes to the improvement of the design. This is possibly the single most valuable source of information for the innovation process.

EVALUATING

This stage is part of the iterative process that delivers specific information to the innovative design stage, and delivers the more general lessons learned to the knowledge-sharing stage. The latter may take the form of evaluation reports available on a central website, recommendations for policy changes, or developed design practice embodied in guidelines or design templates for others to use. The data collected will come from detailed case studies, learning experiments, observation studies and intensive interviews, all designed to challenge the design against use, and to inform the re-design process about learners' needs (see Chapter 10).

Academic management

Administer refereeing process for design of courseware

All universities practise a reviewing and refereeing process for course accreditation, but not for initial course approval. The process needed is similar to the external assessor system in operation at the Open University. Materials are reviewed and commented on against a brief description of the nature of the assessment used on the course. This helps to ensure the acceptability of the design to the discipline area in general, and guards against idiosyncrasies that would damage the wider appeal of the materials produced.

Provide time, resource and support for evaluation

Although formative evaluation is crucial to the success of design, it is not automatically included in the plans of development teams. If it is, it tends to be the final stage, so it is typically ousted to make room for development overruns. The inevitability of this, given the reality of software development, is that management must take responsibility for protecting the quality assurance process by planning for and monitoring the iteration between design and evaluation. Evaluation must be part of the design process, not separate from it. It need not add greatly to the resource needed, and will certainly improve the cost-effectiveness of what is produced. Management must ensure that the evaluation process described in Chapter 10, p.194, is fully supported, and carried through in practice.

Academic teaching

Carry out developmental testing and piloting

The courseware development team is in the best position to contribute to development of the knowledge base of how to use learning technologies effectively. Developmental testing (or formative evaluation) of materials with the target students should provide valuable information from intensive study of how students learn through such media. Designers learn more from watching a small number of students trying to learn from their materials than they ever do from questionnaire studies.

Use evaluation information to define the specification of the course

Part of the value of this iterative process is that it helps to refine the specification of the teaching material – its aims and objectives, the appropriate target students, and the operating conditions under which it succeeds. The observation studies may demonstrate, for example, that some material is inappropriate for the target students, which could lead to a revision of either the content or the target market. Similarly, it may become clear that some materials cannot be stand-alone and need supervision, which affects the specification of the optimal teaching conditions. Output from the evaluation will therefore also feed into the next stage of implementation.

IMPLEMENTING

This is the stage at which the university exploits its innovative products and services for competitive advantage, and where it ensures the success of the innovation by supporting it with new systems and mechanisms as necessary. New technology disturbs the whole environment into which it is introduced. Innovation introduced in one part of the learning–teaching system will not remain contained within it, but will affect all the administrative, technical and management systems that surround teaching. Both management and teaching play a part in ensuring stability in the face of change.

Academic management

Communicate new requirements to students and staff

Academics introducing new technologies must pre-emptively prepare their own students, and other staff for the effects of the innovation if it is not to be undermined by incorrect expectations, lack of preparation, and misunderstanding. Courses never exist in isolation, and others will be affected by new requirements on a course, such as access to equipment, changes in timetabling, changes in support requirements. The responsibility lies primarily with the academic management responsible for a course, which is capable of seeing which existing systems procedures and mechanisms might be disturbed by the innovation, or might need additional resource, at least for an initial period.

Manage marketing of the innovation

The point of innovation in learning and teaching is to improve the university's responsiveness to its students and potential students, so the implementation stage must ensure that this advantage is carried through to the recruitment of students, and indeed staff. A university will attract high quality staff, as well as students, if it has a reputation for innovation in teaching. Prospectuses, websites, advertisements should all help to foster an understanding of the benefits the innovation confers.

Provide support staff for maintenance and administration

Support staff are needed for the maintenance and administration of ICT materials and services, just as they are for print-based materials. The analogy with library staff is close. They have to be institution-based; they have to be responsive to problems and act immediately to correct errors or breakdowns; they have to be able to deal with a range of subjects; they have to be knowledgeable about access to the materials, rather than the details of their content; discipline-based staff are often required to manage the complexity of material and support decisions necessary; they are needed to ensure that the materials are operationally sound. Without this kind of support, students will find it difficult and time-consuming to make proper use of the new technology.

Finalise costs

At this stage, it should be possible to calculate the full cost of development and implementation. Each institution will have its own accounting methods, but they typically work to budget heads that are associated with management units, rather than activities, such as a particular course. It is therefore difficult to cost an innovative course against a traditional course and assess its relative value to the institution. Management will be better able to resource and manage future innovation if they provide access to the costs of activities to all staff involved in resource-related decision-making. The knowledge from this stage would then feed into the overall validation of the innovation, and a better understanding of how to improve the institutional processes.

Ensure that appraisal and promotion procedures reward teaching excellence

In the UK there is a nationally agreed strategy to improve professionalism in teaching via the Institute for Learning and Teaching (see Web References, p.260). This is mirrored in the institutional learning and teaching strategy, where current policy on promotion and reward for teaching excellence is communicated to all staff. If there is seen to be clear support for this at senior management level, backed as it is by national priorities, then the academic community will begin to believe that excellence in teaching has the same status as excellence in research. Without this, innovation in teaching will be confined to the selfless enthusiast, and will not be an integral part of the university's development.

Academic teaching

Provide guidance on the use of learning technology

Guidelines for student briefing were discussed in Chapter 11, which suggested that students using new technology materials will need help in how to make good use of them. The development team must ensure that students and staff are well briefed on what to expect, and how to operate within the new conditions.

Define and support service targets

New technology requires more attention to quality of service than most academic organisations are used to. If students are using their own hardware, there may be problems of lack of standardisation. Software always has bugs, even after rigorous quality control, so a helpdesk service is vital. Students will rely on institutional networks, day and night, and networks can go down as well as up. There are numerous opportunities for system failure of some kind, which could be devastating, for example for students who are desperate to complete an assignment by a deadline. The quality of teaching then becomes dependent on technicians as much as on academics. Academics are responsible for defining the service targets that must be met if students are to receive good quality teaching in this new sense. They must also recognise that failures will happen, and the service targets must be supported by contingency measures within the academic areas, if students are not to suffer. Students can be wonderfully tolerant of difficulties attending innovation, because they delight in the innovation itself, but they are wholly intolerant of breakdowns in the system that adversely affect their learning and assessment.

VALIDATING

This stage completes the full cycle of organisational learning, and provides a reflective account of the innovation in terms of the institutional aims and values. The information from the implementation stage will describe what happened. The validation stage will describe how successful it was in terms of student performance, student satisfaction, staff experience, and cost-effectiveness, as judged in relation to the original intentions.

Academic management

Receive and act on reports

The management committees responsible for resource allocation, curriculum development and staff management need information on how well current implementations are working. These committees will be the recipients of evaluation reports produced by academic development teams on the pedagogical value of course materials and innovations. Together with cost data, and the initial specification for the course or programme, an analysis can be made of the comparison between costs and effectiveness. This makes most sense at university level, because the value of an innovation can go beyond the specific implementation. A poor cost-benefit analysis at one level can be good at another, when other factors are taken into account. Good quality information about the effects of innovations is important for the management making decisions about further areas of development for the university. The value-for-money analysis produced will feed into the origination of developments at the next stage.

Monitor the implementation

Any teaching innovation must be summatively evaluated if it is to contribute to the development of organisational knowledge. This is the key activity at this stage. It requires staff resource: to collect survey data from students and staff, analyse it, and make recommendations for improvements. The staff resource should be an integral part of the way the organisation learns about itself, and is therefore an essential management responsibility.

Academic teaching

Monitor and report on efficiency of institutional procedures

The course team will be the main source of information to the academic management on the successes and failures of the development process. For example, local communications may be ineffective, local administrative procedures may be counter-productive, or inappropriate. With the introduction of new forms of courseware development and academic activities, new administrative procedures or working practices will be necessary. If the project management procedures for a development project include documentation of problems encountered and apparent successes, this will reduce the burden on the course team of providing this information in the validation stage.

Disseminate and publish reports

Each discipline area should take responsibility for developing knowledge about how to teach using new technology in their particular field. The fora created for academic debate and discussion of these issues, such as journals, conferences and websites, need reports on the experience of academics using the new methods. The wide promulgation of the lessons learned will inform that debate. Evaluation reports from individual academics should find a place for discussion and debate outside their own institution, and within the subject field itself. All academic departments should be contributing to the development of knowledge of the teaching of their subject, and their staff should be able to make such a contribution. As this becomes a more accepted part of the role of the professional academic teacher, the quality of information feeding into the next stage of 'expanding knowledge' will improve for everyone.

Analyse use of courseware materials and assignments

The best source of information on the pedagogical value of the learning materials used will come from the way students carry out assignments based on them. Student assignments can be seen as an assessment of teaching as well as learning. If assignment tasks are clearly related to the study materials, then it will be illuminating to treat them as commentary on the success of the design. This data has to be set alongside survey and interview data to achieve a full picture of how students perceived the value of the course, and this triangulation will improve the quality of interpretation of all the data collected.

SUMMARY

An organisational infrastructure for educational technology in higher education must enable the system to learn about itself. The decision-making hierarchy must be in a position to receive feedback on the effects of its decisions at each level in exactly the same way as the student needs feedback on their interactions with the world in order to learn. The full cycle of activities (see Appendix 3), with information flowing through each stage successively and iteratively, will provide for the organisation the same complexity of learning that we have seen is necessary for the individual. This generative methodology for building our knowledge base in learning and teaching was mentioned in the Introduction. With this framework in place, the university will be able to sustain innovation in learning technology, in order to survive in an increasingly competitive educational market.

THE NATIONAL INFRASTRUCTURE FOR INNOVATION IN HIGHER EDUCATION

National higher education systems typically operate a competitive funding environment across universities that depend on government funding. Competition is an essential mechanism for improving quality in industry, but in HE it tends to obstruct the collaboration that is so crucial for developing ICT for learning and teaching.

It is inefficient to promote quality via competition when higher education has necessarily limited income for providing a public service. The already meagre resources are spread even more thinly as academics compete and thereby repeat. There have been attempts in the UK to use national funding for new media developments to encourage collaboration between consortia of universities. Collaboration as an effective means of ensuring quality is as relevant to teaching as it is to research. However, the Dearing Report found evidence of:

> a strong weight of feeling that competitive pressures have gone too far in promoting a climate which is antipathetic to collaboration, even where there would be strong educational or financial grounds in favour of individuals, departments or institutions working together ... Collaboration matters. In some cases, it may make the difference between institutional success and failure.
>
> (Dearing, 1997: 261)

The recommendation was that funding councils should avoid funding arrangements that discourage collaboration, and work to encourage collaboration where appropriate. What might this mean? In this final section, we can extend the principles of organisational learning to consider the implications for a national approach to higher education. If the national HE system is to develop the capacity for adaptation to new technology, then it must be able to investigate, articulate, and share knowledge of learning and teaching. The salient activities for a funding council to address are proposed here.

National academic management

Promote funding of research in student learning

The development process will help to build knowledge about learning through new media if universities adopt the kind of organisational learning strategy argued for in the previous section. However, practitioner knowledge is not sufficient. As in any field, fundamental research is also important. Academics involved in the design of learning media should be able to run longer-term research projects to develop the necessary knowledge about teaching and learning in their subject, as an alternative to research in the subject itself. However, such funds are rarely available. This is the main reason for the dearth of reliable knowledge about learning media. Over the last twenty years, there have been several ICT development programmes in the UK, for example, but only a tiny fraction of public funding has been earmarked for research on the core activity of HE. These programmes have always paid lip service to evaluation, but very little has ever been carried out, as development costs expand to usurp the entire budget. The national funding council can promote research on learning, through the research funding agencies, at a level commensurate with the activity level of teaching and learning in HE. For any other industry to invest so little of its income in research on its core activity would be laughable. For the education industry, it is humiliating.

Provide a national point of contact on learning technology

The UK funding council has now established a Learning and Teaching Support Network (see Web References, p.260), with a Generic Centre to house a database and information network on a long-term basis. Such national networks will always be an essential part of the development process for new technology, as they offer access to existing knowledge. However, it is probably more effective to share this kind of knowledge by embedding it in the basic design of ICT learning formats and systems, just as the optimal design for print material is embedded in the format of a book. Central funds for development of learning materials should be focused on customisable design formats, therefore. Embedding good design helps to migrate good teaching ideas, and avoids costly and wasteful competitive developments. It is contrary to tradition for universities either to use each other's teaching, or to collaborate directly on teaching, whereas this form of collaboration is more feasible. If the funding council recognises the importance of universities sharing the burden of development of courseware, it could fund this form of collaboration by funding the costs of (1) designing materials explicitly for later customisation, and (2) assisting the transfer of 'generic learning activity models' to new contexts.

The UK funding council has supported a project of this kind, which is investigating the conditions for successful transfer, and the feasibility of designing materials to be customisable. Initial findings are that collaboration and transfer are feasible and cost-effective, but conditions for success are highly specialised (Twining *et al.*, 1998). This complex process needs support and promotion: it will not happen naturally.

Promote excellence in teaching alongside excellence in research

If teaching excellence is the aspiration of universities, then it must become the aspiration of individual academics. The sector cannot even begin to build knowledge of learning and teaching without such commitment by the academic community. This means that excellence in teaching must be accorded both the status and rigorous judgemental procedures that research has. The funding councils are responsible for the seriousness with which the academic community regards quality of teaching. Judging the excellence of innovation in ICT methods will involve peer group judgements, e.g. through the adoption by peers of generic learning activity models, and students' judgements of the quality of learning provided.

Establish design standards

Design standards must be established so that re-use by other academics is feasible. Standards should be as minimal as possible so as not to stifle creativity, and should relate to ease of use and production values. Standard descriptors for content, level and teaching style are also under development. A good model would be the standards being developed by the Instructional Management Systems project (see Web References, p.260), supported in the UK by the national funding council. Whatever the standards, they have to be defined centrally. The value of these systems is yet untested, but they should help to underpin the collaborative development needed.

CONCLUSIONS

The blueprint for an organisational infrastructure capable of continual improvement is essential for innovation in learning and teaching. Learning technologies entail a departure from the traditional modes of teaching at university level, which have always provided adequate opportunities for the teacher–student discussion that has been identified as so important for learning at this level. To improve continually, the development of new technology must have the cyclical character of any learning process. To be successful, the implementation must address the full context of the teaching–learning process. To be effective, the design must address all the activities essential for learning. To be applicable to higher education the design process must acknowledge the special nature of academic learning. All these requirements have been built into the organisational infrastructure identified in this chapter.

Higher education is evolving and adapting to new conditions while desperately trying to preserve the traditional high standards of an academic education. I began with the premise that academics must take responsibility for what and how their students learn. Universities have to maintain that responsibility, and not allow their standards to be undermined by new forms of competition. The transient epithets of 'the online university', 'the e-university', or 'the digital university', misconstrue the impact of new technology. A university is not defined by the incidentals of its delivery infrastructure, any more than the traditional university was adequately characterised as 'ivy-leaf', or 'redbrick'. Its character is defined by its role, 'to enable a society to make progress through an understanding of itself and its world' (Dearing, 1997: 72). I have argued throughout the book that the aim of making progress through understanding presupposes a Conversational Framework. At the heart of a university is the iterative dialogue between teacher and learner, nurturing the ideas and skills that constitute understanding. As we imagine the future forms of universities, that dialogue should remain the salient feature, with the delivery infrastructure always in support of it, never in the foreground. In this way, universities preserve the ability to be reflective and adaptive to their students' learning needs: it is not a business model that defines their aims, but the vision of a learning society.

Appendix 1

Extract from Plato's Meno dialogue

This extract shows how Socrates elicits from Meno's slave the proof of a Euclidean theorem. Socrates is setting out to demonstrate that all knowledge is innate, even geometrical knowledge. It would be more accurate, perhaps, to describe Socrates as demonstrating the capacity of the uneducated boy to discern the local logic of the argument being constructed by Socrates. In the extract below, the boy's contribution to the development of the proof is highlighted to emphasise how minimal it is. All the work is done in Socrates' questions. The Socratic method is driven entirely by the teacher, leaving little opportunity for construction by the learner at anything other than the most localised level of the argument structure.

> "To find a square, A, with an area which is double that of another square, B, A has to have a side equal to the length of B's diagonal." Prove this.

Socrates: Mark now, the further development. I shall only ask him, and not teach him, and he shall share the enquiry with me: and do you watch and see if you find me telling or explaining anything to him, instead of eliciting his opinion. Tell me boy, is this not a square of four feet which I have drawn?

Boy: **Yes**.

Socrates: And now I add another square equal to the former one?

Boy: **Yes**.

Socrates: And a third, which is equal to either of them?

Boy: **Yes**.

Socrates: Suppose that we fill up the vacant corner?

Boy: **Very good**.

Socrates: Here, then are four equal spaces?

Boy: **Yes**.

Socrates: And how many times larger is this space than this other?

Boy: **Four times**.

Socrates: But we wanted only one twice as large, as you will remember?

Boy: **True**.

Socrates: Now does not this line reaching from corner to corner bisect each of these spaces?
Boy: **Yes**.
Socrates: And are there not here four equal lines which contain this space?
Boy: **There are**.
Socrates: Look and see how much this space is.
Boy: **I do not understand**.
Socrates: Has not each interior cut off half the spaces?
Boy: **Yes**.
Socrates: And how many such spaces are there in this section?
Boy: **Four**.
Socrates: And how many in this?
Boy: **Two**.
Socrates: And four is how many times two?
Boy: **Twice**.
Socrates: So that this space is of how many feet?
Boy: **Of eight feet**.
Socrates: And from what line do you get this figure?
Boy: **From this**.
Socrates: That is from the line which extends from corner to corner of the figure of four feet?
Boy: **Yes**.
Socrates: And that is the line which the learned call the diagonal. And if this is the proper name, then you, Meno's slave, are prepared to affirm that the double space is the square of the diagonal?
Boy: **Certainly, Socrates**.
Socrates: What do you say of him Meno, were not all these answers given out of his own head?
Meno: Yes they were all his own.

(Jowett, 1953: 282–284)

Appendix 2

Subject teaching journals available on the Web

ERIC

The ERIC database of education literature is now freely available online at:

http://www.accesseric.org/searchdb/dbchart.html

ERIC enables academics to search for articles in their particular area, and in many cases provides abstracts of articles.

A number of ERIC Clearinghouses in specialist areas provide details of journal articles at:

http://www.accesseric.org/sites/barak.html

The British Education Index

The British Education Index, the other main bibliographic database for education, is not freely available to search, but they do provide access to a list of indexed journals, with some links to publishers' websites. At:

http://www.leeds.ac.uk/bei/

follow the 'List of journals indexed' link. Some publishers will provide tables of contents for a selection of volumes of a particular title.

The British Education Index oversees Education Line, which provides access to grey and pre-print literature in education at:

www.leeds.ac.uk/educol/

Education Line is searchable, and there are some full text articles available.

The Social Science Information Gateway

This provides details of journal articles available online, and details of journal titles at:

http://sosig.ac.uk/

under the 'Education' option.

Uncover

The Uncover document delivery database provides details of articles from 18,000 multidisciplinary journals, with the option to order articles online at:

http://uncweb.carl.org

through 'Search UnCover' option. The database can be searched by keyword, and journals are listed alphabetically.

Appendix 3

Summary of activities for an effective organisational infrastructure

This appendix summarises the activities discussed in Chapter 12, grouped by locus of responsibility with academic management and teaching at the institutional level, and then at national level.

Academic management

Expanding knowledge

- Provide access to an institutional database of learning materials and evaluation reports.
- Promote access to national databases of learning materials.
- Provide funding for journals, travel.
- Convene teachers' forum to identify key areas for development.
- Recruit new staff.

Sharing knowledge

- Optimise the deployment of staff resources.
- Optimise the organisation of teaching.
- Encourage use of good materials developed elsewhere.
- Establish a programme of staff development.
- Set up multi-skilled development teams.
- Set up forum for teachers to discuss ideas, experience.

Innovating

- Agree development resources and costing.
- Agree staff time commitment.
- Monitor project management.
- Establish policy on reversioning support, productivity tools, and standards.
- Match innovation in assessment to innovation in learning and teaching.

Evaluating

- Administer refereeing process for design of courseware.
- Provide time, resource and support for evaluation.

Implementing

- Communicate new requirements to students and staff.
- Manage marketing of the innovation.
- Provide support staff for maintenance and administration.
- Finalise costs.
- Ensure that appraisal and promotion procedures reward teaching excellence.

Validating

- Receive and act on reports.
- Monitor the implementation.

Academic teaching

Expanding knowledge

- Analyse course provision at universities.
- Access national database of courseware, courseware reviews.
- Analyse market needs.
- Carry out literature search in key journals.
- Build on previous evaluation reports.
- Analyse exam scripts.

Sharing knowledge

- Share awareness of current developments in design.

Innovating

- Address innovation to the critical aspects of learning the subject.
- Use productivity tools, reversioning principles, and standards.
- Receive and act on evaluation reports.

Evaluating

- Carry out developmental testing and piloting.
- Use evaluation information to define the specification of the course.

Implementing

- Provide guidance on the use of learning technology.
- Define and support service targets.

Validating

- Monitor and report on efficiency of institutional procedures.
- Disseminate and publish reports.
- Analyse use of courseware materials and assignments.

National academic management

- Promote funding of research in student learning.
- Provide a national point of contact on learning technology.
- Promote excellence in teaching alongside excellence in research.
- Establish design standards.

Glossary

Adaptive Refers to learning and teaching activities that enable the student or teacher to adjust their actions in the light of results of previous actions. Describes also media that facilitate this, such as computer programs that give intrinsic feedback (q.v.) on the student's input. Simulations and modelling programs both do this.

Anathemagenic Coined by the author to contrast with 'mathemagenic' (q.v.), to describe learning activities that 'give birth to loathing'.

Approach to learning The umbrella term used to describe what a student brings to learning, including both how they handle the information, and their personal learning intentions.

Asynchronous Contrasts with 'synchronous' (q.v.) to mean 'not at the same time'; applied to forms of communication where interlocutors are not both present at the same time, such as electronic mail.

Audiographics Refers to a form of communication where the audio channel, e.g. a telephone line, or audio on the Web, allows normal conversation, and a data channel allows the interlocutors to exchange data for display on their computer screens at the same time.

Audiovision A term in common use at the Open University to describe a combination of audio and visuals, e.g. an audiocassette talking the student through the visual component displayed in a diagram.

Collaborative learning Means what it says, but is often used to refer to students working on a computer-based learning program that requires them to collaborate by taking different roles, or operating different controls.

Communicative Refers to media that facilitate discussion, or discursive activities (q.v.) between students and teachers. They may be synchronous (q.v.), like the telephone, or asynchronous (q.v.), like email.

Concealed multiple choice question (CMCQ) Describes a version of multiple-choice question (MCQ) (q.v.) that conceals the choices. The program invites open-ended input from the student and compares it, using a matching algorithm, with each choice programmed in. The closest match is taken to be the student's choice. The program thereby knows the student's choice without the disadvantage of suggesting answers to the student, in the manner of the MCQ.

Discursive Describes the learning activity of discussion, or a medium that supports it. The discussion may be between students, or between student and teacher. Each interlocutor must be able to articulate a view, re-articulate it in the light of the other's utterance, ask questions, and reply to questions, though not necessarily synchronously. Thus letter-writing is discursive, whereas lecturing is not. Communicative media (q.v.) support discursive activities.

Evaluation Refers to ways of testing the quality or value of something: in the educational context usually course materials, or teaching methods, but sometimes also students. However, evaluation of students is more usually referred to as 'assessment'. Evaluation methods for course materials include pre- and post-testing of students' knowledge, observation, interviewing, questionnaires.

Experiential knowledge/learning Describes knowledge gained through experience, or learning through experience. Contrasts, and often conflicts with academic knowledge and learning through instruction.

Extrinsic feedback Contrasts with 'intrinsic feedback' (q.v.), and describes someone's evaluation of an action (e.g. applause as a comment on an effective kick in football), where the feedback is generated from a context external to the action itself.

Formative evaluation Contrasts with 'summative evaluation' (q.v.). Describes the evaluation of course materials that provides information for improvement of those materials.

Interactive Often used to refer to user control of a medium, e.g. interactive video allows the user to stop, start, rewind, etc. In this book, it refers to learning activities that enable the student to control and explore a set of resources. Describes also media that facilitate this, e.g. the Web is interactive because users control the sequence and presentation of content.

Intrinsic feedback Contrasts with 'extrinsic feedback' (q.v.), and describes the result of an action, (e.g. a goal as the result of a kick of a football) where the feedback is generated from within the context of the action itself.

Mathemagenic Coined by Rothkopf to describe activities that 'give birth to learning', from the Greek *mathema* meaning 'something learned' and *-genus* meaning 'given birth to'.

Microworld A computer program that embodies rules governing the behaviour of defined objects and their interaction with each other, thus evoking the impression of 'a little world'. The user can manipulate the objects to build something in that world, via a language understood by the program.

Modelling program A program that takes as input descriptions of a system, allowing the learner to create their own model of its behaviour. The program determines the form of the description, and the form of the output (numerical, graphical or text), but uses the learner's definition to generate the system's behaviour.

Multiple-choice question (MCQ) The most common form of interaction offered by computer-based learning programs: the question is put, and is followed by some possible answers, including the correct answer and some plausible distracters, or common incorrect answers. The student selects one, and this is meant to represent their answer. Contrasts with 'concealed multiple-choice question' (CMCQ) (q.v.).

Narrative Refers to a medium that supports the presentation of a linear narrative. Narrative provides a structure that creates global coherence for any text or speech. Narrative media include print, lectures, videos, demonstrations, Web pages, and originally, of course, storytelling.

Pedagogenic error Coined by the author to mean 'teacher-induced error' (from the Greek *paedagogos* meaning 'teacher', and *-genus* meaning 'given birth to'). It is the teaching profession's equivalent of 'iatrogenic disease', meaning 'disease induced by the physician'.

Phenomenography Coined by Marton to mean 'descriptions of the phenomena', specifically, the alternative ways students conceptualise key phenomena; contrasts with the philosophical method of 'phenomenology', which 'studies the phenomena' to develop a fully justified and unitary knowledge of what is.

Productive Refers to a medium that facilitates the student's own production of material. The material could be a text (e.g. via Word), or presentation (e.g. via PowerPoint), or any other combination of audio, visual and software designs.

Reflective Refers to those teaching methods or learning activities that encourage the student to reflect on what they know, or on what they have experienced.

Self-assessed question (SAQ) Used in distance-teaching texts to enable the student to check their answer to the question against a model answer. The answer is usually given at the end of the text.

Simulation A computer program that runs a model of the behaviour of a system, and displays that behaviour in text, numerical or graphical form, e.g. a spreadsheet simulating the cash flow of a business. The user can usually control the initial values of parameters in the model.

Summative evaluation Contrasts with 'formative evaluation' (q.v.). Describes the evaluation of course materials that provides information on the success or otherwise of the implementation of those materials, in terms of the aims of the course, possibly in comparison with alternative teaching methods.

Supplantation Coined by Salomon to describe the way a medium, particularly television, can use special techniques to simulate certain kinds of cognitive processing for the viewer, e.g. a zoom to 'supplant' selective attention to part of a scene.

Synchronous Contrasts with 'asynchronous' (q.v.) to mean 'at the same time'; applied to forms of communication where interlocutors are both present at the same time, such as the telephone, or chat rooms on the Internet.

Teleconferencing Any form of interactive person(s)-to-person(s) communication at a distance, from the Greek *Tele-* meaning 'far off'.

Tutorial program A computer program that presents information, sets exercises for the student, accepts answers in some specified format, and gives feedback on those answers. Some tutorial programs also define the sequence of tasks for a student to achieve specified objectives.

References

BOOKS AND JOURNAL ARTICLES

Barker, P. (1994) 'Designing interactive learning', in Ton de Jong and Luigi Sarti (Eds), *Design and Production of Multimedia and Simulation-based Learning Material*, Dordrecht: Kluwer Academic Publishers.

Barnett, R. (1990) *The Idea of Higher Education*, Milton Keynes: Open University Press.

Bates, A. (1991) 'Third generation distance education: the challenge of new technology', *Research in Distance Education*, 3 (2): 10–15.

Beer, S. (1985) *Diagnosing the System: for Organisations*, Oxford: John Wiley and Sons.

Biggs, John (1999) *Teaching for Quality Learning at University*, Buckingham: SRHE and Open University Press.

Bollom, C.E., Emerson, P.A., Fleming, P.R. and Williams, A.R. (1989) 'The Charing Cross and Westminster interactive television network', *Journal of Educational Television*, 15 (1): 5–15.

Booth, S.A. (1992) 'The experience of learning to program. Example: Recursion', in F. Détienne (Ed.), *5-ème workshop sur la psychologie de la programmation*, Paris: INRIA, 122–145.

Bowden, J., Dall'Alba, G., Martin, E., Laurillard, D., Marton F., Masters, G., Ramsden, P., Stephanou, A. and Walsh, E. (1992) 'Displacement, velocity and frames of reference: phenomenographic studies of students' understanding and some implications for teaching and assessment', *American Journal of Physics*, 60 (3): 262–269.

Bowden, John and Marton, Ference (1998) *The University of Learning: Beyond Quality and Competence in Higher Education*, London: Kogan Page.

Britain, S. and Liber, O. (1999) 'A framework for pedagogical evaluation of virtual learning environments', JTAP Report No. 41, October 1999, http://www.jisc.ac.uk/jtap/word/jtap-041.doc

Brna, P. (1989) 'Programmed rockets: an analysis of students' strategies', *British Journal of Educational Technology*, 20 (1): 27–40.

Brooksbank, D.J., Clark, A., Hamilton, R. and Pickernell, D.G. (1998) 'A critical appraisal of WinEcon and its use in a first-year undergraduate economics programme', *Association for Learning Technology Journal*, 6 (3): 47–53.

Brown, G. and Atkins, M. (1991) *Effective Teaching in Higher Education*, London: Routledge.

Brown, J.S., Collins, A. and Duguid, P. (1989a) 'Situated cognition and the culture of learning', *Educational Researcher*, 18 (1): 32–42.

Brown, J.S., Collins, A. and Duguid, P. (1989b) 'Debating the situation: a rejoinder to Palincsar and Wineburg', *Educational Researcher*, 18 (4): 10–12.

Brown, S.J. and Duguid, P. (1998) 'Organising knowledge', *California Management Review*, 40 (3).

Brown, J.S. and Van Lehn, K. (1980) 'Repair theory: a generative theory of bugs in procedural skills', *Cognitive Science*, 4: 379–426.

Brumby, M. (1984) 'Misconceptions about the concept of natural selection by medical biology students', *Science Education*, 68 (4): 493–503.

Buckingham Shum, Simon, and Sumner, Tamara (1998) 'New scenarios in scholarly publishing and debate', in Marc Eisenstadt and Tom Vincent (Eds), *The Knowledge Web: Learning and Collaborating on the Net*, London: Kogan Page.

Burge, Liz. (1995) 'Electronic highway or weaving loom?', in Fred Lockwood (Ed.), *Open and Distance Learning Today*, London: Routledge.

Calvino, Italo. (1979) *If on a Winter's Night a Traveller*, London: Picador.

Carey, T., Harrigan, K., Palmer, A. and Swallow, J. (1999) 'Scaling up a learning technology strategy: supporting student/faculty teams in learner-centred design', *Association for Learning Technology Journal*, 7 (2): 15–26.

Chalmers, D. and Fuller, R. (1996) *Teaching for Learning at University*, London: Kogan Page.

Chambers, E. and Rae, J. (1999) *Evaluation of the Homer CD-ROM: Final Report*, Institute of Educational Technology, Open University, Milton Keynes, MK7 6AA, UK.

Champagne, A.B., Klopfer, L.E. and Gunstone, R.F. (1982) 'Cognitive research and the design of science instruction', *Educational Psychology*, 17 (1): 31–53.

Collis, Betty (1996) *Tele-learning in a Digital world: The Future of Distance Learning*, London: International Thomson Computer Press.

Dahlgren, L.O. and Marton, F. (1978) 'Students' conceptions of subject matter: an aspect of learning and teaching in higher education', *Studies in Higher Education*, 3 (1): 25–35.

Daniel, J.S. (1991) 'The international role of the Open University', *Reflections on Higher Education*, 3: 15–25.

Dearing, Ron (1997) *Higher Education in the Learning Society*, National Committee of Inquiry into Higher Education, HMSO, ISBN: 1 85838 254 8.

Dobson, M. and McCracken, J. (1997) 'Science, technology, and society: using problem-based learning as a means to evaluate multimedia courseware', in *Educational Multimedia and Hypermedia*, American Association of Computers in Education.

Dobson, M., Hunter, W. and McCracken, J. (2001) 'Evaluation of technology supported teaching and learning: a catalyst to organisational change', *Journal of Interactive Learning Environments*, 9 (2): 143–170.

Duffy, Thomas M. and Cunningham, Donald J. (1996) 'Constructivism: implications for the design and delivery of instruction', in David Jonassen (Ed.), *Handbook of Research for Educational Communications and Technology*, New York: Simon & Schuster Macmillan.

Durbridge, N. (1984a) 'Using audio-vision to teach mathematics', in E. Henderson and M. Nathenson (Eds), *Independent Learning in Higher Education*, Inglewood Cliffs, NJ: Educational Technology Publications.

Durbridge, N. (1984b) 'Developing the use of video cassettes in the Open University', in O. Zuber-Skerritt (Ed.), *Video in Higher Education*, London: Kogan Page.

Durbridge, N. and Stratfold, M. (1996) 'Varying the texture: a study of art, learning and multimedia', *Journal of Interactive Media in Education*, 96:(1), http://www-jime.open.ac.uk/

Edwards, N. (1996) 'Computer-based laboratory simulations: evaluations of students' perceptions', *Association for Learning Technology Journal*, 4 (3): 41–53.

Elton, L. and Middlehurst, R. (1992) 'Leadership and management in higher education', *Studies in Higher Education*, 17 (3): 251–264.

Entwistle, N.J. (1981) *Styles of Learning and Teaching: An Integrated Outline of Educational Psychology*, Chichester: John Wiley.

Entwistle, N.J. and Ramsden, P. (1983) *Understanding Student Learning*, London: Croom Helm.

Eysenck, M.W. and Warren-Piper, D. (1987) 'A word is worth a thousand pictures', in J.T.E. Richardson, M.W. Eysenck and D. Warren-Piper (Eds), *Student Learning: Research in Education and Cognitive Psychology*, Milton Keynes: SRHE and Open University Press.

Gagné, R.M. (1977) *The Conditions of Learning*, New York: Holt Rinehart and Winston.

Gagné, R.M. and Merrill, M.D. (1990) 'Integrative goals for instructional design', *Educational Technology Research and Development*, 38 (1): 23–30.

Gibbons, M., Limoges, C., Nowotny, H., Schwartzmann, S., Scott, P. and Trow, M. (1994) *The New Production of Knowledge*, London: Sage.

Golluscio, R.A., Paruelo, J.M. and Aguiar, M.R. (1990) 'Simulation models for educational purposes: an example of the coexistence of plant populations', *Journal of Biological Education*, 24 (2): 81–86.

Hamilton, W. (1951) *Plato: The Symposium*, translated by W. Hamilton, London: Penguin.

Hawkridge, D. (1993) *Challenging Educational Technology*, London: Athlone Press.

Hawkridge, D. (1998) 'Cost-effective support for university students learning via the Web?', *Association for Learning Technology Journal*, 6 (3): 24–29.

Holland, S. (1987) 'New cognitive theories of harmony applied to direct manipulation tools for novices', *CITE Technical Report* No 17, IET, Open University, Milton Keynes MK7 6AA.

Hounsell, D. (1984) 'Learning and essay writing', in F. Marton, D. Hounsell and N. Entwistle (Eds), *The Experience of Learning*, Edinburgh: Scottish Academic Press.

Hoyles, C., Healy, L. and Sutherland, R. (1991) 'Patterns of discussion between pupil pairs in computer and non-computer environments', *Journal of Computer Assisted Learning*, 7: 210–228.

Ison, R.L. (1994) 'Designing learning systems: how can systems approaches be applied in the training of research workers and development actors?' *Proceedings of the International Symposium on Systems-oriented Research in Agriculture and Rural Development Volume 2: Lectures and Debates*, 369–394. CIRAD-SAR, Montpellier.

Jonassen, D. (1991) 'Hypertext as instructional design', *Educational Technology Research and Development*, 39 (1): 83–92.

Jones, Chris (1999) 'From the sage on the stage to what exactly? Description and the place of the moderator in co-operative and collaborative learning', *Association for Learning Technology Journal*, 7 (2): 27–36.

Jowett, B. (1953) *The Dialogues of Plato. Volumes I to III*, London: Oxford University Press.

Kolb, D.A. (1984) *Experiential Learning: Experience as the Source of Learning and Development*, Englewood Cliffs, NJ: Prentice Hall.

Laurillard, D. (1979) 'The processes of student learning', *Higher Education*, 8: 395–409.

Laurillard, D. (1982) 'D102 Audio-visual media evaluation: interim report blocks 2 and 3', IET, Open University, Milton Keynes MK7 6AA.

Laurillard, D. (1984a) 'Interactive video and the control of learning', *Educational Technology*, 24 (6): 7–15.

Laurillard, D. (1984b) 'Learning from problem-solving', in F. Marton, D. Hounsell and N. Entwistle (Eds), *The Experience of Learning*, Edinburgh: Scottish Academic Press.

Laurillard, D. (1987a) 'The different forms of learning in psychology and education', in J. T. E.Richardson, M. W. Eysenck and D. Warren-Piper (Eds), *Student Learning: Research in Education and Cognitive Psychology*, Buckingham: SRHE and Open University Press.

Laurillard, D. (1987b) 'Evaluation Report on the CADED Project', Queen Mary and Westfield College, Mile End Road, London E1 4NS, UK.

Laurillard, D. (1991) 'Mediating the message: television programme design and students' understanding', *Instructional Science*, 20: 3–23.

Laurillard, D. (1992) 'Phenomenographic research and the design of diagnostic strategies for adaptive tutoring systems', in M. Jones and P. Winne (Eds), *Adaptive Learning Environments*, Berlin: Springer-Verlag.

Laurillard, D. (1997) 'Styles and approaches in problem-solving', in F. Marton, D. Hounsell and N. Entwistle (Eds), *The Experience of Learning* (2nd edition), Edinburgh: Scottish Academic Press.

Laurillard, D. (1998) 'Multimedia and the learner's experience of narrative', *Computers and Education*, 31 (2): 229–242.

Laurillard, D. (1999) 'A conversational framework for individual learning applied to the learning organisation and the learning society', *Systems Research and Behavioural Science, Special Issue: Applying Systems Thinking to Higher Education*, edited by Ray Ison, 16 (2): 113–122.

Laurillard, D., Lindström, B., Marton, F. and Ottosson, T. (1991) 'Computer simulation as a tool for developing intuitive and conceptual understanding', Report No. 1991:03, Department of Education and Educational Research, University of Göteborg, ISSN 0282–2156.

Laurillard, D., Stratfold, M., Luckin, R., Plowman, L. and Taylor, J. (2000) *Journal of Interactive Media in Education*, Open University:
http//:www-jime.open.ac.uk

Laurillard, D. and Taylor, J. (1994) 'Designing the stepping stones: an evaluation of interactive media in the classroom', *Journal of Educational Television*, 20 (3): 169–184.

Lockwood, F. (1992) *Activities in Self-Instructional Texts*, London: Kogan Page.

Luckin, R., Plowman, L., Laurillard, D., Stratfold, M., Taylor, J. and Corben, S. (2001) 'Narrative evolution: learning from students' talk about species variation', *International Journal of Artificial Intelligence and Education*, 12 (1): Special Issue on Analysing Educational Dialogue Interaction (forthcoming).

Lybeck, L., Marton, F., Strömdahl, H. and Tullberg, A. (1988) 'The phenomenography of the "mole concept" in chemistry', in P. Ramsden (Ed.), *Improving Learning: New Perspectives*, London: Kogan Page.

Macdonald, J. (1999) *Appropriate Assessment for Resource Based Learning in Networked Environments*, Unpublished PhD thesis, Open University.

Mandinach, E.B. and Cline, H.F. (1996) 'Classroom dynamics: the impact of a technology-based curriculum innovation on teaching and learning', *Journal of Educational Computing Research*, 14 (1): 83–102.

Manguel, Alberto (1997) *A History of Reading*, London: Flamingo.

Marton, F. (1988) 'Describing and improving learning', in R.R. Schmeck (Ed.), *Learning Strategies and Learning Styles*, New York: Plenum.

Marton, F., Beaty, E. and Dall'Alba, G. (1993) 'Conceptions of learning', *International Journal of Educational Research*, 19: 277–300.

Marton, F. and Booth, S. (1997) *Learning and Awareness*, Marwah, NJ: Lawrence Erlbaum Associates.

Marton, F., Hounsell, D. J. and Entwistle, N. J. (Eds) (1997) *The Experience of Learning* (2nd edition), Edinburgh: Scottish Academic Press.

Marton, F. and Ramsden, P. (1988) 'What does it take to improve learning?', in P. Ramsden (Ed.), *Improving Learning: New Perspectives*, London: Kogan Page.

Marton, F. and Säljö, R. (1976a) 'On qualitative differences in learning I: outcome and process', *British Journal of Educational Psychology*, 46: 4–11.

Marton, F. and Säljö, R. (1976b) 'On qualitative differences in learning II: outcome as a function of the learner's conception of the task', *British Journal of Educational Psychology*, 46: 115–127.

Marton, F. and Wenestam, C-G. (1979) 'Qualitative differences in the understanding and retention of the main point in some texts based on the principle-example structure', in M.M. Gruneberg, P.E. Morris and R.N. Sykes (Eds), *Practical Aspects of Memory*, London: Academic Press.

Mason, Robin (1994) *Using Communications Media in Open and Flexible Learning*, London: Kogan Page.

Mason, Robin (1998) 'Models of online courses', in L. Banks, C. Graebner and D. McConnell (Eds), *Networked Lifelong Learning: Innovative Approaches to Education and Training Through the Internet*, University of Sheffield, 1998.

McCracken, J. and Laurillard, D. (1994) 'A study of conceptions in visual representations: a phenomenographic investigation of learning about geological maps', paper presented at the World Conference on Educational Multimedia and Hypermedia, Vancouver, 25–30: June 1994.

McCracken, J. and Dobson, M. (1999) 'Illuminating learner conceptions: a needs assessment method', in M. Dobson (1999), 'Selecting evaluation methods for technology based learning projects'. Symposium presented to European Association for Research in Learning and Instruction, Gothenburg, Sweden: 1999.

McDermott, L.C. (1991) 'Millikan lecture 1990: what we teach and what is learned – closing the gap', *American Journal of Physics*, 59 (4): 301–315.

McInnes, Craig (1995) 'Less control and more vocationalism', in Tom Schuller (Ed.), *The Changing University?*, Buckingham: SRHE and Open University Press.

McMahon, H. (1990) 'Collaborating with computers', *Journal of Computer Assisted Learning*, 6: 149–167.

McNaught, C. and Kennedy, P. (2000) 'Staff development at RMIT: bottom-up work serviced by top-down investment and policy', in D. Squires, G. Conole and G. Jacobs (Eds), *The Changing Face of Learning Technology*, Cardiff: University of Wales Press.

Moyse, R. (1991) 'Multiple viewpoints imply knowledge negotiation', *Interactive Learning International*, 7: 21–37.

Moyse, R. (1992) 'A structure and design method for multiple viewpoints', *Journal of Artificial Intelligence in Education*, 3: 207–233.

Murison-Bowie, S. (1999) 'Forms and functions of digital content in education', in Anne Leer (Ed.), *Masters of the Wired World*, London: FT Pitman Publishing.

Naughton, John (1999) *A Brief History of the Future: The Origins of the Internet*, London: Weidenfeld and Nicolson.

Neuman, D. (1987) 'The origin of arithmetic skills', *Göteborg Studies in Educational Sciences*, Vol. 62, University of Gothenburg.

Nonaka, I. (1994) 'A dynamic theory of organizational knowledge creation', *Organization Science*, 5 (1): February 1994.

OECD (1987) *Universities under Scrutiny*, Paris: Office of Economic Co-operation and Development, ISBN 92-62-129227.

Ogborn, J. (1990) 'A future for modelling in science education', *Journal of Computer Assisted Learning*, 6: 103–112.

Open University A295 Course (1993) *Homer: Poetry and Society*, Milton Keynes: Open University.

Palloff, Rena M. and Pratt, Keith (1999) *Building Learning Communities in Cyberspace*, San Francisco: Jossey-Bass.

Papert, S. (1980) *Mindstorms: Children, Computers, and Powerful Ideas*, Brighton, Sussex: Harvester Press.

Pask, G. (1976) 'Conversational techniques in the study and practice of education', *British Journal of Educational Psychology*, 46: 12–25.

Perkins, P. (1995) 'The development of computer assisted learning materials for archaeology and art history', *Computers and the History of Art*, 5 (2): 79–91.

Perry, William G. (1959) 'Students' use and misuse of reading skills', *Harvard Educational Review*, 29 (3): 193–200.

Perry, W.G. (1970) *Forms of Intellectual and Ethical Development in the College Years*, NY: Holt Rhinehart and Winston.

Perry, W.G. (1988) 'Different worlds in the same classroom', in P. Ramsden (Ed.), *Improving Learning: New Perspectives*, London: Kogan Page.

Peters, Otto (2000) 'Digital learning environments: new possibilities and opportunities', *International Review of Research in Open and Distance Learning*, 1 (1): 1–19.

Petrie, M., Carswell, L., Price, B., and Thomas, O. (1998) 'Innovations in large-scale supported distance teaching: transformation for the Internet, not just translation', in Marc Eisenstadt and Tom Vincent (Eds), *The Knowledge Web: Learning and Collaborating on the Net*, London: Kogan Page.

Plowman, L. (1996) 'Designing interactive media for schools: a review based on contextual observation', *Information Design Journal*, 8 (3): 258–266.

Prosser, M. and Trigwell, K. (1999) *Understanding Learning and Teaching*, Buckingham: SRHE and Open University Press.

Ramsden, Paul (1992) *Learning to Teach in Higher Education*, London: Routledge.

Ramsden, Paul (1988) (Ed.), *Improving Learning: New Perspectives*, London: Kogan Page.

Ramsden, Paul (1998) *Learning to Lead in Higher Education*, London: Routledge.

Resnick, L. and Omanson, S. (1987) 'Learning to understand arithmetic', in R. Glaser (Ed.), *Advances in Instructional Psychology*, Vol. 3, Hillsdale, NJ: Lawrence Erlbaum Associates.

Reusser, K. (1992) 'Tutoring systems and pedagogical theory: representational tools for understanding, planning, and reflection in problem-solving', in S. Lajoie and S. Derry (Eds), *Computers as Cognitive Tools*, Hillsdale, NJ: Lawrence Erlbaum Associates.

Richardson, John T.E. (2000) *Researching Student Learning*, Milton Keynes: Society for Research into Higher Education and Open University Press.

Robbins, L. (1963) *Higher Education: Report of the Committee*, Cmnd 2154, London: HMSO.

Romiszowski, A. (1988) *The Selection and Use of Instructional Media*, NY: Kogan Page.

Rothkopf, E.Z. (1970) 'The concept of mathemagenic activities', *Review of Educational Research*, 40: 325–336.

Rowntree, D. (1992) *Exploring Open and Distance Learning*, London: Kogan Page.

Ryan, S., Scott, B., Freeman, H. and Patel, D. (2000) *The Virtual University: The Internet and Resource-Based Learning*, London: Kogan Page.

Säljö, R. (1979) 'Learning in the learner's perspective. Some common-sense conceptions', Internal Report, Department of Education, University of Göteborg No. 76.

Säljö, R. (1984) 'Learning from reading', in F. Marton, D. J. Hounsell and N. J. Entwistle (Eds), *The Experience of Learning*, Edinburgh: Scottish Academic Press.

Säljö, R. (1988) 'Learning in educational settings: methods of enquiry', in P. Ramsden (Ed.), *Improving Learning: New Perspectives*, London: Kogan Page.

Salmon, G. (2000) *E-Moderating: The Key to Teaching and Learning Online*, London: Kogan Page.

Salomon, G. (1979) *Interaction of Media, Cognition and Learning*, San Francisco: Jossey-Bass.

Saunders, P. (1991) 'The third edition of the test of understanding in college economics', *Journal of Economic Education*, 22 (3): 255–272.

Scott, Peter and Eisenstadt, Marc (1998) 'Exploring telepresence on the internet: the KMi Stadium Webcast experience', in Marc Eisenstadt and Tom Vincent (Eds), *The Knowledge Web: Learning and Collaborating on the Net*, London: Kogan Page.

Sellman, R. (1991) 'Hooks for tutorial agents: a note on the design of discovery learning environments', CITE Technical Report No 145, IET, Open University, MK7 6AA.

Senge, P.M. (1993) *The Fifth Discipline: The Art and Practice of The Learning Organization*, London: Century Business.

Svensson, L. (1977) 'On qualitative differences in learning III: study skill and learning', *British Journal of Educational Psychology*, 47: 233–243.

Stevens, A., Collins, A. and Goldin, S.E. (1979) 'Misconceptions in students' understanding', *International Journal of Man-Machine Studies*, 11: 145–156.

Tabachneck-Schijf, H.J.M. and Simon, H.A. (1996) 'Alternative representations of instructional material', in Donald Peterson (Ed.), *Forms of Representation*, Exeter: Intellect Books Ltd.

Tait, K. (1994) 'DISCOURSE: the design and production of simulation-based learning environments', in Ton de Jong and Luigi Sarti (Eds), *Design and Production of Multimedia and Simulation-based Learning Material*, Dordrecht: Kluwer Academic Publishers.

Thompson, Ian (1999) *Convergence in Television and the Internet* (2nd edition), London: FT Business Ltd.

Twigg, C. (1994) 'The changing definition of learning', *Educom Review*, 29 (4): http://educom.edu/web/pubs/review/reviewArticles/29620.html

Twining, P., Stratfold, M., Kukulska-Hulme, A. and Tosunoglu, C. (1998) 'SoURCE – software use, re-use and customisation in education', *Active Learning*, 9: 54–56.

Vygotsky, L . (1962) *Thought and Language*, Cambridge, MA: MIT Press.

Wenger, Etienne (1999), *Communities of Practice: Learning, Meaning, and Identity*, Cambridge University Press.

Wertsch, J.V., Minick. N. and Arns, F.J. (1984) 'The creation of context in joint problem-solving', in B. Rogoff and J. Lave (Eds), *Everyday Cognition: Its Development in Social Context*, Cambridge, MA: Harvard University Press.

Whalley, P. (1998) 'Collaborative learning in networked simulation environments', in M. Eisenstadt and T. Vincent (Eds), *The Knowledge Web: Learning and Collaborating on the Net*, London: Kogan Page.

Whelan, G. (1988) 'Improving medical students' clinical problem-solving', in P. Ramsden (Ed.), *Improving Learning: New Perspectives*, London: Kogan Page.

Winer, L., Chomienne, M. and Vazquez-Abad, J. (2000) 'A distributed collaborative science learning laboratory on the Internet', in Michael G. Moore and Geoffrey T. Cozine (Eds), *Web-based Communications, the Internet, and Distance Education*, Readings in Distance Education No. 7, The American Center for the Study of Distance Education, The Pennsylvania State University.

WEB REFERENCES

ILT (Institute for Learning and Teaching) http://www.ilt.ac.uk/
IMS (Instructional Management Systems) http://www.imsproject.com
JIME (Journal for Interactive Media in Education) http://www-jime.open.ac.uk/
LTSN (Learning and Teaching Support Network) http://www.ltsn.ac.uk/
MENO (Multimedia, Education and Narrative Organisation, ESRC project) http://meno.open.ac.uk/meno/
OU/BBC Broadcast programmes support http://www.open2.net/
PLUM (Programme on Learner User of Media, Open University) http://iet.open.ac.uk/plum/evaluation/contents.html
ROUTES (Resources for Open University TEachers and Students) http://routes.open.ac.uk/
Virtual Microscope (Open University) http://met.open.ac.uk

Index